Alkali-Aggregate Reaction in Concrete:
A World Review

T0172797

Alkali-Aggregate Reaction in Concrete: A World Review

Editors

Ian Sims
RSK Environment Ltd, Hemel Hempstead, UK

Alan Poole
Consultant, Oxford, UK

CRC Press
Taylor & Francis Group
Boca Raton   London   New York   Leiden

CRC Press is an imprint of the
Taylor & Francis Group, an **informa** business

A BALKEMA BOOK

Applied for

Published by: CRC Press/Balkema
Schipholweg 107C, 2316 XC Leiden, The Netherlands
e-mail: Pub.NL@taylorandfrancis.com
www.crcpress.com – www.taylorandfrancis.com

First issued in paperback 2020

© 2017 by Taylor & Francis Group, LLC
CRC Press/Balkema is an imprint of the Taylor & Francis Group, an informa business

No claim to original U.S. Government works

ISBN 13: 978-0-367-57333-1 (pbk)
ISBN 13: 978-1-138-02756-5 (hbk)

Visit the Taylor & Francis Web site at
http://www.taylorandfrancis.com

and the CRC Press Web site at
http://www.crcpress.com

Typeset by MPS Limited, Chennai, India

Although all care is taken to ensure integrity and the quality of this publication and the information herein, no responsibility is assumed by the publishers nor the author for any damage to the property or persons as a result of operation or use of this publication and/or the information contained herein.

Library of Congress Cataloging-in-Publication Data

Contents

Preface

In 1992 a new and ground breaking attempt to provide a worldwide review focussing on the problem and implications of alkali-silica reaction (ASR) in concrete structures was published as a collection of contributions from internationally recognised materials experts. These contributions were brought together by the editor, Professor Narayan Swamy, in a single volume. This book, *The Alkali-Silica Reaction in Concrete*, was published by Blackie and Sons in the UK and Van Nostrand Reinhold in America. It provided the civil engineer, materials scientist and other professionals concerned with concrete structures with a unique review of this relatively rare but very costly cause of concrete deterioration that had been identified in a number of countries round the world.

This first review explained the nature of the mechanisms of alkali-silica reaction, which is entirely different from other causes of premature concrete degradation. The review gave first insights into the various international approaches relating to the diagnosis, evaluation and avoidance of this type of deterioration caused by the reaction. Although the countries providing information in this first compilation included the UK, Denmark, Iceland, Canada, New Zealand, Japan and India, it is apparent that many parts of the world remained absent from this list, including the USA where the problem had first been identified.

The contributions in the 1992 edition clearly demonstrated that laboratory research studies alone are inadequate to describe the complex behaviour of this reaction in real concrete structures. The conditions and effects of the deterioration that develop appeared to be different in apparently unpredictable ways in different regions of the world. It came to be realised that factors additional to the materials used in the production of concrete will have an influence on the initiation, progression and on the deleterious effects resulting from the reaction between aggregate and alkalis in the cement. These included variations in atmospheric humidity, temperature fluctuation and ranges, freeze-thaw cycles and rainfall. Consequently, in order to investigate the effect of the reaction on a particular concrete structure fully, all these factors needed to be evaluated. This is not to say that laboratory research studies are irrelevant, it is rather that a combination of both laboratory and field investigation of concrete structures is essential to a proper understanding of the problems caused by the alkali-aggregate reaction.

In the 25 years since 1992 considerable progress has been made in understanding the problem of alkali-aggregate reaction in concrete and how best to avoid or mitigate the premature concrete degradation that it causes. This increase in knowledge has come about in part through intensive research efforts both in the laboratory and on field concrete. Also, there has been a gradual realisation that some concrete structures in many more countries round the world than was first realised are susceptible to this type of decay mechanism, with its serious consequence for important structures such as dams and bridges. This research has been a truly international collaboration with research studies undertaken in many different countries and shared globally through a series of regular international conferences since 1974. The studies have focussed on several different areas, including better understanding of the mechanisms of the reaction, the most effective means of avoiding the use of materials in concrete that are susceptible to initiating the reaction, and options for mitigating or repairing concrete structures that have become damaged by alkali-aggregate reaction.

This current book, like its 1992 predecessor, is a compilation of contributions from many international experts that have been edited into a single volume to provide a state of the art review of the problem of alkali-aggregate reaction as it affects concrete structures in countries around the world. It provides an up to date appraisal of the research conclusions that have been reached and the progress made in understanding the causes and effects of alkali-aggregate reaction in concrete, both at a laboratory research level, as a construction material and in the concrete structures themselves.

A large number of countries world-wide now realise that the alkali-aggregate reaction in concrete is a real and costly problem that may affect their own concrete structures. Many of these countries have assessed the particular aggregate materials and the concrete mix designs they use and have developed their own national test methods for the diagnosis and evaluation of the reaction. A number of these have introduced national standard specifications for concrete, aimed at avoiding susceptible materials and mix designs in new concrete structures.

This book is unique in that it provides not only a review of the current state of research findings relating to the understanding of the reaction and its effects, but also an up to date summary of current national test procedures and specifications that have been adopted in countries round the world. It also provides illustrative case study investigations of alkali-aggregate reaction in concrete structures encountered in these countries. It has already been noted that many countries and regions were missing from the book published in 1992; this new edition has attempted to address and update this incomplete coverage. As can be seen from the table of contents, most regions and countries round the world are represented in the various chapters so that an almost complete global coverage of the reaction and its effects on concrete structures is now available.

The first five chapters of this book attempt to pull together and synthesise current laboratory and field research findings in the principal areas of importance to the understanding of the mechanisms, causes and effects of the alkali-aggregate reaction in concrete and related materials. Chapter 1 is concerned with the chemistry and mechanisms underlying the reaction including the controversial explanations of the effect relating to the sometimes observed variable expansions of the concrete that results from the reaction and referred to as 'the pessimum proportion' effect. The second chapter provides a generalised review and a systematic approach to estimating risk of alkali-aggregate reaction and the measures required to minimise or eliminate its initiation and expansive effects. It also examines the range of tests and specifications for diagnosis and avoidance of the reaction in current use globally.

The possibility of an expansive reaction involving alkaline concrete pore fluids and impure carbonate aggregates is discussed in detail in Chapter 3. Although the development of de-dolomitisation reaction rims around dolomitic carbonate aggregate particles in a concrete is well documented, it has been unclear as to whether the rare examples of expansion of concretes containing these aggregates is due to variants of an alkali-carbonate reaction, or the result of a cryptic alkali-silica reaction arising from siliceous components in the carbonate aggregate. The evidence and research relating to this matter is examined fully in Chapter 3 which presents the latest findings and conclusions concerning this issue.

Chapter 4 deals with the various methods that are in current use as methods of avoiding, or preventing alkali-aggregate reaction in concrete. It provides an overview of the generally accepted approaches to this problem assembled from all the available expertise and experience that has been gathered from countries worldwide. Chapter 5 follows this, with a detailed global appraisal of methods currently available for diagnosing the reaction in an existing concrete structure and an overview of the repair and management options available for dealing with concrete structures that are affected by expansions caused by reaction of this kind.

The remaining chapters from 6 to 16 review and discuss in detail the particular problems of alkali-aggregate reaction in concrete as experienced by countries round the world. The materials and mechanisms of the reaction affected by the particular factors present in the country or region are presented together with the counter measures and testing regimes they adopt. Case histories are presented to provide practical illustrations of the investigations and the measures used to avoid or mitigate the problem of premature deterioration of concrete structures, due to the reaction, that are currently in use in these countries.

In a few countries there is no information available concerning alkali-reactivity in concrete structures. In a number of others, although the civil engineering organisations are aware of the possibility of alkali-silica reactivity in concrete and have test methods and specifications in place to avoid the use of potentially reactive materials in their concrete, no data relating to case histories have been published. This may reflect the validity of the preventative measures they have put in place, alternatively the expansive reaction may not have been correctly identified in structures, or the details of relevant investigations have not been published.

The countries and regions reviewed in these last eleven chapters are: the UK & Ireland, Nordic Europe, Mainland Europe, Turkey & Cyprus, the Russian Federation and neighbouring Asian Countries, North America, South and Central America, Southern and Central Africa, Japan, China and South-East Asia, Australia and New Zealand, the Indian Sub-Continent and the Middle East and North Africa. Between them they cover almost all the developed and developing world and thus represent the first full global review of instances of alkali-aggregate reaction in concrete.

It has been said that editing a book of chapters written by various authors is like trying to herd cats and we understand the feeling, but this book only exists because of the generosity of our many gifted friends and colleagues, who have selflessly and patiently donated their time and expertise to the cause. We are really very, very grateful and hope that they will be proud of the outcome and can forgive us in due course. They are all recognised in the Contributors & Acknowledgements sections.

This current up to date review with its many expert contributions has been generated from a basis of knowledge and wisdom built up over many years by the expertise of scientists such as Thomas Stanton, who introduced the world to the uncomfortable reality of AAR in 1940. There are of course many other internationally acclaimed engineers and scientists who have since influenced our understanding of AAR and our strategies for minimising its risk and managing affected structures. Amongst the many, perhaps we might mention Gunnar Idorn, who brought AAR awareness to Europe, developed the role of petrography in its study and launched the on-going series of International Conferences on AAR (the ICAARs), which have facilitated the global dissemination of knowledge about AAR and solutions for its control.

Finally, we are indebted to Emeritus Professor R Narayan Swamy for arranging publication of the first (1992) edition of this book and for so encouraging and supporting us to achieve this new edition; it is only to be hoped that the updated and greatly enlarged volume is a worthy successor.

Ian Sims & Alan Poole
February 2017

Editors and Authors

EDITORS

Ian Sims
ian@simsdoc.com / isims@rsk.co.uk

Dr Ian Sims is a Director of RSK Environment Ltd in the UK, where he is responsible for Materials Consultancy and Expert Witness Services. He graduated in geology at Queen Mary College (London University) in 1972 and then undertook doctoral research in concrete technology, including AAR in the British Isles, and his PhD was awarded in 1977. Ian joined Sandberg LLP in London in 1975 and gained wide experience with construction geomaterials. In 1996, he moved to STATS Limited, which joined RSK Group PLC in 2008. He has specialised for over 40 years in concrete, its constituents and all aspects of AAR. Between 1988 and 2014, Ian was Secretary of the RILEM Technical Committees on AAR, when he was awarded RILEM Fellowship. As a Fellow of the Geological Society, he was Secretary for four sequential Engineering Group working parties, producing report-books on

Aggregates, Stone and Clay materials and construction in Hot Deserts, also being an editor for the current edition of Aggregates and for Clays; Ian received the Society's Engineering Group Award and later the Coke Medal. He has served on many other committees, including chairing the editorial panel for ICE's journal 'Construction Materials' and currently chairing the British Standards committee on Aggregates. Ian's publications include 'Concrete Petrography: a handbook of investigative techniques', now in its second edition.

Alan B Poole
abpoole@btinternet.com

Dr Alan Poole followed an academic research and teaching career in the University of London until 2000, as senior lecturer, the director of the Geomaterials Masters course and as supervisor to over 20 PhD students. His research interests were principally concerned with the petrology of civil engineering materials and in the last 30 years he specialised in problems associated with concrete and related materials. He became an acknowledged expert in all aspects of alkali-aggregate reaction in concrete. This expertise has led to his appointment as a consultant to numerous governmental and major civil engineering organisations worldwide. He has been actively involved with the series of International Conferences into Alkali-Aggregate Reaction in Concrete (ICAARs) as UK representative on its International Organising Committee until 2004. He chaired the two UK ICAAR conferences (1976 and 1992) and has contributed numerous papers on AAR to these conferences. His researches, consultancy investigations and association with international experts at conferences has led to his authorship of chapters in a number of technical books on constructional materials, over 120 scientific papers and co-authorship of 'Concrete Petrography' (2016), which deals with all aspects of the investigation of concrete. He is secretary to the Geological Society Engineering Group's 'Applied Petrography Group' (APG) and is also a member of British Standards Institution committees concerned with concretes and aggregates.

AUTHORS

Mark Alexander
mark.alexander@uct.ac.za

Dr Mark G Alexander is Emeritus Professor of Civil Engineering in the University of Cape Town, and a Member of the Concrete Materials and Structural Integrity Research Unit at UCT. He has a PhD from the University of the Witwatersrand, Johannesburg, and is a Fellow of the South African Institution of Civil Engineering, the South African Academy of Engineering, RILEM, and the University of Cape Town. He is a registered Professional Engineer in South Africa. He teaches and researches in cement and concrete materials engineering relating to design and construction, with interests in concrete durability, service life prediction, concrete sustainability, and repair and rehabilitation of deteriorated concrete structures. He is active in international scientific circles and publishes in local and international journals. He is Immediate Past President of RILEM. He acts as a specialist consultant on concrete materials problems. He has co-authored 'Aggregates in Concrete' (2005) and 'Alkali-Aggregate Reaction and Structural Damage to Concrete' (2011), and edited a new book 'Marine Concrete Structures – Design, Durability and Performance' (2016).

Özge Andiç-Çakır
ozge.andic@ege.edu.tr

Dr Özge Andiç-Çakır is currently Associate Professor in the Civil Engineering Department at Ege University in Izmir, Turkey, with a PhD in materials of construction. She has been the supervisor of many national projects in the field of building materials and sustainable materials technologies for eight years. Her research interests are functional concrete design and technology, aggregates, bio-inspired materials, energy efficient and sustainable building materials. Özge supervises the Imaging and Microstructural Analysis Laboratory at Ege. She has participated in several national and international projects, publishing more than 70 scientific papers. She also has experience in technology transfer, organization and development of dissemination activities like conferences, seminars and workshops.

Karin Appelquist
karin.appelquist@ri.se

Dr Karin Appelquist has been working at the Swedish Cement and Concrete Research Institute (CBI) since 2010. She has a PhD from Gothenburg University in Mineralogy and Petrology (2010). Her main activities at CBI are aggregate and concrete petrography and she is responsible for ASR expansion methods in Borås. She is also a member of RILEM TC 258-AAA (2014–2019).

Geoff Blight
(sadly deceased)

The late Professor Geoffrey E. Blight obtained his BSc Degree in Civil Engineering from the University of the Witwatersrand (Wits) in 1955, and an MSc (Eng) degree from Wits in 1958. In 1961, he obtained his PhD at Imperial College, London, for work on soil mechanics, and then, in 1975, a DSc(Eng) from London University for work on geotechnical engineering. This was followed by a DSc(Eng) from Wits University in 1985 for innovative work in materials engineering. In 1993, he was awarded a DSc(Eng) from the University of Cape Town for his work on developing design information from *in situ* tests. Wits University awarded him a D.Eng in 2001 for the contribution of his research work in developing and changing civil engineering practice.

Geoff was Professor of Civil Engineering at Wits University from 1969 until his retirement in 2002. From then until his death in 2013, he was an Honorary Professorial Research Fellow in the School of Civil and Environmental Engineering at Wits, and he continued to undertake research and supervision of postgraduate students. Among numerous achievements, his work stands out for contributions in the fields of soil mechanics and geotechnical engineering, and on the structural stability of concrete deteriorated by alkali-silica reaction, both of which represent unique contributions to these fields of study internationally.

Mario de Rooij
mario.derooij@tno.nl

Dr Mario R de Rooij is a Senior Project Manager at TNO in Delft, The Netherlands. He received his training as a materials scientist and engineer at Delft University of Technology. After his Masters (1994) in the field of inorganic chemistry, he specialized in cement chemistry and microscopy during a PhD in Civil Engineering, also at Delft University of Technology (2000) and a Post-Doc period at Northwestern University (2000), Illinois, USA. He continued his career in the field of concrete durability working for TNO in the Buildings Materials Group, focusing on the remaining service life of existing concrete structures. For a period of six years (2002–2008) he was part-time Assistant Professor at Delft University of Technology, teaching microstructure development of cement-based materials and microscopy. It was during this time that he initiated the RILEM Concrete Microscopy Course in Delft for PhD level students. Returning full-time to TNO in 2008, his interest shifted to re-use and adding value to secondary materials for building materials purposes, including alkali-activated materials.

Eduardo Fairbairn
eduardo@coc.ufrj.br

Dr Eduardo M. R. Fairbairn is a Professor in Civil and Environmental Engineering at the Federal University of Rio de Janeiro, Brazil. He has a PhD in Mechanics applied to Constructions from the Université de Paris VI in France. On sabbatical leave, he worked as a researcher at Laboratoire Central de Ponts et Chaussées (now IFSTTAR) and held a senior internship at Arizona State University in the area of nanotechnology. His main interests are in numerical modelling and experimental analysis of concrete structures, from oil well slurries to massive concrete and sustainable cementitious materials. He is currently the chairman of the RILEM technical committee TC-254 'Thermal Cracking of Massive Concrete Structures'. Prof. Fairbairn has served as a consultant for government agencies and private corporations in the fields of electrical energy generation and the oil and gas industry. He has coordinated several projects for numerical modelling and experimental analysis of alkali-aggregate reaction (AAR) and has also supervised several PhD and MSc theses in the use of fibres and pozzolanic materials for AAR mitigation.

Vyatcheslav Falikman
vfalikman@yahoo.com

Professor Vyatcheslav R Falikman is Head of the Sector for Structural Concrete Durability at the Scientific Research Institute for Concrete and Reinforced Concrete after A.A. Gvozdev (NIIZhB), and Professor at the Moscow State University of Civil Engineering (MSUCE) – State National Research University. He is a full member of the Russian Engineering Academy and International Academy of Engineering (IAE). Vyatcheslav is the Russian Federation Government Prize winner in the field of science and technology, The Honorary Builder of Russia and The Engineering Merit Award winner. Member of DAC in RILEM, the Regional Convener of the RILEM in East Europe and Central Asia. He is a RILEM Fellow, also a Member of ACI and Member of ACI Technical committees on material science and nanotechnologies in concrete, and Head of Russian National delegation in fib. Vyatcheslav is Vice-President of IAE, Russtandard representative in ISO committees, and First Vice-President of the Russian Structural Concrete Association. He has authored more than 350 research papers and 70 patents. Research interests include chemical admixtures for concrete and mortars, special binders and concrete, problems of concrete durability, nanotechnologies in construction, and sustainable development.

Isabel Fernandes
mifernandes@fc.ul.pt

Dr Isabel Fernandes holds an MSc degree in Engineering Geology and a PhD in Geology of Alkali-Silica Reaction (ASR) and the evaluation of aggregates for concrete. Currently, Isabel is Assistant Professor at the University of Lisbon in Portugal, teaching courses of Engineering Geology and Rock Mechanics. She is co-author of 3 books, 10 book chapters, and about 100 papers published in scientific journals and international conferences. In addition to her lecturing duties, she supervises a number of MSc and PhD students with theses covering different topics of engineering geology including the evaluation of aggregates for concrete. As a member of RILEM Technical Committee 219-ACS from 2008 to 2014, Dr Fernandes was a leading participant in compilation of the recommendation 'AAR-1.1, *Detection of Potential Alkali-Reactivity – Part 1: Petrographic Examination Method*', included in the 2016 book '*RILEM Recommendations for the Prevention of Damage by Alkali-Aggregate Reactions in New Concrete Structures*'. She was author and editor of the companion volume,

'*Recommended Guidance AAR-1.2, Petrographic Atlas*'. This book, published in 2016, presents the main reactive aggregates used worldwide for which 70 samples were collated from 25 countries in all continents. Isabel also develops R&D activities with commercial companies in the areas of aggregate characterization, diagnosis of concrete deterioration mechanisms by petrographic methods and geologic and geotechnical mapping of rock masses for large dams, reservoirs and tunnels, with about 90 unpublished reports.

Miguel Ferreira
miguel.ferreira@vtt.fi

Dr Miguel Ferreira is an expert on deterioration mechanisms of mineral-based building materials; service life design, assessment, durability modelling of reinforced concrete structures; and ageing management systems. He is an active member in the international Technical Groups: RILEM, FIB, and ICIC, as well as the Finnish Concrete Association. Miguel has authored more than 100 peer-reviewed scientific publications. He has also supervised more than 20 MSc theses and co-supervised 1 PhD thesis.

Kevin Folliard
folliard@mail.utexas.edu

Dr Kevin J. Folliard is a Professor and Austin Industries Endowed Teaching Fellow in the Department of Civil, Architectural, and Environmental Engineer at the University of Texas at Austin, where he has been on the faculty since 1999. Prior to this, he was an Assistant Professor at the University of Delaware from 1997–1999 and a Senior Research Engineer at W.R. Grace & Co. from 1995–1997. Dr Folliard received his PhD in Civil Engineering from the University of California at Berkeley in 1995. His main research interest is in the area of the durability of Portland cement concrete and he teaches courses related to civil engineering materials, concrete technology and concrete durability. Dr Folliard is a Fellow of the American Concrete Institute (ACI), and he received the ACI Young Member Award for Professional Achievement in 2003 and the ACI Wason Medal for Materials Research in 2010 and 2015. He received the Ervin S. Perry Student Appreciation Award four times (2001, 2002, 2003, 2009) while at the University of Texas at Austin, and he also received the College of Engineering Award for Outstanding Teaching by an Assistant Professor in 2004. In 2013, Dr Folliard was honoured with the highest teaching award given by the University of Texas System, the Regents' Outstanding Teaching Award. He has been the Principal Investigator on over $15 million in research projects while on the faculty at the University of Texas at Austin, primarily in the area of concrete durability (especially alkali-silica reaction) and in more recent years on the use of alternative binders for rapid repair applications. Dr Folliard has authored or co-authored over 150 technical publications in his career to date, including more than 70 refereed journal papers.

Benoit Fournier
benoit.fournier@ggl.ulaval.ca

Dr Benoit Fournier is Professor in the Department of Geology and Engineering Geology at Laval University in Québec City (Québec, Canada), from where he obtained his PhD in 1993. From 1990 to 2007, he worked as a research scientist (1990–1998) and manager (1998–2007) for the Advanced Concrete Technology Program of CANMET, Department of Natural Resources (Ottawa, Canada). Benoit's main research interests are in the various aspects of aggregate technology (natural and recycled), recycling and sustainable development in concrete construction, durability of concrete, especially issues related to alkali-aggregate reaction in concrete and concrete incorporating supplementary cementing materials (SCMs). He is currently the Chair of the Canadian Standards Association (CSA) technical subcommittee on Aggregate Reactions in Concrete, and is also a Work Package Leader for RILEM technical committee TC

258-AAA (Avoiding alkali aggregate reactions in concrete – Performance based concept). He was awarded the CSA Order of Merit in 2008.

Sue Freitag
sue.freitag@opus.co.nz

Ms Sue A Freitag is a concrete materials specialist at Opus International Consultants Ltd, New Zealand. She has over 35 years' experience in testing, applied research, and consultancy services relating to concrete materials, durability, asset performance, and compliance issues in New Zealand, including the evaluation of AAR in existing structures and managing the risk of AAR in new structures. In 2009 Sue was awarded the Kevin Stark Memorial Award by Engineers Australia Coastal & Ocean Engineering for a conference paper demonstrating a strong multidisciplinary approach for identifying a significant concrete durability problem in two marine structures. Sue participates in industry forums in New Zealand and Australia, is a member of RILEM, and contributes to the development of New Zealand Standards relating to concrete materials.

Colin Giebson
colin.giebson@uni-weimar.de

Dr Colin Giebson completed his Diploma's degree in Civil Engineering at the Bauhaus-University Weimar (Germany) in 2004. Afterwards, he started as scientific co-worker at the F.A. Finger-Institute (FIB) of the Bauhaus-University Weimar. His work has been highly focused on ASR in concrete. Owing to the large number of ASR-affected concrete pavements since the late 1990s, main topics have been the influence of external alkalis on ASR as well as the development of a realistic ASR performance test method. For his PhD thesis (2013) about ASR in concrete pavements he received the CEMEX (Germany) concrete award 2014. He was author and co-author of many papers, has supervised numerous Bachelor and Master Theses and, since 2006, has been a member of the RILEM Technical Committees TC 219-ACS and TC 258-AAA.

Bruno Godart
bruno.godart@ifsttar.fr

M. Bruno Godart is currently the Deputy Director of the Materials and Structures Department at IFSTTAR (French Institute of Science and Technology for Transport, Development and Networks), a Public Establishment supported by the Ministry of Transportation and the Ministry of Research. Bruno received a CE diploma from the Ecole Nationale des Travaux Publics de l'Etat in France in 1975, and a Master of Science from Stanford University in 1979. The first part of his career was at the Laboratoire Central des Ponts et Chaussées (LCPC – Central Laboratory for Roads and Bridges), where his last position was Technical Director for Bridges. Bruno is the author of many publications in the field of pathology, investigations, repair, strengthening, durability and management of bridges. He leads investigations on structures such as bridges, dams, nuclear power plants, and participates actively in the development of the French and European technical doctrines. He is the co-author of the French Recommendations for preventing disorders in Structures due to AAR (1994) and due to DEF (2007). Bruno is also the President of the French Civil Engineering Association (AFGC), the head of the Prestressing Sector Committee of the French Association for the Qualification of Prestressing and Equipments for Buildings and Civil Works (ASQPE), Senior Lecturer at the Ecole Nationale des Ponts et Chaussées (ENPC) and is involved in several international associations, including IABSE, RILEM and IABMAS.

Paddy Grattan-Bellew
p.grattan-bellew@sympatico.ca

Dr Patrick E. Grattan-Bellew retired in 1998 and is now President of Materials & Petrographic Research G-B Inc. in Ottawa, Canada. Prior to retirement, from 1971, he was a research officer with the materials group in the Institute for Research in Construction, where he conducted research into a wide variety of problems relating mostly to the durability of building materials. However, his main area of interest, for the past 30 years, has been alkali-aggregate reaction (AAR), both alkali-carbonate reaction (ACR) and alkali-silica reaction (ASR). During this time he developed test methods for evaluating the potential reactivity of aggregates and contributed to the development of Canadian Standards relating to AAR. Paddy had graduated in Science from University College, Dublin, Ireland, before obtaining an MSc in Exploration Geology from McGill University, Montreal, Canada and then, in 1968, a PhD from Cambridge, England, on phase transition mineralogy. After Cambridge, he joined the National Research Council of Canada, in Ottawa. Since 1998, Dr Grattan-Bellew has specialized mostly in the petrographic evaluation of concrete from large dams affected by ASR and has developed a method for the quantitative evaluation of concrete affected by ASR, the 'Damage Rating Index' method. He has edited and contributed chapters to various books and, in addition, has over 100 publications in technical journals, has made numerous presentations at international conferences and is a member of the American Concrete Institute.

Bent Grelk
bng@ramboll.dk

Mr Bent Grelk has been involved in work related to buildings, building materials and concrete structures for more than 30 years. This has included work in the field of concrete technology, research in analysis of concrete and aggregate, studies of the impact of climate and aggressive exposures on concrete and the testing of concrete, aggregate and mortar. He is an experienced engineer specialized within the fields of concrete technology, condition surveying of concrete structures, concrete and aggregate testing. He has been involved in a number of EU funded research projects. Bent managed several activities in the 'PARTNER' project (ASR in concrete). Mr Grelk is also an expert witness appointed by the Danish Court of Arbitration in cases concerning problems with natural stone and concrete, as well as being a member of RILEM TC 258-AAA (2014–2019).

Erika Holt
Erika.Holt@vtt.fi

Dr Erika Holt has a PhD in Civil Engineering (2001, University of Washington, Seattle, USA) and has 20 years of research experience at VTT in Finland. Her areas of interest include concrete materials, durability, sustainability practices and applications of concrete for major infrastructure. She currently serves as VTT's coordinator for safe and sustainable nuclear energy research. She has over 60 publications and is active in RILEM, ACI and the Finnish Concrete Association. Her latest concrete projects are for smart city applications of pervious pavements for storm water management.

R. Doug Hooton
hooton@civ.utoronto.ca

A Professional Engineer in the Province of Ontario, Dr R. Douglas Hooton is a Professor, and NSERC/Cement Association of Canada, Senior Industrial Research Chair in *Concrete Durability and Sustainability*, in the Department of Civil Engineering at the University of Toronto, where he has taught for 30 years. Prior to that he was a research engineer at Ontario Hydro where he worked on construction and maintenance of hydroelectric and nuclear power stations. His research has focused on the fundamentals of durability performance of cementitious materials in concrete as well as on performance tests and specifications. Doug is a member of RILEM TC258 AAA Avoiding Alkali-Aggregate Reactions in Concrete, and the previous TC 219 ACS Alkali-Aggregate Reactions in Concrete Structures. He chairs American Concrete Institute committee C233 on Slag Cement, ACI C130A on Sustainability of Materials, is Secretary of ACI C201 on Durability and serves on the ACI Board of Directors. Also, Doug is Secretary of ASTM C09.50 on Risk Management for Alkali-Aggregate Reactions, Chair of C01.29 on Sulfate Resistance and vice-Chair of C01 on Hydraulic Cements. He chairs the CSA A3000 Committee on Cementitious Materials, is vice-chair of CSA A23.1 on Concrete, and a member of the CSA subcommittee on Alkali-Aggregate Reactions since 1986.

Jason H Ideker
jason.ideker@oregonstate.edu

Dr Jason Ideker is an Associate Professor at Oregon State University and Co-Director of the Green Building Materials Laboratory. He holds a B.S. in Civil Engineering from The Georgia Institute of Technology and an M.S.E and Ph.D. from The University of Texas at Austin. Jason's main research areas are in service-life of concrete with a focus on early-age behaviour of high performance cementitious materials, mitigation and test methods for alkali-silica reaction and durability of calcium aluminate cements. He has been working in the research area of alkali-silica reaction since 2001. Dr. Ideker and his group do transformational research, whereby results are implemented into improved test methods and specification development. His work spans micro- to macro- scale, from addressing fundamental reaction and mitigation mechanisms to diagnosis and repair of structures suffering from ASR. He has authored over 75 technical articles and reports. Dr. Ideker is a member of ACI Committees 201, 231 and 236, ASTM C01 and C09, and serves on the Executive Board of C09. He chairs ASTM Subcommittee

C09.50 – 'Risk Management for Alkali-Aggregate Reactions'. Jason is a recipient of the ACI Young Member Award for Professional Achievement, and a 3-time recipient of the PCA Education Foundation Fellowship. He started the international "Corvallis Workshops", bringing together researchers to improve concrete performance. The third workshop will be held in July of 2017 entitled: 'Service-Life Prediction for Concrete'.

Tetsuya Katayama
tetsuya_katayama@taiheiyo-c.co.jp

Dr Tetsuya Katayama is a geologist and petrographer with Taiheiyo Consultants in Japan, working in the field of concrete durability and evaluation of cementing materials in historic mortar and concrete, after working at Kawasaki Geological Engineering for 18 years and Sumitomo Osaka Cement for 16 years. He is a regular participant in the International Conferences on Alkali-Aggregate Reaction (ICAARs) since 1986 and is active in RILEM TC 259-ISR, 'Prognosis of ASR'. Tetsuya received a Master of Science degree with a study of geology and dolomitisation of the Pliocene Shirahama formation, and Doctor of Science with his lifelong petrographic study of alkali-aggregate reactions from the University of Tokyo. He conducted international co-operative studies on AAR with many countries including Australia, Austria, Canada, Iceland, New Zealand, Portugal, Turkey and the USA, plus researches on cementitious binding materials with France and the UK. Dr Katayama's comprehensive investigations into the mechanism of 'so-called ACR' have been internationally accepted. He received the first Gunnar M Idorn award from the 13th ICAAR in Trondheim, Norway in 2008. Tetsuya also received the best paper award from the 8th International Conference on Road and Airfield Pavement Technology, Taipei, Taiwan, 2013. In the past decades, Tetsuya has developed several important methodologies in concrete petrography, including microscopical assessment of the stage of ASR in concrete, quantitative EDS analysis of ASR and CSH gels, so-called 'Katayama plot' of gel compositions on his [Ca/Si]-[Ca]/[Na+K] diagram, and estimation of alkali budgets in concrete. His petrographic method was adopted for the Japanese national guideline by JNES (now NRA) in 2014, and came into use in diagnosing ASR in existing nuclear power plants in Japan.

Ted Kay

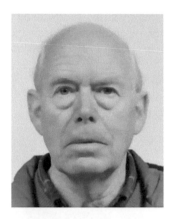

Ted Kay is an Engineer who is now a semi-retired advisor on all matters relating to concrete. Most of his career was spent with Civil Engineering consultants including Halcrow, now subsumed into CH2M Hill Inc., and Travers Morgan, now subsumed into Capita Property and Infrastructure. He also spent a four year spell with the Concrete Advisory Service of the Concrete Society. For most of his career, Ted has specialised in the Middle Eastern region, spending a six year period based in Dubai. During this time, he was part of a team that was one of the first to investigate the concrete problems in the region at that time and to implement the first specifications to include provision for AAR testing of aggregates. Ted Kay was involved with most of the keynote technical reports on developing concrete technology appropriate to the extremely aggressive local environment.

Markku Leivo
markku.leivo@vtt.fi

Dr Markku Leivo is a Principal Scientist at VTT in Finland and has 30 years' experience of building materials research. His main research topics have been concrete material properties, mix-design, durability and use of secondary cementitious materials. Markku has been involved in many EU-funded research projects as participant and/or project manager. He is an author of tens of scientific papers and presentations from 1985 onwards.

Jan Lindgård
jan.lindgard@sintef.no

Dr Jan Lindgård is a Senior Scientist at SINTEF Building and Infrastructure, where he has the professional responsibility for testing and research related to alkali-silica reaction (ASR). Since the early 1990s, he has been involved in research projects related to various ASR issues, as well as assessment of concrete structures affected by ASR. Jan's experience includes field surveys, moisture condition assessment and various laboratory analyses for evaluation of different deterioration mechanisms. For more than 20 years he has been involved in the development of ASR test methods, both on a national and a global scale. He is now the Secretary of RILEM TC 258-AAA (2014–2019). In 2013, Jan defended his PhD project related to reliability of ASR performance test methods and his findings made a significant contribution to the guidance provided by the predecessor RILEM TC 219-ACS. In additional to a wide range of research reports, he has more than 50 publications. Jan also performs peer-review work for several journals.

Toyoaki Miyagawa
miyagawa.toyoaki.6z@kyoto-u.ac.jp

Toyoaki Miyagawa is now Emeritus Professor at Kyoto University in Japan. He originally graduated in civil engineering from Kyoto University in 1973, then obtaining a Master's degree in 1975. He remained at Kyoto University, Department of Civil Engineering, initially as a Research Associate (1975 to 1989), then as a Lecturer (1989 to 1991), Associate Professor (1991 to 1998) and finally Professor (1998 to 2003). Professor Miyagawa then moved to the Department of Civil and Earth Resources Engineering at Kyoto (2003 to 2015), before becoming a Specially Appointed Professor in the Infra-System Management Research Unit at Kyoto (from 2015). Toyo was also a Visiting Professor at Aston University in Birmingham, UK, in 1991 & 1992. His research interests include performance over time, diagnosis, prognosis, sustainability and rehabilitation of concrete structures damaged by ASR and/or chloride induced corrosion. Professor Miyagawa has published many papers about the fracture of reinforcing steel in concrete structures damaged by ASR, plus their assessment and repair. He was also Chairman of the Japan Society of Civil Engineers' Committees on 'Countermeasures for Damage due to ASR' (2003 to 2005) and 'Concrete' (2007 to 2011).

Ajoy K Mullick
ajoy_mullick@rediffmail.com

Dr. Ajoy K. Mullick is former Director General of the National Council for Cement and Building Materials (NCB) in India and taught Civil Engineering at JP University of Engineering and Technology, Guna (Madhya Pradesh). He has spent more than five decades in research, teaching, design and consultancy, devoted to propagation of sustainable concrete practices. At present, he is a private Consultant to various Infrastructure Projects throughout the country. Ajoy graduated in Civil Engineering from Patna University, India, and then obtained Master's and PhD degrees in Civil Engineering, both from the University of Calgary, Canada. Most of his research on AAR in India has been conducted at NCB. Ajoy's other research interests are durability, high performance concrete with ternary binder systems, sustainability in construction, and waste utilization including C&D Wastes. Dr Mullick has authored 170 papers, three books and three book chapters; and is co-inventor of six patents.

Philip Nixon

Dr Philip J Nixon retired from the UK Building Research Establishment (BRE) in 2004 as Technical Director in the Centre for Concrete Construction. During his career there he had overseen several investigations into chemical problems affecting concrete, including alkali-aggregate reactivity (AAR) and the thaumasite form of sulphate attack (TSA). He was also the chairman of the BSI Committee on aggregates and the CEN Committee on aggregates for concrete. Philip joined BRE after completing a doctorate in Mineral Engineering at the Royal School of Mines, which is part of Imperial College, London, when he studied chemical methods of extracting tin from low-grade ores. This PhD research followed his graduation in chemistry from The Royal College of Science, Imperial College. Dr Nixon initiated work on AAR within RILEM, chairing the first RILEM TC on this subject, TC 106, which was set up in 1988 specifically to address the problems arising from the proliferation of test methods for aggregate reactivity. He continued to chair the succeeding committees, TC 191-ARP and TC 219-ACS, which expanded the scope of their work to include the development of integrated assessment, specification, diagnosis and procedures for appraising and managing affected structures. He also headed an EU funded project, PARTNER, which assessed the RILEM methods as the potential basis for European Standard (EN) methods.

Nikolai K Rozentahl

Dr Nikolai Konstantinovich Rosenthal is currently Head of the sector for Concrete Corrosion at the Scientific Research Institute for Concrete and Reinforced Concrete after A A Gvozdev (NIIZhB) in Moscow, Russia, where he has undertaken research into precast concrete and concrete structures since 1962. Nikolai graduated from Moscow State University of Civil Engineering in 1958 and was awarded a PhD Eng from NIIZhB in 1970, and later a DSc Eng in 2005, also from NIIZhB. Dr Rosenthal's research interests include corrosion and protection of steel reinforcement in concrete; corrosion and protection of concrete in aggressive environments; concrete and concrete products; diagnostics of corrosion state of concrete and reinforced concrete in structures and buildings. He is recognized in Russia as especially having long experience and expertise in respect to AAR. Dr Rosenthal is a RILEM senior member. He is the author or co-author of 4 monographs, 293 papers and reports, 12 norms and standards, and 24 patents.

Katrin Seyfarth
katrin.seyfarth@uni-weimar.de

Katrin Seyfarth is currently leading (since 2010) the ASR team at the F.A. Finger-Institute (FIB) of the Bauhaus-University, Weimar, in Germany, which is involved in the work of several German boards that prepare national guidelines and recommendations to prevent ASR. She completed her Diploma's degree in Building Materials and Process Engineering at the Bauhaus-University Weimar in 1985. Katrin spent the following year at the Building Academy in Berlin, before she returned to Weimar and started as scientific co-worker at FIB of the Bauhaus-University Weimar. Her work has been focused on the durability of concrete, especially related to ASR and Delayed Ettringite Formation (DEF). A key aspect of her research has been the development of a realistic performance test method for concrete, suitable for different durability issues. Owing to the large number of ASR-affected concrete pavements since the late 1990s, the activities focused on ASR since 2004. Katrin has been author and co-author of many papers, has supervised numerous Bachelor and Master's theses and, since 2006, has been a member of the RILEM Technical Committees TC 219-ACS and TC 258-AAA.

Ahmad Shayan
ahmad.shayan@arrb.com.au

Dr Ahmad Shayan is a Chief Research Scientist at the Australian Road Research Board (ARRB), where he leads the AAR team. Before joining ARRB in 1997, Ahmad was at CSIRO in Australia from 1980. Ahmad originally graduated in 1968 from the University of Ahwaz, Iran, in Agricultural Engineering, majoring in Soils. He subsequently completed a Post-graduate Diploma in Soil Science at the University of Sydney, Australia, in 1974 and a PhD degree in 1977, followed by three years as Assistant Professor at the University of Technology, Isfahan, Iran. Dr Shayan now has around 37 years of experience in research and consulting on the various aspects of concrete technology, deterioration of concrete structures, and methods of preventing damage to reinforced concrete structures. Ahmad's research interests include AAR, DEF, other degradation mechanisms of concrete structures such as sulfate attack, utilisation of fibre-reinforced polymers to confine AAR-affected elements in concrete structures, steam cured concrete, corrosion of steel reinforcement in concrete under aggressive environments and utilisation of stainless steel, waste materials utilisation in concrete, and geopolymer concrete. Dr Shayan has written over 500 papers and technical reports. He was chairman, Standards Australia Committee CE/12 for over 10 years, has served on the International Organising Committee for the ICAAR conferences, and was chairman of the 10th ICAAR in Melbourne, 1996. Ahmad was a member of the RILEM Technical Committees on AAR (1988 to 2014). He won the prestigious Clunies Ross Medal in 2003 for his work on durability of concrete structures and for contribution to the economic well-being of Australia. Dr Shayan currently holds the Adjunct Professor position in the Departments of Civil Engineering at Monash and Swinburne Universities, both in Melbourne, Australia.

Michael D A Thomas
mdat@unb.ca

Dr Michael Thomas is a Professor and Chair in the Department of Civil Engineering at the University of New Brunswick (UNB), Canada, and a registered Professional Engineer in New Brunswick. He graduated in civil engineering from the University of Nottingham, UK, in 1982 and obtained a PhD from the University of Aston, in Birmingham, UK, in 1987. Michael has been working in cement and concrete research since 1983. Prior to joining UNB in 2002, he had been on faculty at the University of

Toronto since 1994 and previous to this he worked as a concrete materials engineer with Ontario Hydro in Canada and as a research fellow with the Building Research Establishment (BRE) in the UK. Dr Thomas' main research interests are concrete durability and the use of industrial by-products including pozzolana and slag. His studies on durability have included AAR, DEF, sulfate attack, de-icer salt scaling, carbonation, chloride ingress and embedded steel corrosion; he is also active in the areas of service-life modelling and the repair and maintenance of concrete structures. He has authored more than 200 papers and reports, including the book 'Supplementary Cementing Materials in Concrete'; he is also co-author of the service life model, 'Life-365'. He is active on technical committees within the American Concrete Institute (ACI), ASTM, RILEM and the Canadian Standards Association (Awarded Order of Merit in 2010). He was a recipient of the ACI's Wason Medal for Materials Research in 1997, 2009 and 2014, the ACI Construction Practice Award in 2001 and was elected to an ACI Fellow in 2006. Michael is also a Fellow of the UK's Institute of Concrete Technology. Dr Thomas is additionally President of C&CS Atlantic Inc. and provides consulting services and expert testimony on concrete materials, durability and rehabilitation.

Jan Trägårdh
jan.tragardh@ri.se

Jan Trägårdh has been working as a researcher at the Swedish Cement and Concrete Research Institute (CBI) since 1991. His main areas of research have been concrete durability (AAR, sulfate resistance, leaching) and the microstructure of concrete related to mix design. Since 2002, Jan is now leading a research department focusing on sustainable materials in concrete mix design, LCA and durability. He graduated in mineralogy & petrology in 1980, then an MSc in 1982 and a research degree (Phil.lic.) in 1989, all from the University of Lund, Sweden. He has been active in RILEM Technical committees including 'Durability of Self-Compacting Concretes', 'Recycling of Concrete' and 'Modelling of AAR in Concrete'. He has also been involved in EU-projects (CONTECVET, BRITE-EURAM, FISSAC).

Børge J Wigum
borge.j.wigum@ntnu.no

Professor Børge Johannes Wigum is an engineering geologist, appointed as a Senior Project Manager of Development and Application at HeidelbergCement Northern Europe, based in Norway and Iceland. Since 2007, he has also been Adjunct Professor at the Norwegian University of Science and Technology (NTNU), in Trondheim, Norway, where he teaches aggregate production and utilization, in addition to supervising Master's and PhD students, in the Department of Geoscience and Petroleum. Børge finished his own PhD at the same university in 1995, where he studied Alkali-Aggregate Reactions (AAR) in concrete, and the properties, classification and testing of Norwegian cataclastic rocks. During his PhD study he spent one year at Queen Mary and Westfield College, part of London University, UK, where he was supervised by the late Dr William (Bill) J. French. Professor Wigum's major field of interest is the production and utilization of aggregates as building materials in civil engineering structures. Research activities have particularly been associated with AAR in Concrete. This includes research and development of new test methods regarding petrographic analyses and laboratory expansion tests (i.e. mortar bar tests and concrete prism tests), and production of manufactured sand for use in concrete. Børge was a member of the RILEM technical committees on AAR that concluded in 2014, since when he has been Chairman of the current RILEM TC 258-AAA, which is tasked with developing an AAR performance test procedure.

Kazuo Yamada
yamada.kazuo@nies.go.jp

Dr Kazuo Yamada is a Research Fellow in the Radiological Contaminated Off-site Waste Management Section of the National Institute for Environmental Studies, in Fukushima, Japan. He graduated in geology from the University of Tokyo in 1986, from where he also obtained a Master's degree in 1988. Afterwards, Kazuo undertook work within the Central Research Laboratory of Onoda Cement (now Taiheiyo Cement), where he became Manager in 1996. In 2000 he obtained his doctorate from Tokyo Institute of Technology, on the working mechanism of polycarboxylate-type superplasticizers. In 2012, Dr Yamada joined the National Institute for Environmental Studies, initially as a Senior Researcher (Research Fellow from 2014), and moving to Fukushima in 2016.

Acknowledgments

This state of the art review, 'Alkali-Aggregate Reaction in Concrete: A world review', would not have been possible but for the willing co-operation and unstinting efforts of a large number of international experts who have given freely of their time to make this book a reality. The two editors are greatly indebted to all these people whether their particular role has been as authors to chapters, contributors enhancing the content of particular chapters, or as advisors who have assisted both authors and editors with ensuring that the information presented is both accurate and as up to date as possible.

The chapter authors and co-authors are separately listed, with their short biographies, in the preceding section. Clearly, all these chapter authors and co-authors, with their expertise, have been essential in the production of this world review. The editors wish gratefully to acknowledge and thank these authors for their generous and invaluable assistance in compiling this book.

In addition, many other expert scientists and engineers have provided invaluable support and advice, or have supplied additional material, photographs and also case history illustrations. All this support has been freely provided and is gratefully acknowledged by both the chapter authors and the editors. The material they provided has greatly enhanced the coverage and the value of this book, so the editors offer them their sincere thanks and have attempted to compile the following list of those who have assisted them with this enterprise. If the editors have mistakenly missed anyone from this list they most humbly apologise for the omission and hope they may be forgiven.

Dr David Crofts (RSK Environment Ltd, UK)
Prof Sidney Diamond (Lafayette, IN, USA)
Mike Eden (Sandberg LLP, UK)
Prof Isabel Fernandes (University of Lisbon, Portugal)
James Ferrari (RSK Environment Ltd, UK)
Jeremy Ingham (Mott Macdonald Ltd, UK)
Peter Laugesen (Pelcon Materials & Testing, Denmark)
Dr Peter Mason (Consultant, UK)
Dr Chris Rogers (Consultant, Canada)
Dr Ted Sibbick (GCP Applied Technologies Inc. – formerly Grace, Cambridge, MA, USA)
Statkraft UK Ltd (London, UK)
Bradley Staniforth (RSK Environment Ltd, UK)
Prof Michael Thomas (University of New Brunswick, Canada)
Dr Graham West (Retired, ex TRL)

Prof Jonathan Wood (Consultant, UK)

This book could not have been completed without the extensive, patient, practical and technical expertise of Claire Bennett, of RSK Environment Ltd, UK, whose assistance the editors especially need to recognise and for which they wish to express their sincere gratitude. The editors also thank the members of the publishing team at the Taylor & Francis Group, in particular Janjaap Blom and Lukas Goosen. The editors are indebted to all these people for their practical help, advice and encouragement.

Finally, Ian Sims and Alan Poole are aware of the stress and inconvenience that working on book projects imposes on their families and take this opportunity to express their great appreciation for the generous support and tolerance that they have experienced in respect of this book.

Chapter 1

Introduction, Chemistry and Mechanisms

Alan B. Poole

1.1 BACKGROUND

Several varieties of concrete were used as construction materials by the Romans. The development of concrete based on hydraulic cement similar to that used at the present time dates from the 19th Century. Since then a great number of technological improvements have occurred, extending the varieties of cements and hence concretes available today. In one form or another, concrete has become an essential and ubiquitous constructional material across the modern world.

Along with the technological refinements, a basic scientific understanding of the nature of this material has been developed, including the realisation that certain problems can and do occur, some of which can lead to its premature deterioration and failure.

One such problem arises from an injudicious selection of aggregate and cement types when designing and mixing the concrete. Although the concrete may initially meet all the required specifications of strength development and quality, after several years of service, it has been found that a few concretes begin to expand and crack, necessitating expensive remedial work or replacement. The particular problem involved is referred to as Alkali-Aggregate Reaction (AAR) and constitutes the subject of the present volume. This book updates and replaces the first edition entitled 'The Alkali-Silica Reaction in Concrete' (Editor: R.N. Swamy, 1992) and offers a greatly extended world coverage of AAR, including a global review of the problem and some of the possible solutions currently available.

AAR in concrete was first observed and identified in North America during the late 1930s and the first comprehensive scientific investigations describing the reaction and its effects were published by Stanton in December, 1940. Stanton investigated a number of occurrences of expansion and cracking leading to failures in concrete pavements in California, and found that the problem developed only when certain types of mineral components were present in the aggregates, and only when cement alkalis exceeded some minimum threshold percentage concentration. The reaction between the aggregate and the alkali gave rise to the name "alkali-aggregate reaction". His findings encompassed a great deal of detailed information regarding the nature and practical significance of the reactions taking place, and included early versions of test methods that later became standards, and of methods of preventing or at least mitigating the results of the reactions. In a recent appreciation, Thomas (2011) suggested that the impact of Stanton's findings could not be overstated. Prior to his publication, aggregate was considered to be an inert filler, but afterwards it was realized that some aggregates were not suitable for use in concrete without taking special precautionary measures. Of course, Stanton's findings that cement alkalis could influence the durability of concrete caused great concern for the cement manufacturing companies and perhaps the most disturbing implication of Stanton's study was that concrete could *"fail... even if it is exposed to only normal curing and weathering conditions"*.

Since this early work a tremendous volume of research into causes, effects, avoidance and possible solutions to the problem of AAR in concrete has been undertaken. This work has been driven in part by the major costs involved in making good the damage AAR can cause, but also by the world-wide extent of the problem, AAR being found to occur in so many countries (see Figure 1.1). Despite the fact that the actual occurrence of damaging AAR in concrete structures is relatively rare and seemingly spasmodic, the repair or replacement of affected structures involves a major expense for their owners.

Most of the research into this problem has taken place since the Second World War, giving rise to an extensive technical literature numbering several thousand research papers and technical reports. An important source of information concerning this work has been in the form of the proceedings of a series of International Conferences on AAR ('ICAARs'). The first of these was held in Copenhagen in 1974 and the 15th was held in Sao Paolo, Brazil in 2016. In addition, numerous national specifications and guidance documents for avoiding AAR have been published and will be referred to in later chapters of his book.

This second edition offers an up-to-date general overview of research into AAR in concrete. The coverage includes methods of diagnosis and testing (Chapter 2), a discussion of 'alkali-carbonate reaction' (Chapter 3) and conclusions concerning the prevention, mitigation and management of affected concrete (Chapters 4 and 5). The later chapters (6 – 16) then review AAR affected concrete and how the problem is dealt with in the different countries and regions of the world. The world map shown in Figure 1.1 illustrates the countries where AAR has been investigated or reported. Text-Box 1.1 lists the countries and regions where AAR has been investigated. It gives an indication of potentially reactive aggregate types and an indication of the reported cases and types of structures affected in the different regions.

1.2 ALKALI-REACTIVITY IN CONCRETE

1.2.1 The chemical and mineralogical composition of Portland cement

Before considering the alkali-aggregate reaction in concrete in detail it may be useful to briefly review the mineralogy of Portland cement and its hydration.

A generalised chemical and mineralogical composition of Portland cement (PC) is given in Table 1.1. As can be seen cement is composed of four principal minerals with C_3S and C_2S predominating. On addition of water they all hydrate, but at different rates with the C_3A hydrating most rapidly and the C_2S most slowly. The C_3A reacts very rapidly with water and would cause 'flash set' of the cement, so a small percentage of gypsum is added to cement to modify the reaction to form insoluble calcium sulphoaluminate hydrate (ettringite) which only slowly changes to tricalcium

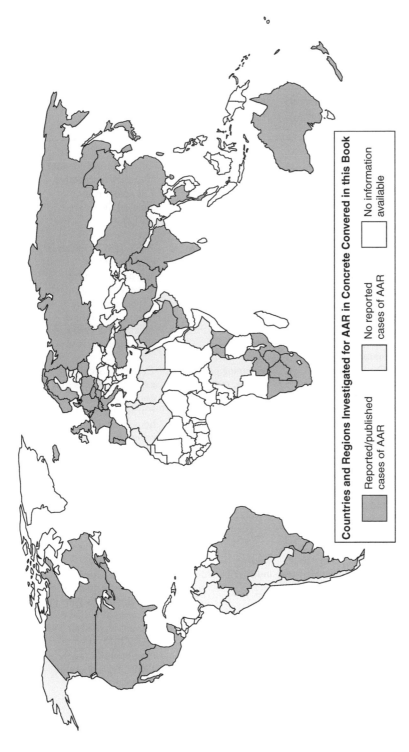

Figure 1.1 A world map indicating countries and areas or regions where cases of AAR in concrete structures have been identified, investigated, reported or where avoidance specifications are in place.

Text-Box 1.1 A summary listing of countries and regions detailed in later chapters where examples of AAR in structures have been reported, or where potential for AAR has been investigated.

Country/ Region	Examples of potentially Reactive aggregate types	Some Reported Case of AAR	Comments and References	Chapter
Europe				6,7 & 8
Austria	Sedimentary sands and gravels	Roads, pavements dams	Quartzite and gneisses identified as reactive	8
Belgium	Quaternary sands and gravels, argillaceous limestone, quartz diorite, dacite	Bridges, bridge decks, hydro structures	150 structures affected Denmars (2012)	8
Cyprus	Crushed opaline siliceous limestone, river and beach gravels	Marine structures buildings	Poole (1975) Siliceous reef limestone aggregate with high water absorption (3.3)	8
Czech Republic	Granodiorite, quartzite, phyllite, acid volcanics, quartz-feldspar tuffs, siliceous limestone	Highway structures, bridges, airport pavements, dams, tunnels	Cryptocrystalline quartz identified as reactive in a range of rock types	8
Denmark	Opaline flint, calcareous flint, microporous flints	Swimming pools Bridges, roads	Cements low alkali, external sources may contribute alkali. Marked pessimum behaviour. Rapid reaction. Nielson et al. (2004)	7
France	Siliceous limestone, quartzite, Rhône gravels, meta-granite and gneiss in Alpine region	Dams, bridges, retaining walls	ASR in over 400 structures, 10 demolished. Most in N. France and Brittany. Some cases complicated by DEF and sulfate attack	8
Finland	Pre-Cambrian gneiss, cataclasites, mylonites	Bridges, housing, industrial buildings	70+ structures identified. Confused with freeze-thaw. Air entrainment & ggbs in common use. Local cement high-alkali. Pyy et al. (2012)	7
Germany	Opaline sandstone, flint, greywacke, siliceous limestone	Bridges, highway pavements, precast slabs	First identified early 1950s, typically rapid reaction time	8

Text-Box 1.1 (Cont.)

Country/ Region	Examples of potentially Reactive aggregate types	Some Reported Case of AAR	Comments and References	Chapter
Hungary	Sands and gravels, carbonates and volcanics	None reported	Only andesite potentially reactive. Common use of ggbs cement replacement	8
Iceland	Sea dredged basalt, andesite, rhyolite, aggregates containing secondary opal	Domestic housing, hydraulic structures, pavements	Early cements are high alkali. Fly ash, pozzalans & silica fume in common use. Wigum et al. (2009/10)	7
Italy	River sands and gravels and carbonates containing chert, jasper, chalcedony	Residential and industrial buildings, pavements	Barisone, G. & Restivo, G. (1992b)	8
Netherlands	River sands and gravels	Bridges, viaducts, tunnel linings	Sea dredged gravel non-reactive common use of pfa. Broekmans (2002)	8
Norway	Sandstones, greywacke, claystone, mylonite, cataclasite, acid volcanics	Bridges, hydraulic structues, precast units, foundations, pipes, dams	Typically slowly reacting, use of fly ash in cement to mitigate ASR. Wigum et al. (2004)	7
Poland	Sands and gravels containing opal and chalcedony, sandstone, siliceous limestone and dolomite	Viaducts, buildings, precast elements	Large producer of aggregates, but variable quality. Góralczyk (2001)	8
Portugal	Granitoid types, basalts	Dams, bridges	Mostly slow reaction ASR (Silva et al., 2016) Dolomite + crypto crystalline silica are potentially reactive.	8
Republic of Ireland	Chert and greywacke	None reported	Some high alkali cements	6
Slovenia	River sands and gravels, opaline breccia	Concrete columns	ASR in floor due to crushed glass contamination	8
Spain	Quartzite, granodiorite, granite, monzonite, schist, slate	Hydraulic structures, dams, bridges	Quartzite and monzonite with micro and crypto crystalline quartz fast reacting others slow	8

Text-Box 1.1 (Cont.)

Country/ Region	Examples of potentially Reactive aggregate types	Some Reported Case of AAR	Comments and References	Chapter
Sweden	Gravels with porous flint, limestone with chert, cataclasites	Floors, pavements, bridges, dams, swimming pools	Most AAR reactions are slow. Recent introduction of higher alkali cements. Aggravated by de-icing salts	7
Switzerland	Fluvio-glacial sands and gravels containing sandstone, siliceous limestone, greywacke, gneisses, granites, schists	Dams, bridges, walls pavements	Several 100 cases reported, reactive mineral cryptocrystalline quartz. Aggravated by chloride ingress into cracks	8
Turkey	Cherts, glassy rhyolites and andesites	Airport aprons, bridges	Crushed limestone replacing reactive aggregates and use of blended cements	8
United Kingdom	Greywacke, gneiss, flint, chert, sandstones, granite, marine and river sands and gravels	Dams, bridges, buildings, pavements	Pessimum proportion of some alkali-reactive cherts from gravels recorded.	6
North America				*10*
Canada	siliceous carbonates, andesites, greywacke, argillites	dams, hydraulic structures, bridges, pavements	Cases in all provinces, ACR case reported. Extensive preventative measures in place	10
USA				*10*
Atlantic Seaboard	Granite gneiss, schists, quartzites, meta greywacke volcanics and chert	Dams, bridges, hydraulic structures, pavements, precast units	Examples of AAR occur in nearly every state. Preventative measures include performance criteria and specification limitations	10
Southern States	Opaline chalcedonic carbonates and shales	Dams, bridges, hydraulic structures, pavements, precast units	Preventative measures include performance criteria and specification limitations	10
Mid-West	Opaline chalcedonic carbonates, shales and sandstones	Dams, bridges, hydraulic structures, pavements, precast units	Preventative measures include performance criteria and specification limitations	10

Text-Box 1.1 (Cont.)

Country/ Region	Examples of potentially Reactive aggregate types	Some Reported Case of AAR	Comments and References	Chapter
Great Plains	Opaline chalcedonic carbonates, shales and sandstones	Dams, bridges, hydraulic structures, pavements, precast units	Preventative measures include performance criteria and specification limitations	10
Basin and Range	Glassy and cryptocrystalline rhyolite, andesites, volcanics and chert	Dams, bridges, hydraulic structures, pavements, precast units	Preventative measures include performance criteria and specification limitations	10
Pacific Coast	Glassy and cryptocrystalline rhyolite, andesites, volcanics, chert and sedimentary rocks	Dams, bridges, hydraulic structures, pavements, precast units	Preventative measures include performance criteria and specification limitations	10
South America				**11**
Argentina	granites, quartzites migmatites, basalts, andesites, rhyolites, tuffs, sandstones, dolomites	highway structures, overpasses, airport ruways pavements buildings,	First cases 1950s Marfil & Maiza (2008) ACR reported, Milanesi et al. (2016)	11
Brazil	granite, granite gneiss, basalt, mylonite, migmatite, quartzites, mine waste.	dams, hydraulic structures, buildings, bridges	Mitigation measures include: pozzalan, low-alkali cement, fibre reinforcement	11
Chile		No reported cases		11
Columbia		No reported cases	Lab. tests indicate potential for AAR. Bolivar (2003)	11
Dominican Republic		No reported cases	Lab tests indicate aggregates with potential for AAR. Silica fume used to mitigate effects	11
Ecuador		No reported cases	Lab. tests indicate potential for AAR. Garcia (2013)	11
Mexico	Crushed limestone, river sand with chalcedony, rhyolite and andesite	Pavements	Olague et al. (2003) Alkalis derived in part from aggregate	11
Paraguay		No reported cases	Lab. tests indicate potential for AAR.	11
Peru		No reported cases		11
Uruguay	Basalt and silica sand	Baygorruia Hydro- electric plant	Patrone, (2008, 2013)	11

Text-Box 1.1 (Cont.)

Country/ Region	Examples of potentially Reactive aggregate types	Some Reported Case of AAR	Comments and References	Chapter
Africa				**12**
Ethiopia	Granites	No reported cases	Volcanic ash used in block making. Low alkali OPC used. Dinku and Bogale (2004).	12
Kenya	Leucocratic and pink granitic gneiss (opaline veins)	Kambura dam spillway	Expansion in 8 years prevented gate 1 operating	12
Namibia	Archaean gneiss and granite	Mast foundations		12
Mozambique	Crushed granite	Cahora Bassa Dam, but ASR not identified	Aggregate sourced from the dam foundations. Some small expansion height increase 11mm in 30 years	12
South Africa	Greywacke, quartzite, coarse granite aggregate, metasediments	Piles, pile caps, dams. Bridges, mast footings, retaining walls, water retaining structures, airport aprons, roads	Prevention by using blended cements and specified limits on alkali content/m^3 concrete	12
Uganda	Various schists and gneisses.	Owen Falls dam	Expansion and cracking adjustable spacers placed under turbines	12
Zaire	Unknown	N'Zilo Dam	Expansion and cracking, not identified as ASR	12
Zambia	Coarse red granite with opaline veins	Dam	Thalow (1983)	12
Zimbabwe	Gneiss coarse aggregate	Kariba dam	Dam remains serviceable. Charlwood et al. (2012)	12
SE Asia, China, Japan				**9**
Republic of China	Sandstones, gneiss, tuff, flint, rhyolite, basalt, slates, andesite, granite, siliceous limestone dolostone	Bridges, dams, hydraulic structures, railway sleepers and structures, airport pavement	Many examples of ASR damage to structures. Reports of expansion due to ACR. Deng et al. (2008)	13
South Korea	Tholeiite with crystobalite, Pre-Cambrian – Palaeozoic rocks with cryptocrystalline quartz	Highway pavements	Slow ASR expansion, but more rapid if chalcedony is present	13

Text-Box 1.1 (Cont.)

Country/ Region	Examples of potentially Reactive aggregate types	Some Reported Case of AAR	Comments and References	Chapter
Taiwan	Sandstones and quartzite with chalcedony or chert	Marine structures		13
Thailand	Granite mylonite with cryptocrystalline quartz	Highway bridge piers and footings		13
Indian Sub-Continent				**15**
India	Granites, glassy and cryptocrystalline basalt, opaline and chalcedonic sandstones	Dams, bridges and hydraulic structures	Reactive aggregates associated with strained quartz	15
Bangladesh	Unknown	No reported cases	Prism testing of USA aggregates for ASR potential . Latifee in Rangaraju et al. (2016)	15
Butan	River gravels	No reported cases	Use of low alkali cement, silica fume and 50% slag cement	15
Pakistan	River gravels, with schist/gneiss, greywacke, quartzite	Dams, hydraulic structures	Slow ASR reaction, high alkali cement used	15
Russian Federation				**9**
	Gravels, quartzite, sandstones, sands, volcanics, siliceous dolomite	Sleepers, house footings, concrete buildings, port facilities, hydraulic structures	ASR risk minimised by aggregate specification, limit on alkali content and use of mineral admixtures	
Australia & New Zealand				**15**
Australia	River sands and gravels, quartzites, gneisses, rhyolite, rhyodacite tuff, meta-sediments, greywacke, siliceous limestone, glassy basalt, chert, hornfels	Dams, spillways and related structures, marine jetties, rail sleepers, harbour structures, water storage tanks	ASR occurs in all parts of the country, Appears to be dependant on opaline or cryptocrystalline silica as a component of the aggregate	14
New Zealand	Andesites, rhyolitic sand, dacite, basalt	Hydro power station, airport pavements, anchor blocks, precast units	ASR minimised by use of low alkali cement and by specification of less than 2.5kg/m3 Na (e) in concrete	14

Text-Box 1.1 (Cont.)

Country/ Region	Examples of potentially Reactive aggregate types	Some Reported Case of AAR	Comments and References	Chapter
North Africa & Middle East				
Arabian Peninsula	Soft limestones with detrital quartz, dolomitic limestones, some cherts	Concrete foundations in the eastern province. ACR-type rim-forming dolomitic limestone; concrete cracking possibly caused by ASR	Katayama (1997) P. Bennett-Hughes (personal communication)	16/13
Bahrain	Soft limestones with detrital quartz, some cherts, clays with gypsum	One structure using Khobar dolomite aggregate	Katayama (1997) ACR and ASR work together, but ASR caused the expansion	13
Egypt	soft limestone with detrital quartz and cherts. Polymictic gravels	No reported cases		16
Iran	Polymictic gravels containing chert, chert-rich gravels, siliceous limestones	Bridges, Dams, pavements	Lab. tests indicate potential for AAR. Extensive use of pozzolanas, mineral admixtures and low alkali cement	16
Iraq	Polymictic gravels and siliceous limestone	No reported cases	Katayama (1997)	16
Israel	Siliceous limestone with amorphous silica	Marine retaining walls. Power station drainage culverts	Deterioration due to ASR often mis-identified	16
Jordan	Impure limestones and dolomites	No published cases	Katayama (1997) Anecdotal only	16/13
Lebanon	Impure limestones and dolomites	No published cases	Katayama (1997) Anecdotal only	16/13
Libya & other North Africa countries	Wadi gravels, including some opaline sandstone and siliceous duricrusts and limestones	No published cases. Extensive efforts to avoid any ASR in major concrete pipeline project[0] (see Chapter 16)	ASR in transmission tower bases, Central Libya, Anecdotal only	16
Oman	Polymictic gravels	Bridge support crossheads	Anecdotal only	16
Yemen	Volcanics, basalt, rhyolite	One structure affected at 3 months post casting	Katayama (1997) Sims and Poole (1980) interstitial glass in basalt and rhyolite thought reactive	13

Table 1.1 Typical chemical and mineralogical composition of Portland cement.

Chemical composition as oxides				Typical percentage ranges by mass
Calcium			CaO	60–68
Silica			SiO_2	17–25
Aluminium			Al_2O_3	3–8
Iron (Ferric)			Fe_2O_3	0.5 – 0.8
Total alkali (estimated as Na_2O + 0.658 K_2O)			Na_2O	0.3–1.2
Sulphate			SO_3	2.5–5.5
Mineralogical composition				
Alite	tricalcium silicate	$3CaO.SiO_2$	$(C_3S)^*$	45–65%
Belite	dicalcium silicate	$2CaO.SiO_2$	(C_2S)	10–30%
Aluminate		$3CaO.Al_2O_3$	(C_3A)	5–12%
Ferrite		$4CaO.Al_2O_3.Fe_2O_3$	(C_4AF)	6–12%

Cement chemists' notation

aluminate hydrate. The principal products of the hydration of the cement are various forms of calcium silicate hydrate gels and calcium hydroxide. These together form the binding agent which holds the aggregate particles together to form the familiar, rigid, dimensionally stable material we know as concrete.

Concrete is basically composed of a mix of coarse and fine aggregates, cement and water. The definitions of aggregate are that particles over 4 or 5 mm in size are coarse aggregate, while fine aggregate consists of the fraction below 4 or 5 mm. Typically, modern concretes will contain of the order of 75–80% of graded aggregate set within a matrix of hydrated cement. The cement itself is today often modified by the inclusion of fly ash (such as pulverised-fuel ash, pfa), ground granulated blast furnace slag (ggbs) or other pozzolanic material. These are referred to as mineral additions (or 'supplementary cementitious materials' [SCMs] in North America). The concrete may also contain very small amounts of chemical admixtures, which serve to modify the rheological and setting characteristics of the concrete. The amount of water added to the mix is normally quoted as a ratio to the quantity of cement in the mix, this is thus referred to as the water/cement ratio (w/c) [or water/binder ratio when there are SCMs in addition to cement]. The w/c ratio is critical to the strength and other properties of the hardened concrete and is usually kept to a minimum consistent with thorough mixing, placing and compaction, so as to maximise concrete strength. Hydration of the cement is rarely complete and even in old concretes a few grains of the original cement will usually remain as unhydrated relicts. These enable a petrographer to identify the type of cement used in a particular concrete.

1.2.2 The basic chemistry of the alkali-aggregate reaction

In simple terms AAR in concrete is a variety of chemical reactions which develop within the fabric of a concrete involving three components. The first is a solid component within the aggregate that is susceptible to attack by the second component, the OH⁻ ions present in the alkali pore solution of the concrete. These OH⁻ ions are usually, but not always, derived from the ionic dissociation of the alkali constituents in the Portland cement used during its hydration. The third component is water, which acts both as a solvent and a carrier for the hydroxyl and alkali ions, and is also required to enable the reaction products to expand.

The term "alkali-aggregate reaction", while universally used, is something of a mis-nomer. The primary reaction is between the susceptible aggregate components and the concentrated ionised alkali hydroxide solution in the pores of the concrete, rather than with the sodium and potassium ions *per se*, though one or both of these become incorporated into the structure of the alkali-silica gel reaction product when it forms. Calcium hydroxide is usually also present in the pore solution and appears to be an important component in modifying the rheological and swelling characteristics of the gel.

The conclusions that have been reached in research studies concerning the reaction have suggested that there may be as many as three types of alkali-reactivity in concrete, though there is some compelling evidence that in reality, the first two are just variants of the third:

1 Alkali-Silicate Reactivity (Gillott, 1975)
2 Alkali-Carbonate Reactivity (ACR; Swenson, 1957; Milanesi *et al.*, 2012; see also Chapter 3)
3 Alkali-Silica Reactivity (ASR; Stanton, 1940)

The alkali-silicate reaction involved reaction between certain argillite, greywacke and other sedimentary and metamorphic aggregates containing disordered forms of cryp-tocrystalline silica contained in high alkali concrete. The reaction is typically slow to develop the hydrophilic gel leading to the expansive forces necessary for cracking to occur. It was originally considered to be different in form from ASR, but this is no longer generally accepted and the subdivision appears to be an arbitrary one. Nevertheless, there is evidence to suggest that in certain circumstances alkali may be leached from aggregates of this kind into the pore fluid and this alkali may contribute to the slow development of ASR in the concrete (Stark & Bhatty, 1986).

The alkali-carbonate reaction (ACR) produces well defined reaction rims at the margins of dolomitic limestone aggregate particles. Several types of rim have been identified (Poole, 1981) but their possible association with a damaging expansion in concrete is unclear and remains a matter for research (Lagerblad, 2012). Katayama (2010) coined the term 'so-called ACR' to describe a reaction involving carbonate aggregates that is actually expansive as a result of ASR. Alkali-carbonate reaction is considered in greater detail in Chapter 3.

The role of the aggregate in ASR arises from the chemical hydrolysis reaction between certain types of disordered, or poorly crystalline silica present as a component of certain types aggregates and the ionised alkali hydroxide pore fluid of the concrete. Provided the concentrations of these reactants are sufficient, reaction will take place to produce an alkali-silica gel product at the silica reaction sites. Typically because alkalis present in cement are sodium or potassium hydroxides, the gels formed also contain these metal ions. Such gels are hygroscopic and absorb water from the surrounding pore fluids and this can lead to swelling pressures sufficient to crack and disrupt the surrounding concrete.

In the simplest terms the reaction can be expressed according to the following idealised equations (Hansen, 1944):

$$4SiO_2 + 2NaOH \rightarrow Na_2Si_4O_9 + H_2O$$
$$3SiO_2 + 2NaOH \rightarrow Na_2Si_3O_7 + H_2O$$

However, the chemical composition of alkali-silica gel is variable and indefinite. Vivian (1951) and others have shown that the key factor in this reaction is the OH^- ion concentration and the metal cation is only relevant insofar as it becomes incorporated into the gel. It is commonly found that other ions are identified in gel analyses and may be of importance in controlling gel properties; the most notable of these is calcium. This observation has also led to the suggestion that there may be a process of ion exchange in gels between sodium and calcium (French, 1980).

Dent-Glasser and Kataoka (1981b) suggest the reaction will proceed in stages, with an acid-base reaction between the silica and hydroxyl followed by disintegration of the silica as the siloxane bridges are attacked. The secondary overlapping stage is the absorption of water by the hygroscopic gel and a consequent expansive pressure is developed by the gel provided calcium ions are also present.

Stage 1. $H_{0.38}SiO_{2.19} + 0.38NaOH = Na_{0.38}SiO_{2.19} + 0.38H_2O$
Stage 2. $H_{0.38}SiO_{2.19} + 1.62NaOH = 2Na^{2+} + H_2SiO_4^{2-}$

This expansion leads to an increase in volume and development of a swelling pressure which may exceed the tensile strength of the concrete and thus be sufficient for cracks to develop. These pressures vary considerably and appear to be related to gel composition. They are also sensitive to temperature, but can reach 6 to 7 MPa (Diamond, 1989).

The alkali-silica gel reaction product if present in a concrete can normally be identified by the examination of thin sections of the concrete using a petrographic polarising optical microscope, provided the concrete has been sampled appropriately. However, the gel may be leached from the concrete by water percolating though cracks and the gel itself is prone to carbonation by atmospheric carbon dioxide. A detailed account of the methods used in the petrographic investigation of alkali-aggregate affected concrete is given in 'Concrete Petrography' (Poole & Sims, 2016).

1.3 ASR IN CONCRETE, THE REACTANTS AND PRODUCTS

1.3.1 Reactive aggregates

The majority of rock types used as aggregate in concrete and mortar contain several mineral components. In order for a given aggregate to be susceptible to ASR it must contain a 'reactive' form of silica. This may only be a minor constituent but is the common factor present in the many types of 'reactive aggregates' referred to in the later chapters of this world review.

Although the majority of rock types are composed of several minerals, a few, such as limestone and dolomite are monomineralic if pure, but many of these also contain small amounts of other minerals, with quartz and clay being the most common. Sandstones and quartzites are also considered monomineralic if pure, but can also contain a variety of minor constituents such as mica and feldspar. Even when composed only of quartz, the silica cementing the grains together is often different in type from the silica of the grains themselves. Greywacke, tuffs and hornfels may in the simplest terms be considered as very impure varieties of sandstones, but they may contain a variety of other mineral, rock and clay species. Again the material cementing the grains together may be different from the silica of the quartz grains that are contained in the rock.

Some sedimentary rocks are largely formed of organic material, with perhaps coal formed from vegetation and limestone from shell fragments being the most familiar examples. Certain organisms use silica to form their hard parts, and important among these are diatoms, which produce outer coatings (tests) of opaline silica, and certain sponges which develop microcrystalline quartz stiffening filaments (spicules). Such siliceous material may form an important component of certain rock types. Alternatively, the silica they contain may be leached and redistributed within the sediment in which they are buried. Such redistribution of siliceous material by groundwater or other means is generally considered to have produced chert, flint, chalcedony, agate and related siliceous materials that are sometimes found in these rocks.

Hornfels, mylonites and other metamorphic rocks are recrystallized versions of sedimentary and igneous rock types. The severity and type of metamorphism will control which new minerals are formed. The common metamorphic minerals which form from clay rich or sandy rocks are micas, chlorites, garnets and feldspars, but during the process the original quartz will partly or wholly recrystallize. This recrystallization may produce cryptocrystalline silica crystals at grain boundaries and tectonic stress may introduce lattice strain in the quartz crystals which can be recognised microscopically as undulose extinction under crossed polar illumination (see also Fernandes *et al.*, 2016).

Igneous rocks such as granites and basalts typically contain several different mineral species. There is usually one or more of the ferromagnesian minerals such as biotite, amphibole, pyroxene and sometimes olivine, together with feldspars and often quartz is also present if the magma from which the rock crystallised is sufficiently siliceous. There are also various other minor mineral constituents such as titanium and iron oxides in nearly all igneous rocks. It should be noted that igneous rocks will have originally crystallised at high temperature and so some of the minerals formed may only be metastable at normal temperatures.

Certain igneous rocks, notably volcanic varieties, cool very rapidly in the atmosphere and may not crystallise fully but contain a glassy phase which is usually interstitial to the early formed crystals. This glassy phase typically contains the most volatile low-temperature melting point components of the original melt. These are likely to be constituents rich in silica and alkali.

As has already been noted that some form of 'reactive silica' is an essential requirement for ASR in concrete to take place. However, the concentration required to produce deleterious effects can be quite small. In certain cases as little as 2% of reactive silica in the aggregate has been reported as causing severe distress in concrete. The majority of the wide range of rock types in use globally as aggregate for concrete will contain forms of silica, with the exception of particularly pure varieties of monomineralic rocks already noted. Although a relatively rare occurrence, any of these might possibly contain a small proportion of reactive silica either as an original, a primary or a secondary mineral. Consequently, it is only on the basis of their service record, or from careful and thorough test result data that the vast majority of these silica containing aggregates can be designated non-reactive even if used in concrete made with high-alkali cement.

It is clear from the above comments that it is incorrect to consider rock type as the criterion for an aggregate's potential for reactivity. Instead attention should be focused on the mineral constituents and textures of the rocks themselves. Although many types of rock may contain silica usually as quartz, the number of reactive forms of silica is quite small. Perhaps the most important factors for a siliceous material to be alkali-reactive are that it should be poorly crystalline, contain lattice defects or alternatively should be amorphous or glassy in character. Surface area is also important, so microporosity and particle size are also relevant considerations. Some of the siliceous materials that meet these criteria are listed in Table 1.2. A suggestion for the possibility of increasing the 'activation' of silica by irradiation has been proposed by Escadeillas *et al.* (2000) which could have implications for storage of nuclear waste in concrete.

Since there is a very wide range of rock types used as concrete aggregate, a petrographic examination forms an important diagnostic method for determining whether a potentially reactive form of silica is present among the other constituent minerals in an aggregate. In some cases, for example, where the siliceous cement between quartz sand grains is suspected as reactive, this may be difficult to confirm so that further investigation using more sophisticated techniques and specific testing such as the 'Gel Pat test' (BS 7943: 1999) or a concrete prism or mortar-bar expansion test (BS 812-123: 1999; Nixon & Sims, 2016) may be necessary. Nixon and Sims (2016) have reported the work of the RILEM technical committees, which suggests that the potential for aggregate reactivity may be classified as low, medium or high based on petrographic, test and case history evidence (see Chapter 2). This classification then dictates what, if any, constraints should be applied to their use in concrete. Further guidance on using petrography to identify reactive aggregates may be found in an excellent pictorial atlas developed for RILEM (Fernandes *et al.*, 2016).

1.3.2 The sources of alkali in concrete

The principal, but not the only possible source of alkali in concrete is derived from the cement used. Portland cements are manufactured by sintering and reacting together intimate mixtures of powdered calcium carbonate (limestone or chalk), together with a smaller proportion of silica/aluminous material usually a clay or shale at temperatures up to 1400°C. Typically the shale or clay contains small amounts of iron, magnesium and sulphate, together with alkalis which are derived from detrital or secondary micas and feldspars.

The proportion of alkali present in a cement will depend on the detailed composition of the raw materials used and the minor components they contain. The mix proportions of these materials, the firing temperatures, feed rates and oxidising/reducing conditions in the kiln are all important controls. These factors together are critical in determining the mineralogical composition of the particular cement produced and the proportion of alkali it will retain (see Table 1.1).

During the firing process the alkalis being the most volatile of the components are partly lost in the flue gases and dust, but the small percentage of alkali in the raw materials has a strong affinity for SO_3 in the liquid phase and crystallize to form a range of alkali sulphates in the cement clinker. Before problems of alkali-reactivity were

Table 1.2 Some siliceous minerals which have been identified as potentially alkali reactive.

Mineral Type	Typical occurrence	Comment
Opaline silica $SiO_2.nH_2O$	Primary and secondary mineraloid, cement between mineral grains in rocks, diatomite, siliceous sinter, wood opal	Water content 3 - 6%, but can range up to 21%. Usually very highly reactive, as little as 1% can cause damaging expansion. An ordered or disordered open array of crypto-crystalline cristobalite spheres
Cristobalite SiO_2	Metastable mineral in volcanic rock glass, high temperature hornfels	As little as 2% can cause damaging expansion in concrete. Normally forms above 1470°C. Open crystal structure, but more dense than tridymite. (ρ = 2.33)
Tridymite SiO_2	In cavities in acid and intermediate volcanic rocks. Constituent of silica brick	More common than cristobalite, forms above 867°C. Open crystal structure. Small percentages can cause expansion in concrete. Less dense than cristobalite. (ρ = 2.26)
Chalcedony, and varieties: e.g Agate Carnelian Jasper Onyx SiO_2	A principal constituent of cherts and flints. As a cementing agent in some quartzites. As nodules or stratified layers in sedimentary rocks and as a secondary void infilling	Contains sub-microscopic pores and is a variable mixture of crypto-crystalline fibrous silica, moganite and amorphous hydrated silica. The names given depend on the type and distribution of colour due to impurity. Moderately alkali reactive in some cases, but dependent on exact composition and structure.
Natural glasses Volcanic glasses, Obsidian and Perlite, Pitchstone	Common interstitial constituent of fine grained volcanic rocks and pyroclastics	Natural glasses e.g., obsidian. In volcanic crystalline rocks compositions close to the bulk rock. Interstitial glasses may contain higher silica than the bulk rock. Alkali reactive in some cases depending on composition, porosity and water content
Synthetic glasses	A wide range of types are manufactured including alkali resistant and low expansion types	Borosilicate glass (Pyrex) is alkali reactive giving large expansions with ASTM C227 mortar bar test.
Calcined flint	Manufactured for use as a decorative constructional material, skid resistant road surfacing and for the ceramic industry	Calcining removes water, carbonates and organic impurities. Converts flint to cristobalite (Weymouth & Williamson, 1951). Alkali reactive in concrete and mortar.

recognised the flue dusts were often trapped and re-cycled into the kiln, but as many countries have now set limiting specifications on the maximum percentage total alkali allowed in Portland cement, the practice has been discontinued. A low-alkali cement is defined in many countries as one containing less than 0.6% equivalent alkali (*i.e.*, $Na_2O + 0.658K_2O$) by mass.

Although the cement used in a concrete is the principal and in many cases the only source of the alkali hydroxides which are present in the pores and capillaries of the cement matrix, other sources have been cited as contributing to the total available alkali. Several possible sources of alkali additional to that derived from the cement include feldspars and micas present in some rock aggregates. Much of this alkali is fixed in the minerals concerned and is not available for leaching even in the alkali conditions pertaining in a concrete. A recent review of alkali release from silicate minerals has been published by Lagerblad (2012).

Although in general the additional alkali derived from the aggregate is usually so small that it may be ignored, it has been suggested that in rare cases an addition of up to 10% of the alkali may be derived from the aggregate (Constantiner & Diamond, 2005). Release of alkalis into the pore solution from an andesitic aggregate for at least 90 days has been demonstrated by Kawamura *et al.* (1989). In some cases glassy material in volcanic rock aggregates may be alkali-reactive (as noted in Table 1.2), but they may also contain alkalis which could similarly contribute to the total alkali in the pore solution.

Mineral additions such as fly ash (pfa) or ground granulated blastfurnace slag (ggbs) are currently very widely used in concrete. As a cement replacement material they have the advantage of reducing the amount of Portland cement used and consequently the available alkali contributed from the cement and thus reducing the possibility of deleterious alkali-aggregate reaction. In some respects this effect is surprising in view of the fact that some of these additions contain significant concentrations of alkali in their glassy phases along with amorphous silica, so they might be expected to contribute to any ASR expansion. The reduction in expansive reactivity of concretes containing them is not fully understood. They typically have a pozzolanic effect (Bennett & Vivian, 1955; Hobbs, 1988) in reducing the calcium hydroxide concentration of the pore fluids and very fine particle size of these materials suggests that they may also contribute to a 'pessimum proportion' effect for the reactive siliceous components by inducing very rapid reaction while the concrete is still setting. Consequently, the reactants become depleted to levels where reaction is greatly reduced, or essentially ceases before the concrete has fully set while any early reaction products formed are accommodated in the still semi-plastic concrete.

Another possible source of additional alkali can be derived from de-icing salts or even sea water. These alkalis will be in solution and consequently will most effectively permeate the concrete if it is already cracked, or is porous. They may then contribute additional alkali hydroxide to the alkali reactions that may have already been initiated in the concrete.

1.3.3 The role of water

Moisture is always present in the pores of a concrete, with the possible exception of a few millimetres of the surface layers, or very thin slabs if the surrounding atmosphere is dry. Consequently the relative humidity (RH) within concrete will typically remain at above 80%.

Water has a dual role in connection with AAR: firstly it acts as a carrier for the alkali and hydroxyl ions allowing reaction to progress, and secondly it is absorbed by the hygroscopic alkali-silica gel reaction product which swells developing the pressures that may be sufficient to crack the concrete. Such cracking then provides additional pathways for moisture to access the interior of a concrete and so promote further reaction.

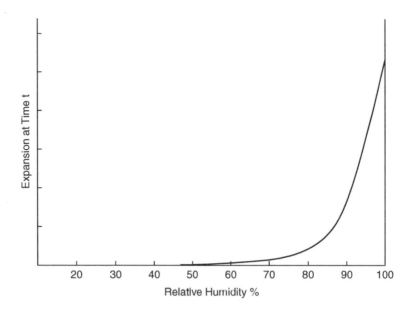

Figure 1.2 The effect of relative humidity on the expansion of concrete due to formation of alkali-silica gel.

Experimental, case study and anecdotal evidence show water to be an essential requirement for the ASR gel to expand. Experimental concrete shows little evidence of ASR expansion if the RH is kept below 80% but increases rapidly, as measured by expansion, if the RH increases beyond this as shown diagrammatically in Figure 1.2.

An interesting case concerns concrete paving in a very dry desert environment. Although the concrete used was potentially alkali-reactive no cracking or other deterioration was observed. After a prolonged period in very dry conditions, some of the slabs were wetted. Expansion and cracking developed within days suggesting that the reaction had in fact occurred, but for a long period there was insufficient moisture available for the alkali-silica gels to swell until the additional moisture became available. A similar observation with reactive mortar bars stored for 40 days in dry air did not expand, but when place in water expanded rapidly (Vivian, 1950).

It is also a common observation that a concrete structure suffering from ASR will show greater evidence of cracking and deterioration on the surfaces facing towards the local prevailing wind and rain direction. The sequence of wetting and drying which will be more extreme on weather faces is thought to introduce a cyclic pumping effect with moisture movement in and out of the surface layers thus accelerating the reaction in that zone.

1.3.4 The gel reaction products

Sodium/potassium silica gels are very variable in chemical composition. Some alkali-silica gel compositions and typical quoted ranges of composition are shown in Table 1.3, which clearly reflects the variation. The calcium oxide content will in part relate to the degree to which the gel sample was carbonated prior to analysis, while the alkali variation in most examples reflect the alkali type available from the original cement and this in turn depends on the raw materials that were used in its original manufacture.

Table 1.3 Some typical Alkali-Silica Gel Analyses and Analysis Ranges. Gels from Concrete Structures.

Oxide	1	2	3	4	5	6	7	8	9
SiO_2	53.9	61.7	52.2	48.6	47.8	56 – 86	60 – 77	36 – 72	27 – 75
Na_2O	12.9	14.9	4.3	0.9	1.1	0.4 – 20	8 – 13	0 – 3.6	0 – 18
K_2O	– 5.2	0.5	14.2	6.8	2 – 8	9 – 18	0 – 2.6	0 – 14	
CaO	2.9	0.6	26.0	27.4	36.2	1 – 28	1 – 18	25 – 57	0 – 35
MgO	0.6 – 10.0	0.0	0.1	–	–	0 – 10	–		
H_2O	–	–	–	–	–	10 – 30	–	–	–

1: Stanton (1940). **2**: Idorn (1967). **3**: Poole (1975): *average of 2.* **4**: Hanson et al. (2003): *mean of 20* (2003). **5**: Hanson et al. (2003): *mean of 12.* **6**: Berube and Fournier (1986). **7**: Davis and Oberholster: *range of 6 crystalline gels (1986).* **8**: Mladenovic et al. (2009): *range of 5 mortar bars.* **9**: Hanson et al. (2003): *'typical ranges'.*

Recent studies using electron microscopy (Katayama, 2012) show that alkali-silica gels will partially crystallise to alkali-rich rosette forms and also needle, rod or blade-like forms, but all appear to have similar chemical compositions related to the first formed gel (see Hanson *et al.*, 2003). Fluid alkali-rich sols can coexist with silica-rich viscous gels in the same concrete and it is suggested that the equilibrium between the two depends on the relative humidity of the concrete, ambient temperature and the incorporation of other ions in the gel (Katayama, 2012). A calculated relationship between a simplified gel composition and its viscosity using data available for the $Na_2O - SiO_2 - H_2O$ system is illustrated in Figure 1.3, with the viscosities given in poise (Moore, 1978). It indicates that the proportion of silica in the gel is a factor in determining its viscosity and this in turn may relate to the swelling pressures developed by the gel.

Observational evidence indicates that a viscous gel generates greater swelling pressure than a more fluid one within the concrete. However, this diagram does not include carbonate, which is present in almost all gels and is likely to affect their swelling characteristics. Also the composition of gels analysed tend to plot on the lower left hand portion of this diagram. The chemical compositions of gels, though variable, do not appear to correlate either with the type of alkali-aggregate reactivity, or the reactive silica type. The expansive pressures exerted by these gels within a concrete are not considered to be the result of the crystallisation of the gel itself, but is dependent on its hygroscopic absorption of water causing the gel to expand (Katayama, 2012).

The high calcium concentrations in most gels (Table 1.2) have been ascribed to the slow carbonation of the gel once formed and French (1991) has suggested that there may be an ion exchange mechanism operating where calcium ions exchange for sodium or potassium ions, thus possibly allowing these alkalis to resupply the alkali-silica reaction.

1.4 INTERNAL AND EXTERNAL FACTORS RELATING TO REACTIVITY AND EXPANSION

1.4.1 Concentration and activity of the reactants

The alkali-silica reactants are the reactive silica within the concrete aggregate and the OH- ions in the pore solution. In a normal concrete the OH- ions arise from the ionisation of

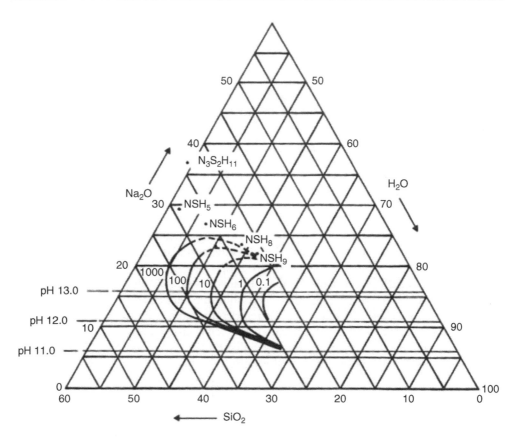

Figure 1.3 The relationship between gel composition and viscosity in poise calculated from vapour pressure values from the $Na_2O - SiO_2 - H_2O$ system. After Moore, 1978.

the alkali hydroxide in the pore solution with the alkali (sodium or potassium) becoming incorporated into the gel product and so maintaining the charge balance.

The importance of the OH- concentration in this reaction is evidenced by experimental work where positively charged tetramethyl ammonium and negative hydroxyl ions will react with opal to produce gel, whereas high concentrations of potassium nitrate or sulphate do not react because no OH- ions are present.

The reaction is one between a solid (a reactive form of silica, see Table 1.2) and a solution so that rate of reaction at a given temperature will depend on the rate at which the solid can be dissolved by the OH- ions in the pore solution. Consequently, the nature of the silica, its internal structure, micro-porosity, specific surface and other characteristics specific to each form of reactive silica are major controlling factors. Another rate controlling factor will be whether the reaction product will itself form a barrier which will then modify the rate of diffusion of the OH- toward the reaction site. The capillary permeability of the concrete cement paste matrix will be yet another control on the rate of diffusion.

If sufficient reactive silica is present in the concrete aggregate, then the other major reaction rate controlling factor is the concentration of OH- ions in the pore fluid of the concrete. High concentration will increase the activity of these ions thus increasing the rate of dissolution of the silica and consequent increase in the formation of gel product. However, as the reaction proceeds, an increasing proportion of the OH- ions get used up in forming the gel so reaction gradually slows down, as is illustrated diagrammatically by Figure 1.4, until it no longer produces additional measurable amounts of gel, or further expansive behaviour.

Further complications concerning the fixing of OH- and alkali ions in the gel reaction product perhaps need further consideration. One of these as noted previously, is the suggestion by French (1991), that positive alkali ions in the gel product may exchange with calcium from the surrounding cement pore solution, thus freeing the alkali ions back into the pore solution and changing the gel composition with time. Also Thomas (2006) and others, have demonstrated that the OH- ions will react with the silica, but the dissolved silica will remain as a fluid sol (Helmuth et al., 1993) unless calcium hydroxide is also available in the pore solution allowing a gel to form with a composition which can swell and exert pressure that can cause expansion of the concrete.

There also appears to be a minimum OH- concentration threshold (in the presence of calcium hydroxide) in the pore solution, below which the reaction slows sufficiently so as to cause no measurable additional expansion or damage irrespective of the amount of reactive silica remaining in the concrete. At this threshold concentration, the reaction may be considered to have reached an equilibrium situation, but in reality the reaction may continue within small and localised areas in the concrete where there is higher than average OH- ion concentration, though the overall expansive effect will be too small to measure. This limiting value concentration provides the rationale for the specification

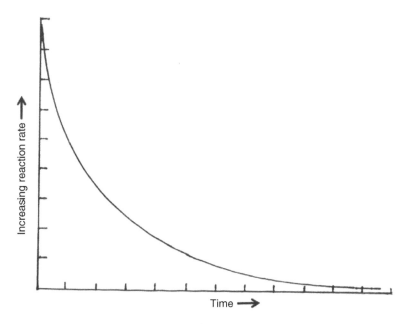

Figure 1.4 Diagrammatic indication of the rate of reaction slowing down as reactant concentrations reduce with time.

standards in many countries limiting the alkali equivalent in the cement or the concrete to below a particular value.

As a summary of the discussion above, the formation of gel (provided $Ca(OH)_2$ and water are present) resulting from alkali-silica reaction will at first be rapid as reactants are initially at their most concentrated and will thus cause expansive pressure. The actual reaction rate will also depend on a number of other factors as well as the concentrations of the reactants. However, with time the reactant concentrations will slowly reduce as they form gel and reaction will consequently slow down, so the deleterious expansive effects of the formation of gel will also reduce. In cases where sufficient accessible reaction sites are available, and permeability is high the alkali and the hydroxyl concentrations in the pore solution will decrease rapidly to below a threshold level and further reaction will become immeasurably slow.

1.4.2 External sources of reactants

As noted in section 1.3.2, some alkali in concrete may be derived from sources other than the cement. There is some evidence that in a few cases this additional alkali hydroxide may be derived from minerals in the aggregate (Kawamura et al., 1989). This is most likely where the aggregate particles have been partially degraded by geological weathering, or where large surface areas are open to attack from the alkali pore fluids. The most common minerals in aggregates which have the potential for releasing alkalis are the micas (K), the feldspars (Na and K), illitic clays (K), degraded volcanic glasses and ash or tuff (K and Na).

As already mentioned, the most commonly quoted potential source of additional alkali is de-icing agents, the commonest being sodium chloride, but organic agents are sometimes used and these include potassium acetate, potassium formate and sodium formate. If such salts are to be effective in increasing the reaction with silica the positive alkali ions must form hydroxyls which can ionise to provide the additional OH- ion concentration in the pore solution. Consequently the negative chloride, acetate or formate group ions must combine with other components of the cement matrix.

1.4.3 Chemical gradients within the concrete

Formation of an alkali-silica gel can only take place at reactive silica sites within the aggregate particles, so the alkali and hydroxyl ions must be able to migrate toward these sites down the chemical gradient produced as the ions will become fixed in the gel product this providing a downward concentration gradient toward each reaction site. The rate of this migration depends on the concentration of alkali hydroxide in the pore solution as discussed above and also on the permeabilities of the reacting silica particle and of the surrounding cement matrix, the availability of water and the temperature of the system.

The ease of ionic interactions within reactive silica particles in the aggregate have an important effect on reaction rate and will be greatest for silica with an open or porous structure. The particular example of opal, which is highly reactive, may be considered an extreme case of a porous material since its internal structure consists of a three dimensional array of silica spheres 150 – 300 nm in diameter with empty or water filled channels between them, which provide ready access to the interior of the mineral allowing chemical attack and dissolution of the silica. The example of opal suggests that other disordered forms of silica with open or disordered crystal structures, less dense crystal structures, or silica varieties containing high pore-water contents are likely to be the most reactive. Table 1.2 lists the more common reactive siliceous minerals, an indication of their relative reactivity may be related in some measure to their structural characteristics as noted in the table.

Reactive forms of silica are typically present as one small component among others within aggregate particles. The pore solution must be able to access these reactive silica particles in order for reaction to occur. If the whole aggregate is permeable to solutions through micro-cracks or along grain boundaries then reaction will be rapid, but if aggregate particles are 'tight' and essentially impermeable, only the reactive silica at aggregate margins will be open to attack, while the reactive silica inside will be protected.

It was noted in Section 1.3.3 that additional external surface moisture and the 'pumping' effect of wetting and drying in the surface layers of a structure assists the rate of movement of alkali and hydroxyl towards reaction sites and also the dispersal of the fluid gel product.

It is difficult to determine the distribution of alkali and OH- ions within the pore structure of the cement matrix. Electron microscopy microanalysis results showing alkali gradients within a few tens of microns of reaction sites are unreliable because specimens are dried prior to analysis with the possibility of redistribution of alkali ions. Also OH- ions cannot be identified by this method. Perhaps the most satisfactory model of the distribution of these ions in the matrix pore solution is of a fairly uniform ionised alkali solution within the pore spaces of the cement matrix in the concrete, but with small local variations on a scale of millimetres, and small regions surrounding reacting particles where alkali and hydroxyl gradients might be expected to occur.

1.4.4 Temperature of reaction

AAR in concrete will be initiated within the reactive silica particles leading to the formation of gel if reactant concentrations are above a threshold level. Provided sufficient moisture together with calcium ions are both present, the hydrophilic gel can swell with the potential to split the reactive silica grain and to widen pathways at grain boundaries and along microcracks in the aggregate particle, unless these pathways are already wide enough to allow the gel to escape into the pore spaces of the surrounding cement matrix. Overall the swelling gel can induce expansive and possibly disruptive pressure within the fabric of the concrete, but key controls will be the viscosity of the first formed gel, the permeability of the aggregate particles where reaction is sited and the capillary permeability of the surrounding cement matrix. If permeability is high and viscosity low, then any gel formed may be absorbed into the pores in the cement matrix without producing expansion.

With endothermic reactions an increase in temperature normally increases reaction speed as the energy added to the reacting system enables the reacting ions to combine more readily. This is well documented for alkali-silica reactions by a number of the test procedures in current use, which employ elevated temperatures of storage for the mortar-bars or concrete prisms in order to accelerate reaction and allow expansions to be monitored over a period of months rather than the years that ASR normally takes to exhibit expansion in actual concrete structures.

Although elevated temperature will speed the reaction and formation of the gel product, the gel itself may also be modified by increased temperature. Several factors are at play here; firstly increased temperature is likely to reduce the viscosity of the newly formed gel, but higher temperature may cause the gel to dry and dehydrate more rapidly. Increased temperature may also induce changes to the composition of the gel itself with the possibility changing its swelling characteristics.

1.4.5 Alkali-silica gel

The alkali-silica gel product may not remain at a constant composition during the whole period of reaction and consequently the expansive characteristics of the gel may also change as the reaction proceeds. The composition and expansive properties of a gel will depend on a number of factors. These include the concentration levels of each of the reactants in the system, the reactivity of the siliceous components in the aggregate, the presence of calcium ions in the pore fluid and factors such as the permeability of the cement matrix and temperature.

As already noted, Thomas (2006) has demonstrated that calcium hydroxide is necessary in a pore solution during reaction to provide a gel composition which can exert a swelling pressure and expansion of the concrete. He was also able to show that an increase in the calcium hydroxide concentration caused the 'pessimum proportion' (see Section 1.5 below) of reactive silica giving the maximum increase in expansion. As noted in Section 1.3.2, the calcium hydroxide requirement may also partially explain the effectiveness of mineral admixtures, such as pfa or ggbs in reducing or eliminating expansion in concrete because their pozzolanic effect removes calcium hydroxide during hydration of the cement. Other workers (Powers & Steinour, 1955; Helmuth *et al.*, 1993) have suggested that the lime/alkali ratio in the gel may control its swelling properties, a low ratio giving expansion and very high ratios no expansion. However, Helmuth *et al.* (1993), Struble and Diamond (1981a) and Dent Glasser and Kataoka (1981a) explored the relationship between gel composition and expansive properties using artificially prepared gels, but were unable to find any consistent relationship equating composition with expansive pressure; Struble and Diamond commenting that little is known about gel composition and its effect on swelling pressures.

The silica/alkali ratio in the gel may also be another controlling influence on the swelling characteristics of alkali-silica gels (Katayama, 2012). It has been suggested by Dent Glasser and Kataoka (1981b) and Taylor (1990) that the swelling characteristics of alkali-silica gel may relate to particular ratios of silica to alkali in the gel itself.

1.5 THE PESSIMUM PROPORTION CONCEPT

1.5.1 Observation of the effect

The presence of a reactive form of silica, with a sufficiently high OH- ion concentration in the pore fluid (normally formed by ionisation of alkali hydroxides in the concrete) and sufficient moisture in local areas within the concrete will lead to the initiation of an alkali-silica reaction.

Studies have also shown that some reactive forms of silica in an aggregate will cause expansion and damage to concretes within the space of a few years, typically 8 to 12 years, while other types designated 'slow reactors' may react more slowly so that damage does not become apparent for many years but then damaging expansion may continue for at least 50 to 70 years. The reason for these differences most probably rests in the relative ease or difficulty with which the OH- ions can access the sites of reaction in the aggregate, as noted in section 1.4.3.

The simple view is that the higher the proportion of the reactive silica present, or the higher the mobile alkali hydroxide in the pore fluid, the longer the reaction can continue which in turn will lead to increasing expansion and consequent damage due to the amount of expansive gel reaction product formed. This appears to be incorrect in some cases, as there are examples where a particular proportion of the reactive siliceous component in the aggregate appears to result in a maximum expansion, while at both a lower or higher concentration of the same siliceous material lower expansions are observed.

This leads to the concept of a 'pessimum proportion' of reactant (usually the reactive silica component in the aggregate) which may be defined as the proportion of reactive silica that has combined with the available hydroxyl to produce a gel of a composition that results in a maximum expansion of the concrete.

This proportion producing a maximum expansion is often found to be between 5 and 10% (Curve A, Figure 1.5) for highly reactive siliceous components within the concrete in the UK (*e.g.*, Hobbs, 1981, 1988), but by contrast, Ozal (1975) demonstrated a pessimum expansion at between 44 and 73% by weight of reactive chert (Curve B, Figure 1.5). Ozal worked with mortar-bars and followed the ASTM test method C227. His series of mortar bars all contained the same alkali content (0.9%), but were made

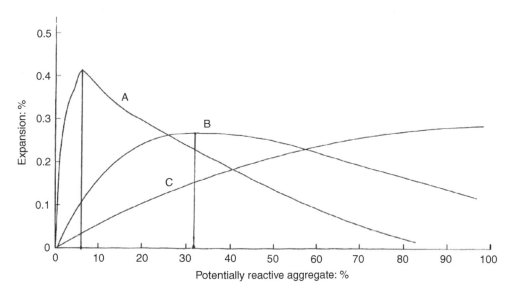

Figure 1.5 Hypothetical examples of curves showing % expansion against % reactive component in a concrete (from Poole, 2010). A: rapidly increasing expansion as the amount of reactive aggregate increases up to a maximum at about 5% (the pessimum proportion), but decreasing as the amount increases further (some UK examples are of this type). B: a more gradual increase to a pessimum at about 30% and then a gradual decline. C: a gradual increase in expansion with increase in reactive aggregate and no pessimum (typical of slowly reacting systems).

with varying proportions of a highly reactive chert aggregate included in the otherwise inert aggregate in a range from 5 to 100%.

1.5.2 Hypotheses to account for decreases in expansion beyond the pessimum

A problem in explaining the pessimum proportion arises from the correlation of degree of reaction and the extent of the expansion observed. The latter is easily measured, but this may be difficult to assess. The graph in Figure 1.5 refers to expansion, which may in fact be very different from degree of reaction. In an ASR-affected concrete clearly if there is only a very small proportion of reactive silica present then the resulting expansion will also be small. Expansion may be expected to increase as the reactive silica proportion in the aggregate increases, but several hypotheses have been proposed to account for the decreased expansion observed when the reactive siliceous material is present in quantities above the pessimum proportion, some suggestions are given below:

1 Because reactive centres become more numerous when silica proportion is increased beyond the pessimum the gel product becomes more thinly distributed in the concrete allowing a more rapid uptake of calcium ions converting it to non-expansive CSH and also allowing rapid carbonation of the gel that has formed (Hobbs, 1981).
2 Composition of the gel formed changes as reaction proceeds leading to gel compositions which have lower expansion potential (Struble & Diamond, 1981b; Taylor, 1990; Helmuth et al., 1993).
3 As the reactive silica surface area increases more gel product is produced thus increasing the expansion. However, at the same time the concentration of alkali hydroxide available reduces rapidly as the gel forms leading to slower reaction rates so that less gel is formed at later stages, thus reducing expansion (Ozal, 1975).
4 Initial reaction and formation of gel is rapid so that a proportion of the early expansion is absorbed by the still semi-plastic concrete before it has fully set (Ozal, 1975; Hobbs, 1988). The reaction then slows because of the much reduced alkali hydroxide concentration remaining.
5 Thinner distribution of the gel may lead to reduced expansive stresses in the concrete (Taylor, 1990).
6 With slowly reacting alkali-aggregate systems (curve C, Figure 1.5), the production and possible change in gel composition is also slow, hence a pessimum proportion is not observed.

The underlying assumptions of these suggestions are that the amount of gel product formed will be proportional to the expansion produced. Also, that the rate of reaction depends on the concentrations of the reactants so will slow down as they get used up in the formation of the gel.

However, if reaction is rapid and the available alkali, OH- (and calcium hydroxide) is used up then this in turn may lead to changes in gel composition and its expansive properties. In Table 1.3 it can be seen that gel compositions are very variable particularly with respect to calcium and the silica/alkali ratio.

Although the reduced expansions observed beyond the pessimum proportion are matters of debate, it would seem that combinations of causes are the most reasonable explanation. Changes in gel composition as expansion proceeds and the more widely distributed reaction sites are perhaps the most likely cause of the reduced expansion. These effects could form the subjects of careful experimental research.

1.6 AVOIDANCE OF ASR

A great deal of global research effort has been directed toward the avoidance of ASR. In summary, avoidance measures are best accomplished if one or more of the reactants can be removed from the concrete, or is reduced to a concentration level that is too low to allow the reaction to initiate.

As is indicated in later chapters, many countries have produced guidance documents and specifications specifically designed to avoid the possibility of damaging ASR in concrete. In addition to other information given in the later chapters relating to the avoidance measures adopted in particular regions, a more detailed general consideration of the standards and specifications relevant to the testing and assessment of materials and of concrete with the objective of identifying and excluding material or mix designs that may induce damaging ASR is covered in Chapter 2.

Chapter 4 specifically addresses the methods whereby ASR can be avoided and the most widely used guidance documents relevant to this objective. It also discusses the general approaches to the manufacture and mix design which may be used to mitigate and minimise the incidence and effects of ASR in concrete.

1.7 THE EFFECT OF ASR ON CONCRETE STRUCTURES

In global terms the incidence of ASR in structural concrete is rare and is by no means as common as, for example, the corrosion of reinforcing steel which is probably the most important mechanism of deterioration of reinforced concrete structures worldwide.

However, where it does occur it is likely to be a major problem for the owner of the structure. Also, because ASR is a reaction that occurs throughout the concrete mass it is very difficult, or impossible to control, repair permanently, or to provide satisfactory remediation once it has been initiated. Furthermore, the cracking it induces allows other processes of deterioration to develop. Consequently it is both difficult and costly to deal with and in many cases eventual replacement of the structure is the option selected.

Chapter 5 considers matters of repair and remediation in detail and examines the options currently available and their effectiveness. It gives advice on how to assess a damaged structure and offers suggestions of the treatment options based on the feedback of past experience. The later chapters of this book give numerous specific case study examples of individual structures from round the world that have been the subject of repair procedures and will provide the reader with an insight as to the relative effectiveness of particular measures.

REFERENCES

ASTM C227 (2010) Standard test method for potential alkali reactivity of cement-aggregate combinations (Mortar-bar method). *The American Society for Testing and Materials*, West Conshohocken, USA.

Barisone, G. & Restivo, G. (1992b) Alkali-silica reactivity of alluvium deposits evaluated using chemical and psammographic methods. In: Poole, A. (ed.) *Proceedings of the 9th International Conference on Alkali-Aggregate Reactivity in Concrete (ICAAR)*. London, UK, pp. 46–52.

Bennett, I.C. & Vivian, H.E. (1955) Studies in cement-aggregate reaction xxii: The effect of fine ground opaline material on mortar expansion. *Aust J Appl Sci.*, 6, 88–93.

Berube, M.A. & Fournier, B. (1986) Products of alkali-silica aggregate reaction in concrete: Study of a case of the Quebec region. *Can Mineral*, 4 (2), 271–288.

Bolivar, I.C.O.G. (2003) *Manual of aggregates for concrete*, Universidad Nacional de Colombia, Facultad de Minas, Escuela de Ingenieria Civil, Medellin. (in Spanish).

Broekmans, M.A.T.M. (2002) The alkali-silica reaction: Mineralogical and geochemical aspects of some Dutch concretes and Norwegian mylonites. *PhD-thesis University of Utrecht, Geologica Ultraiectina* (217). p. 144.

BS 812-104 (1994) *Testing aggregates, method for qualitative and quantitative petrographic examination of aggregates*. London, British Standards Institution.

BS 7943 (1999) *Guide to the interpretation of petrographical examinations for alkali-silica reactivity*. London, British Standards Institution.

BS 812-123 (1999) *Testing aggregates, Part 123, Method for determination of alkali-silica reactivity, Concrete prism method*. London, British Standards Institution.

Charlwood, R., Sciivener, K. & Sims, I. (2012) Recent developments in the management of chemical expansion of concrete in dams and hydro projects – Pat 1: Existing structures. Hydro 2012, Bilbao, Spain.

Constantiner, D. & Diamond, S. (2005) Alkali release from feldspars into the pore solution. *Cement Concrete Res.*, 33, 549–554.

Davies, G. & Oberholster, R.E. (1986) The alkali-silica reaction product: A mineralogical and an electron microscope study. *Proceedings of the 8th International Conference on Cement Microscopy*. Duncanville, Texas I.C.M.A. pp. 303–326.

Demars, Ph. (2012) Les reactions alcalis-granulats dans les géres parale MET, *Unpublished Report*.

Deng, M., Xu, l., Lan, X. & Tang, M. (2008) Microstructures of alkali-aggregate reactivity of Emeishan group basaltic rocks. In: *Proceedings of the 13th International Conference on Alkali-Aggregate Reaction (ICAAR)*, Trondheim, no. 30.

Dent Glasser, L.S. & Kataoka, N. (1981a) Chemistry of the 'alkali-aggregate reaction'. *Cement and Concrete Research*, 11, 191–200.

Dent-Glasser, L.S. & Kataoka, N. (1981b) The chemistry of alkali-aggregate reactions. *Proceedings of the 5th International Conference on Alkali-Aggregate Reaction in Concrete*. Cape Town, South Africa. S525/23. 1–7.

Diamond, S. (1989) Another look at mechanisms. In: Okada, K., Nishibayashi, S., Kawamura, M. (eds.) *Proceedings 8th International Conference on Alkali-Aggregate Reaction*. Kyoto, Japan. pp. 83–94.

Dinku, A. (2005) The need for standardization of aggregates for concrete production in ethiopian construction industry. *International Conference on African Development Archives*. Paper 90.

Dinku, A. & Bogale, B. (2004) Alkali-aggregate reaction in concrete: A review of the Ethiopian situation. *Journal of EEA*, 21, 47–58.

Escadeillas, G., Gibergues, A.C. & Massias, E. (2000) A case study of alkali-silica reaction in high density concrete. In Bérubé, M.A., Fournier, B. & Durand, B. (eds.) *Proceedings of the 11th International Conference, Alkali-Aggregate Reaction in Concrete*. Quebec, QC Canada. pp. 831–839.

Fernandes, I., Ribeiro, A.Md., Broekmans, M.A.T.M. & Sims, I. (2016) *Petrographic atlas: Characterisation of aggregates regarding potential reactivity to alkalis. RILEM TC 219-ACS*

Recommended Guidance AAR-1.2, for use with the RILEM AAR-1.1 Petrographic Examination Method, Springer, Dordrecht, RILEM, Paris, p. 204.

French, J.W. (1991) Concrete petrography: A review. *Q J Eng Geol.*, *24* (1), 17–48.

French, W.J. (1980) Reactions between aggregates and cement paste: An interpretation of the pessimum. *Q J Eng Geol.*, *13* (4), 231–248.

Gillott, J.E. (1975) Alkali-aggregate reactions in concrete. *Eng Geol.*, *9*, 303–326.

Góralczyk, S. (2001) Usefulness of carbonate aggregates to concrete, PhD Dissertation, AGH, University of Science and Technology, Cracow.

García, F.D.F. (2013) *Alkali-silica reactivity potential of combinations of cementious materials and aggregates – Metod of mortar bar.* Ecuela Politéica Nacional, Facultad de Ingeniería civil y Ambiental, Quito, Ecuador, Proyecto previo a la obtención del título de inginero civil. (in Spanish).

Hansen, W.C. (1944) Studies relating to the mechanism by which alkali-aggregate reaction produces expansion in concrete. *J Am Concr Inst.*, *15*, 213–217.

Hanson, K.F., Van Dam, T.J., Peterson, K.R. & Sutter, L.L. (2003) Effect of sample preparation on the chemical composition and morphology of alkali-silica reaction products. TRB *Annual Meeting*, Downloadable CD Rom.

Helmuth, R. Stark, D. Diamond, S. & Moranville-Regourd, M. (1993) *Alkali-silica reactivity: an overview of research.* Strategic Highways Research Council, Washington DC, USA.

Hobbs, D.W. (1981) The alkali-silica reaction – a model for predicting expansion in mortar. *Mag Concrete Res.*, *33* (117), 208–220.

Hobbs, D.W. (1988) *Alkali-silica reaction in concrete.* Thomas Telford Ltd, London. pp. 1–183.

Idorn, G.M. (1967) *Durability of concrete structures in Denmark.* Copenhagen, Danish Technical Press.

Katayama, T. (1997) Petrography of alkali-aggregate reaction in concrete – Reactive minerals and reactive products, East Asia Alkali-Aggregate Reaction Seminar, Supplementary papers A43–59.

Katayama, T. (2010) The so-called alkali-carbonate reaction (ACR) – Its mineralogical and geochemical details, with special reference to ASR. *Cement Concrete Res.*, *40* (4), 643–675.

Katayama, T. (2012) ASR gels and their crystalline phases in concrete: Universal products in alkali-silica, alkali-silicate and alkali-carbonate reactions. In: Drimalas, T., Ideker, J.H. & Fournier, B. (eds.) *Proceedings of the 14th International Conference on Alkali-Aggregate Reaction in Concrete.* Austin, Texas, USA. 030411-KATA-03. pp. 1–12.

Kawamura, M., Koike, M. & Nakano, K. (1989) Release of alkalis from reactive andesitic aggregates and fly ashes into pore solution in mortars. In: Okada, K., Nishibashi, S. & Kawamura, M. (Eds.) *Proceedings of the 8th International Conference on Alkali-Aggregate Reaction in Concrete.* Kyoto, Japan. pp. 271–278.

Lagerblad, B. (2012) Alkali release from silicate minerals and alkali-silica reaction in concrete. In: Drimalas, T., Ideker, J.H. & Fournier, B. (eds.) *Proceedings of the 14th International Conference on Alkali-Aggregate Reaction in Concrete.* Austin, Texas, USA. 052411-LAGE. pp. 1–6.

Marfil, S. & Maiza, P. (2008) Petrographic study of a building deteriorated due to alkali-silica reaction, in Buenos Aires city (*Argentina*). In: *Proceedings of the 13tth International Conference on Alkali-Aggregate Reaction (ICAAR)*, Trondheim. pp. 983–993.

Milanesi, C.A., Locati, F. & Marfil, S. (2016) Microstructural and chemical study on an expansive dolostone from Argentina. In: de Mayo Bernardes, H. & Hasparyk, N.A. (eds.) *Proceedings of the 15th International Conference on Alkali-Aggregate Reaction in Concrete.* Sao Paulo, Brazil 4–7, July 2016 (downloadable from the 15th ICAAR web site).

Milanesi, C.A., Marfil, S., Maiza, P.J. & Batic, O.R. (2012) Expansive dolostone from Argentina: The common dilemma ACR or another variant of ASR? In: Drimalas, T.,

Ideker, J.H. & Fournier, B. (eds.) *Proceedings of the 14th International Conference on Alkali-Aggregate Reaction in Concrete.* Austin, Texas, USA. 030211-MILA. pp. 1–10.

Mladenovic, A., Šturm, S., Mirtic, B., & Šuput, J.S. (2009) Alkali-silica reaction in mortars made from aggregates having different degrees of crystallinity. *Ceramics-Silikáty.*, *53* (11), 31–41.

Moore, A.E. (1978) An attempt to predict the maximum forces that could be generated by alkali-silica reaction. In: Diamond, S. (ed.) *Proceedings of the 4th International Conference on the Effects of Alkalies in Cement and Concrete.* Perdue University, Indiana USA. pp. 363–368.

Nielsen, H.O., Grelk, B. & Nymand, K.K. (2004) ASR – Alkali-Silica Reactions, Dansk Vejtidsskrift no. 2, (in Danish) pp. 10–13.

Nixon, P.J., Hawthorn, F. & Sims, I. (2004) Developing an international specification to combat AAR. Proposals of RILEM TC 191-ARP. In: Tang, M. & Deng, M. (eds.) *Alkali Reactions in Concrete. Proceedings of the 12th International Conference on Alkali-Reaction in Concrete.* International Academic Publishers/World Publishing Corporation. pp. 8–16.

Nixon, P.J. & Sims, I. (eds.) (2016) RILEM recommendations for the prevention of damage by alkali-aggregate reactions in new concrete structures. *RILEM State-of-the-Art Reports, 17,* Springer, Dordrecht, RILEM, Paris, p. 168.

Olague, C., Wenglas, G. & Castro, P. (2003) Influence of alkalis from different sources than cement in the evolution of alkali-silica reaction. *Materiales de Construcción.* [Online] 53 (271–272), 189–198. Available from: http://tecnociencia.uach.mx/numeros/v6nl/data [Accessed 8th May 2014] (in Spanish).

Ozal, M.A. (1975) The pessimum proportion as a reference point in modulating alkali-silica reaction. *Symposium on alkali-silica reaction, preventative measures*, Reykjavik, Iceland. pp. 113–129.

Patrone, J.C. (2008) Effects and Remedy of the expansive concrete of the "Baygorria" dam, *Memorias* [Online], 6, Available from: http://www.um.edu.uy/_upload/_investigacion/web_investigacion_51_Memoria_3_ExpansionHormigon.pdf [Accessed 11 May 2014]. (in Spanish).

Patrone, J.C. (2013) Intervention in the turbines of the Baygorria dam affected by alkali-aggregate reaction, *Congreso Argentino de Presas y Aprovechamientos Hidroeléctricos – CAPyAH 2013*, Comité Argentino de Presas, San Juan. (in Spanish).

Poole, A.B. (1975) Alkali-silica reactivity in concrete from Dhekelia, Cyprus. *Symposium on alkali-silica reaction, preventative measures*, Reykjavik, Iceland. pp. 101–112.

Poole, A.B. (1981) Alkali-carbonate reactions in concrete. In: Oberholster, R.E. (ed.) *Proceedings of the 5th International Conference on Alkali-Reaction in Concrete.* National Building Research Institute of the CISR, Cape Town, South Africa. S252/34. pp. 1–7.

Poole, A.B. (2010) The alkali-aggregate reaction (AAR) damage to concrete. In: Soustos, M. (ed.) *Concrete durability, a practical guide to the design of durable concrete structures.* Thomas Telford, London. pp. 65–69.

Poole, A.B. & Sims, I. (2016) *Concrete Petrography: A handbook of investigative techniques, Second Edition.* CRC Press, Taylor and Francis Group, London & New York.

Powers, T.C. & Steinour, H.H. (1955) An interpretation of some published researches on alkali-aggregate reaction, Part 1 – The chemical reactions and mechanisms of expansion. *Journal of the American Concrete Institute*, Michigan, USA, *51*, 497–516. PCA Research Dept. Bull. 55.

Pyy, H. Holt, E. & Ferreira, M. (2012) Pre-study on alkali aggregate reaction and its existing in Finland. VTT Helsinki, *Report VTT-CR-0054-12/Fi 27s* (In Finnish).

Ranjaraju, P.R., Afshinnia, K., Engugula, R. & Latifee, E.R. (2016) Evaluation of alkali-silica reaction potential of marginal aggregates using miniature concrete prism test (MCPT). In: de Mayo Bernardes, H. & Hasparyk, N.A. (eds.) *Proceedings of the 15th International Conference on Alkali-Reaction in Concret.* Sao Paulo, Brazil 4–7 July 2016 (downloadable from the 15th ICAAR web site).

RILEM (2003) Recommended test method AAR-0: Detection of alkali-reactivity potential in concrete – Outline guide to the use of RILEM methods in assessments of aggregates for potential alkali-reactivity (prepared by Sims, I. and Nixon, P.), *Mater Struct.*, *36* (261), 472–479.

Silva, A.S., Fernandes, I., Sores, D., Custódio, J., Riberio, A.B., Ramos, V. & Medeiros, S. (2016) Portuguese experience in ASR assessment. In: Barfoot, J. *Proceedings of the 15th International Conference on Alkali-Reaction in Concrete.* Sao Paulo, Brazil 4–7 July 2016 (downloadable from the 15th ICAAR web site).

Sims, I. & Poole, A.B. (1980) Potentially alkali-reactive aggregates from the Middle East. *Concrete.*, *14* (5) 27–30.

Stanton, T.E. (1940): Expansion of concrete through reaction between cement and aggregate. *Proceedings ASCE*, 66 (10), 1781–1811.

Stark, D. & Bhatty, M.S.Y. (1986) Alkali silica reactivity: effect of alkali in aggregate on expansion. In: *Alkalis in concrete.* American Society for Testing and Materials, Philadelphia USA. ASTM STP 930, pp. 16–33.

Struble, L. & Diamond, S. (1981) Swelling properties of synthetic alkali silica gels. *J Am Ceram Soc.*, *64* (11), 652–656.

Struble, L. & Diamond, S. (1981) Unstable swelling behaviour of alkali silica gels. *Cement Concrete Res.*, 11, 611–617.

Swamy, R.N. (ed.) (1992) *The Alkali-Silica Reaction in concrete.* Glasgow, London, Blackie and Sons Ltd.

Swenson, E.G. (1957) A Canadian reactive aggregate undetected by ASTM tests. *ASTM Bulletin*, 226, 48–51.

Taylor, H.F.W. (1990) *Cement chemistry.* London, San Diego, Academic Press Ltd.

Thaulow, N. (1983) Alkali-silica reaction in the Itezhitezhi dam project, Zambia. In: Idorn, G.M. and Rostam, S. (eds.) *Proceedings of the 6th International Conference on Alkalis in Concrete.* Copenhagen, Denmark. pp. 471–477.

Thomas, M.D.A. (2006) The role of calcium in alkali-silica reaction. *Materials Science of Concrete – The Sidney Diamond Symposium*, 325–337.

Thomas, M.D.A. (2011) The effect of supplementary cementing materials on alkali-silica reaction: A review. *Cement Concrete Res. 41* (3), 209–216.

Vivian, H.E. (1950) The reaction product of alkalis and opal. *Studies in Cement-Aggregate Reaction. CSIRO Bulletin* 256, Commonwealth Scientific and Industrial Research Organization, Australia, pp. 60–81.

Vivian, H.E. (1951) Studies in cement aggregate reaction. XVI The effect of hydroxyl ions o the reaction of opal. *Aust J Appl Sci.*, *2*, 108–113.

Weymouth, J.H. & Williamson, W.D. (1951) Some physical properties of raw and calcined flint. *Mineralogical Magazine 29*, 557–557.

Wigum, B.J. (2010) *Petrographic examination of state concrete.* Report from Mannvit to the Icelandic Housing Financing Fund.

Wigum, B.J., Bjarnosan, E. & Hólmgeídottir, Þ. (2009) *Steypuskemmdir í púnsumynigrien 20 ára.* (in Icelandic) *(Concrete deterioration in houses younger than 20 years)* Report from Mannvit to the Icelandic Housing Financing Fund.

Wigum, B.J., Haugen, M., Skjølsvold, O. & Lindgård, J. (2004) Norwegian petrographic method – Development and experiences during a decade of service, In: Tang, M. & Deng, M. (eds.) *Proceedings of the 12th International Conference on AAR in Concrete*, Beijing (China), October, 2004, pp. 444–452.

Chapter 2

Assessment, Testing and Specification

Philip Nixon & Benoit Fournier

2.1 INTRODUCTION

As noted in Chapter 1, the Alkali-Silica Reaction (ASR) is a result of the reaction between the alkaline pore solution in concrete and reactive silica in the aggregate. The reaction leads to the formation of a secondary reaction product, a so-called alkali-silica gel, which can absorb water and exert an expansive force on the concrete. In certain conditions, and as a function of the extent of internal/external restraints applying on the affected structure/element, the reaction can lead to damaging expansion and cracking in the concrete. For such damaging expansion to occur, all of the following conditions must be present simultaneously:

- A sufficiently alkaline pore solution
- A critical amount of reactive silica
- A sufficient supply of water

Effective specifications to avoid damage from the reaction are based on ensuring that at least one of these conditions is absent. Additionally, there are interactions with other environmental actions, such as freezing and thawing, application of de-icing salts and exposure to a marine atmosphere, for which allowance must be made.

Such specifications can result in greater costs and in adverse environmental effects, for example by restricting the choice of aggregates or enhancing disposal of alkaline cement kiln dust, so it is also important to tailor the precautions to the nature and service life of the structure. Therefore, the development of the precautions should take the following form:

1. Determination of the necessary level of precaution;
2. Undertaking recommendations according to the level of precaution required.

This chapter is closely based on RILEM Recommendation AAR-7.1 *International Specification to Minimise Damage from Alkali Reactions in Concrete: Part 1 Alkali-Silica Reaction* (included in Nixon & Sims, 2016). This has been developed by international experts based on the most advanced specifications around the world and is now being used widely. References are also given, when relevant, to other existing standard test methods and specifications in current use in North America and in Australia where similar specifications have also been developed over the past few years.

2.2 DETERMINNG THE LEVEL OF PRECAUTION AND THE PRECAUTIONARY MEASURES

The key activities required by RILEM AAR-7.1 are the characterisation of the structural needs, the service life and the characterisation of the environment.

2.2.1 Determining the degree of reactivity of the aggregate

The continuing series of RILEM technical committees on AAR, which started work in 1988, were initially charged with establishing reliable procedures for determining aggregate reactivity on a worldwide basis. Accordingly, an integrated assessment system was developed and the current guidance and methods are described in Nixon and Sims (2016). Basically, it is a three stage scheme (explained in their AAR-0), starting with petrographic examination (designated AAR-1.1), including an optional screening step using accelerated mortar bar tests (designated AAR-2 or 5, for siliceous or carbonate aggregates) and, when necessary, undertaking a concrete prism test (designated AAR-3 or 4.1, for the conventional 38 °C or accelerated 60 °C versions). RILEM has also developed a petrographic atlas (designated AAR-1.2) to assist with the first stage of this assessment scheme (Fernandes *et al.*, 2016a). This RILEM scheme (including a flow chart in Figure 2.1) is explained in more detail in Section 2.4.

In North America, the CSA Standard Practice A23.2-27A (CSA, 2014), AASHTO Standard Practice PP-65 (AASHTO, 2010) and ASTM Standard Guide C1778 (ASTM, 2014), first require determination the degree of reactivity of the concrete aggregates in the process of establishing the level of prevention to implement. This is based on the fact that, when the other conditions for the development of AAR are present, the level of prevention generally increases with the degree of reactivity of the aggregate. Based on the expansions measured in the Concrete Prism [*e.g.*, CSA A23.2-14A (2014) and ASTM C1293 (2008b)] or the Accelerated Mortar Bar [*e.g.*, CSA A23.2-25A (2014) and ASTM C1260 (2014)] tests, the aggregate can be classified into four categories, as shown in Table 2.1 for ASTM Standard Guide C1778 (ASTM, 2014).

Similarly, in the Australian specification SA HB 79: 2015, an aggregate reactivity classification (*i.e.*, non-reactive, slowly reactive and reactive) is used for selecting preventive measures against AAR, in combination with an initial assessment of the risk category of the proposed structure and an evaluation of its in-service environment.

2.2.2 Structural needs and service life

Damaging alkali-aggregate reactions are slow and progressive. Typically, in temperate and cooler climates, they begin to cause visible damage after 5 to 10 years and then may continue for 20 or 30 years or longer; in warmer climates, the reactions are usually accelerated (Fournier *et al.*, 2010). However, some slowly reacting aggregates result in deterioration that takes much longer to develop, but can eventually be more destructive. The damage is evidenced as cracking and expansion of the concrete, occasionally leading to relative displacement between individual elements in a structure. While this has structural consequences, particularly if the reinforcement is insufficient, any deterioration will normally be evident well before there is danger to the integrity of the structure and may cause serious operational issues, especially in the case of hydraulic

Table 2.1 Classification of aggregate reactivity, after ASTM C1778 (2014).

Aggregate reactivity class	Description of aggregate reactivity	14-day accelerated mortar bar expansion (ASTM C1260)	1-year Concrete Prism Expansion (ASTM C1293)
R0	Non-reactive	< 0.10%	< 0.04%
R1	Moderately reactive	≥ 0.10%, < 0.30%	≥ 0.04%, < 0.12%
R2	Highly reactive	≥ 0.30%, < 0.45%	≥ 0.12%, < 0.24%
R3	Very highly reactive	≥ 0.45%	≥ 0.24%

Note: slight differences in the above limits exist between CSA Standard Practice A23.2-27A (CSA, 2014), AASHTO Standard Practice PP-65 (AASHTO, 2010) and ASTM Standard Guide C1778 (ASTM, 2014).

dams, locks and related structures. RILEM 7.1, as well as SA HB 79:2015, indicate three levels of categorization of structure according to the risks associated with any deterioration and the consequent need for precautions (see Table 2.2): S1 – low risk; S2 – normal risk; S3 – high risk.

It is the responsibility of the owner, or authority responsible for the structure, to decide on the appropriate level of risk in co-operation with the designer. This decision will be affected by the economic effects of any failure or deterioration as well as engineering and safety considerations. Other factors to be taken into account are the ease with which any deterioration can be detected, monitored and managed, the importance of the appearance of the structure and likely public perceptions of safety. Table 2.2 presents the criteria which will assist in making this decision, as per RILEM 7.1 and the Australian guidelines SA HB 79:2015; it also includes reference to the North American guidelines where the structures are classified in four categories (instead of the three categories proposed in RILEM 7.1). Examples of structures in each category are also provided in Table 2.2, although some owners may decide to use their own classification system.

2.2.3 Characterisation of the environment

When all the necessary compositional factors are present, the likelihood and extent of damaging alkali-silica reaction is dependent above all on the supply of moisture. In the majority of cases, a supply of moisture extraneous to the concrete itself is necessary. Other aggravating factors, which will influence the likelihood of damage and its severity, include the application of sodium chloride based de-icing salts, exposure to seawater and the synergistic effects of freezing and thawing damage. In ASR-affected concrete roads, the stress variation caused by fluctuating loads may also accelerate the deterioration process. The following three levels of categorisation of environment are therefore appropriate:

E1. The concrete is essentially protected from extraneous moisture
E2. The concrete is exposed to extraneous moisture
E3. The concrete is exposed to extraneous moisture and additionally to aggravating factors, such as sodium chloride based de-icing salts, freezing and thawing or wetting and drying in a marine environment

Table 2.2 Structures classified by risk category [adapted from the following guidelines: RILEM 7.1 (Nixon & Sims, 2016), SA HB 79: 2015, CSA A23.2-27A (2014), ASTM C1778 (2014) and AASHTO PP-65 (2010)].

Guidelines RILEM, SA HB 79: 2015	CSA, ASTM & AASHTO	Category – consequences of damage	Acceptability of ASR damage	Examples
S1	St1, S1, SC1	Safety, economic or environmental consequences of deterioration small or negligible	Some deterioration from ASR is acceptable	• Non load-bearing elements inside buildings • Temporary or short service life structures (likely design life 10 to 20 years; 5 years (CSA); < 50 years (SA HB 79)) • Small numbers of easily replaceable elements (SA HB 79, RILEM) • Most low-rise domestic structures (SA HB 79, RILEM)
S2	St2, S2, SC2	Some safety, economic or environmental consequences if major deterioration	Minor ASR damage is acceptable manageable	• Most building and civil engineering structures • Precast elements where economic costs of replacement are severe; e.g., railway sleepers • Normally designed for service life up to 100 years
		Some safety, economic, or environmental consequences if major deterioration	Moderate risk of ASR is acceptable	• Pedestrian pavements (sidewalks), curbs, and gutters • Service-life < 40 years • Most low-rise domestic structures (CSA)
	St3, S3, SC3	Significant safety, economic, or environmental consequences if minor damage	Minor risk of ASR acceptable	• Pavements, culverts, highway barriers • Rural, low-volume bridges/roads • Large numbers of precast elements where economic costs of replacement are severe • Service life normally 40 to 75 years (74 years (ASTM))
S3	St4, S4, SC4	Serious safety, economic or environmental consequences if any deterioration/minor damage	No significant damage acceptable (ASR cannot be tolerated – CSA, ASTM, AASHTO)	• Long service life (+100 years; +75 years (CSA, ASTM, AASHTO) or highly critical structures/elements where the risk of deterioration from AAR damage is judged unacceptable, such as: • Nuclear installations, dams, tunnels • Exceptionally important bridges or viaducts • Structures retaining hazardous materials • Exceptionally critical elements impossible/very difficult to inspect or replace/repair • Structures where the economic risk of non-serviceability would be unacceptable

Table 2.3 Environmental classes [from RILEM 7.1 (in Nixon & Sims, 2016)].

Environmental class	Description	Environment of concrete (see Note)
E1	Dry environment protected from extraneous moisture	• Internal concrete within buildings in dry (1) service conditions
E2	Exposed to extraneous moisture	• Internal concrete in buildings where humidity is high; e.g., laundries, tanks, swimming pools • Concrete exposed to moisture from the external atmosphere, to non-aggressive ground or immersed in water • Internal mass concrete should be included in this category (2)
E3	Exposed to extraneous moisture plus aggravating factors	• Internal or external concrete exposed to deicing salts • Concrete exposed to wetting and drying by seawater (3) or to salt spray • Concrete exposed to freezing and thawing whilst wet • Concrete subjected to prolonged elevated temperatures whilst wet • Concrete roads subject to fluctuating loads

Notes:
1. A dry environment corresponds to an ambient average relative humidity condition lower than 75% (normally only found inside buildings) and no exposure to external moisture sources.
2. A risk of alkali-silica reaction exists for mass concrete elements in a dry environment because the internal concrete may still have a high relative humidity. Vulnerable mass concrete elements are those with a least dimension of 1 m or more.
3. Concrete constantly immersed in seawater does not suffer a higher risk of ASR than a similar element exposed to humid air, buried in the ground, or immersed in pure water, because the alkali concentration of sea water is lower than the alkali concentration of the pore solution of most concretes, and the penetration of chloride ions is usually limited to a few centimetres.

More detail on the factors affecting the environmental categorisation is given in Table 2.3. A similar approach is used in the North American (CSA, AASHTO, ASTM) and the Australian guideline documents mentioned before. In all cases, the critical impact of moisture on the development of ASR is acknowledged. However, in the AASHTO and ASTM guidelines, the number of the *Environmental classes* increases from 3 to 4, as the E1 class is subdivided into two classes, *i.e.*, for non-massive or massive (least dimension >0.9 m) elements maintained in a dry environment (see Table 2.5). This is actually covered in Note 2 of Table 2.3.

2.2.4 The level of precaution and precautionary measures required

The structural and environmental categorisations are combined into the level of precaution in Table 2.4, where the following four levels of precaution are identified:

P1. No special precautions against AAR
P2. Normal level of precaution
P3. Special level of precaution
P4. Extraordinary level of precaution

Table 2.4 Determination of level of precaution [adapted from RILEM 7.1 (in Nixon & Sims, 2016) and SA HB 79:2015].

Category of Structure (Table 2.2)	Environment Category (Table 2.3)		
	E1	E2	E3
	Level of Precaution (see note)		
S1	P1	P1	P1
	Nil	Nil	Low
S2	P1	P2	P3
	Low	Standard	Standard
S3	P2	P4	P4
	Standard	Extraordinary	Extraordinary

Note: in the table, the *Precaution levels* are included for both RILEM 7.1 (P1 to P4) and Australian standard SA HB 79:2015 (Nil to Extraordinary).

The levels of precaution P1 to P4 will in turn translate into the following precautionary measures M1 to M4 for ASR, which will be described in detail in section 2.2.5:

M1: Measures to restrict the alkalinity of the pore solution
M2: Measures to ensure the use of a non-reactive aggregate combination
M3: Measures to reduce the access of moisture and maintain the concrete in a sufficiently dry state
M4: Measures to modify the properties of any gel such that it is non-expansive

2.2.4.1 Level of precaution P1 (RILEM)

At this level of precaution, appropriate standards and guidance should be followed for the specification of the concrete and good practice employed in its placing and curing but no special precautions against AAR damage are necessary. However, it should be realized that if this level of precaution is adopted, some damage from ASR is possible. Therefore, the structure must be able to withstand this and the level of damage must be acceptable to the owner. The definition of P1 in RILEM 7.1 corresponds to that of the "Nil" level in SA HB 79:2015.

2.2.4.2 Level of precaution P2 (RILEM)

This *normal* level of precaution against AAR damage is appropriate to structures where minor ASR damage is acceptable or damage can be monitored and managed. The definition of P2 in RILEM 7.1 corresponds to that of the "Low" level in SA HB 79:2015. In precaution level P2, one of the RILEM precautionary measures M1, M2, M3 or M4 should be applied in the case of S2 (normal risk) structures, or one of the precautionary measures M1, M2 or M4 in the case of S3 (high risk) structures.

2.2.4.3 Level of precaution P3 (RILEM)

This *special* level of precaution is appropriate where, like P2, minor ASR damage is acceptable or damage can be monitored and managed but where the structure is exposed to aggravating factors such as de-icing salts, freezing and thawing or wetting and drying in a marine environment. The definition of P3 in RILEM 7.1 corresponds to that of the "Standard" level in SA HB 79:2015. In the case of P3, one of the RILEM precautionary measures M1 to M4 should be applied and, additionally, the concrete should be designed to resist the aggravating factor; *e.g.*, it should be freeze/thaw resistant or it should resist the ingress of de-icing salts or seawater.

2.2.4.4 Level of precaution P4 (RILEM)

This *extraordinary* level of precaution (same in SA HB 79:2015) is only needed for structures where the consequences of any deterioration are unacceptable. In general, it will necessitate the combined application of at least two of the RILEM precautionary measures M1 to M4 and, additionally, the concrete in environmental class E3 should be designed to resist any aggravating factors such as freezing and thawing whilst wet, de-icing salts or wetting and drying in a marine atmosphere. This level should not be specified without careful thought and good reason, as it will almost certainly result in increased construction costs.

In the case of some large/remote structures such as dams, where for environmental and/or economic reasons it is necessary to use local materials, it may not be possible to apply two separate precautionary measures. In that case, extra protection can be obtained by the more rigorous application of one of the precautionary measures; *e.g.*, the use of a low alkali limit (see 2.2.5.1) and the inclusion of a fly ash or slag (see 2.3.3), or the use of a non-reactive aggregate combination (see 2.2.5.2) subject to a lower acceptance limit in the testing. Additionally, extra protection can be achieved through design and construction measures, such as detailing of the reinforcement, weather protection of critical elements, drainage or inclusion of expansion joints.

A very similar approach is applied in the North American guidelines [CSA A23.2-27A (2014), AASHTO PP-65 (2010), ASTM C1778 (2014)]. It is also based on the size/environment and a classification of the "criticality" of the structure, but also, as mentioned in section 2.2.1, on the degree of aggregate reactivity (Table 2.5). This risk analysis results in six potential levels of precaution (V to ZZ), covering cases where no specific preventive measures are needed (V≈ P1) to those where exceptional preventive actions (ZZ ≈ P4) are required (Table 2.6).

Table 2.5 Determination of the level of ASR Risk [after ASTM C1778 (2014), AASHTO PP-65 (2010)]

Size and exposure conditions	Aggregate reactivity class (Table 2.1)			
	R0	R1	R2	R3
Non-massive[A] concrete in a dry environment[B]	Level 1	Level 1	Level 2	Level 3
Massive[A] concrete in a dry environment[B]	Level 1	Level 2	Level 3	Level 4
All concrete exposed to humid air, buried or immersed	Level 1	Level 3	Level 4	Level 5
All concrete exposed to alkalis in service[C]	Level 1	Level 4	Level 5	Level 6

[A] Massive element: least dimension >0.9 m.
[B] Dry environment: average ambient RH <60 % (e.g., interior of buildings).
[C] Includes, for example, marine structures exposed to seawater, highway structures exposed to de-icing or anti-icing salts.

Table 2.6 Determination of the level of prevention [after ASTM C1778 (2014)].

Level of ASR risk (Table 2.5)	Classification of structure (Table 2.2)			
	Class SC1	Class SC2	Class SC3	Class SC4
Risk Level 1	V	V	V	V
Risk Level 2	V	V	W	X
Risk Level 3	V	W	X	Y
Risk Level 4	W	X	Y	Z
Risk Level 5	X	Y	Z	ZZ
Risk Level 6	Y	Z	ZZ	Note

Note: it may not be permitted to construct a SC4 structure if the risk of ASR is Level 6. Measures should be taken to reduce the level of risk in these circumstances.

2.2.5 Precautionary Measures

Different precautionary measures will be appropriate in different countries/regions according to their particular materials and practices. Overall, the possible measures can be categorized as M1, M2, M3 or M4 as described hereafter [after RILEM AAR-7.1 (in Nixon & Sims, 2016)].

2.2.5.1 Precautionary measure M1: Limiting the alkalinity of the pore solution

There are a number of ways of achieving this:

- Limiting the alkali content of the concrete
- Using a low alkali cement
- Including a sufficient proportion of a low-lime fly ash, another pozzolana demonstrated to be effective, or ground granulated blastfurnace slag in the concrete

Concrete alkali content: In many respects, limiting the alkali content of the concrete is the most easily applied and monitored of the measures. The principal source of alkalis that control the alkalinity of the concrete pore solution is the alkali content of the Portland cement (Diamond, 1989). This is usually expressed as the equivalent sodium oxide content:

$$\% \ Na_2O \ equivalent \ (Na_2Oeq) = \% \ Na_2O + 0.658 \ \% \ K_2O$$

The alkali content of the concrete is then the equivalent alkali content of the cement multiplied by the cement content of the mix plus any other reactive alkalis that should be included:

$$Concrete \ alkali \ (kg/m^3) = cement \ alkalis \ (\%) \ x \ cement \ content \ (kg/m^3)$$
$$+ \ other \ sources \ of \ alkali$$

The effective application of this measure requires the support of the cement manufacturing and/or supply industry in a particular country as it needs both the quality assured declaration of the average cement alkali level from a particular work and a measure of its variability. This is now done in several countries (*e.g.*, France, UK, Canada & the USA) as part of the control and standardisation process of concrete. Alternatively, users will need to assess cement alkali content by the reliable independent analysis of representative samples of the cement.

Other sources of alkalis: There is no universal consensus on what other sources should be included. The general practice is to include alkalis from these sources:

- Residual salt in marine aggregates
- Cementitious additions such as fly ash and slag
- Admixtures
- Mix water, especially if not from potable sources

The proportion of alkalis that should be included from fly ash and slag is dependent on the proportion used and their reactivity (see 2.3.3). In a blended cement containing fly ash or slag, the manufacturer will need to declare an "effective" alkali content, calculated from the alkali in the Portland cement and the alkalis to be included (if any) from the fly ash or slag on the basis of 2.3.3.

Admixtures, particularly some superplasticisers, can contain substantial alkalis. The calculation of the alkalis in the concrete will need to be based on the effective alkali content declared by the manufacturer.

More problematic is the question of alkalis originating from the body of the aggregate. Many aggregates contain alkalis, but the extent to which these are released and contribute to the pore solution alkalinity remains uncertain and possibly variable (Bérubé & Fournier, 2004; Menéndez et al., 2016; Soares et al., 2016). There is as yet no widely accepted method for assessing the potentially releasable alkali content of aggregates although a method is currently being developed by RILEM (to be designated AAR-8). At present, RILEM 7.1 (in Nixon & Sims, 2016) recommends that, unless there is national evidence to the contrary, this source of alkalis should not be included in the calculation except in concrete in S3 structures. In such structures, the possibility of alkali release by the aggregate should be considered when using aggregates containing significant amounts of altered and/or weathered feldspar, or other minerals capable of releasing alkalis.

2.2.5.2 Precautionary measure M2: Ensuring the use of a non-reactive aggregate combination

In the context of ASR, reactive silica occurs almost exclusively in the aggregate. Therefore, to make use of this precautionary measure, tests are needed to identify "non-reactive aggregate" combinations.

The proportion of silica that can lead to the most damaging reaction will depend on the reactivity of the silica. In some cases, a small amount of highly reactive silica in the aggregate (e.g., opal, cristobalite, volcanic glass) will be most damaging, whereas, if the aggregate contains a high proportion of such highly reactive silica, there may be little damage. If an aggregate containing highly reactive silica is mixed with a non-reactive one, the behaviour of the mix will vary from very damaging to not damaging at all, depending on the proportions of the mix. This feature is known as the "pessimum" effect. Conversely, for aggregates containing forms of silica of relatively lower reactivity (e.g., micro- to cryptocrystalline quartz, strained quartz) or when the silica is not easily exposed to the alkaline pore solution, the worst damage may occur when the greatest amount of silica is present (Thomas et al., 2013a).

Because of this, it is important that the whole aggregate combination is assessed, as amounts of reactive silica that are innocuous in either the fine or coarse aggregate alone may be damaging in the combined aggregate. Conversely, apparently reactive fine or coarse aggregates may be safe when used in combination. When used correctly, controlling this 'pessimum' effect can be an effective way of combating AAR damage and making use of available natural resources.

The testing of aggregates to identify potential reactivity is considered in detail in 2.4.1 below.

2.2.5.3 Precautionary measure M3: Reducing moisture access and maintaining the concrete in sufficiently dry condition

This can be achieved at the design stage by the use of, for example, external cladding or tanking (protection using a completely waterproof barrier). Inclusion of well-designed drainage that can be inspected and maintained is also important. Use of cladding on a concrete structure in a cold and very humid environment may however not sufficiently reduce the moisture content in the air behind the cladding. The use of cladding is regarded as most feasible in a warm and dry environment.

The effectiveness of surface treatments (coatings and impregnations) in preventing (or in arresting) AAR damage has been found to be variable, and so surface treatment is not regarded as a sufficient precautionary measure on its own. If a surface treatment is applied as an extra precaution measure, it should be vapour permeable so that the concrete can dry out when the humidity is low (Thomas et al., 2013b). The maintenance of any coating and/or drainage measures used is vital.

2.2.5.4 Precautionary measure M4: Modifying the properties of the gel such that it is non-expansive.

Inclusion of sufficient, soluble, lithium salts in the concrete mix water can be an effective means of counteracting AAR damage. The mechanisms enabling lithium to be effective are however not fully understood. It is believed by some to operate by modifying the nature of the ASR gel such that it does not absorb water and exert an expansive force (Feng et al., 2010). On the other hand, while Tremblay et al. (2010) found an increased chemical stability of reactive silica in the presence of lithium salts, Leemann (2016) suggested that the ASR-suppressing effect of lithium is solely caused by the formation of a protective lithium-silicate-hydrate layer at reactive sites.

Lithium nitrate is the preferred salt. The recommended dosage levels depend on the alkali level in the concrete and the nature of the aggregate. Recent North American research suggests that lithium salts are not equally effective with all reactive aggregate types (Tremblay et al., 2007). Some aggregates require much higher doses of lithium than others, whilst expansion with other reactive aggregates can sometimes be controlled with lower lithium doses than have previously been recommended. Unlike other measures based on using additions, such as fly ash or ground granulated blastfurnace slag, the lithium dose required does not appear to be related to the degree of aggregate reactivity. Accordingly, it is not possible at present to recommend a single dosage of lithium nitrate that will be effective for all aggregates. Instead, performance testing to evaluate its effectiveness and determine an appropriate lithium dosage is recommended for those considering the use of precautionary measure M4 in new concrete.

North American experience suggests that currently the best means of evaluation is a version of a concrete prism test such as RILEM AAR-3 (in Nixon & Sims, 2016) or

CSA A23.2-28A (2014) with an acceptance criterion of 0.040% after two years (CSA A23.2-28A).

2.3 METHODS OF MINIMISING ALKALI REACTIVITY (REFERENCE TO PRECAUTIONARY MEASURE M I)

2.3.1 Alkali limits and aggregate reactivity

The alkali limits specified for the concrete should be set by national guidance based on experience and practice in the particular country/region.

There is now good evidence that the level of alkali in the concrete necessary to cause a damaging alkali reaction depends on the reactivity of the aggregate (Sibbick & Page, 1992; RILEM AAR-7.1). Therefore, to enable the most efficient use to be made of the local aggregate and cement resources, the alkali limits should be based on the reactivity of the aggregate to be used. Aggregates that are essentially non-reactive can be identified by petrographic methods such as RILEM AAR-1.1. The reactivity of more reactive materials is best defined in terms of the alkali threshold at which damaging ASR is first identified. To establish this threshold, a concrete prism method such as RILEM AAR-3.2 (in Nixon & Sims, 2016), ASTM C1293-08b (2014) or CSA A23.2-14A (2014) can be used. However, the reliable determination of the alkali threshold may be misled due to the leaching of alkalis that occurs under the high temperature and humidity conditions prevailing during concrete prism testing, especially for concrete specimens with relatively small cross-section (*e.g.*, 70-75 mm) (Lindgård *et al.*, 2013).

Alternatively, a particular country can designate the reactivity of particular rock types on the basis of their geology and known behaviour in concrete structures. When classifying the reactivity of an aggregate, it is important to assess the whole aggregate combination in order to allow for any potential "pessimum" effects (see 2.2.5.2).

Three levels of aggregate reactivity can usefully be differentiated (RILEM AAR-7.1):

1 Low reactivity aggregates.

   These correspond to the aggregates identified as "non-reactive".

2 Medium reactivity aggregates.

   These are aggregates that fall neither into the low nor high reactivity categories. They may well be the majority of aggregates in some countries, *e.g.*, the UK (where these medium reactivity aggregates are termed "normal reactivity"), and include the siliceous sands and gravels that arc found widely (see Chapter 6). They will have alkali thresholds of exceeding 4.0 kg/m^3 sodium oxide equivalent (Na$_2$Oeq).

3 High reactivity aggregates.

   This category corresponds to the aggregates shown by tests using concrete specimens to have low alkali thresholds; typically 4.0. kg/m^3 Na$_2$Oeq or less.

N.B. Aggregates containing substantial proportions of opal are likely to be even more reactive than high reactivity aggregates and should either not be used in concrete or used only with special precautions that have been proven effective by trials or performance tests etc.

Table 2.7 Alkali limits and aggregate reactivity. (after RILEM 7.1)

Aggregate reactivity	Alkali limit (kg/m³ Na₂Oeq)
Low	None required
Medium	Typically 3.0 to 3.5 kg/m³
High	Typically 2.5 to 3.0 kg/m³

Examples of using the aggregate reactivity categories to set alkali limits are shown in Table 2.7 [after RILEM AAR-7.1 (in Nixon & Sims, 2016)].

Alkali limits such as those in Table 2.7 allow a "safety margin" compared with the alkali thresholds determined in the laboratory (*e.g.*, if the determined threshold was 4.0 kg/m³, in this example, the limit has been set at 3.0 kg/m³: *i.e.*, a 1.0 kg/m³ safety margin). This is to allow for the known differences between laboratory and field specimens, experimental uncertainty (*e.g.*, from alkali leaching) and site batching variability. National specifications making use of such limits will need to decide on an appropriate "safety margin" from local experience.

Where, in a particular country, the alkali contents of the cements used are within a known narrow range, restrictions on the cement content, according to aggregate reactivity and environment, can be used as an alternative method of limiting the alkali levels in the concrete mix. This can be easier for the concrete producer to apply, but care must be taken as the composition of the cements may change over time because of, for example, changes in raw materials or importation of cements or clinkers, use of alternative fuels, etc.

In the North American approach, a "sliding-scale" is also proposed for the alkali limit used as a preventive measure, which is based on the risk analysis mentioned before (which includes the determination of the aggregate's reactivity level) (Table 2.1). The value for the allowable concrete alkali content ranges from 3.0 to 1.8 kg/m³ Na₂Oeq for mild (W) to strong (Y) prevention levels (Table 2.8). Very strong (Z) and exceptional (ZZ) preventive actions call for either stronger alkali restrictions (Z) or the combination of alkali control and the use of supplementary cementing materials (SCMs) (Z & ZZ – Table 2.12).

Table 2.8 Maximum alkali contents in Portland cement concrete to provide various levels of prevention [after CSA A23.2-27A (2014), AASHTO PP-65 (2010) and ASTM C1778 (2014)].

Prevention level (Table 2.6)	Alkali limit (Na₂Oeq), kg/m³*
(V)	No limit
Mild (W)	3.0
Moderate (X)	2.4
Strong (Y)	1.8
Very strong (Z)	Table 2.12
Exceptional (ZZ)	Table 2.12

*Calculated from the total alkali equivalent for the Portland cement fraction of the binder multiplied by the Portland cement content of the concrete mixture.

2.3.2 Low Alkali Cements

Although less precisely connected with the alkali concentration in the concrete pore solution, this is the longest established and most pragmatic countermeasure. In the case of low alkali Portland cements, an upper limit of 0.60 %Na_2O equivalent is generally applied. The use of such low alkali cements has been found to be effective in some regions in preventing AAR damage, although there are occasional reports of damage despite their use and there is evidence from field trials in North America that with some aggregates damaging expansion can occur in concrete specimens made with low alkali cements (MacDonald *et al.*, 2012; Fournier *et al.*, 2016).

Drawbacks are that cements guaranteed to meet this limit will often be more costly and, to achieve it, the manufacturer may have to discard kiln dust with adverse environmental effects. Moreover, this measure may not be effective in the case of concretes with unusually high cement contents, if there are significant sources of internal or extraneous alkali, or if the passage of moisture concentrates the alkalis in certain parts of the structure.

Some countries have also designated low alkali slag cements. In Germany, for example, two types of low alkali blastfurnace slag cement are defined, the allowable alkali content depending on the percentage of slag. In Austria, there is good experience of using Portland-slag cements, compliant with particular Austrian standards, which have quite high alkali levels. With local experience, such approaches can be an effective way of using these supplementary cementing materials in avoiding ASR damage.

2.3.3 Use of fly ash, other pozzolana, slag and other mineral additions

The use of concrete containing such additions, and of cements in which they are interground or mixed during manufacture, has been the subject of much controversy and research. It is clear that some of these materials can be very effective in combating ASR damage. However, their variability, internationally, makes it problematic to give specific recommendations. The following summarizes the general consensus.

Low lime fly ashes, *e.g.*, to BS EN 450-1 or ASTM C618 class F, and ground granulated blastfurnace slags that are well established as effective cementitious materials or constituent materials of the cement, will provide effective protection against AAR damage provided a sufficient proportion (as a proportion of the total cementitious material) is used (Thomas *et al.*, 2013a). The proportion necessary will depend, amongst other things, on their composition and on the reactivity of the aggregate (Table 2.9).

Table 2.9 Recommended minimum fly ash and slag proportions [after RILEM AAR-7.1 (in Nixon & Sims, 2016)].

Aggregate reactivity	Low lime fly ash (<8% CaO and <5%Na_2Oeq)	Medium lime fly ash (8 to 20% CaO & <5%Na_2Oeq)	Ground granulated blastfurnace slag (<1.5% Na_2Oeq)
	% by mass of total cementitious material		
Low	any	any	any
Medium	25	30	40
High	40	not recommended	50

Table 2.10 Alkali contributions from lower proportions of additions than are recommended in Table 2.9 [after RILEM AAR-7.1 (in Nixon & Sims, 2016)].

Proportion of addition in cement		Proportion of alkali from addition to include in calculation of alkali content of concrete mix
GGBS	25 – 39%	50%
	<25	100%
Fly ash	20 – 24%	20%
	<20	100%

Provided that these minimum proportions are used, and subject to local experience with particular materials, the alkali content of the fly ash or slag needs not be included in the calculation of the "reactive" alkalis in the concrete. If lower proportions of addition are used, some countries (*e.g.*, UK, France, Australia) recommend that a proportion of the alkali content of the ash or slag is included in the calculation of the alkali content of the concrete mix. Based on these, a tentative recommendation is given in Table 2.10. In the Australian guidelines SA HB 79-2015, the recommended fly ash content (with a total alkali content < 3% Na_2Oeq) for ASR mitigation is 25%, while it is 65% for slag (with a total alkali content < 1% Na_2Oeq).

There is also good evidence that other highly active pozzolanic additions, such as silica fume and metakaolin, can be effective in protecting against ASR damage. Typical recommendations are that for concrete containing medium reactivity aggregates, the minimum proportions are as follows: 8% silica fume (> 85% SiO_2); 15% metakaolin (> 45% SiO_2). This approach is not recommended for concrete containing high reactivity aggregates. Also, if such materials are used, it is vital that they are well dispersed in the concrete as agglomerations of silica fume have caused damaging ASR expansions (Maas *et al.*, 2007).

Otherwise, sources of fly ash [*e.g.*, high lime (> 20% CaO)] or slag that are not well established, or other pozzolanic materials or mineral additions, should only be used if their performance has been established by a two year concrete performance test. It is preferable that such performance tests are backed up with long-term outdoor field tests or by site experience since some accelerated laboratory tests, using elevated humidities and temperatures, may give unrealistically optimistic results for the effectiveness of some pozzolanic materials in combating ASR expansion.

A similar "prescriptive" approach is proposed in North America, as recommendations are provided for the minimum levels of supplementary cementing materials (SCMs) when the required preventive level ranges from mild (W) to very strong (Z) [CSA A23.2-27A (2014), AASHTO PP-65 (2010), ASTM C 1778 (2014)] (Table 2.11). Within the above range of prevention levels, and considering some variations between the ASTM, CSA and AASHTO guidelines, the recommended replacement levels range as follows:

- 15 to 40% (ASTM & AASHTO) / 45% (CSA) for fly ash with a CaO content up to 18 % (ASTM & AASHTO) / 20% (CSA), and with an alkali content up to 4.0% (ASTM) / 4.5% (CSA & AASHTO) Na_2Oeq; and
- 25 to 60% (CSA) / 65% (ASTM & AASHTO) for slag with Na_2Oeq < 1.0%.

Table 2.11 Minimum levels of SCM to provide adequate level of prevention [adapted from ASTM C1778 (2014), AASHTO PP-65 (2010) and CSA A23.2-27A (2014)].

Type of SCM	Alkali level of SCM (Na_2Oeq)	Minimum replacement (% by mass of cementitious materials)				
		Level W	Level X	Level Y	Level Z	Level ZZ
Fly ash	CaO ≤ 15% – CSA; CaO ≤ 18% – ASTM & AASHTO	< 3.0 – CSA & ASTM ≤ 3.0 – AASHTO	15	20	25	35	Exceptional requirement (see Table 2.12)
		≥ 3.0 ≤ 4.5 – CSA ≥ 3.0 ≤ 4.0 – ASTM > 3.0 ≤ 4.5 – AASHTO	20	25	30	40	
	CaO 15 – 20% – CSA	< 3.0	20	25	30	40	
		≥ 3.0 ≤ 4.5	25	30	35	45	
GGBS	≤ 1.0 – ASSHTO < 1.0 – CSA & ASTM	25	35	50	60 (CSA); 65 (ASTM, AASHTO)	
Silica Fume (SiO_2 ≥ 85%)[1]	≤ 1.0 – ASSHTO < 1.0 – CSA & ASTM	2.0 x KGA	2.5 x KGA	3.0 x KGA	4.0 x KGA[2]	

[1] The amount of silica fume is calculated on the basis of the alkali loading of the concrete contributed by the Portland cement (in kg/m³ – KGA).

[2] In the case of Prevention Level Z, CSA requires that the minimum cement replacement level by silica fume be determined in accordance with CSA A23.2-28A (2014).

The recommended replacement levels of silica fume (>85% SiO_2) for preventing ASR is calculated on the basis of the alkali loading of the concrete contributed by the Portland cement (in kg/m^3). The levels range from 2.0 to 4.0 times the total alkali content in the mix (Table 2.11). In all cases, the minimum proportion of silica fume, when used as a sole preventive action, must be >7%. Exceptional preventive actions (ZZ) call for a combination of alkali control and the use of SCMs (Table 2.12).

The minimum replacement levels prescribed for any given preventative level are valid for concretes made with cements with alkali content in the range of 0.70 to 1.0% Na_2Oeq. The minimum replacement levels can be reduced or must be increased when cements with lower or higher alkali levels than the above values are used, respectively (Table 2.13).

No "prescriptive" recommendations are given for fly ashes with Na_2Oeq or CaO contents greater than the values mentioned in Table 2.11. The efficacy of high alkali/calcium fly ashes, in addition to all combinations of reactive aggregates and SCMs (binary and ternary systems), can be evaluated through a "performance" approach,

Table 2.12 Options for using SCM and/or limiting alkali loading of concrete to provide highest levels of prevention [adapted from CSA A23.2-27A (2014), ASTM C1778 (2014) and AASHTO PP-65 (2010)].

Prevention level	CSA A23.2-27A	ASTM C 1778 & ASSHTO PP-65
Very strong (Z)	• Control of alkalis only: 1.2 kg/m^3; or SCM only: level Z in Table 2.11; or 1.8 kg/m^3 + SCM (level Y – Table 2.11)	• Control of alkalis only: not allowed SCM only : level Z in Table 2.11; or 1.8 kg/m^3 + SCM (level Y in Table 2.11)
Exceptional (ZZ)	• Control of alkalis only: not allowed SCM only : not allowed 1.2 kg/m^3 + SCM (level Z in Table 2.11)	• Control of alkalis only: not allowed SCM only : not allowed 1.8 kg/m^3 + SCM (level Z in Table 2.11)

Table 2.13 Adjusting minimum SCM level based on the alkali content of the Portland cement. [after CSA A23.2-27A (2014), ASTM C1778 (2014) and AASHTO PP-65 (2010)].

Cement alkali content (% Na$_2$Oeq)	Level of SCM
< 0.70 – CSA & ASTM; ≤ 0.70 – ASSHTO	Reduce the minimum amount of SCM given in Table 2.11 by one prevention level [1]
0.70 to 1.00	Use the minimum levels of SCM given in Table 2.11
>1.00 to 1.25[2]	Increase the minimum amount of SCM given in Table 2.11 by one prevention level
>1.25	No guidance given

[1] The replacement levels should not be below those given in Table 2.11 for prevention level W, regardless of the alkali content of the Portland cement (ASTM & AASHTO).
[2] CSA A23.2-27A (2014) provides further recommendations as a function of the reactivity level of the aggregates. So, for moderately-reactive aggregates, the levels of SCMs in Table 2.11 can be used with a Portland cement of ≤ 1.15% Na_2Oeq, while the level of SCMs needs to be increased by one prevention level for Portland cements of 1.15% to 1.25% Na_2Oeq. In the case of highly- and extremely-reactive aggregates, the levels of SCMs in Table 2.11 can be used only with a Portland cement of ≤ 1.00% Na_2Oeq; however, the level of SCMs shall be increased by one prevention level for Portland cements of 1.00% to 1.25% Na_2Oeq,

using the concrete prism and/or the accelerated mortar bar tests in accordance with test procedures described in CSA A23.2-28A (2014), AASHTO PP-65 (2010) or ASTM C1778 (2014).

2.4 TESTS AND SPECIFICATIONS

Evaluating the potential alkali-reactivity of concrete aggregates is generally considered as the first step in preventing deleterious reaction in concrete.

2.4.1 Accelerated test methods for aggregates

2.4.1.1 General comments

ASR is a slow reaction, typically taking many years to produce visible damage in structures. Therefore, when developing test methods to identify vulnerable aggregates, research workers have almost always made attempts to accelerate the reaction in some way. This can be done by intensifying one of the parameters controlling the speed of reaction; temperature, availability of water, alkalinity around the aggregate or the fineness of the aggregate. By increasing the temperature, ensuring a saturated atmosphere, increasing the alkalinity and crushing the aggregate, rapid reactions can be achieved reducing the testing time to days if necessary. Unfortunately, the more the reaction is accelerated, the less the conditions resemble those actually experienced by the aggregate in concrete and the less reliable the result can be. Usually this will result in false positives; *i.e.*, the identification of aggregates as reactive which can be used safely in practice. Sometimes, however, greatly accelerated tests can fail to identify reactivity in aggregates, which then do cause damage in practice. Usually, this is a problem with slow reacting aggregates which cause damage over very long timescales in structures, perhaps 20 to 50 years.

Ever since AAR was first recognized as a threat to concrete, there have been attempts to develop accelerated tests to identify aggregates that are vulnerable to the reaction. Early research in the USA led to a suite of methods that were standardized by ASTM and became the most widely used methods worldwide for many years. They were the ASTM C227 mortar-bar method, and the ASTM C289 chemical method, of which the latter has now been withdrawn. In C227-10 (ASTM, 2014), mortar-bars made with aggregate and high alkali cement are stored in containers at high humidity and 38°C and the expansion of the bars monitored over a 3 to 6-month period. In the withdrawn chemical method, a 25g sample of the aggregate (150 to 300 μm particle size) was immersed in a 1N NaOH solution at 80°C for 24 hours. The amount of dissolved silica from the aggregate material and the reduction in the alkalinity of the storage solution were then determined and plotted against each other on a graph which was divided into zones which were designated as deleterious, potentially deleterious or innocuous.

These methods could be effective with the types of aggregates that were first recognized as being involved in damaging ASR, *i.e.*, siliceous aggregates containing fast reacting minerals like opal or volcanic glass. However, as the methods became used more and more widely, with very different rock types from those for which they were developed, it became clear that they were not universally reliable. In particular, ASTM C227 failed to identify reactivity in more slowly reacting aggregates (Grattan-Bellew,

1978; Stark, 1980), including a highly-reactive greywacke/argillite aggregate that generated huge ASR issues in the famous Mactaquac dam in New Brunswick, Canada (Moffatt *et al.*, 2016).

In response to this, many countries began to develop tests that were specific to their aggregate types, for example the opaline sandstones in north Germany. Gradually, a proliferation of methods developed, from ultra-fast autoclave methods (Nishibayashi *et al.*, 1987; Fournier *et al.*, 1991; Tang *et al.*, 1983) to methods using quite large concrete specimens (Lindgård *et al.*, 2013). Alternative chemical methods were also proposed, such as the chemical shrinkage method in Denmark (Knudsen, 1987). Typically, these methods gave reliable results when used in the region where they were developed but were somewhat unreliable in other regions with different geologies.

It was because of this proliferation that the first RILEM TC on AAR was formed, RILEM TC 106, *Accelerated tests for aggregate reactivity*. At about the same time, the problem was also being identified in North America and the ASTM and Canadian Standards (CSA) committees began to review their methods. From this work, a consensus has begun to emerge. ASTM, CSA and RILEM all now recommend a methodology based on the following three stages: 1) petrographic examination, 2) screening using an accelerated mortar-bar test, and 3), longer term testing using concrete specimens. Integrated schemes based on the RILEM methods and the North American guidelines are shown in Figures 2.1 and 2.2, respectively.

2.4.1.2 Petrographic examination

This is a powerful method for identifying different types of siliceous phases in the aggregate. It does, however, depend on a knowledge of which types of silica are reactive in a particular region and this has to be developed from a study of concrete from damaged structures or laboratory specimens made with aggregates from the region. It is also a technique dependent on the experience and skills of the petrographer. Petrography can be used quantitatively to determine the proportion of reactive particles and this has been used in a few countries, *e.g.*, Denmark, Norway, Sweden, as the basis of specification. However, studies of the precision of petrography as a quantitative method, such as that carried out by the PARTNER Project (Lindgård *et al.*, 2010), have shown very poor interlaboratory agreement, especially where the personnel are unfamiliar with the particular rock types.

The most complete petrographic methodology for use in identifying alkali reactive aggregates is that produced by RILEM [designated AAR-1.1, in Nixon & Sims (2016)]. RILEM has also produced a petrographic atlas of reactive rock types to assist petrographers who are examining rock types with which they are unfamiliar [designated AAR-1.2 (Fernandes *et al.*, 2013, 2016a)]. Other guides to petrographic examination of aggregates have been published as ASTM C295-12 (2014) and BS 812-104 (1994).

Given the poor precision of the method, it is best used as a preliminary exercise to assist in the choice and effective use of other laboratory methods. This is the approach recommended by RILEM in their overall guidance on the use of the RILEM methods [designated AAR-0 in Nixon & Sims (2016)].

In the North American approach (Figure 2.2), the petrographic examination is used to identify potentially reactive rock types but also to identify aggregates that are

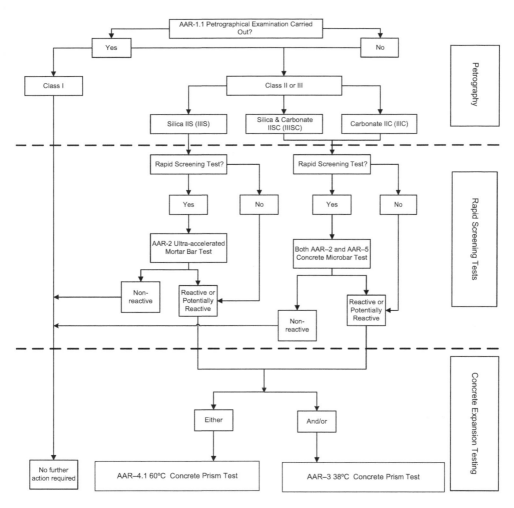

Figure 2.1 Integrated aggregate assessment scheme for the assessment of alkali-reactivity potential of aggregates (RILEM AAR-0).

produced in quarrying operations from carbonate rocks. Such materials are to be subjected to a quick screening chemical method for the detection of potential alkali-carbonate reactivity. This will be discussed in section 2.4.3.

2.4.1.3 Screening using an accelerated mortar-bar test (AMBT)

This is often the second step in the graduated approach for evaluating the potential alkali-silica reactivity of concrete aggregates. The advantage of such methods is that they can give an answer in a few weeks, or in the case of extremely accelerated tests, a few days. However, this high degree of acceleration is also liable to produce unreliable results, usually in the direction of unnecessarily identifying usable aggregates as being reactive (Thomas *et al.*, 2013a).

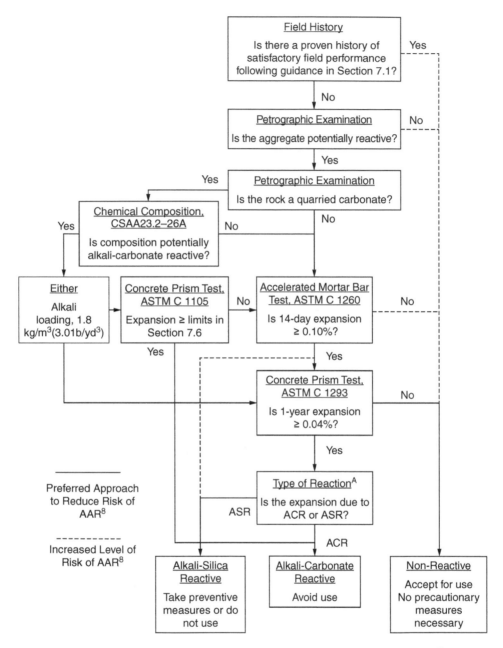

Figure 2.2 Sequence of laboratory tests for evaluating aggregate reactivity [after ASTM C1778 (2014)].

In recent years, a consensus has grown around the use of the accelerated mortar-bar test as the preferred screening method. This method was first developed in South Africa (Oberholster & Davies, 1986) and has been standardized by a number of standards bodies, *e.g.*, ASTM C1260 (2014), AASHTO T303 (2008), CSA A23.2.25A (2014), AS 1141 60.1 (2014) and RILEM AAR-2 (in Nixon & Sims, 2016). In the method, mortar bars made with the graded aggregate are immersed in a 1N sodium hydroxide solution at 80°C for a period of about 14 days (depending on the exact version of the method) and any expansion monitored.

Experience of the test has generally been positive and inter-laboratory trials by RILEM (Nixon & Sims, 2000) and the EU "PARTNER" project (Lindgård et al., 2010) have shown good correlation with practical experience. The expansion limits used in the test vary based on local experience, but generally ranges between 0.10% and 0.15% for non-reactive aggregates. There are suggestions that the method is unreliable with porous flint aggregates and there is some evidence, for example from Argentina, that some slowly reactive aggregates are not detected using the standard criteria. In Australia, expansion limits are used at 10 and 21 days for identifying slowly reactive aggregates (SA HB 79:2015).

The North American experience is such that the accelerated mortar-bar test yields results that will agree in a fair number of cases with the results from ASTM C1293-8b (2014) tests or field experience. However, there is also a wide range of aggregates that will give erroneous and misleading results in the accelerated test, when compared to the more accurate and realistic concrete prism test (Thomas et al., 2013a). In the North American testing scheme illustrated in Figure 2.2, it is recommended that quarried carbonate aggregates be subjected to a quick chemical method [CSA A23.2-26A (2014)] prior to potentially testing them in the AMBT. Testing has indeed shown that some argillaceous dolomitic limestone aggregates from the Kingston/Cornwall area in Ontario (Canada) that have been historically associated with *alkali-carbonate reactivity* and that have caused extensive cracking in field structures actually pass the AMBT (Thomas et al., 2013a). This will be discussed further in section 2.4.3.

2.4.1.4 Longer term tests using concrete specimens (CPT)

These are generally regarded as being the laboratory methods most likely accurately to reflect the behaviour in real structures. A number of standards bodies have published methods, e.g., RILEM as AAR-3 (in Nixon & Sims, 2016), ASTM C1293-8b (2014), CSA A23.2-14A (2014), BS 812-123 (1999), Norway (Norwegian Concrete Association, 2005), Germany (Deutscher-Ausschuss fur Stahlbeton, 2001) and AS 1141 60.2 (2014). The most usual specimen size is 75x75x250-300 mm, though in the German and Norwegian methods, larger specimens are used, e.g., 100x100x500 mm in the Norwegian CPT. Generally, a high content (420 – 440 kg/m^3) of a high alkali cement ($\approx$ 0.8 – 1.3% Na$_2$Oeq, depending on the standard method) is used to produce a high alkali level in the concrete; in some methods [e.g., CSA A23.2-14A (2014), ASTM C1293-8b (2014), AS 114 60.1 (2014)], NaOH pellets are dissolved in the mix water in order to increase the total alkali content in the concrete to 1.25% Na$_2$Oeq by mass of cement. The specimens are then stored at an elevated temperature (generally 38°C) and in high humidity. Any expansion is monitored and typically the test period is one year. Pessimum effects can be identified by varying the proportion of the test aggregate. The expansion limits used in the test vary based on local experience, but generally range between 0.03% and 0.05% for non-reactive aggregates.

The most problematic aspect of these methods has been found to be the balance between ensuring a sufficiently high humidity and minimizing loss of alkali by leaching

(Lindgård *et al.*, 2013). Earlier practices of using fog rooms or wrapping specimens in damp cloth have been found to produce excessive leaching so the consensus now is to store unwrapped specimens in airtight containers above water.

Inter-laboratory trials by RILEM (Nixon & Sims, 2000) and the EU "PARTNER" project (Lindgård *et al.*, 2010) have found good agreement with practice for most aggregates, though identification of slowly reacting aggregates may need either a longer test period or a lower acceptance criteria. Reasonable precision was found by the PARTNER project. However, Fournier *et al.* (2012) have shown that the multi-laboratory variability of the test can be somewhat high, depending on various factors such as the non-reactive fine aggregate used in combination with the coarse aggregate under test.

An accelerated concrete test has been developed by RILEM as AAR-4.1 (in Nixon & Sims, 2016). This is based on a French Performance Test (AFNOR NF P18-454, 2004; Sims & Nixon, 2006) in which concrete specimens are stored in a high humidity reactor at 60^{0}C. The test period is much shorter, 15 weeks in the RILEM method, and good correlation has been found with the 38°C tests (*e.g.*, Fournier *et al.*, 2004a). The precision of the test was found to be reasonable in the PARTNER trials and it was also found that this method was more effective at identifying slowly reacting aggregates.

2.4.1.5 Outdoor exposure sites

Over the past few decades, a number of outdoor exposure sites have been developed in order to establish correlations between rapid laboratory test methods and the long term performance in field conditions, especially in the following two areas: 1) evaluation of the potential alkali-reactivity of concrete aggregates, and 2) evaluation of the preventive effects of supplementary cementing materials and lithium-based admixtures (*e.g.*, Fournier *et al.*, 2004b, 2016; Thomas *et al.*, 2011; MacDonald *et al.*, 2012; Borchers & Müller, 2012). In such experiments, large concrete specimens are stored outdoors to give the most realistic simulation of real structures though the timescales are much longer.

2.4.2 Performance Tests of Concrete

A reliable performance test of the actual concrete mix to be used in a structure is the ideal way of making a decision on its suitability for use. So far, however, the difficulties of getting sufficient acceleration to give an answer in a useful timescale, whilst still giving a reliable assessment of performance, have limited progress in the development of such tests. AFNOR NF P18-454 (2004) have standardized a method based on monitoring the expansion of concrete prisms stored in a reactor at 60°C and this is the basis of the aggregate test [designated AAR-4.1 by RILEM, in Nixon & Sims (2016)] discussed above. One way that this test is used in France is to use it to establish the threshold level of alkalis necessary to give a damaging reaction and then to reduce the alkali level in the actual mix to be used by a safety factor.

RILEM has been working for several years to develop such a Performance Test and current thinking (by the present TC 258-AAA) is that this will be based on the AAR-3 (in Nixon & Sims, 2016) concrete prism test at 38°C (Wigum *et al.*, 2016).

2.4.3 Specification and Assessment of Carbonate Aggregates

Alkali-silica reactive limestone aggregates have been identified in several countries around the world. Research has shown that the reactive phase is micro- to cryptocrystalline quartz finely disseminated in the matrix of those rocks (Fernandes *et al.*, 2016b). In most cases, the deleterious expansion and cracking in concrete incorporating such aggregates, *e.g.*, the world-renowned Spratt's limestone, can be controlled through the use of an adequate quantity of efficient supplementary cementing materials such as slag and class F fly ash (Thomas *et al.*, 2013a; Fournier *et al.*, 2016). Experience in Austria suggests that some dolomitic carbonate aggregates may be usable in concrete in which the cement contains high replacement levels of blastfurnace slag (Thomas & Innis, 1998). On the other hand, the deleterious expansion of concrete incorporating some argillaceous dolomitic limestone aggregates from the Kingston/Cornwall area (Ontario, Canada), and from a few states in the USA, historically qualified as *alkali-carbonate reactive* aggregates, cannot be prevented through the use of low-alkali cements or the use of even large quantities of ground granulated blastfurnace slag or Class F fly ash (Thomas *et al.*, 2013a; Fecteau *et al.*, 2016; Shehata *et al.*, 2016). The exact mechanism involved with the above argillaceous dolomitic limestone aggregates is still the source of controversy, although Katayama (2010) suggested that is corresponds to a form of alkali-silica reaction (Jensen, 2012; Katayama *et al.*, 2016).

So, given the present level of knowledge and understanding, the primary strategy in the case of quarried carbonate rocks is thus focused on identifying and avoiding the use of the deleterious rock facies for which preventive actions will be ineffective. There are some distinct features for such materials that differentiate them from the "typical" alkali-silica reactive limestones, including the nature and composition of the rock, the timeframe of the reaction, the influence of aggregate size, the inability of some standard tests to identify reactive aggregates, the threshold alkali content required to generate expansion, and the impact of preventive measures such as pozzolana, slag, and lithium-based compounds (Thomas *et al.*, 2013a).

In general, the potential alkali-reactivity of carbonate aggregates is best determined through concrete prism testing. However, before running the CPT, RILEM suggests a screening process that combines testing the aggregate through the RILEM AAR-2 AMBT and a concrete microbar test [RILEM AAR-5 in Nixon & Sims (2016)]. The latter uses 40 x 40 x 160 mm specimens, composed of cement paste and aggregate particles of 4 to 8 mm in size. Considering that typical *alkali-carbonate* reactive aggregates generally induce limited expansion when ground to "mortar size fractions" for the AMBT, the comparative testing of the material in the RILEM methods AAR-2 and AAR-5 allows a quick assessment of potential *alkali-carbonate* reactivity. In the North American guidelines (*e.g.*, Figure 2.2), the potential *alkali-carbonate* reactivity of quarried carbonate aggregates can be assessed on the basis of their chemical composition using the test method CSA A23.2-26A (2014). This test involves the determination of the lime (CaO), magnesia (MgO), and alumina (Al_2O_3) content of the rock, and determining where the composition of the rock falls on a plot of CaO/MgO ratio versus the Al_2O_3 content, as shown in Figure 2.3. As mentioned above, confirmation of the potential *alkali-carbonate* reactivity, following any screening tests, can be achieved through the concrete prism tests [*e.g.*, RILEM AAR-3 [in Nixon & Sims (2016), ASTM C1293-8b (2014), CSA A23.2-14A (2014)].

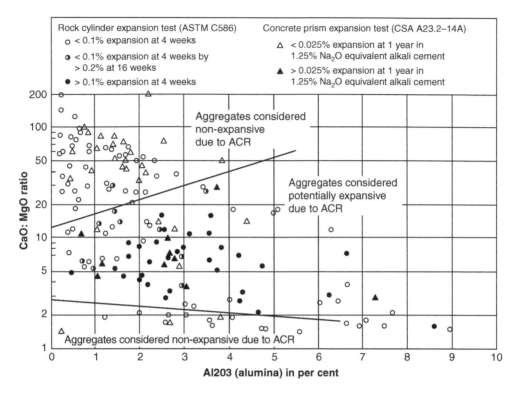

Figure 2.3 Using the chemical composition of the quarried carbonate aggregate to assess alkali-carbonate reactivity potential [after CSA A23.2-26A (2014)].

REFERENCES

AASHTO (2010) Designation PP-65, Standard practice for determining the reactivity of concrete aggregates and selecting appropriate measures for preventing deleterious expansion in new concrete construction, American Association of State Highway and Transportation Officials (AASHTO), Washington, DC (USA), 20p.

AASHTO (2008) Standard Method of Test for Accelerated Detection of Potentially Deleterious Expansion of Mortar Bars Due to Alkali-Silica Reaction. AASHTO T 303, American Association of State Highway and Transportation Officials, Washington, DC, 6p.

AFNOR NF P18-454 (2004) Reactivity of a concrete formula with regard to the alkali-aggregate reaction, Paris, and Interpretation Criteria for NF P 18-454.

AS 1141 60.1 (2014) Methods for sampling and testing aggregates – Potential alkali-silica reactivity – Accelerated mortar bar method. Prepared by the Standards Australia Committee CE-012, Aggregates and Rock for Engineering Purposes.

AS 1141 60.2 (2014) Methods for sampling and testing aggregates – Potential alkali-silica reactivity – Concrete prism method. Prepared by the Standards Australia Committee CE-012, Aggregates and Rock for Engineering Purposes.

ASTM (2014) *ASTM – Annual Book of ASTM Standards 04.02*, ASTM International, West Conshohocken (USA).

- ASTM C227-10. Standard Test Method for Potential Alkali-Silica Reactivity of Aggregates (Mortar Bar Method).
- ASTM C289-07. Standard Test Method for Potential Alkali-Silica Reactivity of Aggregates (Chemical Method).
- ASTM C295-12. Standard Guide for Petrographic Examination of Aggregates for Concrete.
- ASTM C618-15. Standard Specification for Coal Fly Ash and Raw or Calcined Natural Pozzolan for Use in Concrete.
- ASTM C1293-08b. Standard Method for Determination of Length Change of Concrete Due to Alkali Silica.
- ASTM C1260-14. Standard Test Method for Potential Alkali Reactivity of Aggregates (Mortar Bar Method).
- ASTM C1567-13. Standard Test Method for Determining the Potential Alkali-Silica Reactivity of Combinations of Cementitious Materials and Aggregate (Accelerated Mortar-Bar Method).
- ASTM C 1778-14. Standard Guide for Reducing the Risk of Deleterious Alkali-Aggregate Reaction in Concrete.

Bérubé, M.A., & Fournier, B. (2004) Alkalis releasable by aggregates in concrete – significance and test methods. In: *Proceedings of the 12th International Conference on Alkali-Aggregate Reaction (AAR) in Concrete*, Beijing (China), October 2004, Tang and Deng Editors, International Academic Publishers, Beijing World Publishing Corporation, 1, pp. 17–30.

Borchers, I., & Müller, C. (2012) Seven years of field site tests to assess the reliability of different laboratory test methods for evaluating the alkali-reactivity potential of aggregates. In: *Proceedings of the 14th International Conference on Alkali-Aggregate Reaction (AAR) in Concrete*, May 2012, Austin (Texas)).

BS 812-104. (1994) *Testing aggregates. Method for qualitative and quantitative examination of aggregates*. London, BSI Global.

BS 812-123. (1999) *Testing aggregates. Method for determination of alkali-silica reactivity. Concrete prism method*. London, BSI Global.

BS EN 450-1. (2012) *Fly ash for concrete. Definition, specification and conformity criteria*. London, BSI Global.

CSA A23.1. (2014) (Concrete materials and methods of concrete construction) & A23.2 (Test methods and standard practices for concrete), Canadian Standards Association, Mississauga, Ontario, Canada.

- CSA A23.2-14A. Potential expansivity of aggregates (Procedure for length change due to alkali-aggregate reactions in concrete prisms at 38°C), pp. 350–362.
- CSA A23.2-25A. Test method for detection of alkali-silica reactive aggregate by accelerated expansion of mortar bars, pp. 425–433.
- CSA A23.2-26A. Determination of potential alkali-carbonate reactivity of quarried carbonate rocks by chemical composition, pp. 434–438.
- CSA A23.2-27A. Standard Practice to identify degree of alkali-reactivity of aggregates and to identify measures to avoid deleterious expansion in concrete, pp. 439–451.
- CSA A23.2-28A. Standard Practice for laboratory testing to demonstrate the effectiveness of supplementary cementing materials and lithium-based admixtures to prevent alkali-silica reaction in concrete, pp. 452–457.

Deutscher Ausschuss fur Stahlbeton (ed.) (2001) *DAfStb, Vorbeugende Massnahmen gegen Alkalireaktion in Beton: Alklai-Richlinie*. Berlin, Beuth.

Diamond, S. (1989) ASR – Another look at mechanisms. In: Okada, K., Nishibayashi, S., & Kawamura, M. (eds.) *Proceedings of the 8th International Conference on Alkali-Aggregate Reaction*. The Society of Materials Science, Kyoto, Japan. pp. 83–94.

Fecteau, P.L., Fournier, B., & Duchesne, J. (2016) Use of SCMs on ACR-affected concrete: expansion and damage evaluation through the Damage Rating Index (DRI). In: *Proceedings of the 15th International Conference on Alkali-Aggregate Reaction (AAR) in Concrete*, July 2016, Sao Paulo (Brazil), 10p.

Feng, X., Thomas, M.D.A., Bremner, T.W., Folliard, K.J., & Fournier, B. (2010) Summary of research on the effect of $LiNO_3$ on ASR in concrete. *Cement Concrete Res.*, *40*, 636–642.

Fernandes, I., Broekmans, M.A.T.M., Nixon, P., Sims, I., Ribeiro, M.A., Noronha, F., & Wigum, B.J. (2013) Alkali-Silica reactivity of some common rock types: A global petrographic atlas, *Q J Eng Geol Hydroge.*, *46*, 215–220.

Fernandes, I., Ribeiro, M.A., Broekmans, M.A.T.M., & Sims, I. (2016a). *Petrographic atlas: characterisation of aggregates regarding potential reactivity to alkalis. RILEM TC 219-ACS Recommended Guidance AAR-1.2, for use with the RILEM AAR-1.1 Petrographic Examination Method*, Springer, Dordrecht. RILEM, Paris, 204p.

Fernandes, I., Broekmans, M.A.T.M., Ribeiro, M.A., Sims, I. (2016b). Assessment of the alkali-reactivity potential of sedimentary rocks. In: *Proceedings of the 15th International Conference on Alkali-Aggregate Reaction (AAR) in Concrete*, July 2016, Sao Paulo (Brazil), 10p.

Fournier, B., Bérubé, M.A., & Bergeron, G. (1991) A rapid autoclave mortar bar method to determine the potential alkali-silica reactivity of St. Lawrence Lowlands carbonate aggregates (Quebec, Canada). *Cement Concrete Aggr.*, *13* (1), 58–71.

Fournier, B., Chevrier, R., DeGrosbois, M., Lisella, R., Folliard, K., Ideker, J., Shehata, M., Thomas, M.D.A., & Baxter, S. (2004a) The accelerated concrete prism test (60°C): variability of the test method and proposed expansion limits. In: *Proceedings of the 12th International Conference on Alkali-Aggregate Reaction (AAR) in Concrete*, Beijing (China), October 2004, Tang and Deng Editors, International Academic Publishers, Beijing World Publishing Corporation, 1. pp. 314–323.

Fournier, B., Nkinamubanzi, P.C., & Chevrier, R. (2004b) Comparative field and laboratory investigations on the use of supplementary cementing materials to control alkali-silica reaction in concrete. In: *Proceedings of the 12th International Conference on Alkali-Aggregate Reaction (AAR) in Concrete*, Beijing (China), October 2004, Tang and Deng Editors, International Academic Publishers, Beijing World Publishing Corporation, 1. pp. 528–537.

Fournier, B., Ideker, J.H., Folliard, K.J., Thomas, M.D.A., Nkinamubanzi, P.C., & Chevrier, R. (2010) Effect of environmental conditions on expansion in concrete due to ASR. *Mater Charact.*, *60*, 669–679.

Fournier, B., Rogers, C.A., & MacDonald, C.A. (2012) Multilaboratory study of the concrete prism and accelerated mortar bar expansion tests with Spratt aggregate. In: *Proceedings of the 14th International Conference on Alkali-Aggregate Reaction (AAR) in Concrete*, May 2012, Austin (Texas)).

Fournier, B., Chevrier, R., Bilodeau, A., Nkinamubanzi, P.C., & Bouzoubaa, N. (2016) Comparative field and laboratory investigations on the use of supplementary cementing materials (SCMs) to control alkali-silica reaction in concrete. In: *Proceedings of the 15th International Conference on Alkali-Aggregate Reaction (AAR) in Concrete*, July 2016, Sao Paulo (Brazil), 10p.

Grattan-Bellew, P.E. (1978) Study of Expansivity of a suite of quartzites, argillites and quartz arenites. In: *Proceedings of the 4th International Conference on the Effect of Alkalies in Cement and Concrete*. Purdue University, Indiana. pp. 113–140.

Ideker, J.H., Bentivegna, A.F., Folliard, K.J., & Juenger, M.C.G. (2012) Do current laboratory test methods accurately predict alkali-silica reactivity ? *ACI Mater J.*, *109* (4), 395–402.

Jensen, V. (2012) The controversy of alkali-carbonate reaction: state-of-the-art on the reaction mechanisms and behavior in concrete. In: *Proceedings of the 14th International Conference on Alkali-Aggregate Reaction (AAR) in Concrete*, May 2012, Austin (Texas)).

Katayama, T. (2010) The so-called alkali-carbonate reaction (ACR) – Its mineralogical and geochemical details, with special reference to ASR. *Cement Concrete Res.*, 40, 643–675.

Katayama, T., Jensen, V., & Rogers, C.A. (2016) The enigma of the 'so-called' alkali-carbonate reaction. In: *Proceedings of the Institution of Civil Engineers: Construction Materials* (Part 2 of a themed issue on AAR), 169 (CM4), 10p.

Knudsen, T.A. (1987) A continuous quick chemical method for the characterization of the alkali-silica reactivity in aggregates. In: Grattan-Bellew, P.E. (ed.) *Proceedings 7th International Conference on AAR. Noyes publication, Park Ridge (NJ, USA).* pp. 289–293.

Leemann, A. (2016) The influence of lithium on the structure of ASR products in concrete. In: *Proceedings of the 15th International Conference on Alkali-Aggregate Reaction (AAR) in Concrete*, July 2016, Sao Paulo (Brazil), 9p.

Lindgård, J., Nixon, P.J., Borchers, I., Schouenborg, B., Wigum, B.J., Haugen, M., & Akesson, U. (2010) The EU "PARTNER" Project-European standard tests to prevent alkali reactions in aggregates: Final Results and Recommendations, *Cement Concrete Res.*, 40 (4), 611–635.

Lindgård, J., Thomas, M.D.A., Sellevold, E.J., Pedersen, B., Andiç-Çakır, O., Justnes, H., & Rønning, T.F. (2013) Alkali–silica reaction (ASR) – performance testing: Influence of specimen pre-treatment, exposure conditions and prism size on alkali leaching and prism expansion, *Cement Concrete Res.*, 53, 68–90.

MacDonald, C.A., Rogers, C.A., & Hooton, R.D. (2012) The relationship between laboratory and field expansion – Observations at the Kingston exposure site for ASR after twenty years. In: *Proceedings of the 14th International Conference on Alkali-Aggregate Reaction (AAR) in Concrete*, May 2012, Austin (Texas)).

Maas, A.J., Ideker, J.H., & Juenger, M.C.G. (2007) Alkali silica reactivity of agglomerated silica fume. *Cement Concrete Res.*, 37, 166–174.

Menéndez, E., García-Rovés, R., Aldea, B., & Ruíz, S. (2016) Alkali release of aggregates – effectiveness of different solutions and conditions of test. In: *Proceedings of the 15th International Conference on Alkali-Aggregate Reaction (AAR) in Concrete*, July 2016, Sao Paulo (Brazil), 10p.

Moffatt, E.G., Thomas, M.D.A., Hayman, S., Fournier, B., Ideker, J.H., & Fletcher, J. (2016) Remediation strategies intended for the reconstruction of the ASR-induced Mactaquac dam. In: *Proceedings of the 15th International Conference on Alkali-Aggregate Reaction (AAR) in Concrete*, July 2016, Sao Paulo (Brazil), 8p.

Nishibayashi, S., Yamura, K., & Matsushita, H. (1987) A rapid method of determining the alkali-aggregate reaction in concrete by autoclave. In: Grattan-Bellew, P.E. (ed.) *Proceedings 7th International Conference on AAR. Noyes Publication, Park Ridge (NJ, USA).* pp. 299–303.

Nixon, P.J., & Sims, I. (2000) Universally accepted testing procedures for AAR; the Progress of RILEM Technical Committee 106. In: Berube, M.A., Fournier, B., & Durand, B. (eds.) *Proceedings 11th International Conference on Alkali Aggregate Reaction, Quebec. (Canada)*, pp. 435–444.

Nixon, P.J., & Sims, I. (eds.) (2016) RILEM recommendations for the prevention of damage by alkali-aggregate reactions in new concrete structures. *RILEM State-of-the-Art Reports*, Volume 17, Springer, Dordrecht, RILEM, Paris, 168p.

Norwegian Concrete Association (2005) Norwegian Concrete Association, NB, Alkali-aggregate reaction in concrete, Test methods and Requirements to Test Laboratories, 2005, NB Publication no. 32, p. 39.

Oberholster, R.E., & Davies, G. (1986) An accelerated method for testing the potential reactivity of siliceous aggregates. *Cement Concrete Res.*, *16*, 181–189.

RILEM TC 258-AAA – Avoiding Alkali Aggregate Reactions in Concrete – Performance Based Concept. Technical committee for the period 2014–2019.

RILEM Recommendations for the Prevention of Damage by Alkali-Aggregate Reactions in New Concrete Structures, State-of-the-Art Report of the RILEM Technical Committee 219-ACS, RILEM State-of-the-Art Reports, Nixon, P.J. and Sims, I. (eds.), 17, DOI 10.1007/978-94-017-7252-5.

- RILEM Recommended Test Method: AAR-0, Outline Guide to the Use of RILEM Methods in the Assessment of Alkali-Reactivity Potential of Aggregates, pp. 5–34.
- RILEM Recommended Test Method: AAR-1.1, Detection of Potential Alkali-Reactivity – Part 1: Petrographic Examination Method, pp. 35–60.
- RILEM Recommended Test Method: AAR-2, Detection of Potential Alkali-Reactivity – Accelerated Mortar-Bar Test Method for Aggregates, pp. 61–78.
- RILEM Recommended Test Method: AAR-3, Detection of Potential Alkali-Reactivity – 38°C Test Method for aggregate Combinations Using Concrete Prisms, pp. 79–98.
- RILEM Recommended Test Method: AAR-4.1, Detection of Potential Alkali-Reactivity – 60°C Test Method for aggregate Combinations Using Concrete Prisms, pp. 99–116.
- RILEM Recommended Test Method: AAR-5, Detection of Potential Alkali-Reactivity – Rapid Preliminary Screening test for Carbonate Aggregates, pp. 117–130.
- RILEM Recommended Specification: AAR-7.1, International Specifications to Minimise damage from Alkali Reactions in Concrete – Part 1: Alkali-Silica Reaction, pp. 131–146.
- RILEM Recommended Specification: AAR-7.2, International Specifications to Minimise damage from Alkali Reactions in Concrete – Part 2: Alkali-Carbonate Reaction, pp. 147–154.

SA HB 79 (2015) *Handbook – Alkali aggregate reaction – Guidelines on Minimizing the Risk of Damage to Concrete Structures in Australia.* Australia, Sydney, SAI Global Limited. 78p.

Shehata, M., Jagdat, S., Lachemi, M., & Rogers, C. (2016) Alkali-carbonate reaction: Mechanisms and effects of cement alkali and supplementary cementing materials. In: *Proceedings of the 15th International Conference on Alkali-Aggregate Reaction (AAR) in Concrete*, July 2016, Sao Paulo (Brazil), 10p.

Sibbick, R.G., & Page, C.L. (1992) Susceptibility of various UK aggregates to alkali-aggregate reaction. In: *Proceedings of the 9th International Conference on Alkali-Aggregate Reaction in Concrete, Westminster, London*, Conference Papers, Vol. 2, Ref CS 104 (two volumes), The Concrete Society, Slough (now Camberley), UK. pp. 980–987.

Sims, I., & Nixon, P.J. (2006) Assessment of aggregates for alkali-aggregate reactivity potential: RILEM International Recommendations. In: Fournier, B. (ed.) *Marc-Andre Berube Symposium on Alkali-Aggregate reactivity in concrete.* Montreal Canada. pp. 71–91.

Soares, D., Santos Silva, A., Mirão, J., Fernandes, I., & Menéndez, E. (2016) Study on the factors affecting alkalis release from aggregates into ASR. In: *Proceedings of the 15th International Conference on Alkali-Aggregate Reaction (AAR) in Concrete*, July 2016, Sao Paulo (Brazil), 10p.

Stark, D. (1980) Alkali-silica reactions: Some reconsiderations. *Cement Concrete Aggr.*, *2* (2), 92–94.

Tang, M.S., Han, S.F., & Zhen, S.H. (1983) A rapid method for identification of alkali-silica reactivity of aggregate. *Cement Concrete Res.*, *13* (3), 417–422.

Thomas, M.D.A., & Innis, F.A. (1998) Effect of slag on expansion due to alkali-aggregate reaction in concrete, *ACI Mater J.*, *95*, 716–724.

Thomas, M.D.A., Fournier, B., & Folliard, K.J. (2013a) Alkali-aggregate reactivity (AAR) facts book. Federal Highway Administration (FHWA), U.S. Dept of Transportation, FHWA-HIF-13-019, 212p.

Thomas, M.D.A., Folliard, K.J., Fournier, B., Rivard, P., & Drimalas, T. (2013b) Methods for evaluating and treating ASR-affected structures: Results of field application and demonstration projects. Volume I: Summary of findings and recommendations. Report FHWA-HIF-14-002, Federal Highway Administration (FHWA), U.S. Dept of Transportation, November 2013, 70p.

Thomas, M.D.A., Dunster, A., Nixon, P., & Blackwell, B. (2011) Effect of fly ash on the expansion of concrete due to alkali-silica reaction – Exposure site studies. *Cement Concrete Comp.*, *33*, 359–367.

Tremblay, C., Bérubé, M.A., Fournier, B., Thomas, M.D.A., & Folliard, K.J. (2007) Effectiveness of lithium-based products in concrete made with Canadian aggregates. *ACI Mater J.*, *104* (2), 195–205.

Tremblay, C., Bérubé, M.A., Fournier, B., Thomas, M.D.A., & Folliard. K.J. (2010) Experimental investigation of the mechanisms by which $LiNO_3$ is effective against ASR. *Cement Concrete Res.*, *40*, 583–597.

Wigum, B.J., Lindgard, J., Sims, I., & Nixon, P.J. (2016) RILEM activities on alkali-silica reactions: From 1988 to 2019. In: *Proceedings of the Institution of Civil Engineers: Construction Materials (Part 2 of a themed issue on AAR)*, *169* (CM4). 4p.

Chapter 3

So-Called Alkali-Carbonate Reaction (ACR)

Paddy E. Grattan-Bellew & Tetsuya Katayama

3.1 INTRODUCTION

The term 'alkali-carbonate reaction' (ACR) would appear to apply to a reaction between the alkaline pore solution in concrete and any aggregate made from carbonate rocks such as limestones or dolostones (dolomite rocks). However, the term alkali-carbonate reaction is really applied only to the reaction between certain argillaceous dolomitic limestones and the alkaline pore solution in concrete that gives rise to rapid expansion and cracking of concrete. In North America the ACR dolomitic limestones are of Ordovician age, but in the Sichuan Province of China, some are of Triassic age. RILEM considered the use of carbonate-aggregate reaction (CAR) instead of ACR, but the latter was too established.

The argillaceous dolomitic limestones, that are quite rare, form in shallow intertidal or lagoon environments adjacent to the edge of an ancient continent (Tang *et al.*, 2000). The Pittsburg quarry in Kingston, Ontario, Canada, from where the ACR aggregate was first encountered, is in the Ordovician Gull River formation that lies unconformably upon the Pre-Cambrian Canadian Shield and is overlain by limestones formed deeper in the basin. The presence of gypsum in the rock of the Pittsburg quarry is consistent with the formation of the rock in a shallow marine environment. Evidence from thin sections indicates that the deposit was initially laid down as limestone and only later became dolomitized. Fossils are generally rare or absent from the reactive argillaceous dolomitic limestones. In the Pittsburg quarry, the expansivity of aggregates in concrete varies considerably from layer to layer, but the pit-run aggregate from the entire first lift of the quarry is deleteriously expansive.

The term 'alkali-carbonate reaction' (ACR) has been in use since the 1960s, but recent research indicates that the expansion and cracking of concrete caused by this reaction is actually due to the formation of alkali-silica gel by reaction of the alkaline pore solution and cryptocrystalline quartz in the argillaceous dolomitic limestone (Katayama, 1992, 2004, 2006, 2010a, 2011; Katayama & Sommer, 2008; Grattan-Bellew *et al.*, 2009; Katayama & Grattan-Bellew, 2012; Grattan-Bellew & Chan, 2013; Katayama *et al.*, 2016). In 2004 and 2006, Katayama coined the term "so-called ACR" to take into account that the expansion was due to an alkali-silica reaction and not, as originally thought, some other type of reaction involving dolomite, as envisaged by Swenson (1957).

Retention of the term "so-called ACR" to distinguish it from typical alkali-silica reactive limestones may be justified, because of its different behaviour in the mortar-bar test and also because the proportions of supplementary cementitious materials (SCMs)

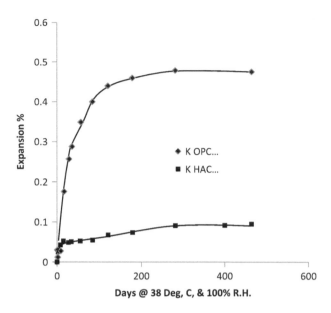

Figure 3.1 Expansion of concrete prisms made with ACR aggregate and a high alkali Portland cement (OPC) and high-alumina cement (HAC).

that are effective in reducing expansion in alkali-silica reactive limestones are not effective in concrete made with the "so-called ACR" dolomitic limestones. This may, in part at least, be due to the formation of additional alkaline hydroxide in the dedolomitization reaction, equations (i, ii).

Reaction rim within aggregate: $CaMg(CO_3)_2 + 2MOH \rightarrow CaCO_3 + Mg(OH)_2 + M_2CO_3$

$$(i)$$

Carbonate halo within surrounding cement paste: $M_2CO_3 + Ca(OH)_2 \rightarrow 2MOH + CaCO_3$

$$(ii)$$

Expansion of typical ASR aggregates in concrete can be prevented by the use of a low alkali-cement, (ASTM 0.60% Na_2O equivalent), but in the case of ACR aggregates such as those from the Pittsburg quarry, the alkali content of the cement has to be less than 0.31% to reduce the expansion of concrete prisms below 0.04%, Swenson (1957). Grattan-Bellew, unpublished results, has shown that the expansion of concrete prisms made with ACR aggregate (Pittsburg) can be significantly reduced by use of high-alumina cement (HAC), Figure 3.1. In these concrete prisms, dedolomitization occurs on the coarse aggregate mixed with HAC (Figure 3.2a), but ASR gel forms when used with high alkali Portland cement (Figure 3.2b).

3.2 HISTORICAL PERSPECTIVES

ACR was first described by Swenson (1957) in a paper entitled 'A Reactive Aggregate Undetected by ASTM Tests'. On the basis of a different response in the mortar-bar test

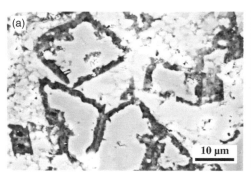

Figure 3.2 SEM micrographs of the concrete prisms depicted in Figure 3.1: a) [left] dedolomitization of the aggregate mixed with high-alumina cement (Katayama & Grattan-Bellew, 2012), but b) [right] ASR gel forming when mixed with a high-alkali Portland cement.

method ASTM C227, Swenson apparently assumed that the reaction was different from that of the classical alkali-silica reaction as described by Stanton (1940). The term ACR was not used by Swenson in his original paper. The ACR aggregate evaluated by Swenson came from the Pittsburg quarry in Kingston, Ontario, Canada which is of Ordovician age. The term 'alkali-carbonate rock reaction' was first used by Swenson and Gillott (1964).

There was considerable research on the problem of the reactivity of carbonate rocks in the U.S.A. in the 1960s. In 1962, Newlon and Sherwood described the occurrence of Alkali-Reactive Carbonate Rock in Virginia, USA. In 1964, Hadley reported on Alkali-Reactive Carbonate Rocks in Indiana and Lemish and Moore, (1964) reported on alkali-carbonate reactions in Iowa, USA. In 1969, Buck reported on the potential alkali reactivity of carbonate rock in Virginia and Kentucky. Subsequently, the ACR problem in the U.S.A., was, for the most part, forgotten apart from a few papers by Ozol (1974, 1994), until a large number of cases of concrete deterioration were investigated by the present authors in Kentucky during the period 2006 to 2008 (Katayama & Grattan-Bellew, 2012). The petrology and expansivity of the Kentucky argillaceous dolomitic limestone is almost identical to that of the Kingston aggregate (Figure 3.3).

Katayama (1992), based on literature reviews, predicted that ACR is a combined phenomenon of harmless dedolomitization and expansive ASR of cryptocrystalline quartz hidden in the dolomitic aggregate. Since then, this hypothesis has been examined in detail by various petrographic methods (polarizing microscopy, SEM observation and quantitative EDS analysis), using i) field ACR concretes from the type locality (Kingston) of ACR (Katayama, 2004, 2006, 2010a, 2011; Gratten-Bellew *et al.*, 2009), ii) field reproduced ACR concretes containing the Kingston ACR aggregate (Katayama & Gratten-Bellew, 2012; Grattan-Bellew & Chan, 2013) and iii) laboratory produced ACR concrete specimens (Katayama, 2004). After all these efforts, however, a paradox was encountered, that the concrete deterioration meeting the original definition of ACR by Swenson & Gillott, that is, deleterious expansion other than typical ASR but associated with dedolomitization, had not been identified, even when typical ACR concretes from the type localities Kingston and Cornwall in Ontario

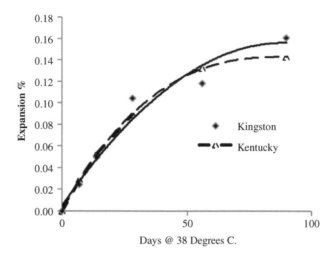

Figure 3.3 Expansion of aggregates from Kingston and Kentucky in the CSA concrete prism test A23.2-14A.

were examined. Consequently, this led to re-defining the classical term ACR as "so-called ACR" in 2004 & 2006 by Katayama (2004, 2006) (Katayama *et al.*, 2016).

ACR was reported by Milanesi *et al.*, in Argentina in 1996. Qian *et al.* (2001) reported the occurrence of ACR aggregate from Jixian, Tianjin, China in 2001. However, before 2004, the mechanism of dedolomitization and expansion were not yet clearly distinguished based on detailed petrographic examination, hence early reports on ACR should be reviewed critically.

3.3 MECHANISM OF SO-CALLED ACR REACTION

3.3.1 General

The deleterious alkali-carbonate reaction occurs in argillaceous dolomitic limestones with porphyrotopic or mosaic textures (Swenson and Gillott, 1960; Buck, 1969; Melanesi *et al.*, 1996). The rocks consist essentially of a matrix of fine grained micritic calcite crystals (2μm), intermixed with clay minerals, mostly illite with minor chlorite and cryptocrystalline quartz (chert), in which are embedded large dolomite rhombohedra (50 μm). According to Hadley (1964) the deleteriously expansive rocks generally contain over 7% clay minerals. Figure 3.4a shows a photomicrograph of a typical thin section of the Kingston dolomitic limestone taken through a petrographic microscope. A photograph taken of a polished section of the Kingston dolomitic limestone taken using SEM is shown in Figure 3.4b. Dolostone (dolomite rock) and dolomitic limestone with a mosaic texture produce marked dedolomitization but are not always expansive. Hence, it is important to distinguish two distinctive phenomena, *i.e.*, reaction (dedolomitization) and expansion, Katayama (2004, 2006, 2010a, 2011).

Hadley (1961) showed that the dolomite content of the deleteriously expansive dolomitic limestones was about 50%. Lemish and Moore (1964) found that expansion

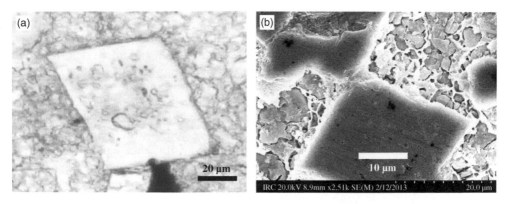

Figure 3.4 Kingston dolomitic limestone: a) [left]: Thin section photomicrograph showing a large dolomite rhomb in a matrix of fine grained calcite. b) [right]: SEM micrograph of dolomite rhombs in a matrix of fine grained calcite.

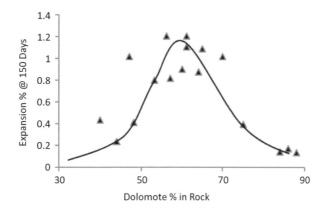

Figure 3.5 Dolomite content of aggregate vs expansion of concrete, adapted from Lemish and Moore (1964).

was greatest at dolomite contents of 60% (Figure 3.5). However, Dolar-Mantuani (1964), using the rock prism test, found most expansive rocks in Ontario contain less than 40% dolomite. Katayama, (1992) reviewed the North American carbonate rocks and indicated that insoluble residue at least 5-10% is the sole common characteristic among expansive carbonate rocks, irrespective of whether it is a limestone or a dolostone. Katayama & Sommer (2008) found that the argillaceous dolomitic limestone in the most expansive layer of the Pittsburg quarry in Kingston (Pt-16) contained 16.8% of insoluble residue and 4.8% of mostly cryptocrystalline quartz by H_3PO_4 extraction (Figure 3.10a).

3.3.2 Dedolomitization

It was initially noticed that the ACR involved dedolomitization of the aggregates in an alkaline medium. Hadley (1961) reported that a large single crystal of dolomite expanded by 0.15% after 100 days in 3M NaOH solution. Hadley concluded that *"the expansion of the de-dolomitizing crystals themselves would be sufficient to account for the expansion of the reactive carbonate rocks"*. However, Gillott and Swenson (1969) failed to reproduce the expansion of a dolomite single crystal and noted that the molar volumes of the products of dedolomitization were less than that of the reactants and hence they concluded that dedolomitization could not account for the expansion of dolomitic limestone aggregates in concrete. Deng and Tang (1993) calculated that the solid volume change on dedolomitization was - 4.32 % (less than zero), however they noted that the products of dedolomitization are fine and enclose many voids. They calculated that the void space was 29.21 % and from this they calculated that the volume change for the dedolomitization reaction was 36.56 %. Walker (1978) concluded that *"Dedolomitization can be an innocuous reaction but in specific kinds of carbonate rocks it can result in expansion of the aggregate particles and destructive expansion of the concrete"*.

In his critical review of the published references on carbonate rock reactions before 1992, Katayama concluded that dedolomitization is an innocuous reaction, and that deleteriously expansive ACR had been confused with dedolomitization. In 2000, Grattan-Bellew and Rogers (2000) found that, after 13 years outdoor exposure, most of the dolomite in the Kingston ACR aggregate was unreacted and so it was most unlikely that dedolomitization could be the underlying mechanism of expansion.

3.3.2.1 Reaction rims

How, then, does dedolomitization proceed in concrete? Decomposition of dolomite crystals, equation (i), forms a reaction rim within the dolomitic aggregate. Figure 3.6a shows a conspicuous dark rim of dedolomitization developed within a reacted dolomitic limestone aggregate in the 6-month old field concrete in Saudi Arabia. At the interface between the dolomite crystals and cement paste, a thin rim of Mg-Al-bearing hydrate resembling hydrotalcite may be formed on the dolomite absorbing Al ions from the cement paste, instead of brucite (Katayama *et al.*, 2008; Katayama, 2010a, 2012a).

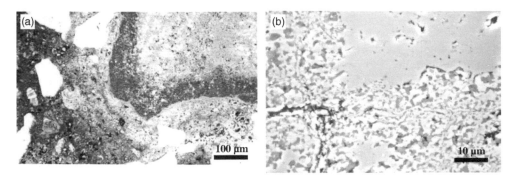

Figure 3.6 Dedolomitization rims around dolomitic limestone aggregate, a) [left] showing a brucite-rich dark zone with an invisibly narrow hydrotalcite layer at the very interface with the cement matrix, Saudi Arabia (Katayama, 2012a), b) [right] showing minute spots of brucite (dark) and calcite (bright) formed by dedolomitization of Austrian dolostone in a RILEM AAR-5 concrete microbar.

By contrast, within the reaction rim of the dolomitic aggregate, or along the cracks extending from the periphery of the reacted aggregate into the interior of the aggregate, minute spots (<3 μm) of brucite and calcite are formed as a result of dedolomitization (Figure 3.6b) (Katayama, 2004, 2010a, 2011; Katayama & Sommer, 2008).

3.3.2.2 Carbonate haloes

In the dedolomitization, secondary calcite forms a carbonate halo, equation (ii), within the cement paste that surrounds the reacted dolomitic aggregate. It should be noted that no radiating crack suggestive of expansion is formed in the cement paste. In sufficiently thin polished thin sections (15μm), under the crossed polars, the carbonate halo has a high birefringence showing a bright yellow colour, characteristic of calcite (Figure 3.7a). This carbonate halo is a porous crust composed mainly of calcite that had precipitated from the interstitial solution and replaces CSH gel in the cement paste (Figure 3.7b). Elemental mapping of the carbonate halo by EDS revealed a zonal distribution of alkalis corresponding to the formula (ii) (Katayama, 2004, 2010a).

3.3.3 Evidence of expansion

Previous research on ACR was likely to be confused by believing that dedolomitization is evidence of expansive ACR. However, according to Katayama (2004, 2006, 2010a, 2011), wherever expansion cracks exist in concrete, ASR gel is always present. Katayama found that evidence of expansion in concrete is demonstrated by the formation of radiating cracks from the reacted particles of the aggregate, irrespective of whether it is typical ACR of argillaceous dolomitic limestone or typical ASR of andesite or another reactive siliceous rock. It has been shown, aided by SEM observations on polished thin sections, that there is no difference in the process of crack formation in concretes between ASR and ACR. Katayama and Katayama and Grattan-Bellew (2012), revealed that the sequence of the crack formation in ACR concretes is: 1) rim formation, 2) internal cracking with ASR gel within the aggregate, 3) cracking from

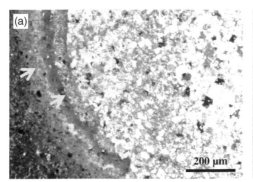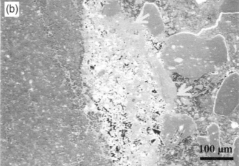

Figure 3.7 Carbonate halo in the cement paste surrounding the dedolomitized aggregate particle: a) [left] showing the distinctive colour indicative of calcite, RILEM AAR-5 concrete microbar with Austrian dolostone, b) [right] showing that the carbonate halo replaces CSH gel in the cement paste in a concrete building in Saudi Arabia.

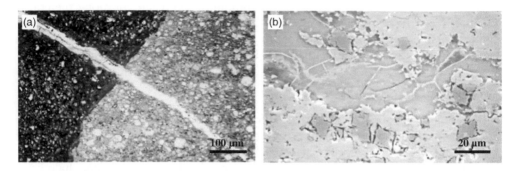

Figure 3.8 Photomicrgraphs from expanded AAR-5 concrete microbars containing dolomitic aggregate: a) [left] ASR gel lining a crack emanating from the expanding particle and into the
surrounding cement matrix, b) [right] veins of ASR gel within the reacted dolomitic limestone from a concrete pavement (sidewalk) in Kingston, Ontario, wherein the dolomite
rhombs are only slightly decomposed (from Katayama & Grattan-Bellew, 2012).

aggregate into the cement paste exuding ASR gel, and 4) ASR gel lining air voids and
cracks distant from the reacting aggregate.

With RILEM AAR-5 (in Nixon & Sims, 2016) concrete microbars, expansion
continues even when brucite disappears, transforming into non-expansive magnesium-
-silicate gel, and exudation of ASR gel becomes prominent along radial expansion
cracks (Katayama, 2006, 2010a; Katayama & Grattan-Bellew, 2011) (see Figure 3.8a).

3.3.4 Formation of alkali-silica gel in the concrete

In 1986, Katayama (1992) collected a piece of concrete from Cornwall, Ontario,
Canada and examined ultra-thin sections (15 µm thick) made from it. He observed
alkali-silica gel filling cracks in the concrete suggesting that ASR had at least a role in
the deterioration of the concrete made with an ACR aggregate. During 2004 to 2006,
Katayama, after examining a number of samples of concrete affected by the 'so-called
alkali-carbonate reaction', including the typical ACR concrete from near Kingston
constructed in the late 1950s, concluded that the *"ACR could be a mixture of deleteriously expansive ASR of cryptocrystalline quartz, and non-expansive dedolomitization
of dolomitic aggregates"*, and that *"ACR is believed to be ASR, influenced more (fine
aggregate) or less (coarse aggregate) by dedolomitization"* (Katayama, 2004).

Katayama identified ASR of crypto- to microcrystalline quartz in the dolomitic
aggregates in several structures from Montreal to Quebec, Canada. He confirmed
that ACR is a combined effect of deleteriously expansive ASR of cryptocrystalline
quartz, and harmless dedolomitization, which produces brucite and carbonate haloes
without accompanying expansion cracks (Katayama, 2004, 2006, 2010a; Katayama &
Sommer, 2008). What is important is that dolomitic aggregates do not develop expansion cracks in the embedding cement paste, unless ASR is involved. Figure.3.8a shows
the typical expansion crack filled or lined with ASR gel in the ACR aggregate in an
AAR-5 concrete microbar (Katayama & Grattan-Bellew, 2012). In the Kingston
experimental pedestrian pavement (sidewalk) at age 22 years, radial expansion crack
is filled with veins of ASR gel within the reacted dolomitic limestone aggregate

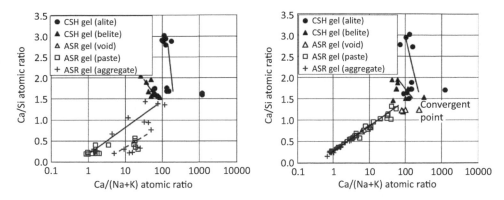

Figure 3.9 Comparison of ASR gel compositions: a) [left] from a typical ACR bridge near Kingston, Ontario (Katayama, 2010a), and b) [right] from a typical example of ASR caused by andesite in Japan (Katayama, 2010b).

(Figure.3.8b). Note the dolomite rhombs are only slightly decomposed into a thin rim, consisting of brucite or magnesium silicate gel, whereas ASR gel is formed abundantly. Lu *et al.* (2006) also observed gel in samples of the ACR aggregate from the Pittsburg quarry autoclaved at 150°C in KOH solution.

3.3.4.1 Compositional trends of ASR gel in ACR concrete

According to Katayama (2010b, 2008) and Katayama *et al.* (2008), ASR gel in typical ASR concretes migrates along cracks from the reacted aggregate into the cement paste, absorbs calcium from cement paste, and approaches the composition of CSH gel. In principle, on the [Ca/Si]-[Ca]/[Na+K] diagram, alkali-silica reaction terminates at a convergent point of CSH gel at around Ca/Si=1.5, Ca/(Na+K)=100-200. It should be noted that there is no difference in the compositional trends of ASR gel between ACR concretes and ASR concretes. Figure 3.9a shows the data of typical ACR bridge near Kingston, while Figure 3.9b is typical ASR of andesite in Japan. In both cases, ASR gel crystallizes into the rosette-like aggregation at Ca/(Na+K) <1. Where leaching of alkalis occurs along cracks in concrete, a short parallel compositional line appears, irrespective of whether it is ACR as shown here (Katayama, 2010a), or ASR (Katayama, 2010b, 2008).

3.3.5 Identification of cryptocrystalline quartz

In the ACR reactive argillaceous dolomitic limestone from Kingston, cryptocrystalline quartz is contained in an isolated form, and not as an aggregation like a chert. This explains why this form of reactive silica is difficult to identify in thin section, and had long been missed in the optical microscopy of the ACR aggregate from Kingston. This cryptocrystalline quartz was extracted from the Kingston rock by the phosphoric acid treatment, Katayama (2004, 2010a) (Figure.3.10a).

In field concretes undergoing ACR, cryptocrystalline quartz in the Kingston rock is difficult to identify, because this quartz reacts very fast (Katayama, 2004, 2010a, 2011;

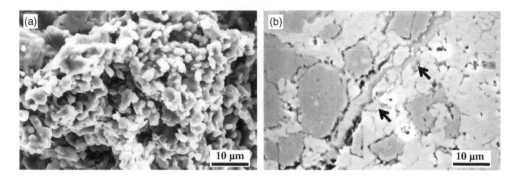

Figure 3.10 Cryptocrystalline quartz from the Kingston dolomitic limestone aggregate: a) [left] extracted by phosphoric acid treatment, b) [right] in some AAR-5 concrete microbars, small amounts of cryptocrystalline quartz remain in isolated form, but most has converted to ASR gel.

Katayama & Sommer, 2008). However, in RILEM AAR-5 concrete microbars, some cryptocrystalline quartz occasionally remains, but mostly converts to ASR gel (Figure 3.10b) or Mg-silicate gel (Katayama, 2004, 2010a; Katayama & Sommer, 2008). In the siliceous argillaceous dolomitic aggregates undergoing ASR in Quebec, crypto- to microcrystalline quartz is abundant and is partly converting to ASR gel as revealed by SEM observation on the polished thin section, Katayama (2006).

3.3.6 Formation of magnesium silicate gel

The magnesium silicate gel (Mg-silicate gel) occurs as an intermediate product between ASR gel and magnesium hydroxide gel or brucite in the reacted dolomitic aggregate that contains reactive silica (Katayama, 2004, 2006, 2010a, 2011, 2012a). It is found within a dedolomitized rim or along cracks in the interior of the aggregate. Mg-silicate gel forms a thin inner rim of the reacted dolomite crystals (Figure 3.11a), or a

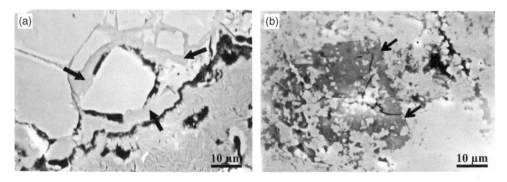

Figure 3.11 Magnesium silicate gel: a) [left] forming a thin inner rim around reacting dolomite particles (adapted from Katayama, 2006), b) [right] forming a pseudomorph after the original dolomite crystals and showing shrinkage cracks.

pseudomorph after the dolomite crystals keeping their original rhombohedral outlines (Figure 3.11b).

To identify the Mg-silicate gel, it is essential to perform a quantitative SEM-EDS analysis on the polished thin section at higher magnifications (*e.g.*, > 1500 x) to obtain a sharp image to distinguish the minute textures composed of dolomite, Mg-silicate gel, ASR gel, etc. In the aged concrete constructed in the 1950s near Kingston containing the ACR reactive Pittsburg aggregate, Mg-silicate gel became more or less crystalline, showing the transformation into sepiolite-like material or flakes of antigorite-like material as estimated by their composition, (Katayama, 2006, 2010a, 2011). Where dolomitic rock is argillaceous (Pittsburg aggregate), Al-Mg silicate gel forms (Katayama, 2006), and on aging this gel converts to a chlorite-like material.

There has been no evidence that Mg-silicate gel generates expansion, because no radiating cracks are found to accompany its formation, except for shrinkage cracks (see Figure.3.11b). The occurrence of the Mg-silicate gel in field concretes has been documented from variously reactive dolomitic aggregates in Ontario (deleterious), Austria (harmless) (Katayama, 2004, 2006, 2010a; Katayama & Sommer, 2008) and Saudi Arabia (harmless) (Katayama, 2012a), as well as in the ASR-reactive aggregates in Quebec and Japan (harmless) (Katayama *et al.*, 2008) based on SEM-EDS analysis, but has not been identified by other researchers in a number of works before around 2000.

3.4 COUNTER ARGUMENTS AGAINST GEL FORMATION AS THE CAUSE OF EXPANSION DUE TO ACR

The arguments on this subject have been reviewed critically by Katayama (2010a, 2011) and Katayama *et al.*, (2016). The main points of the problem are that 1) the conventional methods for preparation and examination of concrete thin sections are inappropriate, 2) general understanding on the expansion behaviours and remedial measures for a wide range of ASR remains stereotypical and has been reproduced by the papers without critical reviewing, and 3) interpretations on the expansion mechanisms related to dedolomitization lack support by petrographic evidence.

3.4.1 Why is gel not observed in thin sections of ACR concrete?

3.4.1.1 Thickness of thin section

The standard thickness of rock thin sections is generally accepted by geologists to be from around 30 µm in Europe and North America to 20 µm in Japan, and published optical properties for minerals usually relate to this range of thickness (although 30 µm is too thick for quartz, because it presents a pale yellow interference colour). However, whilst traditional 30 µm sections are usable for identification of the main mineralogy and types of coarse-grained rocks, they are often too thick for concrete petrography, particularly for the detailed study of AAR. For the petrographic examination of Portland cement clinker, thin sections with thickness of <15 µm have been accepted as standard (Insley & Fréchette, 1955). Similarly, the ideal thin section thickness for fine-grained alkali-reactive rock types (*e.g.*, volcanic rocks, cherts, and carbonate rocks) is 15 µm, because most of the reactive constituents contained in them

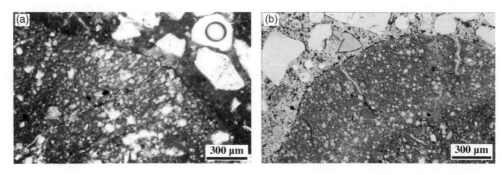

Figure 3.12 Comparison of clarity between traditional 30 and 15 µm thick thin sections: a) [left] near-opaque appearance of cement paste matrix with a traditional 30 µm thick section, and b) [right] clearer resolution of both aggregate and matrix using a 15 µm thick section (adapted from Katayama, 2012b).

(cristobalite, tridymite, volcanic interstitial glass, cryptocrystalline and microcrystalline quartz) are small (mainly <10µm) (Katayama, 2010a, 2012b; Katayama *et al.*, 2016).

The calcite crystals that comprise the matrix of the ACR aggregate are about 2 µm in size (see Figure 3.14b, Grattan-Bellew *et al.*, 2009) and hence, in 30 µm thick thin sections, a petrographer is looking through about 15 layers of calcite that effectively obstructs the view, making it almost impossible to observe fine gel filled cracks. The 30 µm thin section of concrete gives an almost opaque appearance of cement paste (Figure 3.12a), whereas 15 µm section permits detailed observation by giving a transparent appearance of cement paste and argillaceous carbonate matrix of the dolomitic aggregate (Figure 3.12b).

Many papers on ACR, both published and submitted manuscripts to journals, are based on observations of either very thick thin sections, justifying the thickness of 30 µm in the references on petrography of natural rocks, or on the SEM observations on the fracture surface of concrete. In addition, enthusiasm for producing large-area thin sections (*e.g.*, 12 cm by 18 cm, 30 µm thick) prevented preparation of sufficiently thin sections suitable for detailed petrographic examination. This is one of the main reasons why the presence of crypto- to microcrystalline quartz, which is responsible for the deleterious expansion due to ASR in the ACR ('so-called ACR' by the present definition), has long been missed in the petrography of concrete (Katayama, 2010a).

3.4.1.2 Carbonation of ASR gel

In the ACR concretes undergoing dedolomitization, both cement paste and ASR gel are most likely subject to secondary carbonation by the carbonate ions that are liberated from the reacted dolomite through dedolomitization (Katayama, 2010a). Because the carbonate minerals calcite and dolomite constituting the dolomitic aggregates have a high-birefringence, they overlap with fine cracks and ASR gel that lines or fills these cracks, as well as crypto- to microcrystalline quartz in the aggregate, thereby hindering

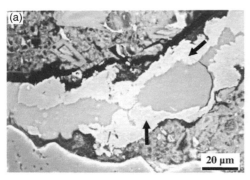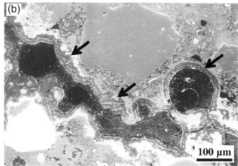

Figure 3.13 Clarity enabled by polished thin (15 μm thick) thin sections: a) [left] vein of combined ASR gel and calcite formed through carbonation during storage of a concrete prism (450 days), b) [right] ASR gel mixed with secondary calcite (from Katayama & Grattan-Bellew, 2012).

the details of the microscopic texture in the transmitted light. In the concretes affected by more or less atmospheric weathering, such as field structures and concrete prisms stored for a long time, secondary calcite forms along the cracks often replacing veins of ettringite and ASR gel (Katayama, (2010a). As a result, ASR gel is likely to be missed by the thin section microscopy.

Although the above alteration processes obscure the presence of ASR gel in the transmitted polarizing microscopy, SEM observation on the polished thin section is able to reveal clearly the presence of the composite vein in which ASR gel is enveloped by its carbonation product calcite vein (Figure 3.13a, Katayama & Grattan-Bellew, 2012), as well as thin films of ASR gel intermixed with secondary calcite (Figure 3.13b). In the thick thin section (~30 μm) with a cover glass, none of fine cracks, ASR gel or cryptocrystalline quartz is detectable. To identify these fine textures and reaction products, combined petrographic examination is recommended, which consists of polarizing microscopy in both transmitted and reflected lights, SEM observation to characterize the reaction sites, and quantitative EDS analysis, using the same polished thin section (15 μm thick) (Katayama, 2004, 2006, 2010a, 2011).

3.4.1.3 Fluorescence method to identify thin films of ASR gel

Subsequent to Katayama, Grattan-Bellew *et al.* (2010) concluded on the basis of laboratory experiments that the formation of alkali-silica gel was the cause of the expansion and deterioration of concrete made with ACR aggregates. Data from Gillott (1961) and Swenson and Gillott, (1964) show that there is a direct correlation between the amount of chert in the insoluble residue of the Kingston aggregate and the expansion of concrete prisms. The fluorescence of alkali-silica gel on the fractured surface of a core, treated with uranyl acetate, from the 23 year old experimental sidewalk in Kingston made with the argillaceous dolomitic limestone and high alkali cement is shown in Figure 3.14a.

Further confirmation for the occurrence of ASR in concrete made with so-called ACR aggregates was presented by Katayama and Grattan-Bellew (2012) and by

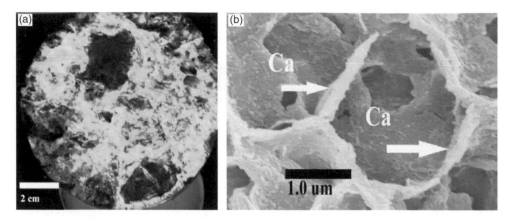

Figure 3.14 Detection of ASR gel in concretes made with ACR aggregate: a) [left] fluorescence of gel after treatment with uranyl acetate, b) [right] thin films of ASR gel between micro-crystals of calcite.

Grattan-Bellew and Chan (2013), showing the presence of the thin films of alkali-silica gel between the small crystals of calcite in the matrix of the rock, Figure 3.14b.

3.4.2 Why does coarse aggregate expand more?

Why do ACR aggregates not cause expansion in mortar-bars if the mechanism of expansion is ASR? This subject has been discussed in detail by Katayama (2010a, 2011). According to Katayama and Sommer (2008) and Katayama (2010a), in mortar-bars the -4 mm size of the aggregate results in more or less complete dedolomitization of the finer particles that produce brucite $Mg(OH)_2$, which reacts with the silica gel, produced by reaction of the crypto-crystalline quartz with the alkaline pore solution, to produce a magnesium silicate gel that is either less expansive or non-expansive and so expansion of mortar-bars is much reduced.

In general, typical ASR proceeds by the expansion generated in the interior of the aggregate. With the coarse aggregate, expansive alkali-rich ASR gel tends to remain within the interior of the aggregate particle, continues expansion, and generates wide cracks filled with ASR gel. By contrast, fine aggregate particles tend to lose the expansive nature rapidly through the replacement of alkali ions in the ASR gel by calcium ions from the surrounding cement paste, thus producing small cracks. In the alkali-immersion tests, the size of the aggregate is important. In the accelerated mortar-bar test (ASTM C1260: 2004), chert particles do not expand significantly due to dissolution of ASR sol into NaOH solution, whereas in the accelerated concrete microbar test [AAR-5 (in Nixon & Sims, 2016), Sommer *et al.*, 2004] they produce expansion due to formation of ASR gel within the coarse aggregate. Nowadays, ASR gel is considered to be a universal product for generating expansion of all the traditionally known varieties of alkali-aggregate reaction, *i.e.*, alkali-silica

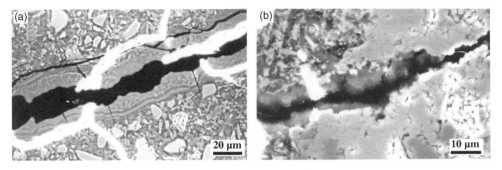

Figure 3.15 a&b Cracking filled or lined with ASR gel in a RILEM AAR-5 concrete microbar made using 50% Portland cement replacement by Japanese ground granulated blastfurnace slag, in which expansion was reduced by 30 % but cracking not prevented (from Katayama & Grattan-Bellew, 2012).

reaction, alkali-silicate reaction, and alkali-carbonate reaction (Katayama, 2012b, 2012c).

3.4.3 Why are supplementary cementitious materials not effective?

3.4.3.1 Blastfurnace slag

Supplementary cementitious materials that are usually effective in minimizing expansion due to ASR are not effective when used in concrete made with ACR aggregates, which may be due to a number of causes. The crystallinity of the cryptocrystalline quartz in the ACR aggregate is lower than in typical ASR limestones, resulting in it being much more reactive and hence more difficult to suppress (Grattan-Bellew & Rogers, 2000). Katayama (2008) reported that replacement of 50% of Portland cement by fine grained Japanese blastfurnace slag reduced expansion by 30%, but that it failed to prevent the formation of cracks filled with ASR gel at later ages (Figure 3.15a,b). Because cryptocrystalline quartz in the Pittsburg aggregate occurs as isolated forms, rather than a massive aggregation like a chert, this quartz is considered to be more reactive than the particles of ground granulated slag and fly ash (Katayama, 2012b). Generally, blastfurnace slag is not always effective in preventing ASR: replacement of 40% of cement by slag could not suppress cracking due to late-expansive ASR in a mass concrete structure in Japan.

3.4.3.2 Fly ash

Fly ash is not effective in suppressing ACR in concrete. In Kentucky, high-calcium fly ash (40% CaO in glass, content 40%) was not effective in preventing cracking by the ACR aggregate and ASR of chert particles (Figure 3.16a, Katayama & Grattan-Bellew, 2012). In the reacted dolomitic aggregate, expansion cracks originated from the pool of ASR gel that had formed surrounding the calcite particles (Figure 3.16b). With typical ASR, even

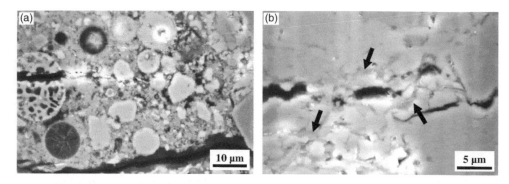

Figure 3.16   Cracking of concrete in Kentucky caused by ACR, a) [left] not prevented by the use of 40% cement replacement by high-calcium fly ash, b) [right] cracks originating from pools of ASR gel surround the calcite particles (from Katayama & Grattan-Bellew, 2012).

low-calcium fly ash is not effective when the aggregate is highly reactive. In a Japanese case, low-calcium fly ash (<4% CaO in glass, replacement of 18%) was effective for the late-expansive ASR of the sandstone pebbles in concrete, but was not effective for the early expansive ASR due to the glassy bronzite andesite aggregate. The average SiO_2 content of the fly ash glass was 58%, while that of the rhyolitic interstitial glass in the andesite was 75%, which explains the low pozzolanic reactivity of this fly ash (Katayama, 2010b).

3.4.3.3 Lithium salt

The addition of lithium salts to concrete made with ACR aggregates is not effective in suppressing expansion (Tang & Deng, 2004). However, this is also the case with late-expansive ASR of orthoquartzite and greenschist in Canada (Durand, 2000). All these may be due to the presence of a dense matrix surrounding the reactive particles that either prevents the lithium ions from reaching them or the lithium is absorbed by clay minerals (Katayama, 2010a).

3.4.4 Dedolomitization-related mechanisms

3.4.4.1 Hypothetical volume-increase dedolomitization

López-Buendia *et al.* (2006, 2008) claimed that dedolomitization itself is a volume-increase reaction, because calcite has a larger d-spacing (d104 = 3.03Å) than that of dolomite (d104 = 2.89Å). To compare the unit cell volumes of dolomite and calcite, mass balance should be considered, then one molar dolomite is equivalent to two molar calcites, expressed as equation (iii) (Katayama, 2011, 2010b). This is a hypothetical dedolomitization, in which complete ion-exchange takes place between Mg in the dolomite and Ca in the cement paste. Mg ions liberated from dolomite into solution do not precipitate as brucite, and more than 10% of volume increase may result (Katayama, 2010b). However, in concrete, such an entire replacement of dolomite by calcite in equation (iii) does not occur, as is evident from the microtexture of Figure

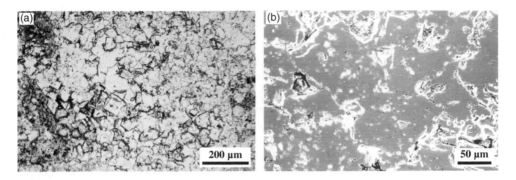

Figure 3.17a) & b) Natural dedolomitization, in which calcite pseudomorphs replace dolomite rhombohedra without producing brucite.

3.6b which follows the equation (i). Actually, concrete specimens of López-Buendía *et al.* (2008) made with a Spanish dolomitic limestone aggregate presented contraction.

For comparison, natural dedolomitization occurs in geological time by the near-surface weathering of dolomitic formations (Katayama, 2010b). In this process, calcite replaces the dolomite crystals, leaving a pseudomorphic texture of the original outlines of the dolomite rhombs (Figure 3.17a,b). This dedolomitization is a dissolution/precipitation process in the open system, and there is no volume change after the precipitation of calcite. Dolomites are dissolved into interstitial solution and secondary calcites are precipitated into the dissolution voids formed after the dolomite crystals. Mg ions liberated from the dolomite are carried away and do not precipitate as brucite, which is different from the dedolomitization in concrete. This reaction without volume change could be written as equation (iv) (Katayama, 2011, 2010b).

$$\text{Hypothetical dedolomitization: } CaMg(CO_3)_2 + Ca^{2+} \rightarrow 2CaCO_3 + Mg^{2+} \qquad \text{(iii)}$$

$$\text{Natural dedolomitization: } CaMg(CO_3)_2 + 0.75Ca^{2+} \rightarrow 1.75CaCO_3 + Mg^{2+} + 0.25CO_3^{2-} \qquad \text{(iv)}$$

3.4.4.2 Uptake of moisture and swelling of dry clay minerals in the dolostone

Gillott and Swenson (1969) concluded that dry clay minerals in dolomite crystals in the aggregate would become exposed to alkalis and moisture due to dedolomitization and the swelling of previously un-wetted, non-swelling clays (illite) could account for the expansion of the aggregates. However, if a swelling clay, typically smectite, is contained in the aggregate, this would produce distinctive shrinkage cracks of cement paste concentrically surrounding the aggregate through the repeated drying and wetting seasonally encountered in the outdoor conditions. Katayama (2011) reported that no such shrinkage cracks have been observed in the Kingston dolomitic aggregate in the typical ACR concretes from Ontario.

3.4.4.3 Expansion due to dedolomitization and formation of brucite in confined space

Tong and Tang (1999) concluded that *"brucite formed in a confined space and generated the force to cause the expansion of rock prisms and aggregate particles in concrete microbars"*. This hypothesis appears rather unlikely as only a small amount of the dolomite limestone de-dolomitizes in the field deteriorated concretes.

The accelerated condition (150°C) of concrete microbars and compacted mortars produces portlandite and brucite in the particles of pure limestone and magnesite (Tong, 1994; Tang & Deng, 2000). They believed that the expansion was caused by a "topochemical reaction" that took place within a confined space to generate a pressure of crystallization of portlandite and brucite, claiming that the expansion of the ACR is due to the formation of brucite through the dedolomitization. However, if this mechanism is accepted, then pure limestone should also be regarded as deleteriously expansive, which contradicts the reality, since the expansion of this limestone aggregate in the autoclave is attributed to the thermal expansion of coarsely crystalline calcite (Mu *et al.*, 1996). Hence, it is unlikely that brucite causes expansion, Katayama (2011).

3.5 TEST METHODS

3.5.1 Rock Cylinder Test (ASTM C 586)

The first reported use of the rock prism test is by Hadley (1961). It was also used by Swenson and Gillott (2009), Dolar-Mantuani (1964), Walker (1978) and intensively by Rogers (1985). ASTM C586 (2006) specifies that the cylinders or prisms be stored in 1N NaOH at a temperature range of 20 to 27.5°C. The overall size of the specimens specified to be 9 mm diameter by 35±5 mm in length. However, Deng *et al.* (1993) stored the prisms at 80°C in 1N KOH. The relatively short length of the cylinders or prisms permits evaluation of individual layers in a quarry either by taking rock hand samples, or by evaluating cores, something that is difficult to achieve using other test methods. Hadley found good correlation between expansion of rock prisms after 14 days in 1M NaOH and the expansion of concrete prisms stored at 38 °C.

A modified miniature rock prism test was developed by Grattan-Bellew (1981) and this method was also used by Tang and Lu (1986), who tested prisms made from the Kingston Rock. They showed that expansion increased with temperatures from 20 to 80°C. It should be noted that laminated argillaceous carbonate rocks have a strong anisotropy in expansion by the direction of bedding. Rock cylinders cut perpendicular to the bedding plane produce expansion larger than those parallel to the bedding: *e.g.*, 20% to 40% larger at age 28 days to 19 months with the Kingston ACR rock, whereas 7 times larger was found with a Japanese dolomitic limestone (Katayama *et al.*, 1996).

This test does not necessarily reproduce ACR as recognized in concrete, because cement is not involved and interaction between the dolomitic rocks and calcium in the cement paste, *e.g.*, formation of the carbonate halo, is thus not generated. Swelling of the clay minerals by water and alkali solution, may occur, but this is not relevant to ACR. In addition, NaOH solution used as the immersion medium tends to dissolve ASR sol before solidifying into ASR gel. On the basis of these differences from the tests using concrete specimens, rock cylinder/prism test is not necessarily suitable for the study of the mechanism of ACR.

3.5.2 Concrete Prism Testing

The concrete prism test [ASTM C1105 (2006), CSA A23.2-14A (2014), French Standard NFP 18-454 (2004), RILEM AAR-3 (in Nixon & Sims, 2016)] is the most widely used method for the evaluation of the potential expansivity of argillaceous dolomitic limestones and probably comes closest to replicating the performance of concrete made with ACR aggregates in the field. However, the use of a specified mixture design of the concrete places some limitations on the use of this test method, for example, to predict the field performance of mixtures with lower cement contents and those using lower alkali cements. The CSA standard specifies the use of a non-reactive fine aggregate in the concrete prism test, however, Fournier *et al.* (2000), showed the importance of evaluating the potential expansivity of combinations of coarse and fine aggregates as the reactivity of the fine aggregate would likely have an effect of the expansion of the prisms. The Canadian Standard CSA A23.2-14A (2014) was developed from the work of Swenson (1957). The specified size of the prisms in the current standard is 75 x 75 x 275 mm or a maximum length of 405 mm (coarse aggregate: max.19 mm). The prisms are stored at 38 °C as for testing for ASR. The specified cement content for the concrete prisms is 420 kg/m^3 and the alkali content of the cement is increased to 1.25% Na$_2$O equivalent by addition of NaOH.

In ASTM C1105 (2006), the specified cross section of the prisms is the same as in CSA, but the storage temperature is 23°C. In the French standard NF P 18-454 (2004) the prisms are 70 x 70 x 282 mm. The prisms are stored in a "reactor" at 60 °C to accelerate the reaction. In RILEM AAR-3 (in Nixon & Sims, 2016), the prisms are 75 x 75 x 250mm, and the storage temperature is 38°C. An accelerated version, RILEM AAR-4, (in Nixon & Sims, 2016) was developed from the French method, which uses the prisms of 75 x 75 x 250mm stored at 60 °C in a double container to avoid drying up and leaching of alkalis from the prisms.

Expansion curves of concrete prism tests often present a plateau at later ages, forming asymptote towards a certain expansion value, but this does not necessarily mean that reactive constituents in the aggregate were entirely consumed. Leaching of alkalis from the concrete prisms occurs during the long-term storage, perhaps in the order of 30-40% in some cases, irrespective of the temperature applied, 23°C or 60 °C. For this reason, late-expansive ASR aggregates present a sigmoidal expansion curve with apparent asymptote, whereas early-expansive ACR aggregates tend to give a simple convex curve with asymptote, rather than a sigmoid with an inflection point. At the same time, secondary carbonation proceeds along the cracks that were formed on the concrete prisms and calcite replaces ASR gel and ettringite in varying degrees (Katayama, 2010a).

3.5.2.1 Correlation Between Expansions in the Concrete Prism Test & in the Field

There are only two papers on the comparison between expansions in the field and in concrete prisms made with ACR aggregates (Grattan-Bellew, 2000; Lu *et al.*, 2008).

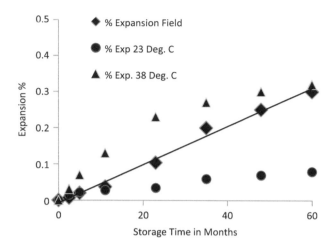

Figure 3.18 Comparison between actual field expansion of the experimental concrete sidewalk in Kingston and that predicted by concrete prism tests at 23 °C and 38 °C.

Both used data from the experimental sidewalk in Kingston, Ontario, Canada. Rogers and Hooton, (1992) noted that after one year the expansion of concrete prisms stored at 23 °C was about the same as concrete slabs exposed outdoors under ambient conditions. However, after 5 years, leaching of alkalis from the concrete prisms stored at 23 °C had reduced expansion significantly below that of the slab exposed outdoors, but in the case of one slab, the expansion of the prisms stored at 38 °C was similar to that of the slab exposed outdoors, Figure 3.18. It is concluded from these results that the expansion of concrete prisms stored at 38 °C and ~100% humidity provides a better prediction of the long term expansion of concrete in the field. The main disadvantage of the concrete prism test is that it takes up to one year to obtain results.

3.5.3 Concrete Microbar Test

In this test procedure [RILEM AAR-5 (in Nixon & Sims, 2016), Tang & Deng, 2004], concrete microbars, 40 by 40 by 160 mm are made with a single size fraction of aggregate, 4.0 to 8.0 mm. The water to cement ratio is 0.33 and the aggregate to cement ratio is 1:1. The length change of the bars is monitored for 28 days in 1 M NaOH @ 80°C. In Tang and Deng (2004), expansions of bars made with ACR aggregates in excess of 0.1% at 28 days are considered to be indicative of potentially deleterious expansion in concrete. Good correlation was found between the expansions of concrete microbars and concrete prisms (Lu *et al.*, 2008). The similar procedure in RILEM AAR-5 is used to compare behaviour with the conventional accelerated mortar-bar test for ASR [AAR-2 (in Nixon & Sims, 2016)], on the basis that reactive carbonates will give greater expansion with the coarser aggregate particle size (4 to 8 mm) used in AAR-5.

During the storage, concrete microbar specimens are not affected by atmospheric carbonation. In this test, all the important features and reaction products characteristic of the so-called ACR, *i.e.*, dedolomitization rims, carbonate haloes, brucite, Mg-silicate gel and ASR gel, can be reproduced. Besides, owing to the larger size of the aggregate than that in the mortar-bar tests (ASTM C1260, RILEM AAR-2), chert aggregates produce deleterious expansion without undergoing excessive dissolution of ASR sol/gel into the NaOH solution (Yamada *et al.*, 2015).

3.5.4 Chinese Autoclave Test

Concrete microbars 20 by 20 by 60 mm are used in this test (Tang *et al.*, 1994). The aggregate size is 5-10 mm, the aggregate:cement ratio is 1 and the water/cement ratio of 0.3. The alkali content of the cement is adjusted to 1.5% Na_2O equivalent by adding KOH to the mixing water. After demolding the bars are steamed at 100°C for 4 hours and then autoclaved at 150 °C. Expansions in excess of 0.1% after 6 hours of autoclaving are considered to be indicative of potentially deleterious expansion of ACR aggregates, (Guangren, 2001). However, this condition (150°C) is too vigorous for well-crystallized carbonate aggregates with distinctive cleavage planes within them because it produces irreversible thermal expansion (Katayama, 2011). Coarsely crystalline, non-reactive limestone from Nanjing produced abnormal expansion, independent of the alkali level of the cement and curing conditions (Mu *et al.*, 1996), that is, coarser aggregate (5-10mm) gave larger expansion (0.09%), comparable to the threshold for deleterious expansion in this test (0.10%). Frequent heat treatment for measuring the length changes at various storage ages was responsible for the larger expansion.

According to Tong (1994), compacted mortar specimens (9 mm in diameter by 30 mm in length) containing the limestone or coarsely crystalline magnesite aggregate also produced significant expansion with the same autoclaving condition (150°C). Calcite and magnesite were decomposed into portlandite and brucite, although the total solid molar volumes should decrease after the reactions (Tang & Deng, 2004). Because the aggregates tested were coarsely crystalline rocks and the heat cycles applied were frequent, irreversible thermal expansion must have occurred with their specimens. Hence, care should be taken in interpreting the expansion data of the autoclave method.

3.5.5 Chemical Screening Test

This test procedure (CSA A23.2-26A: 2014) consists of determining the aluminum oxide, the calcium oxide and the magnesium oxide contents of pulverized carbonate rock samples and plotting a graph of Al_2O_3 versus the CaO:MgO ratio. The potential expansivity of the aggregate is determined from the position of the plotted point on the graph (Figure 3.19). This test procedure was developed by analysis of a large suite of carbonate rocks from Ontario for the Canadian Standards Association by Rogers (personal communication, 2000). This method was also used successfully by the authors to determine the potential expansivity of argillaceous dolomitic limestone from Kentucky, U.S.A.

According to Jensen (2009), about 6% of Norwegian carbonate aggregates tested was *"considered potentially expansive"* and therefore potentially ACR-reactive on this diagram, but ACR has not been diagnosed in Norway. Considering that ASR of

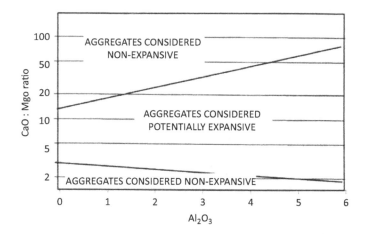

Figure 3.19 Graph showing the chemical composition of expansive and non-expansive carbonate rocks after CSA A23.2 26A (2014).

the cryptocrystalline quartz is the real cause of the deleterious expansion of the so-called ACR aggregates, and that metamorphic carbonate rocks common in Norway contain coarsely crystalline quartz with lower ASR reactivity, this diagram should better be applied to the non-metamorphosed carbonate rocks in North America. Another version of the screening diagram (CaO:MgO and insoluble residue) was proposed by Rogers (1986), which is equivalent to the ratio of calcite-dolomite-insoluble residue on the triangular diagram (Katayama, 1992), meaning that the dolomitic limestones in Ontario with a certain range of impurity are deleteriously expansive.

3.6 CONCLUSION

In 1992, Katayama reported on the occurrence of alkali-silica gel in concrete made with ACR argillaceous dolomitic limestone indicating that the formation of gel was at least a contributing factor in the expansion of the concrete. Up until 1992 ACR was generally considered to have a different mechanism of expansion from that of concrete made with classical ASR aggregates, and a number of mechanisms were put forward to explain the mechanism of expansion, particularly in Canada, the U.S.A. and China. Subsequently, Katayama (2004, 2006, 2010a) obtained more data to support the formation of gel as the underlying mechanism of expansion of concrete containing ACR aggregates, and Katayama's work has been supported by observations of Grattan-Bellew (2009, 2013), Katayama and Grattan-Bellew (2012) and Katayama *et al.*, 2016. It is concluded that the so-called alkali-carbonate reaction (ACR) is a special case of alkali-silica reaction similar to that occurring in siliceous limestones.

The long-standing confusion about ACR is due to the inappropriate methods applied in the sample preparation, observation and analysis, as well as to the misunderstanding that the distinctive chemical reaction (dedolomitization) is directly or indirectly responsible for the deleterious expansion.

REFERENCES

ASTM C586. (2006) Standard Test Method for Potential Alkali Reactivity of Carbonate Rocks as Concrete Aggregates (Rock Cylinder Method), Annual Book of ASTM Standards, 2006 Section 4 Construction, Volume 04.02 Concrete and Aggregates. ASTM International, 100 Barr Harbor Drive, PO Box C700, West Conshohocken, PA 19428–2959, pp. 312–316.

ASTM C1105. (2006) Standard Test Method for Length Change of Concrete Due to Alkali-Carbonate Rock. Annual Book of ASTM Standards, 2006 Section 4 Construction, Volume 04.02 Concrete and Aggregates, ASTM International, 100 Barr Harbor Drive, PO Box C700, West Conshohocken, PA 19428–2959, pp. 600–603.

ASTM C1260. (2004) Standard Test Method for Potential Alkali Reactivity of Aggregates (Mortar-Bar Method), Annual Book of ASTM Standards, 2004 Section 4 Construction, Volume 04.02 Concrete and Aggregates. ASTM International, 100 Barr Harbor Drive, PO Box C700, West Conshohocken, PA 19428–2959, pp. 680–684.

Buck, A.D. (1969) Potential Alkali Reactivity of Carbonate Rock from Six Quarries. Miscellaneous Paper C-69-15, U.S. Army Engineer Waterways Experiment Station Vicksburg, Mississippi, U.S.A.

CSA A23.2 14A. (2014) Potential Expansivity of Aggregates (procedure for length change due to alkali-aggregate reaction in concrete prisms at 38°C. CSA Standards A23.1–09/A23.2–09 Concrete materials and methods of concrete construction/Test methods and standard practices for concrete. Canadian Standards Association, 5060 Spectrum Way Suite 100, Mississauga, Ontario, Canada L4W 5N6, pp. 285–296.

CSA A23.2–26A. (2014) Determination of potential alkali-carbonate reactivity of quarried carbonate rocks by chemical composition. CSA Standards A23.1–09/A23.2–09 Concrete materials and methods of concrete construction/Test methods and standard practices for concrete. Canadian Standards Association, 5060 Spectrum Way Suite 100, Mississauga, Ontario, Canada L4W 5N6, pp. 366–370.

Deng, M. & Tang, M. (1993) Mechanism of dedolomitization and expansion of dolomitic rocks. Cement Concrete Res., 33, 1397–1408.

Deng, M., Han, S.F., Lu, Y.N., Lan, X.H., Hu, Y.L. & Tang M.S. (1993) Deterioration of concrete structures due to alkali-dolomite reaction in China. Cement Concrete Res., 23, 1040–1046.

Dolar-Mantuani, L. (1964) Expansion of Gull River Carbonate Rocks in Sodium Hydroxide." Highway Research Record, No. 45, pp.178–195.

Durand, B. (2000) More results about the use of lithium salts and mineral admixtures to inhibit ASR in concrete. In: Bérubé, M.A., Fournier, B. & Durand, B. (eds.) Proceedings of the 11th International Conference on Alkali-Aggregate Reaction in Concrete (ICAAR), Quebec, Canada, pp. 623–632.

Fournier, B., Bérubé, M.A., & Frenette, J. (2000) Laboratory investigations for evaluating potential alkali-reactivity of aggregates and selecting preventive measures against AAR; What do they really mean? In: Fournier, B. & Durand, B. (eds.) Proceedings of the 11th International Conference on Alkali-Aggregate Reaction in Concrete, Quebec City (Canada), June 2000, CRIB, Université Laval, Québec, pp. 287–296.

Gillott, J.E. (1961) Chemistry of the Kingston Dolomitic Limestone. DBR Internal Report No. 186, National Research Council of Canada, 8p.

Gillott, J.E. & Swenson, E.G. (1969) Mechanism of the alkali-carbonate rock reaction. J Eng Geol., 2, 7–23.

Grattan-Bellew, P.E. (1981) Evaluation of miniature rock prism test for determining the potential alkali-expansivity of aggregates. Cement Concrete Res., 11, 699–711.

Grattan-Bellew, P.E. & Chan, G. (2013) Comparison of the morphology of alkali-silica gel formed in limestones in concrete affected by the so-called Alkali-Carbonate Reaction (ACR) and Alkali-Silica Reaction (ASR). *Cement Concrete Res.*, *47*, 51–54.

Grattan-Bellew, P.E. & Rogers, C.A. (2000) Expansion Due to Alkali-Carbonate Reaction: Laboratory Prognosis Versus Field Experience. In: Fournier, B. & Durand, B. (eds.) *Proceedings of the 11th International Conference on Alkali-Aggregate Reaction in Concrete*, Quebec City, CRIB, Université Laval, Québec, pp. 41–49.

Grattan-Bellew, P.E., Mitchell, L.D., Margeson, J. & Deng, M. (2009) Is Alkali-Carbonate Reaction just a Variant of Alkali-Silica Reaction, ACR = ASR? *Cement Concrete Res.*, *40*, 556–562.

Guangren, Q., Min, D. & Tang, M. (2001) ACR expansion of dolomites with mosaic textures. *Mag. Concrete Res.*, *53*, (5), 327–336.

Hadley, D.W. (1961) Alkali Reactivity of Carbonate Rocks – Expansion and Dedolomitization. *Highway Research Board Proceedings*, *40*, pp. 462–474.

Hadley, D.W. (1964) Alkali-Reactive Carbonate Rocks in Indiana – A Pilot Regional Investigation. *Highway Research Record, No. 45*, pp. 196–221.

Insley, H. & Fréchette, van D. (1955) *Microscopy of Ceramics and Cements*. New York, Academic Press Inc. 286p.

Jensen, V. (2009) Alkali Carbonate Reaction (ACR) and RILEM AAR-0 Annex A: Assessment of Potentially Reactivity of Carbonate Rocks. In: *Proceedings of the 12th Euroseminar on Microscopy Applied to Building Materials*, Dortmund, Germany. pp. 15–19.

Katayama, T. (1992) A Critical Review of Carbonate Rock Reactions – Is their Reactivity Useful or Harmful? In: *Proceedings of the 9th International Conference on Alkali-Aggregate Reaction in Concrete*, London England, July 1992. pp.508–518.

Katayama, T. (2004) How to identify carbonate rock reactions in concrete. *Materials Characterization*, *53* (2–4), 85–104.

Katayama, T. (2006) Modern Petrography of Carbonate Aggregates in Concrete – Diagnosis of So-Called Alkali-Carbonate Reaction and Alkali-Silica Reaction. In: *Proceedings of the Professor Marc-André Bérubé Symposium, Oral Presentation*, 22p.

Katayama, T. (2008) ASR Gel in Concrete Subject to Freeze/thaw Cycles – Composition between Laboratory and Field Concretes from Newfoundland, Canada. In: *Proceedings of the 13th International Conference on Alkali-Aggregate Reaction*, Trondheim, Norway. pp. 174–183.

Katayama, T. (2010a) The so-called Alkali-Carbonate Reaction (ACR) – Its Mineralogical and Geochemical details, with special reference to ASR. *Cement Concrete Res.*, *40*, 643–675.

Katayama, T. (2010b) Diagnosis of Alkali-Aggregate Reaction – Polarizing microscopy and SEM-EDS analysis. In: Castro-Borges, P., Moreno, E.I., Sakai, K., Gjorv, O.E. & Banthia, N. (eds.) *Proceedings of the 6th International Conference on Concrete under Severe Conditions, Environment and Loading (CONSEC'10). Merida, Yukatan, Mexico. Concrete under Severe Conditions*. London, Taylor & Francis Group. pp. 19–34.

Katayama, T. (2011) So-called Alkali-Carbonate Reaction – Petrographic Details of Field Concretes in Ontario. In: *Proceedings of the 13th Euroseminar on Microscopy Applied to Building Materials*, Ljubljana, Slovenia. 15p.

Katayama, T. (2012a) Rim-Forming Dolomitic Aggregate in Concrete Structures in Saudi Arabia – Is Dedolomitization Equal to the So-Called Alkali-Carbonate Reaction? In: *Proceedings of the 14th International Conference on Alkali-Aggregate Reaction in Concrete*. Austin, USA, 10p. Paper 030411-KATA-01.

Katayama, T. (2012b) Petrographic Study of Alkali-Aggregate Reactions in Concrete. Doctoral Thesis (Science), The University of Tokyo, Japan.

Katayama, T. (2012c) ASR Gels and Their Crystalline Phases in Concrete – Universal Products in Alkali-Silica, Alkali-Silicate and Alkali-Carbonate Reactions. In: *Proceedings of the 14th International Conference on Alkali-Aggregate Reaction in Concrete, Austin, USA*. 12p. Paper 030411-KATA-03.

Katayama, T. & Grattan-Bellew, P.E. (2012) Petrography of the Kingston Experimental Sidewalk at Age of 22 Years – ASR as the Cause of Deleterious Expansive, So-Called Alkali-Carbonate Reaction. In: *Proceedings of the 14th International Conference on Alkali-Aggregate Reaction in Concrete*. Austin, USA. 10p. Paper 030411-KATA-06.

Katayama, T. & Sommer, H. (2008) Further Investigation of the Mechanism of So-Called Alkali-Carbonate Reaction Based on Modern Petrographic Techniques. In: *Proceedings of the 13th International Conference on Alkali-Aggregate Reaction in Concrete*, Trondheim, Norway. pp. 850–860.

Katayama, T., Jensen, V., & Rogers, C. (2016) The Enigma of the 'so-called' Alkali-Carbonate Reaction. In: *Proceedings of the Institution of Civil Engineers: Construction Materials*, 169 (CM4). pp. 223–232.

Katayama, T., Ochiai, M., & Kondo, H. (1996). Alkali-Reactivity of Some Japanese Carbonate Rocks Based on Standard Tests. *Proceedings of the 10th International Conference on Alkali-Aggregate Reaction in Concrete*, Melbourne, Australia. pp. 294–301.

Katayama, T., Oshiro, T., Sarai, Y., Zaha, K., & Yamato, T. (2008) Late-expansive ASR Due to Imported Sand and Local Aggregates in Okinawa Island, Southwestern Japan. In: *Proceedings of the 13th International Conference on Alkali-Aggregate Reaction in Concrete*, Trondheim, Norway. pp. 862–873.

Lemish, J. & Moore, W.J. (1964) Carbonate Aggregate Reactions: Recent Studies and an Approach to the Problem. *Highway Research Record, No. 45*, pp. 57–71.

López-Buendía, A.M., Climent, V., & Verdu, P. (2006) Lithological influence of aggregate in the alkali-carbonate reaction. *Cement Concrete Res.*, *36*, 1490–1500.

López-Buendía, A.M., Climent, V., Mar Urquiola, M., & Bastida, J. (2008) Influence of Dolomite Stability on Alkali-Carbonate Reaction. In: *Proceedings of the 13th International Conference on Alkali-Aggregate Reaction in Concrete*, Trondheim, Norway. pp. 233–242.

López-Buendía, A.M., Climent, V., & Verdu, P. (2006) Lithological influence of aggregate in the alkali-carbonate reaction. *Cement Concrete Res.*, *36*, 1490–1500.

Lu, D., Fournier, B., Grattan-Bellew, P.E., Zhongzi, X., & Tang, M. (2008) Development of a universal accelerated test for alkali-silica and alkali-carbonate reactivity of concrete aggregates. *Mater. Struct*, *41*, 235–246.

Lu, D., Mei, L., Xu, Z., Tang, M., & Fournier, B. (2006) Alteration of alkali reactive aggregates autoclaved in different alkali solutions and application to alkali-aggregate reaction in concrete (I) alteration of alkali reactive aggregates in alkali solutions. *CementConcrete Res.*, *36*, 1176–1190.

Melanesi, C.A., Marfil, S.A., Batic, O.R., & Maiza, P.J. (1996) The Alkali-Carbonate Reaction and Its Reaction Products – An Example with Argentinean Dolomite Rocks. *Cement Concrete Res.*, *26*, 1579–1591.

Mu, X., Xu, Z., Deng, M. & Tang, M. (1996) Abnormal Expansion of Coarse-Grained Calcite in the Autoclave Method. In: *Proceedings of the 10th International Conference on Alkali-Aggregate Reaction in Concrete*, Melbourne, Australia. pp. 310–315.

NF P 18-454 (2004) Reactivity of a concrete formula with regard to the alkali-aggregate reaction, performance test, French Association of National Standards, Paris (in French).

Newlon, H.H., Jr. & Sherwood, C. (1962) An Occurrence of Alkali-Reactive Carbonate Rock in Virginia. *Highway Research Board Bull, No. 355*, pp. 27–43.

Nixon, P.J. & Sims, I. (eds.) (2016) RILEM Recommendations for the Prevention of Damage by Alkali-Aggregate Reactions in New Concrete Structures, RILEM State-of-the-Art Reports, Volume 17, Springer, Dordrecht, RILEM, Paris, 168p.

Ozol, M.A. (1994) Alkali-Carbonate Rock Reaction. ASTM STP 169C, pp. 372–387.

Ozol, M.A. & Newlon, H.H. (1974) Bridge Deck Deterioration Promoted by Alkali-Carbonate Reaction: A Documented Example. Transportation Research Record, No. 525, pp. 55–63.

Qian, G., Deng, M. & Tang M. (2001) ACR expansion of dolomites with mosaic textures. *Magazine of Concrete Research*, 53 (5), 327–336.

Rogers, C.A. (1985) Alkali Aggregate Reactions, Concrete Aggregate Testing and Problem Aggregates in Ontario – A Review. Ontario Ministry of Transportation and Communications, Engieering Materials Office, Report EM-41, 5th Revised Edition, 44p.

Rogers, C.A. (1986) Evaluation of the potential for expansion and cracking of concrete caused by the alkali-carbonate reaction. *Cement Concrete Aggr.*, *8*, 13–23.

Rogers, C.A. (2000) Personal Communication.

Rogers, C.A. & Hooton, R.D. (1992) Comparison between Laboratory and Field Expansion of Alkali-Carbonate Reactive Concrete. In: *Proceedings of the 9th International Conference on Alkali-Aggregate Reaction in Concrete*, London, pp. 877–884.

Sommer, H., Grattan-Bellew, P.E., Katayama, T. & Tang, M. (2004) Development and Inter-Laboratory Trial of the RILEM AAR-5: Rapid Preliminary Screening Test for Carbonate Aggregates. In: *Proceedings of the 12th International Conference on Alkali-Aggregate Reaction*, Vol. 1, Beijing, pp. 407–412.

Stanton, T.E. (1940) Expansion of Concrete Through Reaction Between Cement and Aggregate. In: *Proceedings of the American Society of Civil Engineers*, December 1940, pp. 1781–1811.

Swenson, E.G. (1957) A Canadian Reactive Aggregate Undetected by ASTM Tests. *ASTM Bulletin No. 226*, December 1957, pp. 48–51.

Swenson, E.G. & Gillott, J.E. (1960) Characteristics of Kingston Carbonate Rock Reaction. *Highway Research Board Bull. No. 275*, pp. 18–31.

Swenson, E.G. & Gillott, J.E. (1964) Alkali-Carbonate Rock Reaction. *Highway Research Record, No. 45*, pp. 21–40.

Tang, M. & Lu, Y. (1986) Rapid Method for Determining the Alkali-Reactivity of Carbonate Rock. In: Grattan-Bellew, P.E. (ed.) *Concrete Alkali-Aggregate Reactions*, Noyes Publications, Park Ridge, New Jersey, U.S.A., pp. 286–287.

Tang, M. & Deng, M. (2004). Progress on the Studies of Alkali-Carbonate Reaction. In: *Proceedings 12th International Conference on Alkali-Aggregate Reaction in Concrete*, Beijing, China, pp. 51–59.

Tang, M., Deng, M. & Zhongzi, X. (2000) Comparison between Alkali-Silica Reaction and Alkali-Carbonate Reaction. In: Bérubé, M.A., Fournier, B. & Durand, B. (eds.) *Proceedings 11th International Conference on Alkali-Aggregate Reaction in Concrete*, Québec City, pp.109–117.

Tang, M.S., Lan, X. & Han, S.F. (1994) Autoclave method for identification of Alkali Reactive Carbonate Rocks. *Cement Concrete Comp.*, *16*, 163–167.

Tong, L. 1994. "Alkali-Carbonate Rock Reaction," Ph.D thesis, Nanjing Institute of Chemical Technology, Nanjing, China.

Tong, L. & Tang, M. (1999) Expansion mechanism of alkali-dolomite and alkali-magnesite reaction. *Cement Concrete Comp.*, *21*, 361–373.

Walker, H.N. (1978) Chemical Reactions of Carbonate Aggregates in Cement Paste, ASTM Special Technical Publication 169B, Chapter 41, pp.722–743.

Yamada, K., Kawabata, Y., Ogawa, S., & Maruyama, I. (2015). A study on alkali-aggregate reaction in nuclear power related facilities. *Cement Science & Concrete Technology.*, *68*, 457–464, see Figure 2 (in Japanese).

Chapter 4

Prevention of Alkali-Silica Reaction

Michael D.A. Thomas, R. Doug Hooton & Kevin Folliard

4.1 INTRODUCTION

This chapter discusses various strategies for preventing alkali-silica reaction (ASR) in concrete, including avoiding reactive aggregates, controlling the alkali content of the concrete, using mineral additions or 'supplementary cementing materials' (SCMs), and the use of lithium-based compounds. The chapter does not discuss methods for preventing alkali-carbonate reaction (ACR). Strategies for controlling ASR are reported as not generally effective for preventing ACR and, consequently, alkali-carbonate reactive rocks that produce expansion should not be used in concrete. 'So-called alkali-carbonate reaction' is discussed in Chapter 3.

4.2 PREVENTATIVE MEASURES – OPTIONS

As stated in previous chapters, there are three basic requirements for damaging ASR to occur in concrete; these are:

• A sufficient quantity of reactive silica (within aggregates)
• A sufficient concentration of alkali (primarily from Portland cement)
• Sufficient moisture

Elimination of any one of these requirements will prevent the occurrence of damaging alkali-silica reaction. Exclusion of water from civil engineering structures is not practical in most cases and so, from a consideration of the fundamental aspects of ASR discussed in Chapter 1, the most obvious options for preventing expansion due to ASR are the following:

1. Avoid the use of reactive aggregates
2. Minimize the amount of alkalis from the Portland cement and other sources

Two other options are as follows:

1. Use of supplementary cementing materials (SCMs)
2. Use of lithium-based compounds.

In fact, the use of an SCM is a form of Option 2 as these materials consume a portion of the alkalis contributed by the Portland cement into their lower Ca/Si ratio C-S-H and alumina-substituted C-A-S-H (Hong & Glasser, 1999, 2002; Hooton *et al.*, 2010) and reduce their availability for reaction with the aggregate. Lithium compounds work in a different manner by changing the nature of the reaction product. Both of these mechanisms are discussed later in the chapter.

4.3 USE OF NON-REACTIVE AGGREGATES

Using non-reactive, or more correctly, non-deleteriously-reactive, aggregates is certainly a viable method of preventing ASR-induced damage if such aggregates are available. Competent and thorough testing is required to ensure that aggregate sources are non-reactive and this can be achieved through a combination of petrographic examination [ASTM C295 (2012) or RILEM AAR-1.1 (in Nixon & Sims, 2016) & AAR-1.2 (Fernandes *et al.*, 2016)], expansion testing of mortar [ASTM C1260 (2014) or RILEM AAR-2 (in Nixon & Sims, 2016)] or concrete [ASTM C1293 (2008) or RILEM AAR-3 (in Nixon & Sims, 2016)], and field performance. Such testing should be performed on a regular basis to ensure that the composition (and reactivity) does not change within a pit or a quarry. If the aggregate sources can be confirmed to be truly non-reactive, no further precaution is required to prevent ASR.

There is a number of reasons why the option of using non-reactive aggregates is not always feasible; these include:

- Non-reactive aggregates are not available locally and the cost of shipping non-reactive materials from other locations is prohibitive
- Reactive aggregates that are otherwise wholly suitable for concrete are readily and abundantly available at lower cost and reduced environmental impact compared with non-reactive materials
- Lack of confidence in test results (or testing laboratories) or test results are ambiguous (*e.g.*, different test methods do not agree)
- All of the locally-available materials fail the very aggressive accelerated ASTM C1260 (2014) test (regardless of true reactivity).

In such cases, it is necessary to adopt one of the other options for preventing damaging ASR.

Furthermore, some instances warrant extra caution even when using aggregates believed to be non-reactive; examples include the design of critical structures (*e.g.*, prestigious structures or those with an extended design life) and the construction of structural elements exposed to a very aggressive environment (*e.g.*, structures exposed to seawater or deicing salts, which may provide an external source of alkalis).

4.4 LIMITING THE ALKALI CONTENT OF THE CONCRETE

Stanton's (1940) formative work on ASR indicated that expansive reaction is unlikely to occur when the alkali content of the cement is below 0.60% Na_2Oeq, ($Na_2Oeq = Na_2O$ equivalent). However, later research indicated that damaging ASR could occur both in the laboratory and in the field (*e.g.*, Woolf, 1952; Stark, 1980; Blaikie *et al.*, 1996) when low-alkali cements are used. Despite this, the 0.60% value has become the widely accepted maximum limit for cement to be used with reactive aggregates in the United States, and appears in ASTM C150 (2016) Standard Specification for Portland cement as an optional limit when concrete contains deleteriously reactive aggregate.

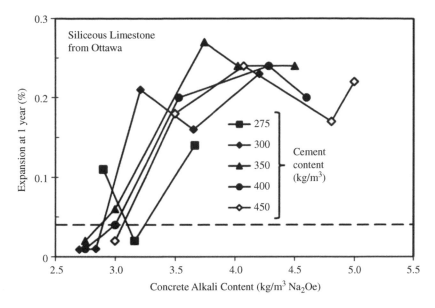

Figure 4.1   Expansion of concrete prisms as a function of alkali content (produced from unpublished data from the Building Research Establishment, U.K.). The dashed line represents the critical 1 year expansion limit (Lindgard *et al.*, 2012).

Appendix X1 of ASTM C33 (2016) Standard Specification for Concrete Aggregates includes "low-alkali cement" (meeting the ASTM C150 limit of 0.60% Na_2O) as a measure *"known to prevent excessive expansion"*.

However, simply limiting the alkali level of the cement takes no account of the cement content of the concrete which, together with the cement alkali content, governs the total alkali content of concrete, and is considered to be a more accurate index of the risk of expansion when a reactive aggregate is used in concrete. Figure 4.1 shows the relationship between alkali content and expansion for concretes produced with a range of Portland cement contents and cements of varying alkali content (Lingård *et al.*, 2012). The figure was produced using previously unpublished data from the Building Research Establishment (Blackwell, private communication, 2000). The relationship clearly shows that it is the product of cement content and cement alkali level (*i.e.*, the alkali content of the concrete) that controls the alkali content rather than the cement alkali level alone.

Figure 4.2 shows expansion of concrete prisms plotted against the alkali content of the concrete for three different reactive aggregates (selected to demonstrate the range of behaviour observed). It can be seen that the threshold alkali content required to initiate damaging expansion in the concrete prism test varies considerably between aggregates, values ranging from approximately 3.0 kg/m³ Na_2Oeq to more than 5.0 kg/m³ Na_2Oeq for the aggregates shown in Figure 4.2.

A number of specifications have employed a maximum concrete alkali content as an option to control expansion in concrete containing reactive aggregates. Nixon and Sims (1992) reported that maximum permissible alkali contents between 2.5 and 4.5 kg/m³ Na_2Oeq have been specified by various countries and agencies, with the allowable alkali contents sometimes varying depending on aggregate reactivity.

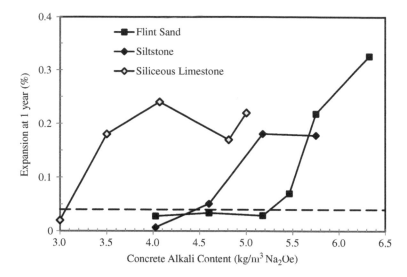

Figure 4.2   Effect of alkali content on expansion of concrete prisms (stored over water at 38°C, 100°F) with different aggregates (data from Thomas et al., 1996).

Unfortunately, traditional concrete prism tests such as ASTM C1293 (2008) tend to underestimate the threshold alkali content for aggregates as a result of alkali being leached from the concrete during exposure. This phenomenon has been discussed in the literature (Thomas *et al.*, 2006; Lindgård *et al.*, 2012) and a further example is shown in Figure 4.3. The figure shows the expansion of concrete blocks (0.38 x 0.38 x 0.71 m) containing reactive (Jobe, Texas) sand and various levels of alkali, and stored on an outdoor exposure site at the University of Texas in Austin, compared with the expansion of concrete prisms from the same mix stored over water (ASTM C1293: 2008). It is evident that the concrete prism test yields a higher threshold alkali content than may be observed for larger elements stored under field conditions.

In 2000, in Canada CSA A23.2-27A (now 2014) introduced a "sliding-scale" for the alkali limit used as a preventive measure, the value varying from 1.8 to 3.0 kg/m^3 Na$_2$Oeq as shown in Table 4.1.

Table 4.1   Concrete alkali limit according to level of prevention (after CSA A23.2-27A: 2000).

Level of prevention required	Alkali limit (Na$_2$Oe)[*] kg/m^3
Mild (W)	3.0
Moderate (X)	2.4
Strong (Y)	1.8
Exceptional (Z)	1.8 + SCM

[*]Calculated from the total alkali equivalent for the Portland cement fraction of the binder multiplied by the Portland cement content of the concrete mixture.

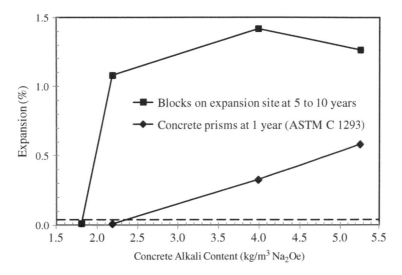

Figure 4.3 Effect of alkali content on the expansion of blocks stored outdoors and prisms stored over water (unpublished data from the University of Texas at Austin).

As the data available from laboratory tests at the time were considered unreliable for the purpose of establishing threshold alkali values, the maximum alkali content limits were based on published data from structures and field experience (*e.g.*, Thomas *et al.*, 2012; Hooton *et al.*, 2000). Thomas (1996a) had previously reported damaging ASR in concrete dams in the U.K. and Canada, where the estimated alkali content was in the range of 2.0 to 2.4 kg/m^3 Na$_2$Oeq. Rogers *et al.* (2000) reported results from a study of 8-year-old concrete blocks (0.6 x 0.6 x 2 m) stored on an exposure site in Ontario, Canada. Specimens produced with high-alkali cement showed very significant expansion and cracking after 8 years. Specimens with low-alkali cement (0.46% Na$_2$Oeq) and a calculated concrete alkali content of 1.91 kg/m^3 Na$_2$Oeq showed significantly less expansion; however, the expansion did exceed 0.04% at 8 years and small cracks were evident. Subsequent investigations (Hooton *et al.*, 2006; MacDonald *et al.*, 2012; Hooton *et al.*, 2013) of these blocks at later ages (14 and 20 years) confirmed the presence of significant ASR-induced damage in the blocks produced with low-alkali cement and measured expansion of approximately 0.08% at 20 years (see Figure 4.4). Based on the information available, a maximum alkali limit of 1.8 kg/m^3 Na$_2$Oeq was selected for Prevention Level Y (*e.g.*, concrete with highly reactive aggregate, exposed to moisture with a service life up to 75 years) and it was decided to require the same limit plus incorporate minimum levels of supplementary cementing material (SCM) for Level Z. This was supported by anecdotal evidence available to the committee preparing the guidelines, there being no known case of ASR in concrete structures with lower alkali contents. In the absence of any other data, the same alkali limits were adopted in AASHTO PP65 (2011) and ASTM C1778 (2016).

Aggregates that are used in concrete with an alkali content below the threshold for expansion may still cause damaging expansion if the alkali content of the concrete increases at some locations during the service life of the structure. This may occur through alkali

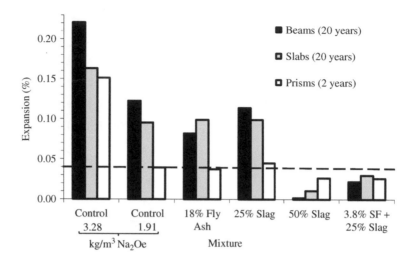

Figure 4.4 Expansion of concrete prisms in the laboratory and concrete blocks and slabs on the Kingston exposure site (data from MacDonald et al., 2012).

concentration caused by drying gradients, alkali release from aggregates, or the ingress of alkalis from external sources, such as deicing salts or seawater (Nixon *et al.*, 1987). Stark (1978) reported increases in soluble alkali from 1.1 to 3.6 kg/m^3 Na$_2$Oeq close to the surface of some highway structures. Migration of alkalis due to moisture, temperature, and electrical gradients has also been demonstrated by laboratory studies (Xu *et al.*, 1993).

Alkali migration can occur very early in the life of concrete and has been implicated as the cause of pop-outs at the surface of slabs (Landgren & Hadley, 2002; Cong *et al.*, 2004). Alkalis may be carried upwards with the bleed water in freshly placed concrete and this migration may exacerbated during and after setting by the evaporation of water from the surface of the slab especially during hot, dry weather. The concentration of alkalis at the surface leads to ASR in the near surface of the concrete where the formation of gel and expansion of aggregates can lead to the fracture of a small conical-shaped mortar fragment (a 'pop-out') overlying the reacting aggregate. Often the damage is restricted to the surface of the concrete where the alkalis are concentrated as there is insufficient alkali in the bulk concrete to cause damage. Some studies have reported surface concentrations that are as much as six times the alkali concentration in the bulk concrete (Nixon *et al.*, 1979). This phenomenon can occur even when low-alkali cement or pozzolans are used (Cong *et al.*, 2004).

A number of workers have demonstrated that many aggregates contain alkalis that may be leached out into the concrete pore solution, thereby increasing the risk of alkali-aggregate reaction. Stark and Bhatty (1986) reported that, in extreme circumstances, some aggregates release alkalis equivalent to 10% of the Portland cement content. A comprehensive review on alkali release from aggregates, including methods for determining the "releasable or available alkali" in aggregates, has been presented by Bérubé *et al.* (2002).

Supplementary cementing materials (SCMs), such as fly ash, silica fume, ground granulated blastfurnace slag and natural pozzolana may also contain significant quantities of alkali. However, with the exception of high-calcium fly ashes with alkali equivalents exceeding 5 kg/m^3 [note that CSA A23.2-27A (2000) requires fly ashes with >4.5 kg/m^3 alkali to be used at a higher cement replacement level], these alkalis generally do not need to be included in the calculation of the concrete alkali content as SCM tend to reduce the alkalis that are available for reaction with the aggregate; this is discussed in the next section.

4.5 USE OF SUPPLEMENTARY CEMENTING MATERIALS

One of the most efficient means of controlling ASR in concrete containing reactive aggregates is the appropriate use of supplementary cementing materials (SCMs). Such materials include pozzolans (*e.g.*, fly ash, silica fume, calcined clay, or shale) and ground granulated blastfurnace slag (GGBS). The potential use of pozzolans to control ASR dates back as far as the discovery of ASR, as reported in the first major publication on the phenomenon (Stanton, 1940). In this paper, Stanton not only demonstrated that damaging reaction would only occur if there was a sufficient quantity of alkalis in the Portland cement and reactive silica in the aggregate, but also that expansion was reduced when a pozzolanic cement was used. Ten years later, Stanton (1950) further demonstrated that partially replacing Portland cement with a sufficient quantity of pozzolan (pumicite or calcined shale) eliminated deleterious expansion whereas replacement with similar quantities of ground quartz (Ottawa) sand did not, indicating that the beneficial action of the pozzolan extended beyond merely diluting the cement alkalis. In the early 1950s, various studies (Cox *et al.*, 1950; Barona, 1951; Buck *et al.*, 1953) showed that other SCMs, namely fly ash and GGBS were also effective in reducing expansion.

Since these early studies, there have been literally hundreds of studies and technical papers dealing with the effects of SCM on ASR and it is now generally recognized that the use of a sufficient quantity of a suitable SCM is one of the more efficient preventive measures for controlling expansion when a deleteriously reactive aggregate is used in concrete (Thomas *et al.*, 2008). Thomas (2011) reviewed selected published works dealing with (i) the mechanisms by which SCM controls ASR, (ii) the effect of SCM composition on its efficacy in this role and (iii) test methods for determining the amount of SCM required to minimise the risk of damaging expansion to an acceptable level.

Thomas (2011) showed that almost any SCM can be used to control ASR provided it is used at a sufficient level of replacement. The amount required varies widely depending on, amongst other things, the following:

- The nature of the SCM (especially mineralogical and chemical composition); more SCM is required as its silica content decreases or as its alkali and calcium content increase,
- The nature of the reactive aggregate; generally, the more reactive the aggregate, the higher the level of SCM required,
- The availability of alkali within the concrete (*i.e.*, from the Portland cement and other sources); the amount of SCM required increases with the amount of available alkali,
- The exposure conditions of the concrete; concrete exposed to external sources of alkali may require higher levels of SCM.

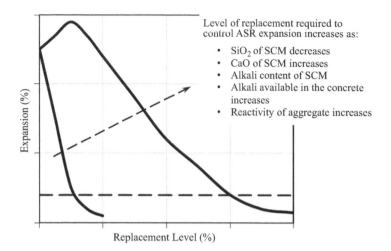

Figure 4.5 Expansion versus SCM Content – Conceptual Relationship (Thomas, 2011).

Figure 4.5 shows the (conceptual) relationship between the (long-term) expansion of concrete and the level of replacement for different SCMs. Generally, as the level of replacement increases with a particular SCM, the expansion decreases and eventually reaches an acceptable level at which no damage occurs. SCMs that are very high in reactive silica, such as silica fume and metakaolin, tend to be very efficient in controlling expansion and are only required at relatively low levels of replacement (*e.g.*, 10 to 15%). On the other hand, SCMs with lesser amounts of silica, such as Class C fly ash and GGBS, need to be used at higher levels of replacement (*e.g.* $\geq 40\%$). Some SCMs (*e.g.*, some Class C fly ashes) may produce a pessimum effect by increasing the amount of expansion (compared to concrete without SCM) if they are used at low levels of replacement, but decreasing expansion at higher levels of replacement.

It is generally considered that the principal mechanism by which SCMs control ASR expansion is by reaction with and consumption of the alkalis in the concrete pore solution, which reduces the alkali available for reaction with the aggregates. However, the use of SCMs also results in a reduction in the availability of calcium (due to the dilution of and consumption of calcium hydroxide) and in a refinement in the pore structure, which leads to reduced ionic and moisture diffusivity; these effects may also be beneficial in terms of minimizing the risk and extent of ASR (Thomas, 2011).

This section begins with a discussion on the effect of SCMs on the composition of the pore solution and the availability of alkalis, and then provides examples on the impact of SCMs on the expansion of concrete. The role of SCM composition, cement alkalis and aggregate reactivity on the expansion of concrete containing SCMs is also discussed.

4.5.1 `Effect of SCM on the Availability of Alkalis

Although all SCMs contain some level of alkali and some may contain significantly more alkali than the Portland cement that they partially replace, the main mechanism by which SCMs reduce expansion due to ASR is by reducing the alkalis that are available to the concrete pore solution. Once the alkalis in the binder phase (Portland cement + SCM) of concrete are "released" by hydration they may be present in one of three ways: dissolved within the pore solution, bound by the hydration products or incorporated in alkali-silica gel. In the absence of reactive aggregate, alkalis will not be consumed by ASR and the partition of the alkalis between the pore solution and the hydrates is largely a function of the composition of the binder.

Numerous workers have shown that SCMs have a significant impact on the concentration of alkalis in the pore solution (see Thomas, 2011). Studies on the effect of fly ash and GGBS on the pore solution of pastes have been reviewed by Thomas (1996b) and studies involving silica fume have been reviewed by Thomas and Bleszynski (2001). These studies show that the incorporation of most SCMs leads to a reduction in the concentration of alkali hydroxides in the pore solution of pastes, mortar and concretes, the amount of reduction increasing with higher SCM replacement levels. Figure 4.6 shows the evolution of the hydroxyl ion concentration of the pore solution extracted from sealed paste samples with w/cm = 0.50 and Figure 4.7 shows the OH- concentration at 2 years as a function of the level of SCM (Shehata et al., 1999; Ramlochan et al., 2000; Bleszynski, 2002; Shehata & Thomas, 2002). Silica fume is the most efficacious SCM in this role, at least initially, followed by metakaolin, low-calcium fly ash and GGBS. High-calcium or high-alkali fly ashes are less effective and have to be used at relatively high levels of replacement to produce a significant reduction in the pore solution alkalinity.

It is interesting to note that in the case of the paste with 10% silica fume, the OH⁻ concentration drops rapidly over the first 28 days but then starts to increase slowly with time beyond 3 months; similar behaviour was observed in pastes containing 5% silica fume (Shehata & Thomas; 2002). This behaviour was not observed for pastes containing any of the other SCMs. As shown in Figure 4.6, the long-term increase in the OH⁻ concentration seems to be prevented in pastes

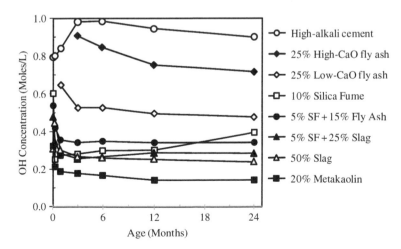

Figure 4.6 Evolution of the Pore Solution in Pastes Containing SCM (Shehata et al., 1999; Ramlochan & Thomas, 2000; Bleszynski, 2002; Shehata & Thomas, 2002).

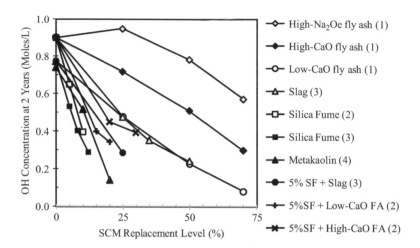

Figure 4.7 Effect of Type and Amount of SCM on Pore Solution Composition (Shehata *et al.*, 1999; Ramlochan *et al.*, 2000; Bleszynski, 2002; Shehata & Thomas, 2002).

containing 5% silica fume by the addition of either slag (25%) or fly ash (15%). It is conjectured that the presence of alumina in the SCM possibly contributes in some way to prevent the long-term release of alkalis back in to the pore solution. Hong and Glasser (2002) showed that introducing alumina into C-S-H, to form C-A-S-H, markedly increases its alkali-binding capacity and they suggest that this partially explains the beneficial effects of aluminous SCM with regards to reducing pore solution alkalinity and the potential for ASR.

Figure 4.8 shows an empirical relationship between the OH⁻ concentration of the pore solution extracted from 2-year-old sealed pastes with w/cm = 0.50 (Thomas & Shehata, 2004; Thomas & Folliard, 2007) and a "chemical index" derived from the chemical composition of the binder. A total of 79 different binders were tested including the following:

- 100% Portland cement at a range of different alkali contents (0.36 to 1.09% Na_2Oeq),
- Binary mixes with 25 to 70% fly ash using 18 different fly ashes with a range of chemical compositions (1.1 to 30.0% CaO, 1.4 to 9.7% Na_2Oeq),
- Binary mixes with 25 to 50% GGBS, 5 to 10% silica fume and 10 to 20% metakaolin; each SCM came from a single source,
- Ternary mixes containing silica fume blended with either slag or fly ash; both low-calcium and high-calcium fly ash were used.

The relationship was derived empirically to find the "chemical index" that was most reliably correlated to the OH⁻ concentration based on a least-square fit. The best-fit index was found to be the product of the equivalent alkalis and calcium divided by the square of the silica content of the binder $(Na_2Oe \times CaO)/(SiO_2)^2$. Although this is an empirical relationship, it makes sense intuitively because the alkalinity of the pore solution can be expected to be a function of the alkalis in the binder and the ability of the hydrates to bind alkalis, which has been shown to be a function of the calcium-to-silica ratio of the binder (see discussion below). The alumina

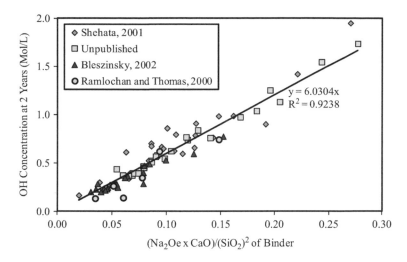

Figure 4.8 Relationship Between Pore Solution Composition and the Chemical Composition of the Binder (Thomas, 2011).

content of the binder was not found to be statistically significant based on the empirical analysis conducted using this dataset, despite the apparent benefit of alumina discussed above.

Analysing the composition of the pore solution extracted from a paste sample only provides one point on the equilibrium curve between bound and free alkalis. If the alkali content of the pore solution in concrete decreases, due perhaps to leaching or reaction with reactive silica in the aggregate, a portion of the bound alkalis may be released to regain equilibrium. It is important to establish what portion of the alkalis in a binder are "available" to a solution at a pH that is just able to sustain the alkali-silica reaction, as it is these alkalis that are available to fuel the reaction. Shehata and Thomas (2006) studied the alkali release characteristics of pastes produced with high-alkali cement and combinations of silica fume and various fly ashes. Paste samples, 1 to 3 years of age, were immersed in solutions of alkali hydroxide at initial molar concentrations of 0, 0.10, 0.25 and 0.40 and the change in concentration was observed to determine how much alkali was leached from the binder. Figure 4.9 shows a selection of the data produced in this study.

When mature paste samples were immersed in distilled water (pH = 7.0), almost all of alkalis present in the binder (80 to 90%) were released regardless of composition. As the alkali concentration of the leaching solution increased, the amount of alkali released from the binder decreased and was strongly dependent on the composition of the binder. Pastes containing 100% Portland cement, 5% silica fume or 25% high-CaO fly ash still released a significant portion of the alkalis present in the binder (50 to 80%), even in the solution of the highest initial alkali concentration (0.40M). Much less alkali (~ 20%) was released from pastes containing 25% low-CaO fly ash or ternary blends containing silica fume with either low-CaO or high-CaO fly ash. A correlation of the data (Shehata & Thomas, 2006) available for 24 different binders showed that the amount of alkali released to a solution with an initial alkali hydroxide concentration of

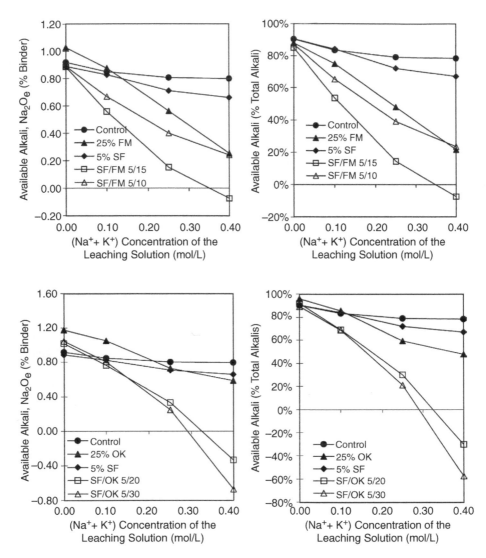

Figure 4.9 Available Alkalis in Pastes with SCM stored in Solutions of Varying Reactivity (Shehata & Thomas, 2006).

0.25M (assumed to be the concentration necessary to sustain alkali-silica reaction) was related to the chemical composition of the binder as represented by the parameter $(Na_2Oeq \times CaO)/(SiO_2)$.

The ability of SCMs to reduce the pore solution alkalinity is linked to their effect on the composition and alkali-binding capacity of the hydrates (especially C-S-H). Bhatty and Greening (1978) found that C-S-H with a low Ca/Si ratio was able to retain more alkali (Na + K) compared with hydrates of higher lime to silica ratios. The addition of fly ash reduces the Ca/Si ratio of the C-S-H hydrates and there is a concomitant increase in the alkali content. Rayment (1982) observed significant differences in the

C-S-H composition of Portland cement and fly ash pastes after just 8 days curing at 20°C. However, Uchikawa *et al.* (1989) found little difference in pastes after 91 days at 20°C, but substantial changes due to the incorporation of fly ash after 60 days at 40°C, indicating the role of the pozzolanic reaction in the C-S-H composition. Thomas *et al.* (1991), reporting results for 7-year-old concretes containing reactive flint sand, showed that the alkali binding capacity of C-S-H hydrates in concretes was increased significantly by the addition of fly ash. Uchikawa *et al.* (1989) showed that GGBS has a similar effect to low-calcium fly ash on hydrate composition.

Glasser and Marr (1985) explain the differences in alkali absorption on the basis of the surface charge on the C-S-H, which is dependent on the Ca/Si ratio. At high ratios, the charge is positive and the C-S-H tends to repel cations. As the Ca/Si ratio decreases the positive charge reduces becoming negative at low Ca/Si ratios, *e.g.*, less than 1.3 (Glasser, 1992). Negatively charged C-S-H has an increased capacity to sorb cations, especially alkalis. Hong and Glasser (1999) confirmed the importance of the Ca/Si ratio on the alkali-binding capacity of synthesized single-phase C-S-H, but subsequently showed that the binding capacity could be greatly increased by introducing alumina into the C-S-H to form C-A-S-H (Hong & Glasser, 2002).

Many of the studies on the alkali-binding of C-S-H have involved microanalysis (*e.g.*, using scanning electron microscopy equipped with energy dispersive X-ray analysis) of the inner-product C-S-H forming around remnant alite and belite grains. However, outer-product C-S-H also forms by reaction between $Ca(OH)_2$ and pozzolans, but this phase is more difficult to identify and analyse separately than the inner-product C-S-H. The pozzolanic reaction is actually very similar to the alkali-silica reaction. The reactive silica in the pozzolan reacts first with the alkali-hydroxides in the pore solution and alkali-silica gel containing small amounts of calcium is formed. Over time, calcium exchanges for alkali in the gel and C-S-H forms with a relatively low Ca/Si ratio compared to that formed in Portland cement paste.

The only substantial differences between this pozzolanic reaction and the alkali-silica reaction is the timescale over which the reactions occur and the absence of any detectable expansion due to the pozzolanic reaction. The lack of expansion can perhaps be explained by the fact that pozzolans are very-finely divided materials and the alkali-silica gel that forms and is subsequently converted to C-S-H is distributed throughout the cement paste, whereas the presence of reactive aggregate particles leads to the accumulation of larger deposits of alkali-silica gel in discrete locations that can become sites of expansion.

The importance of the size and distribution of the reactive silica can be demonstrated in two ways. Firstly, it has been known since the formative work of Stanton (1940) that if a reactive aggregate is ground to sufficient fineness (sub-180 micron in Stanton's studies) expansion is eliminated. Figure 4.10 shows data from Thomas (2011) relating to the effect of ground Vycor glass (sub-100 micron) on the expansion or mortar bars containing sand-sized Vycor glass as a reactive aggregate. The sand-sized Vycor glass behaves as a reactive aggregate causing expansion of the mortar with Portland cement as the only binder. However, the same material, when ground, behaves like a pozzolan when it is used to replace 20% of the Portland cement and prevents expansion. Secondly, if finely-divided pozzolans agglomerate and form sand-sized particles, these particles will behave like reactive aggregates and may result in expansion and cracking. This effect has been observed with agglomerated silica fume both in the field and in the laboratory. Figure 4.11 (from Maas *et al.*, 2007) shows a back-scattered electron image

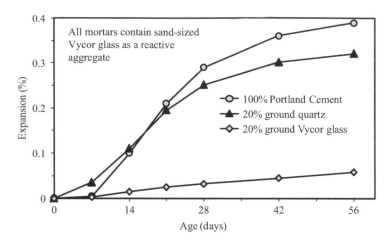

Figure 4.10 Role of Particle Size on the Behaviour of Vycor Glass (Thomas, 2011).

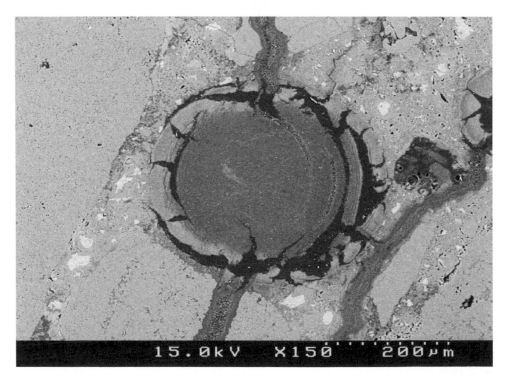

Figure 4.11 Agglomerated Silica Fume Particle behaving as a Reactive Aggregate and Source of ASR Expansion (Maas et al., 2007).

of a mortar bar containing agglomerated silica fume after storage in 1M NaOH solution at 80°C for 14 days (Maas *et al.*, 2007). The mortar, which contained non-reactive sand, expanded during test and the expansion was attributed to the reaction of the agglomerated silica fume.

4.5.2 Effect of SCM on the Expansion of Concrete

Stanton proposed a test method for evaluating the potential for cement-aggregate combinations to expand due to ASR in his first major paper on the subject (Stanton, 1940). This involved the manufacture of 25x25x250 mm mortar-bars and storing them over water in sealed containers. Subsequent modifications included elevating the temperature to 38°C (100°F) and the test was eventually standardized to become ASTM C227 (2010). Pozzolans were often evaluated using a modified version of this test with either the job aggregate or a standard reactive aggregate, Pyrex glass; the test with Pyrex became standardized as ASTM C441 (2011). ASTM C227 is no longer widely used in North America as it fails to detect many slowly-reacting aggregated because the small sample size and test conditions promote the leaching of alkalis from the bars; such a drawback is also a problem for evaluating SCMs (Thomas *et al.*, 2006).

ASTM C441 is a much more rapid test (typically 14 to 56 days) because of the high reactivity of the Pyrex glass and leaching is less significant during the test. However, this test fails to account for the nature of the reactive aggregate which is known to impact the amount of SCM required, and tends to overestimate the amount of SCM required to control expansion with natural aggregates (Thomas *et al.*, 2006). However, the use of Pyrex glass does allow a comparative evaluation of pozzolans and a number of investigators have used the test to observe the reduced efficacy of high-calcium ash compared with low-calcium ash (Dunstan, 1981; Buck & Mather, 1987; Carrasquillo & Snow, 1987; Klieger & Gebler, 1987; Smith, 1988). Dunstan's (1981) work was the most comprehensive with regards to fly ash composition. He reported results from Pyrex mortar bar tests for 17 ashes of varying chemistry and showed a reliable correlation between the calcium content of the ash and the expansion of mortar bars at 14 days.

The test methods most commonly used today to evaluate the efficacy of SCM in controlling ASR expansion are the concrete prism test (ASTM C1293: 2008) and the accelerated mortar-bar test (ASTM C1567: 2013). The accelerated mortar-bar test is by far the most widely used test, however, it is only intended as a screening test and should not be relied upon for phenomenological studies. The test involves the immersion of small mortar-bars in 1M NaOH solution at 80°C and this tends to mask the importance of the alkalis in the system under test. Since SCMs control ASR expansion mainly by reducing the availability of alkalis, providing an inexhaustible supply of alkalis is not desirable as eventually the beneficial effects of the SCM will be swamped by the ingress of alkalis from the storage solution. Furthermore, the very high temperature used in this test is not representative of the conditions that concrete encounters in the field. The review presented here focuses on expansion tests involving concrete exposed either to field conditions or to accelerated conditions (up to 38°C) in the laboratory [*e.g.*, ASTM C1293 (2008) concrete prism and similar tests]. Data from accelerated tests, such as the accelerated mortar-bar test (ASTM C1567: 2013), are not included.

4.4.2.1 Effect of SCM Composition on Expansion

As discussed above and illustrated in Figure 4.5, almost any SCM can be used to control expansion due to alkali-silica reaction provided it is used in sufficient quantity. The amount of SCM required is, of course, dependent on the composition of the SCM, but also on the reactivity of the aggregate, the quantity of alkalis supplied by the Portland cement (and other sources) and whether the concrete will be exposed to alkalis (*e.g.*, seawater, deicing chemicals) during service. On one extreme, a highly efficient pozzolan with a high level of reactive silica and negligible alkali content, may be expected to eliminate damaging expansion with a moderately reactive aggregate when used with a moderate-alkali cement at replacement levels of about 10%; this scenario is represented by the left-hand curve in Figure 4.5. On the other extreme, as represented by the right-hand curve in Figure 4.5, an SCM with a higher alkali and lower silica content might need to be used at a replacement level of 50 to 60% or more with highly reactive aggregate and high-alkali cement.

Figure 4.12 shows the expansion of concretes at 2 years as a function of the type and amount of SCM used; the tests were performed in a single laboratory using a reactive aggregate from a single source (siliceous limestone from the Spratt's quarry in Ontario, Canada) and were, generally, performed in accordance with ASTM C1293 (2008). Silica fume and metakaolin are the most efficient with regards to the replacement levels needed to reduce the expansion at 2 years, followed by low-calcium fly ash. While GGBS works well to mitigate ASR, it typically needs to be used at higher replacement levels (35-60% depending on reactivity of the aggregate). High-calcium fly ash and high-alkali fly ash were less efficient and had to be used at significantly higher replacement levels to control expansion to below 0.040% at 2 years. Generally, the effect of the different SCMs on the expansion of concrete prisms was consistent with their effect on pore solution alkalinity.

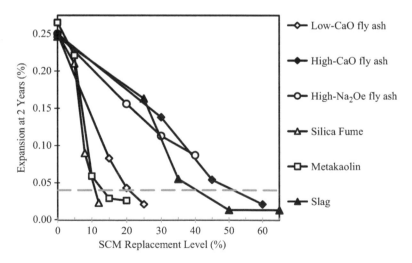

Figure 4.12   Effect of SCMs on Two-Year Expansion of Concrete Containing Siliceous Limestone (Thomas & Innis, 1998; Ramlochan et al., 2000; Shehata & Thomas, 2002).

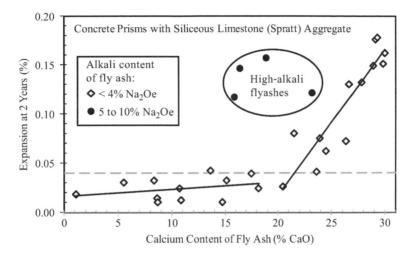

Figure 4.13 Effect of the Calcium Content of Fly Ash on the Two-Year Expansion of Concrete Containing Siliceous Limestone (Shehata & Thomas, 2002; Thomas, 2011).

Figure 4.13 (modified from Shehata & Thomas, 2000) shows the 2-year expansion of concrete containing 25% fly ash and a siliceous limestone (Spratt's) aggregate; fly ashes from 29 different sources are shown. Fly ashes with low to moderate alkali ($\leq 4\%$ Na_2Oeq) and calcium contents ($\leq 20\%$ CaO) are generally effective in controlling expansion below 0.040% at 2 years. As the calcium content increases above 20% CaO, there is a marked increase in expansion with increasing calcium content. In the Canadian CSA A3001 (2006) standard, Class F fly ashes have $\leq 15\%$ CaO, whereas in ASTM C618 (2011), Class F fly ashes have $SiO_2+Al_2O_3+Fe_2O_3 > 70\%$. Fly ashes with high alkali contents (> 5% Na_2Oeq) are not effective in controlling expansion when used at a replacement level of 25%, regardless of the calcium content of the fly ash. High-calcium fly ashes may be effective in controlling ASR expansion when used at increased levels of replacement.

Figure 4.14 shows the expansion of concrete prisms at 2 years plotted as a function of the level of fly ash replacement for Class F fly ashes with less than 10% CaO and Class C fly ashes with more than 20% CaO. Whereas 20 to 25% low-CaO Class F fly ash was sufficient to control expansion, replacement levels of 50% or more were required with the high-CaO Class C fly ashes.

Ternary blends of Portland cement with two SCMs are also effective in controlling ASR expansion. Figure 4.15 shows the expansion of concrete containing 5% silica fume in combination with fly ashes with different calcium contents. Combinations of 5% silica fume with either 10 to 15% low-CaO fly ash or 20 to 30% high-CaO fly ash are effective in controlling expansion ($\leq 0.040\%$ at 2 years). Similarly, Figure 4.16 shows that combinations of moderate amounts of silica fume (2 to 6%) are effective in combination with moderate amounts of GGBS (15 to 35%).

Figure 4.17 shows an empirical relationship between the expansion of concrete at 2 years and a "chemical index" derived from the chemical composition of the total cementing materials to produce 132 different concrete mixes, which were tested in

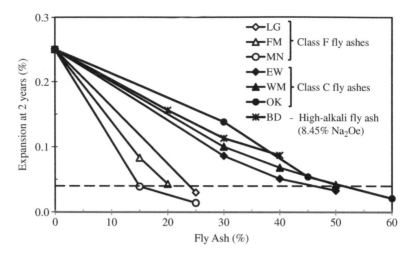

Figure 4.14 Effect of Fly Ash Replacement Level on Expansion of Concrete (Shehata & Thomas, 2000).

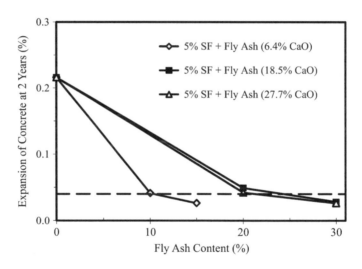

Figure 4.15 Expansion of Concrete with Blends of Silica Fume and Fly Ash (Shehata & Thomas, 2002).

accordance with ASTM C1293 (2008) (Thomas & Shehata, 2004). The cementing materials used to produce these concretes were the same as those used for the pore solution study discussed above. The reactive coarse aggregate was siliceous limestone (Spratt's). The best fit between expansion and chemical composition was found to be with the following index: $[(Na_2Oe)^{0.33} \times CaO]/(SiO_2)^2$. This relationship is not intended as a method for predicting expansion based on the chemical composition of the binder phase, but merely to examine what constituents of the binder tend to influence ASR expansion the most. The relationship is likely quite different if a different reactive aggregate or, even, a different test method is used. However, the

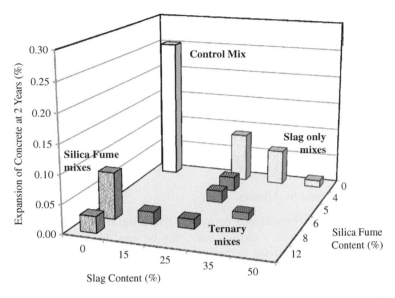

Figure 4.16 Expansion of Concrete with Blends of Silica Fume and Slag (Bleszynski, 2002; Bleszynski *et al.*, 2002)

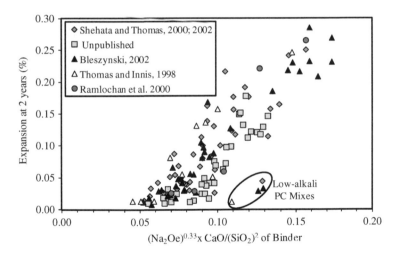

Figure 4.17 Effect of Binder Composition on the Expansion of Concrete Containing Siliceous Limestone.

relationship does indicate that expansion is likely to increase as the alkali and calcium content of the binder increase or as the silica content decreases, and this is somewhat intuitive.

It is interesting that the alkali content of the binder appears to play a less important role in determining expansion compared with the pore solution composition, but this is likely an artifact of the test conditions as significant leaching of alkalis occurs during the concrete prism test and this may reduce the apparent importance of the initial alkali content. This effect can be observed when looking at the expansion data for the concrete mixes produced with low-alkali cement. The expansion is lower than that expected based on the chemical composition. However, it is known that the concrete prism test will likely underestimate the expansion with low-alkali cement because of leaching (Thomas *et al.*, 2006). As with the relationship with pore solution, the alumina content of the binder does not appear significantly to affect the expansion of concrete. The role of alumina, however, is not yet well understood and further study is needed to determine its impact (Chappex & Scrivener, 2012).

4.4.2.2 Effect of Cement Alkalis

One of the drawbacks of the concrete prism test is that significant alkali leaching occurs during the test, which means it cannot generally be used to determine the threshold alkali content required to initiate expansion with a specific aggregate or to determine how the minimum amount of SCM required varies as the alkali content of the cement changes (Thomas *et al.*, 2006). Larger samples exposed under natural conditions should be relatively immune from the effects of alkali leaching, however, much longer testing periods are required.

Figure 4.18 shows the expansion of 300 mm concrete cubes containing a reactive hornfels aggregate and stored outdoors at the National Building Research Institute (NBRI) in South Africa (Oberholster & Davies, 1987; Oberholster, 1989). Two series of mixtures were cast with cementitious contents of approximately 350 and 450 kg/m^3. Within each series 5% or 10% of the Portland cement by mass was replaced with an equal volume of silica fume, resulting in silica fume levels of 3.5% and 7.0% by mass. The "active" alkali content was maintained at a constant level within a given series by addition of alkali hydroxide (using the same Na_2O to K_2O ratio

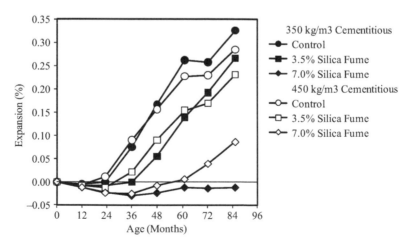

Figure 4.18   Effect of Cement Alkalis and Expansion of Concrete Containing Silica Fume (Oberholster, 1989).

as the cement). The "active alkalis" included the available cement alkalis [using ASTM C311 (2013)] plus the alkali hydroxide, but excluded alkalis in the silica fume. The use of 3.5% or 7% silica fume delayed the onset of expansion and time to cracking in all cases. However, only the mixture at the lower cement content with 7% silica fume failed to expand after just over 7 years field exposure. These data clearly show the effect of alkali content on the efficacy of silica fume in controlling expansion. A replacement level of 7% silica fume appears to have been sufficient to provide long-term prevention of expansion when the "active alkalis" of the mix were just less than 4 kg/m^3 Na$_2$Oeq, but not at the higher alkali content of 5 kg/m^3 Na$_2$Oeq.

Figure 4.19 shows data from Fournier et al. (2004) for concrete blocks stored outdoors in Ottawa, Canada. The blocks contain a reactive greywacke coarse aggregate (Springhill Quarry in New Brunswick) and 420 kg/m^3 of cementing material. High-alkali Portland cement with 0.90% Na$_2$Oeq was used to manufacture the blocks and in some cases the alkali content of the Portland cement component of the concrete was boosted to 1.25% Na$_2$Oeq by the addition of NaOH to the mix water as per ASTM C1293 (2008). The data in Figure 4.19 show that, although the differences in expansion between alkali-boosted and un-boosted blocks is not large, increased amounts of SCM are clearly needed to control expansion to acceptable levels when the alkali content of the mixture is increased.

In recent work by Hooton et al. (2012), Portland cements with up to 1.22% Na$_2$Oeq were used with different levels of SCMs in concrete prisms and outdoor exposure blocks with a reactive siliceous limestone (Spratt's) and reactive greywacke (Sudbury in Ontario) coarse aggregates. Alkalis were boosted by 0.25% Na$_2$Oeq above the alkali content of the cements. Results indicate that with up to 1.10% Na$_2$Oeq, the levels of SCMs required to mitigate expansion are unchanged relative to those for Portland cements with up to 1.0% Na$_2$Oeq. However, above 1.15% alkali content cements, SCM levels need to be increased one level above that normally required in CSA A23.2-27A (2014) [e.g., approximately an additional 10% slag or 5% fly ash].

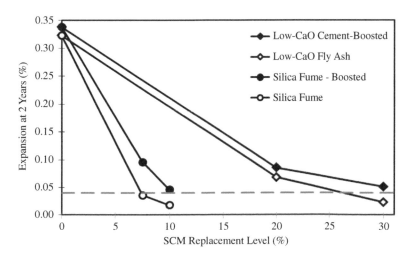

Figure 4.19   Effect of Cement Alkalis on Expansion of Concrete Containing Fly Ash or Silica Fume (Fournier et al., 2004).

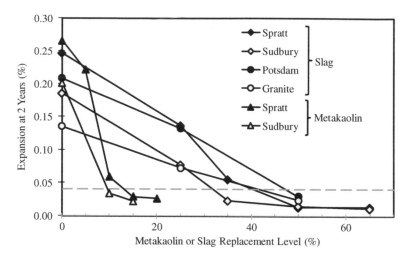

Figure 4.20   Effect of Aggregate Type on the Amount of Metakaolin or Slag Required to Control ASR Expansion (Thomas & Innis, 1996; Ramlochan *et al.*, 2000).

4.4.2.3 Effect of Aggregate Reactivity

Figure 4.20 shows 2-year expansion data for concrete prism tests containing various reactive aggregates and different amounts of GGBS (Thomas & Innis, 1998) or meta-kaolin (Ramlochan *et al.*, 2000). It can be seen that the amount of GGBS required to limit expansion below 0.040% at 2 years varies between 35% and 50% depending on the aggregate type. Figure 4.21 shows similar data for blocks exposed on an outdoor exposure site in Ottawa (Fournier *et al.*, 2004); the blocks contain high-alkali cement (boosted to 1.25% Na_2Oeq), either low-CaO Class F fly ash or silica fume, and aggregates of varying reactivity. For three of the reactive aggregates, which produced expansion levels between 0.097% and 0.219% when tested with 100% Portland cement, a fly ash replacement level of 20% or a silica fume replacement level of 7.5% was sufficient to reduce expansion below 0.040%. For the aggregates that produced an expansion of 0.338% when tested with 100% Portland cement, neither 30% fly ash nor 10% silica fume was quite sufficient to reduce the expansion below 0.040%, although expansions were much reduced (0.051% and 0.046% with 30% fly ash and 10% silica fume, respectively). For the aggregate that produced the largest expansion when tested with Portland cement (0.386%), significant expansion (0.148%) still occurred with 30% fly ash. In this figure, it would appear that the amount of fly ash required increases as the reactivity (as determined by the expansion when tested with Portland cement alone) of the aggregate increases.

4.5.3 Summary on the Effects of SCM on ASR

Supplementary cementing materials (SCMs) are an effective means for controlling expansion due to alkali-silica reaction and most, if not all, SCMs can be used in this role provided they are used at a high enough level of replacement. The level of SCM required generally increases with the following parameters:

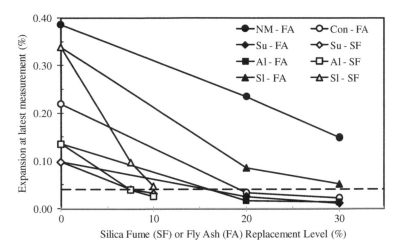

Figure 4.21 Effect of Aggregate Type on the Amount of Silica Fume (SF) or Fly Ash (FA) Required to Control ASR Expansion (Fournier *et al.*, 2004). NM, Con, Su, Al and Sl are different aggregates. FA and SF are fly ash and silica fume, respectively.

- The alkali available from the Portland cement increases
  (if significant alkalis are available from the aggregates – *e.g.*, feldspars, grey-wackes – or from external sources, this will likely also increase the level of SCM required)
- The alkali from the SCM increases
- The CaO/SiO$_2$ ratio of the SCM increases
- The reactivity of the aggregate increases.

The amount of SCM required to prevent damaging ASR expansion generally falls in the ranges shown in Table 4.2.

However, the level of SCM required may exceed these values under exceptional conditions (*e.g.*, extremely reactive aggregate, high alkali availability in concrete – including alkali contribution from aggregates, concrete exposed to high concentrations of alkali in service, and critical structure with extended service life).

Table 4.2 Levels of various types of SCM required to prevent damaging ASR (modified from Thomas & Folliard, 2007; fly ash designations amended as in CSA A3001 (2008).

Type of SCM	Level required (%)
Low-calcium fly ash (< 15% CaO)	20 to 30
Moderate-calcium fly ash (15 - 20% CaO)*	25 to 35
High-calcium fly ash (> 20% CaO)*	40 to 60
Silica fume	8 to 15
Slag (ground granulated blastfurnace slag)	35 to 65
Metakaolin (calcined kaolin clay)	10 to 20

*Higher levels may be required if Na$_2$Oeq levels > 4.5%

4.6 USE OF LITHIUM

The ability of lithium to control deleterious expansion due to alkali-silica reaction (ASR) in mortar and concrete was first demonstrated by McCoy and Caldwell (1951). They showed that, out of more than 100 chemical compounds tested, various salts of lithium (*e.g.*, LiCl, Li_2CO_3, LiF, Li_2SiO_3, $LiNO_3$, and Li_2SO_4) were the most promising and could virtually eliminate the expansion of mortar containing Pyrex glass provided they were used at sufficient levels of replacement. Since then, there have been numerous studies which corroborate this earlier discovery (Building Research Establishment, 2002; Hooper *et al.*, 2004; Feng *et al.*, 2005).

It is somewhat paradoxical that lithium compounds are effective suppressants of ASR as lithium is an alkali metal like sodium and potassium. The precise mechanism by which lithium controls ASR is not known, although many theories have been put forward (Feng *et al.*, 2005). The simplest and most commonly used explanation is that lithium salts will react with reactive silica in a similar way to sodium and potassium salts, but the reaction product is an insoluble lithium-silicate with little propensity to imbibe water and swell. The lithium silicate forms around reactive aggregate particles and protects the underlying reactive silica from "attack" by alkali hydroxides.

The initial work of McCoy and Caldwell (1951) showed that the amount of lithium required to control expansion was a function of the availability of other alkalis (Na + K) in the system and they concluded that the expansion of mortar bars containing reactive Pyrex glass could be effectively suppressed provided that the lithium-to-sodium-plus-potassium molar ratio was greater than 0.74, *i.e.*, [Li]/[Na+K] > 0.74. Since then numerous workers have demonstrated a similar relationship between the amount of lithium required and the amount of alkali available, but the minimum value of [Li]/[Na+K] has been shown to vary depending on a number of issues such as the form of lithium, nature of reactive aggregate and, perhaps, the method of test used (Feng *et al.*, 2005).

Although most lithium compounds have a beneficial effect, lithium nitrate ($LiNO_3$) is considered to be the most efficient form for suppressing ASR (Stokes *et al.*, 1997). Lithium nitrate solution is commercially available from a number of companies in North America being marketed as an *"ASR-suppressing admixture"*. Currently the product is sold as a 30% solution of $LiNO_3$. To achieve a lithium-to-sodium-plus-potassium molar ratio of [Li]/[Na+K] = 0.74 using a 30% solution of $LiNO_3$ requires a dose of 4.6 litres of $LiNO_3$ solution per 1 kg of Na_2Oeq (0.55 gallons of solution per 1lb Na_2Oeq). This has been referred to as the "standard dose" of lithium nitrate solution.

Recent research (Tremblay *et al.*, 2007) has highlighted the influence of aggregate type on the amount of lithium required to suppress expansion due to ASR. Figure 4.22 shows the 2-year expansion of concrete prisms with 12 different reactive aggregates and 1 non-reactive aggregate (NF), and various levels of lithium (standard dose is [Li]/[Na+K] = 0.74). For 6 of the 12 aggregates, 75% to 100% of the standard dose was sufficient to control expansion (≤ 0.040% at 2 years). For 3 of the aggregates, 125% to 150% of the standard dose was required. However, for the remaining 3 aggregates, expansion could not be controlled even at 150% of the standard dose.

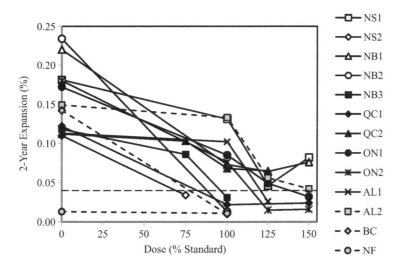

Figure 4.22 Effect of Lithium Dose on the Expansion of Concrete with Different Reactive Aggregates (from Tremblay *et al.*, 2007).

As the effectiveness of lithium appears to be extremely aggregate dependent, it is not possible to prescribe a single dose for controlling ASR, and the minimum dose must be determined by testing lithium with the specific reactive aggregate being considered for use. At this time there is no consensus regarding the appropriateness of accelerated tests for determining the correct lithium dose and it is therefore recommended that the concrete prism test is used for this purpose.

REFERENCES

AASHTO PP65. 2011 (2015) *Standard practice for determining the reactivity of concrete aggregates and selecting appropriate measures for preventing deleterious expansion in new concrete construction.* American Association of State Highway and Transportation Officials, Washington DC, USA. 20p. (Redesignated as AASHTO R80-17 in 2017, 24 p.)

ASTM C33 (2016) *Standard specification for concrete aggregates.* American Society for Testing and Materials, West Conshohocken, USA.

ASTM C150 (2016) *Standard specification for Portland cement.* American Society for Testing and Materials, West Conshohocken, USA.

ASTM C227 (2010) *Standard test method for potential alkali-reactivity of cement-aggregate combinations (mortar-bar method).* American Society for Testing and Materials, West Conshohocken, USA.

ASTM C295 (2012) *Standard guide for petrographic examination of aggregates for concrete.* American Society for Testing and Materials, West Conshohocken, USA.

ASTM C311 (2013) *Standard test methods for sampling and testing fly ash or natural pozzolans for use in Portland-cement concrete.* American Society for Testing and Materials, West Conshohocken, USA.

ASTM C441 (2011) *Standard test method for effectiveness of pozzolans or ground blast-furnace slag in preventing excessive expansion of concrete due to the alkali-silica reaction.* American Society for Testing and Materials, West Conshohocken, USA.

ASTM C618 (2015) *Standard specification for coal fly ash and raw or calcined natural pozzolan for use in concrete*. American Society for Testing and Materials, West Conshohocken, USA.

ASTM C1260 (2014) *Standard test method for potential alkali-reactivity of aggregates (mortar-bar method)*. American Society for Testing and Materials, West Conshohocken, USA.

ASTM C1293. 2008 (2015) *Standard method for determination of length change of concrete due to alkali-silica reaction*. American Society for Testing and Materials, West Conshohocken, USA.

ASTM C1567 (2013) *Standard test method for determining the potential alkali-silica reactivity of combinations of cementitious materials and aggregate (accelerated mortar-bar method)*. American Society for Testing and Materials, West Conshohocken, USA.

ASTM C1778 (2016) *Standard Guide for Reducing the Risk of Deleterious Alkali-Aggregate Reaction in Concrete*. American Society for Testing and Materials, West Conshohocken, USA.

Barona de la O, F. (1951) Alkali-aggregate expansion corrected with Portland-slag cement. *J Am Concrete I.*, 22 (7), 545–552.

Berube, M-A., Duchesne, J., Dorion, J.F. & Rivest, M. (2002) Laboratory assessment of alkali contribution by aggregates to concrete and application to concrete structures affected by alkali–silica reactivity. *Cement Concrete Res.*, 32, 1215–1227.

Bhatty, M.S.Y. & Greening, N.R. (1978) Interaction of alkalis with hydrating and hydrated calcium silicates. *Proceedings of the Fourth International Conference on the Effects of Alkalis in Cement and Concrete, Purdue University*. pp. 87–112.

Blaikie, N.K., Bowling, A.J. & Carse, A. (1996) The assessment and management of alkali-silica reaction in the Gordon River Power Development intake tower. In: Shayan, A. (ed.) Alkali-Aggregate Reaction in Concrete, *Proceedings of the 10th International Conference on Alkali-Aggregate Reaction*, Melbourne. pp. 500–507.

Bleszynski, R.F. (2002) The performance and durability of concrete with ternary blends of silica fume and blast-furnace slag. PhD Thesis, University of Toronto.

Bleszynski, R., Hooton, R.D., Thomas, M.D.A. and Rogers, C.A. (2002) Durability of ternary blend concrete with silica fume and blast-furnace sag: Laboratory and outdoor exposure site studies. *ACI Mater J.*, 99 (5), 499–508.

Buck, A.D., Houston, B.J. & Pepper, L. (1953) Effectiveness of mineral admixtures in preventing excessive expansion of concrete due to alkali-aggregate reaction. *J Am Concrete I.*, 30 (10), 1160.

Buck, A.D. & Mather, K. (1987) "Methods for controlling effects of alkali-silica reaction in concrete." U.S. Army Corps of Engineers, Waterways Experiment Station, Technical Report No. SL-87-6, U.S. Army Engineer Waterways Experiment Station, Vicksburg.

Building Research Establishment (2002) *Minimising the risk of alkali-silica reaction: alternative methods*. Information Paper IP1/02, BRE (CRC), Watford, UK, 8p.

CSA A23.2-27A (2014) *Test methods and standard practices for concrete, 27A. Standard practice to identify degree of alkali-reactivity of aggregates and to identify measures to avoid deleterious expansion in concrete*. Canadian Standards Association, Mississauga, Ontario, Canada.

CSA A3001 (2006) *Cementitious materials for use in concrete*. CSA Technical Committee? Canadian Standards Association, Mississauga, Ontario, Canada.

Carrasquillo, R.L. & Snow, P.G. (1987) "Effect of fly ash on alkali-aggregate reaction in concrete." *ACI Mater J.*, 84 (4), 299–305.

Chappex, T. & Scrivener, K. (2012) Alkali fixation of C–S–H in blended cement pastes and its relation to alkali silica reaction. *Cement Concrete Res.*, 42, 1049–1054.

Cong, D.X., Lawrence, B.L., Deno, D.W. & Patty, T.S. (2004) ASR-induced surface defects. In: *Proceedings of the 12th International Conference on Alkali-Aggregate Reaction on Concrete*, October 15–19, 2004, Beijing China, Vol. 2. pp. 1142–1147.

Cox, H.P., Coleman, R.B. & White, L. (1950) "Effect of blastfurnace-slag cement on alkali-aggregate reaction in concrete." *Pit and Quarry*, 45 (5), 95–96.

Dunstan, E.R. (1981) "The effect of fly ash on concrete alkali-aggregate reaction." *Cement Concrete Aggr.*, 3 (2), pp. 101–104.

Farny, J.A. & Tarr, S.M. (2008) "Concrete Floors on Ground." PCA EB075, Portland Cement Association, Skokie, IL, 239p.

Feng, X., Thomas, M.D.A., Bremner, T.W., Balcom, B.J. & Folliard, K.J. (2005) Studies on lithium salts to mitigate ASR-induced expansion in new concrete: A critical review. *Cement Concrete Res.*, 35, 1789–1796.

Fernandes, I., Ribeiro, A.Md., Broekmans, M.A.T.M., & Sims, I. (2016) *Petrographic atlas: Characterisation of aggregates regarding potential reactivity to alkalis. RILEM TC 219-ACS Recommended Guidance AAR-1.2, for use with the RILEM AAR-1.1 Petrographic Examination Method*, Springer, Dordrecht, RILEM, Paris, 204p.

Fournier, B., Nkinamubanzi, P.-C. & Chevrier, R. (2004) Comparative field and laboratory investigations on the use of supplementary cementing materials to control Alkali-Silica reaction in concrete. In: Mingshu, T. and Min, D. (eds.) *Proceedings of the 12th International Conference Alkali-Aggregate Reaction in Concrete, Vol. 1, International Academic Publishers/World Publishing Corporation*, Beijing, China, pp. 528–537.

Glasser, F.P. (1992) Chemistry of the alkali-aggregate reaction. In: Swamy, R.N. (ed.) *The Alkali-Silica Reaction in Concrete*, Blackie, London. pp. 96–121.

Glasser, F.P. & Marr, J. (1985) The alkali binding potential of OPC and blended cements. *Il Cemento*, 82, 85–94.

Hong, S-Y. & Glasser, F.P. (1999) "Alkali binding in cement pastes Part I. The C-S-H phase." *Cement Concrete Res.*, 29, 1893–1903.

Hong, S-Y. & Glasser, F.P. (2002) "Alkali sorption by C-S-H and C-A-S-H gels Part II. Role of alumina. *Cement Concrete Res.*, 32, 1101–1111.

Hooper, R.L., Nixon, P.J. & Thomas, M.D.A. (2004) Considerations when specifying lithium admixtures to mitigate the risk of ASR. In: *Proceedings of the 12th International Conference on Alkali-Aggregate Reaction in Concrete*, Beijing World Publishing Corporation, Beijing, October, Vol. 1, pp. 554–563.

Hooton, R.D., Donnelly, R. & Clarida, B. (2000) An Assessment of the Effectiveness of Blast-Furnace Slag in Contracting the Effects of Alkali-Silica Reaction. In: Fournier, B. and Berube, M.A. (eds.) *Proceedings, 11th International Conference on Alkali-Aggregate Reaction*, Quebec, pp. 1313–1322.

Hooton, R.D., Fournier, B., Kerenidis, K. & Chevrier, R. (2012) Mitigating Alkali-Silica Reaction When Using High-Alkali Cements. *e-Proceedings 14th International Conference on Alkali Aggregate Reactions*, Austin Texas, 10p. (Paper 030211-Hoot-01).

Hooton, R.D., Rogers, C.A. & Ramlochan, T. (2006) The Kingston Outdoor Exposure Site for ASR – After 14 Years What Have We Learned? In: *Proceedings, Marc-André Bérubé Symposium, Seventh CANMET/ACI International Conference on Durability*, Montreal, May 31–June 2, 2006, 22p.

Hooton, R.D., Rogers, C.A., MacDonald, C.A. & Ramlochan, T. (2013) 20-year field evaluation of alkali-silica reaction mitigation. *ACI Mater J.*, 110 (5), 539–548.

Hooton, R.D., Thomas, M.D.A. & Ramlochan, T. (2010) Use of pore solution analysis in design for concrete durability." *Adv Cem Res.*, 22 (4), 203–210.

Klieger, P. & Gebler, S. (1987) 'Fly ash and concrete durability.' In: Scanlon, J.M. (ed.) *Concrete Durability, Katherine and Bryant Mather International Conference*, ACI SP-100, Vol. 1, American Concrete Institute, Detroit, pp. 1043–1069.

Landgren, R. & Hadley, D.W. (2002) Surface Popouts Caused by Alkali-Aggregate Reactions. Research and Development Bulletin RD121, Portland Cement Association, Skokie, IL, 13p.

Lindgård, J., Andic-Cakir, O., Fernandes, I., Ronning, T.F. & Thomas, M.D.A. (2012) Alkali–silica reactions (ASR): Literature review on parameters influencing laboratory performance testing. *Cement Concrete Res.*, 40, 223–243.

Maas, A.J., Ideker, J.H. & Juenger, M.C.G. (2007) Alkali silica reactivity of agglomerated silica fume. *Cement Concrete Res.*, 37, 166–174.

MacDonald, C.A., Rogers, C. & Hooton, R.D. (2012) The relationship between laboratory and field expansion – observations at the Kingston outdoor exposure site for ASR after twenty years. In: *Proceedings of the 14th International Conference on Alkali-Aggregate Reaction in Concrete*, Austin, TX, USA, May.

McCoy, W.J. & Caldwell, A.G. (1951) New approach to inhibiting alkali-aggregate expansion. *J Am Concrete I.*, 22 (9), 693–706.

Nixon, P.J. & Sims, I. (1992) Alkali Aggregate Reaction ± Accelerated Tests Interim Report and Summary of National Specifications. In: *Proceedings of the 9th International Conference on Alkali-Aggregate Reaction in Concrete*, Vol. 2. Slough, The Concrete Society. pp. 731–738.

Nixon, P.J. & Sims, I. (eds.) (2016) RILEM recommendations for the prevention of damage by alkali-aggregate reactions in new concrete structures. *RILEM State-of-the-Art Reports*, Volume 17, Springer, Dordrecht, pp RILEM, Paris, 168p.

Nixon, P.J., Collins, R.J. & Rayment, P.L. (1979) The Concentration of Alkalies by Moisture Migration in Concrete – A factor influencing alkali aggregate reaction. *Cement Concrete Res.*, 9, 417–423.

Nixon, P.J., Canham, I. & Page, C.L. (1987) Aspects of the pore solution chemistry of blended cements related to the control of alkali-silica reaction. *Cement Concrete Res.*, 17 (5), 839–844.

Oberholster, R.E. & Davies, G. (1987) The effect of mineral admixtures on the alkali-aggregate expansion of concrete under outdoor exposure conditions. In: Grattan-Bellew, P.E. (ed.) *Proceedings of the 7th International Conference on Concrete Alkali-Aggregate Reactions*, Noyes Publications New Jersey, pp. 60–65.

Oberholster, R.E. (1989) Alkali-aggregate reaction in South Africa. Some recent developments in research. In: Okada, K., Nishibayashi, S. and Kawamura, M. (eds.) *Proceedings, 8th International Conference on Alkali-Aggregate Reaction*, Kyoto, pp. 77–82.

Ramlochan, T., Thomas, M.D.A. & Gruber, K.A. (2000) The effect of metakaolin on alkali-silica reaction in concrete. *Cement Concrete Res.*, 30 (3), 339–344.

Rayment, P.L. (1982) The effect of pulverized-fuel ash on the C/S molar ratio and alkali content of calcium silicate hydrates in cement. *Cement Concrete Res.*, 12 (2), 133–140.

Rogers, C.A., Lane, B. & Hooton, R.D. (2000) Outdoor exposure for validating the effectiveness of preventative measures for Alkali-Silica Reaction. In: Marc-André Bérubé, Benoit Fournier and Benoit Durand (eds.) *Proceedings of the 11th International Conference on Alkali-Aggregate Reaction*, Quebec, June, pp.743–752.

Shehata, M.H. & Thomas, M.D.A. (2000) The effect of fly ash composition on the expansion of concrete due to alkali-silica reaction. *Cement Concrete Res.*, 30, 1063–1072.

Shehata, M.H. & Thomas, M.D.A. (2002) Use of ternary blends containing silica fume and fly ash to suppress expansion due to alkali-silica reaction in concrete. *Cement Concrete Res.*, 32 (3), 341–349.

Shehata, M.H. & Thomas, M.D.A. (2006) Alkali release characteristics of blended cements. *Cement Concrete Res.*, 36, 1166–1175.

Shehata, M., Thomas, M.D.A. & Bleszynski, R.F. (1999) The effect of fly composition on the chemistry of pore solution. *Cement Concrete Res.*, 29 (12), 1915–1920.

Smith, R.L. (1988) Is the available alkali test a good durability predictor for fly ash concrete incorporating reactive aggregate. In: Fly Ash & Coal Conversion By-Products: Utilization & Disposal IV, MRS Symposia Proceedings, Vol. 113, Materials Research Society, Pittsburgh, pp. 249–256.

Stanton, T.E. (1940) Expansion of concrete through reaction between cement and aggregate. *Proceedings of the American Society of Civil Engineers*, 66 (10), 1781–1811.

Stanton, T.E. (1950) Studies of use of pozzolans for counteracting excessive concrete expansion resulting from reaction between aggregates and the alkalies in cement. Pozzolanic Materials in Mortars and Concretes, ASTM STP 99, American Society for Testing and Materials, Philadelphia, pp. 178–203.

Stark, D. (1978) Alkali-silica Reactivity in the Rocky Mountain Region. In: *Proceedings of the 4th International Conference on Effects of Alkalis in Cement and Concrete*, Publication No. CE-MAT-1-78, Purdue University, W. Lafayette, Indiana, pp. 235–243.

Stark, D. (1980) Alkali-Silica Reactivity: Some Reconsiderations. *Cement Concrete Aggr.*, 2 (92), 92–94.

Stark, D. & Bhatty, M.S.Y. (1986) Alkali-silica Reactivity: Effect of Alkali in Aggregate on Expansion. In: Alkalis in Concrete, ASTM STP 930, American Society for Testing and Materials, Philadelphia. pp. 16–30.

Stokes, D.B., Wang, H.H. & Diamond, S. (1997) A lithium-based admixture for ASR control that does not increase the pore solution pH. In: Malhotra, V.M. (ed.) *Proceedings, 5th CANMET/ACI International Conference on Superplasticizers and Other Chemical Admixtures in Concrete*, ACI SP-173, American Concrete Institute, Detroit. pp. 855–867.

Thomas, M.D.A. (1996a) "Field studies of fly ash concrete structures containing reactive aggregates." *Mag Concrete Res.*, 48 (177), 265–279.

Thomas, M.D.A. (1996b) Review of the effect of fly ash and slag on alkali-aggregate reaction in concrete. Building Research Establishment Report, BR314, Construction Research Communications, Ltd, Watford, U.K..

Thomas, M.D.A. (2011) The effect of supplementary cementing materials on alkali-silica reaction: A review. *Cement Concrete Res.*, 41, 1224–1231.

Thomas, M.D.A. & Bleszynski, R.F. (2001) The use of silica fume to control expansion due to alkali-aggregate reactivity in concrete – a review. In: Mindess, S. and Skalny, J. (eds.) *Materials Science of Concrete VI*. Westerville, OH, American Ceramics Society, pp. 377–434.

Thomas, M.D.A. & Folliard, K.J. (2007) Concrete aggregates and the durability of concrete. In: Page, C.L. & Page, M.M. (eds.) *Durability of Concrete and Cement Composites*, Cambridge, U.K., Woodhead. pp. 247–281.

Thomas, M.D.A. & Innis, F.A. (1998) Effect of slag on expansion due to alkali-aggregate reaction in concrete. *ACI Mater J.*, 95 (6), 716–724.

Thomas, M.D.A. & Shehata, M. (2004) Use of blended cements to control expansion of concrete due to alkali-silica reaction. In: *Proceedings, 8th CANMET/ACI International Conference on Fly Ash, Silica Fume, Slag and Natural Pozzolans in Concrete*, Supplementary Papers, Las Vegas, pp. 591–607.

Thomas, M.D.A., Blackwell, B.Q., & Nixon, P.J. (1996) Estimating the alkali contribution from fly ash to expansion due to alkali-aggregate reaction in concrete. *Mag Concrete Res.*, 48 (177), 251–264.

Thomas, M.D.A., Nixon, P.J., & Pettifer, K. (1991) The effect of pulverized fuel ash with a high total alkali content on alkali silica reaction in concrete containing natural U.K. aggregate. In: Malhotra, V.M. (ed.) *Proceedings of the 2nd CANMET/ACI International Conference on Durability of Concrete*, Vol. 2, American Concrete Institute, Detroit, pp. 919–940.

Thomas, M.D.A., Fournier, B., Folliard, K., Ideker, J. & Shehata, M. (2006) Test methods for evaluating preventive measures for controlling expansion due to alkali-silica reaction in concrete. *Cement Concrete Res.*, 36 (10), 1842–1856.

Thomas, M.D.A., Fournier, B., Folliard, K.J., Shehata, M., Ideker, J. & Rogers, C.A. (2007) Performance limits for evaluating supplementary cementing materials using the accelerated mortar bar test. *ACI Mater J.*, 104 (2), pp. 115–122.

Thomas, M.D.A., Hooton, R.D., Rogers, C.A., & Fournier, B. (2012) 50 years old and still going strong: Fly ash puts paid to ASR. *Concrete Intl.*, 34 (1), 35–40.

Tremblay, C., Berube, M-A., Fournier, B., Thomas, M.D.A., & Folliard, K.F. (2007) Effectiveness of lithium-based products in concrete made with Canadian reactive aggregates. *ACI Mater J.*, *104* (2), 195–205.

Uchikawa, H., Uchida, S., & Hanehara, S. (1989) Relationship between structure and penetrability of Na ion in hardened blended cement paste mortar and concrete. In: Okada, K., Nishibayashi, S. & Kawamura, M. (eds.) *Proceedings of the 8th International Conference on Alkali-Aggregate Reaction*, Kyoto, pp. 121–128.

Woolf, D.O. (1952) Reaction of aggregate with low-alkali cement. *Public Roads*, August, *27* (3), 50–56.

Xu Zhongxi & Hooton, R.D. (1993) Migration of alkali ions by several mechanisms. *Cement Concrete Res.*, *23* (4), 951–961.

Chapter 5

Diagnosis, Appraisal, Repair and Management

Bruno Godart & Mario R de Rooij

5.1 INTRODUCTION

Being responsible for a structure out in the field, one realizes that there are no museum display cards outlining that this damage is due to AAR and that damage is not. Deterioration and possible loss of serviceability could be the result of different deleterious mechanisms, often acting simultaneously or consecutively, increasing the damage observed. This requires careful diagnosis and appraisal with an open mind for every mechanism that may have contributed to the observed phenomena.

Diagnosis of damage to concrete affected by AAR is not straight forward because the external characteristics of the damage resemble those also caused by other actions such as, for example, frost attack, drying shrinkage or sulfate attack. Furthermore, variations in the development and extent of the damage are caused by a wide variety of aggregate types used in concrete worldwide, by different reactivity and properties of the reactive constituents in the aggregates involved and by a wide range of environmental factors that influence the reaction result. Finally, the different steps in the damaging process are not all understood equally well. All these factors make diagnosis difficult and have sometimes led to subjective or even erroneous interpretations and conclusions.

In this chapter an overview is given of the main steps and possible methods to diagnose and evaluate structures suspected of being damaged by AAR, to suggest how a damaged structure may be treated based on the feedback of past experiences, and to propose a methodology to manage as best as possible a stock of affected structures.

5.2 DIAGNOSIS – FIRST SYMPTOMS

5.2.1 How does AAR show itself?

AAR is an expansive reaction between alkalis (generally from the cement) and aggregates in the concrete. This reaction process takes time. Generalizing over the many types of aggregates and environmental conditions potentially capable of developing AAR, this means that the visible damage due to AAR will typically not show itself in the first five years of the structures' lifetime, except for concrete structures incorporating highly reactive aggregates. Hence, if the damage is already there before that time, it is most likely caused by something other than AAR.

If the AAR shows itself (and note this could also take much longer than the typically mentioned five years), the first symptoms are usually noted during routine visual inspection. The most commonly reported observations are:

- **Variable expansions** causing relative movements, displacements and deformations at different scales between parts of a structure and/or out-of-plane displacement of crack lips, etc. These expansions can also cause attached equipment or installations not to work properly, jammed due to the expansions of the concrete.
- **Development of extensive cracking** (see also Figure 5.1).

  - This cracking is often in a 'map-cracking' pattern where concrete is unrestrained by adjacent structures or by internal reinforcement.
  - In case of a direction of dominant restraint of the expansion, the crack pattern can have a more parallel distribution. This can happen for instance with axially loaded columns where the crack pattern tends to develop longitudinally, as the

a = concrete pavement

b = part of a concrete wall

c = part of the wall of a lock

d = part of the wall of a lock

Figure 5.1 Map-cracking of concrete shown in the surface of various members of non-reinforced concrete structures damaged by AAR. Some of the cracks are filled with carbonated ASR-gel (from Godart *et al.*, 2013; photograph 'b' courtesy of T Katayama, photographs 'c' & 'd' courtesy of TNO).

transverse cracking is suppressed by compression stresses from restraint or loading.

- **Surface discoloration**, particularly along cracks.
- **Gel exudations** from cracks, where they are filled with reaction products.
- **Occasional surface pop-outs** caused by reaction of coarse aggregate particles close to the surface. Similar type damage could also be caused by frost.

It is usually during a routine inspection that observations like the ones noted above are made. However, as AAR is only one of the possible mechanisms that may be responsible for damage to the concrete in a structure, it is necessary to consider all possible mechanisms that may have contributed to the deterioration or damage observed. It is too soon to jump to conclusions towards AAR based on visual observations alone. At this point only hypothesis to be investigated by further research is the correct way to move forward.

In the next section the importance of a well set-up routine inspection is further elaborated, especially because this provides us with the hypothesis for further investigation to determine the cause of the observed damage or deterioration.

5.2.2 The routine inspection

A routine inspection is ordinarily a starting point for any existing structure in a cyclic process to maintain the safety and reliability of the structure. As such it is part of an entire process in which information and data are collected, recorded, reported and compared to previous information available from the structure. Based on the comparison and possible differences between the previous situation and the current situation further actions can be specified. Schematically the process is presented in 5.2.

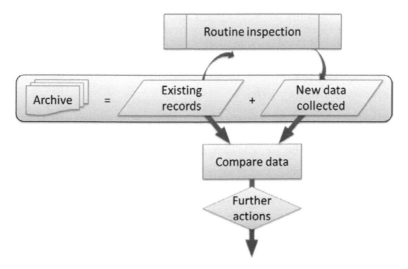

Figure 5.2 Schematic process around a routine inspection to maintain safety and reliability of a structure.

5.2.2.1 Review of existing records

In principle much information is already available from the archives. Prior to any planned inspection, important information from these archives should be reviewed. Important information to look for includes:

- **Archive information** regarding official name of the structure, type and location.
- **Structural design information** regarding design loading and construction details, particularly the position and arrangement of reinforcement or prestressing. Original 'as built' and/or revised scale drawings are very useful to assess this information. Furthermore, for AAR particularly, information on foundations and drainage systems could be of importance.
- **Details of concrete construction procedure** such as the building period including date of completion, mix designs used, sources of materials. Early thermal data and curing procedures may be significant to assess possible causes of cracking.
- **Exposure conditions** to which the structure has been exposed because of its location or function. Also changes in exposure or environmental conditions (temperature, exposure to water, humid air or de-icing salt) from drainage, waterproofing, cladding etc. can contain clues to sudden changes in behaviour or damage propagation in the structure.
- **Previous inspection reports.** Particularly the ones in which the signs of damage or deterioration were first noticed, and when major changes in the rate of the damage development were recorded.
- **Reports of any remedial works,** with details whether major or minor.
- **General status of similar structures** in the area, especially of similar or older age.
- **Access information** to the structure and requirements for access equipment.
- Any other information that may be relevant for the field inspection and the investigation.

Unless it is the very first time a structure is being inspected, based on the review of the existing records, a rough idea of what to expect upon a routine inspection of this structure should be clear. Anything deviating from this expected idea should be given more attention and could require follow up investigation.

5.2.2.2 Personnel and Expertise

A successful routine inspection depends largely on the skills and expertise of the people performing the inspection. A reliable field inspection must therefore be performed by experienced personnel, who have been trained to identify damage, to make careful observations of the external characteristics of damage and to record their findings in a systematic manner. Because of the breadth of the expertise required, this usually involves a multi-disciplinary team. Preferably the personnel involved should consist of qualified civil, structural, geological or materials engineers with a good understanding of AAR and other concrete deterioration mechanisms. The advised minimum is a two-person team composed of a qualified structural engineer, capable of discriminating between cracks and different disorders caused by AAR and those caused by other effects as well as of evaluating the structural integrity, and a materials scientist expert in the chemical and petrographic analysis of concrete affected by AAR. In case of AAR, because of the strong interaction between AAR cracking and the stress fields and underlying reinforcement, the assessment of the structural integrity must be done by a qualified structural engineer experienced in AAR assessments.

5.2.2.3 Field equipment and Materials

Detailed records of the inspection should be kept and written into a notebook or on forms held on a waterproof clipboard (suitable electronic tablet devices have also been developed by some inspectors). Wherever possible, visible signs of deterioration should be recorded on copies of scale drawings. Copies of 'as built' or revised scale drawings showing underlying reinforcement are often useful for making accurate recordings of observations during the inspection.

Next to written notes, a comprehensive photographic record should be made with description of each photograph recorded in a notebook, without forgetting to make overview pictures too. The digital camera should be equipped with a flash gun and must be set for high resolution images. Every photograph should include a scale identifier (*e.g.*, person, pen, coin or hammer).

A minimum list of equipment for field inspection should include:

- Measurement tape (steel, say 4 metres long).
- Crack width gauge or crack width microscope.
- Hand lens or strong magnifying lens.
- Marking crayons.
- Powerful electric torch.
- Mirror.
- Binoculars.
- Small hammer and chisel.

Non-destructive testing equipment may be useful, such as a Schmidt rebound hammer and an electromagnetic cover meter. The cover meter will be essential to check reinforcement locations prior to coring. Flexible endoscopes can be useful to obtain information from areas inside the structure, which are not readily accessible such as those hidden by cladding.

In some cases, the field inspection and sampling may both be carried out at the same time. Then it will be necessary to consider what equipment will be needed for sampling. In any event, the inspector should carry a small selection of sample bags to collect samples and indelible marking pens to code the sample bags.

5.2.2.4 The on-site inspection

The on-site or field inspection should be designed on the basis of the type of structure concerned. There are many well written up case studies of the inspection, testing, and petrographic evaluation of specific types of structures, which provide examples of good practice. These include pavements, (Juliani *et al.*, 2008) bridges, (Eriksen *et al.*, 2008), (Wood *et al.*, 2008), dams (Fernandes *et al.*, 2008), railway sleepers (Silva *et al.*, 2008) and swimming pools (Rodum *et al.*, 2008). Furthermore, the approach for the inspection can be guided by the review of the existing records as mentioned in section 5.2.2.1.

The validity of documentary evidence, particularly earlier test results, can be checked against new on-site observations or sampling.

The aim of the on-site inspection is to acquire information in order to:

- Establish the condition and safety level of the structure
- Formulate hypotheses on the cause or progress of deterioration mechanisms
- Locate representative areas for sampling
- Remove some first (core) samples for laboratory studies.

During the routine inspection, each component of the structure should be examined separately. Observations on the type, extent (severity), and location of the defects should be recorded in a consistent manner. Determination with the structural engineer of the severity and extent of cracking related to stress fields and underlying reinforcement configuration and assessment of the deformations in specific members of the structure are important. These features will be used to decide how further to proceed with the diagnosis and monitoring. Therefore examples of the pattern of cracking and its intensity in the various members should be carefully mapped and described, with the view that after the site inspection a reliable assessment can be made of the extent and severity of each feature in each of the elements inspected.

The process of AAR is in essence an expansive mechanism. Sometimes this can cause significant movements in the structure, Therefore record should be made of any differential displacement, such as closing of joints, relative displacement of adjacent concrete sections, excessive deflection, and twisting or bulging of originally flat surfaces. Simple markers and gauges may be fitted during routine inspection so that further movements will be measured in the future.

AAR typically develops or sustains in concrete elements with internal relative humidity > 80-85%. Therefore, for AAR related damage, particular attention should be given to areas where water is involved such as run-offs, leakages, standing water, groundwater (buried portions of concrete foundations), seawater or condensation. In addition, an overview or record should be made of all the surfaces which may have come into contact with external alkalis. The buried parts of a structure often have large concrete elements with limited reinforcement in consistently damp conditions and are often the most structurally sensitive to AAR. They need to be considered in the preliminary inspection and if necessary be exposed in more detailed further studies.

5.2.3 Magnitude of the problem

At the construction level, the main external evidence of deterioration caused by AAR is usually abnormal cracking. In the early stages, slow developing AAR tends to increase the magnitude of pre-existing cracking before its magnitude reaches a level (0.4 to 0.6 mm/m free expansion) at which the strains from AAR exceed those from other causes of cracking.

Besides cracking, the other mentioned features in this chapter could also indicate the potential of AAR presence in the structure. An initial assessment of the probability that damaging AAR has occurred can be made by considering the apparent damage to the structure and determining whether or not the nature of the damage is consistent with AAR. Table 5.1 lists a number of features that may indicate AAR and classifies their

Table 5.1 Classification system for assessment of probability of presence of AAR based on routine inspection results (modified after CSA A846-00: 2000, from Godart et al., 2013).

Feature	Probability of AAR		
	Low	Medium	High
Cracking and crack pattern	None	Some cracking pattern typical of AAR (e.g., map cracking or cracks aligned with major reinforcement or stress)	Extensive map-cracking or cracking aligned with major reinforcement or stress; misalignment of facings (differential expansion on each side of the crack)
Deformation and/or displacement of elements	None	Some evidence (e.g., closure of joints, spalls, misalignment between structural members	Fair to extensive signs of volume increase leading to spalling at joints, displacement and/or misalignment of structural members
Surface discoloration	None	Slight surface discoloration associated with some cracks	Many cracks with dark discoloration and adjacent zone of light colored concrete
Exudations	None	White exudations around some cracks; possibility of colorless jelly-like exudations	Colorless, jelly-like exudations readily identifiable as ASR gel associated with several cracks
Pop-outs with ASR gel deposit	None	Some	Many
Environment	Dry and sheltered	Outdoor exposure but sheltered from wetting	Parts of component frequently exposed to moisture, e.g., rain, groundwater, water due to natural function of the structure (hydraulic dam)

occurrence as indicating a low, medium or high probability of AAR (modified after CSA A846-00: 2000).

To assess the severity of cracking as an initial guideline, it is recommended to adopt a standard method for describing the width of cracks. A suitable scheme for the structural assessment of the current condition of the structure is shown in Table 5.2. It should be pointed out that also the smaller crack widths mentioned in Table 5.2 can

Table 5.2 Classification of structural crack widths on site (Godart et al., 2013).

Crack width (mm)	Abbreviation	Full name	Remark
< 0.1	F	Fine	Usually present
0.1 – 0.3	N	Normal	To allowed limit for RC
0.3 – 0.5	L	Large	Over limit
0.5 – 1.0	MW	Moderately wide	Record all
2.0 – 5.0	W	Wide	Refer to Engineer
> 5.0 – 10.0	VW	Very wide	Refer to Engineer

Note: for a prestressed concrete structure, the scale has to be adapted towards lower values of crack width.

lead to structural risks over time due to durability related processes that are either increasing the crack width or reducing the function of the reinforcement.

The magnitude of the AAR problem can be further assessed in terms of scale. As concrete structures are made in batch sequence, one pour at a time, it could be that only a few pours of concrete are affected. In those cases special attention is needed towards the location of the damage. Is AAR affecting the main structural integrity of the structure or is the damage confined to areas that are of less vital necessity to the structural function and safety?

More often however, the AAR damage, when present in one structure, may also be present in other structures in the area, when built with the same source material. As the existence of AAR damage is due to reactive aggregates which came from a certain quarry or series of quarries from a similar geological context, it is advised to survey, at a minimum, the records of structures in the neighbourhood. Have the same aggregate sources been used? When were the structures built (to determine development of possible damage over time)?

5.3 PROCEDURES TO GAIN CONFIRMATION OF AAR

5.3.1 Introduction

In the previous section symptoms that may indicate the presence of AAR have been described. Once such symptoms have shown themselves, it is necessary to determine whether or not the damage is actually caused primarily by AAR. This is the topic of the current section. Once the presence of AAR has been confirmed, the assessment of the severity of the AAR damage is the next step in the process. This next step of severity investigation will be dealt with in the following section.

To confirm the presence of AAR beyond doubt requires showing the actual reacted aggregate and preferably the consequences in terms of presence of fractures and gel (see Figure 5.3). The generally accepted method for doing so is through concrete petrography, where it is advised to use optical microscopy complemented by scanning electron microscopy. Other methods that aid to the diagnosis of AAR include core-scanning (taking an image of the cylindrical outer surface of a core, see Katayama *et al.*, 2004), visual and stereomicroscopy examinations, point-counting, X-ray diffraction (XRD) analysis, mapping of elements using Electron Probe Micro Analysis (EPMA) and quantitative EPMA.

Unfortunately, not finding such a reaction in a small thin section sample does not automatically prove the absence of AAR. Due to the heterogeneity of concrete itself the reactive aggregate may not be present in the small sample obtained for the microscopy analysis. In this section a structured approach is provided to increase the reliability as much as possible when making a statement about the presence of AAR.

5.3.2 On-site sample collection

To confirm the presence of AAR within the concrete of a structure, samples of that concrete are necessary. The purpose of on-site sampling is to collect representative cores or suitable materials from specific locations of the structure for further examination and testing in the laboratory. This stage should under no circumstances be taken

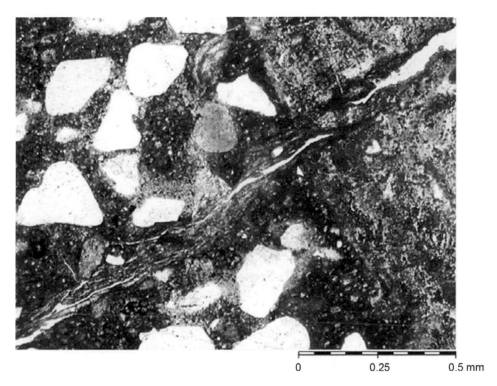

0   0.25   0.5 mm

Figure 5.3 Reacted porous chert aggregate in top right corner, with streaming gel and cracks going to the bottom left hand corner (photomicrograph courtesy of TNO).

lightly, because 'sampling at random' is inappropriate when a thorough answer is expected from laboratory investigation.

Owing to the heterogeneity of most concrete within a large structure, at the very minimum samples should be taken from both sound and damaged concrete areas. If possible the latter is preferably split into typical or moderately damaged and severely damaged concrete. Such a differentiation between damaged areas of concrete will help to establish correlations between the observed damage and the composition and/or characteristic properties of the collected concrete samples.

The number of samples that is preferred is dependent on the type and complexity of the structure, the range of concrete mixes used and the various environments. Almost inevitably, the limitations of access and economics will be deciding factors in the selection of sampling locations and the number of samples to be taken. It should be noted that a more representative assessment is obtained from selected tests from different parts of the structure, as distinct from concentrating a wide range of testing methods on a few samples (CUR, 2005).

5.3.2.1 Cores

Core samples are collected through drilling with a diamond bit. Before coring, any adjacent reinforcement buried in the concrete must be located using information from the drawings, checked with a cover meter. For coring in locations of structurally critical areas, coordination with or direction by the structural engineer in charge is necessary. To avoid excessive damage or fracturing of larger cores, the equipment used for coring should be well fixed to the structure during coring. It is desirable to take cores with a diameter at least three times the maximum size of the coarse aggregate used. To be able to remove the cores from the structure, drilling depth should be more than the diameter of the core. 100 mm diameter cores are often considered, but the use of larger size cores (*e.g.*, 150 mm in diameter) will be beneficial in the case of expansion testing on cores at 38 °C and R.H. > 95 percent, as it will contribute to reducing the leaching of alkalis during the test, thus generating more reliable test data for estimating the potential for future expansion.

The cores taken from the structure should be defined by:

- Sketch showing the location of the core in the structure
- Photograph of the core location
- Indelible marking
- Numbering according to the sample locations in a systematic way
- Defining the top and bottom parts of the cores, including the broken parts
- Showing any other aspects of the cores that may be relevant for the investigation such as orientation.

Concrete cores and fragments should be surface dried if wet cored and then sealed tightly immediately after sampling to prevent desiccation and secondary carbonation. This can be done by wrapping these samples with commercial 'Cling-film' in several layers, after which they should be sealed in polyethylene or vinyl bags for protection. If the cores are long and fragmented, it is recommended to use core boxes for transportation.

After coring, the hole created should be filled or sealed with a suitable cement-based material. To minimize shrinkage this must have a sufficient aggregate content with moderate cement content and a low water/cement ratio to match the concrete (proprietary shrinkage-compensated repair mortars are also available). The mixture should be well rammed home.

Percolation of water into core holes though deck surfacing can lead to severe local acceleration of AAR. If a water proofing layer exists on the top surface of a structure (*e.g.*, a bridge), it is recommended to core from below and upwards (and to stop well before the waterproofing membrane).

5.3.2.2 Fragments and powdered material

Fragments of concrete can be obtained by hammer and chisel from the concrete surface. Because hammering can cause additional cracking (which is misleading) and old surfaces of concrete are more or less weathered, evidence of any AAR can be masked by carbonation of both the cement paste and the gel. Hence, such fragments are not recommended as samples to confirm AAR.

Drilled powered samples of the type used for determining chloride profiles are of little value in investigating AAR, other than as an indication of Na^+ ingress associated with the Cl^-. Drilled powders should be sealed in a small polyethylene bag, expelling the air. Drilled powders should not be used for petrographic examination. It is also not recommended to use powdered drilling to determine the alkali content of the bulk concrete. It has been found (Wood *et al.*, 1996) that there are large spatial variations in alkali content particularly near the surface.

5.3.2.3 Efflorescence and exudations

Identification of efflorescence materials can be performed using X-ray diffraction (XRD). Soft whisker substances forming on the humid surface of concrete can be collected with a brush. They should be stored in a small glass bottle with a tight cap as they can otherwise be dehydrated or carbonated into a different form during transportation and before analysis. Also large bottles are not recommended because moisture from the sample cannot be maintained in the sample in a large space. Hard encrusting types of efflorescence, resembling stalactites, can be collected by scratching with a needle or by chiseling the concrete surface with the efflorescence attached.

Uranyl acetate (uranium) solution reacts with ASR gel and can be used as a screening test for ASR gel in laboratory or on-site. The reaction produces a fluorescing constituent visible in ultraviolet light. However, some gels may not be detected, while others producing a fluorescence are not always ASR gel. It has also to be remembered that ASR gel is water soluble and may have been washed away. Therefore AAR should not be diagnosed by this method alone. Uranyl acetate ($UO_2(C_2H_3O_2)_2.2H_2O$) is an internationally regulated substance, which is radioactive and hazardous to both human health and natural environment. Precautions should be taken for its storage, use and disposal. UV light will damage sight and protective glasses should be used. For further information on the method, see LCPC (1993).

5.3.3 Macroscopic inspection of cores

5.3.3.1 Description of cores 'as-received'

Upon arrival in the laboratory, the cores should be weighed. Keeping record of the weight of the samples at all steps is essential for checking any drying and subsequent water uptake during testing. The cores are first examined in an 'as-received' condition. They should arrive at the laboratory being wrapped with a 'Cling-film' to retain moisture.

The macroscopic inspection of the cores is generally performed with the naked-eye assisted with a magnifying lens (7x to 10x) and/or a stereo-binocular (generally up to 60x). Stereomicroscopy is usually used to examine areas of the sample that may be of interest for subsequent examination by means of polarizing and fluorescent microscopy (PFM) or scanning electron microscopy (SEM).

In addition to observations normally made with core sample descriptions, such as size and distribution of aggregates, compaction and void observations, presence and condition of reinforcement, the following macroscopic features may assist in the diagnostic process of AAR and their presence should be noted:

- **Damp patches** on the concrete surface. Damp or 'sweaty' patches looking dark may be found on the core surfaces around any reacted aggregate particles, which is suggestive of ongoing AAR in concrete. It is desirable to select these active portions of AAR for the preparation of polished thin sections. If the surfaces of the cores are dry, the core should be re-wetted and resealed in a plastic bag or by wrapping film tightly and then stored overnight. On the next morning possible damp patches can provide more information.

- **Reaction/weathering rims** (dark outer / white inner rims) around aggregate particles. The formation of a rim on crushed stone in concrete can indicate AAR has taken place. However, some aggregate particles from gravel have a weathering rim before introduction into concrete. Where dolomitic aggregate is present, any discolored carbonate halo in the cement paste adjoining reacted dolomitic aggregate particles should be examined because this may be suggestive of dedolomitization. ASR has been known to be associated in this rim-forming dolomitic aggregate in concrete, represented by the Canadian Pittsburg aggregate (St. John *et al.*, 1998). Some classical literature on alkali-carbonate reaction (ACR) recommended etching of carbonate aggregate particles in concrete with hydrochloric acid to discriminate the rim, but this method is destructive and therefore not recommended at this stage. Instead combined tests of SEM observation and elements mapping with EMPA analysis is recommended. It should also be noted that reaction rims are not invariably present in AAR damaged concrete. See Chapter 3 for up-to-date information on 'so-called ACR'.
- **Crack patterns** of cement paste and internal cracking of aggregates. Cracking is a common feature in all AAR damaged concretes and should be characterized: cracking location (*e.g.*, around or through aggregate particles), crack width, depth of surface cracking, etc. Cracking possibly caused by AAR is usually widening toward the concrete surface. 'Sub-parallel' cracking developing parallel to the surface, usually close to the plane of the reinforcement, can be a precursor of structural delamination.
- **Presence of any ASR-gel** in relation to cracking, around aggregate particles, inside the aggregates, in the cement paste, or exuding from the core.
- **The pattern/distribution** of the macroscopic feature over the core (through the entire core or only locally).
- Any other signs of irregularities or concrete disintegration.

Photographs of the cores as-received should be taken, if necessary using a series of overlapping views. The logged cores are photographed with a scale bar, a colour scale and a clearly visible core identification (*e.g.*, on a card). The texture of its cylindrical surface may be recorded by a core scanner (slit camera, or panorama camera), if available. This type of camera takes a photograph of the continuous outer surface (360 degrees) of the core (Katayama *et al.*, 2004). Especially for small cores the core scanner provides a wide area for analysis compared to the narrow longitudinal cross-section of the same core. A flat-bed core scanning device has also been developed by True (see Poole & Sims, 2015).

If the aggregate is very active, pop-outs may occur within one week after sampling around a reacting particle on the cylindrical core surface as a result of liberation of strain caused by expansive ASR gel that had accumulated around the particle. In this case a flat conical shaped concrete fragment, attached with a reacting particle on its tip, can be removed from the host concrete.

When alkali-silica reaction occurs in concrete, the interior of concrete expands more than the outer surface of the concrete structure, resulting in the formation of large tension cracks perpendicular to the exposed surface. Therefore, if some of the cracks in the concrete section are filled with ASR gel, and the internal cracks system is typical of that of ASR, then AAR is thought to be a probable cause of the deterioration at this preliminary stage of investigation. Some forms of AAR (*e.g.*, UK cherts in fine aggregate in South West England) produce expansion and cracking of aggregate particles extending out into the concrete, but with little apparent gel even when severe cracking occurs.

5.3.3.2 Selection of specimens for further testing

The specimens that are required for preparing polished- and thin-sections should be selected from areas where the features of the AAR damage are more pronounced. In the laboratory, concrete cores should be cut by a diamond saw to produce samples for further testing.

As concretes undergoing alkali-aggregate reaction are often badly cracked and fragile, they may disintegrate during cutting or coring in a small size. In this case, before cutting, core samples should be reinforced by epoxy resin or plaster filled within a framework, or hardened either by vacuum impregnation with resin or by immersing more permeable samples in cyanoacrylate. Such cores should be dried at temperatures below 50 °C by slow heating, or in the vacuum at ambient temperature in advance using a cold trap. It should be noted that overheating during preparation can produce various misleading artefacts, such as map-cracking within the cement paste and ASR gels, thermally-induced cracks and openings along cement-aggregate boundaries, and alteration of the optical properties of cement hydrates, particularly ettringite (elongation and refractive index).

To cool the diamond saw different options exist: water, oil or coolant air. As water can dissolve parts like certain cement hydrates, ASR gel and efflorescence materials (mainly Na, K, Ca, S and Cl), this can lead to spurious observations in microscopy examinations and chemical analysis. On the other hand, oil is viscous and specimens need sufficient cleaning using an ultrasonic vibrator to remove abrasive powders and dust that fill voids in concrete. Coolant air is not the most effective cooling medium and can raise the level of Si particles in the work environment as a negative effect for the human work conditions. Where multiple examinations are planned using the same core, the design of the core cutting deserves special attention before the actual cutting.

5.3.4 Petrographic examination

5.3.4.1 Introduction

The experienced examination of thin sections will allow positive identification of most of the diagnostic features of AAR, when present. To confirm AAR, thin sections should be prepared from sample areas where there is apparent evidence of AAR, including where possible the surface. To minimize misinterpretation due to other effects within the concrete, also a sample from an apparently sound area of the concrete should be examined.

To be able to identify the various components in a thin section and adequately characterize the microstructure, the petrographic examination should be performed using polarizing and fluorescent microscopy. This is a two-fold method using polarized light microscopy and fluorescent light microscopy (Larbi & Visser, 1999; CSA A846-00: 2000; Poole & Sims, 2015). By means of transmitted light microscopy with magnifications up to 400 times, the various components of concrete and any damage caused by AAR can be determined.

Thin section specimens are impregnated under vacuum at normal temperatures with an epoxy resin containing, where deemed appropriate, a fluorescent dye. The most popular dye has a strong yellowish colour. The strong colour of the dye may hinder details of the optical properties of minerals and gel in concrete under the microscope in polarizing light. The dye presents a greenish fluorescence on excitation by an ultraviolet light. The fluorescence makes it easier to study crack distribution and aspects of micro-porosity in concrete. Furthermore, information can be obtained regarding microstructural aspects, such as changes in the capillary porosity of the cement paste, content of air voids and their fillings.

If possible it is useful to make the polished thin sections thinner than the traditional thickness of around 25-30 μm. This is because the average size of cement and fly ash particles before hydration is 6 μm and 15 μm respectively. Interstitial phases of cement clinker are even smaller. In Chapter 3, it was shown that the true nature of 'so-called ACR' can only be determined using suitably thin and polished thin sections.

To identify opaque constituents or objects smaller than the thickness of a thin section, reflectance microscopy on polished thin sections is a suitable method. Reflecting microscopy is also useful to check the surface roughness in selecting the target area for EPMA analysis. For thin sections to be examined using reflecting microscopy, SEM and/or EPMA, it is necessary to prepare polished thin sections with a mirror surface, finished by diamond paste polishing (grain size 1 – 2 μm) *without mounting a cover glass*. Thus they can be both used for observations under the polarizing microscope and reflecting, SEM or EPMA analysis. Polished thin sections without a cover glass should be stored in a vacuum desiccator to avoid secondary hydration and carbonation of cement particles exposed to the surface.

5.3.4.2 Aggregates

Petrographic examination of concrete thin sections under the polarizing microscope is particularly useful for identifying the rock type(s) of the aggregate particles, especially those that have already reacted, started to crack and produced ASR gel, or have shown another type of unsound behaviour in concrete. It can also be used to determine whether potentially reactive minerals are contained or reacting within the damaged concrete.

Reaction rims, which are usually distinct to the unaided eye on the fracture surfaces of concrete, may paradoxically not be visible when observed in thin sections in transmitted light. Examination of the polished thin section under reflected light can provide details of the rim-forming texture.

Since concrete has a very coarse-grained texture, broadly analogous to a natural conglomerate or breccia, the volumetric ratio of each rock type that constitutes the coarse aggregates in concrete cannot be determined on the area of a regular thin section size. It can be achieved by examining much wider areas by, for instance, point-counting. Such areas can be obtained from the cylindrical outer surface of a core sample either by manual tracing or by core scanning with a panorama camera or adapted flat-bed scanning device.

Point counting on thin sections is usually possible for determining the volumetric composition of the rock and mineral types comprising the fine aggregates in concrete. In some cases only the fine aggregates may be responsible for any AAR, while the coarse aggregate is non-reactive (*e.g.*, chert-bearing sand or volcanic rock-bearing sand with a pure limestone coarse aggregate).

Other deleterious minerals that can cause unsound effects in concrete include smectite, laumontite, iron-bearing brucite, and pyrite. Likewise, some artificial compounds can behave deleteriously in concrete, for example when contaminated by periclase, free lime, anhydrite, and ettringite ($3CaO.Al_2O_3.3CaSO_4.32H_2O$) comprised in an expansive cement (see Katayama & Futagawa, 1997). All these deleterious materials can produce pop-outs on the concrete surface, and should be distinguished from any pop-outs caused by AAR associated with highly reactive opal or hydrated glass.

5.3.4.3 Cement matrix

Petrography can be used to determine the type of cement used. The presence of alkali-aluminate is suggestive of a high-alkali cement being used in concrete. The aluminate occurs as elongated, optically anisotropic alkali-aluminate, which is often a characteristic feature of AAR affected concrete. The abundant presence of periclase can be a cause of delayed expansion of concrete.

In mass concrete, blastfurnace slag and fly ash are often used as a partial replacement (20 – 70%) for Portland cement, to decrease heat of hydration. As a secondary effect they potentially act also as a preventive measure against AAR (although some types of fly ash may not prevent AAR due to their high alkali content).

Ground granulated blastfurnace slag is observed in thin sections as angular glassy shards (colourless, isotropic, refractive index between quartz and alite), often surrounded by one or more hydration rims. Fly ash shows as distinctive spherical particles, ranging from large transparent hollow glass particles ('cenospheres') to small coloured solid glassy (sometimes opaque) grains, or, more rarely, as irregular shaped carbonaceous matter (unburned coal).

5.3.4.4 Cracks

It is important to record the distribution, width and pattern of cracks, and their relationship to any reacting aggregate particles and the core axis relative to the structural stress field. If a fluorescent microscope (UV-light) is available, impregnation by resin containing a fluorescent dye may be helpful in detecting cracks, crack patterns and void structures in the sample.

In concrete where AAR has induced expansion, fine cracks may radiate from reacting particles, with or without reaction rims, into the surrounding hardened cement paste. The rate of cracking varies widely within the structure and within a core, depending on the availability of water, variability of reactive aggregate concentrations and local alkali content. Where the reactivity of aggregate is relatively high, cracks may be filled with abundant ASR gel, and the permeation of ASR gel often darkens the bordering cement paste. Cracks may pass through unreacted aggregate particles, but usually skirt them.

5.3.4.5 ASR gel

ASR gel is mobile in wet condition and consolidates on drying, creating distinctive desiccation cracks on its surface. ASR gel can be found far from reacting particles, having penetrated into air voids and migrated along cracks, as well as in cracks that cut through unreacted aggregate particles. However, it should be possible to trace the ASR gel path through the cracks, back to the reacting aggregate from which it derives. The gel may occur as a layered composite, lining walls of cracks and air voids, which suggest that the gel had migrated periodically into these open spaces in concrete.

In later stages, gel may become crystalline and birefringent, which can often be identified as a rosette-like aggregation of lamellar crystals under SEM. The quantity of ASR gel that is observed exuded from cracks does not necessarily reflect the intensity of deleterious expansion. For instance, late-expansive ASR related to cryptocrystalline or microcrystalline quartz in some rock types typically does not produce abundant ASR-gel, even though internal cracking of concrete is intense.

5.3.5 Additional tests

In addition to what might be considered as the 'petrographic standard' to establish AAR damage to concrete in structures, there is a variety of tests that are intended to assist in the diagnostic process. It is recommended that these tests should not be used in isolation or instead of petrographic analysis, but rather to supplement the primary tests and analysis, noting that SEM may sometimes be the only method to diagnose AAR in a concrete (in case of the finely divided reacting silica inside a limestone aggregate: see also Chapter 3).

5.3.5.1 Scanning Electron Microscopy

When thin section microscopy has difficulty identifying the presence of ASR gel, the next step is scanning electron microscopy (SEM). Using SEM much higher magnifications (1000x to 15000x) can be obtained (see Figure 5.4), which are not possible with transmitted optical microscopy. SEM observations are most commonly made on small pieces of fracture surfaces of concrete including the areas of interest (reaction rims, exudations, gel-pockets, etc.). Another option is using the polished thin section, but this is only possible when no cover glass has been applied in the preparation of the thin section. Photographs of the test specimen taken up front, *e.g.*, under a stereo microscope, can assist in identifying the target area when using SEM. Situations where SEM assistance can be useful include:

- Initial stages of AAR where only reaction rims are visible, *e.g.*, in concrete structures younger than 2 years.
- Scarcity of gel formation in cracks and around aggregate particles, while field and other laboratory evidence strongly suggest the occurrence of AAR.

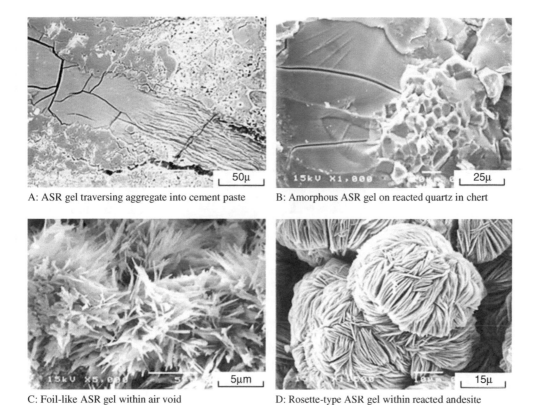

A: ASR gel traversing aggregate into cement paste   B: Amorphous ASR gel on reacted quartz in chert

C: Foil-like ASR gel within air void   D: Rosette-type ASR gel within reacted andesite

Figure 5.4 SEM photographs showing typical morphological types of ASR gel in concrete undergoing ASR. (a) polished thin section, (b), (c), (d) fracture surface. (Godart *et al.*, 2013, photographs courtesy of T Katayama).

- Compositional similarities between products of ASR and pozzolanic reaction.
- Clear-cut boundaries between fine-grained reactive minerals and reaction products.
- Details of AAR products like:

  - Massive-textured amorphous gel with desiccation cracks, which fills air-voids or interstices between reaction rims and cement paste.
  - Clusters of platy to foil-like crystals.
  - Rosette-like aggregations of lamellar to platy crystals, often found in the interior (open spaces formed inside the reaction rim) of the reacted aggregate particles in old concrete.

5.3.5.2 Electron-probe Micro-analyzer

SEMs are often equipped with an analyzing system. Such a system can provide chemical element information to the image that is observed. This option has great merit since all classical wet chemical analyses are based on bulk analysis of the sample and the sample is destroyed in the analysis process.

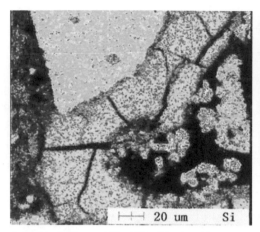 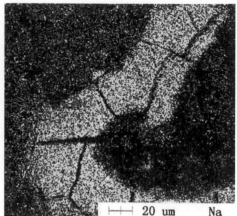

A: Image of silicon within ASR gel in air void   B: Image of sodium distribution within ASR gel

Figure 5.5 Mapping of elements by EPMA (EDS). Polished thin section showing distribution of principal elements of ASR gel in concrete affected by ASR (Godart *et al.*, 2013, images courtesy of T Katayama).

Possible analysis systems come in two types. Often used because it can analyze multiple elements simultaneously is the Energy-Dispersive Spectrum (EDS). The Wavelength-Dispersive Spectrum (WDS) is the other option. This latter system can only cope with one or a few elements at a time. WDS can potentially detect smaller quantities than EDS, but needs a higher beam current (100 times that of EDS) and could damage the sample more.

To use EPMA effectively, a polished sample surface is necessary. Analysis on fracture surfaces should only be used to identify main components. Reliable semi-quantitative analysis on fracture surfaces is not possible. Preferably the polished sample for EPMA analysis should be prepared without the use of UV-hardening type of resin, because this resin is rich in sulfur, which can interfere with the analysis.

EPMA is often used for spot analysis to identify chemical composition of the detail under the analysis spot or for mapping of specific elements over a larger area. Two examples of the latter option are shown in Figure 5.5. Element mapping is useful for following the changing composition of ASR gel as it can vary widely depending on its location relative to the cement paste and its stage of evolution.

Quantitative EPMA analysis of unhydrated cement phases can provide a first estimate of the alkali content in the original cement used. Because this method omits the water-soluble alkalis, the value is a minimum amount of the total alkali, resulting in an underestimation of possibly up to 20 – 30 % (depending on the local condition of the cement production, which may also change with time).

5.3.5.3 Determination of soluble alkali content

By contrast with the EPMA analysis, the bulk wet-chemical analysis of concrete cores almost always overestimates the amount of alkalis of cement, because substantial amounts of alkalis obtainable in this method come from the aggregates contained or from some admixtures, even if the mildest water-extraction is employed. A correction method has not yet been established worldwide, because the alkali-release of aggregates (such as andesite, granite, alkali-basalt, argillaceous sedimentary rocks, etc.) is regionally different.

This being specified, various methods are available for determining the soluble alkali content in concrete (LCPC, 1997b). The amount of soluble alkali in concrete depends on the extraction method used, *i.e.*, diluted acid-soluble or water-soluble, hot or cold, etc. In most countries, a water-extraction is applied. Which method is chosen depends on the circumstances (aggregate type, local experience, equipment, specific purpose of the investigation) of each country. The spatial variability of cement and alkali concentration in site concretes is substantial (Wood *et al.*, 1996). Furthermore, the initial composition is modified by migration and leaching. This needs to be considered both in the selection of samples and in presenting the results which should be of sufficient number to quantify the variability.

In countries like Japan, where the main reactive aggregates are argillaceous rocks and alkali-bearing volcanic rocks and an estimation of the cement alkali amount is important, water-extraction is recommended, since acid-extraction (even HCl with 1/100 dilution) extracts considerable amounts of alkalis from illite, feldspars and feldspathoids. However, in other countries, a light acid-extraction is recommended, as in France where limestone aggregates and quartzite aggregates are the main reactive rock types, which are not particularly alkali-releasing, and where a light acid-extraction (HNO_3 with 1/50 dilution) is applied because this type of extraction is commonly used for the chemical analysis of concrete. The French experience shows that the difference between the alkali content measured by light acid-extraction and that measured by water extraction is smaller than the scattering of the results obtained by several laboratories doing a water extraction during a 'round robin' test. The choice between light acid and water extraction may also be guided by taking into consideration the duration of the period in which the release of alkalis is taking place in the structure, as well as the type of the reactive aggregate.

5.4 APPRAISAL – SEVERITY OF THE SITUATION

5.4.1 Introduction

Once the presence of AAR has been established in the concrete through investigation as described in the previous section, the next step is to analyze the severity of the problem. Going from diagnosis to appraisal, one should keep in mind that it is possible that isolated areas of reaction may have been identified under the microscope, even if it has not caused significant expansion or cracking of the structure. The discovered AAR may be just coming from an isolated batch of aggregates issued from a different quarry, or could be in fact the early beginning of a potentially much more developing AAR problem. Many concretes (including perfectly sound concretes) can present very localised areas showing AAR gel and reactive particles, which are not sufficient to produce significant cracking or damage to the structure. Hence, how severe is the current situation, how fast will it get worse, and how much worse will it become are the main questions addressed in this section.

In principle the same process is being followed as with a routine inspection as sketched in Figure 5.2. The only difference is that now AAR has been identified as a likely cause of the problem at hand. So, data need to be collected and compared with data from existing records. This will provide the necessary information to judge both the immediate structural safety of the structure, by comparing it to the original design calculations, and how fast the situation is changing by comparing it with previous inspection data. The data are collected in two places: on-site and in the laboratory, as presented thereafter.

5.4.2 On-site

5.4.2.1 Cracks

Damage due to AAR presents itself most systematically due to its crack pattern. Hence, the simplest way to determine the severity of the AAR progress is by noting its crack pattern. In the technical literature different approaches are described like the 'Expansion to date' (Institution of Structural Engineers, 1992), the 'LCPC-cracking index' (LCPC, 1997a) or just the 'cracking index method' (Fournier *et al.*, 2010). As the general principle behind all these methods is the same, here they will all be referred to as the 'Cracking Index'.

The Cracking Index (CI) is a crack mapping process in which the widths of cracks within a measurement grid are summarized. In order to generate a statistically representative assessment of the extent of cracking, a minimum of two CI reference grids at least 0.5 m in size should be drawn on the surface of the most severely cracked structural components. For future monitoring of the crack development it is strongly recommended to install 'Demec' or similar gauges as reference corner point or points of the grid.

For typical map cracking patterns, a measurement grid should be drawn consisting of lines perpendicularly (*i.e.*, parallel and perpendicular) to the main restraints on the surface of the concrete element being investigated. In case of oriented cracking, for instance by restraints due to reinforcement, parallel lines should be drawn perpendicular to the principal crack orientation at least 250 mm apart from each other. For documentation, it is advised to take a picture of the reference grid, as well as its location in the structure (overview).

To determine the cracking index, the crack width of each crack at the crossing with the grid lines is measured using *e.g.*, a magnifying lens with internal gradation or a plastic crack size comparator. Each crack is to be measured with at least 0.1 mm accuracy. All the crack widths at the crossings with the grid lines are summarized and then divided by the total length of the grid lines to arrive at a cracking index expressed in mm/m. Thus the cracking index provides an estimated expansion of the concrete. When the cracking is oriented, the method may be applied but the value of the cracking index is not the average of the values obtained, but the maximum value obtained on the axis perpendicular to the cracks.

To evaluate the cracking index, the Institute of Structural Engineers (1992) has proposed a severity rating index ranging from I (< 0.6 mm/m) to V (>2.5 mm/m), see Figure 5.6. Also an LCPC (1997a) rating for the severity of cracking exists, which is slightly more progressive. A comparison between the two is provided in Figure 5.7.

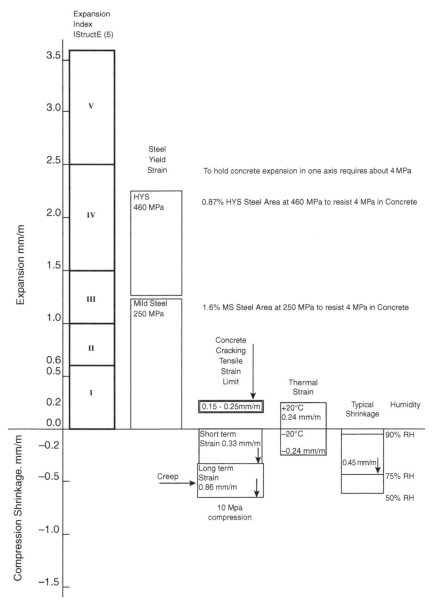

Derived from IStructE 1992 'Structural effects of alkali-silica reaction: technical guidance on the appraisal of existing structures'.

Figure 5.6 Severity rating index of expansion as derived from the Institution of Structural Engineers (1992).

Besides the actual state of cracking as measured to date, the real value of the cracking index is the change over time to monitor the progress of the AAR process. Because of the significant effect of temperature and humidity on crack width, repeated monitoring measurements should be carried out under similar conditions (sun exposure, outdoor

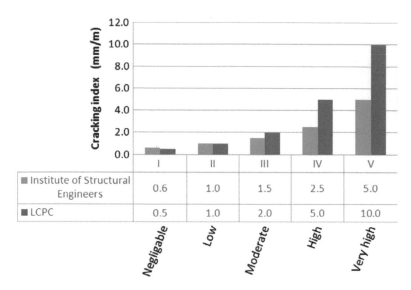

The following data accompanies the chart:

	I	II	III	IV	V
■ Institute of Structural Engineers	0.6	1.0	1.5	2.5	5.0
■ LCPC	0.5	1.0	2.0	5.0	10.0
	Negligable	Low	Moderate	High	Very high

Figure 5.7 Comparison between classification on cracking index from Institute of Structural Engineers and LCPC. Maximum of cracking index is provided per class.

temperature, outdoor humidity conditions). Depending on the rate of the process, measurements are to be repeated every 1 to 5 years as a useful monitoring scheme.

5.4.2.2 Global deformation and stress

The measure of global deformations of structures with time is an important tool for the appraisal of structures, because it can be used directly to adjust some parameters of the model used for the calculation of the structure. Compared with a cracking index, its advantage results from the integration of the expansion of the concrete between cracks.

The expansion process can be monitored by installing 'Demec' points or metallic studs or devices on the surface of the concrete. Ideally, these points or studs should be installed on the same members that have been chosen for crack mapping, to optimize the information generated from the same testing area. Periodic length-change measurements can be taken using extensometers of various shapes and ranges, such as vernier calipers with great capacity, a displacement transducer located at the end of a long steel bar free to slide, a distancemeter with Invar wire, and an infrared distancemeter, etc., as described in an LCPC guide (2009). Fibre optic and vibration wire systems can also be used with deformation measurements being performed and the data transmitted automatically to central servers for further treatment.

Evidence of stress build-up in reinforcing steel and the surrounding concrete resulting from restrained ASR expansion can be obtained from the measurements of stress in reinforcing bars. The reinforcement is partly exposed by cutting away the concrete. Strain gauges are applied on the exposed reinforcement and then the reinforcement is cut to release the tensile stress, which is measured by the strain gauges (Fournier *et al.*, 2010). Stresses in the concrete may also be measured by the relieve stress method using flat jacks inserted in thin slots executed by sawing the concrete.

5.4.2.3 Environment

Humidity and temperature readings can provide useful information on treatment and interpretation of expansion and crack measurements, as both are influenced by these environmental conditions. It is commonly accepted that ASR develops or sustains in concrete elements with relative humidity greater than 80 to 85 %. The relative humidity in concrete structures can be measured over time with depth or laterally in different concrete elements using various techniques such as wooden stick, portable or permanent probes (Stark & Depuy, 1987; Stark, 1990; Jensen, 2000; Siemes & Gulikers, 2000).

Also temperature readings can be made periodically using portable probes, or automatically using *in situ* setups. The temperature readings are often necessary to make any corrections required for temperature variations.

5.4.3 Laboratory investigation

5.4.3.1 Mechanical testing

The development of an AAR in concrete often alters its microstructure and its mechanical characteristics. Measurements of the physical and mechanical properties of concrete can provide information on the severity of damage within the structure. The most important tests are the following:

- Uni-axial compressive strength test where the modulus of elasticity ought also to be measured;
- Tensile strength test;
- Stiffness damage test.

Strength and elasticity

Of course, the measurement of the compressive strength is the most important concrete characteristic to know when assessing a structure. However, for an AAR-affected concrete the classical relations between compressive strength, tensile strength and Young's modulus do not exist anymore, and the return of experience shows that, in most cases, the reduction in Young's modulus and then in tensile strength of concrete affected by AAR is more pronounced than the reduction of compressive strength (Institution of Structural Engineers, 1992; CSA, 2000; Siemes *et al.*, 2002).

Reductions of up to 60 % for compressive and splitting strength, 80 % for direct tensile strength and 60 % for the elasticity modulus have been reported (Institution of Structural Engineers, 1992; CSA, 2000; Siemes *et al.*, 2002; Fournier *et al.*, 2010). Most rapidly affected are the modulus of elasticity and the direct tensile strength, even before significant levels of expansion are attained. Compressive and splitting tensile strengths generally behave similarly, being significantly affected only at relatively high expansion levels.

Standard compressive strength tests on short specimens may not show the loss of compressive strength in concrete with AAR, as friction on the plates restrain the specimen ends. Uniaxial compressive strength on longer specimens (length/diameter > 2) are needed. Furthermore, the compressive strength varies according to the development of the planes of microcracking, as microcracking develops anisotropically in conditions of constraint (Godart et al., 2013).

Considerable reduction in tensile strength and in the Young's modulus should be compared with the results of microscopy studies on thin sections regarding the bonding of cement paste to aggregate. If the bonding is poor, the concrete tends to yield considerably lower tensile strengths (Norris et al., 1990; Siemes et al., 2002).

Care should be taken in the interpretation of tensile strength values determined using splitting tests, especially those performed on cores smaller than 150 mm diameters, since these tend to overestimate the tensile strength. The results are very sensitive to the orientation of splitting. Therefore, splitting type tests are of limited value for evaluating AAR, and direct tensile tests are preferable.

Stiffness Damage Test (SDT)

Changes in stiffness and development of hysteresis (Crouch & Wood, 1988; Chrisp et al., 1993; Smaoui et al., 2004), provide the most sensitive measure of the development of mechanical damage from microcracking that lead to the degradation in tensile strength. These can be measured on cores in the Stiffness Damage Test (Chrisp et al., 1993) as a precursor to other tests.

The test was originally proposed by Chrisp et al. (1989). It is based on the cyclic uniaxial compressive loading of concrete core samples in 5 cycles between 0.5 and 5.5 MPa. The reduction in Young's elastic modulus, the energy dissipated during the load-unload cycles, which correspond to the surface area of the hysteresis loops, and the accumulated plastic strain after these cycles, are associated with the closure of the existing cracks and to a slip mechanism, and thus represent a measure of the damage in the specimen (microcracking) in the direction of the applied stress.

A modified SDT was proposed in 2004 (Smaoui et al., 2004; Berube et al., 2005), changing the maximum load to 10 MPa. The test was found to be useful to estimate the expansion attained to date by the AAR affected concrete using the energy dissipated during the first cycle. It should be noted that the correlation obtained between dissipated energy and expansion due to ASR have been found to be aggregate-dependent. Hence, this requires the use or the establishment of a calibration curve suitable for the particular reactive aggregate present in the concrete under investigation.

5.4.3.2 Residual expansion tests on cores

In considering the severity of the situation, the ongoing and future expansion due to AAR is a critical parameter to consider. To determine this parameter requires monitoring, which then can be extrapolated for estimating the potential for future expansion. While in situ monitoring of the actual structure will generally take 2 – 3 years to yield useful information (which is still influenced by mechanical, thermal and climatic variations), expansion tests on cores can yield first results in a relatively short period of time (6 months to 1 year).

The diameter of the cores should be at least three times the maximum size of the aggregate, the length should be two to three times their diameter. Thicker cores are preferred to minimize the effect of alkali leaching during the test. Furthermore, as this test assesses the mass concrete quality, the first 50 mm from the outer surface of the concrete should be avoided and cut off the test specimen. This outer skin of the concrete is often more macro-cracked, and may vary significantly in alkali content, either due to alkali leaching or alkali enrichment due to evaporation or supply of de-icing chemicals.

The procedures used for testing cores from AAR affected structures vary greatly from one study to another (Berube *et al.*, 2005). Generalizing, the factors to be considered are:

- Elevated temperature, to speed up the reaction process.
- Access to high humidity/water, to allow for the continued swelling.
- Management of the alkali levels, referring either to the possibility of alkali leaching to a water based environment or alkali supply when an alkali solution is used.

Usually it means that three types of tests are considered:

1. **Expansion test in saturated humidity controlled environment** – These tests are usually performed at 38 ± 2 °C and consist of monitoring the changes in length of concrete cores over a period of time (about one year) during which the cores are stored in saturated humidity conditions that are often referred to as "100% Relative Humidity". Two difficulties arise from maintaining a constant saturated humidity and from alkali leaching. In order to obtain homogeneous saturated humidity conditions, the best solution consists of storing cores above a small amount of water in small containers, and then to store the containers in a reactor containing water in the bottom, or in a fog storage room.

2. **Expansion test in water controlled environment** – These tests are usually performed at 38 ± 2 °C. The cores can be stored vertically above water in sealed plastic pails with wickets inside thus resembling conditions used in the concrete prism test (*e.g.*, ASTM C1293: 2008). Alternatively, each core can be sealed in a small inclined container, where the core can absorb water directly from a measured amount of added water, see Figure 5.8. The amount of water is topped up when core measurements for expansion and weight are performed to maintain 10 g of water surplus per kg concrete core.

3. **Expansion test in alkaline solution** – Expansion and expansion rates may be increased in expansion tests when cores are immersed in a 1-molar sodium hydroxide solution. Expansions in alkaline solution are affected by numerous parameters. In particular:

   a. the larger the core diameter (longer delay before the test solution impregnates the core samples),
   b. the lower the alkali content (slower equilibrium between the initial concrete pore solution and the immersion solution), and
   c. the lower the water/cement ratio (lower permeability and slower penetration of the alkaline solution), the lower is the expansion in the short term.

However, in the long term, *e.g.*, after one year, the influence of these three parameters is in general not as important, provided the water/cement ratio is not too low (water/cement ratio in the range of 0.30 or less).

Experience with this test is limited. In some countries where chert and flint are dominant reactive rock types, the aggregates dissolve into alkaline solution rather than producing larger expansion. In other areas where volcanic rocks are the main

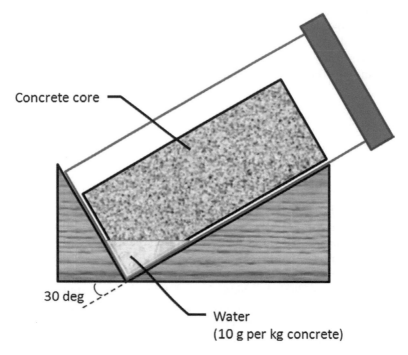

Figure 5.8 Expansion test in inclined container, after Wood (2011).

reactant, the severity of field deterioration of the structures correlates well, with the core expansion rates (diameter 55 mm, length 130 mm) in NaOH solution. When the concrete is tested with its own alkali content, allowing controlled water access and minimizing alkali leaching, the test results are said to be more realistic for evaluating further expansion of AAR-affected concrete (Fournier *et al.*, 2010). When additional alkalis are being supplied through an alkali solution, worst case scenarios of maximizing the 'residual reactivity' of the aggregates in the concrete are pursued. Furthermore, it is important to mention that expansion tests on cores extracted from deteriorating concrete structures will often lead to more pessimistic results than those obtained from *in situ* monitoring of the structures themselves (Fournier *et al.*, 2010), as the cores are not tested under the same stress conditions (possibly counteracting on expansion), nor under similar environmental conditions, such as temperature, humidity, wetting/drying and freezing/thawing.

Concrete in which AAR has caused deleterious expansion will often exhibit rapid expansion during the initial stage of the expansion test. This initial expansion can be due to various mechanisms unrelated to the residual/further AAR expansion:

- Thermal expansion, since the test temperature (38°C) is normally higher than the temperature at which the core samples were stored before being tested;
- Moisture uptake since test humidity conditions are normally higher than in nature;
- Stress release, as a result of lower stress with respect to the conditions prevailing in un-cored situation (when the samples are tested too early after coring).

The expansion and weight are measured at intervals for at least one year from the start of testing. Because the alkali content of field concrete is normally lower than for laboratory concrete made in accordance with the Concrete Prism Test ASTM C1293 (2008), relatively low expansions are usually obtained during residual expansion tests. In order to improve the statistical significance of the results, (1) the tests are often extended over the usual one-year period; (2) the measurements are more frequent than in the standard Concrete Prism Test ASTM C1293 (2008), and (3) linear regression analysis is recommended in order better to assess the annual rate of expansion (Fournier *et al.*, 2010).

A common feature of these tests is that the expansion of a core has to be evaluated with several measurements on the surface because of the large heterogeneity of ASR-expansions. Core locations have to be carefully chosen in order to be representative of the structure (direction of coring compared to casting direction and/or cracking, saturation level within the structure at the coring places, etc.). The whole time-evolution of ASR expansion curves has to be used for assessing the behaviour of damaged structures and for introducing it in the calculation model (Multon *et al.*, 2007).

5.4.3.3 Damage Rating Index (DRI)

Grattan-Bellew (1992, 1995) described a method for evaluating the condition of concrete by counting the number of typical petrographic features of ASR on polished concrete sections (16x magnification) over an area of minimum 200 cm^2 in size. The total number of each type of defect is multiplied by a weighting factor (see Table 5.3), representing the relative importance in the overall deterioration process. Next, these weighted results are summarized. The *Damage Rating Index* (DRI) represents the normalized value of the summation to an area of 100 cm^2.

As Table 5.3 shows, the weighting factor for debonded aggregate is high. Since frost or chemical attack can also cause debonding, the DRI is not suited for analyzing concrete that has also been damaged by these degradation mechanisms.

Furthermore, Bérubé *et al.* (2004, 2005) tried to estimate the amount of expansion reached by concrete cores extracted from structures affected by ASR incorporating siliceous limestone aggregates. The DRI method could not differentiate between concrete affected most and least by ASR; high DRI values were found for all concrete

Table 5.3 Weighting factors for the Damage Rating Index (as modified in Godart et al., 2013).

Petrographic feature	Weighting factor
Coarse aggregate with cracks but not ASR-gel	0.25
Coarse aggregate with cracks and ASR-gel	2
Coarse aggregate debonded from cement paste	3
Reaction rims around aggregates	0.5
Cement paste with cracks	2
Cement paste with cracks and ASR-gel	4
Air voids lined or filled with gel	0.5

investigated. For that reason the report of the Federal Highway Administration (Fournier *et al.*, 2010) does not recommend the DRI method for evaluating the expansion to date of concrete affected by AAR.

5.4.4 Assessment of severity

Until now, very little information has been given on absolute values regarding the range of severity, or threshold values to differentiate between different classes of severity. This is because each structure is different and has its own boundary condition. Hence no generally applicable uniform limit states exist. Therefore, in prioritizing the severity of the situation for a specific structure, only a general approach is given here, without any clear cut numbers. In the general approach described here, it is assumed that AAR has been established (see section 5.3) and that AAR related data from both the on-site investigation as well as the laboratory investigation are available. The type of data for these two types of investigations is discussed below.

The information from the on-site investigation provides data in terms of amount of cracking, which can be recalculated to strain (mm/m) through the cracking index, or of amount of deformations. Furthermore, it gives input on the temperature and humidity conditions (is ample water available?). Finally, it provides information on expansion and misalignment.

For the laboratory investigation, the collection of evidence is firstly obtained on a microscopic level (through polarizing and fluorescent microscopy (PFM) and the Damage Rating Index (DRI), and secondly at the core level (through the reduction in strength and elasticity from mechanical testing, and in terms of expansion based on expansion tests on cores).

When combining the results from on-site with laboratory investigation, Table 5.4 gives guidance to the direction of the severity of the situation and gives also advice on subsequent actions that could be taken.

LL   If neither the laboratory investigation nor the on-site inspection show a significant number of AAR features on any of the indicated parameters, the found AAR is most likely too little to cause any damage yet, if ever. Additional

Table 5.4 Summary of diagnosis directions based on on-site and laboratory investigation findings. The detailed explanation of the four quadrants is given below the table.

		On-site investigation (cracks, expansion, water availability)	
		Low	High
Laboratory investigation *(microscopy results, mechanical reduction, core expansion)*	Low	LL	LH
	High	HL	HH

awareness during the regular inspection regime of the structure is enough until the indications change to features that can be counted and/or monitored.

LH   If the indications from on-site inspection indicate a high probability of AAR, but no evidence of AAR was found on samples investigated in the laboratory, two possibilities exist:

a)   The sampling programme may have missed the severe locations found in the on-site investigation. To solve this problem would require new sampling.

b)   The features observed during on-site investigation, although seemingly consistent with AAR, are the result of other mechanism(s). If this had not been considered before, it would require a new investigation to pursue this direction.

If the on-site inspection does find major degradation features, a monitoring on the progress of the structure degradation is strongly advised.

HL   If significant evidence of AAR has been found on laboratory samples, but only very limited indications have been observed during on-site investigation, it could be that AAR is only a very localised problem in part of the structure. It could also be that the laboratory results show the very early stages of AAR, which have not yet led to major visible features in the structure itself. Yet another option is that AAR has occurred, but that the visual presence on the structure has been masked by other mechanisms. In any of these cases, it is advisable to keep monitoring the structure for progress of the degradation process.

HH   If both the on-site and laboratory investigation show significant numbers of AAR presence, it is clear that the structure has an AAR problem. Monitoring of the structure alone is not enough anymore. A structural assessment of the structure is strongly advised, as well as an investigation programme to predict future development of the AAR. This needs to provide the information to allow for judgment on the safe operation of the structure at all times.

5.4.5 Structural Assessment

A structural assessment is needed, as indicated previously in the case HH. Its objective is to identify significant potential serviceability and safety issues which have an impact on the function of a structure. This structural assessment comprises a detailed structural severity rating and a detailed assessment of critical elements and overall behaviour of the structure.

5.4.5.1 Detailed structural severity rating

The detailed structural severity rating of a reinforced concrete structure is predominantly based on the magnitude of the free expansion to date, and on the effectiveness of the reinforcement configuration in containing the AAR damage in three dimensions. A good illustration of this rating is the approach adopted in the guidance of the Institution of Structural Engineers: 'Structural effects of ASR' from 1992 with its

Table 5.5 Element rating based on detailed structural assessment of reinforced concrete structures (Institution of Structural Engineers, 1992 & its 2010 addendum).

Site environment	Reinforcement detailing class	Expansion index									
		I		II		III		IV		V	
		Consequence of failure (slight or significant)									
		Slight	Sign.	Slight	Sign.	Slight	Sign.	Slight	Sign.	Slight	Sign.
Dry	1	N	N	N	N	N	N	N	N	N	N
	2	N	N	N	N	N	N	N	N	N	D
	3	N	N	N	N	N	N	N	D	D	C
Intermediate	1	N	N	N	D	D	C	D	C	D	C
	2	N	N	D	C	D	C	C	C	C	B
	3	N	D	D	B	C	B	B	A	B	A
Wet	1	D	D	D	C	D	C	C	B	C	B
	2	D	D	C	B	C	B	B	B	B	A
	3	D	C	C	A	B	A	A	A	A	A

addendum of April 2010. This guidance provides a method of classifying each element on a scale:

A: Very Severe, B: Severe, C: Moderate, D: Mild, N: Negligible,

The rating for a structure is that of its most vulnerable element.

The preliminary determination of the rating for each element in a bridge, dam, building or other structure must consider the quality of the reinforcement (when present), the cracking, the moisture availability and the consequences of a failure, as presented in Table 5.5.

An important aspect of the assessment is the consideration of the reinforcement configuration relative to the severity and orientation of the cracking. The reinforcement configuration is ranked on the three following classes:

- Class 1: full containment with hooked or welded laps and links;
- Class 2: conventional reinforcement with corner lapped links;
- Class 3: 2-D surface reinforced with no through thickness steel; light or defective reinforcement.

The comparison of cracks with reinforcement drawings will identify where delamination may occur and where crack development is not controlled by sufficient anchored reinforcement across the cracks. Where the uncontrolled development of the cracking may initiate a failure mechanism (Rating A: Very Severe), a full structural analysis becomes a priority.

The structural analysis of critical elements is mainly based on a calculation inevitably involving numerical modelling due to the complexity of the problem. Sometimes, the numerical modelling is used to analyse the overall behaviour of a structure. In this analysis, the consequences of a local failure of part of the structure and its wider implications need to be considered.

5.4.5.2 Structural assessment based on numerical modelling

As AAR can be a long-term process, the forecasting of further expansion and long-term stability, as well as the prediction of structural responses to various mitigation measures requires the use of numerical models. In 1995 Léger *et al.* (1995) described a number of more or less sophisticated numerical models in a state-of-the-art publication. One of the first comprehensive models used to assess ASR-affected concrete structures was developed by Li and Coussy in 2000. This model represents AAR development as the expansion of a gel in a concrete porous network considered as elastic. Gel swelling is supposed to be in linear relation with the chemical extent ξ and it induces a pressure k.ξ(t) on the solid skeleton. With some simple assumptions, the 1-dimensionnal constitutive equation of AAR-affected concrete can be written as an additive relation:

$$\varepsilon = \varepsilon_e + \varepsilon_\infty \cdot \xi(t),$$

where ε_e represents the elastic part of the strain and ε_∞ is the maximal strain observed on the stress-free concrete. The evolution of ξ(t) is given by Larive's law:

$$\xi(t) = \frac{1 - \exp(-t/\tau_c)}{1 + \exp\left(\frac{\tau_l - t}{\tau_c}\right)}$$

in which the parameters τ_c and τ_l are, respectively, the characteristic and latency times (Ulm *et al.*, 2000).

This model can then be implemented in software such as CESAR-LCPC (RGIB modulus) (Omikrine-Metalssi *et al.*, 2010). Since a high temperature is responsible for lower τ_c and τ_l according to Arrhenius' law, whereas a high value of *h* (moisture content) increases both expansion amplitude ε_∞ and kinetics, the chemo-mechanical computation with RGIB modulus is based on the results of two complementary models: one aimed at assessing the temperature field in the structure, and the other consisting of solving the transient non-linear moisture-diffusion equation governing the evolution of *h* in the porous network of the structure. Fitting of Larive's model is based on experimental data such as the results of the residual expansion test on concrete cores extracted from the structure and the results of the cracking measurement and the global deformation monitoring of the structure.

An example of application of this modeling may be found in Seignol *et al.* (2009). This case-study intends to underline the strong interest of this method to answer practical problems of structure management: in the case of a hydro-power plant, modeling allows for correctly informing a structure owner about stresses generated by the reaction in the structure as well as future disturbance in the equipments due to anchorage displacements, hence the loss of serviceability of the facility. Numerical modeling plays a large role as a long-term managing tool and allows testing of possible treatment techniques.

The FEM model of the structure must consider the real reinforcement detailing of the structure and a good definition of the boundary conditions in terms of mechanical restraints, temperature and humidity states. After calculation, the model is able to give displacements, strains and stresses at every point of the structure. It is therefore possible to check if the ultimate and service limit states are attained or not, whether some bars are yielded, whether cracking is developing, whether concrete is able to sustain compression, etc. Once the FEM calculation has been validated by comparison with the actual behaviour of the structure, it is then possible to predict the future evolution of the structure.

Generally, when numerical analysis of the overall behaviour of a structure is required, the problem should be simplified towards a specific model that is able to simulate the main phenomena of interest. Unfortunately, such simplifications are much less possible in the case of AAR. Modelling AAR involves: real structures with complex geometries and details, mechanical aspects like the initial stress state due to loading conditions and creep, physical processes including temperature and moisture effects, chemical reaction rates causing swelling, and other degradation mechanisms that may also be present in the structure. To complicate things further, the determination of parameters required for most of the models require extensive and long-term experimental tests. Due to this complexity, it is recommended that any simulation model, as sophisticated as it may be, should be calibrated based on the monitoring data and pertinent information obtained from *in situ* and laboratory investigation of the structure under investigation.

5.5 REPAIR – WHAT ARE THE OPTIONS?

5.5.1 Introduction

Once the diagnosis and appraisal phases have concluded that remediation measures are necessary, the owner of the structure has to consider a certain number of criteria before choosing the best measure to be implemented. These criteria may be approached under the following list of important questions:

• What is the purpose of the repair and what is its expected duration? (to cope with safety, serviceability or durability limit state…)
• Is it a critical structure? (strategic importance, risk analysis…)
• What are the past and future exposure conditions of the structure? (temperature, moisture, immersion, freeze-thaw, supply of external alkalis, aggressive ions…)
• What is the internal constitution of the structure? (plain concrete, light reinforcement, well-detailed reinforcement, prestress degree, internal redundancy…)
• What are the external links of the structure? (adjacent structures, degrees of freedom, presence of equipments that need adjustments…)
• What is the accessibility to the structure and what are the operational constraints for the repair work? (traffic intensity for bridges and roads, water level for dams, sea tides for marine structures, etc.)
• What is the budget available for the repair?

This section briefly reviews the "repair" techniques that have been applied worldwide. Here the "repair" methods should be taken in its broad acceptance, more in a sense of treatment, because the experience shows that most of the "repair" methods are rather to mitigate the evolution of the reaction and its structural consequences, than to restore an affected structure to a good condition and even to its initial condition. It tries to identify among them those which are capable of extending the life duration of the structure or reducing deterioration of the structure from additional pathologies such as freeze-thaw damage, carbonation, chloride or sulfate ingress, reinforcement corrosion, etc. For most of these techniques, more details are presented in the following chapters of this book dealing with specific applications in various countries.

To facilitate understanding, this review classifies the techniques among four general principles that are:

- to apply mechanical means to confine the expansion or to release the constraints (section 5.5.2);
- to avoid the penetration of water or moisture in the structures (section 5.5.3);
- to apply chemical or electrochemical means (section 5.5.4);
- to replace whole or part of the structure (section 5.5.5).

5.5.2 Methods based on mechanical means

5.5.2.1 Additional reinforcement

Several methods of repair of a mechanical nature exist; it is possible to enclose, to ring, or to insert reinforcements, which are called 'active' or 'passive' according to whether they are tensioned or not, when installed. These reinforcements are primarily made of steel, but one can use passive reinforcements made of FRP (Fibre Reinforced Polymer), and especially Carbon FRP (CFRP) that has progressed at a rapid rate in recent years to strengthen ASR affected structures (Figure 5.9).

One of the methods to strengthen a structure consists of fixing on its facings reinforcement steels and by embedding them with shotcrete. Under the effect of the structure expansion which continues, the reinforced shotcrete cracks and the reinforcement bars take again a part of the overstresses generated by the expansion of the concrete; thus they relieve the existing reinforcements included initially in the structure.

Strengthening by introducing reinforcement with steel plates and bolts was also found to be effective in providing containment for some ASR-affected concrete members. The case of the Montrose bridge, where the treatment work consisted in introducing reinforcement by straps, plates and tensioned bolts on the top chord and tower top area is a good illustration (Figure 5.9) (Wood & Angus, 1995).

It is also possible to strengthen structures by the addition of tensioned bars or prestressing tendons as it has already been implemented to reinforce the foundation of electric pylons, the upper part of several dams and the piers of bridges.

5.5.2.2 Confinement

The principle of this technique is to introduce a three-dimensional prestressing whose objective is to try to confine the expansion. Taking into account the considerable forces generated by the alkali-aggregate reaction, it seems necessary to apply constraints which are in a range of 3 to 10 MPa, depending on the intensity of the chemical reaction.

However, introducing a 3D post-tensioning on a structure is often difficult. It is possible to apply a 3D post-tensioning in some particular elements, like the buttresses of a dam or a box girder bridge, but generally there is at least one direction where it is impossible to apply a prestress force and this affects the overall confinement. Experience

Strengthening of the crossbeam of the Terenez pylon by CFRP

Pylon of the Terenez bridge reinforced by CFRP.

Strengthening of the top area of the Montrose bridge by plates and tensioned bolts

Figure 5.9 Examples of additional reinforcement on bridges.

shows that, if the concrete expansion continues, the presence of passive or active reinforcements creates an additional prestressing of the concrete, because of additional elongations undergone by the steel reinforcements which are obliged to follow the concrete expansion. It is however not advised to take into account this prestressing in the dimensioning of the strengthening because it can be considered as artificial.

Hence, 1D or 2D additional mechanical "containment" does not constrain the deformations and the cracking in the untreated directions as required. It has a short-term effectiveness in controlling cracking, but it can effectively maintain the structural capacity of the core of the concrete on the long-term. If a confinement is considered as treatment, then a three-dimensional compressive stress state is recommended (see Figure 5.10).

5.5.2.3 Addition of props

The addition of props is seldom used, but in some special cases props may aid in taking vertical load from suspected parts of ASR affected structures like piles or corbels. The props are generally metallic; they must be securely fixed and should be designed to be capable of taking the direct load, even if a substantial part of the load continues to transit via the ASR affected part of the structure. The props should be regularly inspected to check their stability during time.

a) General view of the confined cap beam

b) Detail of the edge of the cap beam with the anchorages of the prestress

Figure 5.10 Example of a 2D confinement by prestressing of a concrete cap beam.

5.5.2.4 Stress release

The stress release is made by sawing whole or part of a structure. This technique is mainly used on gravity dams. The first attempt on large dams was made on the Fontana Dam (USA) by the Tennessee Valley Authority (TVA) in 1970, (Charlwood, 2009). Several attempts were then made on some other dams like the Mactaquac dam (Canada), the Beauharnois dam and its Powerhouse (Canada), and the Chambon dam (France).

Other concrete structures like hydroelectric energy production plants, especially those linked to dams which have been constructed with reactive aggregates, may present deformations or movements that are in excess of the deformations that the generating equipment may support. In dams, sluices or spillway gates could be jammed by the expansion of the concrete. In those cases, stress release in concrete by slot-cutting is also effective (see Figure 5.11).

Experience shows that the cut may close rather rapidly if a high compression exists in the dam due to AAR, and a re-cutting may be necessary. Cuts and released cracks

Cuts in progress on the construction site

View of the cutting line on the downstream face

Figure 5.11 Example of cuts realized on a dam to release stresses.

should be respectively sealed or injected in order to prevent water or moisture from penetrating into the concrete. This could be done by using a low elastic modulus grout or resin, or a special water-stop device at the upstream face of the dam.

Saw-cutting is also used in pavements suffering from AAR to remove sections of concrete near the joints; these sections are then replaced by a sound concrete.

Hence, stress release has a short-term effectiveness, but it often requires renewal of the operation insofar as it does not oppose the continuation of the chemical reaction. It thus remains an applicable solution in quite particular cases, like those concrete structures that are not reinforced.

5.5.3 Methods for avoiding the penetration of water or moisture

5.5.3.1 Cladding

Cladding is composed of screens made of steel or plastics designed to protect a structure from the rain. A gap is maintained between the cladding and the structure to control the humidity, to prevent condensation of water and to encourage drying. A ventilated cladding is designed to produce air currents which may lower the moisture content of the concrete structure and thus reduce the reaction rate. Lowering the humidity in a structure this way could necessitate a rather long time if the structure is massive or if the climate is often very humid. It follows that cladding is more often adapted for thin structures located in rather dry or mild climates.

The inconvenience of such a solution is that cladding may mask the eventual evolution of the disorders and that it changes the aesthetics of the structure, unless it is designed to improve it. If done correctly, this technique is used as a preventive measure to reduce moisture access and to reduce the rate of ASR expansion, but it affects only the surface of the structure. It is more adapted for buildings than civil engineering structures.

5.5.3.2 Waterproofing membranes and water drainage

Bridge parts such as abutments, piers, pile caps, decks, curbs and sidewalks may be subjected to water stagnation that increases AAR. It is therefore strongly recommended to drain water out of the structure to reduce the degree of saturation of a concrete element and to decrease the rate of expansion. It is a simple and generally low cost solution.

When water is entering into the deck of a bridge because the waterproofing membrane is lacking or in a bad condition, it is also strongly recommended to renew the waterproofing in order to avoid the penetration of water into the concrete. It is also a good solution to prevent alkalis coming from deicing salts to exacerbate the reaction.

Drainage flows of water should be collected through pipes, gutters or trenches and expelled out of the structure, towards a water retention basin if required by environmental law. For massive structures such as dams or liquid containment, the drainage is more complex because it necessitates the drilling of internal cores close to the face in contact with water or at existing cracks. For dams, the primary purpose of drainage is to reduce the internal uplift pressure and to improve the structure safety. The use of impermeable coatings or membranes on the upstream face does prevent the buildup of detrimental uplift pressures in damaged joints and cracks, even though reaction and swelling is likely to continue; it is also beneficial for limiting water availability for AAR.

5.5.3.3 Application of penetrating sealers

Penetrating sealers are products that penetrate sufficiently in the voids of a concrete surface and remain stable for a time period. They do not bridge cracks, except for very thin cracks (below 0.1 mm). Their objective is to protect the concrete from water penetration and to allow vapour transmission. They must be able to penetrate to a certain depth, and it is recommended to apply them on a rather dry surface, in any case avoiding application on to moisture saturated concrete.

Most of the sealers are formulated with silicon compounds and include silicates, siliconates, silanes, siloxanes and silicones. Based on experience in several countries, siloxanes and silanes have the better performance characteristics regarding AAR protection. These products are chemically bonded to the concrete surface pores and repel water. Silanes are mainly used to prevent water and chlorides penetrating into concrete structures.

In recent years, silanes have been used to attempt mitigating ASR. It is often difficult to draw general conclusions on their efficiency, because there is a great variety of silane products available, depending primarily on the concentration of silane in the specific formulation (ranging from 20 % to nearly 100%), and on the type of carrier with which silane is combined (either water-based or solvent-based). The depths of penetration of silanes are in the order of a few millimetres, and their action is to seal the concrete surface by forming a barrier preventing water from entering the concrete, whilst allowing internal water vapour to escape.

Sealers must be resistant to ultraviolet radiation and have a good stability in the alkaline environment of concrete, noting that the carbonation of the cover concrete may improve this stability. The durability of the penetrating sealers is rather short in comparison with the life of a structure, and it is necessary to renew the treatment rather often. The FHWA report, (Fournier *et al.*, 2010) mentions that re-applying silanes every five years or so is prudent.

5.5.3.4 Application of coatings

Application of coating systems can be divided into two types:

- Application of a painted system (maximum thickness of about 200 to 300 micron).
- Application of a thicker coating system (a few millimeters).

The application of a painted coating on civil engineering concrete structures constitutes the simplest means to implement coatings. The total thickness of the painted system is rather low (maximum 200 to 300 microns), and their effectiveness to combat the effects of the alkali-aggregate reaction is almost nil insofar as the system is permeable to water vapour and relatively permeable for water. Failure of these systems have been shown many times. The case of the Terenez suspension bridge with its painted pylons is an illustration of a failure in France (Figure 5.12). In Belgium, the policy of painting bridges at the end of the construction did not prevent the occurrence of disorders due

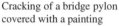

Cracking of a bridge pylon   Cracking reappearing through the painting which is peeling off
covered with a painting

Figure 5.12   Examples of failures of paintings to treat AAR.

to the alkali-aggregate reaction on several bridge piers (Figure 5.12). Hence, the application of a painted system is a solution which has only a very low effectiveness.

The application of a thicker watertight coating having generally a significant thickness (a few millimeters) constitutes another way of protection. An essential element for the choice of such types of coating is their capacity to follow additional expansions of the structures, as well as their characteristic to have a sufficient elasticity at the crack location to be able to accommodate their possible supplementary openings. The installation of a watertight coating can produce a positive result with the provision of utilizing sufficiently tight systems (including tightness against water vapour), provided they are installed correctly and during dry periods.

The application of a coating does not stop the reaction and this latter may continue until the relative humidity in the concrete drops below 80%. The attainment of this degree of humidity depends closely on the structure size, therefore coatings are more efficient on slender structures than on massive structures, where the pore water takes a long time to move from the core to the surface.

As an example, there have been cases reported around the world where bridge beams and other smaller sized structural elements have been dried sufficiently through the application of coatings. In France, several attempts with different types of coatings were made on the soffit of bridges, and after ten years it appears that some coatings have the capability to reduce significantly the measured expansion of the structures: among the best coatings products are a flexible coating composed of a styrene-butadiene resin reinforced by a glass fibre fabric, and then a hydraulic binder modified by a polymer (mixture of cement and acrylic resin). On the other hand, studies of the effects of various coatings on a small dam in Norway (Hunderfossen case dam, see Jensen, 2006) have shown that none of the surface protection systems were able to reduce or stop AAR and that cracking and damage had continued.

Hence, the application of a watertight coating having a sufficient thickness makes it possible to mitigate the expansion of a structure if it is effective in reducing the penetration of water and other aggressive agents in the structure. It is a provisional solution to extend the life of structures, provided that the coating is regularly maintained to avoid any breaching.

5.5.3.6 Injection of cracks

In the past, first attempts made to repair structures affected by the alkali-aggregate reaction consisted of injecting the cracks, generally with epoxy resins. Whatever the country where these attempts took place, almost all showed failures, either because of re-opening of the cracks, or because new cracks occurred close to the injected cracks. It is now well understood that if the resin prevents the water from penetrating into the cracks, it by no means protects the remainder of the facings with respect to the penetration of water or moisture.

Nevertheless, injection of cracks may still be useful in some special cases and may present positive aspects. It is the case when it is required to strengthen a structure by post-tensioning because the compressive forces are better distributed in the structure when the concrete is monolithic. It is also the case when one wants to avoid the ingress of aggressive agents through the cracks, or when the structure expansion is close to its end and the movements of the cracks are nearly stabilized.

An injection with a Portland cement grout is not well adapted because it could increase the alkali content in the vicinity of the cracks and possibly accelerate or reactivate the ASR locally, and because the grout is too viscous to penetrate very fine cracks.

Considering resin grouting for injection of AAR-damaged concrete structures, there is a number of requirements that a resin should meet:

1. it should have a sufficiently low viscosity to be able to penetrate fine hair cracks using very moderate injection pressures;
2. it should however not have such a low viscosity that it becomes absorbed into the pores of the concrete on either side of a crack so that the crack empties when injection ceases;
3. it should have excellent wetting properties so that it can penetrate gel-like reaction products bound to the crack sides and wet the intact concrete forming the crack sides.

Hence, injection and bridging of cracks are inoperative remedies because they do not act on the reaction and they do not improve noticeably the resistance of the concrete to the development of cracking. These techniques simply prevent water from penetrating inside the cracks, provided that the product injected or used in bridging is sufficiently flexible to accommodate the future opening of the cracks. It could nevertheless be necessary to inject cracks if a post-tension is added. It may become appropriate to seal a crack when expansion is finished.

5.5.4 Application of chemical or electrochemical means

5.5.4.1 Effects of lithium

Lithium has been used experimentally as an admixture in concrete in order to prevent AAR, and its required content increases as the amount of alkali in the concrete increases. Although the mechanism of the preventive action of lithium (which is an alkali!) is not well understood, it is generally believed that lithium compounds are incorporated in ASR gels as they form, and change their nature and their behaviour by negating its expansive character. The role of lithium as a measure to prevent AAR is discussed further in Chapter 4.

For existing structures, some laboratory testing has shown that concrete specimens can be treated using lithium-based compounds to slow down the rate of expansion, but the main problem is to achieve adequate lithium penetration, especially if the porosity of the concrete is low. In the USA, many structures have been treated with lithium using either surface (so-called 'topical') application or electrochemical or vacuum impregnation techniques to increase lithium penetration, but up to now, there is no evidence of any general success with such techniques.

Recent field trials conducted under a FHWA Project (Fournier *et al.*, 2010) in Idaho, Massachusetts and Texas have shown that the depths of penetration of lithium have been measured to be quite minimal, with for example only a few millimetres of penetration in the case of an Idaho pavement, even after three treatments. Other research performed under this FHWA project demonstrated that vacuum impregnation was effective in neither the laboratory nor in field structures, with the example of a penetration depth of only 9 to 12 mm in the columns of a bridge. The preliminary findings of a study made on a long-term outdoor exposure site at the Concrete Durability Center at the University of Texas at Austin (Folliard *et al.*, 2012) showed that the application of lithium nitrate solution to the surface of ASR-affected concrete showed no ability to reduce further expansion.

Hence, lithium is not practically effective in reducing AAR-induced expansion in actual structures in the field and, at present at least, cannot be recommended.

5.5.4.2 *Cathodic protection and electrochemical treatment*

Cathodic protection (CP) is one of the various techniques available for stopping the corrosion induced by the ingress of chlorides in many reinforced concrete structures. But CP is likely to increase the alkalinity of the concrete around the cathodically polarised reinforcing steel with the consequential risk that AAR will be induced in concretes containing reactive aggregates. A report of TRL has investigated this problem (Sergi & Page, 1992), describing research aimed at investigating the possibility of a link between CP and ASR. The results of measurements showed that significant expansion in the prisms polarised at -850 mV during a CP treatment occurred in the vicinity of the embedded steel. ASR was identified by petrography in a region up to about 6 mm from the steel cathode and accumulation of alkali ions was detected over a similar region by chemical analysis.

A second TRL report by the same authors (Sergi & Page, 1993) gives results for periods extending to approximately three and a half years. Strain measurements showed that significant expansion and cracking of the concrete around the steel cathodes occurred in four out of the six prisms in which potentiostatic polarisation (-700 mV) was applied. The authors found also that constant potential polarisation necessitating the application of high cathodic current densities (> 100 mA/m^2) during the early stages had a considerable effect in promoting alkali enhancement and ASR around the steel.

Hence, extreme caution must be exercised with the application of electrochemical rehabilitation techniques such as cathodic protection and chloride removal on structures affected by AAR, the latter technique relying on higher current densities over a short period. In any case, realkalisation should not be used for structures presenting reactive aggregates.

5.5.5 Replacement of whole or part of structure

Partial demolition or reconstruction is the safer solution to extend the working life of a structure. However, it could be a very expensive solution when one needs to maintain the structure in operation (like traffic on a suspended bridge) or when one needs to act on the foundations of structures. This solution is therefore seldom applied, especially since, up to the present, no brittle or premature failure of a structure, as a result of AAR, has occurred anywhere in the world.

In most cases, there is an alternative solution to mitigate the effects of an AAR in a structure. Only in a few cases is it necessary partially to demolish or to rebuild a structure, but in these latter cases, there are often other than technical reasons to justify it: it could be for economic reasons, for aesthetic reasons, because the structure will be part of a new development planning, or because the structure already has other initial deficiencies or because its service level is insufficient.

Hence, the demolition and then the replacement of a structure is sometimes an inescapable but expensive solution. Partial demolition followed by a rebuilding can often be considered as a serious alternative. In case of foundations that are not visible and may be deteriorated by AAR, particular care must be taken.

5.5.6 General conclusions on repair options

The first main conclusion that can be drawn is that there does not currently exist a repair or rehabilitation methodology that is sufficiently effective to durably repair AAR-affected structures, or even to stop the evolution of the disorders. As long as the ingredients which are necessary for the development and the continuation of the chemical reaction are present in sufficient quantity within the heart of the structures, it appears rather illusory to stop this type of reaction, which develops considerable swelling forces compared with the modest mechanical restraints achievable by later remedial actions.

However, the second conclusion based on the feedback of superstructures treated by various techniques, is that the first action is to reduce the availability of moisture by upgrading the waterproofing and drainage systems if those are failing. The application of coatings can be effective in slowing down the progression of the disorders.

Thirdly, in some cases for reinforced concrete structures where stresses are only restrained in one or two dimensions, the insertion of additional reinforcement or post-tensioning may provide the best means of making the structure robust by providing a three dimensional restraint on the expansion. In more severe cases, where moisture levels cannot be reliably controlled, for example dams, retaining walls and buried structures, controlled release of the expansion can be applied through slot cuttings or 3D confinement strengthening.

Finally, for the most serious cases, partial or total demolition and replacement may be the most appropriate solution.

5.6 MANAGEMENT OF AAR AFFECTED STRUCTURES

5.6.1 Introduction

In the previous sections of this chapter, the different steps in the process of diagnosis, confirmation, appraisal and repair are presented in a very structured way:

1. Diagnosis: first symptoms; could it be AAR?
2. Confirmation: can we confirm that the main cause of damage is AAR?
3. Appraisal: what is the condition of the structure? Are we at the beginning or at the end? How is it going to evolve?
4. Repair options: what are the possible solutions?

All these actions are integral parts of the management of structures. Very often, AAR affected structures are a part of the set the owner has to manage. The asset manager is therefore confronted with various structures, problems and constraints that are leading him to make prioritizations in order to optimize the global condition of its stock of structures with respect to a minimal safety level. In this framework, the management of the AAR affected structures may only be a part of the various actions for which the asset manager is responsible.

5.6.2 Prioritization

With respect to the structured analyzing process as presented before, the manager is not able, for financial reasons and because of limiting capabilities of the investigation laboratories, to implement the analysis process systematically and in an identical way, on all damaged structures. On the contrary, the manager has to adopt a progressive methodology according to the apparent condition of the structure and the speed of evolution of its disorders. Here the knowledge that AAR swelling is a process, which develops in general rather slowly, can actually help the manager in his prioritizing. The slow process usually allows postponing actions on the less severe cases.

In the most general case, the prioritization process aims to carry out a classification of the damaged (or suspected to be damaged) structures among a patrimony. It allows for detecting the urgencies and planning the investigations to be conducted in the most suitable way to optimize the technical and financial resources. This classification must be established at the end of a specific examination carried out by an AAR specialist engineer and is based on:

- technical criteria characterizing the condition of the structure (localisation of disorders (vital zones affected or not), their intensity and their extent);
- technical criteria characterising the design and the execution, like the degree of redundancy, the intensity of the load effects, the quality of the reinforcement,
- other criteria, such as the severity of the environmental conditions (direct contact with water for example), the strategic or financial importance of the structure (an important bridge in a road network, a plant providing an essential energy source in

the electricity network, or a dyke protecting an important part of the country from flooding).

Several levels of priorities may be considered. For an owner concerned by a great number of structures (for example several hundreds of bridges), the simplest technical prioritization can be based on two classification levels: level 1 for the structures presenting a vital zone affected by intense disorders and level 2 for the other structures. Affected vital zones are defined by either an affected structural element, whose failure would undermine the stability or the integrity of a part of the structure, or an element which can present an immediate risk for the users. Intense disorders correspond to significant visual disorders, even very localised, such as widely opened cracks, concrete delamination or important expansion, etc.

For the structures of the first level, it is necessary to engage simultaneously the monitoring on site and the analysis of cores in laboratory. For the most serious cases of particularly damaged structures, it will be advisable to examine the serviceability and, if necessary, to implement safeguard actions. For the structures classified at level 2, it is enough to initiate at first a monitoring of the structure, with the help of both deformation and cracking survey. It is only if the evolution of the expansion or cracking with time proves to be fast that coring must then be carried out for analyses.

For an owner concerned with a small amount of structures where he has already done laboratory investigations (for example a set of dams), the severity assessment as presented in Table 5.4 in section 5.4.4 provides a base to separate the most severe cases from the slightly less severe cases. Such a prioritisation may be assisted by the classification of the vital role a structure can take in the community. Risk analysis of the consequences upon failure could also be used to provide useful information for the prioritisation exercise.

When preliminary prioritisation has been established in this way, resources can be focused from these structures (global) to part of the structures (more detailed) where risk is the greatest in a similar approach. Each cycle of the prioritisation and rating needs to be progressively refined and followed by a more detailed inspection, or monitoring, or laboratory testing, or numerical assessment, as the investigation demands.

5.6.3 Parallel actions

The prioritization, as has been described so far, is nothing more than a ranking process. After the ranking, complementary actions are necessary to obtain more accurate data or to mitigate the AAR process and strengthen the structure. Again, in practice the asset manager shall often diverge from the textbook step by step approach.

Depending on the available knowledge from similar structures under investigation, or knowledge from similar structures in the same region, the presence of AAR may be more or less likely. The higher the likelihood of AAR being present in the structure, the more time can be saved by planning the different sequential steps as parallel paths in the investigation. In the extreme case of a very damaged structure where safety may be a concern, all the investigation options can be simultaneously applied. For other cases, the asset manager could prefer to apply a methodology to monitor the structure's dimensions or cracking index, rather than the systematic use of destructive cores for elaborated analyses in laboratories.

5.6.4 Frequency of monitoring cycles

Whatever action is taken, whether collecting new data or having (partly) rehabilitated the structure, the result will be that the severity of the structural condition needs to be re-evaluated, leading to a possible new ranking position in the priority listing. And thus a cyclic process is created to maintain and manage AAR affected structures.

The most important aspect from the gathered information is generally the prediction of the progress of the AAR process and, based on that, the time frame frequency until the next action (usually again a monitoring action).

The more significant the (initial) cracking of a structure, the more the recommended frequency for measurement is high. An increase of the frequency of the recommended measurements (*e.g.*, once every 3 months) is justified by the damage of the structure and also the risk of a rapid evolution. However, it should be kept in mind that even if the periodicity retained in this case is 3 months, a reliable interpretation of the measurement results can only be carried out at the end of at least one year period to take into account the seasonal variations. The evolution of deformation and cracking data over at least the period of one year needs to be analyzed. An example of how this can be done, inspired by the methodology used at LCPC (2003) is reported by Godart *et al.* (2004) and is described for two cases a) and b), distinguished by the importance of the width of isolated cracks:

a)   *The reinforced concrete structure **does not** present isolated cracks with significant opening*:

- If the evolution of the Cracking Index (CI) or Deformation measurement (Dm) is > 0.5 mm/m/year, the AAR is still progressive and it is advisable to continue the monitoring with a 3 to 12 month frequency; the higher the evolution the higher the frequency.
- If the evolution of CI or Dm is < 0.5 mm/m/year, the decision is based on the Cracking Index when first measured (CI_0):

   - If CI_0 > 1 mm/year, the structure is considered to have significant cracking, but its evolution is slow, even stabilizing. When the structure in its current state does not prevent any problem of load bearing capacity, nor risk with respect to safety of the users (*e.g.*, by falling of concrete fragments), the forecast of the evolution is rather favourable. The evolution is to be confirmed by programming for new monitoring measurements in the remote term (about 3 years).
   - If CI_0 < 1 mm/m, the structure is considered to have little damage and its evolution is slow. The forecast is also favourable. Specific monitoring frequency can be suspended, but can be carried out within the normal routine maintenance inspections. Specific monitoring frequency can be applied again when routine maintenance data reveals an unfavourable evolution of the visual aspect of the structure.

b) *The structure presents isolated cracks with significant opening*
The criterion on the evolution of CI and Dm is as previously defined, **and** is supplemented by a criterion on the evolution of the opening of the cracks (ΔF):

- If $\Delta F > 0.2$ mm/year, the monitoring should be pursued with a 3 to 12 month frequency.
- If $\Delta F < 0.2$ mm/year, the monitoring can be suspended to programming in the remote term (about 3 years).

For all of the above guidelines, it should be noted that these values are given for normally reinforced concrete structures. They have to be adapted to smaller evolution values and higher monitoring frequencies for prestressed concrete structures.

5.6.5 Forecast of the evolution

The result of the on-site monitoring programme of the structure, but also the residual expansion of the concrete based on cores taken from the structure as described in section 5.4.3.2, all help to establish a forecast on the evolution of the AAR within the structure. The actions decided by the asset manager depend on the results of the tests and the extrapolated forecast. In general these actions will have an effect on the frequency of the monitoring programme, as discussed in the previous section. When the results and/or forecast for a structure, of which a structural part is already strongly affected, could lead quickly to a doubt about its load carrying capacity, it is strongly recommended to proceed to specific recalculation, taking into account the effects of the concrete expansion. As the recalculation is based on chemo-mechanical models coupled to hydrous and temperature diffusion models introduced in a finite element computational software, such calculations are to be carried out by experts. Its implementation requires knowledge of the climatic data history of the structure (temperature, hygrometry), the results of the measured expansion of the structure, and the results of the residual expansion tests on cores.

Further repair actions, limitations to allowed load, or even demolition could be necessary next steps for the asset manager, based on the outcome of the forecast evaluation and computer modelling.

REFERENCES

ASTM (2008) *Standard test method for determination of length change of concrete due to alkali-silica reaction*, C1293-08. West Conshohocken, PA, USA, American Society for Testing and Materials.

Bérubé, M., Smaoui, N., & Côté, T. (2004) Expansion tests on cores from ASR affected structures. In: Tang Migshu & Deng Min (eds.) *Alkali-Aggregate Reaction in Concrete, Proceedings of the 12th International Conference on AAR*, Beijing: International Academic Publishers, Vol II, pp. 821–832.

Bérubé, M., Smaoui, N., Fournier, B., Bissonnette, B., & Durand, B. (2005) Evaluation of the expansion attained to date by ASR-affected concrete – Part III: Application to existing structures. *Can J Civil Eng., 32*, 463–479.

Charlwood, R.G. (2009) AAR in Dams and Hydroelectric Plants. Short Course on Management of Alkali-Aggregate Affected Structures: Analysis, Performance & Prediction, Coyne & Bellier, Paris, France, September 15.

Chrisp, T., Waldron, P., & Wood, J.G.M. (1993) The development of a test to quantify damage in deteriorated concrete. *Mag Concrete Res.*, 247–256.

Chrisp, T., Wood, J.G.M., & Norris, P. (1989) Towards quantification of microstructural damage in AAR deteriorated concrete. In: *Proceedings of the International Conference on Recent Developments on the Fracture of Concrete and Rocks, Elsevier Applied Science,* London, pp. 419–427.

Crouch, R. & Wood, J.G.M. (1988) Damage evolution in AAR affected concretes. In: *Proceedings of the International Conference on Fracture and Damage of Concrete and Rock,* Vienna, Austria, 4–6 July 1988.

CSA (2000) *Guide to the Evaluation and Management of Concrete Structures Affected by Alkali-Aggregate Reaction. General Instruction No.1,* CSA A864-00, Canadian Standards Association, Mississauga, Ontario, Canada.

CUR (2005) *Inspection and assessment of concrete structures attaccked or suspected to be attacked by ASR,* Recommendation 102, CUR, Gouda, The Netherlands.

Eriksen, K., Jansson, J., & Geiker, M. (2008) Assessment of concrete bridge decks with ASR. In: *Proceedings of the 13th ICAAR,* Trondheim, Norway.

Fernandes, I., Silva, A.S., Gomes J.P., Castro A.T. & Noronha F. (2008) Characterization of AAR in Fagilde Dam. In: *Proceedings of the 13th ICAAR,* Trondheim, Norway.

Folliard, K.J., Thomas, M.D.A., Fournier, B., Resendez, Y., Drimalas, T., & Bentivegna, A. (2012) Evaluation of mitigation measures applied to ASR-Affected Concrete elements: Preliminary findings from Austin, TX exposure site. In: *Proceedings of the 14th International Conference on Alkali-Aggregate Reaction in Concrete,* Austin, USA.

Fournier, B., Bérubé, M-A., Folliard, K., & Thomas, M.D.A. (2010) Report on the Diagnosis, Prognosis, and Mitigation of Alkali-Silica Reaction (ASR) in Transportation Structures, Federal Highway Administration, Washington, DC.

Godart, B. (2009) Méthodes de suivi dimensionnel et de suivi de la fissuration des structures: Avec application aux structures atteintes de réaction de gonflement interne du béton, Techniques et méthodes des laboratoires des ponts et chaussées, Guide technique du LCPC, 60p.

Godart, B., de Rooij, M., & Wood, J.G.M. (2013) *Guide to Diagnosis and Appraisal of AAR.* Springer, RILEM.

Godart, B., Mahut, B., & Fasseu, M. (2004) A guide for aiding to the management of structures damaged by concrete expansion in France. In: Tang Mingshu & Deng Min (eds.) *Alkali-Aggregate Reaction in Concrete, Proceedings of the 12th ICAAR, Beijing, International Academic Publishers, China,* 15–19 October, Vol II. pp. 1219–1228.

Grattan-Bellew, P. (1995) Laboratory evaluation of alkali-silica reaction in concrete from Saunders Generator Station. *ACI Mater J,* 92 (2), 126–134.

Grattan-Bellew, P., & Danay, A. (1992) Comparison of laboratory and field evaluation of AAR in large dams. In: *Proceedings of the International Conference on Concrete AAR in Hydroelectric Plants and Dams,* Frederickton, New Brunswick, Canadian Electrical Association.

Haugen, M., Skjolsvold O., Lindgard J. & Wigum B.J. (2004) Cost effective method for determination of aggregate composition in concrete structures. In: *Proceedings of the 12th International Conference on Alkali-Aggregate Reaction in Concrete,* Beijing, International Academic Publishers. pp. 907–911.

Institution of Structural Engineers (1992, including addendum 2010) *Structural effects of alkali-silica reaction.* Technical guidance on the appraisal of existing structures, ISE, London.

Jensen, V. (2000) In situ measurements of relative humidity and expansion of cracks in structures damaged by AAR. In: M. A. Bérubé, B. Fournier & B. Durand (eds.) *Proceedings of the 11th*

Internationaal Conference on Alkali-Aggregate Reaction in Concrete, Quebec City, Quebec, Canada.

Jensen, V. (2006) Surface protection on ASR damaged concrete. Research and state of the art from a postdoctoral project. In: Benoit Fournier (ed.), *Proceedings of the 8th CANMET International Conference on Recent Advances in Concrete Technology, 31 May to 3 June 2006, Marc-Andre' Berube Symposium on Alkali-Aggregate Reactivity in Concrete*, Montreal, Canada.

Juliani, M., Becocci, L., & Carrazedo, R. (2008) Detection of Alkali-Aggregate Reaction in Guarulhos International Airport Aircraft Apron: Methodology. In: Maarten A.T.M. Broekmans & Borge J. Wigum (eds.) *Proceedings of the 13th ICAAR*, Trondheim, Norway.

Katayama, T. (2004) How to identify carbonate rock reactions in concrete. *Mater Charact.*, *53*, 85–104.

Katayama, T. & Futagawa, T. (1997) Petrography of pop-out causing minerals and rock aggregates in concrete – Japanese experience. In: *Proceedings of the 6th Euroseminar on Microscopy Applied to Building Material*, Reykjavik, Iceland.

Katayama, T., Tagami, M., Sarai, Y., Izumi, S., & Hira, T. (2004) Alkali-aggregate reaction under the influence of deicing salts in the Hokuriku district, Japan. *Mater Charact*, *53*, 105–122.

Larbi, J., & Visser, J. (1999) Diagnosis of chemical attack of concrete structures: The role of microscopy. In: Hans S. Pietersen, Joe A. Larbi, Hans H. A. Janssen Eds., *Proceedings of the 7th Euroseminar on Microscopy Applied to Building Materials*. Delft, The Netherlands, Delft University of Technology, *29 June – 2 July, 1999*.

LCPC (1993) *Essai de mise en évidence du gel d'alcali-réaction par fluorescence des ions uranyl*. Method d'essai LCPC No. 36, Ministère de l'Équipement, du Logement, des Transport et du Tourisme, Paris.

LCPC (1997a) *Détermination de l'indice de fissuration d'un parement de béton*. Methode d'essai LCPC No. 47, Ministère de l'Équipement, du Logement, des Transport et du Tourisme, Paris.

LCPC (1997b) *Évaluation de la teneur en alcalins équivalents actifs dans les ciments*. Methode d'essai LPC No. 48, Ministère de l'Équipement, du Logement, des Transport et du Tourism, Paris.

LCPC (2003) *Aide à la gestion des ouvrages atteints de réaction de gonflement interne. (Aid for the management of structures damaged by an internal swelling reaction)*, Guide technique du Laboratoire Central des Ponts et Chaussées, Paris (in French).

Léger, P., Tinawi, R., & Mounzer, N. (1995) Numerical simulation of concrete expansion in concrete dams affected by alkali aggregate reaction: State-of-the-Art. *Can J Civil Eng.*, *22* (4), 692–713.

Multon, S., Barin, F.X., Godart, B., & Toutlemonde, F. (2007) Estimation of the residual expansion of concrete affected by alkali-silica reaction. *J Mater Civil Eng.*, *20* (*Special Issue: Durability and Service Life of Concrete Structures: Recent Advances*), 54–62.

Neville, A. (2003) How closely can we determine the water/cement ratio of hardened concrete? *Mater Struct.*, *36*, 311–318.

Norris, P., Wood, J.G.M., Barr, B. (1990) Torsion test to evaluate deterioration of concrete due to AAR. *Mag Concrete Res.*, *42* (153), 239–244.

Omikrine-Metalssi, O., Le, V.D., Seignol, J.F., Rigobert, S., Humbert, P., & Toutlemonde, F. (2010) Integration of contact elements in RGIB-module of the finite element software "CESAR-LCPC": Application to concrete structures affected by internal swelling reactions. In: Oh, B.H. *et al.* (eds.) *Fracture Mechanics of Concrete and Concrete Structures – Recent Advances in Fracture Mechanics of Concrete*, © Korea Concrete Institute, Seoul, ISBN 978-89-5708-180-8, pp. 568–575.

Poole, A.B. & Sims, I. (2015) *Concrete petrography: A handbook of investigative techniques*, 2nd edition. London, CRC Press (Taylor & Francis). 794p.

Rodum, E., Lindgård, J., Stemland, H., Haugen, M. & Skjolsvold O. (2008) Early damages due to AAR in a swimming pool – Diagnosis and rate of expansion. In: Maarten A.T.M. Broekmans & Borge J. Wigum (eds.), *Proceedings of the 13th ICAAR*, Trondheim, Norway.

Seignol, J-F., Boldea, L-I., Leroy, R., Godart, B., & Hammerschlag, J-G. (2009) Hydro-power structure affected by alkali-aggregate reaction: A case-study involving numerical re-assessment. In: *Proceedings, First International Conference on Computational Technologies in Concrete Structures*, Jeju, Korea, May.

Sergi, G. & Page, C. (1992) *The effects of cathodic protection on alkali-silica reaction in reinforced concrete*, TRL Report CR310, 53p.

Sergi, G. & Page, C. (1993) *The effects of cathodic protection on alkali-silica reaction in reinforced concrete: Stage 2*, TRL Report PR62, September, 22p.

Siemes, T. & Gulikers, J. (2000) Monitoring of reinforced concrete structures affected by alkali-silica reaction. In: *Proceedings of the 11th International Conference on Alkali-Aggregate Reaction in Concrete*, Quebec City, Quebec, Canada.

Siemes, A., Han, N., & Visser, J. (2002) Unexpected low tensile strength in concrete structures, *Cement*, 7, 475–482.

Silva, A., Gonçalves, A. Pipa, M. (2008) Diagnosis and prognosis of Portuguese concrete railway sleepers degradation – a combination of ASR and DEF. In: Maarten A.T.M. Broekmans & Borge J. Wigum (eds.) *Proceedings of the 13th ICAAR*, Trondheim, Norway.

Smaoui, N., Bérubé, M-A., Fournier, B., Bisonette, B., & Durand, B. (2004) Evaluation of expansion obtained to date by ASR-affected concrete. In: *International Proceedings of the 12th International Conference on Alkali-Aggregate Reaction in Concrete*, Beijing.

Smaoui, N., Bérubé, M-A., Fournier, B., Bissonnette, B., & Durand, B. (2004) Evaluation of the expansion attained to date by ASR-affected concrete – Part I: Experimental Study, *Can J Civil Eng.*, 31, 826–845.

St. John, D.A., Poole, A.B., & Sims, I. (1998) *Concrete Petrography: A handbook of investigative techniques*, Edward Arnold & John Wiley, London and New York. See Poole & Sims 2015 for 2nd edition.

Stark, D. (1990) *The moisture condition of field concrete exhibiting alkali-silica reactivity*. Halifax: CANMET.

Stark, D. & Depuy, G. (1987) Alkali-Silica Reaction in five dams in southwestern United States. In: *Proceedings of the Katherine and Bryant Mather International Conference on Concrete Durability*, Atlanta, USA.

Ulm, F.J., Coussy, O., Li, K., & Larive, C. (2000) Thermo-Chemo-Mechanics of ASR-expansion in concrete structures. *J Eng Mech-asce.*, *ASCE*, 126 (3), 233–242.

Wood, J., Grantham, M., & Wait, S. (2008) Tuckton Bridge, Bournemouth. An investigation of Condition after 102 years. In: *Proceedings Concrete Platform 2007*, Belfast, Northern Ireland, UK, Ed. Queen's University of Belfast, 19–20 April 2007.

Wood, J., Nixon, P., & Livesey, P. (1996) Relating ASR structural damage to concrete composition and environment. In: Ahmad Shayan (ed.) *Proceedings of the 10th International Conference on Alkali-Aggregate Reaction in Concrete*, Melbourne, Australia, 19–23 August 1996.

Wood, J. 2011. Personal communication with M.R. de Rooij based on internal document 'Water Supply Expansion Testing of Concrete Cores at Laboratory Ambient Temperatures', Structural Studies & Design Ltd., Surrey, UK, 1997.

Wood, J.G.M., & Angus, E.C. (1995) Montrose Bridge: Inspection, assessment and remedial work to a 65-year-old bridge with AAR, In: M.C. Forde (ed.), *Proceedings of the 6th International Conference on Structural Faults and Repair 1*, London, Engineering Technics Press.

Chapter 6

United Kingdom and Ireland

Ian Sims

This chapter was originally completed in the early 1990s, when the UK was only starting to come to terms with alkali-silica reaction ('ASR'). In the quarter century since, the UK has grown accustomed to living with the risk of ASR and largely developed confidence in its national measures for minimising that risk. Meanwhile, as this book demonstrates, the experience and understanding of alkali-aggregate reaction ('AAR') worldwide has developed greatly, benefiting where applicable from some of the earlier UK findings, but also encountering significant variety. The early sections of this chapter continue to chart the story of the realisation of ASR in the UK, whilst all the later sections are brought up to date. A new separate section (6.9) explains the slightly different approach to ASR in the Republic of Ireland. Finally, although British and Irish engineers and scientists have encountered apparently reactive carbonate aggregates overseas, there have been no UK or Irish examples, apart from the silicified components of limestone, so that this chapter concentrates only on ASR (see Chapter 3 for a discussion on the 'so-called alkali-carbonate reaction').

6.1 INTRODUCTION

When ASR was discovered by Stanton in America in 1940 (a & b), Britain was otherwise engaged. During the war years and to the end of the 1940s, American researchers established most of the basic parameters of ASR and its consequences (Tuthill, 1982). The first ASR tests on British aggregates were carried out in the USA in 1947 (see 6.3.2). In their 13th Annual Report, the Cement and Concrete Association (1948) in the UK announced an initial research programme on 'tests to determine whether the expansive reaction between aggregates and cement so often reported from the USA is a possible cause of disintegration in Britain'. It is not clear to what extent that programme was ever pursued at that time, but certainly a major programme of research had commenced at the Building Research Station (BRS) in 1946, under the supervision of F.E. Jones, and continued through most of the 1950s. The preliminary findings, as National Building Studies (NBS) Research Papers (RP), were published in 1952 (Jones, 1952a, 1952b; Jones & Tarleton, 1952) and the final conclusions and recommendations, also as NBS Research Papers, were issued in 1958 (Jones & Tarleton, 1958a, 1958b).

The reassurance provided by the BRS/NBS Research Papers, which was arguably more apparent than real (see 6.3.3), led to a decade of near complacency, when the UK concrete industry was generally content to regard ASR as 'a foreign problem'. In January 1971, however, a mass concrete dam in Jersey, Channel Islands, was found to exhibit displacements and cracking which were in due course attributed to ASR (Cole & Horswill, 1988), and this finding reactivated British awareness of ASR. The reawakening was rapid, with the Cement and Concrete Association and Queen Mary College (University of London) jointly hosting the Third International Conference on ASR in September 1976 (Poole, 1976). In December 1976, unexplained cracking was found to be affecting concrete bases at an electricity substation in Plymouth, Devon, and by early 1977 this was known to be the result of ASR (Sandberg, 1977). Over the years since that discovery, hundreds, possibly thousands, of concrete structures have been identified on the UK mainland that are affected, to a greater or lesser extent, by ASR.

Like a disease for which the cure is uncertain, the alarm and concern generated by ASR was generally out of proportion to the low incidence of occurrence and the typically limited scale of structural damage involved. However, the degree of interest in ASR enabled good progress to be made in the UK in formulating measures to minimise the risk of ASR in new concrete work (Hawkins, 1983; Concrete Society, 1987), in establishing agreed procedures for the diagnosis of ASR (British Cement Association, 1988) and in providing guidance for the structural assessment of affected structures (Institution of Structural Engineers, 1988). When the 50th anniversary of the discovery of ASR was reached in 1990, it seemed less likely that future UK concrete would suffer ASR and that even already affected structures would mostly prove to be repairable. These expectations would largely be justified, but not without complications and new discoveries along the way.

6.2 INITIAL REASSURANCE: STUDIES AT THE BRS

6.2.1 1952 – review and initial findings

The BRS research (Jones, 1952a, 1952b; Jones & Tarleton, 1952) commenced with a detailed review of the experiences to date in the USA, including references to specific structures, an assessment of test methods and an appreciation of 'corrective' measures to avoid ASR in new works (Jones, 1952a). In his prefatory note to this RP 14, F.M. Lea, then Director of Building Research, warned: *"In the USA, many serious and extensive failures of concrete have since (1940) been attributed to alkali-aggregate reaction. Fortunately, no evidence has yet been obtained of large-scale failures in this country arising from this cause; nevertheless, some aggregate deposits which may be reactive in some degree do occur, and unfamiliar aggregates may need examination before use"*.

The second NBS Research Paper, No. 15 (Jones, 1952b), provided a general preliminary consideration of British Portland cements and British aggregates. The total alkali content of British cements, as analysed at the BRS since 1928, was found to range from 0.5 to 1.0 % $Na_2Oeq.$ and to be typically higher than the 0.6 % maximum level suggested by American experience for preventing ASR. Furthermore, extraction tests indicated that all of the acid-soluble Na_2O and K_2O in British cements would become available for reaction in concrete. In British aggregates, flint in the south-east England gravels was recognised to be 'slightly reactive', probably as a result of microstructure rather than the presence of opaline silica (Midgley, 1951) and *"the acknowledged freedom from trouble in practice"* was thought to be *"associated with the use of whole flint aggregate"*. Thus, even in 1952, the potential importance of the flint 'pessimum' was understood and led to the cautionary statement in RP 15: *"There is a possibility that if flints are used in admixtures with inert aggregates, sufficient expansion to cause trouble might occur"*. 'Malmstone' (an opaline sandstone from the Upper Greensand [Cretaceous] of south-eastern England) and some glassy acid to intermediate volcanic rocks were also identified as potentially reactive materials, but these were not considered to be important sources of aggregates in the UK (Malmstone is used for building and paving stone).

The expansion bar test (a mortar-bar test), which had been identified in the review as seeming to be 'the most generally useful' procedure currently available, was then applied to a limited range of British aggregates using a cement of 'medium' alkali content (0.7 % $Na_2Oeq.$) and employing storage at 20 °C. The results up to 4 years were given in NBS Research Paper No. 17 (Jones & Tarleton, 1952), where the interpretation was based upon comparison with the results obtained in the same test series for two opal-bearing aggregates imported from the USA (Californian siliceous magnesian limestone – 10 % in Thames Valley flint sand - and Nebraska sand). The American 'control' samples gave results of 0.38 % and 0.03 % respectively at 6 months, with 0.72 % and 0.18 % at 4 years, whereas the highest value obtained for any of the British natural aggregates was not more than 0.03 % up to 4 years (Table 6.1 and Figure 6.1). This seemingly hopeful outcome was confirmed by the statement that *"the conclusions regarding the rock types concerned are reassuring for aggregates in*

Table 6.1 Selected mortar-bar results (percentage expansion) from BRS/NBS Research Papers 17 & 20 (Jones & Tarleton, 1952, 1958a). All results are for 'low quality' mortar-bars (higher water/ cement ratios and greater porosities) stored at 20 °C.

Aggregate	Cement type (alkali content)							
	'Medium' alkali (0.7% Na_2O eq.) Test age (years)				'High' alkali (1.2% Na_2O eq.) Test age (years)			
	1	3	4	7	1	3	4	7
American 'control' aggregates								
Siliceous magnesian limestone (10% in Thames Valley sand)	0.51	0.68	0.72	–	0.71	0.72	0.74	0.74
Nebraska Sand	0.10	0.18	0.18	–	0.16	0.25	0.27	0.30
British natural/crushed rock aggregates								
Clee Hill basalt	0.02	0.02	0.02	–	0.02	0.03	0.03	0.03
Quartzite	0.02	0.02	0.02	–	0.02	0.03	0.03	0.03
Carboniferous limestone	0.01	0.02	0.02	–	0.02	0.02	0.02	0.03
Darley Dale sandstone	0.02	0.03	0.03	–	0.03	0.03	0.04	0.04
Thames Valley flint sand	0.01	0.01	0.01	–	0.02	0.03	0.03	0.03
Cambridge flint gravel	–	–	–	–	0.02	–	0.02	0.02
Cambridge flint sand	–	–	–	–	0.02	–	0.05	0.06
Rhyolite (50% in Clee Hill basalt	–	–	–	–	0.02	0.02	–	–
Andesite (50% in Clee Hill basalt)	–	–	–	–	0.02	0.03	–	–

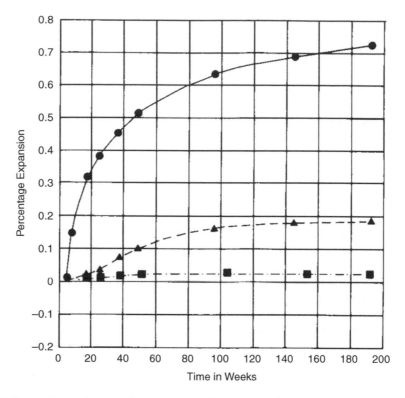

Figure 6.1 Comparison of mortar-bar expansions for reactive American 'control' aggregates and Thames Valley flint sand: ■ Thames Valley sand; ▲ Nebraska sand; ● siliceous magnesian limestone 10 % replacement in graded river sand. All results are for 'medium' alkali cement (0.7 % Na_2Oeq.), 'low-quality' mortar-bars (higher water/cement ratios and greater porosities) stored at 20 °C. Redrawn and adapted from Jones & Tarleton (1952). See data in Table 6.1.

this country", although this view was conditioned by the need to be mindful of the presence of any opaline silica.

With the benefit of hindsight, it is perhaps significant that low-power microscopical examination of broken mortar-bar surfaces after testing identified small amounts of alkali-silica gel deposits associated with the Thames Valley flint sand sample, similar to those found occurring within the 'control' specimens made with American aggregates. This would seem to indicate that ASR had occurred with Thames Valley flint sand, but that the tests had not produced expansions. However, this observation was not emphasised and the impression left was that British aggregates were not expansively reactive, in contrast to some of those found overseas.

6.2.2   1958 – conclusions and recommendations

The series of experiments described in RP 17 (see 6.2.1) were repeated using 'high-alkali' cement, actually the 0.7 % Na_2Oeq. cement with added NaOH to reach an equivalent 1.2 %, and the findings reported in Research Paper No. 20 (Jones &

Tarleton, 1958a). Also, further British aggregates were tested, including representatives of those materials initially identified in RP 14 (Jones, 1952a) as being potentially reactive: flint, 'Malmstone' and acid to intermediate volcanic rocks. Some limited studies were carried out into aggregate mixtures and 'dilutions'.

In the 'high-alkali' cement repeat expansion bar tests, the American control aggregate exhibited higher expansions than were recorded for the earlier 'medium alkali' cement series, especially at 6 months (Table 6.1). After 4 years, none of the British natural aggregates exhibited any substantial expansion, the highest value obtained being only 0.035 %. An extension of testing to 7 years did not significantly alter the conclusion, with the highest value moving up only to 0.042 %.

Tests were carried out on additional flint aggregates from four sources, two in the Thames Valley, one in Essex and another in Cambridgeshire. None of the flint gravels exceeded 0.02 % in these tests, even after 7 years, but the single flint-bearing sand (composed of quartz and flint) exhibited rather more but still comparatively low expansivity, reaching 0.05 % after 4 years and 0.06 % after 7 years. Once again the expansion bar specimens made with flint aggregates exhibited some evidence of ASR, in the form of gel deposits visible to low-power microscopical examination after testing.

The inherent reactivity of flint, notwithstanding the lack of expansion in these tests, was demonstrated by some chemical tests reported in RP 20 (Jones & Tarleton, 1958a): the pat test and rapid chemical test described in RP 25 (Jones & Tarleton, 1958b, see below). A series of 'dilutions' using Clee Hill basalt (considered inert) with Thames Valley flint aggregate did not produce any significant expansions in the tests at 20 °C, although earlier work in the USA had suggested that at least some British flint could be induced to cause expansion, up to 0.14 % at 12 months, when present as around 20 % of the total aggregate, at least at the elevated temperature of 38 °C (McConnell *et al.*, 1947).

A range of acid to intermediate volcanic rock materials, including rhyolites and andesites, was tested using the same regime, variously as 'whole aggregate' or as 50:50 mixtures with Clee Hill basalt (considered inert). None of the aggregates or combinations tested produced any expansions greater than 0.03 %, even after 3 years. Chemical tests also showed a 'complete lack of reactivity' for these volcanic rocks.

Malmstone was subjected to extensive testing as it had been found to contain opaline silica, although it was not in practice a source of aggregate for normal concrete. As expected, the chemical tests indicated considerable reactivity of Malmstone, but the expansion bar tests produced no excessive movements, even in various 'dilutions' with flint sand and Clee Hill basalt. This apparent anomaly was explained in RP 20 (Jones & Tarleton, 1958a) as being caused by the highly porous nature of Malmstone, rather than by any inherent inadequacy in the expansion bar test procedure.

Jones and Tarleton (1958a) considered that the findings reported in RP 20 confirmed those given 6 years earlier in RP 17 (Jones & Tarleton, 1952) 'on a firmer and wider basis', so that it could be concluded that *"the normal British aggregates so far tested, when used as whole aggregates, are not expansively reactive with high-alkali cements at normal temperatures"*. Crucially, however, they added some words of caution in respect of flint aggregates, perhaps partly because of the observation of evidence of reaction in the absence of harmful expansion in the tests: *"There remains the possibility therefore that some flints may be encountered which, under adverse conditions of 'dilution', alkali content, water content and temperature, may cause trouble. So far as present evidence goes, however, this is considered to be rather unlikely"*.

The final NBS Research Paper, No. 25 (Jones & Tarleton, 1958b), comprised two parts, both concerned primarily with test procedures and, as such, accessory to the substantive results and conclusions of RP 20 (Jones & Tarleton, 1958a), but nevertheless influential over future British testing practice. In the first part (actually part V of the NBS series on alkali-aggregate reaction), the apparent deterioration in bending strength of mortar-bars was found to be 'useful supplementary evidence', but was not considered to be a preferable alternative to expansion testing. Also, the observation of gel deposits either within or on the outer surfaces of mortar-bars was thought to provide 'at least valuable evidence of potential expansive reactivity', although it was conceded that such observations had not always been accompanied by expansion in the BRS test programme.

The second part of RP 25 (actually part VI of the NBS series) discussed the test methods used at the BRS and gave recommended test procedures. The pat test (or gel pat test), developed from a method earlier described by Stanton *et al.* (1942), was recommended as 'a useful rapid test for reactive material' and was perhaps used subsequently more in the UK than elsewhere. A rapid chemical test was recommended as 'a useful acceptance test', albeit a modified version of the ASTM C289 (1987) method, and a maximum dissolved silica – alkalinity reduction ratio of 1.5 was suggested, in contrast to the maximum ratio of 1 given in ASTM C33 (1986).

Finally, the expansion bar test was recommended as *"the standard reliable test for determining the degree of expansive reactivity of an aggregate in various combinations with cements"*. A BRS method for the expansion bar test was described, which differed in several important details from the ASTM C227 (1987) method; the cement alkali content was standardised at 1.2 % $Na_2Oeq.$, the aggregate/cement ratio was 2 rather than 1.25 and the preferred storage temperature was 20 °C rather than 38 °C. Using this BRS test, the aggregate was to be classed as 'expansively reactive' if measured expansion exceeded 0.05 % at 6 months or 0.1 % at 12 months, in contrast to the revised ASTM C33 similar limits at 3 and 6 months, respectively. Some mention was made of the possibility of applying the expansion test to concrete specimens, but it would be more than 20 years before such a procedure was developed at the BRS (which became the BRE).

6.2.3 Significance and effect

At the time that Dr Jones was finalising his BRS research programme, there were no known examples of ASR affecting structures in the UK, and this will inevitably have been regarded as an important factor bearing upon the emphasis of the conclusions to be drawn from the laboratory results. The laboratory findings were seemingly consistent with the experience in practice, that is, British natural aggregates were 'not expansively reactive'. To their considerable credit, Jones and Tarleton recognised that flint aggregates, in particular, had at least a hypothetical potential for expansive reaction in certain circumstances, notwithstanding the lack of expansion produced for flint aggregates in the BRS tests. However, the prophetic cautions contained within the conclusions to the BRS work in respect of British flint aggregates were generally superseded in the readers' appreciation by the overall reassurance which appeared to emerge from an apparently thorough survey.

Petrographically, flint (a distinctive chert of Upper Cretaceous age) is a potentially alkali-reactive form of silica, even without the presence of any opaline material. As early as 1947, flint from Dover cliffs had been shown to be expansively reactive in American tests when forming around 20 % of an otherwise non-reactive total aggregate (McConnell *et al.*, 1947). In the BRS test programme, flint aggregates had shown clear evidence of ASR within expansion bar specimens, as well as in separate chemical tests, although no significant expansions had been recorded. For some reason, the expansion testing approach, modelling normal concrete with small mortar specimens of high cement content, appears to have been regarded as beyond reproach, despite its failure to cause expansion even with opal-bearing 'Malmstone' material. Subsequent events have shown justification for Jones and Tarleton's cautious appraisal of British flint aggregates, but it was to be nearly 20 years before the UK concrete industry would accept that ASR was anything other than a 'foreign problem'. Later work on UK flint aggregates will be described in subsection 6.5.3.

6.3 FALSE SECURITY: COMPLACENCY IN THE 1960s

6.3.1 British Standards

The British Standards Institution ('BSI') is an independent chartered body devoted to the assurance of uniformity and quality of materials and products in the UK. Published British Standards ('BS's) are not legally binding, except when they are called up by a contract specification or similar document, but they are accepted throughout the construction industry as being dependable and authoritative guidance. The constitution of BSI ensures that all published standards are compiled by representative committees and that they are regularly reviewed and amended or revised when necessitated by new knowledge, recent developments or changed circumstances. BSs, including Codes of Practice ('CP's), are therefore important and fairly reliable indicators of the contemporary views within the construction industry at the times of their publication, although they do represent a consensus and may not adequately cover particularly contentious matters. In more recent years and increasingly, purely British Standards have now largely been superseded by European Standards (ENs, published in the UK as 'BS EN's), which have a different legal status in relation to European law, but the discussion in this section relates to BSs and CPs.

The materials used in the manufacture of concrete were covered in the 1960s, and until the 1990s, by BS Specifications, notably BS 12 (1978) for Portland cement (ordinary and rapid-hardening), BS 882 (1983) for aggregates (natural) and BS 3148 (1980) for the mix water. Test methods were given for cement in BS 12 (1978) and BS 4550 (1970, 1978) and for aggregates in BS 812 (1984, 1985). Guidance for the making and use of concrete for structures was given in CP 114 (1969), CP 115 (1969) and CP 116 (1969), later superseded by CP 110 (1972), and in turn by BS 8110 (1985), whilst test methods for concrete were given in BS 1881 (1983, 1986). With one small exception, none of these BSI publications made any direct or allusive reference to ASR until the 1980s, and even then only BS 8110 contained any substantive guidance. The single exception was BS 3148 (1980), in which it was suggested (in an appendix) that the alkali content of some mix waters might be of significance in the presence of reactive aggregates and cement, although it was further stated that, *"So far, no naturally occurring expansively reactive aggregates have been found in the United Kingdom"*.

The 1954 edition of BS 882, the specification for natural aggregates for concrete, was extensively revised and a new edition published in 1965. A comparison of the 1954 and 1965 editions shows that the standard was rearranged and, in places, completely rewritten. The final results of the BRS research programme had been published in 1958 (see 6.2.2) and the findings must have been familiar to all members of the revising committee; indeed, although the full membership of that committee is no longer available, it is known that the BRS was represented by Teychenné (1987). Therefore, it seems certain that a conscious decision was taken by the BSI committee to omit any reference to ASR in the concrete aggregates specification. One surviving member of that committee has indicated that the matter was discussed and that, because BRS research had only recently concluded that ASR was unlikely to be a problem with British aggregates and no actual examples had been reported, it was considered inappropriate or even unhelpful to raise concern over ASR in the national specification (Teychenné, 1987). Even in the 1983 revision of BS 882, mention of ASR is restricted to a non-mandatory note within an appendix, and that is only a cross-reference to the Code of Practice, BS 8110.

From the end of the Second World War to the discovery after 1971 that ASR had damaged the Val de la Mare Dam in Jersey, and by 1977 that concrete on the UK mainland was also affected, there was virtually no reference to the possibility of ASR in British Standards relating to concrete and concrete materials. Therefore, any supplier or user dependent upon the provisions of the relevant British Standards for assuring the quality of concrete would not have been caused to make any allowance whatever for the possibility, however remote, of ASR occurring at some time after construction. It has been argued, albeit retrospectively, that the 'broad-brush' quality requirements of BS 882 (1983), including the provision that *"aggregates...shall not contain deleterious materials in such a form or in sufficient quantity to affect adversely the strength at any age or the durability of the concrete"*, could be considered to have embraced ASR, even though ASR was not one of the specific examples of 'such deleterious materials' listed in the standard to clarify the above-quoted requirement. The validity of such an argument depends upon the extent to which alternative authoritative guidance, other than BSI documents, was available in the UK to advise suppliers and users of the possibility of ASR.

6.3.2 Other authoritative guidance and advice

In the UK, apart from BSI, authoritative national guidance on construction materials and practice is available from the Building Research Station (later the Building Research Establishment, and now just 'BRE') and, where appropriate, from the Cement and Concrete Association ('C&CA', now renamed the British Cement Association, 'BCA') and the Transport and Road Research Laboratory (TRRL, now just 'TRL'). Occasionally similarly authoritative and useful guidance may be produced on specific topics by the relevant professional institutions, learned societies or even trade associations. Academic publications and certain textbooks can also provide useful guidance, but, by their nature, are not necessarily representative of the consensus viewpoint.

After the publication of Jones and Tarleton's final NBS reports in 1958 (see 6.2.2), the BRS (BRE) did not produce any further official guidance until 1982, when Digest 258 was issued (BRE, 1982), some 6 years after the identification of ASR on the UK mainland. However, BRE researchers had been involved in the investigation into the Val de La Mare Dam in Jersey (see 6.4) and Midgley (1976) described a technique arising from this investigation at the 3rd International ASR conference in London. The next publication on ASR by BRE staff did not appear until Gutt and Nixon (1979). It is thus apparent that for the whole of the 1960s, and also for a good part of the 1970s, the seeming reassurance provided by the earlier 1946–1958 BRS research programme remained unchallenged. Indeed, BRE Digest 126 (1971), on the subject of concrete appearance, contained the following statement: *"Fortunately, there are no known deposits of aggregate in the United Kingdom that have any significant degree of alkali--aggregate reactivity, but a designer of concrete buildings for other parts of the world might run into serious difficulties if he does not get advice on the suitability of the available aggregates"* (the present author's underlining emphasis). Clearly, ASR was still to be regarded as 'a foreign problem'.

Although the C&CA declared an intention to carry out research into ASR in 1948, there is no available evidence that any such work was carried out, and the earliest C&CA publication on the subject appeared as late as 1977 (Palmer, 1977), with their first research work appearing a year later (Hobbs, 1978). A C&CA advisory note on the subject of aggregate 'impurities' was published in 1970 (Keen, 1970) and made some reference to 'alkali-reactive minerals', but concluded confidently, *"In this country, these minerals are fortunately not present in the aggregates used for making concrete but, in any event, commercially available cements seldom contain enough alkalis to be troublesome. In practice then, the question of aggregate-alkali* (sic) *reaction concerns only cements and aggregates used abroad'* (present author's underlining emphasis). Roeder (1975) carried out some limited research at C&CA into flint reactivity as part of a wider study in 1974–75, but the results were not formally published and he was careful to state that, *"More recently interest has centred on the possibility of reaction between the aggregate and the alkalis in the cement but no examples of such reaction are known in this country"*. Again, therefore, until the late 1970s, the clear impression given was that ASR, when mentioned at all, was only 'a foreign problem'.

The Sand and Gravel Association of Great Britain (SAGA) organised an important symposium in 1968 on the subject of 'sea-dredged aggregates for concrete', which were enjoying a rapidly increasing rate of use. The various authors, for example Shacklock (1969), were primarily concerned with the possible influences of sea salt and shell material, and ASR was only mentioned briefly by two of the contributors. Haigh (1969), from a firm of consulting engineers, stated that, *"Alkali-reactive aggregates, apparently not a problem with aggregates in this country, are to be found frequently overseas"* (present author's underlining emphasis). Van de Fliert (1969), referring to some Dutch concerns over the use of flint aggregates, said that *"About 10 years ago, simultaneous research in Great Britain* (presumably that at BRS) *and in the Netherlands showed that this expansion reaction with crypto-crystalline quartz stone (flint) in concrete need not be feared"* (present author's explanatory parentheses).

Most subjects become covered by numerous textbooks over a period of time, but inevitably some take on a particular authority, either in deference to the eminence of the author and/or his affiliations, or because of popular demand and the consequent reprinting and periodic revising of the work. For illustrative purposes, just three such books will be examined here. Neville's celebrated book, *'Properties of Concrete'*, was first published in 1963; a second edition was produced in 1973, and the third edition was dated 1981. In the 1963 and 1973 editions, a thorough account of ASR is concluded by the statement, *"The alkali-aggregate reaction of the type described has fortunately not been encountered in Great Britain, but is widespread in many other countries"* (present author's underlining emphasis). The 1981 edition is reworded slightly and makes brief reference to examples identified in Great Britain in 1978, but, for nearly 20 years (Neville, 1963, 1973, 1981), this popular and influential book had appeared to support the notion that ASR was 'a foreign problem'.

Lea's important specialist textbook, *'The Chemistry of Cement and Concrete'*, was reissued as the third edition in 1970, being a revision of the second edition by Lea and Desch published in 1956. Since Lea had been Director of Building Research throughout the period of Jones' work at BRS in the 1950s, his rather cautious summary of ASR in respect of the UK is now seen to be significant. Lea (1970) states: *"A survey of the common British aggregates has failed to reveal any containing alkali-reactive constituents, but care is needed when using aggregates of geological types which might contain such constituents and of which no previous service experience is available. It might have been expected that flint, which is such a large constituent of the Thames Valley gravels, would be reactive. Long experience with this aggregate has shown no such troubles, but it appears that it consists of a microporous mass of silica and does not contain opal."*

Although Lea does not imply that ASR is exclusively an overseas problem, and advises caution with some new and untried sources of aggregate, he is nevertheless fairly unequivocal in his assessment of UK flint gravel as a non-reactive aggregate.

Finally, the Road Research Laboratory (later TRRL and now just TRL) published the well-known *'Concrete Roads, Design and Construction'* in 1955, and it remained available for purchase until 1975. The advice given therein in respect of ASR is consistent with the findings emerging from the BRS, and the interpretation embodied by the BSI and other authorities: *"Some aggregates, notably those containing opaline silica or chert, may combine with the alkalis in the cement to produce disruption...This trouble has not been observed in Great Britain, although it is met with in the USA and elsewhere"* (present author's underlining emphasis).

6.3.3 Signs of scepticism

It is apparent from the preceding sections that the overwhelming impression gained by the UK construction industry during the 1960s and the early 1970s was that ASR was an interesting but, fortunately, a foreign problem. Geologically, this perception was always tenuous, since the inherent reactivity of certain mineral and rock varieties does not change at political boundaries, and in fact proper residual concern was expressed to some extent both by Jones and Tarleton (1958a) and by Lea (1970) in respect of those aggregate materials that were not included in the BRS research programme. By the late 1960s and early 1970s, there are some signs that not everyone in the industry was completely reassured, especially those who had experienced alkali-reactivity in other parts of the world.

Gunnar Idorn was invited to present a lecture to the Concrete Society in London in May 1968, on the subject of 'the durability of concrete' (Idorn, 1969). He had recently completed a substantial doctoral study of ASR and other threats to durability of concrete structures in Denmark (Idorn, 1967), and much of his lecture to the Concrete Society concerned ASR. Whilst it is undoubtedly significant that Idorn did not presume to suggest in his lecture that ASR was likely to become recognised as a problem in the UK, it is equally apparent that the materials he had studied were generally similar to many of those found and used in the UK. It was this general similarity with Danish and German flint aggregates, and the uncertain nature of the distribution and provenance of marine aggregates in the North Sea, that caused Fookes to recommend the regular examination and ASR testing of the sea-dredged aggregates to be stockpiled for use in concrete for the prestigious Thames Barrier project in London, preparations for which commenced in the early 1970s (Fookes, 1984).

Applied geological research into ASR and other aspects of concrete commenced at Queen Mary College (QMC), London University, at the end of the 1960s, inspired by Peter Fookes (a visiting lecturer and later visiting professor) and also promoted by the growing awareness of special concrete problems being encountered in the unprecedented Middle East construction boom (Fookes *et al.*, 1977). Doctoral research commenced at QMC in 1972 to include some study of concrete from the Val de la Mare Dam in Jersey and also the alkali-silica reactivity of British flint aggregates (Sims, 1977). Rapid experience was gained by the investigation of reactive or potentially reactive materials from overseas, including Cyprus (Poole, 1975), Bahrain (French & Poole, 1974) and elsewhere (Sims & Poole, 1980). The Geomaterials Group at Queen Mary College is now rightly regarded as having been one of the foremost authorities on ASR in the UK and they started out by being unwilling to accept that it was safe to dismiss ASR as being exclusively 'a foreign problem'.

6.4 FIRST WARNING: THE VAL DE LA MARE DAM IN JERSEY

6.4.1 Background

The Val de la Mare (Figure 6.2) mass concrete gravity dam at St Ouen's in Jersey was built between 1957 and 1961 (Coombes *et al.*, 1975; Coombes, 1976; Sims, 1977; Cole & Horswill, 1988). The structure, which is 29 m maximum height above foundation level and extends for 168 m straight across the valley, retains 950 million litres out of the island's total water storage capacity of 1500 million litres. Construction was in vertical bays or 'blocks', each 6.7 m wide and made up of lift sections each 1.2 m high. The concrete was made using UK Portland cement, together with local aggregates and water from the stream that was to be dammed, and the majority was placed in 1959–1960.

The Jersey New Waterworks Company first reported in 1971 that four of the concrete sections had become discoloured, dampened and cracked, and the concrete crest handrail was in some places displaced by up to 13 mm (Figures 6.3 and 6.4). This led to investigations, including sonic speed measurements (Sherwood *et al.*, 1975), petrographical examinations and aggregate tests, and remedial action including the installation of anchor bars, which was completed by 1975. Monitoring over the following 13 years demonstrated a slowly continuing expansion at a rate of 50–75 microstrains/year and the failure of an anchor bar, caused by seismic shock rather than

Figure 6.2 Val de la Mare Dam, Jersey, in April 1989 (*cf.* Figure 6.7).

ASR, but confidence has grown that *"the dam can now enjoy a working life comparable with other concrete gravity dams"* (Cole & Horswill, 1988).

6.4.2 Reactive constituents

The coarse aggregate comprised mainly crushed diorite or granodiorite from a large quarry at Ronez on the north coast of Jersey, mixed with a smaller proportion of beach gravel containing rounded pebbles of 'Jersey shale', chert and various igneous rocks including dacite. The fine aggregate was a fossil beach sand from a pit at St Ouen's Bay,

Figure 6.3 Part of the downstream face of the Val de la Mare Dam, Jersey, viewed from the parapet walkway, showing map cracking and exudations (from Sims, 1977).

Figure 6.4 Displacement by up to 13 mm of the concrete parapet of the Val de la Mare Dam, Jersey (from Sims, 1977).

comprising quartz, quartzite and feldspar, with small amounts of chert, shell and various igneous materials. Although the beach aggregates were found to contain some potentially reactive constituents, principally chert and dacite (acid volcanic rock), there is little evidence that these materials reacted within the dam concrete. The reactive silica was present as secondary hydrothermal veining and vugh fillings within the crushed dioritic rock aggregate source. The veining material, which is associated with shear zones and lamprophyre dyke zones within the quarry, comprised variable mixtures of chalcedony and opal. These vein and chalcedonised lamprophyre materials showed reaction within a few days at 20 °C in the gel pat test (see Figures 6.5 and 6.6), whereas no such rapid reaction could be obtained with the chert and dacite particles from the beach deposits (Sims, 1977).

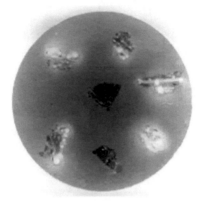

Figure 6.5 Gel pat test of chalcedony/opal vein material, similar to the reactive constituent present in concrete from the Val de la Mare Dam, Jersey, after 14 days of the test at room temperature (from Sims, 1977).

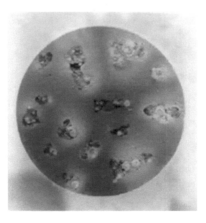

Figure 6.6 Gel pat test of chalcedonised lamprophyre dyke rock, similar to the reactive constituent present in concrete from the Val de la Mare Dam, Jersey, after 14 days of the test at room temperature (from Sims, 1977).

Although the chalcedony/opal veining affected much of the Ronez rock mass exposed in the quarry at that time, it clearly forms much less than 1% of the rock overall, and contemporary work by BRS considered it to be less than 0.1% (Cole & Horswill, 1988). The presence of some beach aggregate in the coarse fraction will have diluted this content of chalcedony/opal still further in the concrete. The 'pessimum' for opal is known to be very low (Hobbs, 1978), say 1%-5%, but expansive reaction seems unlikely with opal contents as low as those computed for the Ronez veining material. It must therefore be presumed that variations during quarrying will occasionally lead to higher concentrations of reactive silica in certain batches of concrete.

The BRS carried out tests on aggregate samples using the mortar-bar method, which had been the basis of their NBS research programme in the 1950s. The test details are not available, but it is reported that none of their results exceeded the ASTM C33 (1986) guidance criteria at 6 months (Cole & Horswill, 1988) and that results for the Ronez and beach aggregates were similar. In view of the reliance placed upon mortar-bar test results in providing reassurance about UK aggregates, this must have been a worrying outcome. It seems likely, from more recent experience, that expansions would only have been produced in the laboratory by the controlled addition of the opal veining material to the aggregate, and perhaps only then using comparatively high alkali contents and concrete rather than mortar specimens.

The source (or sources) of the cement in the UK is not known for certain, and it is possible that cements from various works were supplied over the period of construction of the dam. Approximate total alkali values for two concrete samples have been quoted as 0.74 % and 0.96 % $Na_2Oeq.$ by mass of cement (Cole & Horswill, 1988), and Coombes found that cement supplies during placement of the most severely damaged concrete (June-August 1960) most probably averaged 0.95 % $Na_2Oeq.$ (Coombes *et al.*, 1975). Thus, the most affected portions of concrete might have contained higher than typical amounts of opaline vein material in the aggregate and also relatively high concentrations of alkalis from the cement.

Additional reactive alkalis could have been derived from the rock aggregates themselves, or from seawater salt contamination of the beach aggregates, and in any water-retaining structure there is always the possibility of alkali migrations to form localised concentrations.

6.4.3 Features and damage attributed to ASR

The visible damage and correspondingly low sonic velocities were almost entirely restricted to concrete lifts cast during the critical period June to August 1960 (Figure 6.7) and block 6 was worst affected. The downstream concrete surface of the dam was stained with heavy carbonate exudation from between the sections, but the most serious discoloration was mainly concentrated into those areas cast in the critical period (Figure 6.3). The overall discoloration in the affected areas was from an original grey-buff to a relatively bright orange-brown. Affected areas also displayed random map cracking, these cracks being frequently bordered by orange-brown discoloration of slightly smoother surface texture than the surrounding concrete. There were white, sometimes porcellaneous, exudations from these cracks, especially those orientated horizontally or nearly so, and these deposits included some alkali-silica gel. No surface 'pop-outs' were observed, but there was evidence of significant expansion in the form of

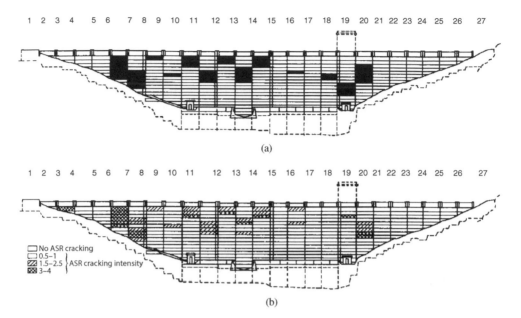

Figure 6.7 Downstream elevation of the Val de la Mare Dam, Jersey, showing (a) the lifts cast June-August 1960 and (b) their cracking intensity (from Cole & Horswill, 1988).

large-scale misalignment (Figure 6.4). Good correlation was found between sonic velocities, date of casting and the visible condition of the concrete. Nearly all of the sonic velocity results for 47 concrete lifts cast during the critical period were less than the overall average of 4.56 km/s for the 77 unaffected lifts measured (Sherwood *et al.*, 1975) and, between 1972 and 1983, the maximum apparent 'reduction' in sonic velocity for the affected concrete was 14 %.

Apart from measurement of the obvious handrail displacement, the early reports (Coombes *et al.*, 1975; Coombes, 1976) do not indicate any attempt to quantify the amounts of any expansive movement which had occurred up to 1971. The BRS is reported to have carried out some tests for latent expansion using core samples, obtaining results of only about 0.01 % after 1 year at 20 °C and 100% relative humidity (RH), and similarly after 3 months at 38 °C and 100% RH. Although these results are low (BCA, 1988; ISE, 1988) and might indicate that most of the potential for reaction and expansion had been exhausted before drilling and testing of the cores, there is no certainty that the samples were kept moist prior to testing in accordance with modern practice, and also it is not clear whether any of the tests were pursued for storage periods longer than those reported. The petrographic examinations carried out by the BRS do not appear to have been published, although it is generally believed that microscopic evidence of ASR was identified.

Sims (1977) obtained four small samples of the concrete and carried out petrographic examinations; he was able to identify some gel deposits within the concrete, variously infilling voids or lining cracks and fracture surfaces, and sporadic potentially reactive aggregate particles, including chalcedony, chert and rhyolite, but he could not find any definite reaction sites in the four samples available. Sims also reported less micro-cracking in the samples than might have been expected for expansive ASR and suggested that perhaps the main damaging reactions had occurred deep within the innermost parts of the mass concrete, perhaps there promoted by conditions of higher temperatures and increased pore water pressures. This suggestion appeared to be supported by the composition of the gel deposits, in which relatively reduced alkali contents and the presence of calcite and sulphates indicated considerable migration away from the reacting centres.

It was further suggested that the conditions prevalent at the site of reaction might have favoured the production of a comparatively fluid gel, able to migrate, rather than a viscous swelling gel. If so, the observed cracking and expansion in the structure may have resulted primarily from the initial reaction, with continued gel-producing reaction not promoting any significant further expansion.

6.4.4 Remedial measures

Although it was assumed that reaction would continue, the investigations in 1971–74 indicated that the concrete would not become incapable of taking the required compressive loads, but that expansive cracking could lead to instability because of increasing internal uplift pressures (Cole & Horswill, 1988). Remedial action was therefore restricted to measures to maintain stability of the dam against increased internal uplift, and also it was decided that only the worst-affected parts would be dealt with initially, other areas being subjected to a similar form of repair as and when further deterioration occurred.

Three 40mm Macalloy high tensile steel anchor bars were installed through the upstream side of the worst-affected block 6 and into the underlying rock; these were post-tensioned to produce a minimum factor of safety against overturning of 1.7 for hydrostatic internal uplift pressure of 100 % in the upstream face and zero on the downstream face (Figure 6.8). Since drilling equipment was available on site, it was decided also to drill pressure-relief drainage holes vertically at 3.4 m centres through the central blocks 10-19, to intercept any seepage flow and provide some control over the development of uplift pressures.

A grouting trial was undertaken on block 6 to try to seal any ASR cracking on the upstream side to inhibit water seepage. Although a low-viscosity oil-based chemical grout was used (Polythixon 60/40 DR), very little penetration was achieved; it is not clear whether this was because of some inadequacy in the grouting system or instead because the concrete interior contained fewer cracks and microcracks than had been expected.

As well as the actual remedial measures, facilities were established, especially on block 6, for the future monitoring of movements (Table 6.2). This involved the instrumental monitoring of block 6, including vibrating-wire anchor bar load cells, electrically operated piezometers and 24 mechanical strain gauges across lift and vertical joints. However, as with most condition monitoring, much reliance is still placed upon regular visual inspections, supported by crack surveys and *in situ* sonic velocity surveys. A further 44 electric piezometers were installed on selected lift joints in 1981, and later removed and reinstalled in 1983 after recalibration and treatment to reduce creep effects, but a new design has proved less effective than the earlier instruments in these conditions, and by 1988 only a third were giving reliable readings (Cole & Horswill, 1988).

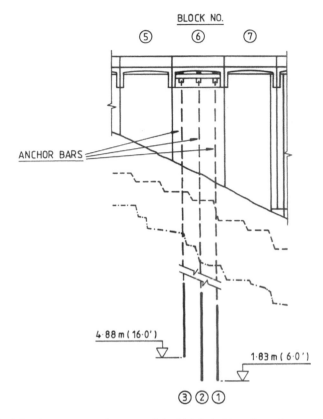

BLOCK NO.

⑤   ⑥   ⑦

ANCHOR BARS

4·88 m (16·0')

1·83 m (6·0')

③②①

Figure 6.8 Detail of the downstream elevation of the Val de la Mare Dam, Jersey, showing the anchor bar installation to worst-affected block 6. The 'failure point' refers to the 1977 anchor bar fracture due to seismic shock (from Cole & Horswill, 1988).

Table 6.2 Val de la Mare Dam, Jersey: main monitoring activities and frequencies (after Cole & Horswill, 1988).

Monitoring activity	Frequency of records 1975–1988
Piezometer readings	2 weeks
Relief drain flows	Monthly
Anchor bar load cells	Monthly
Crest movement gauges	3 months
Visual inspection and survey or face cracking	3 months (later 6 months)
Other block 6 instrumentation	3 months (later stopped)
Sonic velocity survey	3 years (later 4 years)
All readings plotted on graphs and reviewed	Monthly, unless significant changes occur which are notified immediately
Review reports with recommendations for immediate future work	3 years
Inspection by independent panel alkali reactivity engineer	As engineer considers necessary

6.4.5 Experience to date and the future

Around fourteen years after monitoring of the Val de la Mare Dam commenced, the findings generally indicated that 'there can be *"life after ASR" for mass concrete gravity dams'* (Cole & Horswill, 1988). Concern had arisen in 1977, when it was discovered that one of the three anchor bars had fractured at a depth of 18.6 m below the top of the dam. However, an intensive investigation revealed that the failure could only have been caused by a high-velocity tensile dynamic load, probably from a seismic shock, and the anchor was not replaced. The two remaining anchors have continued to perform satisfactorily, and in fact act as large extensometers together with the vibrating-wire load cells. Over 7 years (1975–82) the load on anchor 1 increased by 2.5% and, even after controlled de-stressing in 1982, the load has continued slowly to increase at a rate of around 2.5 kN per annum.

The flows from the pressure-relief drains drilled in 1974 have, with one exception, remained small, and encrustation with calcium carbonate has been the only problem, necessitating periodic cleaning. The exception was a drain in block 12, which did develop unexpectedly high flows in 1980, as a result of leakage through a particular lift joint. This was corrected by the installation of an epoxy-sealed neoprene rubber gasket into the suspect joints on the upstream face of the dam.

There now seems no reason to question the integrity of the Val de la Mare Dam for the foreseeable future, but inevitably it remains of special interest as the first example of ASR recognised in the British Isles. It seems logical that some other concrete structures on Jersey must also be affected by ASR, and a limited survey has been carried out (Fookes *et al.*, 1984; Cole & Horswill, 1988). Concretes made using the granodiorite aggregate source (with sporadic chalcedony/opal veins) exhibited worse cracking at all ages than concrete made with only beach aggregates (Figure 6.9); indeed only one example with beach aggregates was more than a case of 'suspected' ASR. In the case of the granodiorite concretes, the water-retaining structures, including Val de la Mare Dam, were more seriously affected than the 'dry' concretes. Cole & Horswill (1988) claim that, for Jersey concretes, ASR damage does not worsen significantly after an age of 15-20 years (the dam was 9 years old when ASR was discovered and is now (in 2015) more than 50 years old), although further deterioration caused by other destructive processes, including corrosion of embedded steel in reinforced structures, remains a possibility.

Later concrete works for the Queen's Valley Reservoir in Jersey (the dam itself being a rockfill construction) used a different source of coarse aggregate on the island, a crushed granite, which is free of the highly reactive opaline constituent (Bridle *et al.*, 1991).

6.5 FEAR OF EPIDEMIC: THE GROWTH IN MAINLAND EXAMPLES

6.5.1 Plymouth—the first examples

For a while it was possible to regard the dam in Jersey as something of 'a special case': caused by hitherto unsuspected opal-bearing secondary veins in the otherwise well-proven aggregate source and also possibly relating to the quarrying of an unusually intensely veined body of rock and/or the use of an unusually high-alkali cement. Between 1976 and 1978, any such residual complacency was overcome by the

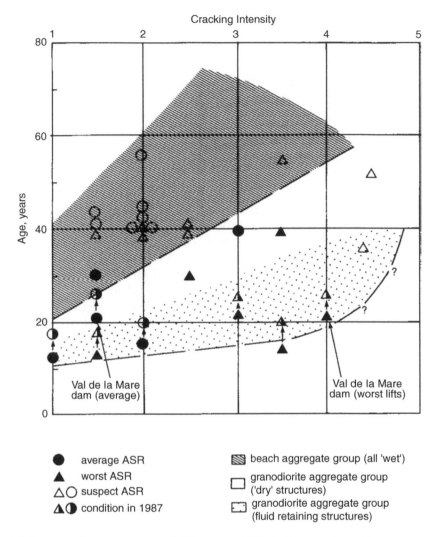

Figure 6.9 Survey of Jersey concretes (1982, updated 1987), showing that ASR damage is largely restricted to concretes made using granodiorite coarse aggregate and especially water-retaining structures. The 'cracking intensity' scale is defined in Fookes *et al.* (1984) (from Fookes *et al.*, 1984; after Cole & Horswill, 1988).

discovery of ASR on the UK mainland, first in the south-west of England and then in the English Midlands and South Wales (Palmer, 1978).

The first UK mainland example was found at an electricity substation at Milehouse in Plymouth, Devon, affecting unreinforced concrete bases set in the ground with their upper surfaces exposed to the weather (Figure 6.10). A number of organisations were associated with the investigation into severe cracking of these bases and the identification of ASR, including Queen Mary College, the C&CA, Messrs Sandberg (now Sandberg LLP) and Plymouth Polytechnic (now Plymouth University) (Sandberg,

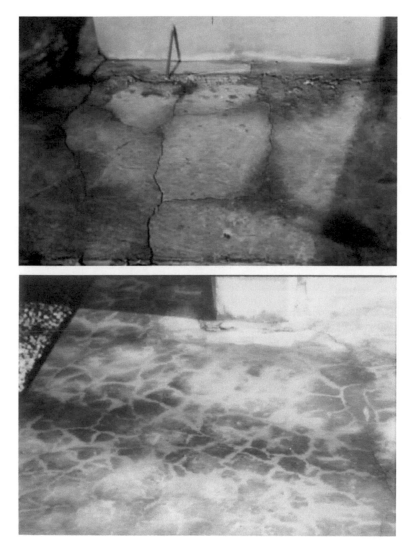

Figure 6.10 Concrete bases at Milehouse electricity substation, Plymouth, showing map cracking (from Sandberg, 1977).

1977; Palmer, 1978). Before long, the Central Electricity Generating Board had identified a number of other installations in the south-west and South Wales exhibiting broadly similar random cracking patterns, concrete compositions and exposure conditions.

At Milehouse, the concrete comprised crushed limestone coarse aggregate, a sea-dredged fine aggregate containing a subordinate proportion of chert and a locally produced high-alkali cement (probably around 1.0-1.2 % $Na_2Oeq.$). Reaction sites were observed in the concrete, involving brown and white chert particles, with the formation of reaction rims and associated gel deposits (Figure 6.11). Although some

Figure 6.11 Detail of the concrete from a cracked base at Milehouse electricity substation, Plymouth, showing reactive chert particles (Sandberg, 1977).

other causes of cracking were apparent (Sandberg, 1977), it has become generally accepted that ASR was the principal mechanism responsible for the damage. It is now considered that the particular conditions to which these concrete bases were exposed, being buried in the ground except for the upper horizontal surface, which was subject to wetting and drying, was an important factor in the promotion of ASR, but some earlier investigations considered that stray electrical currents might have been involved (Moore, 1978).

In Palmer's 1978 review, concrete associated with two reservoirs in the Midlands and another in South Wales were mentioned as examples of ASR, as well as two rather atypical cases respectively involving east coast wartime anti-tank defences and possible reactivity with greywacke aggregates in Scotland (see 6.5.6). However, it was probably the identification of ASR in the Charles Cross Multistorey Car Park in Plymouth which was mainly responsible for causing practising structural engineers in the UK to become acutely aware of the potential threat of ASR. The car park was built in 1970, using a helical ramp design, and was exhibiting slight to serious cracking of the concrete at an age of just 7 years. Some beams exhibited severe surficial cracking, the patterns of which were influenced by the underlying reinforcement and stress fields, and some column heads exhibited substantial cracking and fragmentation (Figure 6.12). The ground-level retaining walls, some parapet walls and pavements on the top level exhibited classic map-cracking patterns. Laboratory examination of the concrete again revealed the presence of reactive chert particles in a sea-dredged fine aggregate, together with a non-reactive coarse aggregate (variously crushed limestone or crushed granite) and a high-alkali Portland cement. The car park has been used for a thorough study of concrete condition classification and crack mapping (Fookes *et al.*, 1983).

a)

c)

b)

d)

Figure 6.12 Charles Cross Car Park, Plymouth, showing its spiral ramp (a), severely cracked beam (b) and column head (c) elements and a photomicrograph (d) from a thin section of concrete from the car park, exhibiting a site of expansive ASR involving a flint particle in the coarse fraction of the sand.

Load tests were carried out on various concrete units at the car park in 1981, 1982, 1985 and 1986 (Wood *et al.*, 1987). The results confirmed that the beams *"were still able to carry their design load, and their actual service load plus a significant margin safely"*. Consequently, the main *in situ* structural frame was strengthened in 1983 and the car park then remained in service for more than 20 years, being monitored and the structural condition being periodically reviewed. It was demolished in 2004, as part of a general redevelopment of that part of Plymouth city centre.

The reactive combination of a chert (or flint)-bearing fine aggregate, a non-reactive (often crushed limestone) coarse aggregate and a high-alkali Portland cement was becoming established as the principal cause of damaging ASR in the UK and, with only a few exceptions, this continued to be the case as more and more examples became identified. Chert or flint had always been recognised as a potentially reactive material (see 6.3.1) and it was the comparatively small proportion of chert in the total aggregate combination which appears to have approached the 'pessimum' for chert, especially in the presence of high alkali concentrations. The combination for concrete of local crushed limestones and crushed granites with sea-dredged sands, imported from further east in the English Channel, was a new and unfortunate development in the south-west of England in the late 1960s or early 1970s, to meet increased construction demand in an area traditionally short of good-quality land-based sand. It was the dilution of the overall chert or flint content of the aggregate by the presence of non-flint coarse aggregate, as opposed to the flint gravel which had more usually been combined with such flint sands in south and south-eastern England, which coincidentally created the reactive mixture. There is no reason to suppose that the use of sea-dredged aggregates, instead of land-based aggregates of similar mineralogical composition, had any adverse influence on the concretes, except of course for any cases where inadequate washing of the marine aggregate might have left a residual content of sea salt (which can then contribute to the overall sodium content of the concrete).

6.5.2 Geographical controls – aggregates and cements

Apart from the Jersey ASR caused by opaline veination of an aggregate source and some other possible cases involving greywacke aggregate (see 6.5.4), nearly all UK examples of ASR have involved chert or flint as the reactive constituent, and then only when present in subordinate 'pessimum' proportions and also in the presence of high concrete alkali contents (Table 6.3). Since most concretes are made using locally

Table 6.3 Some examples of reactive aggregate combinations identified in UK concretes affected by ASR.

Coarse aggregate	Fine aggregate	Structure locations	Reactive minerals
Granodiorite	Quartz beach sand	Jersey	Chalcedony/opal in granodiorite
Limestone or granite	Sea-dredged quartz/flint sand	SW England	Flint/chert in fines
Limestone	Sea-dredged quartz/flint sand	S Wales	Chert in fines
Trent Valley gravel	Trent Valley sand	Midlands	Flint, chert and possibly quartzite and rhyolite
Granite	Essex quartz sand with flint	Midlands	Flint
Limestone, glass and chert	Limestone, glass and chert	NW England	Glass
Lightweight aggregate	Quartz sand with chert	SE England	Chert
Flint gravel	Quartz sand and siliceous limestone	SE England	Silica in limestone
Greywacke sandstone and argillites	Natural sand mainly quartz and quartzite	Wales, SW England and N England	Cryptocrystalline quartz in greywacke/argillite

derived aggregates and cements, it is therefore not surprising that certain regions of the country have tended to yield most of the definite or relatively serious cases of ASR. The mixture of imported flint-bearing marine sand with local limestones and granites in south-west England has already been mentioned. In South Wales, somewhat similarly, sand dredged from the Bristol Channel contains small proportions of chert (probably Carboniferous) and is commonly blended with local crushed limestone aggregate. In the cases of both south-west England and South Wales, at least some of the locally made Portland cements were high-alkali in character.

In the English Midlands, especially in the East Midlands, concretes are frequently made using the abundant fluvioglacial sands and gravels commonly termed 'Trent Valley' aggregates. These materials are dominated by quartzites, but both chert and flint are typically present, in varying ratios and proportions, in both the coarse and fine fractions. Again, at least one of the Portland cements made in the region was a high-alkali type, and ASR has sometimes occurred when the local aggregates and cement have been used together. These sands and gravels are somewhat polymictic, and other types of potentially alkali-reactive constituents have also been recognised, including some metamorphic quartzites and glassy rhyolite, although it is uncertain to what extent, if any, such materials have actually caused or contributed to concrete damage.

In other parts of the UK, to date, local combinations of aggregate and cement have not generally proved to be susceptible to ASR, and only sporadic cases of structures possibly affected by ASR have been reported. However, it would not be sensible to regard any geographical region as being completely free from the possibility of ASR and, in particular, the nationwide distribution and use of precast concrete units must always be borne in mind. General accounts of UK aggregates may be found in Fookes (1992), Sims and Brown (1998), Smith and Collis (2001) and Poole and Sims (2003), whilst specific considerations of potentially reactive UK aggregates are usefully given by Rayment et al. (1990), French (1992), Hobbs (1992a, 1992b, 1992c) and Sibbick and Page (1992a, 1992b).

As well as the presence of a reactive aggregate combination and of a sufficient amount of alkalis, a supply of moisture is necessary for damaging ASR to occur in concrete. The degree and frequency of wetting has an important environmental control over ASR, and it is believed that cyclic wetting and drying may be the most conducive condition (BCA, 1988). Consequently the incidence and severity of ASR may sometimes depend upon geographical aspect; for example, Fookes et al. (1983) found a sympathetic relationship between exposure and deterioration at the Charles Cross Car Park in Plymouth (Figure 6.13). In some circumstances, regular moisture migration might cause localised concentrations of alkalis, so enabling ASR to occur in affected parts of a structure (Nixon et al., 1979).

6.5.3 Flint aggregates

In the late 1970s and into the 1980s (see subsection 6.5.5), it became apparent that flint was the principal cause of ASR damage in the UK. Urgent research was carried out in order to understand the nature of this constituent and its reactivity (Scott, 1991) and also to use expansion tests to study behaviour of the aggregates and to identify parameters that would later assist with devising preventative strategies (Rayment et al., 1990; Rayment, 1992; Rayment & Haynes, 1993, 1996).

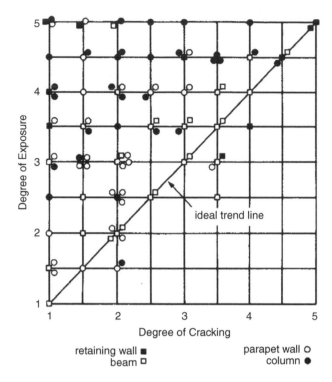

Figure 6.13 Diagram relating the degree of exposure to the degree of cracking at Charles Cross Car Park, Plymouth, showing that exposure helps deterioration to occur. The scales for degrees of exposure and cracking are defined in Fookes *et al.*, 1983 (from Fookes *et al.*, 1983).

A thorough applied geological review of flint and chert in the UK was carried out for the Mineral Industry Research Organisation ('MIRO') by a team at the University of Leicester (Scott, 1991). They noted that primary chert is mainly formed and found in most limestone formations in the UK, being initially re-precipitated from dispersed silica as opal-CT (disordered α-cristobalite: Jones & Segnit, 1971), which usually recrystallises later to form microquartz; 'flint' was the name given to the chert so-formed in the Upper Cretaceous Chalk. Derived flint and other types of chert are then important constituents of terrestrial and marine sand and gravel deposits across and around the UK, when the flint and chert particles have often been through more than one cycle of erosion and deposition.

The study for MIRO (Scott, 1991) included mineralogical laboratory work on about 150 flint and chert samples from primary and superficial sources. Although these materials were mainly low-porosity, they did find general correlation between porosity and the familiar colours, with dark grey-blue-lack particles being least porous, brown-stained particles being relatively more porous and the white or visibly porous materials have the highest porosity. Mineralogically, the cherts and flints were dominated by microcrystalline quartz ('microquartz'), with variable contents of chalcedonic (fibrous) quartz and 'megaquartz' crystals (> 30 μm); small amounts of opal-CT had been found rarely in some primary flint and chert, but not in derived aggregate particles. Importantly for the evident alkali-reactivity of flints and cherts, it was found that there is a variable sub-microscopic porosity associated with dislocations in the microquartz crystals, where water is probably present and may lead to a variably active crystal surface. It was thought that the extent of these microquartz dislocations could relate to the ASR susceptibility of chert and flint; the present chapter author had separately postulated a similar causation in his PhD research: Sims (1977). In their work at the BRE (see next paragraph), Rayment *et al.*, 1990 described a broadly similar reactive element of flint as being 'microchalcedony, which occurs interstitially between quartz lepispheres', so that varieties comprising 'densely packed lepispheres with very little interstitial chalcedony' were less reactive.

Following the increasing occurrences of ASR damage on the UK mainland, the BRE undertook a major programme of expansion testing of British concreting aggregates, using sand and gravel samples from 227 locations and samples of volcanic rock from 16 quarries (Rayment *et al.*, 1990; Rayment, 1992). The work was carried out using a concrete prism test devised at the BRE, which would later form the basis of the current British Standard method (see subsection 6.6.5). As had been observed from the affected structures, it was found that fine aggregates (sands) containing flint and chert rarely showed expansion when tested with the coarse aggregates from the same locations, which in eastern and southern England were usually also almost wholly composed of flint and chert. However, quite frequently the same flint-bearing sands would exhibit expansion when blended with an inert limestone coarse aggregate; mixtures of 30 % flint-bearing sand and 70 % crushed limestone often produced expansions of > 0.2 % or even > 0.3 % (Figure 6.14). It was evidently an example of 'pessimum' behaviour, with tested coarse-fine aggregate combinations exhibiting total flint contents in the range 54 % to 79 % or more usually showing no expansion.

It was further evident to Rayment *et al.* (1990) that the inherent reactivity of flint varied, there being no correlation in their 30:70 (flint sand:limestone) tests between the flint content of the sand and the overall expansion. The 30:70 tests yielding the highest expansions (> 0.4 %) had flint contents in the sand component as different as 5 %, 12 % and 22 %. Also, in some tests on mixtures of 70 % chert/flint gravel and 30 % inert limestone, most showed no expansion, but two showed expansion with a total flint of content of 60 to 65 %, in contrast to other mixtures of similar flint contents from different sources. Finally, flint-rich deposits from a few locations produced test expansions in the 30:70 mixtures, irrespective of whether the 30% component was flint-sand or inert limestone, suggesting an unusually reactive variety of flint. These examples of apparently anomalous behaviour are further discussed in the next paragraph. Overall, the findings from this BRE programme would collectively be used to refine the guidance in UK specifications for minimising the risk of ASR (see subsections 6.6.2 & 6.8.2); where previously the guidance had suggested that flint must be > 60 % or < 2 % for the aggregate combination to be deemed non-reactive, this would change such that when flint is > 60 %, the flint content of the sand must also be at least 5 %.

The BRE team continued their research work, especially concentrating on the causes of the apparently 'anomalous' occurrences of expansion when the flint content was outside the pessimum range indicated by the majority of the tests and experience with structures (Rayment & Haynes, 1993, 1996). Nodules of primary flint within limestone are frequently surrounded by a porous white zone, known as the 'cortex', and the

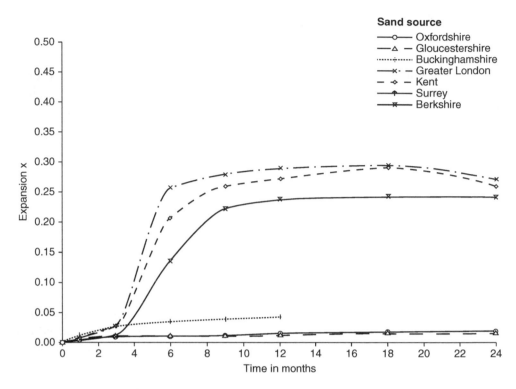

Figure 6.14 Expansion of concrete prisms made using ex-UK 30% flint-bearing sand and 70% inert limestone coarse aggregate (after Rayment *et al.*, 1990).

mode of formation of this cortex remains uncertain, but might be a weathering feature. However, whatever its cause, it is the case that a proportion of derived flint sand and gravel deposits variably comprises cortex material, either remaining whitish or sometimes stained brown; additionally some individual flint gravel particles seem to have developed external cortex zones during their cycles of erosion and deposition, perhaps supporting the notion of a weathering formation for cortex. Further work at BRE suggested that the cortex element was the main reactive part of flint, so that it was the proportion of cortex that determined overall reactivity, rather than the total content of flint. It was thus found that the very small number of mixtures that recorded test expansions despite having total flint contents exceeding the pessimum threshold (say above 60 %) corresponded with flint materials that were unusually low in cortex, so that reactivity and thus pessimum level would be reduced.

It was consequently proposed by Rayment & Haynes (1993, 1996) that water absorption values could be used to identify flint aggregates potentially susceptible to anomalous ASR expansion, with 10-minute absorption values of > 2.0 % indicative of suitably high porous cortex contents, whereas values of < 1.0 % might indicate a risk of anomalous expansion (Figure 6.15). It will be seen (see subsection 6.8.2) that this additional complication to UK specification for minimising the risk of ASR was finally not needed, as it was eventually decided that flint and certain other aggregates would

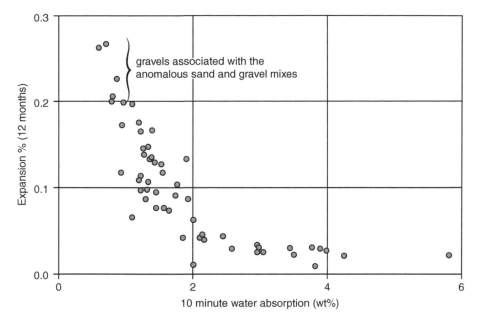

Figure 6.15 Relationship between the porosity of flint-rich gravels (as shown by water absorption) and their expansion at 12 months in the concrete prism test (70% flint gravel & 30% inert limestone sand) (from Rayment & Haynes, 1993).

simply be accepted as exhibiting 'normal' reactivity in the UK, with a range of counteractive criteria thus being required.

6.5.4 Greywacke aggregates

Although Palmer (1978) had reported a case of AAR involving greywacke aggregates, in a small 1936 dam at Muckburn in southern Scotland, this was initially regarded as an unusual occurrence, as the aggregate had been locally obtained for the project from a nearby tunnel excavation, and most national attention had focused on the reactive flint-bearing concrete structures. However, gradually it became apparent that some, often older, concrete structures in the UK were exhibiting evidence of a slower variety of ASR, involving greywacke aggregate (Figure 6.16). A definitive diagnosis of ASR was carried out for the former Maentwrog Dam in North Wales (Figure 6.17) by French (1986), who observed substantial quantities of gel both within the slightly metamorphosed greywacke aggregate particles, forming within cracks reflecting sedimentary rock structures, and within cracks and voids in the concrete. He opined that this ASR had been taking place since before a programme of extensive repairs that was carried out in 1958–59 (Blackwell & Pettifer, 1992). French (1986) also observed that much of the gel in the concrete had been recrystallized and calcified, which would later lead him, after also examining other Welsh cases, to suggest that alkalis might eventually be recycled, as the first-formed alkali-silica gel was altered in this way (French, 2005). This old dam, built in 1926–28, was later replaced by the current larger structure downstream, more to enlarge Trawsfynydd lake than as a result of ASR damage.

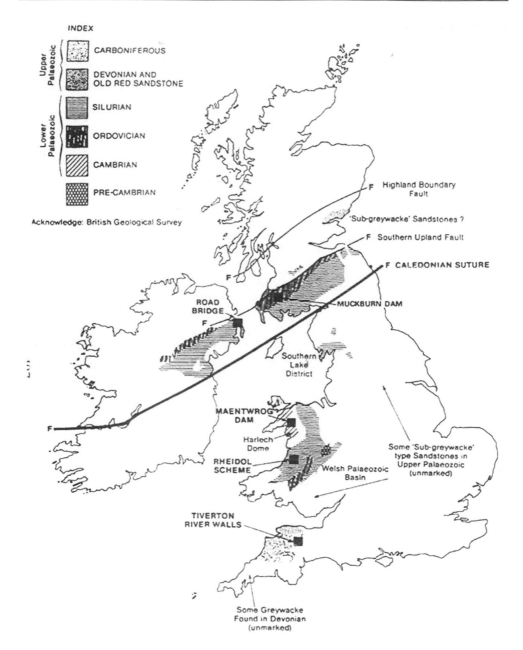

Figure 6.16 Map of the UK showing strata by geological age where greywacke occurs, also marking some of the main reported examples of affected structures (from Blackwell *et al*., 1996).

Further separate work ascertained that these greywacke rocks were closely similar to the varieties that had once been described in Canada (Gillott *et al*., 1973) as being 'alkali-silic*ate* reactive' (but which are now acknowledged to be a particular form of ASR), with the reactivity originating in the fine-grained recrystallized matrix

a)   b)

Figure 6.17 Maentwrog Dam, North Wales (a), showing cracking caused by ASR, but also suffering long-term water leaching and localised freeze-thaw damage (b).

(Blackwell & Pettifer, 1990, 1992), comprising 10–30 µm chlorite and sericite laths, with an interstitial silica cement in 25–50 µm areas. Zhang *et al.* (1990) demonstrated, using Transmission Electron Microscopy (TEM) that this silica generally exhibited good crystallinity, but also had regions with a high density of dislocations caused by metamorphism. This high density of silica dislocations was thought to relate to the potential reactivity of greywacke.

An early study by Blackwell *et al.* (1996) reported that around half of UK greywacke materials were reported to be potentially reactive using concrete prism testing, with those varieties not being geographically restricted. As shown in Figure 6.16, other examples included a bridge near Belfast in Northern Ireland, involving greywacke stratigraphically similar to that in the Muckburn Dam in Scotland, dams and related structures in the Rheidol Scheme in central Wales (Figure 6.18) and the river wall at Tiverton in southwest England (Devon).

Although it had been found that only about half of the greywacke samples studied were potentially reactive, there was no easy means of distinguishing these apparently reactive and non-reactive varieties, beyond long-term expansion testing. Petrographic examination could identify greywacke and even the presence of a recrystallized matrix, but the research by Xhang *et al.* (1990) had shown that the reactive silica in a rock sample could only be directly identified by sophisticated TEM examination, which could not be applied to generalised aggregate samples. Additionally, work at the BRE had found that *crushed* rock greywacke was more reactive than the same greywacke in a natural gravel-sand deposit, leading to a recommendation that natural gravel-sand aggregates containing less than 10 % crushed greywacke would not be classed as 'highly' reactive (Building Research Establishment, 1999) on the basis of the greywacke content.

Accordingly, a BSI sub-committee prepared a protocol for the sampling, testing and interpretation of crushed rock greywacke or natural aggregates containing more than

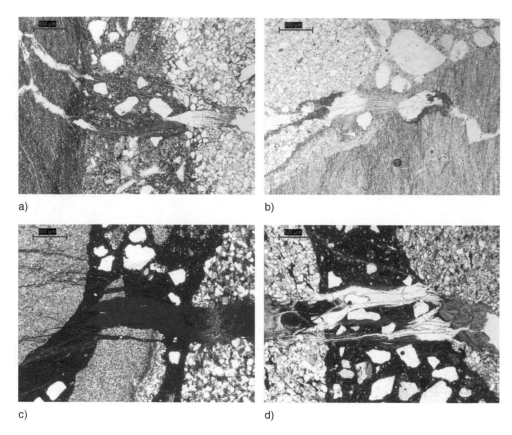

a)   b)

c)   d)

Figure 6.18 Photomicrographs showing ASR associated with greywacke aggregates in concrete from Dinas Dam, part of the Rheidol Scheme in central Wales (images courtesy of RSK Environment Ltd, reproduced with permission of Peter Mason and Statkraft).

10 % crushed greywacke from individual sources (British Cement Association, 1999). Broadly, the then-draft BS concrete prism test was adapted, using four mixes with different total alkali (Na$_2$Oeq) contents, from 3.5 to 5.0 kg/m^3 and considering expansions greater than 0.08 % after 24 months to be 'significant'; the results were used to identify the maximum total alkali content (up to a maximum limit of 3.5 kg/m^3) for a concrete in which the tested aggregate could be safely used. It is not thought that this protocol has been widely used in practice, probably because of the impractical two-year duration of the testing and also because it has become apparent (Blackwell *et al.*, 1996) that the greywacke expansion can be successfully controlled by cement replacement using fly ash (such as pfa) or ground granulated blastfurnace slag (ggbs). Thomas *et al.* (1992) and Thomas (1996, 2011), in particular, have been able to demonstrate over a period how greywacke reaction in the Rheidol Scheme has affected those structures, such as the Dinas Dam (Figures 6.19 & 6.20), that were made with concrete that did *not* contain pfa, whilst comparable concrete containing pfa, such as the Nant-y-Moch Dam, remain unaffected despite containing similar greywacke aggregate (Thomas *et al.*, 2012).

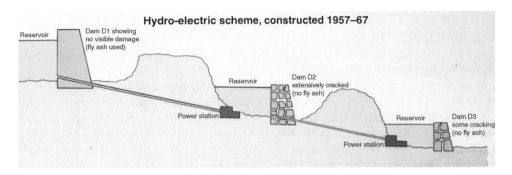

Figure 6.19 Schematic diagram of the Cwm Rheidol hydroelectric system, including the Nant-y-Moch Dam (D1) with no visible damage [pfa used] and the Dinas Dam (D2) with extensive damage [no pfa used] (from Thomas *et al.*, 2012).

Figure 6.20 ASR damage to the Dinas Dam, as seen in 2010, here showing severe cracking of the gravity abutments (from Thomas *et al.*, 2012).

6.5.5 The spread of discoveries and allegations

It is difficult to quantify the number of ASR-affected structures in the UK, because of client report confidentiality. It was hoped that the examples identified in the mid-1970s, such as the electricity substation cases, would prove to be 'special cases' and that ASR would continue to be 'an extremely rare occurrence in the British Isles' (Palmer, 1977), but the number of cases of ASR had risen to 30 by 1982 (BRE Digest 258) and to 170 by 1988 (BRE Digest 330, see subsections 6.6.2 & 3). A majority of these cases were civil engineering structures, rather than buildings, perhaps because these are subject to more rigorous inspection and assessment.

Nevertheless, it did appear that *"the incidence of damage diagnosed as being caused by ASR in the UK is very small when compared with the total amount of concrete construction carried out"* (Concrete Society, 1987). One major study of a sample of the nearly 6000 highway bridges for which the Department of Transport is responsible in England apparently identified signs of ASR in between a half and a third of the structures examined (Montague, 1989; Wallbank, 1989). However, no more than about 5 to 10% overall were damaged to some extent by ASR; moreover highway structures are liable to be particularly vulnerable to ASR because of the conditions to which they are exposed. It is also important to distinguish between those few cases of ASR in which structural damage has resulted and those more numerous cases where any surficial damage is only cosmetic, or which at worst could provide a facility for later deterioration by other agencies, such as frost, if left untreated.

It had also been recognised (BCA, 1988) that microscopic evidence of the presence of ASR in concrete is not necessarily adequate proof that any damage has been caused or even accentuated by ASR, so that many statistics regarding the scale of the effect of ASR in the UK, or indeed elsewhere, must be treated with some caution. In an admirably thorough review of his observations over a decade, for example, French (1987) reported some 40% of 300 concrete structures to be exhibiting evidence of ASR, but these reactions ranged from 'just detectable to serious destructive processes', but he provided no clear indication of the proportion considered to exhibit the latter degree of severity. There is no easy and unequivocal means of quantifying such degrees of damage caused by ASR and, in any case, it is likely to be time-dependent; a minor case of ASR identified today could possibly develop into a more serious case over a period of time. In one appraisal of a group of 93 UK structures (ISE, 1988), using a core expansion test as a measure of severity, only 32% exhibited expansions greater than about 0.10% and just 6% exhibited expansions greater than 0.15% (Figure 6.21).

It has been noticed that the concrete structures found to be affected by ASR in the UK were built during a period from the 1930s to the 1970s, but the number of affected structures appears to show a preponderance in the 15 years 1960 to 1975, which interestingly post-dates the BRS research work of the 1950s (Figure 6.22). Some authors have suggested that this apparent upsurge in ASR in the UK might be attributed, in part at least, to changes in the composition and nature of Portland cement being manufactured in the UK (Buttler *et al.*, 1980). Changes to the cement manufacturing process and greater controls over the emissions into the atmosphere from cement works might have caused the alkali contents to rise, and mineralogical modifications to achieve greater rates of strength gain might have led to the formation of a more permeable hydrated product. There seems to be little definite scientific evidence to support these allegations and, although alkali content records before 1960 are not exhaustive, there do not appear to have been any substantial changes in either the range or the weighted average alkali content of UK cements (Corish & Jackson, 1982). A more probable reason for the apparent upsurge of ASR between 1960 and 1975 is that this period also corresponds to a major growth period in UK construction, most

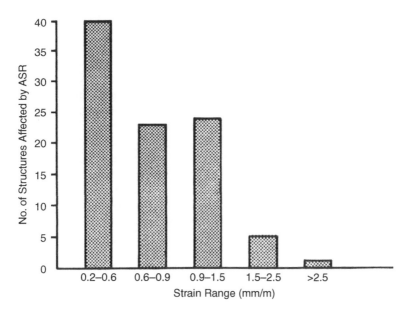

Figure 6.21 Expansion of concrete cores taken from 93 UK structures affected by ASR (from ISE, 1988).

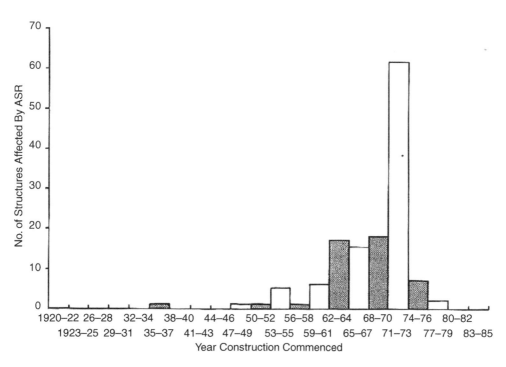

Figure 6.22 Diagram showing the frequency of the identified occurrence of ASR in construction of UK structures between 1920 and 1985. The histogram is not weighted to reflect the variations in intensity of construction (from ISE, 1988).

notably the rapid expansion of the motorway system. Also, like any other new phenomenon to be discovered, it might be assumed that some earlier examples were misdiagnosed and either successfully repaired or demolished.

Whatever the number of affected UK structures may be, very few case studies have been published, even in part form, and most information continues to remain confidential or on restricted access. Some of the best-publicised examples are structures along the A38 trunk road in Devon, including the Marsh Mills Viaduct on the outskirts of Plymouth, an underbridge at Plympton and the Voss Farm bridge further east. The Marsh Mills Viaduct, which was built in 1969–70, consisted of two separate three-lane carriageways, each supported on a series of T-shaped piers. The deck beams were precast in East Anglia and were not affected by ASR, but the piers, cross-beams, deck slab and pile caps were found to be damaged to varying extents by ASR (Mott Hay & Anderson, 1986). As with other cases in the Plymouth area, the reactive combination consisted of crushed limestone coarse aggregate, a sea-dredged sand containing a subordinate proportion of chert and the local high-alkali Portland cement from Plymstock works. Areas of concrete subjected to severe wetting were most seriously affected, including the cross-beam ends beneath deck joints, but most notably the buried pile caps, which were perpetually wet in the marshy ground (Figure 6.23).

Loading on critical elements was reduced by permanently reducing the traffic flow from three to two lanes on each carriageway, and steel stanchions were installed beneath the cantilever heads of the piers (Figure 6.24), to reduce reactions on the bearings on the ends of cantilevers and to provide a more direct load path to the piles to limit shear stresses in the badly cracked pile caps. In 1992, the ASR-affected viaduct

Figure 6.23 Marsh Mills Viaduct, near Plymouth, showing an excavation to reveal a severely cracked concrete pile cap, largely below the local water table.

Figure 6.24 A pier of the Marsh Mills Viaduct, near Plymouth, showing the steel stanchions being installed as a measure to support the cantilever heads and modify load transference.

was largely replaced by an altogether larger construction commensurate with greatly increased traffic requirements.

The Voss Farm overbridge, also built around 1970, was found to exhibit severe cracking attributable to ASR in the columns, the abutments and wing walls and the foundations (Figure 6.25). The concrete mix was similar to that found to be reactive elsewhere in the region, with a minor proportion of chert being the active constituent. The circular columns were heavily reinforced and the crack pattern was wholly orientated according to the stress field, so that only vertical cracks were apparent with no trace of the map cracking often considered typical of ASR (Figure 6.26). Corrective action included the replacement of the columns, and the ground anchoring and refacing of the abutments and wing walls.

Of the cases of ASR in the English Midlands, probably the Smorrall Lane overbridge on the M6 near to the Corley services has been most widely reported (Wood *et al.*, 1987). The two-span bridge, which was built in 1969, exhibited some severe cracking of the central pier-top beam, the deck beams and the soffit of the deck slab; cracking patterns on the beams were reminiscent of flexural and shear cracking (Figure 6.27). Petrographic examinations (Sandberg, 1984) established the presence of ASR (Figure 6.28), with the reactive constituent being subordinate chert present in both the coarse and fine aggregates, which were otherwise dominated by quartzite. Structural engineers considered that *"the two simply supported spans could become an unpredictable hazard within 5 years"* (Wood *et al.*, 1987) and, consequently, the

Figure 6.25 Voss Farm Overbridge, Devon, showing a severely cracked wing wall.

concrete decks were replaced by new steel constructions; the old concrete beams were removed and retained for continued monitoring and further testing.

Another example of ASR in the Midlands has been described in detail by Nixon and Gillson (1987). In 1982, cracking of concrete bases at an electricity substation were once again reported, this time at Drakelow Power Station, near Burton-on-Trent. The main reactive constituent was thought to be chert, present at around 3% of the otherwise sandstone and quartzite sand and gravel aggregate, although some particles of rhyolite may also have reacted (Sandberg, 1986). There were indications that the concrete alkali contents, which were not originally unduly excessive (see 6.6), might have been enhanced by alkalis derived from the aggregates and also by alkali migration into the upper parts of the slabs caused by moisture movements.

This review of publicised cases of ASR in the UK, as with earlier such reviews (Palmer, 1978; Allen, 1981), remains incomplete, unavoidably eclectic and obviously now outdated. Identified cases in south-west England now include foot, road and rail bridges, car parks, office, shop, college and hospital buildings, and concrete blocks at

a)   b)

Figure 6.26 Voss Farm Overbridge, Devon, (a) showing vertical cracking in a heavily reinforced circular column (steel straps installed as temporary precaution). (b) Photograph taken after removal of the column, showing the development of some horizontal cracking in addition to the dominant vertical cracking. (Photograph by courtesy of G.V. Walters of ECC Quarries Limited, Exeter, Devon).

substations, TV transmitters and microwave towers. Jetty, bridge, river wall, reservoir and electricity substation structures have all been found to be affected in South Wales. In the Midlands, identified cases have included foot and road bridges, motorway junction complexes, reservoir structures, colliery shafts and electricity substation bases. Sporadic or 'special' cases (see 6.5.6) have been reported from other areas and, even in 2015 (when this chapter was updated), other cases probably remain to be discovered. It is important to remember that these identified cases still represent a small proportion of the total stock of concrete structures in the UK and also that the severity of any damage is frequently not critical. Moreover, as will become clear in later sections, the occurrence of some ASR in existing UK structures is now handled as a routine matter of identification, assessment and then repair and/or management.

6.5.6 Special cases and circumstances

Apart from the Val de la Mare Dam in Jersey, UK cases of ASR have predominantly involved chert as the reactive constituent, and the physical manifestations of reaction have usually been restricted to surficial cracking, some surface discolorations and, occasionally, small displacements of the concrete. As already noted in subsection 6.5.4 some greywacke aggregates have also emerged as particularly, if slowly, alkali-silica reactive. Some other 'special' exceptions and circumstances are described in this section, with the question of manufactured and recycled aggregates being considered separately in the succeeding section (6.5.7).

As early as 1978, a small dam in Scotland, built in 1936, was found to exhibit evidence of ASR involving a greywacke aggregate derived from an excavation on site

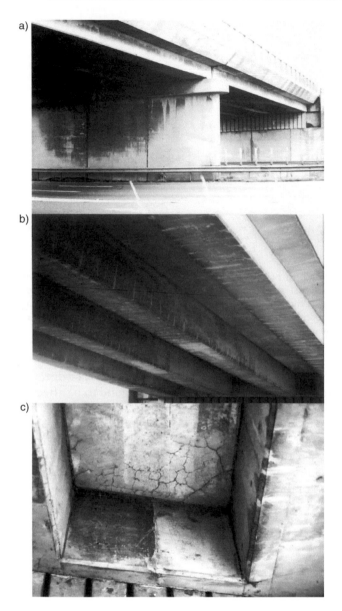

Figure 6.27 Smorrall Lane Overbridge, Warwickshire, (a) general view, (b) showing regular tensile-style cracking of deck beams across the eastbound carriageway and (c) map cracking on soffit of deck (from Sandberg, 1984).

and apparently not used for any other construction (Palmer, 1978). It appeared that the reaction had not caused any disruption and it was considered that the reaction had proceeded to completion, so that no remedial action was required. More recently, a number of structures in Wales and the north of England, also made using greywacke or similar aggregates, have been identified as exhibiting ASR. The old Maentwrog Dam, in

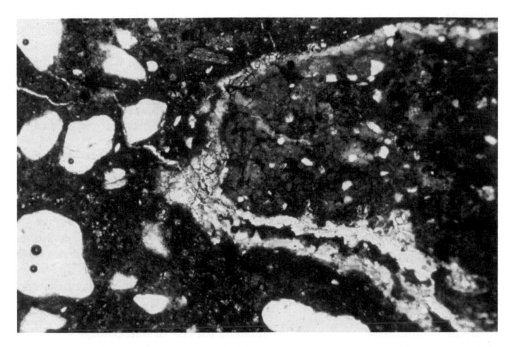

Figure 6.28 Photomicrograph of concrete from Smorrall Lane Overbridge deck slab, showing a reaction centre involving chert. Pln. Pols. (approx. x 15 magnification, from Sandberg, 1984).

North Wales, which was completed in 1927 and has now been superseded by a larger dam, exhibited extensive cracking and heavy white exudations on the downstream face and had apparently experienced several phases of repair during its lifetime. An 'alkali-greywacke' form of ASR was observed in the concrete under the microscope (Anon, 1986; BCA, 1988), as explained in subsection 6.5.4, wherein other cases of greywacke ASR are also identified.

It was noted in subsection 6.5.5 that the apparent upsurge of ASR in UK concrete structures probably coincided with the period of intense highway construction in the years from about 1960, when the motorway system started to be rapidly developed. By the mid-1980s, many concrete bridge and similar highway structures had been identified as exhibiting ASR, yet concrete road pavements were also likely to be at high risk, being wet for long periods and having de-icing salts routinely applied to their surfaces during the winter. Accordingly, the Transport Research Laboratory (TRL) commenced a survey in 1986, the results of which were summarised by West & Sibbick (1988, 1989) and have since been very well described by West (1996). They found that only about 10 % of the concrete roads they examined showed some evidence of ASR and concluded that ASR was not a major problem in existing roads, when the reactive aggregate particles were most usually porous white flint, albeit in the coarse aggregate (whereas the reactive constituent in the bridge examples had typically been in the coarser fraction of the sand). However, a key finding was that ASR could be facilitated by deeper water penetration into damaged and cracked areas of a susceptible concrete road pavement, especially in the vicinity of joints and/or when de-icing salts were in use.

One particular case study by Sibbick & West (1992) concerned the Padiham Bypass in North-west England, which was severely cracked and was considered similar in materials and construction to a similarly damaged part of the nearby M6 motorway. There were two layers of concrete, both of which exhibited ASR damage, by chance caused by different aggregates. A lower layer contained a siliceous limestone aggregate, whilst an upper layer contained a siltstone aggregate, microcrystalline quartz being the reactive constituent in each material. Sibbick & West's distinctive and effective diagrammatic method for demonstrating the reaction evidence is exemplified in Figure 6.29.

It had been found by Swamy (1988), experimenting with highly reactive types of aggregate, that externally derived alkalis, especially in the form of de-icing salts, seemed to *"have the quickest and most devastating effects on hardened concrete undergoing ASR expansion"*. The later TRL work similarly found that de-icing solutions based on sodium chloride (halite) had exacerbated the ASR damage, though the mechanism remained unclear. Further work by Sibbick & Page (1996, 1998) confirmed that ASR expansion could be induced by sodium chloride solutions for concretes made using potentially reactive aggregate combinations but using alkali contents that would not otherwise cause such expansion; reaction was even induced for some aggregates of lower reactivity. Dunster & Crammond (2003) have reported cases of ASR damage with a concrete made using 'normally' reactive aggregate (see subsection 6.6.2) and otherwise conforming with measures to minimise the risk of ASR damage, caused by de-icing treatment leading to ponding on crosshead beams and resultant concentrations of saline solution.

Research and experience has also confirmed that, whilst the reactive alkali content in the concrete mix is probably the most important factor controlling the likelihood and magnitude of ASR in a concrete containing a potentially reactive aggregate combination, those alkalis can be augmented in service. We have already seen that saline solutions deriving from de-icing measures can exacerbate ASR. It has also become clear that the rocks and minerals in some aggregates can 'release' alkalis within the concrete over a long period, which was thought to be a factor with the UK greywacke concrete cases, whilst long-term recrystallisation of ASR gel might also eventually 'recycle' alkalis for further reaction (French, 2005). Page *et al.* (1992) demonstrated that even the zones of enhanced alkalinity associated with cathodic protection could give rise to localised ASR. Al-Kadhimi & Banfill (1996), however, found that the re-alkalisation process, to reverse the loss of protection to reinforcing steel caused by carbonation, did not necessarily lead to enhanced ASR in potentially susceptible concretes, possibly owing to the absence of calcium hydroxide and reduced hydroxyl concentration in the carbonated concrete.

In other parts of the world, surface 'pop-outs' have sometimes been a common symptom of ASR, but this has rarely been the case in the UK, where neither 'pop-outs' nor copious gel exudations are typical features of concrete surfaces on structures diagnosed as being affected by ASR (BCA, 1988). Sims and Sotiropoulos (1989), however, have described some 'special' cases in which surface 'pop-outs' and associated gel exudations are the only defects (Figure 6.30), with no cracking, expansion or other damage being apparent. In these cases, of which several different locations were investigated, the reactive particles comprised chert which formed only a very minor percentage of an otherwise wholly oolitic limestone aggregate. Localised ASR was able

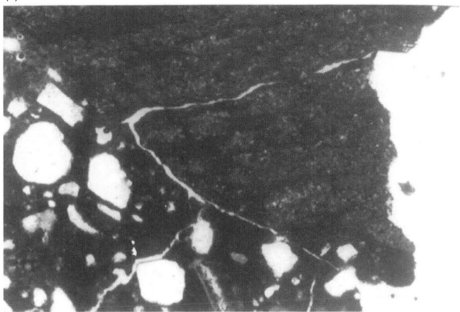

(a)

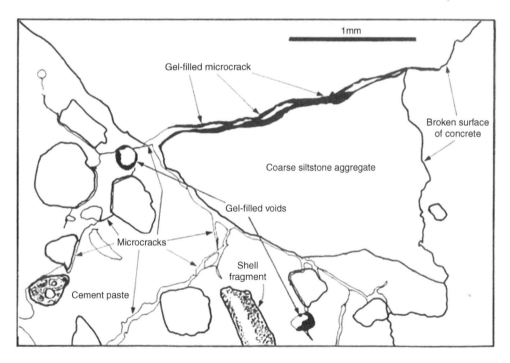

1mm

Gel-filled microcrack

Broken surface
of concrete

Coarse siltstone aggregate

Gel-filled voids

Microcracks

Shell
fragment

Cement paste

Figure 6.29 Photomicrograph and explanatory drawing, showing a) alkali-reaction of siltstone aggregate
in concrete from the Padiham Bypass; a crack in the siltstone is partially filled with ASR gel
and the overall microcracking pattern is 'stepped' to include the gel-filled crack in the
aggregate, and b) a large flint aggregate exhibiting a well-developed branching system of
microcracks with associated gel (from West, 1996).

b)

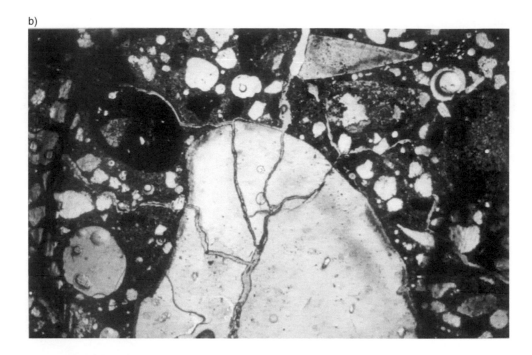

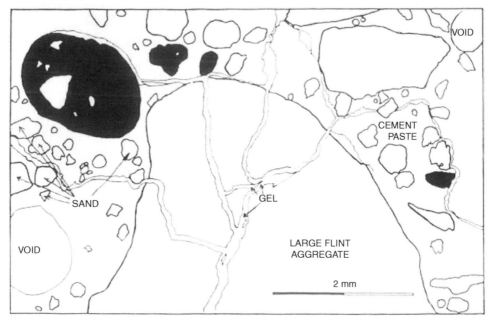

Figure 6.29 (Cont.)

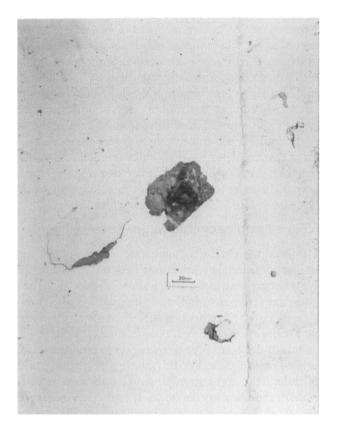

Figure 6.30 Unusual UK example of ASR in which surface gel exudations and pop-outs are the only detrimental features. Chert particles react in a surface zone which is greatly enriched in alkalis (from Sims & Sotiropoulos, 1989).

to occur because of the very substantial concentration of alkalis that was brought about by the continuous migration of moisture though basement retaining walls (Table 6.4), the back sides of which were in direct contact with wet ground and the front sides of which were exposed to an internal drying environment. The alkali content was found to

Table 6.4 The concentration of alkalis ($Na_2Oeq.$, kg/m^3) in the internal surface zone of concrete retaining walls as the result of the continuous migration of moisture; the surface zone is greatly enhanced in alkalis, whilst the centre of the wall is probably depleted in alkalis (after Sims & Sotiropoulos, 1989).

Structure location	Core sample	Centre of concrete wall	Internal surface zone of concrete wall[*]
A	1	1.4	10.1
	2	1.4	12.6
B	1	0.8	15.8
	2	2.0	18.0
	3	0.7	13.1

[*] Approximately 0–50 mm in case of location A and 0–25 mm in the case of location B.

be enhanced by up to five times by this process, and it was thought possible that sealing or painting of the front concrete surface might have exacerbated the effect.

It had been observed at the Maentwrog (old) Dam (see subsection 6.5.4) that some concrete had been damaged by freeze-thaw action, but it was unclear whether or not the long-term ASR damage had caused the concrete to become more susceptible to freeze-thaw action. However, a programme of work by Bolton (1989), concerned with the risk to more than 3500 concrete railway bridges, of which only about 12 had thus far been confirmed with ASR damage, confirmed that ASR made the structural concrete (at least C50 strength grade) more susceptible to freeze-thaw action; resultant expansion was greater than that recorded for ASR alone and led to disintegration. It is now understood that combinations of ASR with other deteriorative mechanisms, such as freeze-thaw or delayed ettringite formation, can behave synergistically.

6.5.7 Manufactured and recycled aggregates

The possibility that some 'manufactured' aggregates in use in the UK might prove to be alkali-reactive was first considered by Jones and Tarleton (1952, 1958a), who included some air-cooled blastfurnace slag in their BRS survey. Jones and Tarleton also utilised Pyrex glass as a known reactive material, and glass has long been recognised as being potentially reactive with Portland cement; glassfibre reinforcement of concrete can only be successful using alkali-resistant glass (Poole & Sims, 2015/2016). Figg (1981) has described a case in which decorative glass aggregate in concrete cladding panels for a building in north-east London caused ASR, resulting in cracking, distortion and spalling of the units. Although the non-structural panels had been made using a low-alkali white Portland cement, the glass aggregate itself had contained sufficient reserves of available alkalis to perpetuate the expansive reaction.

Although many of the aggregates that are manufactured or derived from industrial waste products are both siliceous and poorly crystalline or even glassy, there appear to have been no reports of any concrete structures in the UK being damaged as the result of ASR involving manufactured aggregates, other than the decorative glass already mentioned. There have also been very few published attempts to assess the ASR potential of such manufactured aggregate materials. Sims and Bladon (1984) carried out a limited programme to investigate the ASR potential of sintered pulverized-fuel ash (PFA), which is widely used as a lightweight aggregate ('Lytag'). No significant expansion of mortar-bar and small concrete specimens was recorded, and yet microscopical examination of the concrete at various ages suggested that some gel production might have occurred, impregnating the cement paste matrix but seemingly not giving rise to expansive forces. It is also possible that the voided nature of such lightweight materials might allow a considerable degree of ASR to be accommodated without any disruptive stresses being generated within the mortar or concrete. Further research remains needed before manufactured aggregates can be safely regarded as being incapable of expansive ASR in concrete over a prolonged period of service in conducive exposure conditions.

Until recently, there has only been limited use of recycled aggregates in new concrete in the UK (Dhir & Paine, 2007), but this situation is changing, especially with potential use of coarse 'recycled concrete aggregate' (RCA) being recognised in BS EN 8500-2 (2006), where it is permitted for use in concrete up to strength Class C40/50 and with certain durability classes. In terms of AAR risk, concern over the use of RCA has usually focused on the possibility of additional alkalis being derived from the residual cementitious component of the recycled concrete and/or unsuspected reactive constituents being present in the older aggregate. However, an extensive research programme for WRAP (Waste and Resources Action Programme), carried out at the University of Dundee (McCarthy *et al.*, 2009), found encouragingly low expansions in AAR tests. In the current version of BRE Digest 330 (2004), RCA is accepted as having 'normal reactivity' (see subsection 6.8.2), whereas it was previously deemed to have 'high reactivity' as a precaution.

6.5.8 Urgent search for remedies

As the number of confirmed cases of ASR built up in the UK and it became clear that the problem was no longer restricted to 'foreign' or 'special case' occurrences, it became urgent to identify the main causal factors and so to formulate measures for avoiding the reaction in future construction. At the initiative of the C&CA, an 'Inter-Industry' Meeting was held on 10 June 1981, at which it was agreed to invite Michael Hawkins, as the engineer of the county then most afflicted with cases of ASR (Devon), to form an independent working party of specialists to co-operate in the production of some guidance notes for the practising engineer. The first report of this 'Hawkins Working Party' (Hawkins, 1983) was published (see 6.6.1) and later updated, as a Concrete Society (1987) technical report (see 6.6.2). BRE Digest 258 (1982) was published shortly before the first 'Hawkins Report' and heralded a number of the concepts shortly to be endorsed by the Hawkins Working Party, including the naming of aggregate group classifications ('trade groups') considered 'very unlikely to be susceptible to attack by cement alkalis' and the notion of using a threshold limit on concrete alkali content of 3 kg/m^3, as an alternative to the use of 'low-alkali' cement in an endeavour to avoid ASR in new construction.

Cement alkalis could be established by analysis and were, in any case, usually obtainable upon request from the manufacturer. By contrast, BRE Digest 258 (1982) maintained that, *"There are as yet no British Standard tests for the susceptibility of aggregates to attack by alkalis and ASTM tests have not been found to predict the reactivity of UK aggregates accurately"*. This statement is significant in view of the reliance that had previously been placed on the mortar-bar test by the BRE/BRS in the 1950s, during their previous investigations into the ASR potential of British aggregates. Appropriately, the baton was taken up by the BSI in 1981, when the aggregates committee, CAB/2, formed a new working group, WG1O, to evaluate methods for assessing the ASR potential of aggregates (see 6.6.4). The following Section 6.6 charts the initial and developing approach to minimising the risk of ASR in the UK, but it will be seen in later Section 6.8 that this was gradually refined by research and increased experience to the established and apparently successful scheme in place at the present time.

6.6 COUNTERMEASURES: MINIMISING THE FUTURE RISK OF ASR

6.6.1 Hawkins Working Party – I

The 'Hawkins Report' was published in 1983 (Hawkins, 1983) and emphasised that the incidence of ASR in the UK was 'small', but the importance of 'minimising the risk of ASR' in future work was embodied by the warning that, *"When ASR has been diagnosed in a structure, there are no methods which can be reliably recommended at present for either preventing further damage or carrying out effective and lasting repairs"*. The report was presented as practical guidance, based on contemporary information and intended only to apply to materials, conditions and practice encountered in the UK. This last caution was considered to be particularly important in the Republic of Ireland at that time, where it was maintained that AAR was less likely to occur (see section 6.9).

It was considered that ASR causes damage only if the three following factors are all present: sufficient moisture, sufficient alkali and a critical amount of reactive silica in the aggregate. The recommendations given in the Hawkins guidance notes were thus based on ensuring that at least one of these factors was absent. The risk of ASR for a particular concrete in a particular environment could be assessed by reference to these factors (Figure 6.31) and, if a risk was deemed thereby to exist, the same factors provided a range of possible optional precautions to be implemented. Two levels of precaution were envisaged: a level of 'normal concrete construction' which it was supposed would apply in the majority of cases, and a level for 'particularly vulnerable forms of construction', such as concrete 'buried in water-logged ground with the top surface exposed' (*e.g.* electricity substation bases and some structure foundations).

A sufficiency of moisture for damaging ASR to occur was considered to be any exposure to moisture, including relative humidity levels exceeding 75%. In practice this meant any externally exposed concrete was in an environment deemed conducive to ASR.

The amount of 'reactive alkali' available was considered in three main ways according to: (a) the alkali content of the cement, (b) the alkali content of a blend of cement with a certain minimum content of a mineral addition, such as ground granulated blastfurnace slag (GGBS) or pulverized- fuel ash (PFA), or (c) the alkali content of the concrete. In respect of (a) above, the Hawkins Report stated, *"The use of Portland cement with an alkali content of 0.6% or less is accepted worldwide as the best means of minimising the risk of damage due to ASR"*. Only sulphate-resisting Portland cement (SRPC) was then available in the UK with a guaranteed alkali content of not more than 0.6%, but 'ordinary' Portland cement (OPC) could be found to be 'low-alkali' by examination of manufacturer's analytical data making due allowance for variations, or be rendered 'low-alkali' by the controlled replacement of part of the cement by GGBS or PFA, which were assumed to contribute 'no reactive alkali to the concrete'.

In respect of (b) above, a combination of GGBS and any UK Portland cement which contained 50% or more GGBS was considered equivalent to a 'low-alkali' cement. Similarly, a combination of PFA and any UK Portland cement which contained 25% or more PFA was considered equivalent to a 'low-alkali' cement, with the added condition that the alkali content of the concrete provided by the Portland cement in the combination was not more than 3.0 kg/m^3.

Finally, in respect of (c) above, it was considered that damage from ASR was unlikely to occur if the concrete had an alkali content of 3.0 kg/m^3 or less, which was calculated from the alkali content of the cement and the maximum expected cement content of the concrete, making due allowances for future variations in cement alkali contents and concrete batch weights.

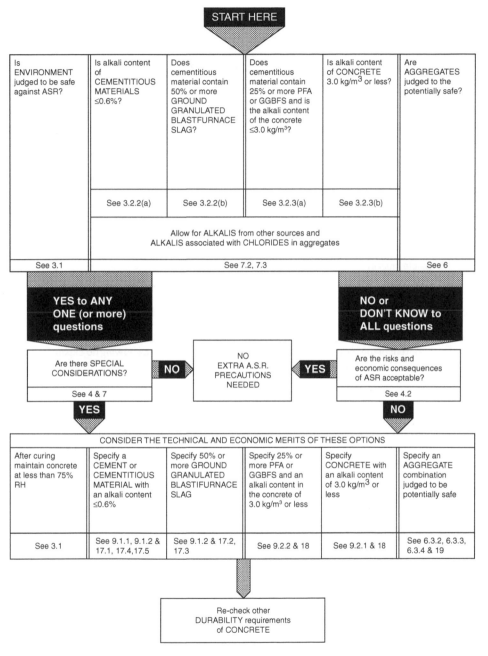

Is ENVIRONMENT judged to be safe against ASR?	Is alkali content of CEMENTITIOUS MATERIALS ≤0.6%?	Does cementitious material contain 50% or more GROUND GRANULATED BLASTFURNACE SLAG?	Does cementitious material contain 25% or more PFA or GGBFS and is the alkali content of the concrete ≤3.0 kg/m³?	Is alkali content of CONCRETE 3.0 kg/m³ or less?	Are AGGREGATES judged to the potentially safe?
	See 3.2.2(a)	See 3.2.2(b)	See 3.2.3(a)	See 3.2.3(b)	
	Allow for ALKALIS from other sources and ALKALIS associated with CHLORIDES in aggregates				
See 3.1	See 7.2, 7.3				See 6

START HERE

YES to ANY ONE (or more) questions

NO or DON'T KNOW to ALL questions

Are there SPECIAL CONSIDERATIONS?	**NO**	NO EXTRA A.S.R. PRECAUTIONS NEEDED	**YES**	Are the risks and economic consequences of ASR acceptable?
See 4 & 7				See 4.2

YES

NO

CONSIDER THE TECHNICAL AND ECONOMIC MERITS OF THESE OPTIONS

After curing maintain concrete at less than 75% RH	Specify a CEMENT or CEMENTITIOUS MATERIAL with an alkali content ≤0.6%	Specify 50% or more GROUND GRANULATED BLASTIFURNACE SLAG	Specify 25% or more PFA or GGBFS and an alkali content in the concrete of 3.0 kg/m³ or less	Specify CONCRETE with an alkali content of 3.0 kg/m³ or less	Specify an AGGREGATE combination judged to be potentially safe
See 3.1	See 9.1.1, 9.1.2 & 17.1, 17.4,17.5	See 9.1.2 & 17.2, 17.3	See 9.2.2 & 18	See 9.2.1 & 18	See 6.3.2, 6.3.3, 6.3.4 & 19

Re-check other DURABILITY requirements of CONCRETE

Figure 6.31 Simplified flow chart for the assessment of risks and precautions against ASR, according to the guidance given by the Hawkins Working Party. From Technical Report No. 30 (Concrete Society, 1987), which developed from the 1983 version and should be consulted for full details. See subsection 6.6.2 for description of Concrete Society (1987).

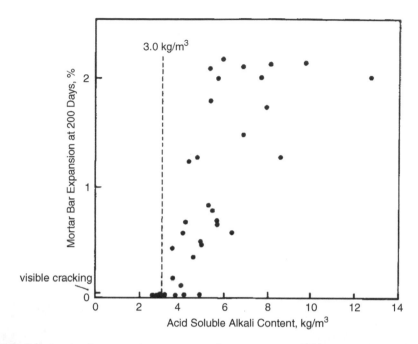

Figure 6.32 Relationship between the expansion of mortar bars at 20°C and the acid-soluble alkali content of the mortar, using opal as the reactive constituent, showing no expansion at alkali levels lower than 3.5 kg/m³. After Hobbs (1988).

This approach was not considered adequate for 'particularly vulnerable construction'. The value of 3.0 kg/m³, which was coincidentally equivalent to using a high-alkali cement of, say, 1.0% Na_2Oeq. in a typical structural concrete of, say, 300 kg/m³ cement content, seems largely to have been based upon early research by Hobbs (1980) at the C&CA (Figure 6.32), using Beltane opal and mortar-bars, although it was also considered that the results agreed with earlier findings in Germany and elsewhere (Palmer, 1981; Hobbs, 1988).

The 1983 Hawkins Report was not able to be particularly helpful in the guidance on aggregates. It was stated that, *"There are at present no British Standard tests for the alkali-reactivity of aggregates"*, but the American chemical (ASTM C289: 1987) and mortar-bar (ASTM C227: 1987) methods *"commonly used in other countries"* were considered not to be *"a practical basis for the specification of British concrete materials"*. Modifying the group classification approach put forward in BRE Digest 258 (1982), the Hawkins Report stated that, *"At present, it seems that aggregates consisting wholly of rock types in the Basalt, Gabbro, Granite, Gritstone, Hornfels, Limestone, Porphyry and Schist trade groups, would be classified in the UK as 'unlikely to be reactive' "*. The word 'wholly' was explained as meaning both coarse and fine aggregates, and that the source was not contaminated by reactive silica (such as the secondary opal veining in the granodiorite source in Jersey). Consideration of the composition of an aggregate could therefore enable an aggregate to be adjudged 'unlikely to be reactive' or 'contains constituents which are sometimes found to be reactive', but, in assessing the amount of reactive silica in the latter case, the 1983

Hawkins Report warned that, *"the critical proportion is not capable of easy prediction and in any case will vary with the aggregate combination"*.

In addition to the three main controlling factors (moisture, alkali content and reactive aggregate), the possibilities of alkali migration through moisture movement and the contribution of alkalis from sources other than cement (from other constituents or from external sources) were recognised, although no additional precautions could be recommended, save by tanking the concrete or by the use of definitely non-reactive aggregates. The working party was also anxious to ensure that zealous efforts to achieve the avoidance of ASR did not lead to the neglect, or even act to the detriment, of other considerations: *"It is essential that adjustments which may be made to the mix in order to avoid ASR should never lead to the use of materials, cement contents or water/cement ratios which will be inadequate to ensure general durability."*

The first Hawkins Report was well received in the industry, but it was only guidance and, in the absence of a BS specification, it was clear that a wide range of approaches to 'minimising the risk of ASR' would arise from different specifiers, including extreme cases where all of the 'optional' precautions would be required to apply instead of just one of the options chosen according to circumstances. The Hawkins Working Party therefore remained in being, now affiliated to the Concrete Society, and commenced work on 'model specification clauses'.

6.6.2 Hawkins Working Party – II

The second substantive edition of the Hawkins Report was published as a Concrete Society Technical Report, No. 30 (Concrete Society, 1987), although an earlier consultation document had been published in 1985. The new edition was styled 'Guidance Notes *and Model Specification Clauses*' (chapter author's emphasis) and the opportunity had been taken to revise and update the guidance notes, which were otherwise based upon the same principles and followed the same general approach as the original Hawkins Report.

In his Foreword to the new edition, Michael Hawkins reaffirmed that damage by ASR affects only a small proportion of concrete construction undertaken in the UK, but added, *"Nevertheless, when it does occur the cost of remedial work can in some cases be very high"*. The report still accepted that ASR tended to occur most frequently in certain regions, but warned that, *"The increasing transportation over long distances of both materials and concrete products means that the potential for damage from ASR will not be restricted to particular regions"*. For this reason, and others, the simple UK map depicting the locations of reported cases of ASR, which had been included in the first edition, was omitted from the revision. The model specification clauses were based upon the revised guidance notes and will not be separately described in this summary, except where convenient to clarify the advice.

The main revisions in the guidance notes concerned the contributions of reactive alkalis from GGBS and PFA, the derivation of additional alkalis from sodium chloride and other sources, and some changes and improvements to the guidance about aggregates. In the first Hawkins Report, the GGBS and PFA components were, for the purposes of that report, considered to contribute no reactive alkali to the concrete system. In the new report, the reactive alkali content of GGBS and PFA is defined as being the water-soluble alkali content, when determined using the BS 812 (1984, 1985)

Site combinations of Portland cement with either ground granulated blastfurnace slag (GGBFS) or PFA

When cement to BS 12 is combined on site with either GGBFS or PFA the reactive alkali content of the concrete is calculated from:

$$A = \frac{(C \times a) + (E \times d)}{100}$$

where A = reactive alkali content of concrete (kg/m³)
       C = target mean Portland cement content of concrete (kg/m³)
       a = reactive alkali content (%) of cement
       E = target mean content of either GGBFS or PFA in the concrete (kg/m³)
       d = average reactive alkali content (%) of either the GGBFS or the PFA as provided by the manufacturers.

Figure 6.33 Calculation of the 'reactive alkali content' of concrete in accordance with the Concrete Society TR30 guidance, one of several equation options given in the report. After Concrete Society (1987).

extraction method for aggregates and the BS 4550 (1970, 1978) analysis method for cements. In consequence, when the concrete alkali content was being calculated, the GGBS or PFA reactive alkali contents were to be taken into account (Figure 6.33).

This approach was disputed, and some authorities maintained that the reactive alkali content of these additions should be based upon a notional proportion of the acid (rather than water) soluble alkali determination: 50% in the case of GGBS and 17% (or one-sixth) in the case of PFA (BRE, 1988). The Hawkins Working Party established a technical subcommittee to carry out further research into this controversy and later the notional acid soluble criteria would be adopted, albeit in the context of a new binder classification approach (see subsection 6.8.2).

In the first Hawkins report, it had been recognised that sodium chloride in sea-dredged and some other aggregates might be a source of additional alkalis, but it was only suggested that the CP 110 (BSI, 1972) limits on chloride for reinforced concrete should be applied. By the time of the 1987 revised edition, it had become clear from experiences in Denmark and from work at the BRE (Nixon *et al.*, 1988) that the addition of sodium chloride to the concrete mix increased the hydroxyl ion concentration and could promote greater ASR expansion. Consequently, provision was made, in the calculation of concrete reactive alkali content, for the additional sodium associated with any chloride present in the coarse and fine aggregates. No recommendations were made in respect of any reactive alkalis which could possibly be released from some rock types found in aggregates, although the opinion was expressed that, *"No significant amounts of reactive alkali will be derived...(chlorides excepted) from natural aggregates in the UK"*.

The guidance on aggregates was amended in two important ways: the classification of aggregates 'unlikely to be reactive' was improved and extended, and a new concept was introduced regarding aggregates comprising largely flint or chert. The list of rocks and minerals considered 'unlikely to be reactive' was revised (Table 6.5) according to the updated BS 812-102 (1984), in which the 'trade groups' had been replaced by a more

Table 6.5   Rocks and minerals 'unlikely to be reactive', according to the
guidance notes given in Technical Report No 30 (Concrete
Society 1987). The list of rock names was based upon that given in
BS 812-102 (1984).

Andesite	Feldspar[*]	Microgranite
Basalt	Gabbro	Quartz[*†]
Chalk[‡]	Gneiss	Schist
Diorite	Gneiss	Slate
Diorite	Granite	Slate
Dolerite	Limestone	Syenite
Dolomite	Marble	Trachyte
		Tuff

* Feldspar and quartz are not rock types but are discrete mineral grains
   occurring principally in fine aggregates.
† Not highly strained quartz and not quartzite.
‡ Chalk is included in the list since it may occasionally be a minor
   constituent of concrete aggregates.

straightforward list of common rock types found in aggregates, and an appendix was included to provide some guidance on those rock types which had not been included on that tabulated list. Feldspar and (unstrained) quartz were included on the list of constituents 'unlikely to be reactive', enabling many fine aggregates to be assessed more readily.

A dilemma was presented by chert and flint, because although these constituents had been involved with most of the UK examples of ASR, they are extremely common in UK aggregates, and in the highly built-up south-east of England they form the major part of most aggregate combinations. The 1987 revised Hawkins Report advised that, *"Experience to date from cases of ASR involving sands and gravels indicates that up to a maximum of 5% by mass of flint, chert or chalcedony taken together in the combined aggregate can be tolerated"*. However, at the other end of the range, *"It is currently considered that a combination of fine and coarse aggregate which contains more than 60% by mass of flint or chert is unlikely to cause damage due to ASR"*. In other words, in the absence of any other potentially reactive constituent, aggregates either containing no more than 5% or alternatively containing more than 60% chert and flint are to be regarded as 'unlikely to be reactive'. The value of 60% was based partly upon observation, whereby concretes made with predominantly flint aggregates were generally not found to be affected by damaging ASR, and partly upon limited laboratory experiments at the BRE, the results of which have since been confirmed by further work at the BRE and elsewhere (Figure 6.34).

No 'detectable' opal was permitted for aggregates 'unlikely to be reactive' and it is important to realise that opal is considered by the report to be unacceptable under any circumstances: *"Sources (of aggregate) known to contain opaline silica should not be used even when the alkali content of either the cementitious material or the concrete is being controlled."* This was because there is some evidence that small amounts of opal in concrete can cause ASR damage even when the cement alkali content is less than 0.6% or the concrete alkali content is less than 3.0 kg/m^3.

No specific guidance is given in respect of alkalis derived from any sources other than the cementitious or aggregate materials, but the report requires that, *"If alkalis in excess of 0.2 kg/m^3 of concrete come from other sources they must be taken into*

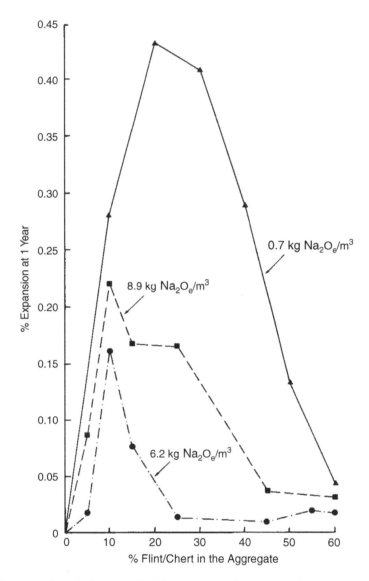

Figure 6.34 Some relationships between flint/chert content of aggregate and concrete prism expansion, indicating a flint/chert 'pessimum' of 10-30% for flint-limestone combinations and no significant expansions for mixtures containing more than 60% flint. Redrawn and adapted from Nixon *et al.* (1989) & Walters (1989).

account". In effect this meant that many engineers would require exhaustive alkali audits to be carried out for all the concrete mixes proposed for a given project.

6.6.3 Department of Transport and other specifiers

BS 8110 (1985) superseded CP 110 (BSI, 1972) as the Code of Practice for 'Structural Use of Concrete'. Whereas CP 110 had not mentioned ASR, BS 8110 (in its clause

6.2.5.4) included recommendations that followed the first Hawkins Report (1983) and BRE Digest 258 (1982). Other national authorities maintained concrete specifications which were widely recognised and respected, including *inter alia* the Property Services Agency (PSA), British Rail, the Central Electricity Generating Board (CEGB) and the British Airports Authority (BAA), but perhaps the most important in this ASR context was the Department of Transport (DTp). Concrete and concrete materials were (and remain, through the Highways Agency today) included in the DTp *Specification for Highway Works* (1986), which superseded the earlier *Specification for Road and Bridge Works* (1976).

The published 'Specification for Road and Bridge Works' had not contained any reference to ASR. However, in July 1982 a new Clause 1618 was issued by the DTp to its regional offices to amend the specification to include '*Measures to control alkali-aggregate reaction in respect of all structural concrete*' (DTp, 1982). The amendment was generally similar to the guidance which appeared in the BRE Digest 268 (1982) and the first Hawkins Report (1983). The total alkali content deriving from cement was not permitted to exceed 3 kg/m^3 of concrete unless both the coarse and fine aggregates were 'not susceptible to alkali-aggregate reaction' according to criteria stipulated. Low-alkali SRPC or blends of cement and GGBS (in which GGBS was at least 50%) were permitted as means of reducing alkali contents that would otherwise exceed 3 kg/m^3. The concrete alkali content was to be calculated from cement manufacturer's data, adding 0.15% to allow for possible future variations, or alternatively adding twice the standard deviation where that would amount to less than 0.15%.

The criteria in Clause 1618 governing the AAR susceptibility of aggregates were very restrictive, being based upon the classification given in BRE Digest 258 (1982), rather than the slightly modified version given in the first Hawkins Report (1983). Only aggregates in which all of the 'significant' rock components (*i.e.* > 5% by mass of the coarse or fine aggregate) were in the 'trade groups': granite or gabbro or basalt (except andesites) or limestones or porphyry (except dacites, rhyolites and felsites) or hornfels, were to be considered 'not susceptible'. Of the commonest UK coarse aggregates, only the limestones could satisfy these requirements, and almost none of the natural fine aggregates could be assessed as 'not susceptible' according to such criteria. Again the presence of any opal rendered an aggregate 'susceptible', whatever the main composition of the aggregate.

The DTp (1986) later published an extensively revised *Specification for Highway Works*, and this included clauses for the 'control of ASR', which were broadly similar to those contained within the 1985 draft revision of the Hawkins Report. The requirements are essentially similar to those provided in the model specification clauses of the revised Hawkins Report later issued as Concrete Society TR 30 (1987) (see subsection 6.6.2), although the DTp was perhaps slightly more restrictive.

Although the 1986 DTp specification allowed that, "*When the proportion of chert or flint is greater than 60% (by weight) of the total aggregate it shall be considered to be non-reactive providing it contains no opal, tridymite or cristobalite*", the lower limit for chert, flint or chalcedony taken together was only 2% by weight instead of the 5% implied by the Hawkins Report in its guidance notes. Also, whereas the revised Hawkins Report considered that the use of the 'certified average alkali content' of cements eliminated the need for any deliberate variation factors in the calculation of the reactive alkali content of concrete, the DTp specification continued to require that twice the standard deviation was added to the average of 25 daily determinations of cement alkali content by the manufacturer and also that 10 kg should be added to the intended cement content of the concrete to allow for possible batching errors. In practice the difference could sometimes be important, with the Hawkins Report recommendations allowing concrete alkali contents to range up to 3.75 kg/m^3, whilst the DTp requirements sought to make 3.0 kg/m^3 an absolute maximum limit.

The 1986 DTp specification agreed with the 1987 revised Hawkins Report (and its 1985 draft) in requiring only the water-soluble alkali contents of any GGBS or PFA to be taken into account. However, the DTp specification would later be amended in line with BRE Digest 330 (1988), in which 50% and 17% proportions of the *acid*-soluble alkali contents were preferred (see subsection 6.6.2). It will be seen in subsection 6.8, that these slightly differing criteria for minimising the risk of AAR damage to UK concrete structures would eventually converge into consistent guidance.

6.6.4 Application of petrography

UK research into AAR in the 1950s (see subsection 6.2) had recognised that aggregates containing potentially alkali-reactive aggregates were in common usage, even if examples of resultant damage had not been reported or recognised. Later, in the 1970s and 1980s, once cases of ASR damage were admitted and being identified, it was becoming clear that certain aggregate types and combinations were most commonly implicated (see subsection 6.5). As strategies for minimising future risk of ASR damage were developed (Hawkins, 1983), avoiding potentially reactive aggregates had seemed a logical option, with lists of apparently non-reactive geological classifications being suggested and the other options (reducing reactive alkalis and avoiding exposure to wet or humid conditions) being difficult to control or achieve and apparently less decisive. However, apart from a growing recognition of the complexity of using geological (or 'petrographical') examination for predicting reactivity with varying forms and proportions of flint and/or greywacke in aggregates (see subsections 6.5.3 & 6.5.4), Sims (1987) also noted the problem of ensuring representative samples for reliable examination; he thought that petrographic examination was a valuable first stage in an assessment of aggregates that helped to select the appropriate further tests (subsection 6.6.5).

Development of a systematic method for the petrographical examination of aggregates had been started by a BSI working group as early as 1981, although this was not specific to ASR and had only recently been issued as a 'draft for public comment' at the time that this book was first published in 1992. The relatively low proportion of potentially reactive constituents that can sometimes be most critical for ASR in an aggregate (some UK specifications were requiring contents of flint or crushed greywacke to be limited to 5 % or even lower in some cases) produced profound problems in formulating a practicable standard petrographical examination method. Statistical requirements to achieve adequate precision, based on established mineral assaying principles, could necessitate the analysis of prohibitively large samples and also impose complicated statistical controls, both likely to cause difficulties in commercial application. The method was finally published as BS 812-104 (BSI, 1994), including qualitative and quantitative options, and providing detailed statistical guidance and requiring the routine analysis of duplicate sample portions in respect of the latter. This BS method remains current at the time of writing, but is being revised and updated by the Applied Petrography Group, which is affiliated to the Engineering Group of the Geological Society. A 'simplified' method for petrographic description was later published as a European Standard, BS EN 932-3 (BSI, 1997d), but this is not adequate for the detailed assessment of AAR potential.

Once the BS 812-104:1994 method was available, it was necessary to provide some guidance on how the findings could be interpreted in respect of the AAR potential of UK aggregates and this was published as BS 7943 (BSI, 1999c; Sims *et al.*, 2000). Following the ASR experience with small amounts of opaline material in a quarry in Jersey (see subsection 6.4), the guidance in BS 7943:1999 is predicated on the basis that the aggregate contains no detectable opaline silica, with a modified version of the old 'gel-pat' test being adapted (and described in an annex) for the purpose of establishing its absence. By the time that BS 7943 was published, the reactivity of opal-free UK aggregates had been simplified into three levels: 'low', 'normal' and 'high' (see subsection 6.8 & BRE, 1997), based on the acceptance that many UK aggregates offered a degree of reactivity that should be regarded as the 'normal' occurrence, in the absences of any opal and of any definitive evidence to show that a particular material or combination exhibited either 'low' or 'high' levels of reactivity. The guidance in BS 7943:1999 is based on these principles; it also accepts that the microscopical measurement of 'undulatory extinction angles' is *not*, after all, a helpful indicator of quartz reactivity (Grattan-Bellew, 1992; Smith & Dunham, 1992). More recently, petrographic atlases have been published, including some reactive UK aggregates (Figure 6.35), which provide detailed, practical, pictorial guidance for petrographers seeking to assess aggregate reactivity (Lorenzi *et al.*, 2006; Fernandes *et al.*, 2012, 2015); in particular, Fernandes *et al.* (2015) is a tour de force.

The 1986 DTp specification introduced an additional requirement for aggregate sources to be examined as well as processed aggregate samples, so that the prospects of future variations could be geologically assessed. Although the BSI working group undertook an appraisal of such source examination procedures, in practice most considerations of ASR potential have continued to be based upon the examination of representative samples of processed aggregate. Some practical research has suggested that the variations in petrographic composition of processed aggregates over a period of production might be less than expected (Sims & Miglio, 1989).

6.6.5 Quests for tests

The test methods adopted for the BRS research in the 1950s (see 6.2) had all been derived from American experience, and in the early 1950s several of these had evolved into ASTM standard methods (ASTM, 1987a, 1987b). In the UK, presumably because of the apparent reassurance arising from the BRS research programme, there were no similar moves to create any British Standard methods for testing concrete constituents for ASR. Once examples of ASR had been identified on the UK mainland in the mid-1970s, a new urgency was created for finding a predictive test method which would enable the problem to be avoided in new construction. At first it was assumed that the mortar-bar expansion test (ASTM C227: 1987b) was reliable but that the practical problem concerned the long duration of the test: Palmer (1977) said that, *"there is no satisfactory short-term test... Only long-term tests on the aggregate can show with any*

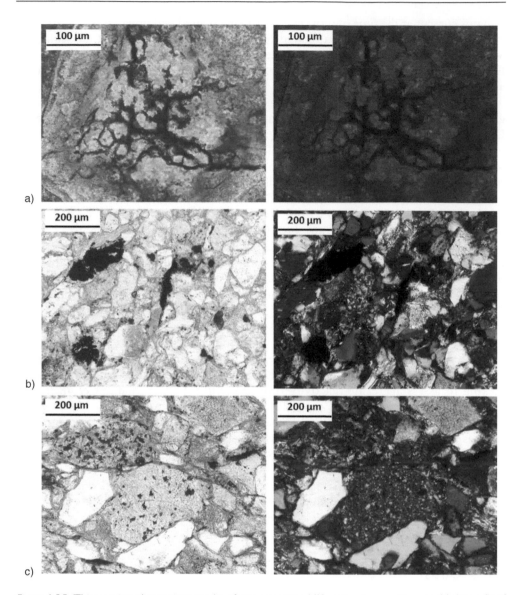

Figure 6.35 Thin section photomicrographs of some reactive UK aggregate constituents: (a) dense flint/ chert comprising cryptocrystalline silica and chalcedony, stained with limonite [left: plane polarised, right: gypsum plate], (b) greywacke with a matrix largely comprising microcrystalline to cryptocrystalline quartz [left: plane polarised, right: cross polarised], and (c) sandstone, including lithic fragments and rock cement containing microcrystalline and cryptocrystalline quartz [left: plane polarised, right: cross polarised] (from Fernandes *et al.*, 2016).

certainty whether there is any danger in its use". Gutt and Nixon (1979) stated, "the mortar-bar method is the most generally accepted method of assessment available and if the limitations of time and sampling are borne in mind it has been shown to give a good guide to the reactivity of the aggregate".

In a thorough review, Sims (1981) emphasised the indicative, rather than definitive, nature of the various tests for ASR and stressed the importance of petrographical examination: *"A knowledge of the nature of the materials being considered is central to the assessment programme and some form of petrographic examination should never be omitted. In some cases the petrographic appraisal may obviate the need for any further testing, and in most cases the most pertinent indicative testing sequence will be identified."* The inadequacy of the mortar-bar test for UK reactive aggregate combinations had not yet been realised, but Sims did warn that, *"In a few cases, mortar-bars produce anomalous results that are difficult to explain"* and noted that *"the flint-bearing aggregates of south-east England invariably produce mortar-bar results that give no cause for alarm, and yet the deleterious alkali-reactivity in the south-west of England has mostly concerned similar materials which are blended with other rock types and are thereby present in very much smaller proportions"*.

It was then considered that testing actual cement-coarse aggregate-fine aggregate combinations, rather than the individual aggregate constituents separately, might avoid such a misleading outcome. However, by 1986, Sims (1987) could report that: *"In the UK the mortar-bar test has not predicted any aggregate combination to be potentially alkali-reactive, including those involved in known cases of ASR. It is possible that the mortar-bar test…is insensitive to those circumstances encountered in the UK which depend upon the occurrence of a critical combination of constituents and conditions."* A concrete prism test was considered to be more reliable (Figure 6.36).

A BSI working group (WG10) was established in 1981, and began its work by comparing a range of existing procedures (Nixon & Bollinghaus, 1983; Nixon, 1986), using some of the most widely used UK aggregates. There was *"an almost complete lack of correlation between the results of the different methods"* (Nixon, 1986). The working group considered it necessary to develop a standardised method for petrographical examination because, *"In the absence of other agreed test methods, the specifications against ASR in the UK are still having to rely on lists of 'safe' rock types and identification of suspect minerals"* (Nixon, 1986); see subsection 6.6.4. Initial work by the group also indicated that a test method using concrete stored at 38 °C and 100 % RH would identify pessimum combinations of aggregates.

Accordingly, a concrete prism test method was developed by the BSI working group, based on experience at the BRE, and was issued as a draft for public comment (BSI, 1988b). Thereafter, a great deal of laboratory work was undertaken, notably by Hobbs (1992a, 1992b, 1993) and also by Livesey (1992), to assess the reliability of the proposed test for UK aggregate combinations. Hobbs (1992a, 1993) found that the laboratory test results only correlated with field concretes of the same mix proportions and high alkali content, which contrasted with BRE experience, in which the expansion of ASR-damaged structures could only be reproduced in the laboratory by raising the test concrete alkali content to 7 kg/m^3. Livesey (1992), on the other hand, found good correlation between the draft test and reported field performance of UK aggregates; similarly for some other tests including the accelerated mortar-bar test (Oberholster & Davies, 1986).

Hobbs also reported that, in his tests, storage at 38 °C led to deleterious expansion taking place at lower alkalis than for similar concretes stored at 20 °C or exposed externally. Importantly for measures to minimise the risk of ASR damage in future construction, Hobbs and Livesey both concluded that concrete alkali levels in excess of

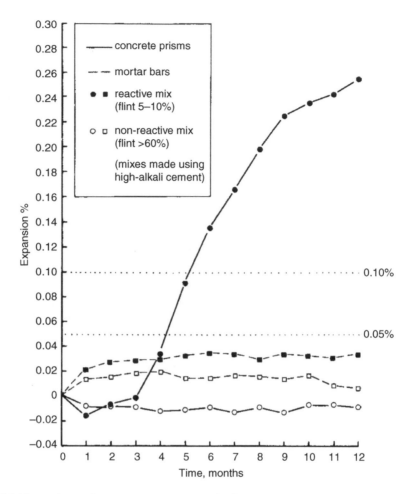

Figure 6.36 Mortar-bar and concrete prism test results for comparable reactive and non-reactive UK aggregate combinations (after Sims, 1987).

5 kg/m^3 were required for deleterious expansion with UK aggregates. Hobbs (1996) continued his laboratory work and later calculated, for concretes stored at 38 °C, 20 °C and external conditions, systematic approximate ratios for the induction of 'abnormal' expansion (1:4:7) and for the rate of expansion after induction (7:4:1). Perhaps more controversially, he also suggested that, for the UK aggregates and concretes he had studied and these three storage conditions, the ages at which expansion was complete were probably 9 to 15 months, 4 to 7 years and 8 to 15 years, respectively. Hobbs' programme does not seem to have included any greywacke aggregates.

The concrete prism test method was subsequently reissued as a Draft for Development, BS DD 218 (BSI, 1995b). It employed four 75 x 75 x 200 mm prism specimens cast from a concrete mix made using the aggregate combination under test and a high cement content (700 kg/m^3), in which the cement alkali content was standardised at 1.0% Na$_2$Oeq. Normal test storage conditions were 38°C and nearly

100% RH, utilising a closely specified regime of wrapping and storage in separate containers. A precision trial was undertaken and the test method was eventually published as BS 812-123 (BSI, 1999b), which remains current. No criteria for interpretation have ever been agreed, however, in the 1997 version of Digest 330 (BRE, 1997), tentative guidance is provided on using the concrete prism test to distinguish 'low', 'medium' and 'high' levels of reactivity (see subsection 6.6.4). The method was also adopted for the greywacke aggregate protocol (BCA, 1999).

In addition to the BS 812-123 concrete prism test, the BSI working group also issued a Draft for Development, BS DD 249 (BSI, 1999a), concerning the accelerated mortar-bar test that had first been suggested in South Africa (Oberholster & Davies, 1986) and also used as the basis for ASTM C1260 (2007). This method, based on storing mortar-bars in a sodium hydroxide solution at 80 °C for 14 days, has never been confirmed as a British Standard (and the DD is long expired).

In 1988, a RILEM international technical committee (TC) was established, under British leadership (Dr Philip Nixon as Chairman and Dr Ian Sims as Secretary), to investigate the wide range of existing and new tests for ASR. Over the nearly 50 years since ASR had been discovered, many had sought to devise a universally reliable test to predict ASR potential, but this 'quest for a test' was looking increasingly like the earlier search for El Dorado. Over 26 years and 3 consecutive TCs, plus a European research programme, 'PARTNER', along the way (Nixon et al., 2008; Lindgard et al., 2010), a great deal of progress has been made with harmonising the worldwide approaches to AAR testing, diagnosis and specification (Godart et al., 2013; Nixon & Sims, 2016; Wigum et al., 2016). A new, fourth TC was formed in 2014, under Norwegian leadership (Prof Borge Wigum as Chairman & Dr Jan Lindgard as Secretary), to take this work forward and especially to develop performance testing for particular concrete mixtures.

6.6.6 Mineral additions (supplementary cementitious materials)

It had been known since the earliest published identification and research on ASR in North America that expansion could be reduced or prevented by using a pozzolanic cement or by replacing part of the cement component with a natural pozzolanic mineral addition (Stanton, 1940b, 1950; Thomas, 2011). Continued research in the 1950s established that the waste/by-product materials, fly ash (now including pulverised-fuel ash, PFA) and ground granulated blastfurnace slag (GGBS), could be similarly effective. Thomas (2011) provides an excellent review of this subject and the factors controlling the efficacy of various types of addition or 'secondary cementitious material' (SCM).

Once the existence of ASR had been accepted in the UK, it was soon realised that the country had an abundance of Class F (low calcium) fly ash (available as PFA), mainly from its coal-fired power stations, and also GGBS from its steel plants (and one in nearby Dunkirk), which were already of interest as partial cement replacement constituents for concrete.

Thorough research into the effectiveness of this PFA in combating ASR expansion with UK aggregates was carried out at Sheffield University (Swamy & Al-Asali, 1990), the BRE (Blackwell et al., 1992; Thomas & Blackwell, 1996) and the BCA (Hobbs, 1994). Hobbs (1994) found that the effectiveness of cement replacement by PFA in

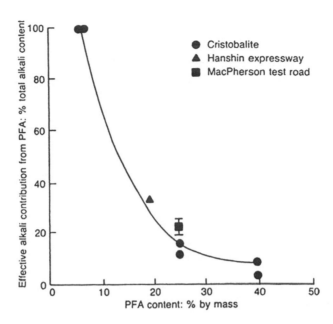

Figure 6.37 Reduction in effective alkali contribution from PFA as replacement level increases (from Hobbs, 1994).

reducing ASR expansion was dependent on the alkali content of the Portland cement, the portion of cement being replaced and the alkali content of the PFA, in a complex interaction, wherein overly low PFA replacement levels could even increase expansion; however, PFA replacements of at least 40% reduced expansion in most circumstances (see Figure 6.37). When the European Standard, BS EN 450 (BSI, 1995a), was introduced for fly ash, it permitted a wider particle size range [up to 40 % retained on a 45 μm sieve] than had the British Standard for PFA, BS 3892-1 (BSI, 1997c) [up to 12 % retained on a 45 μm sieve], and there was some resultant concern that BS EN 450 compliant fly ash might be less effective at controlling ASR. However, investigations at the BRE (Hooper *et al.*, 2004) determined that the BS 3892-1 and BS EN 450 additions could be expected to be equally effective; the alkali content of the ash was found to be more important than its coarseness.

Similar research into the effectiveness of GGBS was carried out, including programmes by or for the GGBS producers (Sims & Higgins, 1992; Connell & Higgins, 1992, 1996) and cement manufacturers (Lumley, 1992). There had been debate about the means by which GGBS suppressed ASR expansion, with some maintaining that the main influence of GGBS replacement of cement was its dilution of the concrete alkali content; however, the work by Sims & Higgins (1992) demonstrated that, at 50 % replacement, the GGBS remained effective even when alkalis were added to compensate for this dilution. Continued work by Connell & Higgins (1996) confirmed these findings and led them to question the validity of taking a proportion of the GGBS alkali content into account when calculating the overall 'reactive' alkali content of a concrete mix (see subsections 6.6.1, 2 & 3).

This extensive UK research into the use of PFA and GGBS with UK aggregates has informed the various schemes for minimising the risk of ASR damage in concrete structures (see subsection 6.8).

6.6.7 Alternative preventative options

As well as fly ash and GGBS, consideration has been given in the UK to some alternative methods for minimising the risk of damaging ASR, as reviewed by a working party organised by the BRE (2002). Two alternative mineral additions were silica fume (also see Dunster, 2009), which was thought to be effective when comprising more than 85 % amorphous silica and using at least 8 % by mass of the total binder, and metakaolin (see also Walters & Jones, 1991; Jones et al., 1992), when the material appeared effective if it comprised more than 45 % silica and using a 15 % by mass of total binder.

Lithium salts were investigated by some of the earliest ASR researchers, as a means of minimising damage, and were included in the BRE (2002) working party review, which provided some tentative guidance levels for the use of lithium hydroxide monohydrate solid and lithium nitrate solution, when the total alkali content of the concrete does not exceed 5.0 kg/m^3 (see also Blackwell et al., 1997; Hooper et al., 2001, 2004). More recently, international work has established that the required dosage of lithium varies with aggregate type (see RILEM AAR-7.1 in Nixon & Sims, 2016), complicating guidance on its use. Finally, the working party looked at another early prospect for controlling ASR expansion, air-entrainment, arising from the observation that non-expansion in mortar-bars was sometimes related to gel forming within voids. However, it was found that the method could not be recommended, because the effectiveness of air-entrainment in reducing ASR expansion was comparatively low, with only a small proportion of entrained air space becoming filled with gel; Hobbs (1998) estimated about 10 % in concrete made with reactive UK aggregates.

Overall, none of these alternative preventative measures have become widely used in the UK, though silica fume (or microsilica) and occasionally metakaolin are used in some concrete, but not necessarily as an intentional measure against ASR risk.

6.7 TAKING STOCK: DIAGNOSIS, PROGNOSIS AND STRUCTURAL MANAGEMENT

6.7.1 Diagnosis of ASR

It was suggested earlier (see 6.5.5) that some cases of ASR in the UK might have been misdiagnosed prior to the mid-1970s, with the cracking and any other damage being wrongly ascribed to a variety of different mechanisms, perhaps including frost damage, sulphate attack and excessive drying shrinkage. Since the 1970s, it has probably also been the case that some examples of concrete cracking have been just as wrongly ascribed to ASR, when they were in fact caused either by more familiar processes or by delayed ettringite formation (DEF) that was not really recognised much before the 1990s (Poole & Sims, 2015/2016).

Reliable diagnosis of ASR can be complicated, partly because the characteristic map pattern of macrocracking caused by ASR is also a common symptom of several other mechanisms, including drying shrinkage, and some of these mechanisms are statistically more likely to occur than ASR; moreover, in stressed locations on structures, macrocracking caused by ASR can completely lack the random map pattern too often regarded as diagnostic and instead exhibit a strongly systematic pattern aligned to the stress field. Sims (1996) has also explained that AAR can occur in various guises and stages, not all of which represent a significant or continuing risk of damage, also (Sims, 2004) that in practice concrete often exhibits combinations of mechanisms. However, the accurate diagnosis of ASR is quite possible and actually involves the careful and coordinated assessment of both site observations and laboratory findings and, even then, it can be difficult to establish with certainty a causal link between any evidence of ASR and any damage which might have occurred to the concrete structure.

In 1984, the C&CA (later BCA) organised a working party of specialists, from a range of firms and organisations actively involved in the diagnosis of ASR, in order to formulate a standardised approach to diagnosis, and the resultant report was published four years later (BCA, 1988). It had been recognised previously that, *"The only certain evidence of alkali-silica reaction is provided by microscopic examination of the interior of concrete to identify positively the presence of gel and of aggregate particles which have reacted"* (Gutt & Nixon, 1979; BRE, 1982). Sims (1987, 1992) termed these 'reliably diagnostic features', and the recognition of such evidence is an essential part of the BCA diagnosis guidance. This first publication dealing solely with ASR diagnosis, which consequently received worldwide attention, was formulated at a time of rapid growth in understanding, so that a second (and still current) edition was issued after four years (BCA, 1992). There was also liaison between the BCA working party and a team at the Institution of Structural Engineers that was working on the effect of ASR on structures and methods for their appraisal (see subsection 6.7.3).

Most considerations of ASR as a possible cause of distress in concrete structures arise because of the presence of surface cracking and the consequent desire or need to establish the cause of cracking prior to any remedial action being implemented. As with any investigation to establish the cause of cracking in concrete, the age at which cracking first appeared is an important piece of evidence, especially to distinguish between early-age phenomena such as shrinkage and longer term damage caused by ASR, but in practice this information is rarely either available or reliable. The BCA report lists the features possibly indicative of ASR to be observed on site, including environmental conditions, cracking, discoloration, exudations, pop-outs and any displacements or deformations, and requires each feature to be classified according to extent and severity (Table 6.6).

The BCA report devoted a lot of attention to the range of laboratory techniques considered appropriate in the UK for the diagnosis of ASR, usually using core samples drilled from those parts of a structure in which any ASR is considered most likely to have occurred. Direct visual identification of features 'reliably diagnostic' of ASR is regarded as the only positive evidence, and this is generally achieved by the microscopical examination of thin-sections and/or sawn sections. Both 'alkali-silica gel deposits' and 'sites of expansive reaction' can usually be recognised if they are present in a concrete sample (Figures 6.38 and 6.39). The second edition of the BCA report provides helpful flow charts, appends more guidance on the examination procedures and generally provides greater assistance with interpretation. Other supporting information can be derived from chemical analyses of the concrete (including determination of alkali content). Hobbs (1988) had suggested that the microcracking pattern resulting

Table 6.6 Classification scheme for assessing the extent and severity of each
feature possibly indicative of ASR to be observed during a site inspection
(after BCA, 1992).

Classification	Index
Extent of feature	
Not significant	1
Slight, up to 5% of area or length as appropriate	2
Moderate, 5 to 20%	3
Extensive, more than 20%, but not all	4
Total, all areas affected	5
Severity of feature	
Insignificant	1
Minor features of a non-urgent or purely cosmetic nature	2
Unacceptable features requiring attention	3
Severe defects or problems requiring immediate attention	4
Structurally unsafe	5

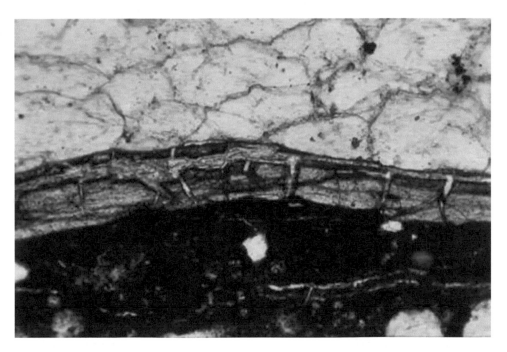

Figure 6.38 Photomicrograph of concrete showing desiccated alkali-silica gel infilling a micro-crack
peripheral to a quartzite coarse aggregate particle. Pln. Pols., approx. x 35 magnification
(from Sandberg, 1984).

from ASR can be distinctive and described an assessment procedure, using impregnation of sawn slices of concrete by fluorescent resin and their examination in ultraviolet
light; this approach was also incorporated in to the second edition of the BCA report
(see Figure 6.40).

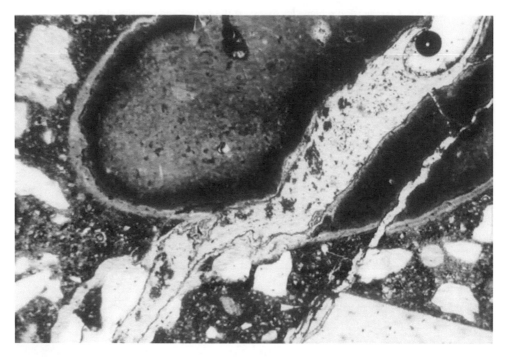

Figure 6.39 Photomicrograph of concrete showing a site of expansive reaction, involving a chert particle. Pln. Pols., approx. x 45 magnification.

The expansion testing of core samples has been used extensively in the UK as a method for diagnosis and also for evaluating any residual capacity for future expansion (BRE, 1982; Wood, 1985). Some consultants have devised their own versions of this test, but the first (1988) BCA report included a standard method of test, using cores of 70-100/150 mm diameter and with a length/diameter ratio of at least 2. Storage of the core was in a sealed container and the conditions were controlled to ensure a temperature of 38 ±2°C and a relative humidity as close as possible to 100%. The measurement of any expansion is achieved by using a mechanical strain gauge periodically to record any length changes between studs affixed systematically to the sides of the core. This core expansion test was significantly modified in the second (1992) edition, following more experience, including wrapping of the cores within the containers in an endeavour to control alkali leaching [similar to the concrete prism test that would be finalised later by BSI (1999b), as explained in subsection 6.6.5]. Additionally, alternative length-change procedures were introduced: Method 1 using end studs and a measuring frame, and Method 2 using side studs and a 'Demec' strain gauge (as for the 1988 version, but using a minimum 100 mm gauge length, instead of 50 mm); some later amendments (BCA, 1993) included a preference for using Method 2, to avoid premature termination of tests by the detachment of an individual stud.

There had been some evidence (Wood *et al.*, 1987) that certain types of concrete can exhibit ultimately greater expansions at storage temperatures lower than 38°C (say 20° C or even 13°C), but the rate of expansion was significantly reduced in such cases and

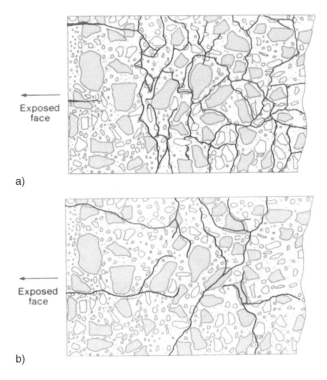

Exposed
face

a)

Exposed
face

b)

Figure 6.40 Two examples of the diagrams provided in BCA, 1992 for assisting with the interpretation of internal cracking patterns: a) caused by ASR, when the reactive silica is in the sand fraction [a common occurrence in the UK] and b) caused by ASR, when the reactive silica is in the coarse aggregate fraction; diagrams are also usefully given in the report for cracking patterns indicative of other mechanisms (from BCA, 1992).

the prolongation of the test would usually be unacceptable for diagnosis purposes. However, both the 1988 and 1992 versions of the BCA core test accommodated alternative storage temperatures or either 38 or 20 °C (though the 1993 amendments include a preference for 38 °C) and the later report included precision data.

Curves of expansion against time of storage in the core test (Figure 6.41) are typically asymptotic, providing moisture loss does not occur, and the maximum expansions recorded at various ages range from about 0.02% (200 microstrains) up to more than 0.25% (2500 microstrains). The comparatively rapid expansion within the first few weeks of the test can help to indicate or confirm the presence within the concrete of existing alkali-silica gel capable of reabsorbing moisture lost and/or absorbing further moisture, causing early expansion. Any expansion at later ages in the test might indicate a potential within the concrete for further, or as yet unrealised, reaction and resultant movement. Tentative guidance is provided in the 1992 report, suggesting that expansions of 700 microstrain (0.07 %) or less in a concrete tested at 38 °C would indicate that damage or further damage was unlikely, with the risk of damage being progressively increased for expansions greater than 700 microstrain (0.07 %), with values up to and occasionally higher than 4000 microstrain (0.4 %) having been recorded for UK concretes.

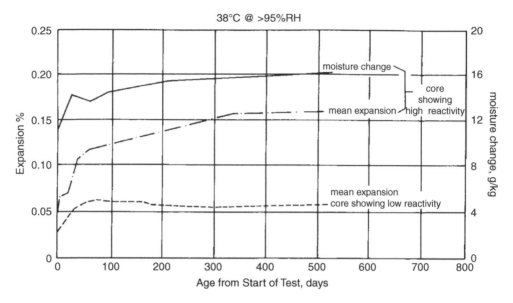

Figure 6.41 Typical levels of expansion recorded at 38°C in tests carried out on cores drilled from structures affected by ASR (redrawn and adapted from BCA, 1988).

The boundary between diagnosis and prognosis is not clear-cut, and the identification of the most likely cause of cracking is of limited usefulness without also assessing the risk of continued expansion and cracking in those areas already affected, also the risk of expansion and cracking in areas currently unaffected. The BCA report introduced a novel 'risk classification system', in which the overall risk can be appraised (low, medium or high risk) from the 'materials risk', in turn assessed from the site observations and laboratory findings, and the 'environmental risk' assessed from the exposure conditions involved. In seeking to utilise this scheme, it is important to recognise that the core expansion test is used as one factor to assist with assessing 'materials risk'; it is not intended as a method for predicting or quantifying expansion within a structure, when the same concrete will be under restraint, probably greatly reducing any expansion (Hobbs, 1999).

Although the BCA (1992) guidance on diagnosis remains valid and useful, many sophisticated techniques, such as scanning electron microscopy (SEM) with associated microanalysis facilities are now more readily and economically available. An early UK example of the usefulness of these techniques was given by Laing *et al.* (1992) at Imperial College, London, and further details on the application of SEM can now be found in ASTM C1723 (2010), Winter (2012) and Godart *et al.* (2013).

6.7.2 Effect of ASR on structural properties and performance

Once it became apparent that ASR was affecting a sizeable number of structures and buildings in the UK, many of them publicly owned, it became necessary to adopt a systematic approach to the appraisal of structural integrity and future performance. Initially, such engineering appraisals were frustrated by the paucity of research, in the

Table 6.7 Percentage reduction (compared with control at same age) in engineering properties of laboratory concrete specimens after expansion caused by ASR involving either opal or fused silica as the reactive constituents. Compiled and adapted from Swamy & Al-Asali (1988).

Concrete property	Concrete expansion (%)			
	0.05		0.10	
	4.5% opal	*15% fused silica*	*4.5% opal*	*15% fused silica*
Compressive strength	9	12	11	11
Modulus of rupture	—	30	—	48
Indirect tensile strength	—	27	—	29
Dynamic modulus of elasticity	16	11	37	20
Ultrasonic pulse velocity	7	2	11	6

UK or elsewhere, concerning the influence of ASR on engineering properties, although it was starting to become apparent from overseas experience that many structures, severely cracked and disfigured by ASR, had not actually suffered any engineering failure even over a long period of time (Tuthill, 1982).

Research results on the effects of ASR on the mechanical properties of concrete were initially limited and variable, but useful work had been conducted in the UK at Sheffield University. Swamy and Al-Asali (1986, 1988), using opal and fused silica reactive constituents, found that tensile and flexural strengths were appreciably more affected by ASR than compressive strength (Table 6.7), with the ratio of tensile to compressive strength being reduced by about 45% at an ASR expansion of 0.1%, and by about 60% at an expansion of 0.3%. There was also a substantial reduction in modulus of elasticity as ASR expansion increased, possibly because of microcracking.

In broad terms, once ASR had been diagnosed as a major cause of damage, the structural engineer had to arrive at two judgements: the existing engineering condition of the structure, and the projected future performance and serviceability. The consequential course of action, whether to demolish, support, repair or monitor further, would depend upon the conclusions reached from structural analyses and a variety of laboratory and site tests. In the UK, the Institution of Structural Engineers (ISE) took the lead in formulating guidance on the appraisal of structures for the effects of ASR and this will be described in sub-section 6.7.3.

As an essential precursor to being able to formulate such engineering guidance, a programme of practical research was funded by the British Government through the Transport and Road Research Laboratory (since privatised as TRL). On the principle of 'not re-inventing the wheel', especially as ASR had been identified around 30 years before it was widely recognised in the UK, one of the earliest contributions to this programme was a thorough critical review of existing international literature pertaining to the structural implications of ASR (Clark, 1989, 1990). This study established that, in practice, ASR expansion could reduce the compressive strength, tensile strength and elastic modulus properties of concrete material, by up to 25 %, 50 % and 60 %, respectively. However, importantly, in structures and structural members, relatively small amounts of reinforcement and/or applied compressive stress could significantly reduce the expansion, depending on the rate of free expansion. Even in reinforced or post-tensioned structural elements exhibiting severe surface cracking attributed to ASR, the service load behaviour was usually little impaired; although circumstances were identified in which ASR might have more effect, including structural elements or areas in which reinforcement anchorage was inadequately restrained by links or transverse bars. Clark's key (1989) study also identified topics in need of further practical research.

Similar conclusions were reached by Hobbs (1990) from another literature review, leading him to warn that visually apparent ASR damage did not necessarily correlate with impairment of structural performance: *"...load tests on affected structures and load tests to failure on affected beams indicate that visually severe ASR cracking in concrete members is deceptive and that the expansion and cracking which ASR induces may not have an unacceptable (sic) adverse effect upon the performance of reinforced or prestressed concrete members"*. Clayton *et al.* (1990) at BRE, using full-scale prestressed beams made using alkali-reactive concrete, found that shear reinforcement could enable elements to recover initially lost shear capacity as expansion continued, an average confining stress from shear reinforcement of about 4 N/mm^2 being necessary to prevent ASR macrocracking.

One major research programme into the effects of ASR on bending, shear, torsional and compressive modes of structural failure was conducted at the BCA, focusing on post 1950 concrete bridge design practice in the UK. The work was conducted and published in two parts: Phase I was aimed at establishing appropriate techniques and obtaining some basic information (Chana & Korobokis, 1991a), whilst Phase II addressed poorly detailed members liable to brittle failure, including columns in compression and members with curtailed tension reinforcement or poor anchorages (Chana & Korobokis, 1992). The findings of Phase I essentially corroborated the indications of Clark's (1989) review. In Phase II, it was found that reduction in shear strength was 'modest' (maximising at 30 % in the tests for beams made using round bars and at 23 % for those made with ribbed bars), even for beams with poor anchorage (Chana & Thompson, 1992); the authors suggested that the ASR expansion might be providing additional anchorage. A similar conclusion was reached by Cope & Slade (1992) from parallel work at Plymouth University (then Polytechnic South West), who suggested that *"AAR, perhaps by pre-stressing the concrete, enhances shear capacity"*.

BCA also carried out research into the important aspect of residual bond strength of reinforcement in ASR-affected concrete structures, especially the anchorage in support regions and also regions in which the reinforcement is lapped (Chana & Korokobis, 1991b). The findings were complicated by the difficulty of realistic testing and possible reinforcement slip as ASR develops, but generally ASR reduced bond strength, especially for plain bars in bottom cast positions: 40 % reduction with 'medium' levels of ASR expansion (microstrain 450 restrained or 2500 unrestrained), rising to 60 % for 'high' expansion (microstrain 1300 restrained or 4500 unrestrained). Ribbed bars were less affected, bond strength reduction in bottom cast positions being 20 % and 40 % for 'medium' and 'high' expansion, respectively. Lapped specimens without links showed similar changes in strength in response to ASR expansion.

Further investigations were undertaken at Queen Mary College (QMC, London University) into variations in the ultimate average bond stress of reinforcement bars in beams made using ASR reactive concrete (Rigden *et al.*, 1992a). It was found that, whilst ASR reduced the bond stress, compared to a control mix, the most expansive (highest alkali, Na_2Oeq: 12 kg/m^3) mix exhibited a much smaller reduction than the

Table 6.8 Effect of ASR expansion on bond stress failure in reinforced concrete beams (after data in Rigden et al., 1992a).

(a) Mix proportions used for concrete beams

Mix No	20 mm Aggregate (kg)	10 mm Aggregate (kg)	Sand (kg)	Fused silica (kg)	Cement (kg)	Water/ cement ratio	Na_2O equivalent (kg/m^3)
1	–	1060	300	240	550	0.41	12.0
2	–	1060	300	240	550	0.41	7.0
3	1060	–	300	240	550	0.37	7.0
4	–	1060	540	–	550	0.41	7.0
5	–	1060	540	–	550	0.41	–
6	–	1060	540	–	550	0.74	–

(b) Average expansion of beams in different directions

Mix No	Average horizontal expansion top edge (mm/m)	Average horizontal expansion bottom edge (mm/m)	Vertical expansion (mm/m)	Radius of curvature (m)
1	14.4	7.45	18.0	10.2
2	11.8	6.71	13.2	13.9
3	12.5	7.72	14.2	14.8

(c) Mechanical properties of the beams

Mix No	Mean compressive (equivalent cube) strength (N/mm^2) (f_{cm})	Failure load = 2P (kn)	Ultimate mean tensile stress $(N/mm^2)(f_s)$	Ultimate mean bond stress $(N/mm^2)(f_b)$	$(f_s)/(f_y)$ $f_y=375\%$	$f_{bu}= 1.4\beta\sqrt{f_{cu}}$ $\beta=0.28$ $f_{cu}=f_{cm}$ (N/mm^2)
1	32.7	4.92 4.23	147.5	4.4	39.3	2.2
2	37.7	2.58 2.57 2.54	82.6	2.5	22.0	2.4
3	35.2	2.29 2.54 2.41	77.8	2.3	20.7	2.3
4	57.2	8.54 7.00	250.5	7.5	66.8	3.0
5	69.4	9.48 10.2	317.3	9.5	84.6	3.3
6	35.9	9.21 7.81 8.64	275.8	8.3	73.5	2.3

two comparable reactive mixes with lower alkali contents (Na_2Oeq: 7 kg/m^3), possibly because of increased adhesion (see Table 6.8). It was also observed that the beams also displayed curvature as ASR expansion progressed, owing to the restraint from bottom reinforcement and slippage of reinforcement (Figure 6.42).

Further valuable work was carried out at QMC, by casting large blocks of ASR-reactive concrete and subjecting them to different stress levels during the year after manufacture,

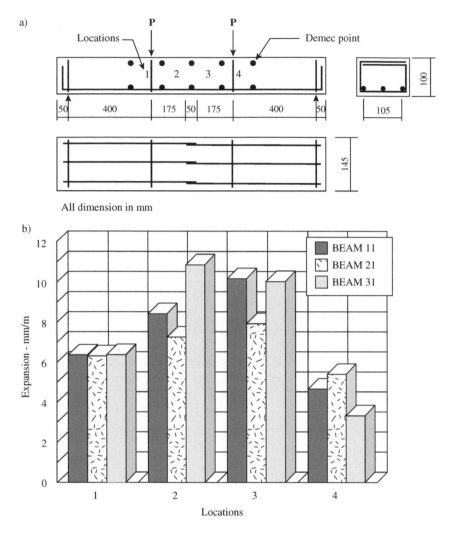

Figure 6.42 Expansion and curvature of reinforced beams made with ASR-reactive concrete: (a) typical loading arrangement of beams for investigation of bond failure, (b) bar chart illustrating reinforcement slippage (from Rigden *et al.*, 1992).

then measuring the effect on mechanical properties, using cores drilled from the blocks in their three principal directions (Rigden *et al.*, 1992b). The effect of ASR on these properties varied considerably according to stress levels and orientation, indicating that indications from testing randomly drilled cores from structures might be misleading. May *et al.* (1992, 1996) similarly recognised the need to allow for reduction in expansion by applied stress in attempting to model the effects of ASR on structural members.

At the time of the British Government funded structural research programme in the late 1980s and early 1990s, several hundred ASR-affected bridge structures had been identified, mainly constructed in the period 1930 to 1975, with a high proportion between 1968 and 1973. Many reinforced concrete bridge and viaduct slabs in the UK were designed and constructed prior to 1973 with little or no shear reinforcement, whereas later codes required shear reinforcement. Accordingly, investigations were carried out at Birmingham University into the effects of ASR on the punching shear strength of reinforced concrete slabs (Clark & Ng, 1992; Ng & Clark, 1992). It was found that generally ASR expansion and cracking did not reduce the punching strength of a slab, but did increase its ductility. However, at 'extremely high' levels of free ASR expansion (say, >6000 microstrain), delaminations could occur parallel to the reinforcement planes and this in turn reduced the punching shear capacity of so-affected slabs.

6.7.3 Assessing structures and the future risks

Alongside a growing understanding of the implications of ASR within concrete for the structures comprising or containing concrete components, as described in subsection 6.7.2, there was a concerted effort in the UK to guide owners and engineers on the appraisal of the large existing stock of concrete structures. The issues to be decided may be simply stated (Hobbs, 1988; Wood, 1985; Somerville, 1985):

(1)   Does any ASR damage significantly impair the strength, stiffness, stability and serviceability of the structure; is the structure safe?
(2)   Is expansion continuing and will damage continue to develop; will the reaction reach an equilibrium or can the reaction be arrested; what will be the ultimate expansion and damage?
(3)   Can the damage lead to secondary deterioration caused, for example, by frost or corrosion of reinforcement? and
(4)   Does the structure need to be repaired or strengthened, need attention from the point of view of its appearance, or need to be monitored and managed?

These considerations were the subject of an *ad hoc* committee set up by the Institution of Structural Engineers in 1986, under the chairmanship of David Doran, and the first full report of the 'Doran Committee' was published in late 1988 as 'Interim technical guidance on the appraisal of existing structures' (ISE, 1988). This interim guidance was soon replaced by an updated version (ISE, 1992; Wood & Doran, 1992), which appears to have been effective, as it remains current, with an Addendum only being issued some 18 years later (ISE, 2010, also Wood, 2008). The UK approach to issues (1) to (3) above will now be briefly considered in accordance with the ISE (1992) guidance; issue (4) is covered in 6.7.4 and 6.7.5. The later addendum (ISE, 2010) will be explained in subsection 6.8.1.

As shown in subsection 6.7.2, when appraising a concrete structure, the effect of ASR on the strength or load-bearing capacity of a structural member is much more important than the properties of the concrete material alone and, for this, much is dependent upon the restraints derived from internal steel reinforcement or from externally applied stresses related to the loading patterns. Some testing of reinforced concrete members in Japan and Denmark had indicated that, in some circumstances, the induced compressive stress arising from the restraint of ASR expansion by reinforcement can lead to unimpaired or even increased shear capacity (Hobbs, 1988). According to Hobbs, quoting Courtier, *"overall expansion"* of structural components is the dominant parameter, with expansions less than about 0.12% or 0.15% having the effect of *"prestressing the member without significantly affecting its load-carrying capacity"*. In practice, the structural adequacy of any particular construction at a particular time can probably only be ascertained by full-scale proof-loading, and any deterioration in the reserves of stiffness and strength can be determined by repeated proof-loading wherever this is practicable (Wood *et al.*, 1987).

One thorough preparatory study, covering the estimation of concrete expansion caused by ASR, the effect of ASR on the structural properties of that concrete and a methodology for assessing the load bearing capacity of ASR-affected structures, was carried out by W S Atkins Consultants Ltd, as part of the Government-funded programme through TRL (McLeish, 1990). An excellent and insightful summary account of this work was published by Courtier (1990), who had also had practical experience of assessing ASR-affected structures in the UK: *"Assessment can be made using an interactive process of assessment of expansion and calculation of the consequences. Strength parameters must be measured, and judgement made concerning local effects"*. At that time, UK engineers had experimented with mechanical tests that would best characterise the microstructural damage caused by ASR, including especially the 'Stiffness Damage Test' (SDT, Chrisp *et al.*, 1989) and also a torsion test (Norris *et al.*, 1990); advice on strength assessment for concrete highways structures was issued by the Highways Agency (1994). Courtier (1990) added: *"...assessment can be improved by monitoring actual performance, thus corroborating estimates. If such measurements are made they should concentrate on overall behaviour of the structure and not be over-concerned with cores"*. One approach to the integrated material and structural assessment, in preparation for rehabilitation, was suggested by Swamy (1997).

The approach to appraisal and the determination of subsequent actions adopted by the ISE guidance (1992), which is complementary to the overall ISE appraisal procedure (ISE, 1996), is based upon a 'structural severity rating' scheme, which in turn utilises assessment of 'expansion index', reinforcement detailing, the site environment, and the consequence of further deterioration (Table 6.9). An arbitrary scale of expansion indices (I-V) is defined, being computed from determinations of 'current expansion' based upon crack width measurements, plus 'potential additional expansion' based upon core expansion testing including an allowance for long-term enhancement. Three classes (1, 2 and 3) of reinforcement detailing are defined, according to whether the cage is one-, two- or three-dimensional and also to the nature of the anchorage; attention must also be paid to particularly sensitive structural details. Three levels of site environment are defined (dry, intermediate, wet) and are broadly equivalent to the three conditions of exposure described in the BCA (1992) report on diagnosis.

Each 'structural severity rating' is then separately considered according to whether the consequence of further deterioration is deemed 'slight' or 'significant'. The consequential action or 'management procedure' is then defined by the ISE guidance according to the determined 'structural severity rating', ranging in five steps from 'negligible' ('n' in Table 6.9) requiring only routine inspections at normal frequency, up to 'very severe' ('A' in Table 6.9) demanding immediate action and detailed investigation. The scheme has merit in standardising the approach and decision-making, but it seems vulnerable to the imprecision of the 'expansion index', which combines two determinative methods (crack width measuring and core expansion testing) that are either suspect or at least not yet capable of reliable interpretation. These challenges are recognised in the ISE guidance, which indicates a preference for core expansion tests using lower storage temperatures (20 or 25 °C) and direct 'water supply' moisture conditions for assessing expansion potential, in contrast to the test procedure described in BCA (1992) for ASR diagnosis purposes.

Table 6.9 Scheme for establishing the 'structural severity rating'. The 'expansion index' is computed from crack width measurements and core expansion test results. The site environment and reinforcing detailing classes are defined in the Institution of Structural Engineers publication: *Structural effects of alkali-silica reaction: technical guidance on the appraisal of existing structures* (ISE, 1992).

Site environment	Reinforcement detailing class	Expansion index									
		I		II		III		IV		V	
		Consequence of failure									
		Slight	Significant	Slight	Significant	Slight	Significant	Slight	Significant	Slight	Significant
Dry	1	n	n	n	n	n	n	n	n	n	n
	2	n	n	n	n	n	n	n	n	n	D
	3	n	n	n	n	n	n	n	D	D	C
Intermediate	1	n	n	n	D	D	C	D	C	D	C
	2	n	n	n	C	D	C	C	C	C	B
	3	n	d	n	B	C	B	B	A	B	A
Wet	1	D	D	D	C	D	C	C	B	C	B
	2	D	D	D	B	C	B	B	B	B	A
	3	D	C	C	A	B	A	A	A	A	A

Structural severity ratings	n – negligible	D – mild	C – moderate	B – severe	A – very severe

Following the publication of the ISE (1992) guidance, Birmingham University undertook a three-year research programme to address many of the remaining uncertainties identified in section 11 of that guidance, and a summary of the findings was published by Jones & Clark (1996). Reassuringly, it was found that the ISE appraisal scheme generally overestimated deterioration in relation to expansion, although it adequately predicted the deterioration in elastic modulus. The research also further investigated the relationship between applied stress and expansion, finding significant variations according to orientation of the reinforcement, relative to casting direction, and specimen size.

Whatever the effect on structural strength, ASR leads to cracking of the concrete surface, and this raises concerns over the possible susceptibility to damage from other causes, including frost attack, carbonation and reinforcement corrosion. There seems to be some empirical evidence (Somerville, 1985) that corrosion may not occur as readily as might otherwise be expected when the concrete cover is cracked by ASR. However, the ISE (1992) guidance rightly warns that the, *"relationship between cracking and corrosion in ASR-affected concrete is not clear"*, but then tentatively suggests that, *"it is reasonable to assume that the effect of ASR on corrosion rates should not be significant"* when *"concrete quality and the cover comply with current codes of practice and surface crack widths do not exceed 0.3 mm"*.

6.7.4 Monitoring and management

Internationally, it is extremely rare for structures affected by ASR to require replacement or substantial repair, and in most cases 'management' by monitoring and selective remedial actions is an adequate response. According to Wood *et al.* (1987), for example, *"for many of the structures we investigate, the robustness of the reinforcement and the mildness of the reaction enable them to be managed on the basis of routine inspections and some action to reduce the moisture and salt supplied to the structure"*. A good example of such structure management, the Val de la Mare Dam in Jersey, has already been described (see 6.4). In those cases where ASR can be shown definitely to have terminated, or reached an equilibrium, it might be possible to consider a lasting remedial solution that will not require long-term management and monitoring.

The scope of monitoring and management varies according to the 'structural severity rating' as determined according to the ISE (1992, 2010) guidance (see 6.7.3). In one early study of some 63 structures (Wood *et al.*, 1987), using the rating that would be given in the ISE (1988) interim guidance, only 10% were rated 'very severe', 24% were 'mild' or 'very mild' (now termed 'negligible') and just less than half were rated as 'serious', which suggests annual engineering inspection, crack monitoring and occasional core testing. Schemes for the inspection of structures for the possibility of ASR damage are described in the BCA (1992) report on diagnosis, by Fookes *et al.* (1983) and more recently by RILEM in their AAR-6.1 international guidance on diagnosis and appraisal (Godart *et al.*, 2013); complementary RILEM recommendations on appraisal and structural management will shortly be published as AAR-6.2 (Godart *et al.*, 2017). In the case of repetitive inspections of a given structure, it is essential to ensure that reliable comparisons can be made between inspections carried out at different times and perhaps even by different engineers; good-quality notated photography and/or scale drawings can assist.

Crack monitoring involves plotting the evolution of crack patterns, as well as measuring fluctuations in crack widths or local displacements, and it is important to monitor initially uncracked locations as well as the more obviously affected areas. Various instrumental techniques are available for measuring movements across cracks or elsewhere on a structure, but it is important properly to control the veracity of such readings and to make appropriate corrections for background environmental factors. Some good examples of the crack monitoring of UK bridge structures have been described by Wood (2004). In the case of the Exe Bridge in Exeter (Figures 6.43 & 6.44), built in 1970 and comprising a combination of high-alkali cement with a

a)

b)

Figure 6.43 Exe Bridge in Exeter: (a) general view and (b) cracking of abutment concrete in 1992.

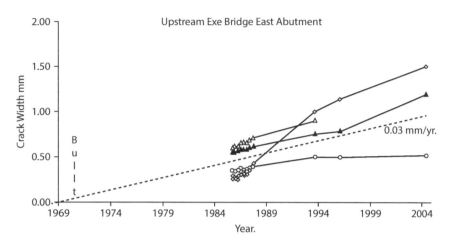

Figure 6.44 Typical crack growth measured on the Exe Bridge, east abutment, upstream (from Wood, 2004).

Figure 6.45 Montrose Bridge in Scotland (image courtesy of Dr Jonathan Wood).

partially crushed, quartzitic sand and gravel aggregate, with a small content of chert, crack monitoring from 1985 had shown a continued linear increase in crack widths for all the elements monitored. An older bridge at Montrose in Scotland (Figures 6.45 & 6.46), built in 1930, was monitored from 1997 to 2002 and had still exhibited measurable crack growth after 72 years (Wood & Angus, 1995).

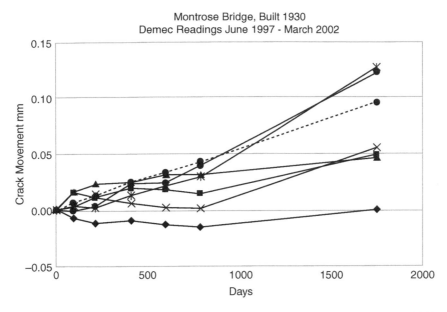

Figure 6.46 Crack growth data from Montrose Bridge, Scotland, June 1997 to March 2002 (from Wood, 2004).

6.7.5 Prospects of treatment and/or repair

It used to be thought that ASR was the cause of an irreversible deterioration of concrete, ultimately leading to structural failure unless action was taken to replace or to 'manage' the affected units. Some researchers considered that attempted repairs could, in some circumstances, make the damage worse, for example by sealing cracks that would otherwise have provided a non-expansive escape route for newly generated gel. It has been agreed for some time that protection of the concrete from exposure to moisture must have some beneficial influence, at least in reducing the rate of reaction, even if the reaction could not be completely arrested, but the performance of coating systems had been extremely variable. It was these uncertainties over the prospect of successful repair that had emphasised the importance of implementing the recommendations of the Hawkins (and later Concrete Society) Working Parties in the UK (see 6.6).

In the UK, at the time of the first edition of this book, most 'repairs' had consisted of substantial replacements or extensive reconstruction, such as at the Val de la Mare Dam, Charles Cross car park, Voss Farm Bridge and Smorrall Lane Bridge (see 6.4.4 and 6.5.5), or temporary support works prior to planned replacement in the foreseeable future, such as the Marsh Mills Viaduct (see 6.5.5). To the best of the author's knowledge, to date, no UK structures have been demolished wholly or only as a result of structural concerns over ASR damage, although several publicised cases have now been replaced as part of overall redevelopment, including the Devon and Exeter Hospital in Exeter, which was the subject of the only ASR-related litigation in the UK (in the mid-1980s), the Marsh Mills Viaduct as planned and, more recently, the

Charles Cross Car Park in Plymouth. Otherwise, further cases of ASR damage have been identified by regular inspection and investigation, then subject to routine monitoring and management, as required and following the ISE (1992, 2010) guidance.

Measures to deflect or re-route water migration, or physically to protect concrete surfaces with shelters or temporary claddings, have usually been regarded as only emergency measures to retard deterioration whilst investigations and structural appraisals are completed. There had been little British experience to that date, and only discouraging international experience, to suggest that surface coating systems can be other than negligibly effective at preventing further ASR and cracking from occurring, although some success has been claimed for silanes (Idorn, 1988), especially if they are periodically reapplied. Swamy and Tanikawa (1992) found some success for an acrylic rubber coating in preventing water ingress and reducing strength loss caused by ASR, especially by comparison with an epoxy coating that blistered and cracked in the same range of conditions; however, they found that the coating was unable to prevent further expansion arising from moisture already within the concrete. Internationally, some interest has been shown in using lithium compounds to arrest ASR damage in existing structures and some success has been shown in treating concrete pavements where the worst damage occurred in the surface region (Thomas & Stokes, 2004), but the technique is less practicable with structures.

The most hopeful experience in the UK remains the observation, over a period of some years of repetitive re-examination of mainly highways structures and buildings, that a level of stability of ASR damage can be reached at some stage, often at between 10 and 15 years after construction, after which further crack development and movements become negligible and very slow. It is not clear why this should occur, but in most cases it seemed more likely to represent a physical equilibrium rather than an exhaustion of one of the chemically reactive constituents. Limited trials suggested that crack sealing and some coating types might prove to be long-term effective treatments for affected concretes, which can be shown to have reached such a stage of physical equilibrium. Further guidance on repair options is expected in forthcoming RILEM AAR-6.2 (Godart et al., 2016).

6.8 NEW REASSURANCE: LIVING WITH ASR

6.8.1 Back in proportion

At the height of the publicity over ASR discoveries in south-west England, one enterprising journalist coined the term 'concrete cancer' to describe ASR to lay readers (Coates, 1983). Although the term was misleading, because ASR is a scarring rather than a potentially terminal 'disease', and engineers rightly deprecated the use of such emotive terminology, it was interesting, because it revealed the level and nature of the public response to the apparently epidemic spread of ASR damage to concrete. Like all human fears of 'diseases' or other phenomena of unknown cause and uncertain remedy, the public response to ASR exaggerated the actual incidence of occurrence and the severity of damage to affected structures. However, the causation has now become much better understood, confidence has grown that future construction should be largely free of ASR, and it has become realised in the UK that most structures disabled by ASR can remain in service, with many being able to satisfy the originally intended lifespan.

It has already been explained (see 6.7.3) that most UK structures found to be affected by ASR can be 'managed' by monitoring and selective repair; demolition and re-provision is rarely a necessary option in the absence of other prejudicial factors. Considerations other than ASR that could lead to a decision prematurely to dismantle an ASR-affected structure would include, *inter alia*, the discovery of inadequate structural detailing in critical positions leading to an unacceptable reduction in the safety margin, secondary deterioration such as reinforcement corrosion, or simply an economic aspect whereby the cost of replacement might be less than the cost of long-term management, especially if replacement also enables desirable improvements to be achieved. Few structures affected by ASR in the UK have needed to be replaced or substantially reconstructed unless other non-ASR factors are involved. Even then, it appears that only a very few cases have been demolished for purely engineering reasons. In short, most structures have survived a discovery of ASR.

The ISE appraisal procedures described in subsection 6.7.3, introduced in 1988 and then published in substantive form in 1992, have continued in use ever since, save for a five-page Addendum issued by the ISE in 2010. This Addendum provides a valuable brief summary of relevant international developments and experience gained with applying the ISE guidance, but does not significantly amend the recommended procedures. A forthcoming RILEM guide (Godart *et al.*, 2017) will build on this ISE scheme for international application and include more advice on remedial options.

6.8.2 Performance specification to minimise risk of ASR damage

By 1990, the UK had established principles for minimising the risk of ASR damage in new concrete construction (see subsection 6.6), with outline guidance in BRE Digest 330 (1988) and more detailed advice in Concrete Society TR30 (1987), including model specification clauses. A summary at that stage was prepared by Nixon (1990) and the Specification for Highway Works had been updated to reflect the current guidance (Department of Transport, 1990). As the new decade proceeded, these principles were gradually developed in the direction of an integrated performance specification that would eventually be included in the British Standard for concrete (BS 8500-1 & 2: 2002, 2006, 2015), which is complementary to the European Standard on concrete (EN 206) that does not itself cover AAR.

A new version of Concrete Society TR30 was issued in 1995, largely just updating the established guidance of the previous version (Concrete Society, 1987), although the recommendations relating to any content of chert and flint in the aggregate had become more complicated in deference to research at the BRE (see subsection 6.5.3). When the content of chert and flint in the total aggregate exceeded 60 % by mass, it could only be considered indicative of non-expansive behaviour if at least 5 % of that chert and flint was present in the sand fraction. It was recognised that there remained uncertainty over the extent to which the alkalis in any ggbs or pfa in the binder should be included in the total alkali content of the concrete, with proportions of one-half and one-sixth, respectively, being considered appropriate, pending results from a technical sub-committee.

Two years later, a completely new version of Digest 330 was published by BRE (1997), now presented in four separate parts; a summary was provided by Blackwell (1997). This new Digest introduced a new approach to specifying the total alkali limit of a concrete, depending on the alkali content of the binder type (Low, Moderate or High) and the reactivity classification of the aggregate. The latter was especially novel, replacing questionable previous attempts to identify and quantify rock and mineral constituents with a broader classification based on UK experience and some research into alkali thresholds. Thus, most UK aggregates were considered to fall into a so-called 'normal' reactivity class, effectively accepting that some degree of alkali-reactivity was likely, whilst a smaller proportion could be regarded (or shown by testing) to have 'low' reactivity and a yet smaller group (primarily crushed greywacke: see subsection 6.5.4) would be regarded as having 'high' reactivity (in due course, any opaline and similarly reactive materials would become labelled as having 'extreme' reactivity). A simplified version of the resultant matrix of recommendations is shown in Table 6.10.

Thus, in the mid-1990s, the two principal sources of AAR guidance on specification in the UK (CSTR 30 & BRE Digest 330) were offering slightly differing guidance. However, by the end of the decade, harmony had been re-established, with the issue of both a 3rd edition of CSTR 30 (Concrete Society, 1999) and a new edition of Digest 330 (BRE, 1999) and a later revision (BRE, 2004) that incorporated new information on some alternative methods (see subsection 6.6.7 & BRE, 2002). In particular, CSTR 30 was able to use the findings from its own technical sub-committee and the recommendations were then also drawn up to ensure correspondence with those given in the BRE Digest 330.

Essentially this meant that the new approach to aggregate reactivity ('low', 'normal' or 'high': see earlier in this subsection) was adopted and similarly the classification of binder alkali contents ('low', 'moderate' or 'high'), based on the content of Portland cement in the binder (*i.e.* the Portland cement clinker component of the increasing used composite cements). It had been found to be unnecessary

Table 6.10 Simplified version of the recommendations introduced by BRE Digest 330, based upon the 2004 edition; see Digest 330 (2004) for a complete version with its accompanying notes.

Aggregate type or combination	Alkali content of the CEM I-type component of the cement (Table 6) or the CEM I component of a combination with ggbs or pfa		
	Low alkali (guaranteed ≤0.60% Na_2O eq on spot samples)	Moderate alkali (declared mean ≤0.75% Na_2O eq)	High alkali (declared mean >0.75% Na_2O eq)
Low reactivity	Self-limiting: no mix calculation needed[†]	Self-limiting: no mix calculation needed[†]	Limit: ≤5.0kg Na_2O eq/m^3 [‡◊]
Narmal reactivity	Self-limiting: no mix calculation needed[#]	Limit: ≤3.5kgNa_2O eq/m^3 [◊$]	Limit: ≤3.0kg Na_2O eq/m^3 [‡◊]
High reactivity	Limit: ≤2.5kgNa_2O eq/m^3 [♦]	Limit: ≤2.5kgNa_2O eq/m^3 [♦]	Limit: ≤2.5kgNa_2O eq/m^3 [♦]

◊ † ‡ # $ ♦ **Notes to table** : These important explanations and qualifications to the recommended limits are given in detail in BRE Digest 330 (2004).

to try to allow for a portion of the alkalis within ggbs or pfa, providing recommended minimum proportions of those additions were used. Overall, this enabled a more flexible approach to the control of concrete total alkali content as a measure for minimising the risk of ASR, which could now range from 2.5 to 5.0 kg Na_2Oeq/m^3, with the common combination of 'normal' reactivity aggregate and 'moderate' binder alkali content yielding a limit of 3.5 kg Na_2Oeq/m^3, compared with the default value of 3.0 kg Na_2Oeq/m^3 that had been in use since the earliest Hawkins (1983) report. This still current CSTR 30 scheme is shown in a flow chart (Figure 6.47), whilst the criteria are summarised in Table 6.11 that is restructured from that in Digest 330 (see Table 6.10). In the UK, it was thought possible to rely upon cement mean alkali contents declared by their manufacturer, as a level that would not be exceeded without prior notice; suppliers of ggbs and pfa would provide similar data. This has largely worked well, although an exception has occurred (see subsection 6.8.3).

6.8.3 Standardised specification and one hiccup

In the mid-1990s, the British Standard for concrete specification was BS 5328, first published in 1991 to replace parts of the BS 8110 code of practice ('Structural Use of Concrete') and then issued in a new edition (BSI, 1997a, 1997b). In 1999, Amendment No 1 was issued, which, as the Foreword to the amended BS 5328-1 explained, "takes account of the recent consensus reached by experts on provisions to resist damaging ASR in the UK. These recommendations are published in the BRE Digest 330:1999 (see subsection 6.8.2). The technical content of this amendment has been derived from these recommendations". There was a short section on ASR in BS 5328-1, then the amendment drew attention to requirements in BS 5328-2 (actually clause 1.4) for minimising the risk of damaging ASR, which "applies to all types of concrete (designed, prescribed and designated)". This clause was essentially based upon BRE Digest 330 (1999), but subclause 1.4.5 allowed a satisfactory track record to be relied upon in some circumstances and subclause 1.4.6 referred to the protocol for testing greywacke (see subsection 6.5.4 & BCA, 1999); subclause 1.4.7 included reference to CSTR 30 (1999).

Of course, at this time British Standards were gradually being withdrawn in favour of appropriate European Standards (ENs) and in 2000, BS 5328 was replaced by BS EN 206-1 (BSI, 2000), with BS 5328 finally being withdrawn near the end of 2003 (Harrison, 2003). Prior to the final withdrawal of BS 5328, guidance on specification to BS EN 206-1 was issued as 'complementary' BS 8500 (BSI, 2002). A little confusingly, BS 8500 was issued in two parts: BS 8500-1 was intended as guidance for the specifier, whereas BS 8500-2 was prepared for use by a producer. As measures for minimising the risk of ASR had been relegated by CEN to national practice, it was necessary for BS 8500 to include this provision and, accordingly, BS 8500-2:2002 provided clause 5.2 on 'Resistance to alkali-silica reactions'. Unsurprisingly, this also followed the guidance in Digest 330 (BRE, 1999) and was generally similar to the ASR coverage in the 1999 amendment to BS 5328-2; there was additional guidance in respect of the risk of ASR with recycled concrete aggregate. When BS 8500-2 was

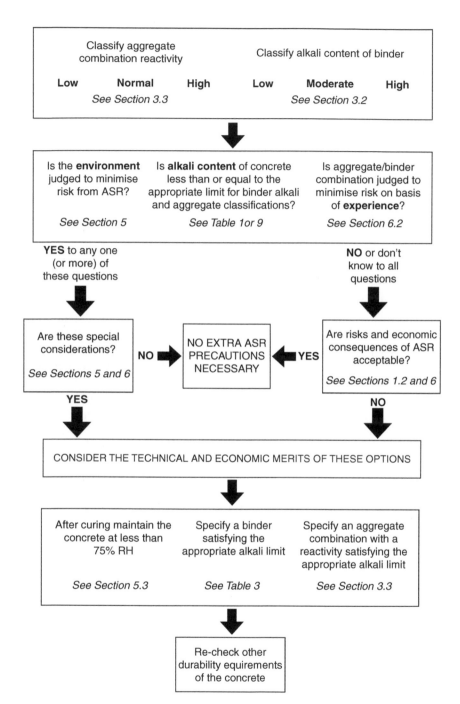

Figure 6.47 Flow chart showing the CSTR 30 assessment scheme (from Concrete Society, 1999): consult CSTR 30 (1999) for full details.

Table 6.11 Recommended CSTR 30 total alkali content limits for concrete (after Concrete Society, 1999). See CSTR 30 for full details of the cited Notes.

Aggregate classification	LOW REACTIVITY			NORMAL REACTIVITY			HIGH REACTIVITY		
Binder alkali classification	Low	Moderate	High	Low	Moderate	High	Low	Moderate	High
Initial alkali limit for concrete (kg Na$_2$Oeq/m^2)	Self-limiting Note 1	self-limiting Note 1	≤ 5.0 Notes 2 and 5	Self-limiting Note 3	≤ 3.5 Notes 4 ond 5	≤ 3.0 Notes 2 and 5	≤ 2.5 Note 6	≤ 2.5 Note 6	≤ 2.5 Note 6

If alkali from other sources exceeds 0.6kg Na$_2$Oeq/m^3 this table does not apply and no guidance is given.
Notes 1 to 6 : These important explanations and qualifications to the recommended limits are given in details in CSTR 30 (1999).

revised (BSI, 2006), subclause 5.2 was greatly reduced in favour of directly citing Digest 330 (BRE, 2004).

As around 20 years had elapsed since BRE Digest 330 first appeared in 1988 and more than 10 years since the 4-part version had started, Livesey (2009a, 2009b) produced a thoughtful and thorough assessment of the guidance, both in the light of experience with construction and performance using the ASR provisions and also taking account of continued scientific research and understanding. In his milestone two-part paper, he cautiously suggested thirteen recommendations in respect of revising the guidance based on Digest 330. This included (as his recommendation 13) the suggestion that the revised guidance on UK ASR specification might be presented as a 'normative annex' to BS 8500-2.

After several years, this proposal has finally been adopted in the recent third edition of BS 8500 (BSI, 2015a, 2015b), wherein, aside from brief introductions in BS 8500-1:2015 (clause A.8.1) and BS 8500-2:2015 (clause 5.2), the guidance on minimising the risk of damaging ASR in concrete is concentrated into a 5-page normative Annex B (the concept was actually first introduced as an Amendment in 2012). These requirements are to be used, according to clause 5.2 in BS 8500-2, for all concretes except prescribed mixes or where a specifier has specified particular provisions for resisting ASR. The specification in the new Annex B is completely restructured and rewritten from the earlier versions of BS 8500-2, deliberately presented as a specification rather than guidance and without any extant reference to BRE 330 or CSTR 30 (though there is reference to Livesey, 2009a, 2009b), but it is still clearly based on the fundamental principles arising from the *"consensus reached by experts"* to which the Foreword to amended BS 5328-1 referred in 1999 (see earlier in this subsection), albeit modified in places as Livesey had recommended. Neither is there any reference to the international work undertaken by a series of RILEM technical committees under British leadership, between 1998 and 2014, including the preparation of separate specifications for mini-mising the risk of damage in the cases of ASR [AAR-7.1], ACR [alkali-carbonate reactivity, AAR-7.2] and large dams and other hydro structures [AAR-7.3] (Nixon *et al.*, 2004; Nixon & Sims, 2016).

The UK practice of relying upon alkali contents declared by cement manufac-turers, for the purpose of assessing project concretes against the ASR guidance in BRE Digest 330, CSTR 30 and/or BS 8500-2, has seemingly worked satisfactorily on the whole. However, as a result of illegal activity by individuals during the period September 2002 to December 2004, which the manufacturer publicised as soon as it was discovered, one works did produce large quantities of cement classified as 'high' alkali (> 0.75 % Na_2Oeq), but labelled and declared as being of 'moderate' alkali ($\leq$ 0.75 % Na_2Oeq). Consequently, some concrete mixes produced with cement from this works did not, in fact, comply with the deliberately conservative guidance for minimising the risk of ASR, contrary to assessments based on the incorrectly declared data. Accounts of this unfortunate occurrence, and some of the investigations carried out to assess the longer-term risk to structures made using these concretes, have been reported by Sims *et al.* (2008) and by Jones & Cather (2013).

Sims *et al.* (2008) described an assessment procedure that, whilst being sepa-rately detailed to suit each particular project, was essentially based on a

Figure 6.48 An office building in southern England, in which the concrete columns and slabs had been inadvertently constructed using a 'high' alkali class cement, instead of the 'moderate' alkali class that had been incorrectly declared (from Sims *et al.*, 2008).

fundamental audit of the actual reactive alkali content of the concrete, plus an informed appraisal of the reactivity potential of the aggregate combination, based upon quantitative petrography. They provided three case study examples, including a recently opened business park structure (Figure 6.48), a food distribution centre and a water treatment works. Overall, in most cases, it had been possible to demonstrate either that the alkali content was within the required limits for an aggregate of 'normal' reactivity and/or that detailed appraisal could establish that the aggregate combination was of 'low' reactivity. Jones & Cather (2013) report similar procedures, but were able to provide a longer perspective and included some expansion testing to corroborate assessments or further investigate uncertain outcomes.

6.8.4 A more confident future and some continued concerns

When the first edition of this book was published, it was nearly 50 years since ASR had first been discovered, but less than about 15 years since it started to become apparent in the UK that ASR was not wholly a 'foreign problem' and that many British concretes were damaged to some extent by the reaction. In that comparatively short time, the British construction industry had identified the means of prevention, established the means of reliable diagnosis and made considerable progress in the development of measures for managing affected structures. The Hawkins Working Party (see subsections 6.6.1, 6.6.2 & 6.8.2) had initiated the principles for an eventual performance specification, now available in a British Standard, and there is ample evidence that the appropriate precautions are being routinely specified and implemented in normal concrete production and construction projects. The guidance provided by the BCA Working Party on Diagnosis (see subsection 6.7.1) ensured that, over the following more than 20 years, the occurrence of ASR in the existing stock of UK concrete structures has been correctly identified and assessed; the principles remain central to the guidance on diagnosis recently published by RILEM (AAR-6.1, Godart *et al.*, 2013) for international application. Similarly, in the case of prognosis and structural assessment, the practical guidance developed by the ISE (see subsections 6.7.3 & 6.7.4) has remained effective for more than 20 years and will feature in the forthcoming RILEM guidance on appraisal and repair (AAR-6.2, Godart *et al.*, 2017) for international application.

There is thus good reason to suppose that, now more than 75 years since ASR was first reported and around 40 years since its occurrence in the British Isles was accepted, the potential problem of AAR is sufficiently understood for the risk of damage to be minimised in routine good practice in the UK and many other countries. However, inevitably there remain possible risks arising from factors that we have not yet recognised or fully understood. A case study is presented in the **Text Box**, which describes the premature deterioration of some concrete sea defences that complied with the contemporary and current requirements of BRE Digest 330, in respect of minimising the risk of AAR, but did not make any additional allowances for the exceptional exposure conditions.

Much of the evidence on which we base our criteria for specification, diagnosis and assessment has been derived from experience with relatively small-scale structures and the current range of concrete constituents, or larger structures comprising interaction between smaller such elements, all with modest design lives cited in decades (say up to about 50 years). There is some continuing concern (Wood, 2000; Charlwood *et al.*, 2012) that larger concrete structures and/or those designed to be in service for 100 years or more, such as some dams, plus some especially sensitive types of construction, such as major tunnels or nuclear installations, might behave differently and will require particular attention. Work has been initiated by RILEM, in coordination with the International Commission on Large Dams (ICOLD), to determine a specification (as AAR-7.3) for large dams and other hydro structures (Sims *et al.*, 2012; Nixon & Sims, 2016).

Text Box   Case study of ASR-affected concrete protection units exposed to severe coastal conditions. Kindly provided by Jeremy Ingham[1], David Collery[1], Ioannis Sfikas[1] & Mike Badger[2] (based on Ingham *et al.*, 2016).

INTRODUCTION

The Clacton Coastal Defence Works (Essex, UK), constructed in 1993, included a 1.4 km length of sloping revetment formed of precast concrete armour units (see Figure 6.49). Each unit comprised a hexagonal head on top of a central hexagonal spine; three legs connected to the head. The spine and legs were connected by a base which exhibited three square perforations. The units contained no reinforcement bars. The revetment was designed for a service life of greater than 50 years, but severe cracking of the armour units (and breakages) necessitated its entire replacement after only 22 years, when the armour units had exhibited deterioration for a considerable period.

THE INVESTIGATION

To diagnose the causes of cracking, an investigation was undertaken, involving the following:

Figure 6.49 General view of the precast concrete coastal defences at Clacton–on-Sea.

[1] Mott MacDonald Ltd
[2] Tendring District Council

- Desk study of existing documentation;
- Site works to undertake a visual inspection and to obtain concrete core samples;
- Laboratory examination and testing of the concrete core samples;
- Finite element analysis of concrete temperature achieved during curing.

The desk study provided information regarding the concrete mix specified for the armour units (see **Table 6.12**).

Table 6.12 Summary of concrete mix details specified for the precast units.

Property	Stated Value
Strength Grade	C50 (cube strength of 50 N/mm^2 at 28 days)
Cement Type	Rapid Hardening Portland Cement (RHPC)
Coarse Aggregate Type	5-20 mm natural gravel coarse aggregate consisting chiefly of quartzite and vein quartz with minor proportions of sandstone, basalt and flint
Fine Aggregate Type	Natural sand consisting chiefly of quartz and quartzite with minor proportions of sandstone and basic igneous rock types
Cement Content	400 kg/m^3
Coarse Aggregate Content	1150 kg/m^3
Fine Aggregate Content	697 kg/m^3
Maximum Water/Cement Ratio	0.45
Calculated Alkali Content for Mix	2.46 kg Na$_2$O eq/m^3

From visual inspection, cracking was frequently observed in units that were within the intertidal zone. In particular, the lower three rows of units frequently exhibited large cracks (typically 1-3 mm wide) concentrated in the head of the unit and narrowing to termination in the spine and legs (see **Figure 6.50**). The units located above the splash zone (above the intertidal zone) appeared to be in generally good condition, with no cracking other than occasional surface crazing.

Six concrete core samples were taken from different units for laboratory testing; four from cracked units in the intertidal zone and two from uncracked units from splash zone. The samples were subjected to the determinations shown in Table 6.13.

Petrographic examination identified that all of the samples were similar in terms of composition and that they had all been manufactured from the specified concrete mix. All of the samples were found to be suffering from alkali-silica reaction (ASR) involving chert and quartzite particles of both the coarse and fine aggregate. The microscopical scale observations of ASR severity correlated well with the macro scale observations of cracking, with ASR being notably worse for the samples from units in the intertidal range. ASR had also initiated in all samples from units in the splash zone, even though these units had no visible cracks.

At the time of construction, the nature of the aggregate had been confirmed by petrography and the alkali content determined for the concrete mix, thus facilitating comparison with the current version of BRE Digest 330 (BRE, 2004). In this way, the aggregate combination for the concrete samples was

Figure 6.50 Cracking in the head of a unit caused by alkali-silica reaction (ASR).

Table 6.13 Examination, analysis and test procedures undertaken.

Test Name	Standard
Petrographic examination	Applied Petrography Group SR2 and ASTM C856
Sulfate content	BS 1881-124: 1988
Alkali content	BS 1881-124: 1988
Compressive strength	BS EN 12504-1:2009 and BS EN 12390-3:2009

classified as being of 'normal' reactivity. For normal reactivity aggregate combinations, BRE Digest 330 (2004) recommends that alkalis be limited to ≤ 3.0 kg Na_2O eq/m^3 (where high alkali cement is used). The calculated alkali content of the concrete mix at the time of construction (2.46 kg Na_2O eq/m^3) complied with this recommendation. Alkali content analysis of the six core samples gave results ranging from 2.25 to 3.38 with a mean value of 2.55 (kg Na_2O eq/m^3) and, for all but one, the values were below the recommended upper limit.

The ASR observed in all core samples during petrographic examination could not have occurred without the availability of sufficient alkali hydroxides in the pore solution of the cement paste. However, the alkali content results were typically within the BRE Digest 330 recommended limit. This raised the question of how alkali levels were ever sufficient to cause ASR. It is suggested that the probable answer is the action of seawater leaching through the concrete, combined with surface evaporation that caused alkalis to migrate and accumulate in the upper (and other) sections of the units. Is should be noted that the geometry of the armour units may have made them relatively susceptible to leaching, with wicking action driving alkali accumulation in the unit head. Also, the units had a significantly higher surface area to volume ratio than other unit designs, such as cubes or grooved cubes).

It is thus probable that the seawater percolation and alkali migration caused the alkali content to exceed the BRE Digest 330 recommended limits in certain parts of the concrete units (especially the heads but also other parts of the units). Most of the alkalis involved in the reaction would still have been those cast-in at the time of construction (from the cement).

Microscopical examination showed that the core samples exhibited deposits of ettringite in voids and cracks that were most likely caused by the action of leaching. However, the possibility that some minor degree of delayed ettringite formation (DEF) had occurred could not be completely ruled out, though the pattern of secondary ettringite occurrence was not typical of DEF. The sulfate content analysis gave results ranging from 3.2 to 3.7 with a mean of 3.4 (% by mass of cement). These values would be considered normal for Portland cement concrete.

When specifying concrete curing temperatures to prevent DEF, it is common to limit the maximum temperature to 65°C (Ingham, 2012). When investigating concrete structures that are suspected of suffering from DEF, it is usually found that much higher curing temperatures are required to cause the reaction. Quillin (2001) provides the following DEF risk classification scheme:

<60°C = No risk of DEF
60 to 69°C = Very low risk of DEF
70-79°C = Low risk of DEF (providing that the Na_2Oeq of the cement is less than 0.85%)
>85°C = Increased risk of DEF

A modelling exercise was undertaken using finite element analysis software to estimate the maximum temperature that the precast concrete units may have reached during early-age curing (up to 3 days). Thermal simulation indicated that the curing temperatures reached were unlikely to have exceeded the temperature at which a high risk of DEF would exist (>85°C). However, they could potentially have reached temperatures where DEF is possible (70-79°C), but the associated risk would be classified as 'low'. Further discussion of thermal modelling techniques can be found in Sfikas et al. (2016).

The thermal simulation showed that it is possible that relatively high unit core temperatures could have been generated, possibly resulting in temperature differentials (to surface temperature) that could have caused some early-age thermal cracking to have occurred. This is likely to have involved relatively narrow cracks; that have apparently passed quality control inspections prior to installation and/or some hidden cracks in the core. Any such early-age thermal cracking may have facilitated seawater penetration into the core of the units to accelerate the initiation and early development of expansive reactions.

In terms of strength, the concrete samples were found likely to have been compliant with the specified 28-day cube strength of 50 N/mm^2 in all cases, with the mean estimated in-situ cube strength being 71 N/mm^2. Moreover, the compressive strength test specimens were specifically prepared to avoid any of the cracks caused by expansive reaction.

OUTCOME AND CONCLUSIONS

After considering all of the available evidence, the cracking observed in the armour units located in the intertidal zone was ascribed to an expansive deleterious reaction. The type of expansive reaction was most likely to have been ASR alone, although the possibility of a combination of ASR and DEF (with ASR being by far the most dominant) could not be ruled out.

For the units located in the intertidal zone, the expansive reaction is likely to have initiated a few years after the structure was commissioned. In most cases of ASR, the expansive effects of the reaction could reasonably be expected to be largely complete at around 15 years after initiation (once the alkalis are consumed). In this case, there was a likelihood of on-going alkali migration within the units. Hence, it was likely that the concrete units or parts of units exposed to variably cyclic seawater wetting would continue to react expansively (and deteriorate) in response to on-going exposure.

The units not directly exposed to seawater (in the splash zone, above the normal tidal range) were in relatively good condition. However, petrographic examination indicated that they had a similar composition to the cracked units that are exposed to seawater and also, that ASR had already been initiated.

In light of the investigation, the ASR-affected concrete armour units were removed from the revetment and a new coastal protection scheme was constructed involving fishtailed groyne structures (rock armour) with beaches of dredged marine aggregate in between. To reduce carbon usage, the removed concrete armour units were crushed down and recycled within core material being used for the fishtailed groyne structures. The groynes were designed to accommodate further ASR-related deterioration of the concrete material once exposed to seawater within the groynes, by blending with sound rock materials to mitigate any effects.

It is notable that the failed concrete armour units complied with the current UK guidelines [BRE Digest 330 (2004)], in terms of the aggregate combination used and the alkali content of the mix. As such, this case study demonstrates the importance of considering the exposure conditions, including the potential action of fluctuating seawater percolation when designing concrete structures to minimise the risk of ASR damage. In particular, as well as ensuring a concrete mix suitable for the exposure conditions, designers of marine or coastal structures should be cautious of using concrete elements with complex geometry and/or relatively low surface area to volume ratios.

6.9 REPUBLIC OF IRELAND

6.9.1 Recognising a history without ASR

Geographically, the British Isles is an archipelago dominated by two islands: Great Britain, including the nations of England, Scotland and Wales, and Ireland, including Northern Ireland (part of the UK, together with Great Britain) and the independent Republic of Ireland. There is naturally considerable geological continuity between Great Britain and Ireland, but the distribution of cases of ASR damage has appeared to differ significantly. Indeed, when the Hawkins Report was first published in the UK (Hawkins, 1983), it was considered that there had been no occurrences of ASR anywhere in the Republic of Ireland. Consequently, after discussions with the Hawkins Working Party, which stated in their report that it was intended to be applied within the UK, but obviously could not prevent it from being referred to in the Republic of Ireland, the Institution of Engineers of Ireland and the Irish Concrete Society jointly formed a parallel working party in 1988. As a result, ASR guidance that was particular to the Republic of Ireland was published in 1991 (de Courcy, 1991).

6.9.2 Guidance from the Institution of Engineers in Ireland – I

A useful summary of the 1991 Irish recommendations was published by de Courcy (1992). Although the considerations and approach were reminiscent of the Hawkins Report in the UK, the elegantly written guidance made especial reference to certain aspects of concrete and concrete construction in the Republic of Ireland up to that time. It was acknowledged that Portland cements of Irish manufacture typically had relatively high alkali contents, ranging from 0.70 to 1.00 % Na_2Oeq and then being stabilised at about 0.85 % Na_2Oeq, but also asserted that there were generally few other sources of alkalis in Irish concrete practice, for example de-icing salts were rarely applied and ggbs or pfa had negligible use.

It was also maintained that Irish aggregates largely comprised constituents that would not be expected to be reactive, with the crushed rock materials being dominated by Carboniferous limestone and natural sand and gravel materials excluding marine-dredged sources (which had been associated with some cases of ASR in South-West England, see subsection 6.5.1); even shore-based sands and gravels were said by de Courcy (1992) to be *"firmly discouraged"*. One controversial consideration was that, unlike the 'flint' type of chert originally derived from Cretaceous rocks that had been associated with many UK examples of ASR, the chert in Irish sands and gravels was *"almost wholly associated with the older Carboniferous limestone rather than the younger Cretaceous limestone"*. Strogen (1993) would later argue for a geological basis for his explanation that the Carboniferous cherts have a higher quartz crystallinity index (QCI) than Cretaceous flints. In fact, this assertion was at least questionable at the time, especially as the premise that UK reactivity was only associated with the flint type of chert was incorrect; also it has now been recognised that QCI is an unreliable guide to silica reactivity (Marinoni & Broekmans, 2013).

In addition to considerations of limited alkali levels and non-reactive aggregates, the 1991 Irish guidance permitted reliance on an 'acceptable' history of use of the aggregates. A flow chart is given in the guidance (similarly in de Courcy's 1992 summary), which shows five separate routes to 'No further precautions necessary', notably including when the cement alkali content is less than 1.0 % Na_2Oeq and when 'the history in use' of aggregates is 'acceptable'. When none of these routes can be achieved, the alkali content of the concrete is limited to 4.0 kg/m^3, which was higher than the 3.0 kg/m^3 limit in use in the UK at the time, or even 4.5 kg/m^3 if the reactive aggregate material was Carboniferous chert.

6.9.3 Guidance from the Institution of Engineers in Ireland – II

As already explained in subsection 6.8.3, the European Standard (EN 206-1) on concrete was published in 2000, which would in due course replace all of the separate national standards on that subject. However, the matter of ASR was left as a matter for national practices, partly in recognition of the way in which the experience of ASR differed greatly between countries and doubtless partly because there was thus little enthusiasm for trying to seek a harmonious approach between all those European countries. In the Republic of Ireland, it was decided by the Institution of Engineers of Ireland and the Irish Concrete Society to take this opportunity to revisit their joint guidance, taking account of further research into the reactivity of Carboniferous chert and also greywacke aggregates, plus the various contributions to concrete alkali content and various international developments. A new edition of the Irish ASR guidance was published (Richardson, 2003). In an explanation of the new guidance, Richardson (2004) was still able to state that *"No cases of damaging alkali-silica reaction (ASR) have been reported in the Republic of Ireland to date"*.

In keeping with its predecessor, the new guidance is clearly written and admirably concise. It is closer to the UK approach in principle, but less confused with detail. Compliance with the overall requirements can be achieved by one of the three familiar basic routes: keeping the concrete effectively dry, using a non-reactive aggregate combination or restricting the total alkali content (or 'load') of the concrete. In the case of the non-reactive aggregates route, the 'history of use' option remains available and testing is also permitted (reference to the RILEM methods is included in an earlier section of the guidance). The 'alkali load' route can be satisfied by using a low-alkali cement or otherwise by controlling the concrete total alkali content, when all the possible alkali contributors, now including ggbs and pfa, have to be taken into account; compared with the 1991 guidance, the alkali limit is raised to 4.5 kg/m^3, except for concrete made with greywacke aggregates (not mentioned in the 1991 guidance), when the limit is 3.5 kg/m^3.

6.10 CONCLUDING REMARKS

ASR continues to be a matter of intense, and perhaps sometimes disproportionate, concern around the world. In the British Isles it had first been regarded as an exotic, rare and, above all, foreign problem; later the exaggerated response to the discovery of ASR in UK structures became reflected in the perception of a 'concrete cancer' epidemic, as portrayed to the general public by the media. The occurrence of ASR is now accepted by most UK engineers as one amongst many durability threats to concrete construction, but the risk of any resultant damage being of serious structural concern is now recognised to be comparatively small. In the Republic of Ireland, ASR damage seems to have remained a remote risk, though its effects in the UK and elsewhere are well recognised. Nevertheless, it is undeniable that damage caused by ASR can occasionally be very significant, and reliable means of arrest and repair cannot always be applied with confidence, so that, for the foreseeable future, critical or particularly vulnerable structures will need to be designed to negate the risk of ASR and precautions will continue to be taken with most normal construction at least to minimise the risk of ASR. The search for dependably predictive performance tests will go on, and hopefully knowledge and understanding will improve so that the likelihood of ASR occurring in a given concrete in a particular location can be more accurately evaluated. It would now seem that most of the existing stock of concrete structures will be able to survive the experience of ASR.

REFERENCES

In the case of British and American Standards, derived from the first edition, the modern edition is cited and reference to any earlier versions is made in parentheses.

Al-Kadhimi, T.K.H., & Banfill, P.F.G. (1996) The effect of electrochemical re-alkalisation on alkali-silica expansion in concrete. In: Shayan, A. (ed.) *Alkali-Aggregate Reaction in Concrete, Proceedings of the 10th International Conference on Alkali-Aggregate Reaction in Concrete*, Melbourne, Australia, 18–23 August 1996, pp. 637–644.

Allen, R.T.L. (1981.) Alkali-silica reaction in Great Britain – a review. In: Oberholster, R.E. (ed.) *Proceedings of the 5th International Conference on Alkali-Aggregate Reaction in Concrete*, S252/18, National Building Research Institute of the CSIR, Pretoria, Republic of South Africa.

American Society for Testing and Materials (1986) *Standard Specification for Concrete Aggregates*. ASTM C33-86 (ASTM C33-55T, ASTM Standards, 1955, Part 3, 1145–1150 and Revised in ASTM Reprint 12, 1956).

American Society for Testing and Materials (1987a) *Standard Test Method for Potential Reactivity of Aggregates* (Chemical Method). ASTM C289-87 (ASTM C289-52T, ASTM Standards, 1952, Part 3, 943).

American Society for Testing and Materials (1987b) *Standard Test Method for Potential Alkali-Reactivity of Cement-Aggregate Combinations* (Mortar-Bar Method). ASTM C227-87 (ASTM C227-52T, ASTM Standards, 1952, Part 3, 44–51).

American Society for Testing and Materials (2010) *Standard guide for examination of hardened concrete using scanning electron microscopy*, ASTM C1723, ASTM, West Conshohocken, PA, USA.

American Society for Testing and Materials (2013) *Standard practice for petrographic examination of hardened concrete*, ASTM C856, ASTM, West Conshohocken, PA, USA.

Anon (1986) Full inspection follows Welsh dam AAR alert. New Civil Engineer, 13th November, 8.

Applied Petrography Group (2010) *A code of practice for the petrographic examination of concrete*, SR2. The Geological Society of London.

Blackwell, B.Q. (1997) BRE Digest 330 – How to avoid ASR, *Concrete, 31* (7), 18–19.

Blackwell, B.Q. & Pettifer, K. (1990) In Confidence Note 6/90, Building Research Establishment, Garston, Watford, UK (cited in Zhang et al., 1990).

Blackwell, B.Q. & Pettifer, K. (1992) Alkali-reactivity of greywacke aggregates in Maentwrog Dam (North Wales), *Mag Concrete Res., 44* (161), December, 255–264.

Blackwell, B.Q., Thomas, M.D.A., Nixon, P.J., & Pettifer, K. (1992) The use of fly ash to suppress deleterious expansion due to AAR in concrete containing greywacke aggregate, In: *Proceedings of the 9th International Conference on Alkali-Aggregate Reaction in Concrete*, London, Conference Papers, *Volume 1*, 102–109, The Concrete Society, Slough (now Camberley), UK.

Blackwell, B.Q., Thomas, M.D.A., Pettifer, K., & Nixon, P.J. (1996) An appraisal of UK grey-wacke deposits and current methods of avoiding AAR, In: Shayan, A. (ed.) *Alkali-Aggregate Reaction in Concrete, Proceedings of the 10th International Conference on Alkali-Aggregate Reaction in Concrete*, Melbourne, Australia, 18–23 August 1996, pp. 492–499.

Blackwell, B.Q., Thomas, M.D.A., & Sutherland, A. (1997) Use of lithium to control expansion due to alkali-silica reaction in concrete containing UK aggregates, In: Malhotra, V.M. (ed.) *Proceedings of the 4th International Conference on Durability of Concrete, Detroit*, USA, American Concrete Institute, SP 170-34, pp. 649–663.

Bolton, R.F. (1989) *An investigation into the freeze-thaw behaviour of ASR affected concrete*, Technical Memorandum, Report Ref. TM CES 127, British Rail Research, Derby, UK, 48p.

Bridle, R., Horswill, P., & Fawcett, S. (1991) Concrete at Queen's Valley Reservoir, Jersey, *Concrete (J Concrete Society)*, 25 (4), 20–23.

British Cement Association (1988) *The Diagnosis of Alkali-Silica Reaction. Report of a Working Party*. British Cement Association, Ref. 45.042, Wexham Springs, Slough, UK.

British Cement Association (1992) *The Diagnosis of Alkali-Silica Reaction, Report of a Working Party*, 2nd edition, Ref. 45.042, BCA, Wexham Springs, Slough, UK, 44p.

British Cement Association (1993) *The Diagnosis of Alkali-Silica Reaction: Further Information and Recommendations for BCA Publication*, Insert to Ref. 45.042, 2nd edition, January 2003, BCA, Crowthorne, UK, 1 page.

British Cement Association (1999) *Alkali-silica reaction, Testing protocol for greywacke aggregates*, Protocol of the BSI B/517/1/20 ad hoc group on ASR, Ref. 45.044, BCA, Crowthorne, UK, 8pp.

British Standards Institution (1978) BS 12:1978 (BS 12:1958, BS 12-2:1971). Specification for Ordinary and Rapid-Hardening Portland Cement.

British Standards Institution (1983) BS 882:1983 (BS 882:1954, BS 882 & 1201:1965, BS 882,1201-2:1973). Specification for Aggregates from Natural Sources for Concrete.

British Standards Institution (1980) BS 3148:1980 (BS 3148:1959). *Methods of Test for Water for Making Concrete* (Including Notes on the Suitability of the Water).

British Standards Institution (1970 and 1978) BS 4550-2: 1970 & BS 4550-0 & 1 & 3 to 6:1978. *Methods of Testing Cement*.

British Standards Institution (1984 and 1985) BS 812-101 and 102:1984, BS 8l2-103, 105, 106 and 119:1985 (BS 812:1960, BS 812:1967, BS 812-1 to 3:1975, BS 812-4:1976). *Testing Aggregates*.

British Standards Institution (1969a) CP 114-2:1969 (CP 114:1957). *The Structural Use of Reinforced Concrete in Buildings*.

British Standards Institution (1969b) CP 115-2:1969 (CP 115:1959). *The Structural Use of Prestressed Concrete in Buildings*.

British Standards Institution (1969c) The Structural Use of Precast Concrete. CP 116-2:1969 and Addendum No. 1, 1970 (CP 116:1965).

British Standards Institution (1972) CP110-1 to 3:1972. *The Structural Use of Concrete*.

British Standards Institution (1985) BS 8110-1 to 3:1985. *The Structural Use of Concrete*.

British Standards Institution (1983 and 1986). BS 1881-101 to 122:1983, BS 1881-125 and 201 to 203 and 205 to 206:1986 (BS 1881:1952, BS 1881-1 to 5:1970, BS 1881-6:1971). *Testing Concrete*.

British Standards Institution (1988a) BS 1881-124. *Testing Concrete, Part 124: Methods of analysis of hardened concrete*. London, BSI.

British Standards Institution (1988b) BS 812-123. *Draft for public comment. Testing aggregates. Alkali-Silica reactivity: Concrete prism method*. BSI Document Ref. 88/11922 DC, June.

British Standards Institution (1994) BS 812-104:1994. *Testing aggregates, Part 104. Method for qualitative and quantitative petrographic examination of aggregates*. London, BSI.

British Standards Institution (1995a) BS EN 450:1995. *Fly ash for concrete – Definitions, requirements and quality control*. London, BSI.

British Standards Institution (1995b) *Testing aggregates, Method for determination of alkali-silica reactivity: Concrete prism method*. DD 218:1995. London, BSI, 16p.

British Standards Institution (1997a) BS 5328-1:1997. *Concrete, Part 1. Guide to specifying concrete.* incorporating Amendment No 1, 1999. London, BSI, 25p.

British Standards Institution (1997b) *Concrete, Part 2. Methods for specifying concrete mixes,* BS 5328-2:1997, incorporating Amendment Nos 1 & 2 and Corrigendum, 1997 & 1999. London, BSI, 20p.

British Standards Institution (1997c) BS 3892-1:1997. *Pulverized-fuel ash, Part 1. Specification for pulverized-fuel ash for use with Portland cement.* London, BSI.

British Standards Institution (1997d) BS EN 932-3:1997. *Tests for general properties of aggregates, Part 3: Procedure and terminology for simplified petrographic description.* London, BSI.

British Standards Institution (1999a) DD 249:1999. *Testing aggregates, Method for the assessment of alkali-silica reactivity, Potential accelerated mortar-bar method.* London, BSI, 10p.

British Standards Institution (1999b) BS 812-123:1999. *Testing aggregates, Method for determination of alkali-silica reactivity: Concrete prism method.* Confirmed 2011, London, BSI, 18p.

British Standards Institution (1999c) BS 7943:1999. *Guide to the interpretation of petrographical examinations for alkali-silica reactivity,* London, BSI (see also reconfirmed version in 2009).

British Standards Institution (2002) BS 8500-2:2002. *Concrete – Complementary British Standard to BS EN 206-1, Part 2: Specification for constituent materials and concrete,* incorporating Amendment No 1, 2003, London, BSI, 34p.

British Standards Institution (2006) BS 8500-2:2006. *Concrete – Complementary British Standard to BS EN 206-1, Part 2: Specification for constituent materials and concrete,* London, BSI, 40p.

British Standards Institution (2009) BS 7943. *Guide to the interpretation of petrographical examinations for alkali-silica reactivity.* London, BSI (reconfirmed version of 1999 edition).

British Standards Institution (2009) BS EN 12504-1. *Testing concrete in structures. Cored specimens. Taking, examining and testing in compression.* London, BSI.

British Standards Institution (2009) BS EN 12390-3. *Testing hardened concrete. Compressive strength of test specimens.* London, BSI.

British Standards Institution (2009) BS EN 12390-7. *Testing hardened concrete. Density of hardened concrete.* London, BSI.

British Standards Institution (2015) BS 8500-1:2015. *Concrete – Complementary British Standard to BS EN 206, Part 1: Method of specifying and guidance for the specifier.* London, BSI, 62p.

British Standards Institution (2015) BS 8500-2:2015. *Concrete – Complementary British Standard to BS EN 206-1, Part 2: Specification for constituent materials and concrete.* London, BSI, 42p.

Building Research Establishment (1982) Alkali – Aggregate Reactions in Concrete. BRE Digest 258. HMSO, London.

Building Research Establishment (1988) Alkali – Aggregate Reactions in Concrete. BRE Digest 330. Department of the Environment, London.

Building Research Establishment (1997) *Alkali-silica reaction in concrete,* BRE Digest 330, Parts 1 – 4, Construction Research Communications (CRC), BRE, Watford, UK.

Building Research Establishment (1999) *Alkali-silica reaction in concrete,* BRE Digest 330, Parts 1 – 4, Construction Research Communications (CRC), BRE, Watford, UK.

Building Research Establishment (2002) *Minimising the risk of alkali-silica reaction: alternative methods,* Information Paper IP1/02, BRE (CRC), Watford, UK, 8p.

Building Research Establishment (2004) *Alkali-silica reaction in concrete,* BRE Digest 330, Parts 1 – 4, HIS/BRE, Watford, UK.

Building Research Station (1971) Changes in the Appearance of Concrete on Exposure. BRS Digest 126. HMSO, London.

Buttler, F.G., Newman, J.B., & Owens, P. (1980) Pfa and the alkali-silica reaction. *Consult Eng*, November, 57–62.

Cement and Concrete Association (1948) Thirteenth Annual Report. 30th June.

Chana, P.S. & Korobokis, G.A. (1991a) *Structural performance of reinforced concrete affected by alkali silica reaction: Phase I*, Contractor Report 267, Transport and Road Research Laboratory (Department of Transport), Crowthorne, UK, 77p.

Chana, P.S. & Korobokis, G.A. (1991b) *Bond strength of reinforcement in concrete affected by alkali silica reaction: Phase II*, Contractor Report 233, Transport and Road Research Laboratory (Department of Transport), Crowthorne, UK, 54p.

Chana, P.S. & Korobokis, G.A. (1992) *The structural performance of reinforced concrete affected by alkali silica reaction: Phase II*, Contractor Report 311, Transport and Road Research Laboratory (Department of Transport), Crowthorne, UK.

Chana, P.S. & Thompson, D.M. (1992) Laboratory testing and assessment of structural members affected by alkali silica reaction, In: *Proceedings of the 9th International Conference on Alkali-Aggregate Reaction in Concrete*, London, Conference Papers, Volume 1, pp. 156–166, The Concrete Society, Slough (now Camberley), UK.

Charlwood, R., Scrivener, K.L., & Sims, I. (2012) Recent developments in the management of chemical expansion of concrete in dams and hydro projects – Part 1: Existing structures, presented at *Hydro 2012: Innovative Approaches to Global Challenges*, Bilbao, Spain, 17pp.

Chrisp, T.M., Wood, J.G.M., & Norris, P. (1989) Towards quantification of microstructural damage in AAR deteriorated concrete, *International Conference on Recent Developments on the Fracture of Concrete and Rock, The University of Wales, Cardiff.* 9pp.

Clark, L.A. (1989) *Critical review of the structural implications of the alkali silica reaction in concrete*, Contractor Report 169, Transport and Road Research Laboratory (Department of Transport), Crowthorne, UK, 89pp.

Clark, L.A. (1990) Structural effects of alkali-silica reaction, One-day Seminar at BCA, 25 October 1990: *Alkali-Silica Reaction, an update on structural and material effects*, Ref TD50960.

Clark, L.A., & Ng, K.E. (1992) Prediction of the punching shear strength of reinforced concrete slabs with ASR, In: *Proceedings of the 9th International Conference on Alkali-Aggregate Reaction in Concrete*, London, Conference Papers, Volume 1, pp. 167–174, The Concrete Society, Slough (now Camberley), UK.

Clayton, N., Currie, R.J., & Moss, R.M. (1990) The effects of alkali-silica reaction on the strength of prestressed concrete beams, *The Structural Engineer*, 68 (15) August, 287–292.

Coates J. (1983) Concrete cancer plague spreading, The Sunday Times, 6th February

Cole, R.G., & Horswill, P. (1988) Alkali-silica reaction: Val de la Mare Dam, Jersey, case history. Proc. Instn Civ. Engrs., Part 1 84 (December), 1237–1259 (Paper 9372).

Concrete Society (1987) *Alkali-Silica Reaction: Minimising the risk of damage to concrete, Guidance notes and model specification clauses*, Technical Report No. 30, second edition, Concrete Society, London (now Camberley, UK). See Hawkins (1983) for the first edition of this guidance.

Concrete Society (1995) *Alkali-Silica Reaction: Minimising the risk of damage to concrete, Guidance notes and model specification clauses, 1995 Revision*, Technical Report No. 30, 1995 Revision (of second edition), The Concrete Society, Slough (now Camberley), UK, 47pp.

Concrete Society (1999) *Alkali-Silica Reaction: Minimising the risk of damage to concrete, Guidance notes and model clauses for specifications, Report of a Concrete Society working party*, Technical Report No. 30, third edition, The Concrete Society, Slough (now Camberley), UK, 72pp.

Connell, M.D., & Higgins, D.D. (1992) Effectiveness of GGBS in preventing ASR, In: *Proceedings of the 9th International Conference on Alkali-Aggregate Reaction in Concrete*, London, Conference Papers, Volume 1, pp. 175–183, The Concrete Society, Slough (now Camberley), UK.

Connell, M.D., & Higgins, D.D. (1996) Effectiveness of granulated blastfurnace slag in preventing alkali-silica reaction, In: Shayan, A. (ed.) *Alkali-aggregate reaction in concrete, Proceedings of the 10th International Conference on Alkali-Aggregate Reaction in Concrete, Melbourne, Australia,* pp. 530–537.

Coombes, L.H. (1976) Val de la Mare Dam, Jersey, Channel Islands. In: *Proceedings of a Symposium on the Effect of Alkalis on the Properties of Concrete,* London. Cement Concrete Ass., Wexham Springs, Slough. UK, pp. 357–378.

Coombes, L.H., Cole, R.G., & Clarke, R.M. (1975) Remedial measures to Val de la Mare Dam, Jersey, Channel Islands, following alkali-aggregate reactivity. Symposium, British National Committee on Large Dams (BNCOLD), University of Newcastle upon Tyne, Proceedings Paper 3.3, pp. 1–10.

Cope, R.J., & Slade, L. (1992) Effect of AAR on shear capacity of beams, without shear reinforcement, In: *Proceedings of the 9th International Conference on Alkali-Aggregate Reaction in Concrete,* London, Conference Papers, Volume 1, pp. 184–191, The Concrete Society, Slough (now Camberley), UK.

Corish, A.T., & Jackson, P.J. (1982) Portland cement properties - past and present. Concrete (J. Concr. Soc.) 16(7), 16–18.

Courtier, R.H. (1990) The assessment of ASR-affected structures, *Cement Concrete Comp.,* 12, pp. 191-201.

de Courcy, J.W. (chairman) (1991) *Alkali-silica reaction – General recommendations and guidance in the specification of building and civil engineering works – a report prepared by a joint working party,* The Institution of Engineers of Ireland, Dublin and The Irish Concrete Society, Stillorgan, Co. Dublin, 36pp.

de Courcy, J.W., & Ryan, N.M. (1992) Managing ASR in the Republic of Ireland, In: *The 9th International Conference on Alkali-Aggregate Reaction in Concrete,* London, Conference Papers, Volume 1, 240–250, The Concrete Society, Slough (now Camberley), UK.

Department of Transport (1976) *Specification for Road and Bridge Works,* 5th Edn (and Supplement No. 1, 1978). HMSO, London.

Department of Transport (1982) Control of alkali-silica reaction. Interim amendments to specifications for all the Department's contracts that include structural concrete usage. Clause 1618 addition to Specification for Road and Bridge Works (see Ref. 85), July.

Department of Transport (1986) Specification for Highway Works, Parts 1–7 and Notes for guidance, Parts 1–6, HMSO, London.

Department of Transport (1990) Sixth CHE [Chief Highway Engineer] Interim Amendment (superseding fourth CHE interim amendment), Control of alkali-silica reaction, Clause 1704.6, Specification for Highway Works, 6th edition, Department of Transport, London, 6pp.

Dhir, R.K., & Paine, K.A. (2007) *Performance related approach to use of recycled aggregates,* WRAP Project AGG0074, WRAP (Waste & Resources Action Programme), Banbury, UK.

Dunster, A. (2009) *Silica fume in concrete,* BRE Information Paper IP5/09, Building Research Establishment, Watford, UK, 12pp.

Dunster, M., & Crammond, N.J. (2003) *Deterioration of cement-based building materials: lessons learnt,* BRE Information Paper IP 4/03, (referring to BRE Report BR 441, Avoiding deterioration of cement-based materials and components, Lessons from case studies: 4), Building Research Establishment, Watford, UK.

Fernandes, I., Broekmans, M.A.T.M., Nixon, P.J., Sims, I., Ribeiro, M.D.A., & Noronha, F., Wigum, B.J. (2012) Alkali-silica reactivity of some common rock types: A global petrographic atlas, In: Drimalas, T., Ideker. J.H., & Fournier, B. (eds.) *Proceedings of the 14th International Conference on Alkali-Aggregate Reactivity in Concrete,* Austin, Texas, USA.

Fernandes, I., Ribeiro, M.D.A., Broekmans, M.A.T.M., & Sims, I. (2016) *Petrographic Atlas: Characterisation of aggregates regarding potential reactivity to alkalis, RILEM TC 219-ACS*

Recommended Guidance AAR-1.2, for use with the RILEM AAR-1.1 Petrographic Examination Method, Springer for RILEM,

Figg, J.W. (1981) Reaction between cement and artificial aggregate concrete. In: Oberholster, R.E. (ed), *Proceedings of the 5th International Conference on Alkali-Aggregate Reaction in Concrete.* S252/7, National Building Research Institute of the CSIR, Pretoria, Republic of South Africa.

Fookes, P.G. (1984) Discussion on the design and construction of the Thames Barrier coffer dam. *Proc. Instn. Civ. Engrs*, Part 176 (May), 567–568.

Fookes, P.G. (1992) Geology and concrete: a review of British aggregates, The 5th Sir Frederick Lea Memorial Lecture, *20th Annual Convention of the Institute of Concrete Technology*, 13–15 April 1992, De Vere Hotel, Coventry, UK. 70pp.

Fookes, P.G., Cann, J., & Comberbach, C.D. (1984) Field investigation of concrete structures in South-West England, Part 3. Concrete. J. Concr. Soc., *18*(11), 12–16.

Fookes, PG., Collis, L., French, W.J., & Poole, A.B. (1977) Concrete in the Middle East (reprints of five articles from Concrete). Cement Concrete Ass., Viewpoint Publication, Ref. 12.077, Wexham Springs, Slough, U.K.

Fookes, PG., Comberbach, C.D., & Cann, J. (1983) Field investigation of concrete structures in South-West England, Parts 1 & 2. Concrete. J. Concr. Soc., *17* (3 and 4), 54–56 and 60–65.

French, W.J. (1986) *Research report on two samples of concrete from Maentwrog Dam for Binnie and Partners*, Geomaterials Research Services Ltd, Billericay, UK, Ref. 1424, 2 September 1986, 30pp (not published).

French, W.J. (1987) A review of some reactive aggregates from the United Kingdom with reference to the mechanism of reaction and deterioration. In: *Proceedings of the 7th International Conference on Concrete Alkali-Aggregate Reactions*, pp. 226–230.

French, W.J. (1992) The characterization of potentially reactive aggregates, In: *Proceedings of the 9th International Conference on Alkali-Aggregate Reaction in Concrete*, London, Conference Papers, Volume 1, The Concrete Society, Slough (now Camberley), UK, pp. 338–346.

French, W.J. (2005) Presidential address 2003: Why concrete cracks – geological factors in concrete failuire, *Proceedings of the Geologists' Association*, 116 (2), 89–105.

French, W.J., & Poole, A.B. (1974) Deleterious reactions between dolomites from Bahrain and cement paste. *Cement Concrete. Res., 4*, 925–937.

Gillott, J.E., Duncan, M.A.G., & Swenson, E.G. (1973) Alkali-aggregate reaction in Nova Scotia. IV. Character of the reaction. *Cement Concrete Res.*, *3*, 521–535 (also National Research Council of Canada, RP 567).

Godart, B., de Rooij, M., & Wood, J.G.M. (2013) *Guide to diagnosis and appraisal of AAR damage to concrete in structures, Part 1 Diagnosis (AAR-6.1)*, RILEM State-of-the-Art Reports, Volume 12, Springer, Dordrecht, pp RILEM, Paris.

Godart, B., Wood, J.G.M., & de Rooij, M. (2017) *Guide to diagnosis and appraisal of AAR damage to concrete in structures, Part 2 Appraisal and repair (AAR-6.2)*, RILEM State-of-the-Art Reports, Springer, Dordrecht, pp RILEM, Paris. *Expected completion for publication in 2017.*

Grattan-Bellew, P.E. (1992) Microcrystalline quartz, undulatory extinction & the alkali-silica reaction, In: *Proceedings of the 9th International Conference on Alkali-Aggregate Reaction in Concrete*, London, Conference Papers, Volume 1, pp. 383–394, The Concrete Society, Slough (now Camberley), UK.

Gutt, W., & Nixon, P.J. (1979) Alkali-aggregate reactions in concrete in the UK. Concrete (J. Concr. Soc.) *13*(5), 19–21.

Haigh, I.P. (1969) General principles of control of aggregates. Proceedings of a Symposium on Sea Dredged Aggregates for Concrete, 9th December 1968. Sand and Gravel Association of Great Britain, Ref. 97.101, London, pp. 47–50.

Harrison, T. (2003) *The new concrete standards – getting started, an introductory guide to the new standards for concrete BS EN 206-1 and BS 8500*, The Concrete Society, Crowthorne (now Camberley), Quarry Products Association, Ref. CS 149, 20pp.

Hawkins, M.R. (ed.) (1983) Alkali-Aggregate Reaction: Minimising the Risk of Alkali-Silica Reaction, Guidance Notes. Report of a Working Party. *Cement Concrete Ass.*, Ref. 97–304, Wexham Springs, Slough, UK.

Highways Agency, The. (1994) The assessment of concrete structures affected by alkali silica reaction, Insert BA 52/94, *Design Manual for Roads and Bridges*, 3a Highways Structures: Inspection and Maintenance (4 Assessment) Part 10, HMSO, London.

Hobbs, D.W (1978) Expansion of concrete due to alkali-silica reaction: an explanation. *Mag. Concr. Res.*, *30*(105), 215–220.

Hobbs, D.W. (1980) *Influence of Mix Proportions and Cement Alkali Content upon Expansion due to the Alkali-Silica Reaction*. Cement and Concrete Association, Technical Report 534, C&CA Ref. 42.534, Wexham Springs, Slough, UK.

Hobbs, D.W. (1988) *Alkali-Silica Reaction in Concrete*. Thomas Telford, London.

Hobbs, D.W. (1990) Cracking and expansion due to the alkali-silica reaction: its effect on concrete, *Structural Engineering Review*, *2*, 65–79.

Hobbs, D.W. (1992a) Deleterious alkali-silica reactivity of a number of UK aggregates, *Concrete* (The Journal of the Concrete Society), *26*(3), 64–70.

Hobbs, D.W. (1992b) *Deleterious alkali-silica reactivity of a number of UK aggregates and an examination of the draft BS concrete prism test*, Ref. C/13, Mineral Industry Research Organisation, British Cement Association, Slough, UK, 88pp.

Hobbs, D.W. (1992c) Deleterious reactivity of a number of UK aggregates and an examination of the draft BS concrete prism test, In: *Proceedings of the 9th International Conference on Alkali-Aggregate Reaction in Concrete*, London, Conference Papers, Volume 1, pp. 451–460, The Concrete Society, Slough (now Camberley), UK.

Hobbs, D.W. (1993) Deleterious alkali-silica reactivity in the laboratory and under field conditions, *Mag Concrete Res.*, *45* (163) June, 103–112.

Hobbs, D.W. (1994) The effectiveness of PFA in reducing the risk of cracking due to ASR in concretes containing cristobalite, *Mag Concrete Res.*, *46* (168) September, 167–175.

Hobbs, D.W. (1996) Long term movements due to alkali-silica reaction and their prediction, In: Shayan, A. (ed.) *Alkali-aggregate reaction in concrete, Proceedings of the 10th International Conference on Alkali-Aggregate Reaction in Concrete, Melbourne, Australia*, pp. 316–323.

Hobbs, D.W. (1998) Effect of air-entrainment upon expansion induced by ASR, In: *Proceedings of the 8th BCA Annual Conference on Higher Education and the Concrete Industry (Concrete Communication Conference '98)*, 9-10 July, *Southampton, UK*, pp. 59–71.

Hobbs, D.W. (1999) Suitability of cores for determining expansion rate and future expansion in concretes adversely affected by ASR, In: *Proceedings of the 9th BCA Annual Conference on Higher Education and the Concrete Industry (Concrete Communication Conference '99)*, 8-9 July, *Cardiff, UK*, pp. 129–140.

Hooper, R., Morlidge, J., Lardner, K., & Thomas, M. (2001) *Use of lithium compounds to prevent damaging ASR in concrete*, BRE BR 426, Building Research Establishment (CRC), Watford, UK.

Hooper, R.L., Matthews, J.D., Nixon, P.J., & Thomas, M.D.A. (2004) The introduction of BS EN 450 fly ash and mitigating the risk of ASR in the UK, In: Tang M., & Deng M. (eds.) *Proceedings of the 12th International Conference on Alkali-Aggregate Reaction in Concrete, Beijing, China*, Volume 1, pp. 544-553.

Hooper, R.L., Nixon, P.J., & Thomas, M.D.A. (2004) Considerations when specifying lithium admixtures to mitigate the risk of ASR, In: Tang M., & Deng M. (eds.) *Proceedings of the 12th International Conference on Alkali-Aggregate Reaction in Concrete, Beijing, China*, Volume 1, pp. 554-563.

Idorn, G.M. (1969) The Durability of Concrete. Concrete Society Technical Paper PCS 46, London.

Idorn, G.M. (1967) Durability of Concrete Structures in Denmark, a Study of Field Behaviour and Microscopic Features. DSc Thesis, Technical University of Denmark, Copenhagen.

Idorn G M (1988) Concrete durability in Iceland, Concrete International – Design and Construction *10*(11) 41.

Ingham, J.P. (2012) Briefing: Delayed ettringite formation in concrete structures, *Proceedings of the Institution of Civil Engineers – Forensic Engineering, 165* (FE2) pp. 59–62, The Institution of Civil Engineers, London.

Ingham, J.P., Collery, D., Sfikas, I., Badger, M. (2016) Failure of coastal protection armour units caused by alkali-silica reaction, *Proceedings of the Institution of Civil Engineers – Maritime Engineering*, 169 (ME3), 115–123, The Institution of Civil Engineers, London.

ISE: Institution of Structural Engineers, The. (1988) *Structural Effects of Alkali-Silica Reaction – Interim Technical Guidance on Appraisal of Existing Structures*, Institution of Structural Engineers, London, 31pp.

ISE: Institution of Structural Engineers, The. (1992) *Structural Effects of Alkali-Silica Reaction – Technical Guidance on the Appraisal of Existing Structures*, Institution of Structural Engineers, London, 45pp.

ISE: Institution of Structural Engineers, The. (1996) Appraisal of existing structures, Second edition, Institution of Structural Engineers, London, 106pp. (nb 1980 First edition was in the process of being revised when the ISE (1992) 'Structural effects of alkali-silica reaction' was published).

ISE: Institution of Structural Engineers, The. (2010) *Structural Effects of Alkali-Silica Reaction - Technical Guidance on the Appraisal of Existing Structures*, Addendum, Institution of Structural Engineers, London, 5pp.

Jones, A., & Cather, R. (2013) The risk of alkali-silica reaction in concrete made with non-conforming cement, *Mag Concrete Res., 65* (6), 377–385.

Jones, A.E.K., & Clark, L.A. (1996) A review of the Institution of Structural Engineers report: "Structural effects of alkali-silica reaction (1992)", In: Shayan, A. (ed.) *Alkali-Aggregate reaction in concrete, Proceedings of the 10th International Conference on Alkali-Aggregate Reaction in Concrete, Melbourne, Australia*, pp. 394–401.

Jones, F.E. (1952a) *Reactions Between Aggregates and Cement. Part I, Alkali-Aggregate Interaction: General*. National Building Studies Research Paper No. 14, DSIR/BRS. HMSO, London.

Jones, FE. (1952b) Reactions Between Aggregates and Cement. *Part II, Alkali-Aggregate Interaction: British Portland Cement and British Aggregates*. National Building Studies Research Paper No. 15, DSIR/BRS. HMSO, London.

Jones, F.E., & Tarleton, R. D. (1952) Reactions Between Aggregates and Cement. Part III, Alkali-Aggregate Interaction: The Expansion Bar Test and its Application to the Examination of Some British Aggregates for Possible Expansive Reaction with Portland Cements of Medium Alkali Content. National Building Studies Research Paper No. 17, DSIR/BRS. HMSO, London.

Jones, F.E., & Tarleton, R.D. (1958a) Reactions Between Aggregates and Cement. Part IV, Alkali-Aggregate Interaction: The Expansion Bar Test and its Application to the Examination of Some British Aggregates for Possible Expansive Reaction with Portland Cements of High Alkali Content. National Building Studies Research Paper No. 20, DSIR/BRS. HMSO, London.

Jones, F.E., & Tarleton, R.D. (1958b) Reactions Between Aggregates and Cement. Part V, Alkali-Aggregate Interaction: Effect on Bending Strength of Mortars and the Development of Gelatinous Reaction Products and Cracking. Part VI, Alkali-Aggregate Interaction: Experience with Some Forms of Rapid and Accelerated Tests for Alkali-Aggregate Reactivity: Recommended Test Procedures. National Building Studies Research Paper No. 25, DSIR/BRS. HMSO, London.

Jones, J.B., & Segnit, E.R. (1971) The nature of opal, 1. Nomenclature and constituent phases, *J Geol Soc Aust.*, *18*(1), 57–68.

Jones, T.R., Walters, G.V., & Kostuch, J.A. (1992) Role of metakaolin in suppressing ASR in concrete containing reactive aggregate and exposed to saturated NaCl solution, In: *Proceedings of the 9th International Conference on Alkali-Aggregate Reaction in Concrete*, London, Conference Papers, Volume 1, pp. 485–496, The Concrete Society, Slough (now Camberley), UK.

Keen, R.A. (1970) Impurities in Aggregates for Concrete. Cement and Concrete Association, Advisory Note No. 18, Wexham Springs, Slough, U.K.

Laing, S.V., Scrivener, K.L., & Pratt, P.L. (1992) An investigation of alkali-silica reaction in seven-year old and model concretes using SEM and EDS, In: *Proceedings of the 9th International Conference on Alkali-Aggregate Reaction in Concrete*, London, Conference Papers, Volume 2, pp. 579–586, The Concrete Society, Slough (now Camberley), UK.

Lea, F.M. (1956 and 1970) The Chemistry of Cement and Concrete, 2nd Edn (with Desch, C.H.) and 3rd Edn, Edward Arnold, London.

Lindgard, J., Nixon, P.J., Borchers, I., Schouenborg, B., Wigum, B.J., & Haugen, M., Akesson, U. (2010) The EU Partner project – European standard tests to prevent alkali reactions in aggregates, *Cement Concrete Res.*, *40*, 611–635.

Livesey, P. (1992) Alkali susceptibility of UK aggregates, In: *The 9th International Conference on Alkali-Aggregate Reaction in Concrete*, London, Conference Papers, Volume 2, pp. 614–621, The Concrete Society, Slough (now Camberley), UK.

Livesey, P. (2009a) BRE Digest 330: Alkali-silica reaction in concrete – the case for revision Part I, *Concrete*, *43* (6, July), 42–44.

Livesey, P. (2009b) BRE Digest 330: Alkali-silica reaction in concrete – the case for revision Part II, *Concrete*, *43* (7, August), 35–36.

Lorenzi, G., Jensen, J., Wigum, B.J., Sibbick, R., Haugen, M., Guedon S., & Akesson, U. (2006) *Petrographic atlas of the potentially alkali-reactive rocks in Europe (PARTNER programme)*, Professional Paper 2006/1-N.302, Geological Survey of Belgium, Brussels.

Lumley, J.S. (1992) The ASR expansion of concrete prisms made from cements partially replaced by ground granulated blastfurnace slag, In: *Proceedings of the 9th International Conference on Alkali-Aggregate Reaction in Concrete*, London, Conference Papers, Volume 2, pp. 622–629, The Concrete Society, Slough (now Camberley), UK.

Marinoni, N., & Broekmans, M.A.T.M. (2013) Microstructure of selected aggregate quartz by XRD, and a critical review of the crystallinity index, *Cement Concrete Res.*, *54* (December) 215–225.

May, I.M., Wen, H.X., & Cope, R.J. (1992) The modelling of the effects of AAR expansion on reinforced concrete members, In: *Proceedings of the 9th International Conference on Alkali-Aggregate Reaction in Concrete*, London, Conference Papers, Volume 2, pp. 638–647, The Concrete Society, Slough (now Camberley), UK.

May, I.M., Cope, R.J., & Wen, H.X. (1996) Modelling of the structural behaviour of AAR affected reinforced concrete members, In: Shayan, A. (ed.) *Alkali-aggregate reaction in concrete*, In: *Proceedings of the 10th International Conference on Alkali-Aggregate Reaction in Concrete, Melbourne, Australia*, pp. 434–441.

McLeish, A. (1990) *Structural implications of the alkali silica reaction in concrete*, Contractor Report 177, Transport and Road Research Laboratory (Department of Transport), Crowthorne, UK, 63pp.

Midgley, H.G. (1951) Chalcedony and flint. Geol. Mag. 88 (May-June), 179–184.

Midgley, H.G. (1976) The identification of opal and chalcedony in rocks and methods of estimating the quantities present. Proceedings of a Symposium on The Effect of Alkalis on the Properties of Concrete, London, pp. 193–201.

McCarthy, M.J., Halliday, J.E., Csetenyi, L.J., & Dhir, R.K. (2009) *Alkali-silica reaction guidance for recycled aggregate in concrete*, Aggregate Research Programme/Final

Report, WRAP Code MRF 108-001, WRAP (Waste & Resources Action Programme), Banbury, UK.

McConnell, D., Mielenz, R.C., Holland, WY., & Greene, K.T. (1947) Cement-aggregate reaction in concrete. J. Am. Concr. Inst. *19*(2), 93–128.

Montague, 5. (1989) Most DTp bridges in danger from chloride. New Civil Engineer, 2nd March, 8 (referring to a report The Performance of Concrete in Bridges: a survey of 200 highway bridges. Published by DTp, HMSO, London 1989). See Wallbank now.

Moore, A.E. (1978) Effect of electric current on alkali-silica reaction. In: *Proceedings of the 4th International Conference on the Effects of Alkalis in Cement and Concrete*, pp. 69–72.

Mott, Hay & Anderson (1986) Marsh Mills Viaduct Appraisal and Recommendations. Report on Full Appraisal and Recommendations for Future Management of Marsh Mills Viaduct. Vol. 1 Report, Vol. 1 Appendices. Department of Transport, London.

Neville, A.M. (1963, 1973 and 1981) Properties of Concrete, 1st, 2nd and 3rd Edns. Pitman, London.

Ng, K.E., & Clark, L.A. (1992) Punching tests on slabs with alkali-silica reaction, *The Structural Engineer*, *70* (14) July.

Nixon, P.J. (1986) Testing the alkali-silica reactivity of UK aggregates. Chemistry and Industry 21st July, 488–489.

Nixon, P.J. (1990) *Review of United Kingdom specifications, test methods and guidance for avoidance of damage from alkali aggregate reactions*, Document PD 44/90, Building Research Establishment (BRE, then Department of the Environment), Watford, UK, 8pp.

Nixon, P.J., & Bollinghaus, R. (1983) Testing for alkali-reactive aggregates in the UK. In: *Proceedings of the 6th International Conference on Alkalis in Concrete, Research and Practice*, pp. 329–336.

Nixon, P.J., Collins, R.J., & Rayment, P.L. (1979) The concentration of alkalis by moisture migration in concrete - a factor influencing alkali-aggregate reaction. *Cement Concrete Res.*, 9, 417–423.

Nixon, P.J., & Gillson, I.P. (1987) An investigation into alkali-silica reaction in concrete bases at an electricity substation at Drakelow Power Station, England. In: *Proceedings of the 7th International Conference on Concrete Alkali-Aggregate Reactions*, pp. 173–177.

Nixon, P.J., Hawthorn, F., & Sims, I. (2004) Developing an international specification to combat AAR – proposals of RILEM TC 191-ARP, In: Tang Mingshu & Deng Min (eds.) In: *Proceedings of the 12th International Conference on Alkali-Aggregate Reaction in Concrete, Beijing, China*, Volume 1, pp. 8–16.

Nixon, P.J., Lindgard, J., Borchers, I., Wigum, B.J., & Schouenborg, B. (2008) The EU "PARTNER" Project – European Standard tests to prevent alkali reactions in aggregates – Final results and recommendations, In: Broekmans, M.A.T.M., & Wigum, B.J. (eds.) *Conference Proceedings of the 13th International Conference on Alkali-Aggregate Reaction*, 16–20 June 2008, *Trondheim, Norway*. Proceedings on CD, Paper 132.

Nixon, P.J., Page, C.L., Canham, I., & Bollinghaus, R. (1988) Influence of sodium chloride on alkali-silica reaction. *Advances Cement Res.*, *1*(2), 99–106.

Nixon, P.J., Page, C.L., Hardcastle, J., Canham, I., & Pettifer, K. (1989) Chemical studies of alkali-silica reaction in concrete with different chert contents. In: *Proceedings of the 8th International Conference on Alkali-Aggregate Reaction*, Kyoto, Japan, pp. 129–134.

Nixon, P.J., & Sims, I. (eds.) (2016) *RILEM Recommendations for the prevention of damage by alkali-aggregate reactions in new concrete structures, State-of-the-Art report of the RILEM Technical Committee 219-ACS*, RILEM State-of-the-Art Reports, Volume 17, Springer pp RILEM, Paris, 184pp.

Norris, P., Wood, J.G.M., & Barr, B. (1990) A torsion test to evaluate the deterioration of concrete due to alkali-aggregate reaction, *Mag Concrete Res.*, *42* (153) 239–244.

Oberholster, R.E., & Davies, G. (1986) An accelerated method for testing the potential alkali reactivity of siliceous aggregates, *Cement Concrete Res., 16* (2), 181–189.

Page, C.L., Sergi, G., & Thompson, D.M. (1992) Development of alkali-silica reaction in reinforced concrete subjected to cathodic protection, In: *Proceedings of the 9th International Conference on Alkali-Aggregate Reaction in Concrete*, London, Conference Papers, Volume 2, pp. 774–781, The Concrete Society, Slough (now Camberley), UK.

Palmer, D. (1977) Alkali-Aggregate (Silica) Reaction in Concrete. Cement and Concrete Association, Advisory Note, Ref. 45-033, Wexham Springs, Slough, UK.

Palmer, D. (1978) Alkali-aggregate reaction, Recent occurrences in the British Isles. In: *Proceedings of the 4th International Conference on the Effects of Alkalis in Cement and Concrete*, Purdue University, Lafayette, Indiana, USA, No. CE-MAT-1-78, 5–7 June 1978, pp. 285–298.

Palmer, D. (1981) Alkali-aggregate reaction in Great Britain - the present position. Concrete (J. Concr. Soc.) *15*(3), 24–27.

Poole, A.B. (1975) Alkali-silica reactivity in concrete from Dhekelia, Cyprus. In: *Proceedings of a Symposium on Alkali-Aggregate Reaction Preventive Measures*, Icelandic Building Research Institute, Reykyavik, Iceland, 113–130.

Poole, A.B. (ed.) (1976) Proceedings of a Symposium on the Effect of Alkalis on the Properties of Concrete, London. Cement Concrete Ass., Wexham Springs, Slough, UK.

Poole, A.B., & Sims, I. (2003) Geology, aggregates and classification, Chapter 5 in: Newman, J. & Ban Seng Choo (eds.), *Advanced Concrete Technology – Constituent Materials*, 5/3-5/36, Elsevier, London, & Burlington, USA.

Poole, A.B., & Sims, I. (2015/2016) *Concrete petrography: a handbook of investigative techniques*, 2nd edition, CRC Press (Taylor & Francis), London, 794pp.

Quillin, K. (2001) *Delayed ettringite formation: in situ concrete*. BRE Information Paper, IP11/01, Building Research Establishment, Watford, UK.

Rayment, P.L. (1992) The relationship between flint microstructure and alkali-silica reactivity. In: *Proceedings of the 9th International Conference on Alkali-Aggregate Reaction in Concrete*, London, Conference Papers, Volume 2, pp. 843–850, The Concrete Society, Slough (now Camberley), UK.

Rayment, P.L., Haynes, C. (1993) Investigation into the anomalous pessimum behaviour of a minority of flint-rich aggregates and the implications for specifications, BRE Client Report CR200/93, Building Research Establishment, Watford, UK. © Crown copyright 1993.

Rayment, P.L., & Haynes, C.A. (1996) The alkali-silica reactivity of flint aggregates, In: Shayan, A. (ed.) *Alkali-aggregate reaction in concrete, Proceedings of the 10th International Conference on Alkali-Aggregate Reaction in Concrete, Melbourne, Australia*, pp. 750–757.

Rayment, P.L., Pettifer, K., & Hardcastle, J. (1990) *The alkali-silica reactivity of British concreting sands, gravels and volcanic rocks*, Contractor Report 218, Transport and Road Research Laboratory, Department of Transport, Crowthorne, UK, 66pp.

Richardson, M.G. (chairman) (2003) *Alkali-silica reaction in concrete – General recommendations and guidance in the specification of building and civil engineering works – a report prepared by a joint working party*, The Institution of Engineers of Ireland, Dublin & The Irish Concrete Society, Drogheda, Co. Louth, 24pp.

Richardson, M.G. (2004) Alkali-silica reaction in the Republic of Ireland: recent research and revisions to national guidance, In: Tang Mingshu & Deng Min (eds.) *Proceedings of the 12th International Conference on Alkali-Aggregate Reaction in Concrete, Beijing, China*, Volume 2, pp. 1120–1129.

Rigden, S.R., Majlesi, Y., & Burley, E. (1992a) Bond stress failure in alkali-silica reactive reinforced concrete beams, In: *Proceedings of the 9th International Conference on Alkali-Aggregate Reaction in Concrete*, London, Conference Papers, Volume 2, pp. 859–864, The Concrete Society, Slough (now Camberley), UK.

Rigden, S.R., Salam, J.M., & Burley, E. (1992b) The influence of stress intensity and orientation upon the mechanical properties of ASR affected concrete, In: *Proceedings of the 9th International Conference on Alkali-Aggregate Reaction in Concrete*, London, Conference Papers, Volume 2, pp. 865–876, The Concrete Society, Slough (now Camberley), UK.

Road Research Laboratory (1955) Concrete Roads, Design and Construction. DSIR. HMSO, London.

Roeder, A.R. (1975) Some Properties of Flint Particles and Their Behaviour in Concrete. Unpublished individual project, Advanced Concrete Technology Course, 1974–1975, *Cement Concrete Ass.*, Wexham Springs, Slough, UK.

Sandberg, M. (1977) 132/33 kV substation, Milehouse, Plymouth. Investigation of Cracking in Concrete Bases. Confidential report to South Western Electricity Board. Messrs Sandberg Ref. MWO/60.

Sandberg, M. (1984) Examination of Concrete for Alkali-Reactivity and Chloride Content, Smorrall Lane Overbridge, M6 Motorway. Confidential report to Warwickshire County Council. Messrs Sandberg Ref. L/4045/G/5.

Sandberg, M. (1986) Examination of Aggregate and Concrete Samples. Drakelow Substation, Alkali-Silica Reactivity Study. Confidential Report to the Central Electricity Generating Board. Messrs Sandberg Ref. L/7067/G.

Scott, P.W. (1991) *The chemistry, mineralogy and distribution of flint and chert in the United Kingdom*, Final report to the Mineral Industry Research Organisation (MIRO) of Project RC55: Study of UK Flint and Chert (additional contributions to the report: Dunham, A.C., Cousens, J.M), October 1991, Department of Geology, University of Leicester, UK.

Sfikas, I., Ingham, J.P., Baber, J. (2016) Simulating the thermal behaviour of concrete structures using FEA: a state-of-the-art review, *Proceedings of the Institution of Civil Engineers – Construction Materials*, DOI: 10.1680/jcoma. 15.00052.

Shacklock, B.W. (1969) Durability of concrete made with sea-dredged aggregate. Proceedings of a Symposium on Sea Dredged Aggregates for Concrete, 9th December 1968. Sand and Gravel Association of Great Britain, Ref. 97.101, London, pp. 31–33.

Sherwood, D.E., Marriott, M., & Smith, J. (1975) Non-destructive testing of concrete dams by sonic speed measurement. Symposium, British National Committee on Large Dams (BNCOLD), University of Newcastle upon Tyne, Proceedings Paper 3.2, pp. 1–7.

Sibbick, R.G., & Page, C.L. (1992a) Susceptibility of various UK aggregates to alkali-aggregate reaction. In: *Proceedings of the 9th International Conference on Alkali-Aggregate Reaction in Concrete*, London, Conference Papers, Volume 2, 980–987, The Concrete Society, Slough (now Camberley), UK.

Sibbick, R.G., & Page, C.L. (1992b) Threshold alkali contents for expansion of concretes containing British aggregates, *Cement Concrete Res.*, 22(5), 990–994 (& Discussion in 23 (2), 1993, 495–499).

Sibbick, R.G., & Page, C.L. (1996) Effects of sodium chloride on the alkali-silica reaction in hardened concretes, In: Shayan, A. (ed.) *Alkali-aggregate reaction in concrete, Proceedings of the 10th International Conference on Alkali-Aggregate Reaction in Concrete, Melbourne, Australia*, pp. 822–827.

Sibbick, R.G. & Page, C.L. (1998) Mechanisms affecting the development of alkali-silica reaction in hardened concretes exposed to saline environments, *Mag Concrete Res.*, 50 (2), June, 147–159.

Sibbick, R.G. & West, G. (1992) *Examination of concrete from the Padiham Bypass, Lancashire*, Digest of Research Report 304, Transport Research Laboratory, Crowthorne, UK

Sims, I. (1977) *Investigations into Some Chemical Instabilities of Concrete*. PhD Thesis, University of London (Queen Mary College).

Sims, I. (1981) Application of standard testing procedures for alkali-reactivity, Parts I and 2. *Concrete (J. Concr. Soc.)* 15(10 and 11), 27–29 and 29–32.

Sims, I. (1987) The importance of petrography in the ASR assessment of aggregates and existing concretes. In: Grattan-Bellew, Patrick. E. *Proceedings of the 7th International Conference on Concrete Alkali-Aggregate Reactions.*, Noyes Publications, Park Ridge, New Jersey, USA. pp. 358–367.

Sims, I. (1992) The assessment of concrete for ASR, *Concrete (The Concrete Society Journal)*, 26 (2) March/April, 42–46.

Sims, I. (1996) Phantom, opportunistic, historical and real AAR: Getting diagnosis right, In: Shayan, A. (ed.) *Alkali-aggregate reaction in concrete, Proceedings of the 10th International Conference on Alkali-Aggregate Reaction in Concrete, Melbourne, Australia,* pp. 175–182.

Sims, I. (2004) Whatdunnit? Forensic petrography and AAR diagnosis, In: Tang Mingshu & Deng Min (eds.) *Proceedings of the 12th International Conference on Alkali-Aggregate Reaction in Concrete, Beijing, China,* Volume 2, pp. 995–1004.

Sims, I., & Bladon, S. (1984) An exploratory assessment of the alkali-reactivity potential of sintered pfa in concrete. In: *Proceedings of the 2nd International conference on Ash Technology and Marketing.* CEGB (Central Electricity Generating Board), London. 16–21 September 1984.

Sims, I, & Brown, B.V. (1998) Concrete aggregates, Chapter 16 In: Hewlett, P.C. (ed.), *Lea's Chemistry of Cement and Concrete*, 4th edition, 903–1011. A new edition of this chapter and book is in preparation, with probable publication in 2017.

Sims, I., & Higgins, D.D. (1992) The use of GGBS to prevent ASR expansion caused by UK flint aggregates, In: *Proceedings of the 9th International Conference on Alkali-Aggregate Reaction in Concrete*, London, Conference Papers, Volume 2, pp. 988–1000, The Concrete Society, Slough (now Camberley), UK.

Sims, I., & Miglio, B.F. (1989) Compositional uniformity of some UK sand and gravel aggregates for concrete. The Institution of Geologists/The Geological Society/The Institution of Mining and Metallurgy/Minerals Industry Research Organisation, Joint Conference, Extractive Industry Geology, Birmingham, April (submitted for publication in Quart. J. Engng. Geol.).

Sims, I., Nixon, P.J., Blanchard, I.G., & Bennett-Hughes, P. (2008) Assessment of concrete in service in the UK – remembering the value of practical petrography, In: Broekmans, M.A.T. M., Wigum, B.J. (eds.) *Conference Proceedings of the 13th International Conference on Alkali-Aggregate Reaction, Trondheim, Norway, 16–20 June 2008*, Paper 167 on DVD.

Sims, I., Nixon, P.J., & Godart, B. (2012) Eliminating alkali-aggregate reaction from long-service structures, In: Drimalas, T., Ideker. J.H., Fournier, B. (eds.) *Proceedings of the 14th International Conference on Alkali-Aggregate Reactivity in Concrete, Austin, Texas, USA.*

Sims, I. & Poole, A.B. (1980) Potentially alkali-reactive aggregates from the Middle East. Concrete (J. Concr. Soc.) 14(5), 27–30.

Sims, I., Smart, S., & Hunt, B. (2000) Practical petrography – the modern assessment of aggregates for AAR potential, In: Berube, M.A., Fournier, B., Durand, B. (eds.) *Alkali-Aggregate Reaction in Concrete, Proceedings of the 11th International Conference, Quebec City, Canada,* June 2000, pp. 493–502.

Sims, I. & Sotiropoulos, P. (1989) A pop-out forming type of ASR in the United Kingdom, Eighth International Conference on Alkali-Aggregate Reaction, Kyoto, Japan, poster-session paper (submitted for publication to Cem. Concr. Res).

Smith, A.S. & Dunham, A.C. (1992) *Undulatory extinction of quartz in granites and sandstones,* Transport and Road Research Laboratory (Department of Transport), Contractor Report 291, Crowthorne, UK.

Smith, M.R. & Collis, L. (2001) *Aggregates: Sand, gravel and crushed rock aggregates for construction purposes* (3rd edition, revised by Fookes, P.G., Lay, J., Sims, I., Smith, M.R., & West, G.), The Geological Society, London (earlier editions 1985, edited by Collis, L., Fox, R.A., & 1993).

Somerville, G. (1985) *Engineering Aspects of Alkali-Silica Reaction*, Cement and Concrete Association, Interim Technical Note 8, Wexham Springs, Slough, UK.

Stanton, T.E. (1940a) Influence of cement and aggregate on concrete expansion. *Engng. News Rec.*, 124 (5), 59–61.

Stanton, T.E. (1940b) Expansion of concrete through reaction between cement and aggregate. *Am. Soc. Civ. Engrs.*, Pap., December 1781–1811.

Stanton, T.E., Porter, O.J., Meder, L.C., & Nicol, A. (1942) California experience with expansion of concrete through reaction between cement and aggregate. *J. Am. Concr. Inst.*, 13 (3), 209–236.

Strogen, P. (1993) *Cherts in Irish aggregates and their use in concrete*, The Institution of Engineers of Ireland, Paper presented to a joint meeting of The Civil Division, Institution of Engineers of Ireland, and the Irish Concrete Society, 17th February 1993.

Swamy, R.N. (1988) Alkali-silica reaction – sources of damage, Highways and Transportation The Institution of Highways and Transportation., 35 (12), 229.

Swamy, R.N. (1997) Assessment and rehabilitation of AAR-affected structures, *Cement Concrete Comp.*, 19, 427–440.

Swamy, R.N., & Al-Asali, M.M. (1986) Influence of alkali-silica reaction on the engineering properties of concrete. *Alkalies in Concrete*, ASTM STP 930, Am Soc TestMater., Philadelphia, PA. pp. 69–86.

Swamy, R.N. & Al-Asali, M.M. (1988) Engineering properties of concrete affected by alkali-silica reaction. *American Concrete Institute (ACI) Mater J.*, 85(5), 367–374.

Swamy, R.N. & Al-Asali, M.M. (1990) Control of alkali-silica reaction in reinforced concrete beams, *ACI Mater J.*, Title 87-M6, 87 (1) January-February, 38–46.

Swamy, R.N. & Tanikawa, S. (1992) Acrylic rubber coating to control alkali silica reactivity, In: *Proceedings of the 9th International Conference on Alkali-Aggregate Reaction in Concrete*, London, Conference Papers, Volume 2. pp. 1026–1034, The Concrete Society, Slough (now Camberley), UK.

Teychenné, D.C. (1987) Personal communication.

Thomas, M.D.A. (1996) Field studies of fly ash concrete structures containing reactive aggregates, *Mag Concrete Res.*, 48 (177), December, 265–279.

Thomas, M.D.A. (2011) The effect of supplementary cementing materials on alkali-silica reaction: A review. *Cement Concrete Res.*, 41 (12), 1224–1231.

Thomas, M.D.A. & Blackwell, B.Q. (1996) Summary of BRE research on the effect of fly ash on alkali-silica reaction in concrete, In: Shayan, A. (ed.) *Alkali-aggregate reaction in concrete, Proceedings of the 10th International Conference on Alkali-Aggregate Reaction in Concrete, Melbourne, Australia.* pp. 554–561.

Thomas, M.D.A., Hooton, R.D., Rogers, C.A., & Fournier, B. (2012) 50 years old and still going strong – fly ash puts paid to ASR, *Concrete International*, January, 35–40.

Thomas, M.D.A., Nixon, P.J., & Pettifer, K. (1992) Suppression of damage from alkali-silica reaction by fly ash in concrete dams, In: *Proceedings of the 9th International Conference on Alkali-Aggregate Reaction in Concrete*, London, Conference Papers, Volume 2, pp. 1059–1066, The Concrete Society, Slough (now Camberley), UK.

Thomas, M.D.A. & Stokes, D. (2004) Lithium impregnation of ASR-affected concrete: preliminary studies, In: Tang Mingshu & Deng Min (eds.) *Proceedings of the 12th International Conference on Alkali-Aggregate Reaction in Concrete, Beijing, China*, Volume 1. pp. 659–667.

Tuthill, L.H. (1982) Alkali-silica reaction – 40 years later. Concrete International, 4 (4), 32–36 [and Discussion (1983) 5 (2), 65-67].

Van de Fliert, C. (1969) Experience with sea-dredged aggregates in the Netherlands. Proceedings of a Symposium on Sea Dredged Aggregates for Concrete. 9th December 1968. Sand and Gravel Association of Great Britain, Ref. 97.101, London, pp. 57–60.

Wallbank, E.J. (1989) *The performance of concrete in bridges – a survey of 200 highway bridges* (Report prepared for Department of Transport by G Maunsell & Ptns), Department of Transport, HMSO, London, 96p.

Walters, G.V. (1989) ECC Quarries Limited, Exeter, Devon. Personal communication.

Walters, G.V. & Jones, T.R. (1991) Effect of metakaolin on alkali-silica reactions (ASR) in concrete manufactured with reactive aggregates, In: Malhotra, V.M. (ed.) *Proceedings of the 2nd International Conference on the Durability of Concrete, Montreal, Canada*, Volume II, pp. 942–947, American Concrete Institute.

West, G. (1996) *Alkali-aggregate reaction in concrete roads and bridges*, Thomas Telford, London, 163p.

West, G. & Sibbick, R.G. (1988a) Alkali-silica reaction in roads. *Highways*, 56 (1936), 19–24.

West, G. & Sibbick, R.G. (1988b) Alkali-silica reaction in roads, Part 2. *Highways*, 57 (1949), 9–14.

Wigum, B.J., Sims, I., Lindgard, J., & Nixon, P.J. (2016) RILEM activities on alkali-silica reactions: from 1988 to 2019 & beyond, In: Sims, I. (ed.) Themed issue on alkali-aggregate reactions: part II, *Proceedings of the Institution of Civil Engineers: Construction Materials*, 169 (CM4) 233-236. ICE Publishing, London.

Winter, N.B. (2012) *Scanning electron microscopy of cement and concrete*, Microanalysis Consultants Ltd, Woodbridge, UK. 192p.

Wood, J.G.M. (1985) *Engineering assessment of structures with alkali-silica reaction.* One day conference, Alkali-Silica Reaction, New Structures – Specifying the Answer, Existing Structures – Diagnosis and Assessment, London, November. The Concrete Society, London.

Wood, J.G.M. (2000) Comparison of field performance with laboratory testing: how safe and economic are current AAR specifications? In: Berube, M.A., Fournier, & B., Durand, B. (eds.) *Alkali-Aggregate Reaction in Concrete, Proceedings of the 11th International Conference, Quebec City, Canada*, June 2000, pp. 543–552.

Wood, J.G.M. (2004) When does AAR stop: in the laboratory and in the field? In: Tang, M., & Deng, M. (eds.) *Proceedings of the 12th International Conference on Alkali-Aggregate Reaction in Concrete, Beijing, China*, Volume 2, pp. 1016–1024.

Wood, J.G.M. (2008) Improving guidance for engineering assessment and management of structures with AAR, In: Broekmans, M.A.T.M. & Wigum, B.J. (eds.) *Conference Proceedings of the 13th International Conference on Alkali-Aggregate Reaction, Trondheim, Norway*, 16–20 June 2008, Paper 180 on DVD.

Wood, J.G.M., & Angus, E.C. (1995) Montrose Bridge, In: *Proceedings of Structural Faults and Repair - 95 Conference*, Volume 1, London.

Wood, J.G.M., & Doran, D.K. (1992) Revision of the Institution of Structural Engineers Report – Structural Effects of Alkali Silica Reaction, In: *Proceedings of the 9th International Conference on Alkali-Aggregate Reaction in Concrete*, London, Conference Papers, Volume 2, pp. 1107–1112, The Concrete Society, Slough (now Camberley), UK.

Wood, J.G.M., Johnson, R.A., & Abbott, R.J. (1987) Monitoring and proof loading to determine the rate of deterioration and the stiffness and strength of structures with AAR. Institution of Structural Engineers/Building Research Establishment, Seminar, Structural Assessment - Based on Full and Large-Scale Testing, 6–8 April.

Wood, J.G.M., Johnson, R.A., & Norris, P. (1987) Management strategies for buildings and bridges subject to degradation from alkali-aggregate reaction. In: *Proceedings of the 7th International Conference on Concrete Alkali-Aggregate Reactions.* pp. 178–182.

Wood, J.G.M., Young, J.S., & Ward, D.E. (1987) The structural effects of alkali-aggregate reaction on reinforced concrete. In: *Proceedings of the 7th International Conference on Concrete Alkali-Aggregate Reactions.* pp. 157–162.

Zhang, X., Blackwell, B.Q., & Groves, G.W. (1990) The microstructure of reactive aggregates, *Br. Ceram. Trans. J.*, 89, 89–92.

Chapter 7

Nordic Europe

Jan Lindgård, B Grelk[1], Børge J Wigum[2], J Trägårdh[3], Karin Appelqvist[3], E Holt[4], M Ferreira[4] & M Leivo[4]

[1] assisted with Denmark
[2] assisted with Iceland
[3] assisted with Sweden
[4] authored Finland

7.1 INTRODUCTORY OVERVIEW

Nordic Europe includes Denmark, Iceland, Norway, Sweden and Finland. The history of alkali-aggregate reactions and the geology are quite different in these countries (Table 7.1). The same is valid for the cement types used (see later). However, the environmental conditions are relatively similar (see later). Furthermore, to the author's knowledge, only ASR is documented in structures (*i.e.*, no cases of ACR are reported). Neither has lithium been used in any of the countries to mitigate ASR.

In general, the mean temperature in Nordic Europe is relatively low compared with countries in more southern Europe. However, the temperature and the precipitation may vary widely, both internally in each country and between the countries [see for example the climate classification by Köppen (https://en.wikipedia.org/wiki/Köppen_climate_classification)].

Nevertheless, in all the countries a huge number of concrete structures might be exposed to many freeze-thaw cycles and to de-icing salt and/or sea water. As a consequence, several concrete structures are damaged by more than one deterioration mechanism. For example, ASR can induce cracks that are easily penetrated by chlorides, leading to reinforcement corrosion? The de-icing salt might also contribute to ingress of alkalis that can enhance the development of ASR, for example as experienced in several Danish bridges (see 7.2).

Due to surface cracking and the absorption of water by the ASR gel, ASR will in general lead to an increased moisture state in the affected concrete structures. Thus, if the "critical water content" of the concrete is reached, freeze-thaw damage might be introduced locally, *i.e.*, a synergy between ASR and freeze-thaw. To avoid freeze-thaw damage, air entraining agents are frequently used in all the countries.

In the subsequent 'national sections' 7.2–7.6, information about national research activities on ASR and the following ASR topics are given: type of reactive aggregates; binder types used; any special features of ASR; preventative and mitigation measures (regulations, test methods and mitigation alternatives); diagnosis and appraisal of structures; and protection, repair and protection of structures. Furthermore, in the first edition of this book (Swamy, 1992), more detailed information can be found regarding the ASR history (including some results from selected research projects) in Denmark and Iceland.

In the two European standards EN-12620: (2004) and EN-206-1: (2013), it is stated that alkali-reactivity shall be assessed in accordance with the provisions valid in the

Table 7.1 Brief overview of the history of ASR and the geology in the Nordic countries.

Country	First reports on ASR damage	When was focus put on ASR?	First ASR regulations introduced	Last update of the ASR regulations[1]	Main types of alkali-reactive aggregates
Denmark	1951[2]	1954[3] I: 1960's II: 1970's	1961[4] 1987[5]	2004	Most aggregate sources contain reactive components in form of highly reactive opaline flint, calcareous opaline flint and more slowly reactive porous chalcedonic flint. Pessimum behaviour of opaline and opaline flint limestone. A content > approx. 2 % by volume in the sand fraction of highly reactive microporous flint is the main problem
Iceland	Early 1960s[6] (1976)	Early 1960s[6]	1979[7]	2012	Several highly reactive volcanic, porous aggregates, mainly basalts. Rhyolite is also reactive. The finest fraction is more severe
Norway	1978[8]	About 1990[9]	1993[10] 1996[11]	2004	Slowly reactive aggregates are spread over most of the country. The coarse fraction has in particular given rise to ASR damage
Sweden	1975[12]	Early 1990s[13]	2008[14]	2015[15]	Aggregates containing opaline flint are locally present in the south-western part (Scania area with similar geology as in Denmark). Different slowly reactive aggregates are located in many regions
Finland	Late 1990s[16]	About 2010[16]	No[16] regulations	No regulations	Slowly reactive aggregates, similar as in Sweden and Norway, are assumed to be present in many regions in Finland (geological mapping of potential reactive aggregates is lacking)

1 More details are given in the various chapters for each of the countries
2 (Nerenst, 1952)
3 The first Danish R&D project on ASR was initiated in 1954 by ATV (Academy of Technical Sciences) based on in-situ investigations of Danish concrete structures. Extensive research work was later carried out in two phases, denoted I and II.
4 (Plum, 1961): Preliminary guidelines
5 ("BBB", 1987): Requirements to aggregates and maximum allowed alkali content
6 (Swamy, 1992, 1st edition): Awareness of potential reactive aggregates in the early 1960's. However, the first proof of serious ASR damages in Iceland was not obtained until 1976 when drilled cores were taken from the exterior walls of a concrete house
7 (Kristjánsson, 1979)
8 (Kjennerud, 1978)
9 (Jensen, 1993): Dr thesis. Extensive research work was performed on a national basis from the early 1990's
10 (Lindgård et al., 1993): Norwegian ASR test methods
11 (Norwegian Concrete Association, 1996): ASR regulations (voluntary until 2001)
12 (Nilsson & Peterson, 1983)
13 (Lagerblad & Trägårdh, 1992)
14 (SS137003: 2008): Concrete – Application of EN 206-1 in Sweden
15 (SS-EN 12620: 2008; SS 137003: 2015)
16 (Pyy et al., 2012a)

place of use. In other words, only national requirements for ASR exist in Europe. However, where aggregates are imported across national boundaries, the purchaser should take account of experiences in the country of origin.

As part of the EU 'PARTNER' project (Lindgård *et al.*, 2010), an overview of various European national ASR standards and requirements was prepared (Wigum *et al.*, 2006). All the Nordic Europe countries, except Finland, were included in this review.

7.2 DENMARK

7.2.1 Brief history of national research

ASR was discovered for the first time in Denmark in 1951 by Poul Nerenst after a visit to the USA to study concrete technology. In 1951, the Danish National Institute of Building Research (DNIBR) started a preliminary investigation of approximately 200 concrete structures in Denmark, to see whether observed phenomena of deterioration could be traced to reactions between alkalis and reactive silica of the aggregate. The result of field observations clearly supported this hypothesis. This discovery led to a large co-ordinated concrete research effort on the concrete field, for the first time in Denmark.

The research project was initiated in 1954 by the Academy of Technical Sciences (ATV). A joint committee was appointed by ATV and DNIBR, which would look into the technical and practical problems of ASR. Three phases of research work were initiated: 1) Field investigations, 2) Investigation of aggregate and 3) Laboratory investigations. The main part of the work was done at DNIBR. The research was continued at the Concrete Research Laboratory in Karlstrup.

The results were published in 23 reports (references are included in Swamy, 1992). The main outcomes were identification and classification of reactive aggregate types, verification of the 'pessimum proportion' (see Chapter 1) and recommendation of preventative measures: use of fine aggregates containing less than 2% reactive component (porous calcareous opaline flint in the sand) or a low-alkali Portland cement or a pozzolana. At this stage, the basic assumption was that the alkali content of concrete was determined by that of the cement (*i.e.*, assumed no alkali contribution from other sources) and that the alkalis were evenly distributed in the concrete. The culmination of the research was Idorn's doctoral dissertation in 1967 (Idorn, 1967).

The second phase of extensive R&D on ASR, carried out in the 1970s until the middle of the 1980s, started with an investigation of concrete roads, which had deteriorated within 4 years of their construction. This investigation showed that extensive ASR had taken place, although the coarse aggregate was non-reactive granite and the Portland cement contained about 0.6% Na_2O_{eq}. It was the result of an interaction between the de-icing salt and reactive silica (in porous calcareous opaline flint) present in the sand. The explicit assumption of this phase of research was that the structures often received alkalis from outside sources (de-icing salt, sea water and salt water in swimming pools) and that there is nearly always a concentration gradient of alkalis in any structure. This finding was the background for the development of the TI-B 51 mortar bar method (see 7.2.4).

In the period after about 1985, little research on ASR was carried out in Denmark. However, due to the extensive ASR damage discovered on many old concrete bridges

(7.2.4), today some new research activities have been initiated focusing on the load bearing capacity of ASR-damaged structures.

7.2.2 Aggregates, binders and admixtures

In the research projects in the 1960s, opaline flint and calcareous opaline flint were identified as the most prevalent reactive aggregate types in Denmark. The reactive compounds were found in both the fine (sand) and coarse (gravel) fractions of many aggregate sources (land based and sea dredged materials). Later research has resulted in the following overview of Danish alkali-silica reactive aggregates:

- Porous opaline flint, especially in the western part of Denmark
- Porous calcareous opaline flint in all parts of Denmark, except the western parts
- Opaline sandstone in the southern parts of Denmark

Dense flint is considered to be non-reactive, porous flint to be reactive. In Denmark, about 90–95 % of all cases with ASR-damaged concrete structures are caused by porous opaline or calcareous opaline flint in the fine (sand) fraction. ASR due to opaline or calcareous opaline flint can occur very quickly (<5 years) under severe conditions [ie exposure to water, external alkalis (for example from de-icing salt) and a critical amount of reactive particles in the aggregates]. A maximum content of 2 volume-% potential reactive particles in the sand fraction is considered to be a safe upper limit to avoid ASR damage.

The aggregate deposits situated in Zealand, Fyn and the eastern part of Jutland do often contain so much reactive porous flint particles (>2 vol % in the sand) that the aggregates extracted might lead to ASR damage (Nielsen *et al.*, 2004). In western Jutland and some other locations, the content of reactive particles (opaline and calcareous opaline flint) is too low to cause ASR damage. However, many sea dredged deposits often contain less than 2 % reactive particles (opaline or calcareous opaline flint) or the less reactive porous chalcedonic flint.

Before about 1990, the alkali content of some of the most commonly used Danish types of cement (rapid, super-rapid and 'ordinary' Portland cement) varied between 0.5% and 0.9% by weight of cement, expressed as Na_2Oeq. However, also two special cements with low alkali content were available: a low alkali sulfate-resisting Portland cement (less than 0.4% alkalis) and a white Portland cement (less than 0.2% alkalis). Blended cement containing about 20–25% fly ash was also available in the period from 1978 to 1995.

Today, the most used cement types today in Denmark are shown in Table 7.2.

The low alkali sulfate resisting cement has, within the last 25–30 years, been the most important cement type in Denmark for infrastructure constructions like bridges and tunnels.

The use of fly ash began in about 1978 because of the oil crisis (the largest power plants then replaced oil with coal). At that time, Aalborg Portland began producing blended cement with about 20–25% fly ash. In 1995, Aalborg Portland stopped the production of this blended cement type. However fly ash is still being used in concrete today, as a separate pozzolanic additive / supplementary cementitious material (SCM).

Silica fume was introduced in Denmark in the late 1970s. It was typically used in relative low quantities around 5%. In the 1980s, it was relatively common to

Table 7.2 Overview of the most used cement types in Denmark today.

Cement type	EN 197-1	Alkali content (%)
Basic cement (ABC cement)	CEM II/A-LL	0.5–0.6
Rapid cement[1]	CEM I	0.5–0.6
Low alkali sulfate resisting cement[2]	CEM I	0.2–0.4
Aalborg White[3]	CEM I	0.1–0.2
Element cement[4]	CEM I	0.5–0.7
Mester cement[4]	CEM I	0.5–0.6

[1] Very frequently used for the last 40–50 years. Also today it is a very common used cement type for all purposes, except bridges and tunnels.
[2] This cement types is the most common used cement for bridges and tunnels in Denmark for the last 25–30 years. Also very frequently used today.
[3] This cement is normally primary being used for white concrete purposes due to aesthetical reasons.
[4] Element and Mester cement have for the last 5–10 years primarily being used by smaller craftsmen for house building projects.

use microsilica in Danish infrastructure concrete projects. However, also triple blends (microsilica, fly ash and Portland cement) have characterized several smaller and larger projects in Denmark in the 1980s and 1990s, for example the Great Belt link. Microsilica is still a common pozzolanic additive or SCM for concrete in Denmark.

7.2.3 Special features of ASR

Most ASR problems in Denmark are caused by porous opaline and calcareous opaline flint that also shows a clear pessimum effect (7.2.2). The reaction might occur quickly under severe conditions (< 5 years). This is in contrast to the other Nordic countries, where only slowly reactive aggregates are present [except in a local area in the south western part of Sweden where porous flint can also be found (see 7.5.2)].

As for the other Nordic countries, ASR has been documented in Denmark in almost all types of various moisture exposed outdoor concrete structures, as well as in swimming pools. However, what is special for Denmark is that severe ASR damage on many bridges built in the 1960s and the 1970s are connected to the intrusion of de-icing salts like NaCl that has increased the alkali content significantly. Probably the most contributing factor is that the successive reaction of the highly reactive porous flint (during the intrusion of de-icing salts) has 'opened up' the concrete and thus increased the rate of alkali ingress (additionally, the water/cement ratio was quite high in several of the bridges). However, this occurs only on bridges and/or in areas where the bridge deck membranes are defective and leaking. As a consequence, many Danish bridge decks are heavily delaminated/cracked due to ASR (see an example in Figure 7.1).

The use of de-icing salt in Denmark is around 300–500.000 tons per year (2010–2011), and the Danish Road Directorate was distributing > 1 kg salt per m^2 road in Denmark in the same period (Ingeniøren, 2012).

Figure 7.1 Typical crack pattern due to ASR on a bridge near Ølstykke, Denmark.

7.2.4 Preventative and mitigating measures – Danish approach

Even though extensive research on ASR was performed in the 1950s and 1960s, the knowledge gained in these R&D projects was not sufficiently implemented in the Danish concrete business sector. Preliminary ASR guidelines were indeed presented in 1961 (Table 7.1), but no formal requirements for ASR were introduced in Denmark until 1987, when the Basic Concrete Specification for Building Structures was published by the National Building Agency [BBB (Basis Beton Beskrivelsen), 1987 in Danish]. Hence, before the second half of the 1980s, most concrete structures were built without taking ASR into consideration. Furthermore, parts of the concrete business sector probably misunderstood the conclusions from the previous research reports. It seems that they wrongly believed that the ASR problems were mainly connected to the coarse aggregate fraction. By replacing coarse reactive aggregate with non-reactive ones [like crushed granite from Sweden, Norway and Bornholm (Denmark)], and keep on using the reactive sand with porous flint, the ASR problems even increased due to the pessimum effect (Nielsen *et al.*, 2004). Consequently, ASR is still a very serious problem in Denmark, owing to the large number of structures built in the period before 1987, including a lot of bridges built in the 1960s and 1970s.

BBB (1987) gave requirements for aggregates and maximum permitted alkali content in the concrete. It applied to ordinary building structures supported by authorities only. However, these requirements were used for almost all concrete structures built after 1987. The test methods referred to in BBB are TI-B51 and TI-B52 (see later), which still are part of the Danish ASR regulations.

The following sections sum up the current Danish requirements for classification of the alkali reactivity of aggregates, for calculation of the 'effective alkali content' of the concrete and for composition of the concrete depending on the environmental class. Additionally, a brief overview of the present Danish ASR test methods for assessment of the aggregate reactivity is given.

7.2.4.1 Requirements for classification of the alkali reactivity of aggregates

The aggregates are classified in 4 groups:

- Class P for use in a passive environment
- Class M for use in a moderate environment
- Class A for use in an aggressive environment
- Class E for use in an extra aggressive environment

The classification of sands with respect to alkali-silica reactive components is described in DS 2426 (2004), see 7.2.2. The methods used are the chemical shrinkage method (TK 84: 1989), the petrographic thin section point counting method (TI-B 52: 1985), the mortar-bar expansion test (TI-B 51: 1985) and the accelerated mortar-bar test (ASTM C1260: 2014) – see later. The DS 2426 standard distinguishes between sand which contains microporous flint and sand which does not. For the first group, the demands for either chemical shrinkage, volume of reactive flint or the mortar-bar expansion test have to be fulfilled. For the second group, the additional demand for the accelerated mortar-bar expansion test has to be fulfilled, as well. If the sand material is crushed, testing results from another batch is allowed if the geological origin is the same.

The classification of coarse aggregates with respect to alkali-silica reactive components is also described in DS 2426 (2004). Several requirements are given, as shown in Table 7.3. The amount of reactive material in the aggregates to be used in the environmental classes M and A is limited by the maximum allowable percentages of particles with a density below 2400 kg/m^3 (lighter particles most likely containing porous flint). For environmental class E, a more restrict requirement is set. Furthermore, there is a maximum limit of water absorption of any flint with density greater than 2400 kg/m^3. This value is determined on those 10% of the flint particles with highest water absorption. The standard distinguishes between coarse aggregates within the borders of Denmark (or within the Danish sea territory) and coarse aggregates from abroad. For the first (national) group, only the requirement to the critical water absorption has to be fulfilled. For the second (abroad) group, an additional requirement for accelerated mortar-bar expansion also has to be fulfilled. If the coarse material is crushed, testing results from another batch is allowed if the geological origin is the same.

7.2.4.2 Requirements for calculation of the 'effective alkali content' of the concrete

The alkali content of the concrete should be calculated as the sum of the acid soluble equivalent Na_2O content in the cement (obtained from the cement supplier), the water soluble equivalent Na_2O content in the sand and the coarse aggregate (obtained from the aggregate supplier) and the equivalent Na_2O content in the admixtures (obtained from the admixture supplier). The alkali content from fly ash (PFA) and microsilica is not included in the calculation of the 'effective alkali content'.

Table 7.3 Classification of Danish aggregates (from DS 2426: 2004, Table 2426–4).

Classification of aggregates

Reference to DS/EN 12620	Property	Method	Class M	Class A	Class E
5.7.3	Alkali reactivity for fine aggregate[5] One of the four alternative methods shall be documented	TK-84 Chemical shrinkage[2, 6]	Max. 0.3	Max. 0.3	Max. 0.2
		TI-B 52 Volume reactive flint (vol. %)[2, 6]	Max. 2	Max. 2	Max. 1
		TI-B 51 Mortar bar expansion (%)[2, 6]	Max. 0.1 after 8 weeks	Max. 0.1 after 8 weeks	Max. 0.1 after 20 weeks
		ASTM C 1260 Accelerated mortar bar expansion (%)[3]	Max. 0.2 after 14 days	Max. 0.1 after 14 days	Max. 0.1 after 14 days
	Coarse aggregate One of the three alternative methods shall be documented	TI-B 75 Critical absorption[1, 2] (%)	Max. 2.5	Max. 1.1	Max. 1.1
		Alkali-Richtlinje 1997 Reaktionsfähiger flint[4]	Max. 10	Max. 3	Max. 3
		ASTM C 1260 Accelerated mortar bar expansion (%)[3]	Max. 0.2 after 14 days	Max. 0.1 after 14 days	Max. 0.1 after 14 days
	The content of light weight particles shall be documented for coarse aggregates containing micro porous flint[7]	DS 405.4 Particles with a density below 2400 kg/m^3 (%)	Max. 5.0	Max. 1.0	–
		DS 405.4 Particles with a density below 2500 kg/m^3 (%)	–	–	Max. 1.0

[1] Critical absorption for the 10 % of the flint particles with the highest absorption and a density above 2400 kg/m^3.
[2] This method shall only be used on aggregates with microporous flint.
[3] This method can't be used on aggregates with microporous flint
[4] This method can be used on (sea dredged) aggregates from the North Sea.
[5] It is the mean value of the latest 3 test results that shall fulfil the requirements
[6] Particles greater than 4 mm shall not be included in the test. The content of particles with a grain size > 4 mm must not be greater than 5 %.
[7] Microporous flint includes porous chalcedonic flint as well as porous opaline and calcareous opaline flint.

7.2.4.3 Requirements for composition of the concrete depending on the environmental class

The requirements for composition of the concrete depending on the environmental class (in Table DS 2426-F.1 of DS 2426: 2004) are based on DS EN 206-1 (2004).

7.2.4.4 Overview of the present Danish ASR test methods for assessment of the alkali-silica reactivity of fine aggregates (sand)

The following test methods are used for fine aggregates:

1 *Alkali Silica Reactivity of Sand – TI-B 51: (1985):*
   Principle: Prisms (40 x 40 x 160 mm) made from mortar, consisting of one portion by weight of cement and three portions by weight of the sand to be tested. The water/cement ratio is 0.50. The prisms are water cured for 28 days and then put into a saturated sodium chloride solution at a temperature of 50 °C. The linear expansion of the prisms is measured for 8 weeks (exposure class M and A), or for 20 weeks (exposure class E), after they have been put into the saturated sodium chloride solution.

   Use: The method should be used for the relative comparison of the alkali-silica reaction of different sands tested, making it possible to choose the sand that will result in the smallest expansion.

   Test result: Mortar prism expansion.

2 *Petrographical Investigation of Sand – TI-B 52: (1985):*
   Principle: Fluorescent impregnated thin sections made from epoxy encased sand, which is to be tested, are analysed using a polarizing microscope to determine the mineralogy of the individual grains of sand.

   Use: The method is used to determine the mineralogical composition of sand for the specific purpose of determining the quantity of alkali-reactive material, *i.e.*, porous opaline flints and/or calcareous opaline flints.

   Test Result: Distribution of rock minerals and content of reactive materials.

3 *Accelerated mortar-bar expansion test – ASTM C 1260: (2014) [DS405.16]*

4 *Chemical shrinkage – TK 84: (1989).*
   Principle: The method is based upon that reactive flint is dissolved in NaOH. A sample of the sand is subjected to hot NaOH and the shrinkage due to the dissolution of flint is measured.

   Use: The method is only used on sand containing microporous flint.

   Test result: Dissolution per kg material.

7.2.5 Diagnosis and appraisal of structures

Signs of premature deterioration in concrete structures that could be related to ASR can generally be detected during routine site inspections. Most people performing such inspections regularly in Denmark have long experience. Thus, they are with acceptable accuracy able to identify whether an observed crack pattern, often combined with gel exudation, might be due to ASR (see example in Figure 7.1). If no signs of ASR are observed, a structure analysis on drilled cores is often performed. This analysis includes a macroscopic analysis of the cores and the prepared plane polished sections, in addition to a microscopic analysis of thin sections. Such thin section analysis is included in most routine inspections and field surveys of concrete structures performed by a number of Danish laboratories and consultants. The thin section analysis can efficiently detect whether ASR is already occurring or if the concrete has a potential to develop severe ASR damage.

In some cases, also residual expansion is measured on drilled cores. Then the cores are exposed to 50°C and either 100 % RH or submerged in saturated salt water (similar as in TI-B 51: 1985). These measurements, giving information about the expansion potential when exposed to 1) unlimited moisture, or 2) unlimited moisture and external alkalis, are used in the planning of any repair actions.

Within the last 5–10 years, inspection personnel have also started to use non-destructive testing such as ultrasonic pulse velocity, impact echo, acoustic methods, etc. to determine the extent of internal cracking or deterioration of the inspected structures.

7.2.6 Protection, repair and rehabilitation

During recent years, more focus is put on the load carrying capacity of Danish bridges with ASR. In 2013, it was realised that about 600 concrete bridges owned by the Danish Road Administration and the 'Banedanmark, Rail Net Denmark' have a potential to develop ASR. The Danish Road Administration has prepared a so-called 'ASR list' of most of their bridges, where each bridge is marked with a 'Smiley' imo ☺ in green if no ASR potential, in yellow if uncertain and in red if a potential risk for development of an ASR damage exists.

Even if the concretes in these 600 or so bridges have a potential to develop ASR damage, the current condition of the bridges is generally good. The reason is that precautions are taken by minimizing the access to moisture/water and alkalis (de-icing salts). The aim is to replace the bridge deck membranes before any leakage occurs. To be able to act in time, routine inspections of most bridges and tunnels in Denmark are performed about every fifth year. Then, all the surfaces of the bridges are surveyed visually and a 'condition grade' is given.

Despite the ASR problems in Denmark being comprehensive (see 7.2.3), still only a few bridges have been demolished due to ASR damage. One reason for this is the increased knowledge about the influence of the ASR cracking on the load carrying capacity. During the last five years, only two rather large bridges have been replaced primarily due to ASR. In 2013, large *in situ* tests of the load carrying capacity were

Figure 7.2 Photos from beneath of the bridge deck after the *in situ* load carrying capacity test on the Vosnæsvej bridge, Denmark (left). Photo after the load carrying capacity test on a beam cut from the pile supported slab bridge slab, Lindenborg Pæledæk, Denmark.

performed on one heavily ASR damaged bridge owned by the Danish Road Administration, situated on Vosnæsvej at Løgten (Figure 7.2). The results from these tests were that the demolition of this bridge has been postponed by at least 25 years or even more without any major repair work. The repair work on the Vosnæsvej bridge consisted of replacing the existing bridge deck water proof membrane (standard main-tenance/repair work conducted on concrete bridges in Denmark).

Encourage by the good results from the Vosnæsvej bridge, in 2014 similar compre-hensive load carrying capacity tests were performed on 18 beams cut from a 312 m long pile supported slab bridge (Lindenborg Pæledæk) located south of Aalborg. This bridge was visually more damaged by ASR than the Vosnæsvej bridge. Once again the results from these load carrying capacity tests were very positive, reliable and consistent. The results were in line with the tests conducted on the Vosnæsvej bridge. The demolition of the Lindenborg bridge is expected to be postponed by more than 25 years. The repair work, including strengthening of the existing bridge slab is expected to be carried out in 2017–2018. The Danish Road Administration (bridge owner) is expected to save between 5 and 10 million euros by repairing the two existing ASR damaged bridge slabs instead of replacing the bridge slabs.

As mentioned above, the main action performed on Danish bridges with on-going ASR is to minimize the access to moisture/water and alkalis (de-icing salts). Furthermore, large in-situ and/or laboratory tests of ASR damaged bridges have been conducted. Until 2016, the demolition of two large ASR damaged bridges has been postponed due to consistent load carrying capacity test results.

7.3 ICELAND

7.3.1 Brief history of national research

No ASR problems have been detected in Iceland on concrete structures built before the early 1960s, since before that period they mostly used non-reactive aggregates in the capital area (Reykjavik) where the majority of the building activities took place. However, when a new aggregate source was discovered in 1962, a sea dredged reactive basalt, the situation changed. This material was used increasingly in concrete houses in the period 1962–1979. At first, they believed that the moisture content in such structures was not high enough to cause ASR damage. By 1978, the Icelandic Building Research Institute (IBRI) had proved that ASR was a widespread problem in exterior walls of concrete houses in Iceland.

Knowing of the potential danger of ASR damage in concrete in Iceland, a Concrete Committee was established in 1967 to facilitate research on durability of concrete. The committee had representatives from the main public concrete users and IBRI, which had the chair and carried out the research. From the beginning, the emphasis was on field inspections of constructions, research on aggregates and cement and the effect of pozzolanas on ASR expansions, as well as those of temperature and moisture, seawater and freeze–thaw cycles.

In 1971, IBRI published a report on ASR in concrete (Guðmundsson, 1971). This report was the first comprehensive report on the subject published in Iceland. It included research results obtained at IBRI on reactivity of Icelandic aggregates, test methods used, field inspections on constructions, effect of different types of cements and climatic conditions in Iceland. In 1975, the third international conference on AAR (ICAAR) was held in Iceland.

That year, three research projects were published dealing with ASR in Iceland; the effect of pozzolanas on ASR and geological prospecting for pozzolanic materials in Iceland (Guðmundsson, 1975; Guðmundsson & Ásgeirsson, 1975) and Sæmundsson, 1975).

The first proof of serious ASR damage in Iceland was obtained in 1976 when drilled cores taken from the exterior walls of a concrete house were classified with extensive ASR (Thaulow, 1976). A wide ranging field survey as well as research on drilled cores followed. The results were published in January 1979, stating that ASR was most common in houses in Reykjavik built in 1962–1964. 17% of them had some damage caused by ASR (Kristjánsson, 1979). In July the same year, preventative measures were taken (see 7.3.4).

The quick response was possible because considerable research on preventative measures had been carried out since the establishment of the Concrete Committee in 1967, especially on the effect of silica fume and other pozzolanas on ASR and other concrete properties. The first extensive field survey was followed by such surveys, though in a more limited scale, more or less every third year until 1992. The results were published in IBRI reports (Kristjánsson et al., 1979–1987).

In the 1980s, considerable research was carried out on how to stop pre-existing ASR in structures (houses) and how to maintain the structures. Full-scale experiments were made using claddings, mortar-insulation systems, hydrophobic impregnation, other surface treatments and shotcrete, Olafsson (1983), Ólafsson and Iversen (1982), Ólafsson (1986), Thordarson (1991), Guðmundsson and Sveinsdóttir (1994). In the late 1980s, a research project looked at the interaction between ASR and freeze-thaw resistance, Olafsson (1989). Later publications on this issue are Sveinsdóttir and Guðmundsson (1994) and Guðmundsson (1996). Around 2005, ASR research was once more focused on by the consultant company Mannvit. The main issue of interest in these later research activities was test methods for aggregates and concrete (see 7.3.4).

7.3.2 Aggregates, binders and admixtures

Various types of basalts are the dominating rock groups in Iceland. They are dark-coloured and fine-grained mafic extrusive rocks. They are mainly composed of augite and plagioclase with opaque minerals. Basalts can be subdivided into tholeiitic basalts and alkali olivine basalts regarding the presence of accessory olivine or/and absence of quartz and low-Ca pyroxenes. As far as the alkali-silica reaction is concerned, the tholeiitic basalt type is the most interesting case, as tholeiite is oversaturated with silica, which may be present as cristobalite, tridymite and quartz within the glassy groundmass. It is often impossible to detect these very fine-grained minerals by optical microscopy, thus X-ray diffraction (XRD) analysis is required.

Rhyolite is an acid volcanic rock with phenocrysts of quartz and alkali feldspar embedded in a glassy or cryptocrystalline groundmass. Rhyolite can have a high proportion of glass. It is the extrusive equivalent of granite. The alkali-reactive components can be the glass and cryptocrystalline quartz.

In Iceland, little has been done in attempts to classify the reactive constituents of alkali-reactive aggregates. However, from experience it is known that certain rock types participate in the reaction. Helgason (1990) recognises the lack of research. However, he makes some general conclusions. He claims that the most common types of reactive particles are of intermediate to acidic composition, i.e., andesite (icelandite), dacite and rhyolite (frequently with a more or less glassy groundmass).

In an earlier study by Helgason (1982), rock samples from the tholeiitic series were collected (except for icelandite) and tested by the ASTM C227 (2010) method with high alkali cement. Unaltered, porous to fine porous olivine basalt was applied as innocuous aggregate, and other rock types were added to test their reactivity and pessimum proportion. Not only the tested dacite and rhyolite exhibited high reactivity, but also a tholeiitic basalt. The reactivity of the quartz normative basalt tested was believed to be due to reactive silica minerals and interstitial glass. Rhyolite was the only rock type tested that did not exhibit a pessimum behaviour in this test. The other tested rock types of dacite and tholeiitic basalt appeared to show a 'pessimum' content of 20–30%.

According to Katayama *et al.* (1996), it is believed that the high reactivity of some Icelandic basalts is due to the presence of secondary opal or chalcedony, in combination with primary cristobalite and rhyolitic glass. Moreover, basaltic rock is usually non-reactive when it is glassy and contains fresh basaltic glass, but may show a potential reactivity when it is highly crystalline and contains rhyolitic interstitial glass as a residual melt.

Contradictory results are observed for the effect of the reactive particle size of aggregates (sand vs. gravel). High amounts of reactive sand appears to be the governing factor for high expansion in the concrete prism tests (Wigum, 2012), whereas the opposite seems to be the case for cubes at the outdoor exposure site after 8 year of exposure. For these cubes, the amount of reactive gravel is the governing factor for high expansion (Wigum, 2016).

The mention of ASR as a possible cause for damage to concrete structures in Iceland is first to be found in an IBRI report from 1954. Testing of various basaltic aggregates showed that some of them were classified as reactive. Knowing of ASR in USA and later Denmark, the common belief was that only structures in constant contact with water such as dams, harbours and bridges were in danger of such damage. In Iceland, precautions were taken in such constructions, prescribing non-reactive aggregates or/ and cement with low alkali content.

The extensive ASR damage in domestic houses in the Reykjavik region was to a great extent due to a sea dredged natural aggregate extracted from the Hvalfjord region. This type of aggregate contains reactive constituents of basalt, andesite and rhyolite. Addition of alkalis from the sea water is also part of the problem.

In 1958, an Icelandic cement factory started to produce Icelandic cement. The only lime available in Iceland was huge shell deposits in the seabed. This raw material as well as rhyolite used for the production were the reasons for the high alkali content of the Icelandic cement with $Na_2Oeq.$ of 1.5%.

Considering the high alkali content of Icelandic cement, the way to combat deleterious ASR in Iceland has mostly been by using pozzolanas. For regular structures such as houses, bridges and harbours, silica fume blended cement has been used. For dams and hydroelectric power stations, an Icelandic pozzolanic cement was developed and has been used since the Sigalda power station was erected in 1975. This cement had 25% replacement of pozzolana, finely ground rhyolite. In a power station project built later, with reactive aggregates, 10% of silica fume was additionally intermixed.

The interest in silica fume in Iceland dates back to 1972. Then the first silica samples were tested when a ferrosilicon plant was conceived in the country. Silica fume proved to be a very effective suppresser of alkali-silica expansion (Kristjánsson, 1979). In line with this, in 1979 the State Cement Works started to produce only cement intermixed with silica fume, replacement at first being 5%, but increased in 1983 to 7.5%.

Prior to silica fume, finely ground rhyolite pozzolana was intermixed with OPC, starting in 1973 with 3% and increasing to 9% in 1975–79.

The situation during recent years is however more complicated. Since the import of low alkali cement from Aalborg Portland started in 2000, there are various types of cement on the market. The original 7.5% addition of silica fume in the Icelandic cement was reduced to 4–6% depending on type of cement. A special Icelandic cement with 10% silica fume and 25% ground rhyolite is no longer available. Production of Icelandic cement was terminated in 2013, and the Icelandic cement producer started to import Norwegian low alkali cement and cement with 20% fly ash.

7.3.3 Special features of ASR

As previously mentioned, the main ASR problem in Iceland is connected to domestic houses built in the Reykjavik region in the period 1962–1979, due to use of a sea dredged natural aggregate extracted from the Hvalfjord region. Due to a lot of freeze-thaw cycles (annual about 80 in Reykjavik) and driving rain, it is also common to encounter combined ASR and freeze-thaw damage. No serious ASR damage is observed on structures subjected to a high degree of moisture, such as harbours, dams and bridges, because precautions were taken early (7.3.4). As a curiosity, serious ASR damage observed in storage tanks for excess water from power plants should be mentioned. This excess water is almost boiling, contributing to rapid development of ASR.

7.3.4 Preventative and mitigating measures – Icelandic approach

7.3.4.1 Experience from use of silica fume to mitigate ASR

The research on preventative measures, starting already in 1967 (7.3.1), made it possible to respond quickly once the ASR problems in houses had been identified. In July 1979, the following preventative measures were taken (Kristjánsson, 1979). Firstly, high quality silica fume was intermixed with the cement in the grinding process, first 5% replacement and then 7.5%. Additionally, the criteria for reactivity according to the ASTM C 227 mortar-bar method (2010) was changed to 0.1% expansion after 12 months instead of 6 months, as recommended in the original method. Lastly, all sea dredged materials were to be washed and criteria set on the residual chloride content. No serious damage caused by ASR has been found in concrete structures since these measures were introduced. However, it is still debated whether addition of these quantities of silica fume is sufficient to mitigate ASR for the most reactive Icelandic aggregates.

In 1987, six trial concrete walls (Figure 7.3) were cast at the Icelandic Building Research Institute (IBRI), now the Icelandic Innovation Centre (NMÍ). In combination with different types of cements, the aggregate used was the well-known alkali-reactive aggregate from Hvalfjörður. In a study by Wigum (2010), concrete cores from the walls were collected and examined by thin sections and plane polished sections. It was found, as documented in many cases earlier, that the aggregate was alkali-aggregate reactive in combination with the high alkali HP cement, containing no silica fume.

Figure 7.3 Outdoor exposure site of trial concrete walls at NMÍ, seen from North West. Photo by Børge J. Wigum.

It was however, quite unexpected to discover that samples from two of the walls were cracked and exhibited some amount of ASR gel, even though both contained the VP cement with 7.5% silica fume and 3% ground rhyolite. The degree of expansion of these concretes is not known. Nevertheless, this finding revealed a need to test potential highly reactive aggregates, even when they are used with cements with additives designed to mitigate AAR. Using the reactive Hvalfjörður aggregate with the PP cement (containing 10% silica fume and 25% ground rhyolite) appeared effective to mitigate the development of AAR. This study provided important information on the state of Icelandic concrete after more than 20 years of exposure, as damage due to ASR was observed in concrete with VP cements that was assumed to be able to mitigate ASR for all Icelandic aggregates (Kristjánsson, 1979).

In a study presented by Wigum (2016), the effect of silica fume on ASR was examined by use of the concrete prism tests RILEM AAR-3 and RILEM AAR-4, along with measurements of concrete cubes exposed for almost 8 years on an Icelandic outdoor exposure site. It was shown that the inter-milled amount of 4% and 6% silica fume appears to hinder the expansion of concrete prism in the AAR-3 method and 6% appears sufficient to hinder expansion in the AAR-4 method. However, concrete cubes with cement containing both 4% and 6% are showing expansion more than 0.10% after almost 8 years at the outdoor exposure site.

7.3.4.2 Building regulations – accelerated test methods – field exposure Site

The principal test method used in Iceland for evaluating ASR has been the well-known ASTM C227 (2010) mortar-bar method. The ASTM 6-months criterion of 0.1% expansion was at first used for guidance. However, in 1979, the Icelandic Building Code demanded a stricter criterion of 0.05% after 6 months and 0.1% after 12 months. Other methods such as the ASTM quick chemical test, ASTM C289 (2007, but now withdrawn) and petrographic analyses have not been found suitable for Icelandic aggregates. These methods were, however, previously used for preliminary information prior to the mortar-bar testing.

Another building code restriction introduced in 1979 was that unwashed sea-dredged reactive aggregates were banned. The reason for this ban was that NaCl from the sea water may exchange ions with $Ca(OH)_2$ liberated during the cement hydration and form NaOH, which increases the alkalinity of the concrete and potentially its alkali reactivity.

In order to avoid deleterious ASR, it has recently (from about 2005) been found necessary to take up research again in Iceland, particularly in accordance with the new test methods developed internationally and the competence of these to mirror the effects of various types of cements and pozzolanas, particularly silica fume (performance testing). This work has been managed by the laboratory of Mannvit. Particular attention has been paid to the two concrete prism methods developed by RILEM: AAR-3 (storage at 38°C) and AAR-4.1 (storage at 60°C) (both in Nixon & Sims, 2016). Other test methods included were the mortar bar test ASTM C 227 (2010) and the accelerated mortar-bar test (RILEM AAR-2, in Nixon & Sims, 2016), which is similar to ASTM C1260 (2014).

The research has shown that the accelerated mortar-bar test (AAR-2 or C1260) exhibits significantly higher expansion than the other test methods examined. Hence, it is proposed that results from RILEM AAR-2 should only be used for assessing the reactivity of aggregates, and not as an assessment of the effects of additives and various cements. For testing Icelandic materials, it is believed that the concrete prism test (AAR-3), using unwrapped prisms as in the final version of the test (in Nixon & Sims, 2016) is the best method to reflect field behaviour[1]. However, based on the shorter testing period for the accelerated concrete prism test, RILEM AAR-4.1 (in Nixon & Sims, 2016), this test is considered likely to be the most suitable test in the future. It is an ambition to continue the development of this latter test in Iceland.

As part of this improvement, research is continuing in order to better correlate the critical limits of the laboratory tests with what is happening in real concrete structures. Thus, a field exposure site was established at Mannvit in 2007, where large concrete cubes (300 x 300 x 300 mm) are exposed outdoors. Today, this exposure site holds more than 30 different concrete mixes where the field expansion is compared with results from accelerated laboratory testing. The exposure site also includes various samples from different parts of the world for comparison.

A new version of the Icelandic Building Regulation (no. 112/2012) was approved in 2012, where it is stated that aggregates for concrete shall be tested with regards to alkali aggregate reactivity. The accelerated mortar bar test (AAR–2) and the concrete prism method (AAR-3), albeit in its earlier draft version (with wrapped prisms)[1], are introduced into this new version, along with the continued use of ASTM C 227 (2010) mortar-bar method.

Aggregates shall be considered innocuous if the expansion of mortar-bars, cast with high alkali cement (for example pure Icelandic Portland cement), is less than; a) 0.05% after 6 months or 0.1% after 12 months, according to the test method ASTM C 227 (2010), or; b) 0.20% after 14 days according to test method RILEM AAR-2.

---

[1] *The final version of the RILEM AAR-3 concrete prism test, with 38°C storage and unwrapped prisms, is used by several Icelandic laboratories, even though the Icelandic Building Regulations (no. 112/2012) currently refer to the earlier AAR-3 draft version published in 2000, which suggested wrapped prisms.*

If an aggregate is classified as reactive after testing, its use can be allowed if one of the following criteria is fulfilled: a) expansion of mortar-bars, cast with the actual type of cement to be used, is less than 0.05% after 6 months or 0.1% after 12 months according to the test method ASTM C227; however, in cases where cement with silica fume is applied (silica fume > 5%), the expansion limit is 0.08% after 12 months, or b) expansion of concrete prisms, cast with the actual type of cement to be used and aggregates, is less than 0.05% after 12 months according to the test method RILEM AAR-3 (*using wrapped prisms*[1]).

Producers of aggregates, and when relevant concrete producers, are obliged to test their materials on a regular basis by an independent and recognized laboratory, providing a written certificate stating whether the aggregate is classified as reactive or innocuous according to the testing described above.

7.3.5 Diagnosis and appraisal of structures

Recent field surveys in Iceland have included sampling of sprayed concrete in tunnels (Wigum *et al.*, 2005), concrete samples from a 20 year exposure site and concrete samples from domestic houses, (Wigum *et al.*, 2009) and (Wigum, 2010). These samples have been examined by plane section in UV-light and by thin section under a petrographic microscope.

7.3.6 Protection, repair and rehabilitation

Since 1980, considerable emphasis has been on research for stopping ASR in constructions and for repair of damages caused by ASR. The best results have been obtained by reducing the moisture content of the concrete in walls in ASR damaged houses. The use of ventilated claddings, mortar insulation systems and impregnation with hydrophobic silanes has been most effective. The research has been reported in various forums; (Ólafsson & Iversen, 1982), (Ólafsson *et al.*, 1988), (Ólafsson & Gestsson, 1988), (Ólafsson, 1983) and (Ólafsson, 1988). On the bases of the research results, IBRI has published recommendations in this field. None of these methods have proven to be effective when used on other types of construction, since consistent temperature gradient through the exterior walls of the houses is an important factor in decreasing the relative humidity in the concrete.

7.4 NORWAY

7.4.1 Brief history of national research

An overview of the Norwegian ASR history, starting about 1990, with focus on research activities, development of test methods and national regulations have previously been summarised by Wigum (2006) and Wigum and Lindgård (2008). Some details are given in the following sub-sections. The web page of "FARIN" (a nationwide 'Forum on Alkali-Reactions In Norway', established in 1999, http://farin.no/) contains a collection of most of the published Norwegian literature on ASR; originating from research projects, post-graduate programmes and doctoral studies. For many of these, pdf copies may be downloaded.

ASR is still a hot topic in Norway. The research activities during the last ten years have mainly focused on reliable performance testing, see for example, Lindgård (2013a). In order to calibrate results from accelerated laboratory testing to field, a field exposure site was established at SINTEF (Trondheim) in 2004 as part of the European 'PARTNER' project (Lindgård et al., 2010). During the COIN programme (2007–2014, www.coinweb.no), this field exposure site was extended with more than 25 additional concrete mixtures. Furthermore, SINTEF have established a research cooperation with LNEC (National Laboratory for Civil Engineering) in Lisbon, Portugal. When new field exposure cubes are produced at SINTEF, parallel cubes are transported to Lisbon for exposure at the LNEC field exposure site. Thus, any expansion on these cubes may be revealed earlier due to the warmer climate in Portugal.

In the years to come, the research activities will most likely focus more on assessment and repair of ASR damaged concrete structures. However, the research on reliable performance testing methods and corresponding critical limits (laboratory/field correlation) will continue. The latter issue is the main research topic of a new Norwegian research project (2014–2018) that is closely linked to the current work in RILEM Technical Committee 258-AAA (2014–2019), with its Norwegian leadership (Prof Børge Wigum is Chairman, whilst Dr Jan Lindgård is Secretary).

7.4.2 Aggregates and binders

The first significant existence of ASR in Norwegian concrete structures was documented in the SINTEF managed research project 'AAR in Southern Norway' (1990–1993). The project primarily focused on mapping the occurrence of ASR and the identification of reactive rock types by petrographic examinations of cores, fluorescence impregnated polished half cores and thin sections from structures (Jensen, 1993). It was found that ASR in Norwegian structures is caused by the coarse fraction of slowly reactive siliceous aggregates. Cataclastic rocks, for example cataclasite and mylonite, were observed as deleteriously alkali reactive in about 50% of all the investigated structures. On this basis, a list of potential reactive Norwegian rock types was included in the first description of the Norwegian petrographic method (Lindgård et al., 1993). The fact that primarily the coarse aggregate fraction has given rise to ASR damage has later been demonstrated in an extensive field survey (Lindgård et al., 2004).

The classification list of Norwegian rock types was slightly revised during the research programme 'NORMIN 2000' (1998–2000) and later published by Wigum et al. (2004). As part of the research programme, a detailed petrographic atlas with micrographs of the various Norwegian rock types was prepared (Wigum, 1999: available on http://farin.no/). The up-to-date rock classification chart, shown in Table 7.4, is included in the Norwegian ASR regulations (Norwegian Concrete Association, 2005).

Norcem, which is now the only Norwegian cement producer (now part of the Heidelberg cement group), started to produce cement about 125 years ago. They have only produced cements based on Portland clinker, having rather similar properties at all the cement plants due to roughly equal properties of the raw materials. During the 1940s, the consumption and production was low, and in the early 1950s, Norway also had to import cements. Some blended cements containing pozzolanas and ground granulated blastfurnace slag (GGBS) were for example used in massive structures such as concrete dams. However, the Norwegian cement production increased

Table 7.4 Classification chart for the alkali-reactivity of Norwegian rock types (Norwegian Concrete Association, 2005).

ALKALI REACTIVE ROCK TYPES (Documented in concrete structures)	AMBIGUOUS ROCK TYPES (In some cases observed to react in concrete structures)	INNOCUOUS ROCK TYPES
1. SEDIMENTARY ROCKS • Sandstone • Arkose • Quartz sandstone • Claystone (including shale) • Siltstone (including shale) • Marlstone (including schistose and/or metamorphic) • Greywacke (also metamorphic) Sedimentary features should be observed 2. MYLONITE/CATACLASITE (Containing free quartz) • Mylonites • Cataclasites • Mylonite gneiss 3. ACIDIC VOLCANIC ROCKS • Rhyolite • Quartz keratophyre 4. OTHER ROCK TYPES • Microcrystalline quartzite • Phyllite • Quartz schist	5. AMBIGUOUS (examples) • Quartzite/quartz schist • Rock types with quartz (Modal quartz >20%) • Limestone with contaminations (contaminated with disperged fine grained quartz) • Hornfels • Mylonites low in free quartz (1–5%)	6. MAFIC ROCK TYPES • Basalt • Greenstone • Gabbro • Amphibolite (including all types of mafic rock types, also metamorphic) 7. ROCK TYPES CONTAINING QUARTZ • Granite/Gneiss • Quartzite/quartz schist • Mica schist 8. FELDSPATHIC ROCK TYPES 9. OTHER/UNIDENTIFIED • Limestone (pure) and marble • Other non-reactive (also single crystals) • Porphyry • Quartz-free mylonites
Typical grain size of quartz < 60 μm (exception: Sandstone)	Typical grain size of quartz 60–130 μm	Typical grain size of quartz > 130 μm, or quartz not present

Rock type nomenclature; Gjelle og Sigmond, 1994: Rock classification and production of maps. Geological Survey of Norway, Publication 113 (in Norwegian).

intensely during the 1950s and kept on rising in the period from 1960 to 1980. From the 1950s up to the early/middle 1990s, Norcem primarily produced high-alkali CEM I cements (alkali content in the range of 1.0–1.3% Na_2Oeq). Thus, the majority of Norwegian concrete structures are built with high-alkali Portland cement. However, the alkali content was lower (about 0.6% Na_2Oeq) in the CEM I cements used to produce the concrete oil platforms in the 1970s and 1980s.

In the middle of the 1990s, Norcem started to produce CEM I cements with lower alkali content (in the range of 0.6–0.9% Na_2Oeq). They also introduced fly ash cement on the Norwegian market from the middle of the 1980s. After the middle of the 1990s, the Norcem fly ash cement with 17–20% of class F fly ash (CEM II/A-V), manufactured by co-grinding clinker and fly ash, has more and more taken over the Norwegian cement market. In addition to environmental reasons (use of less clinker), mitigation of ASR has been an important factor for the increased use of this fly ash cement. Based on performance testing (see sub-section 7.4.4), this Norcem CEM II/A-V fly ash cement is documented to mitigate ASR also when using the most reactive Norwegian aggregate types. Norcem is also introducing alternative fly ash cements on the Norwegian market, including some with higher fly ash contents.

Silica fume was introduced in Norway in the 1970s. It was used in rather high quantities until the middle of the 1980s (for economic reasons in order to fulfil strength requirements with high water/binder ratios). Today, it is quite common to use up to 3–5% of silica fume in the concrete. Since fly ash and silica fume were used in the production of concrete in Norway before being aware of any ASR problems (about 1990), Norway have most likely mitigated ASR in many concrete structures even if this was not the intention.

Today, imported cements constitute roughly 20% of the Norwegian market. Primarily two cement types are imported, low alkali CEM I cement and a blended slag cement (CEM II/B-S according to the European Standard, NS EN 197-1: 2011). This 'second period of cement import' started in the period 2003–2005.

7.4.3 Special features of ASR

ASR has been documented in most moisture exposed outdoor concrete structures, for example concrete dams, bridges, foundations, retaining walls, gravity retaining walls (precast concrete stones), buried concrete pipelines and a huge number of railway sleepers. Severe ASR has also been found indoors when exposed to moisture, for example in power plants (massive foundations for turbines), swimming pools and foundations and roofs of paper mills.

7.4.4 Preventative and mitigating measures – Norwegian approach

The research projects in the first half of the 1990s strongly focused on test methods and corresponding criteria for the prediction of ASR as observed in Norwegian concrete structures. On this basis, the Norwegian ASR test methods were firstly published in 1993 (Lindgård et al., 1993) (Table 7.1). Since 2001, these have been the formal national ASR test methods. The test methods were slightly modified in 2005 and published in the Norwegian Concrete Association publication no. 32 ('NB32')

(Norwegian Concrete Association, 2005, with an English version available since 2010). This publication also gives requirements to laboratories. Presently, only a very few laboratories are approved to perform such ASR testing. Strict requirements with respect to education, experience and approval is also required of petrographers performing petrographic analysis (see later).

In 1996, the Norwegian Concrete Association prepared publication no. 21 ('NB21'), giving guidelines for handling ASR in Norway (Norwegian Concrete Association, 1996). Based on technological knowledge obtained after 1996, and the fact that the publication in 2001 became the harmonised normative reference document to the new European concrete standard, NS-EN 206-1, a revised version of the publication was published (Norwegian Concrete Association, 2004, an English version has been available since 2008). This publication refers to 'NB32'.

As did the 1996-version, the 2004 edition of 'NB21' states that concrete shall be considered secure against ASR if at least one of the following conditions is fulfilled (Dahl et al., 2004):

• The aggregate is documented to be non-reactive
• The binder has a composition that prevents ASR at the actual alkali content (in these calculations, no alkali contribution from aggregates or from the environment is included)
• The concrete will serve under dry exposure conditions.

'NB21' is a key element in the Norwegian system for preventing ASR. The system, which is considered to represent the same security level as corresponding systems for securing against durability damage from reinforcement corrosion and freeze/thaw, also includes:

• Descriptions of the actual test methods ('NB32')
• Certification of producers of aggregates and concrete
• Certification of test laboratories
• Round robin tests for test laboratories
• A committee of experts to be used for interpretation of the regulations in cases of uncertainty or dispute.

The 2004 edition of 'NB21' is divided into two major parts. Part 1 is, in formal terms, describing the mandate and the use of the publication, and how concrete constituents and concrete recipes shall be tested and evaluated with respect to potential ASR, including critical limits for the various test methods (see later). As a first step, individual aggregates must be evaluated by the petrographic method, similar to RILEM AAR-1 (current version in Nixon & Sims, 2016). The good experience with this method during a decade of service was summarised by Wigum et al. (2004). The evaluation based on results from the petrographic analysis can be reassessed by the Norwegian ultra-accelerated mortar-bar test, similar to RILEM AAR-2 (current version in Nixon & Sims, 2016), but the prism size used in Norway is the 40x40x160mm option. Finally, the Norwegian 38°C concrete prism test (CPT) can be used to reassess the evaluation from any of these tests. The CPT is comparable with RILEM AAR-3 (current version in Nixon & Sims, 2016) and ASTM C1293 (2008), but the Norwegian prisms are larger, 100x100x450 mm (Lindgård et al., 2013b). For the evaluation of binders and concrete compositions, i.e., performance testing, only the Norwegian CPT can be used. The current critical limits for the various test methods are shown in Table 7.5 (Norwegian Concrete Association, 2004).

Table 7.5 Overview of critical limits for test methods for documentation of alkali-reactivity of single aggregates or blends of aggregates (Norwegian Concrete Association, 2004).

Documentation of	Critical limits for the laboratory test methods [1]		
	Petrographic analysis (converted results)[2]	Norwegian accelerated mortar bar method [3]	Norwegian concrete prism method [4]
Fine aggregate and blend of fine	20.0 %	0.14 %	0.040 % [5]
Coarse aggregate and blend of coarse	20.0 %	0.08 %	0.040 % [5]
Fine/coarse aggregate	20.0 %	0.11 %	(not applicable)
Blend of a fine- and coarse aggregate, where the fine or coarse is alkali-reactive	20.0 % [6]	0.11 %	0.050 %

[1] A single aggregate or a blend of aggregates shall be classified as innocuous if the values obtained are lower than the specified critical limits.

[2] The comparative value (Sv) shall be compared with the critical limit. Sv is calculated as a weighted average of the previous petrographic analyses (up to 6 analyses) added a security margin. The method for calculation of Sv is presented in (Norwegian Concrete Association, 2004) and (Dahl et al., 2004).

[3] The measured expansion after 14 days of exposure shall be compared with the critical limits.

[4] The measured expansion after 1 year of exposure shall be compared with the critical limits.

[5] A fine aggregate or a blend of fine shall be tested with a coarse non-reactive reference aggregate. A coarse aggregate or blend of coarse shall be tested with a fine non-reactive reference aggregate. Procedure for testing is given (Norwegian Concrete Association, 2005). The binder used shall have an alkali content of 5.0 kg/m^3 Na_2O eq.

[6] A maximum of 15% of the calculated value (Sv) is allowed to come from the coarse aggregate.

Part 2 of 'NB 21' gives advisory guidelines for how the concrete industry can fulfil the requirements given in the Part 1 specifications. More complementary descriptions of the test methods (with reference to publication no. 32) and corresponding evaluation criteria are presented and illustrated by examples. Part 2 also provides a survey of binders and corresponding alkali contents documented to be suited for production of ASR resistant concrete containing all types and amounts of Norwegian reactive aggregates. This 'binder approval testing' is performed with the most reactive Norwegian aggregate types that are presently known ('worst case reactive aggregates'). The list of approved binders, published on the web page of the Norwegian Concrete Association, is continuously updated whenever new binders have obtained satisfactory documentation. Thus, concrete producers do not have to perform separate performance testing with their alkali-reactive aggregates, as long as the cement content and the alkali content are lower than the approved binders.

Recently, (Lindgård et al., 2010) have reviewed 15 years of experience with performance testing in Norway. Despite the long testing time required (1–2 years), one main conclusion is that the Norwegian system for performance testing has proven to be an advantageous and flexible tool to document critical alkali limits for binders and aggregates. Yet, no specific 'safety factor' (beyond a 'general safety factor') is included in the Norwegian regulations to compensate for any alkalis that may leach out from the prisms during the exposure period. However, some extra security is already built into these regulations when testing binders, since most alkali reactive aggregates in common use in Norway are far less reactive than the reference 'worst case' alkali reactive aggregate combinations applied for performance testing of binders. In addition, most commercial concretes containing SCMs will normally contain less alkali than the critical alkali limits documented for various binders. The cement producer (Norcem) has also declared 'upper alkali limits' for their cements to be used for calculating the concrete alkali content according to the Norwegian regulations.

Irrespective of this, the possible influence of alkali leaching on the measured expansions in the Norwegian CPT cannot be neglected, even though rather large concrete prisms are used. The large prism cross section (100x100 mm) has proven to give far less alkali leaching than prisms with a smaller cross section, for example as used in the RILEM CPTs and in the ASTM C 1293 (2008) version (Lindgård et al., 2013c). However, the alkali leaching will most likely be accounted for more restrictively during the next revision of the Norwegian ASR guidelines.

7.4.5 Diagnosis and appraisal of structures

No general national regulations or recommendations exist regarding diagnosis and appraisal of structures with possible ASR. Even after about 25 years with ASR experience in Norway, many owners and consultants lack knowledge and skills in assessing ASR.

With respect to laboratory testing of drilled cores, only a very few experienced laboratories exist. For such examinations, analysis of plane polished sections and thin sections have proven to be a necessary and promising analysing tool, used by SINTEF for about 25 years to diagnose ASR and to assess the degree of damage (Lindgård et al., 2012). SEM analysis is also used, but not regularly. For several structures, the moisture state is measured too, either as water content, degree of capillary saturation (DCS) and/or relative humidity (RH). The relationship between DCS and observed damage on about 50 concrete structures was discussed by Lindgård et al. (2006). They showed that ASR primarily occurs on concrete structures with DCS higher than 90%. More rarely, structural parameters such as compressive strength, tensile strength and dynamic E-modulus are included in the test programme.

Only a limited number of concrete structures are monitored to follow the progress of expansion. In two cases, the expansion has been measured in drilled holes in massive concrete foundations for turbines, by use of a Sliding Micrometer (high position strain meter). Each measurement was taken over a distance of 1 m. The main experience is that the concrete has expanded linearly in each location over the measuring period from 1987 to 2007. However, the expansion varies over the depth (0–6 m), resulting in maximum expansion from zero to 2.6 ‰ during the 20 years measuring period.

For a few structures, the LCPC-cracking index method is being used (LCPC, 1997). For other, the length change between measuring points with distance of 200 mm is measured by a mechanical strain gauge. Three point measurements of crack opening with a strain gauge have also been used (Jensen & Merz, 2008).

Even if monitoring of RH in the field is notoriously uncertain due to various reasons, such measurements have been performed regularly by SINTEF with various sensors (primarily HumiGuard; www.industrifysik.se). Another method used is by means of wooden sticks of the species Ramin (Jensen, 2006).

7.4.6 Protection, repair and rehabilitation

The experience in Norway with protection, repair and rehabilitation of ASR-affected concrete structures is limited. No general recommendations for repair strategy exist. Neither has there been any national research activity on the topic.

Some owners have tried various types of surface protection in order to try to reduce the moisture level in some concrete structures, for example bridges and dams. However, it is very difficult (perhaps impossible?) to reduce the internal moisture state in relatively massive concrete elements/structures, documented to have rather stable moisture levels deeper than 20–50 mm (Lindgård et al., 2006; Relling & Sellevold, 2006). Logically, successful rehabilitation projects are lacking. Another problem experienced is that the degree of damage on concrete structures with incipient freeze/thaw damage might increase significantly after a surface coating has been applied. The reason is accumulation of moisture in the surface layer behind the coating (Rodum et al., 2001).

However, for a few structures it has been reported that various surface treatment systems and products have been able to reduce the RH in the concrete near the surface when the rainwater has been the main source of moisture supply (Jensen, 2003).

On the other hand, some surface protections might be able to reduce or stop any ingress of chlorides from de-icing salts, and thus have a benefit (avoiding reinforcement corrosion) even if the general moisture level is not reduced significantly to reduce the rate of ASR. Thus, parts of some concrete bridges have been surface protected by the owners, in several cases in connection with test sites where various products have been applied.

To the authors' knowledge, only a very few Norwegian concrete structures have been demolished or replaced due to damage caused by ASR. For example, a limited part of an airfield pavement was replaced to avoid that any pieces of the concrete loosened and thus were sucked into the engines of the jet fighter planes. A smaller suspension bridge was also removed partly due to ASR.

However, a large number of heavily cracked concrete railway sleepers have been replaced after about 20–25 years in service (the intended lifetime was at least 50 years), and many more will follow in the future. About five to six million railway sleepers have been produced without being aware of the potential ASR problem. Due to the pre-stressing and no use of stirrup reinforcement, the ASR led to longitudinal cracking of the railway sleepers. Additionally, the bond strength might be reduced when the ASR expansion increases. As a consequence, the load capacity and the fatigue resistance is reduced (Jensen et al., 2000).

Three cases of comprehensive rehabilitation of ASR damaged Norwegian concrete structures are worth mentioning. 'Pillar walls' of a road bridge from 1950 were heavily cracked due to ASR and a lack of stirrup reinforcement at the end of those walls. These pillar ends were strengthened in 2002 by use of bolted steel cladding filled with repair

Figure 7.4 End face of a cracked pillar (left; *photo by Aas-Jacobsen*). One repaired pillar with steel cladding (right; *photo by Knut Grefstad, Norwegian Public Road Administration*).

mortar (Figure 7.4). The repair action seems to have worked well after about 10 further years in service.

The beams of the 200 m long Elgeseter bridge in Trondheim, built around 1950, had expanded about 1 ‰ (200 mm) during about 50 years, leading to closure of the single expansion joint at the end of the bridge. Another consequence was that the topmost part of the slim columns closest to the expansion joint also had moved up to 200 mm, resulting in unacceptable restraint forces. As reported by Jensen (2004), the main rehabilitation work in 2003 (except repairing the expansion joint) was to move the most stressed columns back to a vertical position. To support the beams before demolishing the topmost part of each of these columns by 'micro blasting', a special steel construction was prepared (Figure 7.5). Subsequently, the reinforcement was cut and removed, each column was moved back to vertical position, new reinforcement was welded to the beam and finally new concrete was poured into the mould. The repair actions worked as intended. On other columns of the bridge, a test field with strengthening of the columns (that all have longitudinally cracks due to ASR) with carbon fibre has been established.

A 230 m long slab concrete dam (Votna 2) with a maximum height of 20 m was built in about 1965. The thickness of the slabs varies from 0.30 to 0.80 m (thickest in the bottom part). The distance between the pillars supporting the slabs is 5 m. The development of ASR during 40 years in service and the repair actions taken were presented by Larsen *et al.* (2008). Text extracted from this paper is given as follows: "*During a normal dam visit in May 2003 the dam owner discovered an alarming happening in the corner of the slab concrete dam, where the dam has a change of direction of about 22°. At the pillar in the corner and the two neighbour pillars the front slab had been lifted away from the pillars. At the top of the dam there was a gap of about 50 mm between the pillar and the front slab. This gap decreased towards zero at the bottom of the dam. The cause of this movement was*

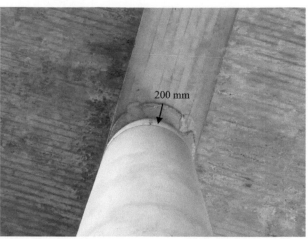

Figure 7.5 A column at Elgeseter Bridge in Trondheim after blasting off the topmost part (left). One rehabilitated column – the topmost parts have been moved about 200 mm (right; *photo by Børge J. Wigum*).

expansions of the slab. Without being noticed by the owner, the slab had expanded due to ASR leading to closed joints between the slab elements (each with a length of about 10 m). Thus, there was no more space for expansion of the slabs when the temperature of the concrete increased on a warm sunny day with a low water filling of the magazine (about 15 m below the maximum approved water filling level)."

In the following years (2003–2006) condition surveys, laboratory investigations of drilled concrete cores, dam safety evaluations, and planning and execution of rehabilitation work of the slab concrete dam were performed. On this basis it was decided to strengthen the front slab of the dam. Another influencing factor was sabotage consideration. After evaluation of different repair methods, the following permanent rehabilitation works were carried out during the summer 2006 (Figure 7.6):

Figure 7.6 Rehabilitation work on the Votna 2 dam, upstream face (left). Upstream face of the dam after the rehabilitation (right). *Photos by Sweco.*

- The old slab had expanded so much due to ASR that most of the joints between the slab elements had been closed over the years. Thus all the joints in the old slab were cut open from top to bottom by use of a diamond saw (to a width of approximately 45 mm), to give room for expansion during temperature rise in the summer season, in addition to allow more expansion due to further development of the ASR in old slab concrete.
- A new concrete front slab was concreted upstream of the old slab over the total length of the dam. The new slab was divided into slab elements with length 10 m separated with expansion joints located at every second supporting pillar (the joints between the old slab elements were located between the pillars). To seal the joints a "water stop" sealing was used.
- To allow free movement between the old and the new slab (except in the bottom part of the dam height where bond is required), a smooth membrane was placed between the slabs.
- A drainage system was established by drilling several holes through the bottom part of the old slab (underneath the lower part of the membrane). Thus any water leaking through the new slab is allowed to drain through the old slab.
- To create a good bond between the two slabs in the bottom part of the dam, steel bolts were placed between the new and the old slab. In addition, the surface of the old slab was prepared by general chiselling and sawing of horizontal slots prior to casting.
- The old railing (concrete wall) at the top of the dam was removed, and a new one was created.

7.5 SWEDEN

7.5.1 Brief history of national research

In 1975, ASR was discovered in several concrete floors ('pop outs' caused by porous flint) in the south-western part of Sweden. To cope with this problem, some research activities on this special topic were initiated in the second half of the 1970s (Nilsson & Peterson, 1983) (see 7.5.3). However, research on general ASR topics was not initiated until the early 1990s (Lagerblad & Trägårdh, 1992a). The reason for not recognizing ASR as a potential deterioration mechanism in Swedish concrete structures until about 1990 is that most known alkali-reactive aggregates in Sweden are slowly reactive and that larger constructions until about the mid 1970s were primarily made with low to moderate alkali Portland cements. Then cement with high alkali content was introduced (see 7.5.2).

Today ASR is recognized as being fairly common in some areas, although not severe, and then primarily in structures built in the 1960s and 1970s. It has also been recognized in some old bridges in which the reaction is probably the result of de-icing salts in combination with frost action and long-term ASR (Trägårdh & Lagerblad, 1996). The research activities regarding ASR today primarily focus on correlations between different test methods, development of a performance test and correlations between laboratory and field testing.

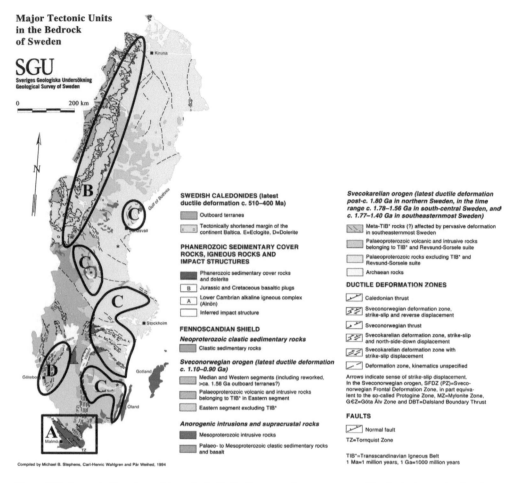

**Major Tectonic Units
in the Bedrock
of Sweden**

SGU
Sveriges Geologiska Undersökning
Geological Survey of Sweden

0   200 km

SWEDISH CALEDONIDES (latest
ductile deformation c. 510–400 Ma)

　Outboard terranes

　Tectonically shortened margin of the
　continent Baltica. E=Eclogite, D=Dolerite

PHANEROZOIC SEDIMENTARY COVER
ROCKS, IGNEOUS ROCKS AND
IMPACT STRUCTURES

　Phanerozoic sedimentary cover rocks
　and dolerite

B Jurassic and Cretaceous basaltic plugs

A Lower Cambrian alkaline igneous complex
　(Alnön)

　Inferred impact structure

FENNOSCANDIAN SHIELD

Neoproterozoic clastic sedimentary rocks

　Clastic sedimentary rocks

*Sveconorwegian orogen (latest ductile deformation
c. 1.10–0.90 Ga)*

　Median and Western segments (including reworked,
　>ca. 1.56 Ga outboard terranes?)

　Palaeoproterozoic volcanic and intrusive rocks
　belonging to TIB* in Eastern segment

　Eastern segment excluding TIB*

Anorogenic intrusions and supracrustal rocks

　Mesoproterozoic intrusive rocks

　Palaeo- to Mesoproterozoic clastic sedimentary rocks
　and basalt

*Svecokarelian orogen (latest ductile deformation
post-c. 1.80 Ga in northern Sweden, in the time
range c. 1.78–1.56 Ga in south-central Sweden, and
c. 1.77–1.40 Ga in southeasternmost Sweden)*

　Meta-TIB* rocks (?) affected by pervasive deformation
　in southeasternmost Sweden

　Palaeoproterozoic volcanic and intrusive rocks
　belonging to TIB* and Revsund-Sorsele suite

　Palaeoproterozoic rocks excluding TIB* and
　Revsund-Sorsele suite

　Archaean rocks

DUCTILE DEFORMATION ZONES

　Caledonian thrust

　Sveconorwegian deformation zone,
　strike-slip and reverse displacement

　Sveconorwegian thrust

　Svecokarelian deformation zone, strike-slip
　and north-side-down displacement

　Svecokarelian deformation zone with
　strike-slip displacement

　Deformation zone, kinematics unspecified

Arrows indicate sense of strike-slip displacement.
In the Sveconorwegian orogen, SFDZ (PZ)=Sveco-
norwegian Frontal Deformation Zone, in part equivalent to the so-called Protogine Zone, MZ=Mylonite Zone,
GÉZ=Göta Älv Zone and DBT=Dalsland Boundary Thrust

FAULTS

　Normal fault

TZ=Tornquist Zone

TIB*=Transscandinavian Igneous Belt
1 Ma=1 million years, 1 Ga=1000 million years

Compiled by Michael B. Stephens, Carl-Henric Wahlgren and Pär Weihed, 1994

Figure 7.7 Areas in Sweden with typical alkali-silica reactive aggregates. Geological map by Stephens *et al.* 1994, ©SGU. The zones and labels are explained in the text.

7.5.2 Aggregates and binders

There are many different rock types in Sweden, and some of these are alkali-silica reactive. An overview of areas with different aggregate types is given in Figure 7.7. A compilation of the most commonly used concrete aggregate types in the Nordic countries is given in a Nordtest report from 1990 (Brandt & Schouenborg, 1990). Different aggregate types in Sweden are further discussed in Lagerblad and Trägårdh (1992b) and Appelquist *et al.* (2014).

Some of the Swedish aggregates are very reactive, for example those in Scania, in the south western part of Sweden (labelled A in Figure 7.7). In this region, parts of the bedrock are of the same type as in Denmark and in northern Germany, *i.e.*, limestone with flint (chert) and sometimes porous flint. Consequently, in this part of Sweden the problems with ASR have been the same as in Denmark and northern Germany. North of the Scania area, siliceous (generally innocuous) gneisses, dominate the bedrock. Therefore, most of the aggregates to be used in aggressive environments are taken from parts to the north of the area with flint bearing limestone.

Along the Swedish-Norwegian mountain chain, the Caledonides (labelled B in Figure 7.7), the bedrock consists mainly of orogenic metamorphic rocks. The known alkali-reactive rock types, regarded to be medium reactive, are grey-wackes, fine grained mylonites and some altered sparagmite sandstone. The experience of aggregates from the Caledonides is limited in Sweden. In general however, the aggregates are similar to those in Norway, facts not too peculiar considering they were derived from 'Norway' or the same area during the Caledonian orogenesis.

In the areas labelled C in Figure 7.7, slowly reactive aggregates occur, such as porphyries and other fine grained metavolcanic rocks, along with fine grained granite, quartzite and fine to medium grained metasediments. Cataclasite is also present along east-west faults in the middle of Sweden. The reactivity of some of these cataclasites has recently been discovered to be more deleterious than previously recognised. This has prompted new research in order to evaluate critical amounts of these rock types.

South of Lake Vänern, along the NE-SW trending Mylonite Zone, in the area labelled D in Figure 7.7, medium reactive mylonites and cataclasites are common. These rock types also occur in N-S trending shear zones and faults of the Protogine Zone south of Lake Vättern.

With a high alkali content of the concrete, it may take up to 15 years for the expansion to reach such a level that the cracking becomes evident (except in the Scania area, labelled A). These medium to slow reactive aggregates rarely cause severe damage when the concrete aggregates are made up of glaciofluvial gravel, although as the years go by new cases appear. The glaciofluvial gravel is generally a mixture of different types of rocks, which typically have a 'pessimum' content of 100% (ie higher contents of reactive material cause greater expansion), whereas the fast reactive flint-containing aggregates usually have a pessimum of 5–10 % flint (Lagerblad & Trägårdh, 1992a, 1992b). However, if the medium to slow reactive types of rocks are used as crushed rocks in combination with high alkali cement, then the mix is very likely to become deleterious.

For a long time, Portland cements with low to moderate alkali content (in the range of 0.4–0.85% Na_2Oeq) have been available in Sweden. These were normally used for larger structures, also before the low alkali CEM I (SS EN 197-1: 2011) Portland cement 'Anläggningscement' was introduced in the early 1980s. For example, many concrete dams were built with a CEM I cement containing only 0.4% Na_2Oeq alkalis. Fortunately, this has limited the number of ASR damaged structures. However, in the mid 1970s a high alkali cement was introduced ($Na_2O_{eq.}$ ~1.2 %), owing to a change in the production process from the so-called 'wet method' to the 'dry method', which resulted in some ASR effects on structures where alkali-reactive aggregates were used. In the late 1990s, a limestone blended CEM II A-LL (SS EN 197-1: 2011) cement was introduced on the Swedish market which decreased the alkali contents in house building concretes. In 2013, a new composite cement was introduced, containing about 17 % fly ash, with the objective of reducing the risk for future ASR damage in Sweden.

In some cases, pozzolanic materials or GGBS were used in concrete dams to limit the heat development, which also have contributed to the prevention of ASR. In dam structures where other binders were used, severe damage is frequently exhibited. After 50–60 years in service, many of these dams are reaching the end of their service life.

7.5.3 Special features of ASR

Severe ASR damage in Sweden has been found in several concrete dams and bridges, for example on edge beams on bridges, which have been exposed to aggravating de-icing salts. ASR damage has also been detected on structures such as balconies, swimming pool complexes and in parking decks. The cements used in the latter structures are mainly CEM I with about 1.1% Na_2Oeq alkalis or CEM II/A-L with about 0.9% Na_2Oeq alkalis.

A particular type of damage, 'pop-outs' on concrete floors in the Scania area, was detected already in 1975, as reported by Nilsson and Peterson (1983): *"The problems with alkali-silica reactions coincide with the introduction of a new cement in the south of Scania. The earlier used cement, however, was a low alkali cement and no alkali-silica reactions where known before. The problems have only been reported on 'moist indoor structures' that have a certain history of moisture changes. The most frequently affected structure is concrete slabs on the ground. More than 9 out of 10 problems with pop-outs due to ASR are to be found on such a structure. Relative humidity of the concrete close to 90%, at different temperatures, has been shown to be 'pessimal', and cause most pop-outs. A test method for pop-outs has been developed using the 'moisture history' as an aid to reproduce the damage in the laboratory. The effect of some inhibitors has been studied in the test method, and has produced promising results."*

7.5.4 Preventative and mitigating measures – Swedish approach

Historically, different specifications regarding ASR have been available in Sweden:

- In recommendations for certification of pre-mixed, dry mortar containing flint in the sand fraction, it was stated in 1992 that the mortar should be initially type tested for alkali reactivity (to avoid pop-outs). The specified method was NT BUILD 295 (NORDTEST, 1985) in a slightly modified version (Lagerblad & Trägårdh, 1992b).
- In 1998, the Swedish Concrete Association issued the second edition of a report called 'Durable Concrete Structures' (Swedish Concrete Association, 1998). In this report, it was stated that concrete exposed to moisture should be made with cement containing less than 0.6 % alkali oxides. In addition, 'Aggregates suspected of containing alkali- reactive particles should be investigated as regards its reactivity'. The recommended test method for sand 0/8 mm with glaciofluvial origin was NT Build 295 (NORDTEST, 1985) that prescribes submersion in a saturated NaCl solution at 50°C. For the purpose of evaluating aggregates with D_{max} 16 mm, the NT Build 295 was also modified to test concrete prisms (Lagerblad & Trägårdh, 1992b). For crushed rocks, both the Canadian concrete prism test (38°C), CSA

A.23.2 14A (now CSA, 2014) and ASTM C1260: (now 2010) accelerated mortar-bar test (80°C, 1N NaOH) were recommended. For rock types like sandstone and phyllite, the ASTM C1260: (2010) was also the recommended test method.

- One 'specification' concerning ASR is included in the Swedish Concrete Regulations ('BBK 04': 2004). It only states that the aggregate may not contain deleterious components in such amounts that the quality and performance of the concrete or reinforcement is decreased. However, there is no definition of acceptable amounts.
- In previous 'steering/governing documents' by the Swedish Road Association, e.g., Swedish Bridge Code of 2004 ('Bro 2004') it was mentioned that aggregates in bridges in contact with sea water should not contain ASR aggregates. The previous Swedish Road Code of 2005 (ATB VÄG: 2005) also gave reference to SS-EN 12620: (now 2008) and SS 137003: (now 2015), but today none of the replacing documents 'TRVK Bro' and 'TRVK Väg' say anything about ASR. Instead, the Swedish Road Association, in their procurement instructions, makes regulations/recommendations in SS 137003: (now 2015) to specific demands within TRVAMA.
- In the Construction Product Regulations (FPC) for aggregates (Swedish Concrete and Aggregate Certification Ltd., 2009), it is stated that the producer shall present a petrographic analysis every third year. The petrographic analysis shall be carried out by a petrographer with documented competence in the area. In addition, the report of the analysis shall comment on the suitability of the aggregate for the intended use. The regulations refer to SS-EN 12620: (now 2008) and SS 137003: (now 2015).
- SS-EN 12620: (2008) states that, when required, aggregates shall be assessed in accordance with the provisions valid in the place of use and results declared. It gives some general guidance on precautions necessary to avoid ASR. Regulations regarding testing for ASR states that *"when necessary or in doubt"*, at least a petrographic description should be made every third year.
- Regulations within the new Swedish standard SS EN 137003:2015 "Concrete – Application of SS-EN 206 in Sweden" are stated below.

The primary petrographic method used today is RILEM AAR-1: (2001)[2], which is required for all concrete aggregates according to SS EN 137003 (2015), except those used in dry environments. It is stated that a quantitative petrographic analysis shall be carried out on thin-sections and at least 1000 points per fraction shall be counted (numbers of thin sections depending on the fraction to be analysed). Fluorescent petrography is used by some laboratories, and is crucial for detection of porous flint.

According to the requirements in SS EN 137003: 2015, the alkali-reactivity shall be evaluated if fast reactive aggregates are present or if the amount of slowly reactive or potentially reactive components is >15%. If fast reactive aggregates are present, the alkali-reactivity shall be tested with NT Build 295 (NORDTEST, 1985). If the amount of slowly reactive material is >15% and the aggregate is still to be used, its reactivity shall be tested with either the RILEM AAR-2 (2016)

---

[2] *These RILEM methods AAR-1 (now AAR-1.1), AAR-2 and AAR-3 (now AAR 3.1) have all been updated in the RILEM compilation edited by Nixon & Sims (2016) and differ from the superseded versions.*

method (mortar-bars at 80°C in 1N NaOH) or the RILEM AAR-3 (2000, now 2016)[3] (unwrapped concrete prisms at 38°C). For aggregates to be considered as low reactive, the expansion shall be below 0.25% at 28 days in RILEM AAR-2 and below 0.050 % in RILEM AAR-3. The limits are the same as recommended for ASTM C1260 (ASTM, 2010) and CSA A.23.2 14A (CSA, 2014). However, if the mortar-bars in the RILEM AAR-2 method still expand after 28 days and is slightly below the critical limit, the aggregate is considered reactive.

If a concrete aggregate is considered alkali reactive according to the test methods described above, the aggregate may still be used in a specific job mix, if tested and evaluated according to the CBI method No. 1 (Lagerblad & Trägårdh, 1992b). In Sweden, this modified NT Build 295 method using concrete prisms has been used since the early 1990s in order to evaluate opaline and porous flint in glaciofluvial deposits from Scania.

To avoid ASR in new Swedish concrete structures, CBI recommend to use a so-called 'deemed-to-satisfy' method. According to this procedure, one should select one or more of the following mitigation alternatives:

- **Limit the alkali content in the cement**: a normal requirement for maximum alkali content of CEM I cement is 0.6 % Na_2Oeq. This requirement is linked to the requirement to maximum alkali content of the concrete of 3.0 kg Na_2Oeq. per m^3. If de-icing salts are present, the maximum allowed alkali content is lowered to 2.5 kg Na_2Oeq. per m^3. Indirectly these requirements are put to the maximum allowed cement content in the concrete; maximum 500 kg/m^3 without de-icing salts and maximum 414 kg/m^3 when de-icing salts are present.
- **Limit the content of reactive aggregates**: the amount of slowly reactive material should be below 15%. If higher, a mortar-bar expansion test according to RILEM AAR-2 (2000, now 2016) can be performed. If still not approved, a third step can be to perform a concrete prism test according to RILEM AAR-3 (2000, now 2016). A concrete performance test is also being developed in Sweden (in communication with RILEM TC 258-AAA).
- **Use a blended Portland cement containing fly ash, GGBS or silica fume**: an example is the new blended fly ash cement introduced in 2013 (see 7.5.2).

Most of these CBI recommendations are included in the revised Swedish ASR regulations SS EN 137003 (2015). Finally, for many major construction projects, special requirements are prepared by consultants.

7.5.5 Diagnosis and appraisal of structures

If ASR is a possible cause of damage of a concrete structure, the extent of cracking is mapped and the crack widths are measured. Occasionally, the crack widths are 'summed' to get a measure of the expansion up to date. Subsequently, cores are drilled for laboratory analyses.

CBI normally performs the following laboratory analyses on the drilled cores:

- Visual inspection.
- Preparation of plane polished sections and thin sections.

---

[3] The former version published in 2000 suggested wrapped prisms, while the final version published in 2016 describes use of unwrapped prisms.

- Point-counting in thin sections in order to determine the amount and type of reactive aggregates.
- Observations of cracking in the plane polished sections.
- Analysis of the alkali content of the cement paste by ion chromatography.

Based on the field survey and the laboratory analyses, an estimation of the future expansion potential might be given.

7.5.6 Protection, repair and rehabilitation

Sweden has little experience in repair and rehabilitation of structures affected by ASR. The common approach in severe cases has been to remove and replace the concrete cover. In some cases, for example bridge edge beams, the entire construction member has been replaced, where ASR was only one of many influencing factors. Currently the efficiency of silanes in mitigating ASR is being studied in a case object in Stockholm. A severely damaged structure has been treated with silanes and RH is monitored.

It is well known that changing the internal moisture state in the concrete can affect several processes including ASR. According to the literature, the critical internal humidity in concrete required to initiate ASR is in the range 80–85%. A water repellent treatment (for example silanes) might change the internal humidity, but it is important to know the source of the moisture before the water repellent agent is applied in order to predict the outcome. As shown by Johansson et al. (2008), different situations can affect the outcome of the treatment. The example illustrates the difference between the north and the south side of the building, which can be as much as 15–20% RH. On the south side, a dry-out effect is seen.

However, the risk for entrapment of water is also present. As reported by Sandin (1998) and Wittmann (1998), the drying is slowed down by a hydrophobic treatment, which means that with a moisture source behind the hydrophobic layer, the humidity level can increase. The importance of knowing the source of the moisture is also pointed out by Verhoef (1998).

7.6 FINLAND

7.6.1 Brief history of national research

Within the Finnish building industry, it is widely believed that there is no risk of ASR in Finland. This results from the belief that the commonly used aggregates originate from high-quality, non-reactive granitic rock and there have not been documented cases of ASR in the past. ASR-related information has traditionally been omitted from the academic curriculum, so that Finnish engineers and concrete practitioners have been less aware of the potential issue. In addition, the visual manifestations of ASR degradation mechanisms can be very similar to those of freeze-thaw attack, which is a very common form of concrete degradation in Finland, and thus ASR may historically have been erroneously undiagnosed (Pyy et al., 2012a) or obscured by freeze-thaw evidence.

Historically there have been no well-publicised cases to contradict the general belief that ASR does not occur in Finland, however, several cases that have been identified provide evidence of its occurrence. There is no understanding yet of the extent of potentially damaged infrastructure at a national level. Much of the infrastructure in Finland has been built during and after the 1970s, and these structures are now coming to the age where ASR has been found. Pyy *et al.* (2012a) suspect that a mixture of extreme climate and favourable geology have helped keep ASR undetected for so long. The ever aging infrastructure is beginning to show more and more signs, and there is the possibility that ASR will become a problem in the near future because of changing materials, such as binders with higher alkalis and the use of alternative aggregate sources, but also because of the increased awareness of this problem.

The first study addressing ASR in Finland was undertaken in 2011, with the main objectives to identify the locations and extent of occurrence of ASR in Finnish concrete structures, and identify the needs proactively to address ASR in Finland in the near future (Pyy *et al.*, 2012b). The study was based on the results of a survey conducted among the Finnish laboratories with thin section microscopy analysis capabilities. The study revealed over 50 cases of concrete structures with obvious signs of ASR confirmed over the previous 15 years. These structures have come from a wide range of geographic areas of Finland and from various types of structures. An additional 20 or more cases have been documented since the initial study, and the increasing number of annual occurrences is expected to increase as Finland's ASR understanding improves and infrastructure ages.

7.6.2 Aggregates, binders and admixtures

Finland is known to have strong, non-reactive aggregates. The bedrock is mainly made of Precambrian plutonic and metamorphic rocks. The occurrence of sedimentary rocks is marginal and these rocks are also geologically old. The metamorphic rocks include different metavolcanics, gneisses, schists and quartzites, and they are often highly metamorphosed and folded. The bedrock is covered by soil from the latest ice age, about 10 000 years ago, so there is a wide time gap between these two.

The soil well represents the composition of the underlying bedrock. Granites and granodiorites are the most common rock types in the bedrock and soil. These rock types cover about 60% of the average composition of a sporadic aggregate. The rest is mainly made of gneisses and schists. In Eastern Finland, quartzites play a more significant role.

From a geological point-of-view, there is very much in common between Finland and most of Sweden. In both countries, the bedrock is made mainly of Precambrian rocks (see the map in Figure 7.8). This is, in itself, quite an important aspect, because Sweden acknowledged the existence of AAR more than 20 years ago (see 7.5) and has addressed this problem accordingly. From the map it can be seen that the bedrock in both most of Sweden and Finland represents the same Precambrian association made of metamorphic, cataclastic and mylonitic schists and gneisses, cut by plutonic rocks like granite and gabbro.

Several types of ASR-suspect Finnish rock types have been studied and presented in the RILEM Petrographic Atlas (AAR-1.2, Fernandes *et al.*, 2016). As an

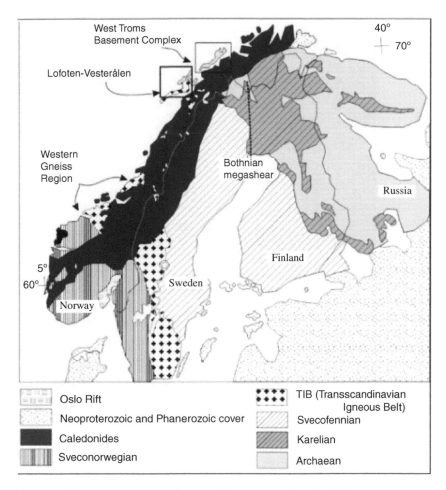

Figure 7.8 A simplified geological bedrock map of Fennoscandia (Bergh, 2012).

example of such studied rocks, Figure 7.9 presents a thin section image of a gneiss composed of quartz, plagioclase, K-feldspar, muscovite, biotite, chlorite, sericite and rutile. The rock exhibits gneissic texture in which biotite-rich domains are segregated out of the quartz-rich domains. The phases suspected of causing reactivity are the stretched quartz with undulatory extinction and inclusions trails. Pyy *et al.* (2012b) suspect that, in Finland, ASR is more connected to certain rock types than to certain geographical areas, though this should be further clarified in the future.

Finnsementti Oy is the only producer of cement in Finland, with operations starting 100 years ago and production of approximately 2 million tons annually. 40% of the buildings in Finland are made of concrete, and approximately 80% of the cement used is produced by Finnsementti Oy. A typical composition of the Finnish clinkers is given in Table 7.6 (Finnsementti, 2012).

Figure 7.9 Thin section image of a Finnish gneiss composed of quartz, plagioclase, K-feldspar, muscovite, biotite, chlorite, sericite and rutile; strained quartz textures are thought likely to be alkali-reactive (Fernandes *et al.*, 2016).

Table 7.6 Finnsementti Oy clinker composition, as percentages (Finnsementti, 2012).

	Normal	Sulphate-resisting	White
CaO	64	63	69
SiO_2	21	21	24
Al_2O_3	5.2	3.0	2.1
Fe_2O_3	3.0	4.0	0.4
MgO	2.7	2.6	0.6
$K_2O + Na_2O$	1.5	1.0	0.2
Other	2.6	5.4	3.7
Total (%)	100	100	100

On average over the past few years, the annual amount of cement used in Finland is approximately (using EN 197-1 designations):

- 45% Yleis cement (CEM II/A-M (S-LL) 42,5 N),
- 35% Rapid cement (CEM II/A-LL 42,5 R),
- 10% Pika cement (CEM I 52.5R),
- 5% Plus cement (CEM II B-M (S-LL) 42,5 N),
- 2% Sulfate-resisting cement (CEM I 42,5 N) and
- 2% Aalborg White cement (CEM I 52,5 R) (Finnsementti, 2012).

All of these, except the Danish-made White cement, are based on Finnish Normal or Sulfate-resisting clinker, where the equivalent alkali contents are over the traditionally recognised level of 0.6% to avoid AAR.

Nowadays, the Finnish cements are typically blended with limestone and secondary cementitious materials, such as fly ash and blastfurnace slag. Historically, large infrastructures such as power plants and dams have been constructed using high amounts of blastfurnace slag, *i.e.*, as 70% of the binder content. Chemically, most historic building constructions were produced with cements of type CEM I, corresponding to 'Rapid cement' from 'Normal' clinker in Table 7.6. With regards to ASR occurrence, Finnish concrete practice has probably benefited from the historic widespread use of blastfurnace slag and fly ash in concrete mixtures. This could partly explain why ASR damage has not been widely detected.

For decades it has also been common practice in Finland to use air entraining admixtures (AEA) in almost all outdoor exposed concretes. From the 1960s, AEA was used in structures in contact with water, and from the 1970s in housing and infrastructures. According to some scientists, the entrained air voids provide space for the gel formation and thus may also contribute to reducing ASR-induced cracking.

7.6.3 Special features of ASR

In the 2011 study mentioned previously, the survey documented a total of 56 cases of obvious ASR damage in Finnish concrete structures that have been investigated over the past 15 years (Pyy *et al.*, 2012b). An obvious ASR case meant that gel could be detected in the cracks and/or aggregates. All of these confirmed ASR cases were listed in confidential reports, and thus the full disclosure of information is not provided here.

The types of structures where the ASR was reported are shown in Figure 7.10, where the majority of cases were in bridges (41%) and houses (34%). There was no

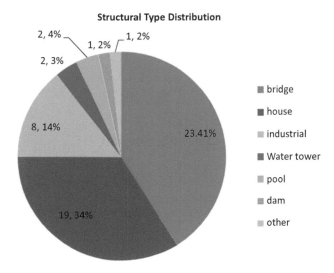

Figure 7.10 Types of concrete structures where ASR has been found based on survey results (note labelling: number of structures, percentage) (Pyy *et al.*, 2012b).

geographical concentration of where the ASR affected structures were located. The aggregate origin and concrete type was not studied in detail for these identified cases, though it should be the focus of future investigations.

7.6.4 Preventative and mitigating measures – Finnish approach

Finnish geological experts should give a statement about the suitability of the aggregates for concrete. ASR should be one of the issues considered, yet there are no Finnish guidelines about how aggregate reactivity should be evaluated. Overall, there are generally no comments on ASR risks when evaluating aggregate performance.

Finland does not have any ASR tests for either aggregate, binder, concrete or structures specified in any codes, standards or guidelines. Normal Finnish building practice also does not refer to other international ASR guides. It is expected that within the next few years these types of test guidelines and practices will be improved based on recent ASR findings and better awareness.

7.6.5 Diagnosis and appraisal of structures

Condition assessments of Finnish structures are typically a well-documented and thorough process, though they are not tailored for addressing ASR. For instance, the Finnish Road Administration has maintained a detailed national bridge registry for decades. Cracking extent is a common assessment, though the cause of damage may not always be clarified. Sampling cores are often taken from the surface, only extending a few centimetres. It is suspected that ASR cracking damage may be misdiagnosed, for instance as attributed to freeze-thaw damage. There are no specific ASR field diagnosis methods in Finland.

There are no established ASR laboratory test methods specified or standardized in Finland. It is common practice to use thin section microscopy for concrete assessments, yet most Finnish petrographers are not trained in ASR detection.

As the number of recognized cases is still limited, Finland does not yet have established monitoring systems for following the progress of ASR.

7.6.6 Protection, repair and rehabilitation

Defining structural damage attributed to either ASR or freeze-thaw attack has not been separately well defined in Finland. The protection, repair and rehabilitation methods have been the same for both of these deterioration types.

In some cases of ASR affected bridges, waterproofing has been used to try to stop the damage from progressing. In a couple of cases, the solution has been to remove the affected cracked area and replace the concrete. Figure 7.11 gives examples of typical visual appearances of Finnish bridges that have been affected by ASR. Their repair methods have been focused on applying waterproofing.

Figure 7.11 Examples of four Finnish bridges affected by ASR (*photos courtesy of Huura Oy*). Views of the (a) underside and (b) edge beam of Mäntylä Bridge near Turku, where edge beams needed to be demolished and recast. (c) Underside of an overpass near Joensuu where extensive cracking and leaching is visible. (d) Bridge on Paltanmäki with ASR penetrating through deck slab (unrepaired).

REFERENCES

Appelquist, K., Döse, M., Göransson, M., & Trägårdh, J. (2014) Petrographic examination of mortar bars of Swedish aggregates exposed to RILEM AAR2. In: Lollino, G., Manconi, A., Guzzetti, F., Culshaw, M., Bobrowsky, P., & Luino, F. (eds.) *Engineering Geology for Society and Territory – Volume 5: Urban Geology, Sustainable Planning and Landscape Exploitation*, Springer, pp. 97–100.

ASTM C1260–14 (2014) *Standard Test Method for Potential Alkali Reactivity of Aggregates (Mortar-Bar Method)*. ASTM International, West Conshohocken, PA, USA.

ASTM C1293-08b (2008) *Standard Test Method for Determination of Length Change of Concrete Due to Alkali-Silica Reaction*, ASTM International, West Conshohocken, PA, USA.

ASTM C227-10 (2010) *Standard Test Method for Potential Alkali Reactivity of Cement-Aggregate Combinations (Mortar-Bar Method)*, ASTM International, West Conshohocken, PA, USA.

ASTM C289-07 (2007) *Standard Test Method for Potential Alkali-Silica Reactivity of Aggregates (Chemical Method)*, ASTM International, West Conshohocken, PA, USA.

BBK 04. (2004) *Swedish Concrete Regulations* (in Swedish).

Bergh, S.G. *et al.* (2012) Was the Precambrian basement of Western Troms and Lofoten-Vesterålen in Northern Norway linked to the Lewisian of Scotland? A comparison of crustal components, tectonic evolution and amalgamation history. In: Sharkov, E. (ed.) *Tectonics – Recent Advances*, ISBN: 978-953-51-0675-3, In Tech. July 18, 2012. doi: 10.5772/48257.

Brandt, I., Schouenborg, B.E. (1990) Guidelines for petrographical micro analysis of aggregates for concrete. Nordtest project no. 826 89, Danish Technological Institute, 110p.

Brandt, I., & Schouenborg, B.E. (1990) *Guidelines for Petrographical Micro Analysis of* BRO 2004: *Bridge Code* (in Swedish).

Byggestyrelsen (1987) *Basisbetonbeskrivelsen for Bygningskonstruktioner "BBB"* (*Requirements for aggregates and maximum allowed alkali content*, in Danish).

CSA (2004) *CSA A23.2-14A-00, Potential Expansivity of Aggregates (Procedure for Length Change due to Alkali–Aggregate Reaction in Concrete Prisms at 38 °C), Methods of Testing for Concrete*, Mississauga, Ontario, Canada, Canadian Standards Association. pp. 246–256.

Dahl, P.A., Lindgård, J., Danielsen, S.W., Hagby, C., Kompen, R., Pedersen, B., & Rønning, T.F. (2004) Specifications and guidelines for production of AAR resistant concrete in Norway. In: Tang, M., & Deng, M. (eds.) *Proceedings of the 12th International Conference on AAR in Concrete*, October 2004, Beijing (China). pp. 499–504.

Danish Standard (1999) DS481, *Concrete – Materials*.

Danish Standard (2002) DS/EN 206-1 (05-09-2002), *Concrete – Part 1: Specification, performance, production and conformity*.

Danish Standard (2004) DS/EN 12620 (02-04-2004), *Aggregates for concrete*.

Danish Standard (2004) DS 2426 (10-05-2004), *Concrete – materials – Rules for application of EN206-1 in Denmark*. European Standard EN 206-1 (2013) *Concrete – Part 1: Specification, performance, production and conformity*.

Danish Technological Institute (1989) *Chemical shrinkage*, TK 84, DTI, Copenhagen. (in Danish).

Fernandes, I., Ribeiro, M.d.A., Broekmans, M.A.T.M. & Sims, I. (eds.) (2016) *Petrographic Atlas: Characterisation of Aggregates Regarding Potential Reactivity to Alkalis, RILEM TC 219-ACS Recommended Guidance AAR-1.2, for use with the RILEM AAR-1.1 Petrographic Examination Method*, Springer, Dordrecht. pp RILEM, Paris, 205pp.

Finnsementti Oy. (2012) *Suomalainen Sementti-Opas* (in Finnish) (*Finnish Cement Guide*). [Online] Available from: http://www.finnsementti.fi, January 2012.

Guðmundsson, G. (1971) *Alkali Efnabreytingar í Steinsteypu* (in Icelandic). The Icelandic Building Research Institute.

Guðmundsson, G. (1975) *Investigation on Icelandic Pozzolans*, Symposium on AAR-Preventive Measures. The Icelandic Building Research Institute.

Guðmundsson, G. & Ásgeirsson, H. (1975) Some Investigation on Alkali Aggregate Reaction, *Cement Concrete Res.*, 5, 211–220.

Guðmundsson, G. & Sveinsdóttir, E.L. (1994) *The condition and characteristics of concrete in Reykjavik – thin section analyses on drilled cores* (in Icelandic). The Icelandic Building Research Institute.

Helgason, Þ.S. (1982) *Islandske vulkanske bjergarter og alkalikiselreaktioner i beton*. Paper presented at the Winter Meeting of Nordic Geologists. Rannsóknastofnun Byggingariðnaðarins (in Danish).

Helgason, Þ.S. (1990) *Characteristics, properties and quality rating of Icelandic volcanic aggregates*. Rannsóknastofnun Byggingariðnaðarins. Report number: 90-08, 8p.

Idorn, G.M. (1967) Durability of concrete structures in Denmark: A study of field behaviour and microscopic features. Teknisk Forlag, Copenhagen, 208.

Ingeniøren (2012), no. 49, 2 (in Danish). Available from: http://ing.dk/mediehuset.

Jensen, V. (1993) *Alkali Aggregate Reactions in Southern Norway*, Doctor Technical Thesis, The Norwegian Institute of Technology, University of Trondheim, 262p + Appendices.

Jensen, V. (2000) In-situ measurements of relative humidity and expansion of cracks in structures damaged by ASR. In: Bérubé, M.A., Fournier, B. & Durand, B. (eds.), *Proceedings of the 11th International Conference on Alkali-Aggregate Reactions in Concrete*, June 11–16, 2000, Quebec, Canada.

Jensen, V. (2004) Measurement of cracks, relative humudity and effects of surface treatment on concrete structures damaged by Alkali Silica Reaction. In: Tang, M., & Deng, M. (eds.), *Proceedings of the 12th International Conference on Alkali-Aggregate reaction in Concrete*, October 15–19, 2004, Vol. II. International Academic Publishers – World Publishing Corporation, pp. 1245–1253.

Jensen, V. (2006) Surface protection on ASR damaged concrete. Research and state-of-the-art from a postdoctoral project. In: Fournier, B. (ed.), *Proceedings of the Eighth CANMET/ACI International Conference on Recent Advances in Concrete Technology/Marc-André Bérubé Symposium on Alkali-Aggregate Reactivity in Concrete*, May 31–June 3, 2006, Montréal, Québec, Canada. pp. 335–355.

Jensen, V. & Merz, C. (2008) Alkali-aggregate reaction in Norway and Switzerland. Survey investigations and structural damage. In: Broekmans, M.A.T.M., & Wigum, B.J. (eds.), *Proceedings of the 13th International Conference on Alkali-Aggregate Reactions in Concrete*, June 16–20, 2008, Trondheim, Norway. pp. 785–795.

Johansson, A., Nyman, B. & Silfwerbrand, J. (2008) Decreasing humidity in concrete facades after water repellent treatment. In: De Clercq, H. (ed.), Proceedings, Hydrophobe V, Fifth International Conference on Water Repellent Treatment of Building Materials, Royal Institute for Cultural Heritage (KIK-IRPA), Aedificatio Publishers, Brussels, Belgium, April 15–16, 2008, Belgium. pp. 379–386.

Katayama, T., Helgason, T.S., & Olafsson, H. (1996) Petrography and alkali-reactivity of some volcanic aggregates from Iceland. In: Shayan, A. (ed.), *Proceedings of the 10th International Conference on Alkali-Aggregate Reaction in Concrete*, August 1996, Melbourne, Australia. pp. 377–384.

Kjennerud, A. (1978) *Alkaligrusreaksjoner påvist i Norge: Skadene mer vanlig enn antatt?* (in Norwegian), (*Alkali-Aggregate reactions documented in Norway: Are the damages more common than assumed?*). Norwegian Building Research Institute (NBI), offprint 259, p. 3.

Kristjansson, R. (1979) *Steypuskemmdir – Ástandskönnun. Keflavík* (in Icelandic). Rannsóknastofnun Byggingariðnaðarins. Report number: 2.

Lagerblad, B. & Trägårdh, J. (1992a) Slowly reacting aggregates in Sweden - Mechanism and conditions for reactivity in concrete. In: Poole, A.B. (ed.), *Proceedings 9th International Conference on Alkali-Aggregate Reaction in Concrete*, London. Concrete Society Publication CS. 104p.

Lagerblad, B. & Trägårdh, J. (1992b) *Alkalisilikareaktioner i Svensk Betong* (in Swedish) (*Alkali-silica reactions in Swedish concrete*). CBI report 4:92. ISSN 0346-8240. 74 pp.

Larsen, S., Lindgård, J., Thorenfeldt, E., Rodum, E. & Haugen, M. (2008) Experiences from extensive condition survey and FEM-analyses of two Norwegian concrete dams with ASR. In: Broekmans, M.A.T.M. & Wigum, B.J. (eds.), *Proceedings of the 13th International Conference on Alkali-Aggregate Reactions in Concrete*, June 16-20 2008, Trondheim, Norway. pp. 914–923.

LCPC. (1997) *LCPC Techniques et méthodes Alcali-réaction du béton. Essai d'expansion résiduelle sur béton durci. Techniques et méthodes des LPC*. Project de Methode d'essai LPC No. 44. Ministère de l'Équipement, du Logement, des Transports et du Tourisme, p. 11.

Lindgård J., Dahl P.A., & Jensen, V. (1993) *Rock composition – reactive aggregates: Test methods and requirements to laboratories* (in Norwegian), SINTEF report no. STF70 A93030, Trondheim, 9 pp.

Lindgård, J. 2013a: *Alkali-silica reaction (ASR) – Performance testing*, Doctoral theses at NTNU, 2013-269, October 2013. Available from: http://ntnu.diva-portal.org/smash/get/diva2:665704/FULLTEXT01.pdf.

Lindgård, J. et al. (2010) The EU "PARTNER" Project - European standard tests to prevent alkali reactions in aggregates: Final results and recommendations. *Cement Concrete Res., 40* (4), 611–635.

Lindgård, J. et al. (2012) Advantages of using plane polished section analysis as part of microstructural analyses to describe internal cracking due to alkali-silica reactions. In: Drimalas, T. Ideker, J.H., & Fournier, B. (eds.), *Proceedings of 14th International Conference on Alkali-Aggregate Reactions in Concrete*, May 2012, Austin, Texas.

Lindgård, J., Pedersen, B., Bremseth, S.K., Dahl, P.A. & Rønning, T.F. (2010) Experience using the Norwegian 38°C concrete rism test to evaluate the alkali reactivity of aggregates, concrete mixes and binder combinations. *Nordic Concrete Res.*, publication, 2, 42, pp. 31–50.

Lindgård, J., Rodum, E., & Pedersen, B. (2006) Field experience from alkali-silica reactions in concrete – relationship between water content and observed damage on structures. In: Malhotra, V.M. (ed.), *Proceedings of the 7th CANMET/ACI International Conference on Durability of Concrete*, Montreal (Canada).

Lindgård, J., Sellevold, E.J., Thomas, M.D.A., Pedersen, B., Justnes, H., & Rønning, T.F. (2013b) Alkali-silica reaction (ASR) – Performance testing: Influence of specimen pre-treatment, exposure conditions and prism size on concrete porosity, moisture state and transport properties. *Cement Concrete Res.*, *53*, 145–167.

Lindgård, J., Skjølsvold, O., & Haugen, M. (2004) Experience from evaluation of degree of damage in fluorescent impregnated plane polished sections of half-cores based on the "Crack index method." In: Tang, M., & Deng, M. (eds.), *Proceedings of the 12th International Conference on AAR in Concrete*, October 2004, Beijing (China). pp. 939–947.

Lindgård, J., Thomas, M.D.A., Sellevold, E.J., Pedersen, B., Andiç-Çakır, Ö., Justnes, H., & Rønning, T.F. (2013c) Alkali-silica reaction (ASR) – Performance testing: Influence of specimen pre-treatment, exposure conditions and prism size on alkali leaching and prism expansion. *Cement Concrete Res.*, *53*, 68–90.

Nerenst, P. (1952) *Concrete technical studies in USA – Report from ECA Technical travel November 14th 1950 to February 15th 1951* (in Danish). the Danish National Institute for Building Research, Copenhagen, Series SBI Study No. 7.

Nielsen, H.O., Grelk, B., & Nymand, K.K. (2004) ASR – Alkali-Silica Reactions, Dansk Vejtidsskrift no. 2 (2004), (in Danish) 10–13.

Nilsson, L.O., & Peterson, O. (1983) *Alkali-silica reactions in Scania, Sweden – a moisture problem causing pop-outs in concrete floors*. Division of Building Materials, Lund Institute of Technology. Report TVBM-3014.

Nixon, P.J. & Sims, I. (2016) *RILEM Recommendations for the Prevention of Damage by Alkali-Aggregate Reactions in New Concrete Structures, State-of-the-Art Report of the RILEM Technical Committee 219-ACS*, Volume 17, Springer (Dordrecht) pp RILEM (Paris), 184pp.

NORDTEST NT Build 295. (1985) *Sand-alkali-silica Reactivity Accelerated Test*. Finland, Nordtest.

Norwegian Concrete Association. (1996) *Durable concrete containing alkali reactive aggregates (NB21)* (in Norwegian), 5+27 pages including appendices.

Norwegian Concrete Association. (2004) *Durable concrete containing alkali reactive aggregates (NB21)*, 22+12 pages including appendices (English version available in 2008).

Norwegian Concrete Association. (2005) *Alkali aggregate reactions in concrete. Test methods and requirements to laboratories (NB32)*, 13+26 pages including appendices (English version available in 2010).

Olafsson, H. (1983) Repair of vulnerable concrete. In: Idorn, G.M., & Rostam, S. (eds.), *Proceedings of the 6th International Conference on Alkalis in Concrete*, June 1982, Copenhagen, Denmark.

Ólafsson, H. (1986) The effect of relative humidity and temperature on alkali expansion of mortar bars. In: Grattan-Bellew, P.E. (ed.), *Proceedings of the 7th International conference Concrete Alkali-Aggregate Reactions*. Noyes Publications, Far Ridge, NJ, pp. 461–465.

Ólafsson, H. (1988) Hydrophobing agents for protecting of low quality concrete, *Nordic Concrete Research Publication 7*, 1988, The Nordic Concrete Federation.

Ólafsson, H. & Gestsson, J. (1988) *Áhrif vatnsfæla á steinsteypu* (In Icelandic). (*Effect of hydrophobic materials on concrete*). The Icelandic Building Research Institute. Reykjavik, Iceland.

Ólafsson, H. & Iversen, K. (1982) *Viðgerðir á alkalískemmdum á steinsteypu* (in Icelandic). (*Repair due to ASR*). The Icelandic Building Research Institute. Reykjavik, Iceland.

Ólafsson, H., Iversen, K., & Ankerfeldt, P. (1988) *Udvendig Facadeisolering – Demonstration i Island* (in Danish). The Icelandic Building Research Institute. Reykjavik, Iceland.

Plum, N.M. (1961) *Alkaliutvalgets Vejledning. Foreløpig Vejledning i Forebyggelse af Skadelige Alkali-Kiselreaktioner i Beton* (in Danish). The Danish National Institute for Building Research, Copenhagen, Series SBI.

Pyy, H., Ferreira, R.M., & Holt, E. (2012a) Assessing the extent of AAR in Finland. In: Drimalas, T., Ideker, J.H., & Fournier, B. (eds.), *Proceedings of14th International Conference on Alkali-Aggregate Reactions in Concrete*, May 2012, Austin, Texas. 7p. (in CD-ROM).

Pyy, H., Holt, E., & Ferreira, M. (2012b) *An initial survey on the occurrence of alkali aggregate reaction in Finland*, VTT Technical Research Centre. VTT-CR-00554-12.

Relling, R.H., & Sellevold, E.J. (2006) In situ moisture state of coastal bridges, In: Alexander, M.G. (ed.), *Proceedings of the International conference: Concrete, Repair, Rehabilitation and Retrofitting, Cape Town, South Africa, 2006*, Taylor & Francis, London, ISBN: 0415396549.

RILEM TC 106-AAR (2000) 'Alkali aggregate reaction' A. TC 106-2 – Detection of potential alkali-reactivity of aggregates – The ultra-accelerated mortar-bar test B. TC 106-3-Detection of potential alkali-reactivity of aggregates-method for aggregate combinations using concrete prisms. *Mater Struct.*, *33*, 283–293.

RILEM TC 191-ARP (2003) RILEM Recommended Test method AAR-1: Detection of potential alkali-reactivity of aggregates – Petrographic method. *Mater Struct.*, *36*, 480–496.

RILEM TC 219-ACS (2006) 'Alkali–silica reactions in concrete structures': RILEM AAR-4.1— Detection of potential alkali-reactivity of aggregates: accelerated (60 °C) concrete prism test. (unpublished draft).

RILEM TC 219-ACS (2011) 'Alkali-silica reactions in Concrete Structures': RILEM recommended test method: RILEM AAR-3 - Detection of potential alkali-reactivity - 38°C test method for aggregate. (Unpublished draft). (*Comment: In the previous version published in 2000, wrapped prisms were used. In this version revised by RILEM TC 219-ACS in 2011, the exposure conditions are similar to ASTM C 1293 (ASTM, 2008), i.e. using unwrapped prisms*).

Rodum, E., Lindgård, J., Skjølsvold, O., Sellevold, E.J., & Gundersen, H. (2001) Repair of frost damaged concrete foundations exposed to severe climate. In: Banthia, N., Sakai, K., Gjørv, O.E. (eds.), *Proceedings of CONSEC*, June 2001, Vancouver, Canada, pp. 2079–2087.

Sæmundsson, K. (1975) Geological Prospecting for Pozzolanic Materials in Iceland, Symposium on AAR-Preventive Measures. The Icelandic Building Research Institute, Reykjavik.

Sandin, K. (1998) Surface treatment with water repellent agents. In: *Proceedings, Surface Treatment of Building Materials with Water Repellent Agents*, November 9–10, Delft, Netherlands. Aedificatio Publishers, Paper 18.

Stephens, M.B., Wahlgren, C.H., & Weihed, P. (1994) Karta över Sveriges berggrund (Bedrock map of Sweden). Sveriges Geologiska Undersökning (The Geological Survey of Sweden).

Swamy, R.N. (ed.) (1991) *The Alkali Silica Reaction in Concretes*, chapter 6 Alkali silica reaction Danish experience.

Swedish Concrete and Aggregate Certification Ltd. (2005) *Certification rules for aggregates*, doc. no. CB7-CE.

Swedish Concrete Association (1998) *Durable Concrete Structures* (in Swedish). Concrete Report No. 1 (edition 2), ISBN 91-971755-9-5, Stockholm.

Swedish Standard SS 137003 (2015) *Concrete – Application of EN 206-1 in Sweden.*

Swedish Standard SS-EN-12620 (2008) *Aggregates for concrete.*

Test Method TI B51 (1985) *Alkali Silica Reactivity of Sand* (in Danish). Technological Institute, Copenhagen.

Test Method TI B52 (1985) *Petrographical Investigation of Sand* (in Danish). Technological Institute, Copenhagen.

Thaulow, N. (1976) *Undersögelse af Beton Borekærne fra Reykjavík* (in Danish). Aalborg Portland.

Thordarson, B. (1991) *ASR Reactivity in Concrete – a Field Survey* (in Icelandic). The Icelandic Building Research Institute.

Trägårdh, J., & Lagerblad, B. (1996) Influence of ASR expansion on the frost resistance of concrete. In: Shayan, A. (ed.), *Proceedings of the 10th International Conference on Alkali-Aggregate Reaction in Concrete*, August 1996, Melbourne, Australia.

Verhoef, L.G.W. (1998) Analysis of Facades. In: *Proceedings, Surface Treatment of Building Materials with Water Repellent Agents*, November 9–10, Delft, Netherlands. Aedificatio Publishers, Paper 20.

Wigum, B.J. (ed.) (1999) *NORMIN-2000, Alkali Aggregate Reaction in Concrete – Petrographic Atlas* (in Norwegian). Oslo, 14p. (English version available at www.this.is/ergo/efarin/bergartsliste.htm).

Wigum, B.J. (2006) Alkali Aggregate Reactions (AAR) in Concrete. Testing, Mitigation & Recommendations. The Norwegian approach during 15 years of research. In: Fournier, B. (eds.), *Proceedings of the 8th CANMET International Conference on recent Advances in Concrete Technology/Marc-André Bérubé Symposium on Alkali-Aggregate Reactivity in Concrete*, May 2006, Montréal, Canada. pp. 111–128.

Wigum, B.J. (2010) *Petrographic examination of state of concrete.* Report from Mannvit to the Icelandic Housing Financing Fund.

Wigum, B.J. (2012) Assessment and development of performance tests for alkali aggregate reaction in Iceland. In: Drimalas, T., Ideker, J.H., & Fournier, B. (eds.), *Proceedings, 14th International Conference on Alkali-Aggregate Reactions in Concrete*, May 2012, Austin, Texas.

Wigum, B.J., Bjarnason, E., & Hólmgeirsdóttir, Þ. (2009) *Steypuskemmdir í húsumyngri en 20 ára* (in Icelandic). (*Concrete deterioration in houses younger than 20 years old*). Report from Mannvit to the Icelandic Housing Financing Fund.

Wigum, B.J. & Einarsson, G.J. (2016) Alkali-aggregate reaction in Iceland. Results from laboratory testing compared to field exposure site. In: Bernandes, H.d.M and Hasparyk, N.P. (ed.), *Proceedings, 15th International Conference of Alkali-Aggregate Reaction in Concrete*, Sao Paolo, Brazil.

Wigum, B.J., Guðmundsson, G., Loftsson, M., Sveinbjörnsson, S., & Harðarson, B.A. (2005) *Ástandskönnun sprautusteypu. Samantekt, lokaskýrsla* (in Icelandic). (*Field conditions of sprayed concrete. Summary, final report*). VGK-Hönnun Report.

Wigum, B.J., Haugen, M., Skjølsvold, O., & Lindgård, J. (2004) Norwegian petrographic method – development and experiences during a decade of service. In: Tang, M., & Deng, M. (eds.), *Proceedings of the 12th International Conference on AAR in Concrete*, October 2004, Beijing (China). pp. 444–452.

Wigum, B.J., & Lindgård, J. (2008) AAR: testing, mitigation & recommendation. The Norwegian approach during two decades of research. In: Broekmans, M.A.T.M., & Wigum, B.J. (eds.), *Proceedings of the 13th International Conference on Alkali-Aggregate Reactions in Concrete*, June 16–20, 2008, Trondheim, Norway. pp. 1322–1333.

Wigum, B.J., Pedersen, L.T., Grelk, B., & Lindgård, J. (2006) State-of-the-art report: Key parameters influencing the alkali aggregate reaction. PARTNER Report 2.1. SINTEF report no. SBF52 A06018, 55p.

Wittmann, F.H. (1998) Influence of a water repellent treatment on drying of concrete. In: *Proceedings, Surface Treatment of Building Materials with Water Repellent Agents*, November 9–10, Delft, Netherlands. Aedificatio Publishers, Paper 31.

Chapter 8

Mainland Europe, Turkey and Cyprus

*Isabel Fernandes, Özge Andiç-Çakir, Colin Giebson &
Katrin Seyfarth*

8.1 HISTORICAL INTRODUCTION AND CASE STUDIES

The present chapter aims at providing an overview on the situation in the countries here described as 'Mainland Europe', with the enlargement of the definition to include all of Turkey and Cyprus, in what concerns alkali-silica reactions, their development, prevention, standards and measures for mitigation. The Nordic European countries are similarly covered in Chapter 7 and the UK and Ireland in Chapter 6.

This group includes a wide range of different countries and the heterogeneities in the amount and quality of information available are enormous. In some countries, such as France, Germany and the Netherlands, a lot of research has been developed, mainly since the 1970s, in consequence of the diagnosis of an increasing number of deteriorated structures due to AAR. This fact is closely related with the variable geology of the respective regions. In Western and Central Europe, the use of alluvial deposits of sands and gravel containing rapidly reactive aggregates seems to be the reason that AAR was first identified as a cause of severe concrete deterioration in these countries, when compared to Southern and Eastern Europe where igneous rocks and limestones are exploited. In addition, the recent history of Europe and the boom in the construction industry that took place on the 1950s and 1960s, including large public works such as bridges, dams and highways, are closely associated with the occurrence and acceptance of AAR mostly from the 1970s to the 1980s.

'Mainland Europe, Turkey and Cyprus' includes countries with quite different geology, raw materials and regulations. When trying to summarise the procedures and standards used in European countries, it can be concluded that although European standards (EN) are well known and accepted, technicians involved in standardisation tend to preserve at least partially the national recommendations which are referred to in notes or annexes in the local versions of the European Standard.

In what concerns the gathering of information, it is important to recognise that this group of countries have different language roots (Celtic, Germanic, Greek, Romance, Slavic, Turkic and Uralic) and in most of them the information available is published in the local language. The case studies are usually reported in confidential documents of/ for governmental institutions or private companies, which dramatically limits access to information.

On a global approach, when focusing on the history of AAR, the countries of Mainland Europe, Turkey and Cyprus accepted the existence of damaged structures due to AAR only quite recently, when compared with the process in North America (from 1940) (Stanton, 1940) or New Zealand (from 1943) (St. John, 1992). In fact, most countries showed some resistance to admitting that AAR was a worldwide problem that should be faced and studied (Broekmans, 2002).

8.2 AGGREGATES, BINDERS AND ADMIXTURES

8.2.1 Aggregates for concrete

The economic growth in any country or region is closely related with the increase of demands on aggregates. Aggregates (crushed stone, sand and gravel) are one of the most important mining products (Menegaki & Kaliampakos, 2010). The aggregates industry differs from most other mining activities, since by their nature, aggregates are high-bulk, low unit value commodities that derive much of their value from being located near the market. Transportation is a major cost and a critical parameter for the market prices. Based on data available from the British Geological Survey, for the years 2001–2005, the main European importers and exporters of aggregates were the Netherlands, Belgium and Germany. The mass balance between imports and exports shows that Germany, the United Kingdom and Norway export much greater quantities of aggregates than they import, while the Netherlands is mainly an importer. The five main producers of aggregates in Europe are Germany, Spain, France, Italy and the United Kingdom, with an annual aggregates production over 200 million tons (Menegaki & Kaliampakos, 2010).

Europe is part of the Eurasian continent and comprises the geological units extending from the Urals and Caucasus mountains, which make the Eastern limit in contact with Asia, to the margins of the Atlantic Ocean in the west, the Mediterranean Sea in the south, and the Barents Sea in the north. The European continent was formed by several major steps of evolution including the Cadomian, Caledonian, Variscan, Cimmerian and Alpine orogenic belts. The formation of sedimentary basins postdating each of these orogens is partly related to the opening of Tethyan and Atlantic oceans, *e.g.*, North and South German, Paris, Aquitan and North Sea basins as well as the Moesian platform. Two main sectors can be identified: the Archean/Early Protorezoic basement with a Middle Proterozoic to Tertiary cover in Eastern Europe; and the Palaeozoic orogens in Central/Western/Southern Europe. These sectors are separated by the Caledonian thrust front and the Tornquist-Teisseyre fault (Neubauer, 2003).

The geology of Europe is varied and complex and produces a wide variety of outcrops and landscapes. Europe's most significant feature is the dichotomy between highland and mountainous regions in Southern Europe and a vast, partially underwater, northern plain ranging from France in the west to the Ural Mountains in the east. Mesozoic and Cenozoic rocks are dominant in Central Europe while Palaeozoic and Precambrian rocks outcrop in limited areas (*e.g.*, geological map 1:5000000, Asch, 2005). Sedimentary rocks occur mainly in the Netherlands, Belgium and France and include sandstones, cherts and various types of limestones, some with a high content of silica (siliceous limestones).

Sedimentary rocks are exploited as aggregates for concrete in all European countries, namely from alluvial deposits of some of the major rivers (*e.g.*, Seine, Rhone, Meuse, Po). The dominant lithology in the origin of AAR in Northern and Central Europe is chert, dense or porous, which is common in France, Belgium, Netherlands and Italy.

Crystalline rocks, such as metamorphic and igneous varieties have been used as aggregates in Austria, Germany, Italy, Portugal, Spain, Switzerland and Turkey. Strained metamorphic rocks such as gneisses and tectonites are common in the mountains of Central Europe (*e.g.*, Switzerland and Alpine Italy) whilst granitic rocks are widely used as aggregates in Spain, Portugal and Germany, and volcanic rocks have been used in Germany, Hungary and Turkey. For some of the crystalline rocks the performance as aggregates is highly influenced by the geological history related mainly to the various episodes of deformation that defined the European geologic structure.

A much simplified summary of the main reactive aggregates identified in the different countries is presented in Table 8.1. In addition, the approximate year (or decade) when AAR was identified and started being studied in each country is shown according to the literature available. In some countries the phenomenon was recognised as a problem long before being published.

The examples found in the literature refer to ASR (alkali-silica reaction) and only seldom (in Poland and Spain) to ACR (alkali-carbonate reaction). Therefore, in the present chapter ASR is considered, just briefly referring to ACR. An up-to-

Table 8.1 Type of aggregates used in the European countries and which proved to be reactive in field structures and/or in laboratory expansion tests.

Country	First study	Potentially reactive aggregates
Austria	1990s	Quartzite and gneiss
Belgium	Early 1984	Mainly porphyries, siliceous limestones and sand with silex
Czech Republic	Late 1990s	Chert-rich limestones, quartz-rich sediments, meta-sediments and other metamorphic rocks (greywacke, meta-greywacke, meta-sediment, quartzite)
France	1980s	Chert, siliceous limestone
Germany	1965	Opaline sandstone, siliceous limestone, chert (flint, radiolarite), greywacke, rhyolite, porphyry, granodiorite, granite, andesite, gneiss
Hungary	?	Andesite
Poland	1981	Dolomite rocks and porous limestone, weathered igneous rocks (melaphyre, porphyry, basalt, tuff)
Italy	1981	Reactive aggregates are mostly from alluvial deposits of the Adriatic seacoast, consisting of siliceous and fossiliferous limestones, with variable amounts of chert, chalcedony and opal. Occasionally, potentially reactive aggregates, containing deformed and/or microcrystalline quartz, come from the Padana Valley.
Netherlands	1991	Porous chert, chalcedony, impure sandstones, greywacke, sericitic sandstone, quartzite, schist/slate
Portugal	1991	Quartzite, granite, sand and gravel from alluvial deposits
Romania	1990s	Quartzite, gneiss, schist, flint and chalcedonic silica, siliceous dolomite
Slovenia	1998	Opal breccia
Spain	1975	Granite, quartzite, granodiorite, dolerite, gneiss, schist and slate
Switzerland	1995	Siliceous sandstones and limestones, greywacke and weakly deformed granites and gneisses (greenschists facies retrograde)
Turkey	1995	Sand-gravel particles containing glassy rhyolite and andesite

date worldwide review of the 'so-called alkali-carbonate reaction' is given in Chapter 3.

8.2.2 Binders and admixtures

European cement and concrete industries contribute to the economy with a total production value of €74 billion, a value added of €22 billion and about 366 thousand jobs. The industry plays a vital role in generating growth, considering that investment in infrastructure is one of the highest multiplier effects on the economy. Over the past 30 years, the cement industry has witnessed an accelerated internationalisation. European cement companies, albeit operating predominantly through regional markets, have positioned themselves as key actors throughout the world (BCG, 2013).

The first standards produced by CEN (the European Committee for Standardisation) regarding the components of concrete date from 2004. At present, the approved standards are in large number and concern all the components of concrete (aggregates, cement, water, additions and admixtures).

The production and use of cement in accordance with EN 197-1(2011) has been accepted by a large number of European countries: Austria, Belgium, Bulgaria, Croatia, Cyprus, Czech Republic, Denmark, Estonia, Finland, France, Germany, Greece, Hungary, Iceland, Ireland, Italy, Latvia, Lithuania, Luxembourg, Malta, the Netherlands, Norway, Poland, Portugal, Romania, Slovakia, Slovenia, Spain, Sweden, Switzerland, Turkey and the United Kingdom. The standard defines three standard strength classes at 28 days (32.5, 42.5 and 52.5 MPa). In addition, three early strength classes are included for each standard strength class: one for low early strength, one for ordinary early strength and the third for high early strength. The 27 products are grouped into five main cement types which are listed below:

- CEM I Portland cement (clinker > 95%)
- CEM II Portland-composite cement (65%–94% clinker)
- CEM III Blastfurnace cement (5%–64% clinker)
- CEM IV Pozzolanic cement (45%–89% clinker)
- CEM V Composite cement (20%–64% clinker)

This standard and the one that covers the production of concrete (EN 206-1, 2007) summarise the conclusions of the research that has been developed in the different participating countries. Both replace the local specifications creating a unified code. Regarding the significant energy, the CO_2 savings and durability benefits over the lifetime of buildings, it is encouraged to produce low clinker cements. It is possible to reduce the clinker content of cement by the use of other main constituents ('Supplementary Cementitious Materials' – SCMs), such as limestone, blastfurnace slag, coal fly ash, and natural pozzolanic materials, based on regional availability. In addition, the cement industry is continuously developing more sustainable production processes and products. Figure 8.1 shows the total cement delivery record by CEMBUREAU between the years 2000 and 2010, as classified by the five main cement types. The chart demonstrates the reduction in the clinker content of cement. Regarding ASR, the consequence of this decrease is that the type of binder is influencing the composition of the concrete pore solution, which is mainly dependent on the alkalis available in the cement clinker.

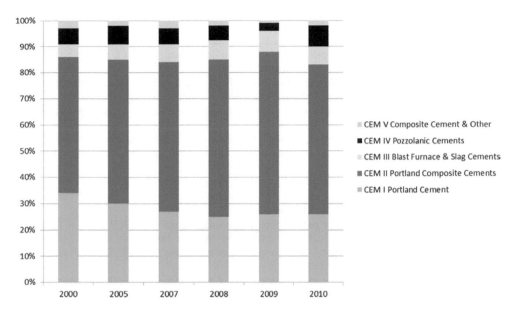

Figure 8.1 The evolution in cement composition, clearly demonstrating the reduction in clinker content (data from www.cembureau.eu)

8.3 CEN REGULATIONS

In 1988, European countries signed the Construction Products Directive (CPD), which marked the start of the work to develop common European Standards, by CEN, across the construction sector. The aim of CPD was the removal of barriers to trade between the participating countries by a unification of the existing national technical requirements.

The main changes in relation to the existing regulations were the terminology, the product descriptions, the standard sieve sizes for aggregates, the grading presentation, the test methods and the establishment of the Factory Production Control (FPC) and a CE Marking requirements. In consequence, more than 30 test methods were published as standards.

In what concerns specifically ASR, the European standards are very poor and comprise just rules regarding concrete and aggregate properties and very basic principles in order to prevent ASR (*e.g.*, composition of concrete and amount of alkalis allowed). A summary of the standards and reports anyhow related to ASR is presented in Table 8.2, including standards regarding the manufacture of concrete, sampling and petrographic characterization of aggregates, additions and admixtures.

Although there is not a specific standard dedicated to ASR, there are particular references in some of the published standards. In EN 206-1, clause *5.2.3.4 Resistance to alkali-silica reaction* is the only one in which attention is directed to the aggregates and it is recommended to follow the CEN CR 1901 (1995) report for the mitigation measures in different countries. In EN 12620, short references in *5.7.3 Alkali-silica reactivity* and *Annex G* are included. However, it is recommended that local experience of use of the aggregates is considered and that reactivity should

Table 8.2 EN standards and relevant reports applied in most European countries, which include indications concerning AAR.

CEN TR 16349:2012	Framework for a specification on the avoidance of a damaging alkali-silica reaction (ASR) in concrete
CEN CR 1901:1995	Regional Specifications and Recommendations for the avoidance of damaging alkali-silica reactions in concrete
EN 197-1:2011	Cement Part 1: Composition, specifications and conformity criteria for common cements
EN 206-1:2007	Concrete: Specification, performance, production and conformity
EN 450-1:2012	Fly ash for concrete — Part 1: Definition, specifications and conformity criteria
EN 932-1:1996	Tests for general properties of aggregates – Part 1: Methods for sampling
EN 932-2:1999	Tests for general properties of aggregates – Part 2: Methods for reducing laboratory samples
EN 932-3:2000	Tests for general properties of aggregates – Part 3: Procedure and terminology for simplified petrographic description
EN 932-5:2012	Tests for general properties of aggregates – Part 5: Common equipment and calibration
EN 932-6:1999	Tests for general properties of aggregates – Part 6: Definitions of repeatability and reproducibility
EN 934-2:2009+A1:2012	Admixtures for concrete, mortar and grout – Part 2: Concrete admixtures – definitions and requirements
EN 12620:2002+A1:2008	Aggregates for concrete
EN 13263-1:2005 +A1:2009	Silica fume for concrete — Part 1: Definitions, requirements and conformity criteria
EN 15167-1:2006	Ground granulated blast furnace slag for use in concrete, mortar and grout — Part 1: Definitions, specifications and conformity criteria

be evaluated according to procedures in national specifications. The recent technical report CEN TR 16349 (2012) states the basic principles to avoid ASR in concrete, but mainly forwards the measures to be adopted to the local/national guidance and specifications. The working group recognises the difficulties in providing pragmatic and economic unified European specifications and concludes that *"unless there is any sound scientific explanation of damaging ASR which can be used uniformly all over Europe, it is premature to have harmonised classes for alkali-reactivity of aggregates and provisions for avoiding a damaging ASR on European level. Additionally, safety margins are determined at national level and are related to the reliability at which damaging ASR will not occur"*. Nevertheless, the report defines three categories of environment and states that specifications for avoiding a damaging ASR in concrete start with the definition of the environmental category, and further by undertaking recommendations for precautionary measures appropriate to concrete depending on the environmental category.

8.4 RILEM RECOMMENDATIONS

In addition to CEN activities, most European countries have representatives of industry, research institutions and academia in RILEM technical committees and are adopting the test methods developed in the scope of their activities. The investigation developed by the members is reflected in the recommendations prepared by the technical committee and results from fruitful discussions and agreements in order to create codes that, as far as possible, include the practices found to be adequate in different regions under different conditions.

The RILEM technical committee, TC 219-ACS (Alkali-aggregate reactions in Concrete Structures: performance testing and appraisal, 2007–2014), which continued the work of the earlier TC106 and TC 191-ARP (1988–2007), was concerned with the development of international guidelines to avoid AAR in new structures, tests for the assessment of the reactivity of the aggregates, alkali release from the aggregates, performance tests for concrete mixtures, mitigation procedures, guidance on diagnosis of AAR in structures and prediction by using computer modelling. Table 8.3 lists the guidelines, test methods and reports that have been published as a result of the committee work.

Regarding the environmental conditions of exposure, there is some uniformity of opinions for most countries and standards. A simplied version is presented in RILEM AAR-7.1 (in Nixon & Sims, 2016), which is similar to the one included in CEN TR 16349 (2012) (see Table 8.4).

Classes of exposure have been adopted in the different countries and, when justified, also changed according to national needs and conditions. In France, the three classes are based on EN 206-1 (2007) (XAR1, XAR2, XAR3) and compared with the categories of structures to obtain the class of risk. In the Netherlands, there are three classes (in accordance with Article 4.1 of EN 206-1 (NE-EN 206-1) and Appendix A of NEN 8005, 2008), but the wet environment with aggravating factors is sub-divided into three sub-classes. In Switzerland, the classes of exposure result from regrouping those

Table 8.3 List of Recommendations developed by RILEM TC 219-ACS (Nixon & Sims, 2016).

AAR-0	Outline guide to the use of RILEM methods in assessments of alkali-reactivity potential of aggregates
AAR-1.1	Detection of potential alkali-reactivity – RILEM petrographic examination method
AAR-1.2	Petrographic Atlas – Characterization of aggregates regarding potential reactivity to alkalis (Fernandes *et al.*, 2016)
AAR-2	Detection of potential alkali-reactivity – Accelerated mortar-bar test method for aggregates
AAR-3	Detection of potential alkali-reactivity – 38° C test method for aggregate combinations using concrete prisms
AAR-4.1	Detection of potential alkali-reactivity – 60° C test method for aggregate combinations using concrete prisms
AAR-5	Rapid preliminary screening test for carbonate aggregates
AAR-6.1	Guide to diagnosis and appraisal of AAR damage to concrete structures, Part 1: diagnosis (Godart *et al.*, 2013)
AAR-6.2	Guide to diagnosis and appraisal of AAR damage to concrete structures, Part 2: Appraisal and management (Godart *et al.*, expected 2017)
AAR-7.1	International specification to minimise damage from alkali reactions in concrete; Part 1- Alkali-silica reaction
AAR-7.2	International specification to minimise damage from alkali reactions in concrete; Part 2- Alkali-carbonate reaction
AAR-7.3	Preliminary international specification to minimise damage from alkali-aggregate reactions in concrete; Part 3: Concrete dams and other hydro structures

Table 8.4 Classes of exposure of concrete structures suggested in CEN TR 16349 (2012) and in AAR-7.1 (Nixon & Sims, 2016).

CEN		RILEM	
Class	Environmental conditions	Class	Environmental conditions
E1	Concrete is essentially protected from extraneous moisture	1 Low	Dry to moderate moisture penetration – design life: max. 15 years
E2	Concrete exposed to extraneous moisture	2 Moderate	Concrete exposed to weather conditions – moderate to highly moisture penetration – surface temperature: always < 20 °C
E3	Concrete is exposed to extraneous moisture and additionally to aggravating factors, such as de-icing agents, freezing and thawing or wetting and drying in a marine environment or fluctuating loads	3 High	External alkali supply (e.g., de-icing agents) – moderate to highly moisture penetration – surface temperature: alternating and maximum temperature > 25 °C – dynamic loading

in EN 206-1 (SIA Merkblatt 2042, 2012). Austria used to follow three classes of exposure, but recently approved a document where just two classes are included. In Germany, three classes of exposure are used as well (DAfStb, 2013), but there are additional regulations for pavements (Bundesministerium, 2013).

8.5 TEST METHODS AND PREVENTIVE MEASURES

8.5.1 Test methods for aggregate assessment

The evaluation of ASR potential should always be an extension of the routine assessment of the suitability of an aggregate, which includes a wide range of test methods (e.g., strength, shape, roundness, size distribution, etc.). However, this evaluation is very specific and implies demanding skills and tests. According to RILEM AAR-0 (Nixon & Sims, 2016), it is possible to consider three main categories of aggregate tests for ASR usually plotted in a flow chart procedure. Similar flow charts are found in national standards (e.g., AFNOR FD P 18-542; CUR Recommendations 102; LNEC E 461):

– initial and essentially qualitative screening (visual method). The use of petrography is extremely useful for an initial indication of the likelihood of ASR to occur within a given aggregate source, by identifying possibly alkali-reactive phases;
– indicator tests. This category refers mainly to accelerated mortar-bar tests (e.g., ASTM C 1260 or RILEM AAR-2, Nixon & Sims, 2016);
– long-term concrete expansion tests (e.g., concrete prism test as RILEM AAR-3 and accelerated concrete prism test as RILEM AAR-4). These tests are used to assess the reactivity performance of particular aggregate combinations, albeit using standardised proportions.

This sequence of tests has been followed in most of the European countries in the last two decades. The petrographic assessment in EN 932-3 (1996) is too 'simplified' regarding ASR; it is just intended to enable aggregate composition to be reasonably categorised. Therefore, there is local guidance for the petrographic study of aggregates, such as LNEC E 415 (1993), SN 670 115 (2005), CUR Recommendation 89 (2008), UNI 11530 (2014) and AFNOR NF P 18-543 (2014). The petrographic analysis of aggregates is mainly performed by the study of thin sections under the polarising optical microscope, ocasionally complemented by other methods such as SEM-EDS and chemical analysis of bulk rock. The component rocks and minerals are identified, the texture described and other features such as alteration, porosity and cracks registered. RILEM has recently published its recommendations for petrographic examination, AAR-1.1 (in Nixon & Sims, 2016) and the atlas in AAR-1.2 (Fernandes *et al.*, 2016).

The content of any potentially reactive forms of silica is quantified, usually by pointcounting. As an example, the forms of silica considered as potentially reactive in CUR Recommendation 89 (2008) are described as: *"the most important, potentially alkalireactive components of aggregates and/or individual aggregate grains are opal, chalcedony, moganite, cristobalite, tridymite, cryptocrystalline quartz, (porous) flintstone (silex / chert / flint), impure sandstone (greywacke, siltstone), siliceous limestone and certain types of volcanic rock (due to the glass present in it)"*. However, other forms of silica can be considered to be unstable in alkaline environment, such as those present in deformed rocks, as listed in AFNOR FD P 18-543 (2014):

- Strained quartz with undulatory extinction; quartz with polygonisation;
- Microcrystalline to cryptocrystalline quartz;
- Rhyolitic quartz with reaction rims; spherolitic textures;
- Secondary micro-quartz (usually generated by dynamic recrystallization);
- Siliceous devitrified glass, with microcracks;
- Siliceous glass;
- Occurrence of tridymite, cristobalite, chalcedony, opal, jasper;
- Altered feldspars and mica (as potential sources of releasable alkalis).

Regarding the expansion tests, although the main principles are similar in the different national standards, there are variations in the procedures, namely in the sizes of the prisms and the conditions of exposure. As an example, in Table 8.5 a comparison is made of the concrete prism tests used in different countries, including the one proposed by RILEM (AAR-3, in Nixon & Sims, 2016). It is worth mentioning that French standards have been followed in the last decades in other countries such as Belgium and Switzerland.

8.5.2 Mitigation alternatives

As stated in CEN TR 16349 (2012), the type of mitigation measures as well as the limits and levels in a measure itself are actually defined on a local level, because they depend on the national safety margin, the experience in building practice, the geology and the climate/exposure conditions. The goal of any mitigation measure is based on the need to act on one or more of the components involved in ASR, namely reactive aggregates, available alkalis and/or relative moisture availability. The measures differ among CEN member countries and depend on the reactivity of the aggregate combination and on

Table 8.5 Concrete prism test. Example of local choices in the procedures.

	Size of the concrete prisms (mm)	Exposure conditions
Austria	100 × 100 × 400	38 °C, 52 weeks
France	70 × 70 × 282	60 °C, 12, 20 weeks or 8 months (duration depends on the concrete mixture)
Germany	100 × 100 × 500	40±2 °C, 9 months
Netherlands	75 × 75 × 250	38 °C, 52 weeks
Spain	75 × 75 × 275	38 °C, 52 weeks
RILEM AAR-3	75 × 75 × 250	38 °C, 52 weeks

the national determined safety margin. However, the suggested precautionary measure options in CEN TR 16349 (2012) are:

- no measure necessary,
- use of a non-reactive aggregate combination,
- limiting the alkalinity of the pore solution by the use of cement with a low effective alkali content, the use of an adequate proportion of slag, fly ash, silica fume or other pozzolana (in cement or as an addition) or conforming to a numerical limit on the effective alkali content of the concrete,
- verification of the suitability of a concrete mix in a performance test.

Some aggregates that are classified as potentially reactive by field experience and/or laboratory tests are not allowed for use in concrete in some areas, such as in Austria, where siliceous limestones cannot be used as aggregates. However, for economic reasons related to the costs of transport for construction in remote areas, sometimes the available aggregate has to be used. In this case, alkalis have to be accurately evaluated.

According with CUR Recommendation 89 (2008), the first precaution against ASR damage is the use of low alkali cements in order that the alkalis in the concrete are below 3.0 kg/m^3. For cements that comply with EN 197-1 (2011), it is assumed that deleterious ASR will not occur if the recommendations are followed.

In France and Belgium a very tight criterion is applied in regard to alkali content of the concrete and it covers both Portland (CEM I), blastfurnace slag (CEM III) and composite (CEM V) cements with limits for alkali content in each case.

The release of alkalis from aggregates is accepted to be important for some types of rocks. As such, research has been carried out in the countries where these rocks are used as aggregates, namely in France where a guideline has been developed and published (AFNOR FD P 18-544, 2014). A new test method for the evaluation of alkalis release is also under development in the scope of RILEM activities (it is an objective of the current AAR committee, TC 258-AAA).

In addition to the limits of alkalis in cement and the possible contribution of alkalis from the aggregates, the total active alkalis in concrete is calculated in the French standard (AFNOR FD P 18-464, 2014), by weighting the alkalis in all the components of concrete, including additions, admixtures and water.

8.6 DIAGNOSIS AND APPRAISAL OF STRUCTURES

8.6.1 Field survey

The site inspection of structures in Europe follows the principles indicated in BCA (1992) and similar recommendations prepared in different countries (*e.g.*, CUR Recommendation 102, 2008; LCPC, 1999). Also in Leemann and Griffa (2013) a good presentation is made about the methodology to follow in the assessment of dams for the diagnosis of ASR. The RILEM guidance on diagnosis is provided in their AAR-6.1 (Godart *et al.*, 2013). See Chapter 5 for a general discussion on diagnosis.

CUR Recommendation 102 (2008), in addition to CUR Recommendation 72 (2008), constitute very detailed documents in which the assessment is made by sequential levels of study, divided into four classes developed according to a flow chart:

- Class I – this inspection is intended to find out whether ASR is present or the cracks have a different cause,
- Class II – this study is to confirm if ASR is indeed present and how seriously the damage has affected the mechanical characteristics of the concrete,
- Class III – in this study more information is collected in a targeted manner,
- Class IV – this study evaluates the structural safety by: manual inspection, numerical inspection and total structural assessment.

Similar documents have been produced in other countries. One of the most widely used in Europe is the French approach, which includes site inspection and laboratory tests (LCPC, 2003). The visual identification performed during the site inspection includes: the characteristic orientation of the cracks; the occurrence of stronger damage in the areas where water flows; the disappearance of mosses and lichens on either side of the cracks; the occurrence of pop-outs; the presence of gel exudations; a tan colour in concrete; the age of the structure at the moment of appearance of damage (+/- 7 years). To be sure that ASR is the cause of damage, drilled concrete cores must be collected for examination in the laboratory. The cores are wrapped in 'Cling film' and sealed in polythene bags. The inspection of the concrete cores aims at: defining the depth of carbonation, detecting 'sweaty' spots after some days of preservation in wet environment and/or white and black halos around some aggregates, and observing fluorescence spots after the uranyl acetate test. The determination of the depth of the cracks is also performed. It is frequently observed that open crack apertures located near the exposed surface (5 mm) become closed at shallow depth (say, 50 mm).

According to CUR Recommendation 102 (2008), the final report can be presented in the following levels, depending on the classes above referred:

- Level A: Reporting of data;
- Level B: Reporting of data and immediate measures to be taken;
- Level C: Reporting of data and control measures to be taken.

8.6.2 Laboratory tests

The site inspection may be followed by sampling of concrete obtained by drill coring for further study in the laboratory.

The study usually starts with the description of the characteristics of the cores under the scale of hand specimen, followed by petrographic analysis for the identification of the phenomena generating the deterioration of the concrete. In many European countries, fluorescence impregnated thin sections are used for determining the cause of damage by performing concrete petrography. Fluorescent dye is added to the epoxy resin used as mounting media in order to highlight voids, porosity and cracks when ultraviolet light is applied. The thin sections also allow the study of aggregates used in old concretes and the identification of gel in cracks in the cement paste or crossing the aggregate particles and filling voids. The study under optical microscope is often complemented by SEM/EDS examination which provides information on the texture and composition of the deterioration products (Figure 8.2).

In Leemann and Griffa (2013), detailed information can be found on the preparation of samples for concrete petrography as well as the techniques that can be used in the

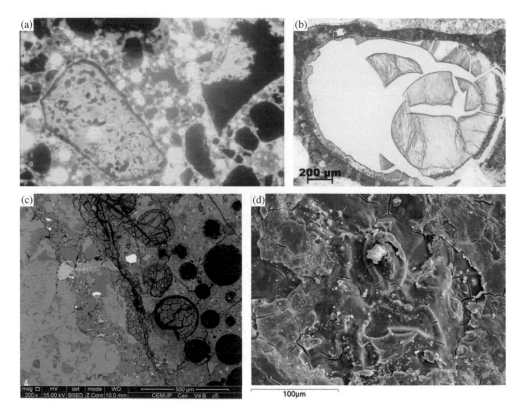

Figure 8.2 Example of images obtained by concrete petrography study: a) and b) Fluorescent yellow dye added to the resin highlights the microcracks and the voids, mainly when observed under UV light [(a) photo by PELCON]; c) image obtained by SEM on a polished thin section of concrete [obtained in CEMUP]; d) alkali-silica gel observed in a fragment of concrete (LNEC).

identification of the ASR indicative signs and characterization of the ASR products (polarized light microscopy, fluorescent light microscopy, SEM/EDS, EPMA).

The determination of residual expansion is part of the French standards, in an attempt to predict the development of the expansion in damaged structures (LCPC no. 44, 1997). Concrete cores, 76 mm diameter and 160 mm in length, are tested, eventually with an external supply of alkalis. In this method the specimens, previously wrapped in absorbent paper and polyethylene sheet, are placed in a closed metal container, with water or 1M NaOH solution in the bottom, and stored in a climatic chamber at 38 °C. The specimens are placed in a vertical position with the lower end of the cores approximately 25 mm above the surface of the water or NaOH solution in the containers. The measurements are made at various intervals over the period of one year. At the end of these tests, the specimens can be crushed for SEM/EDS analysis to confirm the occurrence of ASR.

8.6.3 Monitoring

By monitoring a structure, both the development of the expansion and the crack widths may be measured. Monitoring systems are well known, mainly in large dams, and are designed in accordance to guidelines (*e.g.*, ICOLD, 1991). The monitoring equipment allows the evaluation of the magnitude of ASR pathology by using interpretation techniques on the data obtained. The devices more commonly used measure displacements and strains and consist of plumb-lines, geodetic instrumentation, joint meters, rod extensometers and strain meters, particularly stress-free strain meters (Lopes Batista & Piteira Gomes, 2012). Strains can compromise the serviceability conditions of the structures, namely related with gate operation in dams, and can introduce damage that affects the durability of the concrete and the structural safety (Ramos *et al.*, 1996).

In case of the detection of early ASR signs, the following actions include site inspection, as above referred, laboratory testing for ASR evaluation (petrography, chemical and expansion tests) and laboratory tests to evaluate the reduction of the mechanical properties of concrete (compressive and tensile strengths, Young modulus and creep); also, the measurement of concrete stress can be performed, using flat jacks and over-coring techniques. Leeman and Griffa (2013), and references included, give further information on other non-destructive tests used in the diagnosis of ASR in structures. The integration of all the results obtained should be considered for the use of appropriate structural modelling (Piteira Gomes, 2008).

8.7 PROTECTION, REPAIR AND REHABILITATION

When a structure presents manifestations of possible ASR, the first measure will be the examination of the concrete in order to make the correct diagnosis. It would be important to know the composition of the concrete, namely the type of aggregates and the cement composition. However, this information is usually not available for old structures.

The products and systems for the protection and repair of concrete structures are defined in the European Standard EN 1504, published in 10 sub-documents (publication dates given in parentheses):

Part 1: Definitions. (2005)
Part 2: Surface protection systems for concrete (2004)
Part 3: Structural and non structural repair (2004)
Part 4: Structural bonding (2004)
Part 5: Concrete injection (2004)
Part 6: Anchoring of reinforcing steel bar (2006)
Part 7: Reinforcement corrosion protection (2006)
Part 8: Quality control and evaluation of conformity (2004)
Part 9: General principles for the use of products and systems (2008)
Part 10: Site application of products and systems and quality control of the works (2005)

The workflow defined in the standard can be summarized with steps given below:

- assessment of the condition of the structure,
- identification of the causes of deterioration,
- selection among the appropriate protection and repair methods,
- identification of maintenance requirements.

In the standard, ASR is referred to as a chemical reaction and included in the 'concrete defects' class, within the same classification as the effect of aggressive agents, *i.e.*, sulphates, soft water, salts and biological activities. When deciding the appropriate action for protection and/or repair of the affected structure, the following options are considered:

- doing nothing for a certain time,
- re-analysis of the structural capacity, bearing in mind the downgrading of the function of the concrete structure,
- test loading the structure components,
- prevention or reduction of further deterioration,
- improvement or strengthening of the construction,
- reconstruction of total or a part of the structure,
- demolition of the structure.

In order to protect or repair the damaged structures, the following principles are applied separately or together: protection against ingress (PI), moisture control (MC), concrete restoration (CR), structural strengthening (SS), physical resistance (PR), resistance to chemicals (RC), respectively. Among these, MC and CR are mostly suggested for rehabilitation of ASR damage (Bartuli *et al.*, 2012). In extreme cases, the structure has to be demolished and replaced by a new one.

8.8 CASE HISTORIES

8.8.1 Dams

A large number of dams have been diagnosed with expansive alkali-silica reactions. In this section, three cases are presented.

Chambon dam is a 293 m-long and 90 m-high concrete gravity dam built from 1929 to 1935. The concrete thickness varies from 70 m at the base to 5 m at the crest and it is composed of 150 to 250 kg/m^3 of cement (Chulliat *et al.*, 2012). Damage was identified in 1958 with opening of the concrete joints and cracking on the upstream and downstream walls, as well as in the gallery. The deterioration mainly resulted in important vertical cracking of the structure, likely to affect its integrity under earthquake action. Monitoring indicated that there was a dislocation of the crest upstream in the half-left of the dam.

Tests performed from 1967 to 1996 confirmed the occurrence of ASR and rehabilitation works carried out in the 1990s included the grouting of the cracks, the placement of a waterproofing membrane on the upper 40 m of the upstream face, and vertical diamond wire slot cuts of 32 m depth and 11 mm diameter. During the investigation undertaken from 2007 to 2010, drilling was performed as well as mechanical tests, investigation of the contact between the dam body and the foundation and evaluation of the *in situ* stress. New reinforcement works are in progress and include (i) installation of horizontal prestressed tendons crossing the structure from upstream to downstream, designed to accommodate further expansion, supplemented with a carbon fibre composite net on the upstream face, (ii) realization of seven 16 mm in diameter vertical diamond wire slot cuts, the highest measuring 42 m and (iii) replacement of the existing sealing geomembrane (Chulliat *et al.*, 2012).

Pracana dam is generally accepted as the first Portuguese concrete structure to be diagnosed with ASR. The structure was constructed between 1948 and 1951 and is composed of a 60 m high concrete buttress dam, with a crest length of 245 m, twelve buttresses 13.0 m wide and three massive blocks at each abutment. The total concrete volume of the dam is 129,000 m^3 (WGPNCOLD, 2003a, 2003b). When in March 1952 the normal water level (NWL) was reached, a crack was detected on one of the buttresses. In 4 years new fissures developed above the initial crack. In 1962 several cracks were detected in some buttresses and, in 1964, the seepage through the dam body suddenly increased. The development of these cracks went on together with a progressive upward movement of the crest and a downstream movement of the dam.

During the emptying of the reservoir, which occurred in the summers between 1971 and 1973, the main cracks on the upstream face of the dam were filled with cement grouts. However, it was recognised that the work was not effective and the constraints of the water level were maintained. In 1980 it was decided to empty the reservoir again. Concrete petrography studies were developed by Silva (1992) and Cotelo Neiva *et al.* (1995), allowing the identification of the aggregates as well of the deterioration products which included alkali-silica gel and ettringite. Among the potentially reactive siliceous material, the presence of quartzites, meta-mudstones with chalcedony and meta-pelites has been identified. The rehabilitation works were performed from 1988 to 1992. The structural rehabilitation of the dam comprised (WGPNCOLD, 2003b): the placement of a waterproofing membrane on the upstream face, the grouting of the cracks (Figure 8.3), the construction of foundation buttress downstream struts and upstream plinth, the consolidation of the foundation and the execution of a new grout curtain.

Another damaged structure, the Salanfe dam in Switzerland, is a gravity dam located at an altitude of 1,295 m on the Salanfe Plateau in the Swiss Alps. The dam is equipped with several monitoring systems, which allowed the detection of disorders. It has constantly been deformed by the hydrostatic load, seasonal thermal effects and, since the 1970s, the effects of swelling of concrete have been evident. The measurements

Figure 8.3 Paracana dam: the cracks are lined by the white product used for the filling and sealing during rehabilitation works.

recorded showed irreversible displacements of the crest upstream (the total irreversible downstream-upstream displacement is about 35 mm) and towards the right bank for one of the concrete blocks. This drift started soon after the dam was built but it accelerated in the 1970s.

In 2001 the presence of ASR was confirmed by petrographic analysis and SEM. In 2009, detailed research was undertaken and it was found that the concrete contained 4 kg/m^3 of alkalis (Leroy *et al.*, 2011). There have been rehabilitation works in order to re-establish an acceptable level of stress and deformation to guarantee the safety of the dam. These works included slot cutting of the upper part of the dam using a diamond wire. The rehabilitation project is considering further works to prevent future development of ASR, perhaps including the injection of some cracks in the galleries.

8.8.2 Pavements

Pavement concrete is one of the most stressed types of concrete. The placing must be done under given weather conditions, curing is delicate and in Central Europe, pavements are exposed to changing temperature and moisture conditions, including freeze-thaw cycles and the application of de-icers. Moreover, the permanent traffic loads create 'pre-damages' (micro cracks) that facilitate the progress of deterioration processes (Breitenbücher & Sievering, 2012).

In 2009, an evaluation revealed that about 10 % of the nearly 4000 km of concrete highway pavements in Germany were either proved or suspected to suffer from ASR (Deutscher Bundestag, 2009). The increasing number of ASR-damaged pavements and the fact that other structures with the same aggregates were not necessarily affected, suggested that ancillary conditions had exacerbated the ASR in the pavements. The typical indication of ASR starting in pavements are dark discolorations along the joints, especially at the cross points (Figure 8.4). Later, map cracking appears and the deterioration moves towards the centre of the slab (Figure 8.5). It was assumed that water

Figure 8.4 Typical discolorations along the joints, especially at the cross points, of a concrete pavement indicating the beginning of ASR.

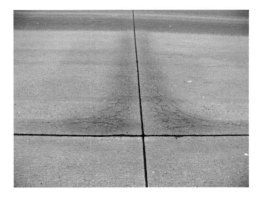

Figure 8.5 Map cracking at the cross points of a concrete pavement.

and de-icer solutions (external alkalis) enters the concrete along the joint regions and triggers the ASR (West, 1996; Stark *et al.*, 2006a, 2006b). The initial map cracking opens the concrete for a further ingress of water and it is the de-icer solution that finally accelerates the deterioration process.

It was confirmed in laboratory studies that alkali-containing de-icers used during snow and ice control operations are indeed highly able to trigger and accelerate ASR in concrete with reactive aggregates (Figure 8.6). Airfield de-icers based on alkali acetates and formates turned out to be much more aggressive than sodium chloride, owing to a special mechanism that results in an increase of the pH in the concrete pore solution (Stark & Giebson, 2008; Giebson *et al.*, 2010a; Giebson, 2013). The cement alkali content takes a back seat position here, since even low alkali Portland cements (CEM I) could only delay but not prevent a deleterious ASR (Figure 8.7).

In Germany, concrete pavements are designed to last for about 30 years. If a pavement concrete has an ASR potential, first signs for an ongoing reaction usually appear after 8 to 15 years in service. Between the appearance of the first signs

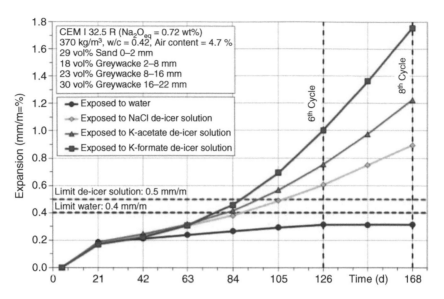

Figure 8.6 Expansion during the climate simulation concrete prism test (CS-CPT) for pavement concrete with reactive greywacke aggregates, exposed to different de-icer solutions and water (control).

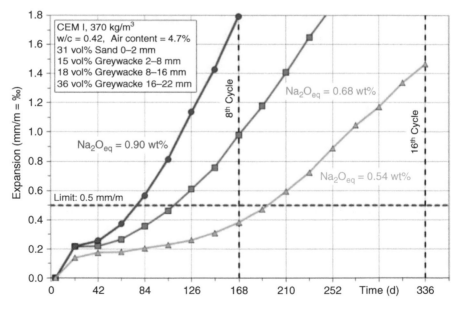

Figure 8.7 Expansion during the climate simulation concrete prism test (CS-CPT) for three tested airfield pavement concretes with reactive greywacke and different cement Na_2O_{eq} exposed to K-acetate de-icer solution.

(discolorations along the joints) and more clear indications (map cracking) several years may elapse. However, severe cracking and even spalling can occur subsequently within one more year only (Vertreter des Bundes und der Länder, 2012).

Since the standard test methods were unable to consider the impact of de-icer solutions, a new test method, the so called 'climate simulation concrete prism test (CS-CPT)', was introduced as an ASR performance test on concrete prisms exposed to de-icer solutions (see section 8.13.3). This test method, developed at the Finger-Institute, turned out to be effective in evaluating the influence of different de-icer solutions on ASR in concrete mixtures of different composition, *i.e.*, it is used for real job mixtures.

8.9 AUSTRIA

8.9.1 Historical introduction and case studies

In Austria, ASR was not considered as a major cause of damage in a number of concrete structures as the general opinion was that there were just a few reactive aggregate sources in the country. However, in 2001 Sommer *et al.* reported two major cases of ASR: a concrete road surface built in 1990 started showing signs of deterioration in 1994, which were attributed to reactive carbonate aggregate, and a dam built in 1942, which showed severe damage and in 1993/94 had to be replaced by a new structure.

Later on, Fischboeck and Harmuth (2009) examined the damage observed in two motorways, A2 and A9, located in the south of the Austrian province of Stryra. The A2 motorway was built in 1976/78 and reconstructed in 2000. Its sub-concrete was made by using the recycled aggregate obtained from the previous structure. The A9 was built in 1985 and showed the first signs of map cracking 13 years after construction. The authors examined the concrete cores taken from both motorways by using both direct techniques (*i.e.*, visual examination, determination of damage rating index (DRI), and performance of concrete petrography) and indirect methods (*i.e.*, ultrasonic pulse velocity, resonance frequency and determination of Young modulus). The investigated cores revealed that quartzite and gneiss aggregates were the cause of reactivity. A connection between the damage degree and the tectonic stress rate of the aggregates was also established. Ettringite formation inside the air voids revealed the mutual occurrence of alkali-silica gel and ettringite formation.

Aggregates that are used for concrete construction in Austria mostly show a minor content of reactive components. The sand and gravel types usually arise from the sedimentary deposits of rivers (Krispel, 2008). However, due to the geological variety of the Austrian aggregates it is not predictable which type of aggregate would cause ASR (Wigum *et al.*, 2006).

8.9.2 Test methods and preventive measures

Some construction elements built several decades ago and found to show characteristic ASR damage, such as air voids filled with gel and gel in and around aggregate particles (Krispel, 2008), indicated a need to study AAR in Austria.

The standard ÖNORM B 3100 (2008) (*Assessment of alkali-silica reactivity in concrete*) was prepared as a guideline particularly adapted to the Austrian aggregates

Table 8.6 Classification of concrete components according to the level of exposure (Krispel, 2008).

	Level of exposure	
	Level 1	Level 2
Classification of cement	Cement for particular concrete specifications according to ÖNORM B 4710-1	Cement "VD" (reduced expansion according to ÖNORM B 3327-1)
Classification of admixtures	Admixtures for particular concrete specifications according to ÖNORM B 4710-1	Admixtures according to ÖNORM EN 934-2; declared alkali content < 1%
Classification of additives	Additives for particular concrete specifications according to ÖNORM B 4710-1	Additives according to ÖNORM B 3309 (1-3)
Measures of design and execution of building components	No additional measures required	Prevention of water-penetration Measures of drainage Adequate and satisfactory curing

which could possibly contain a small amount of reactive components. This standard was aimed at minimizing the residual risk of ASR and at preventing possible damage, based on the consideration that ASR depends on local determining factors. The standard defines the classes of exposure of the concrete structures regarding alkali-silica reactivity and includes recommendations for the design of structural elements. According to this standard, two exposure classes are defined (Table 8.6):

1. Low to medium exposure class; applies to structural elements which cannot be assigned to Class 2.
2. High exposure class; applies to concrete road pavements. Characteristic for this class are strong environmental conditions, such as external alkali supply, medium to high wetting, changing or high surface temperatures of the structural element and cyclic loads.

For exposure Class 1, the aggregate can be used without further testing if structural elements containing the same aggregate showed no visible damage due to ASR for 7 years. Also, any cement and any applicable concrete admixtures and concrete additives according to the exposure class of ÖNORM B 4710-1 (2007) can be used for exposure Class 1. For exposure Class 2, the time span for visible damage is 20 years. The European Standard, *Aggregates for Concrete* (ÖNORM EN 12620) states that petrographic description has to be carried out in accordance with the methodology described in *Tests for General Properties of Aggregates – Part 3: Procedure and Terminology for Simplified Petrographic Description* (ÖNORM EN 932-3) and repeated every 3 years to ensure no changes in geologic structure and petrography. The use of aggregates is based on satisfactory long-term operating experiences. In the event that there is no experience and the laboratory tests identify the aggregate as potentially reactive, this aggregate cannot be used in structures in the most severe class of exposure. If the aggregate fails the petrographic analysis, a concrete prism test one year long has to be carried out.

Concretes for Class 2 exposure have to meet cement and additive requirements so that, in case of the occurrence of reactive silica within the aggregate, there is reduced expansion, which is regulated by ÖNORM B 3327-1 (2005). For concrete pavements the use of cement CEM II/A-S over decades is definitely a reason for the absence of serious ASR damage in Austria (Krispel, 2008). According to ÖNORM EN 934-2, concrete admixtures being added in a quantity of over 1% of the binder's mass must have alkali content under 1% for concrete of exposure Class 2. Only concrete additives according to ÖNORM B 3309 (parts 1,2,3) may be used for exposure Class 2. In addition, structural elements for exposure Class 2 have to be planned and constructed so that minimum water penetration can be assured. If necessary, suitable drainage systems have to be planned and efficient curing has to be performed.

The guideline is also aimed at evaluating the reactivity of new aggregates for which there is no long-term experience in concrete structures and recommends two types of expansion test methods:

- Mortar-Bar Test (similar to RILEM AAR-2): mortar-bars (40x40x160 mm) are prepared with a special CEM I 42.5 R (Austrian "Einheitszement"; mix of all Austrian CEM I cements), aggregate with a defined grading curve and tap water.
- Concrete Prism Test (similar to RILEM AAR-3): the test is performed with concrete prisms (100x100x400 mm) prepared with Einheitszement CEM I 42.5 R (mix of all Austrian CEM I cements), aggregate with a defined grading curve and tap water. The prisms are stored for one year in 1M NaOH solution at 38 °C.

8.10 BELGIUM

8.10.1 Historical introduction and case studies

The first case of damage due to ASR in Belgium was identified in 1988 in a bridge (B51) over the E19 built in 1975 in Kontich (De Ceukelaire, 1988; Elsen *et al.*, 2003). This road bridge, built with lightweight concrete, displayed wide cracks and was replaced in 1989. Shortly after, twelve damaged concrete structures were described by Soers (1988) and 25 more by Soers and Meyskens (1990). The number of affected structures identified increased fast. From the 3500 bridges managed by the Ministry of Equipment and Transportation (actually Public Service of Wallonia) in the Walloon region, ASR was diagnosed in about 50 structures up to 1989 and 75 by January 2007 (Figure 8.8). Besides bridges, ASR was identified in a sluice, a monument, in two motorway sections and in lighting masts. The situation in the Walloon region is now stabilized, with no new cases diagnosed in the last 10 years.

The damaged structures were aged from <10 to 25 years old. The cement used in these structures was Portland (CEM I) and the chemical analyses indicated that the concrete had Na_2O_{eq} of 1.1 to 2.0%. Minor cracking was found in structures manufactured with blended cements (Portland cement+ggbf+slag). Regarding the Flemish region, from a total of 2400 structures, up to 2003 over 80 were found to be affected by ASR, all of them built between 1967 and 1985. In Belgium' the total of structures giving serious concern is actually about 150.

Since 1995, a lot of bridges in Belgium have begun to show serious concrete degradation of their deck slab. Macroscopic manifestations, such as potholes, crumbling of the

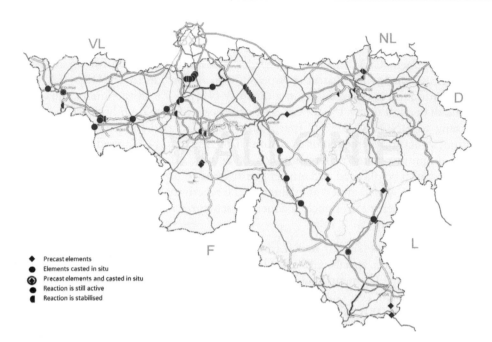

Figure 8.8 Location of the structures affected by ASR in Walloon region (Demars, 2012).

concrete into a gravel consistency on the upper surface of the deck, the perforation of the deck, and patches at the lower surface of the slabs, all indicated a defect of the waterproofing, leading to the so-called *"decay of the deck slabs"*. It has been shown that this degradation mechanism is really an interaction between alkali-silica reaction, chloride pollution and freeze-thaw cycles (Demars *et al.*, 2008).

8.10.2 Aggregates, binders and admixtures

The geological sources for aggregates for concrete used in Belgium range from Palaeozoic to recent formations. The following aggregates, listed in order of importance, have been identified as reactive by petrographic investigation of affected structures (Soers & Meyer, 1990; Soers, 1995; Elsen *et al.*, 2003; Lorenzi *et al.*, 2006):

- Gravels and sands of the Quaternary, recent alluvial deposits of the Meuse river in Limburg (Belgium and the Netherlands) and materials dredged from the North Sea. These materials contain micro- to cryptocrystalline and amorphous silica. The most reactive components are porous chert and opaline grains. They vary from 1 to 4% in Meuse deposits and the 'pessimum' content is of about 2%. In North Sea aggregates, the reactive pebbles may reach 30% and the 'pessimum' content is estimated to be about 10 to 13%.
- Siliceous and argillaceous limestones (Tournaisian Formation) – micritic limestones containing phyllitic minerals and small amounts of chert and cryptocrystalline quartz. This aggregate was studied by Guédon-Dubied *et al.* (2000), who

identified two sub-units in the limestone. The underlying unit consisted of micritic limestone, clayey in places, impregnated with diagenetic silica and containing abundant quantities of fossil fragments (bioclasts): crinoids, bryzoa, brachiopods, gastropods, bivalves, nautiloids and trilobites. The upper unit is composed of clayey micrites also impregnated with diagenetic silica and contains pseudomorphs of gypsum, anhydrite and black siliceous nodules. These nodules are composed of chalcedony and opal. Both the nodules and the diagenetic silica (microscopic) can be the cause of the rock reactivity. Tournaisian limestone containing disseminated reactive microcrystalline silica does not show a 'pessimum' effect due to the small concentration of reactive silica in the otherwise inert limestone.

- Crushed porphyry from Quenast area (Caledonian Massif of Brabant). This is a subvolcanic rock of quartz-dioritic, dacitic composition and porphyric texture. Reactivity might be due to the microcrystalline to amorphous groundmass.
- Sericitic fine grained sandstones quarried in Cambrian and Devonian formations or part of the alluvial gravels mentioned above. These rocks are described by Broekmans and Jansen (1997) as diagenetically altered impure sandstone pebbles with detrital mica and interstitial neogenic phyllosilicates. The examples of ASR with these rocks are rare in Belgium, but common in the Netherlands (see section 8.17).

The first precaution against ASR damage is the use of low alkali cements according to the Belgian standard NBN B 12-109 (2015), which covers both Portland (CEM I), blastfurnace slag (CEM III) and composite (CEM V) cements and defines the limits of alkalis content. There are also limits established for the total content of alkalis in the concrete mix if it is not exposed to other external or internal sources of alkalis. The external source of the alkalis is due to the de-icing salts (mainly NaCl), used every winter on the roads and bridges. The internal sources of alkalis are mainly due to additives or to the aggregates.

The Na_2O_{eq} content of the cements (at the present time) ranges from 0.2 to 0.8%, but in the past it was sometimes up to 1%, while the total amount of alkalis in the concretes ranges from 1.7 kg/m^3 (CEM I) up to 3.2 kg/m^3.

8.10.3 Test methods and preventive measures

The Belgian concrete standard NBN B 15-001 (2012), which is the national supplement to EN 206-1 (NBN EN 206-1, 2001), gives the following guidance for the prevention of ASR (Circulaire 42-03-06-05(01), 2006): use of a composition for the concrete considered non reactive by the expansion tests; application of a cement with limited alkali content (according to NBN B 12-109, 2015); use of mineral additions; use of non-reactive sands and aggregates; limit the degree of saturation of the concrete, for example by using an impermeable membrane.

Regarding aggregates, two main test methods are applied in Belgium for the assessment of potential reactivity: point counting according to the RILEM AAR-1, 2003 (now AAR-1.1, in Nixon & Sims, 2016) petrographic method for aggregates composed mainly of silica (quartz, quartzite, chert, chalcedony, opal) – aggregates containing more than 2% reactive silica are classified as potentially reactive; the accelerated mortar-bar test for rocks containing disseminated reactive microcrystalline silica

(Tournaisian limestone and porphyry) in which an expansion of more than 0.10% is indicative of potential reactivity.

Mineral admixtures (ground granulated blastfurnace slag, fly ash, silica fume) can contribute more or less effectively to the prevention of damage by ASR. The maximum recommended amounts of these admixtures are as follows: slag, 50 to 60%; fly ash, depending on their chemical composition; silica fume, 5 to 7.5%.

8.10.4 Diagnosis and appraisal of structures

The methods of ASR diagnosis in structures include visual examination (site inspection), examination of concrete cores and residual expansion tests. The site investigation is made according to the most common international recommendations and include the identification of the characteristic orientation of the cracks, the occurrence of more damage in the area with water flows, the disappearance of mosses and lichens on either side of the cracks, the occurrence of pop-outs, the presence of exudations of gel (not common in Belgium), a tan colour in concrete and registration of the age of the structure at the moment of appearance of damage (before/after 7 years). In order to overcome the limitations of the human memory, traces are prepared on places representative of the cracking of the structure in squares, with areas of 500 mm square, in which all the cracks are outlined with indelible ink. The aperture and depth of some of the cracks are measured. The results obtained using the monitoring frameworks lead to the determination of a crack index according to LCPC n° 47 (1997). This parameter allows the quantification of the cracking and also the modelling of its evolution. In addition, the determination of ultrasound velocity can be performed which gives some information on the distribution and evolution of the cracking.

Visual examination of the concrete cores includes recording the detection of the depth of carbonation, the occurrence of 'sweaty' spots after some days of preservation in a wet environment, the presence of fluorescence spots after the Cornell fluorescence test (LCPC n° 36, 1993), the identification of the type of cement applied (only Portland cement, CEM I, shows important ASR), and the presence of white and black halos around some aggregates. Concrete petrography is performed using both the petrographic microscope and SEM/EDS in order to identify the type of product of ASR found.

Most of the remediation techniques are based on avoiding water in the system, such as painting concrete at early ages, cladding, and use of waterproof materials on the concrete (by silane or a composite of polymers). Injection of the cracks proved not to be effective.

8.11 CYPRUS

8.11.1 Historical introduction and case studies

The island of Cyprus in the Eastern Mediterranean lies on the southern margins of the Eurasian and Anatolian tectonic plates. The collision of these plates with the African plate has led to the Cyprus Arc and the uplift of the island of Cyprus. It has been seismically active with numerous earthquakes since the Miocene.

The island can be divided into four major bedrock units as illustrated in Figure 8.9. The first, the Keryneia (Beşparmak), is a sequence of sediments including limestones of

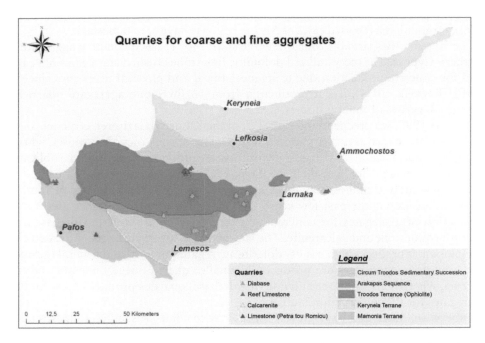

Figure 8.9 Sketch map of Cyprus showing the four major lithological units (provided by Cyprus Geological Survey, 2016).

Permian to Recent age. The second is the Troodos Ophiolite, which occupies the central region and contains lithosphere material, dykes, pillow lavas and deep water shales, radiolarites and mantle material that is largely altered to serpentinite. This is surrounded by the Circum Troodos unit, shallow marine sediments of Upper Cretaceous to Pleistocene age. The forth unit in the south western part of the island is referred to as the Mamonia Complex, which is an assemblage of volcanic, sedimentary and metamorphic rocks of Middle Triassic and Upper Cretaceous age and was thrust onto the southern margin of Cyprus in the Cenozoic.

The first published account of alkali–silica reaction in concrete in Cyprus was by Poole (1975), and related to the rapid deterioration observed in a marine jetty at Dhekelia in Larnaca Bay. The aggregates used were polymictic beach gravels containing sandstones, cherts, igneous rocks and siliceous limestones. The reactive component in this aggregate was the siliceous limestone which contained opaline siliceous diatom skeletons (tests). The jetty was constructed in 1966 and had been the subject of several investigations because of its rapid deterioration but has more recently been abandoned as unsafe.

8.11.2 Aggregates, binders and admixtures

As a consequence of the varied geology of Cyprus, a range of rock types are potentially available as sources of aggregate. Fluvial and marine gravels and sands are likely to be polymictic in character as is evidenced in the Dhekelia jetty concrete.

Components of concrete aggregates that may be considered to have a potential for alkali-aggregate reactivity will include the siliceous limestones, cherts and radiolarites. Although the potential for alkali-silica reactivity had been noted in southern Cyprus (Nixon & Sims, 1992), no additional reported cases of AAR in concrete have been published.

The Kyrenia (Beşparmak) zone offers the source for crushed concrete aggregate for northern Cyprus. The recrystallized dolomitic limestone crushed aggregates have been used for concrete production due to its mechanical and physical properties since the 1990s (Rezaieh, 2014). Total production from 16 limestone aggregate quarries in northern Cyprus is 1 million tonnes.

Prior to 1990, sea dredged sand-gravel from the Kumköy-Gaziveren coastline and/or river gravel were used as primary source of aggregate for construction purposes, whereas use of fine and coarse crushed aggregate for concrete production later became obligatory (Alkaravlı, 2007, Salihoğlu et al., 2014, Eren, 2017, personal communication).

Approximately 6 million tonnes of aggregate are produced annually in southern Cyprus, considering the past few years. The most important rock types used for the production of aggregates for concrete and road making are dolerite (diabase), reef limestone/limestone and calcarenite. The largest aggregate quarry zone is located close to Limassol. It produces high quality dolerite aggregates for local and general European use. Limestone aggregates are produced in smaller quarries near Limassol and near Nicosia, where both limestone and dolerite (diabase) quarries produce coarse and fine aggregates for the construction industry.

Otherwise, beach and river gravels together with some crushed reef limestone and dolerite (diabase) form the main aggregate sources in southern Cyprus.

Since 1984, the majority of quarries of natural aggregates ceased to operate except for those producing natural sand. In the early 1990s, the production of fine sand from crushed calcarenite was introduced, which progressively replaced the production of natural sand.

Almost the entire 'Sheeted Dyke Complex' of the Troodos Ophiolite (Upper Cretaceous) is basaltic to doleritic in composition. The dolerite (diabase) occurs in the form of dykes that are parallel and contiguous, with thicknesses varying from 0.30 m to more than 3 m. After emplacement, the dykes have undergone hydrothermal alteration resulting in local modification of the original composition of the dolerite. The varieties that have been recognized are quartz-dolerite, epidote-dolerite (diabase), amphibole-dolerite and albite-dolerite. Therefore, the quality of the dolerite used for the production of concrete aggregate depends on the degree of alteration that the sheeted dykes have undergone.

Currently, there are ten active and one inactive dolerite (diabase) quarries of fine and coarse aggregates producing about 4 million tonnes, corresponding to approximately 66% of the total aggregate production in Cyprus. These quarries are located in Lefkosia, Larnaca and Limassol districts. Figure 8.10 shows the Parekklishia Quarry zone, one of the most important in southern Cyprus. It is estimated to have over 25 years reserves.

The limestone used for fine and coarse aggregate is principally obtained from the Koronia Member (Upper Miocene), formed of recrystallized bioclastic, bioherm and biostrom reefs, and the Terra Member (Lower Miocene) of the Pakhna Formation. Four active and two inactive quarries are located in the Mitsero area, Lefkosia district, three in the Xylophagou area, Larnaca district and three in the Antrolykou area in Pafos district. In addition, limited quarrying of crushed aggregates is undertaken in one active quarry in the limestone of the Petra tou Romiou of the Mamonia Complex.

This dolomitic or magnesium-rich reef limestone is generally hard, massive and relatively porous and locally it may be bedded, brecciated and contain conglomeratic

Figure 8.10 Part of the Parekklisia Quarry Zone 2016 (picture kindly provided by Mr Christodoulos Hadjigeorgiou, Cyprus Geological Survey).

and/or sandy facies. Currently, there are two active zones in the Koronia Member, producing fine and coarse aggregates for concrete and road making.

The Terra Member represents a first phase of reef growth occurring in the lower part of the Pakhna Formation. The reefs were cemented by Mg-calcite and botryoidal aragonite during and shortly after growth. The reef limestone is generally hard, massive and relatively porous.

Currently, there are total of eleven active and two inactive limestone quarries for fine and coarse aggregates in southern Cyprus. The production of aggregates from this type of rock is approximately 1.5 million tonnes and corresponds to some 26% of the total production of aggregates in southern Cyprus.

Calcarenite for the production of fine sand is obtained from outcrops of the Athalassa Formation (Pliocene) which are located on the Agios Sozomenos plateau and Kellia area in Lefkosia and Larnaca districts respectively. There are three operating quarries, producing sand for concrete. The production of fine sand of this type represents approximately 0.5 million tonnes, and corresponds to approximately 8% of the production.

Very little cement is imported into southern Cyprus and the principal source of Portland cement is limited to a single gas fired semi-dry cement plant located at Vasilikos on the south coast of the island, approximately midway between Limassol and Larnaca. It has an output of up to 600 tonnes/day and produces a range of cements including composite cement (S-L), pozzolanic cement, sulphate-resisting, low alkali types and white cement as well as ordinary Portland cement. The cement types produced conform to EN 197-1 (2000).

One cement factory provides the cement for northern Cyprus where imported clinker is ground with gypsum and other admixtures in the mills. Gypsum is found in the Karpaz region of the island and ground granulated blastfurnace slag is imported from Turkey. The types of cements available in the northern Cyprus markets are; CEM I, CEM III/A, CEM IV/B (P) and CEM II/B-M- (S-L) types conforming to EN 197-1 and BPC52.5R (white cement) conforming to relevant Turkish standard, TS 21, respectively.

8.11.3 Test methods and preventative measures

Southern Cyprus adopts European Norms as well as ISO International standards (CYS EN, CYS ISO and CYS EN ISO) and in the northern part of the island Turkish adopted versions of European Norms are also used (TS-EN).

Other standards such as RILEM AAR-2 and AAR-5 (Nixon & Sims, 2016), BS, TS, ASTM and DIN test procedures have also been used for testing and specification purposes for concrete structures and laboratories in Cyprus. Rezaied (2014) presented a limited series of tests on two aggregate types, dolerite (diabase) and limestone, using the accelerated mortar-bar test method AAR-2 proposed by RILEM (2000, now in Nixon & Sims, 2016). These tests indicated that the aggregates tested showed little potential for alkali-reactivity.

The quality of aggregates produced in Cyprus are specified particularly in terms of their particle size distribution, physical, mechanical and chemical properties by the EN 12620:2002+A1:2008 Aggregates for concrete.

The aggregates are sampled using the Standard EN 932-1:1996 *Tests for general properties of aggregates – Part 1: Methods for sampling* and, in terms of AAR, are tested by the EN 932-3:1996 *Tests for general properties of aggregates – Part 3: Procedure and terminology for simplified petrographic description* and ASTM C289, *Standard Test method for Potential reactivity of Aggregates (Chemical method)*, as necessary.

In view of the fact that the components of some aggregates have the actual or potential risk for being alkali-aggregate reactive, specifications for new concrete structures will include requirements for appropriate aggregate testing and other specification requirements, such as the use of low alkali cement and/or the use of mineral admixtures in accordance with the guidance provided by European Standards (CEN).

8.11.4 Case examples of alkali-reactivity in concrete structures in Cyprus

As noted earlier (8.11.1), the Dhekelia Jetty concrete is the only reported case of ASR (Poole, 1975). There are unpublished cases of ASR damaged concrete structures that have been mainly the concern of government authorities in southern Cyprus; these are mostly in refugee settlements built after 1974 on the breakwater in New Port, Limassol. In all cases natural aggregates from river and beach deposits were used.

According to the Cyprus Geological Survey database, market surveillance for aggregate quality identifies several quarries of natural sand and gravel in southern Cyprus that contain chert that may cause ASR in concrete. Currently only crushed aggregates that quality control shows to be innocuous are used as concrete aggregate. Similarly, the crushed dolomitic limestone provided from the Beşparmak mountain quarries was tested by the laboratories of Eastern Mediterranean University and no reactivity potential has been indicated (Eren, 2017, personal communication).

8.12 CZECH REPUBLIC

8.12.1 Historical introduction and case studies

The alkali-silica reaction has been accepted as a serious problem of concrete deterioration in the Czech Republic since the late 1990s, due to extensive deterioration observed on the highway D11 from Prague to Poděbrady (Pertold *et al.*, 2002). After this case, an interdisciplinary programme was started to test a wide range of aggregate sources, including igneous, metamorphic and sedimentary rocks. As a result, an important database on Czech aggregates was established, which includes rock types, their locations, and their ASR potential. In addition, a database of cements used in the Czech Republic including their chemical composition and Na_2O_{eq} values was created (Modrý *et al.*, 2003). Czech former standards ČSN 721179 (1967), ČSN 721153 (1984), and ČSN 721180 (1968) for the characterization of aggregates were updated by the inclusion of selected international standards. The final summary was published as the Czech technical regulation TP 137 (2003), the first Czech regulation specifying ASR in concrete and its petrographic investigation.

Since 2003, ASR has been detected in some deteriorated concrete pavements (Lukschová *et al.*, 2010; Pertold *et al.*, 2010; Šachlová *et al.*, 2011), bridges (Lukschová, 2008; Lukschová *et al.*, 2009b), airport pavements (Škarková, 2005, 2006), dams (Lukschová *et al.*, 2009c) and tunnels (Gregerová & Všianský, 2008). In the following, some examples of damaged structures are presented:

- **Highway D11** was the first identification of ASR in the Czech Republic. The highway D11 was constructed in 1984–1985 leading from Prague to Poděbrady. The main parts of the highway were reconstructed in 1993 and 1999. Intensive systems of cracks in deteriorated pavement surfaces were firstly investigated in 1999. ASR was assessed to be the main degradation mechanism caused by highly alkali-silica reactive volcano-sedimentary quartz-and-feldspar rich rocks (*e.g.*, quartz-and-feldspar-rich tuff, tuffitic greywacke, and tuffitic siltstone) (Pertold *et al.*, 2002).
- **Road constructions in the NW part of Prague** indicated ASR caused by silica-rich limestones. The road construction forming the NW part of the Prague circumferential highway was constructed in 2000. In 2007, cracks were observed in deteriorated crash barriers. ASR was confirmed as a main degradation mechanism based on microscopical studies (Pertold & Lukschová, 2007, 2008).
- **Vrané nad Vltavou dam** shows a combination of ASR and DEF (Lukschová *et al.*, 2009a). The mutual occurrence of DEF and ASR was investigated in the concrete dam, the residual expansions recorded and the concrete petrography carried out. It was concluded that both reactions had caused deterioration in concrete samples. The concrete samples that were exposed to periodic contact with water have shown the highest extent of cracking due to ASR and DEF.
- **Concrete pavements of D1, D11 and D5 highways** stimulated development of a new methodology, combining quantitative macroscopic and microscopic identification of ASR in concrete pavements (Lukschová *et al.*, 2011; Šachlová *et al.*, 2011). Various alkali reactive rock types were identified by polarizing microscopy, petrographic image analysis and SEM/EDS methods.

8.12.2 Aggregates, binders and admixtures

Lukschová (2008) and Lukschová *et al.* (2009b), examined the core samples taken from 13 different affected bridges and a crash barrier by *in situ* observation, by laboratory methods and, on the microscale, by petrographic methods. The microscopical analysis was also conducted on specimens after mortar-bar and gel-pat tests. According to this analysis, the chert aggregates had shown a high degree of reactivity, whereas the quartz-rich aggregates have shown a medium degree of reactivity. The aggregates composed of magmatic rocks, *e.g.*, granitoids, serpentinite and basic volcanic rocks, showed little or no reactivity. This study also showed that very small new quartz grains, originating during a low degree recrystallization mechanism, exhibited important ASR, whilst well recrystallized coarse aggregates exhibited little or no ASR. Moreover, the grain size parameters were important for the reactivity degree of quartz-rich aggregates.

Seidlová *et al.* (2012) investigated volcanic rocks (13 basalts, 5 spilites, 4 melaphyres, 2 phonolites, 1 rhyolite, and 1 dolerite) widely used in the Czech Republic as crushed aggregate. Experimental methods, *i.e.*, accelerated mortar-bar test, gel-pat test and microscopical techniques (*i.e.*, polarizing microscopy, SEM/EDS and image analysis) were conducted. Some volcanic rocks, such as basalt, dolerite and phonolite, showed a low degree of ASR under laboratory conditions, as well as in real concrete samples, whereas acidic volcanic rocks, such as rhyolite and melaphyre, showed a medium degree of ASR. Interestingly, spilite samples were found to exhibit either a low or high degree of reactivity depending on the level of spilitization.

Šťastná *et al.* (2012) combined standard microscopical techniques (polarizing micro-scopy and SEM/EDS method) with cathodoluminescence (CL), with the aim to verify source materials of reactive aggregates in concrete. Quartz coming from different volca-nic (blue CL colour), plutonic (blue to violet CL colour), and metamorphic (brown to dark CL colour) parent rock was identified. The highest alkali-reactivity was exhibited by quartz–feldspar-rich pyroclastic and epiclastic rocks (mainly tuffs, tuffaceous sandstones and tuffaceous siltstones), containing significant amounts of very fine-grained quartz. Some plutonic rock types (*e.g.*, biotite granodiorite), as well as some polycrystalline quartz aggregates affected by various deformation mechanisms (with a brown to dark CL colour), showed a medium degree of ASR. Monocrystalline and polycrystalline quartz aggregates (with red and blue to violet CL colour), aggregates of carbonatized volcanic rocks (with an intense orange CL colour of calcite) and undeformed granitic aggregates rarely showed features of ASR. The different CL characteristics pointed to various sources, as well as the deformation mechanisms of the silica materials, thus helping to evaluate the alkali–silica reactivity of the aggregates studied.

Šachlová (2013) combined microscopical techniques (polarizing microscopy and SEM/EDS method) and petrographic image analysis with the aim to find microstruc-tural parameters affecting ASR of various quartz-rich aggregates. The highest degree of ASR (expansion of mortar-bars exceeding the 0.10% limit, measured according to the standard ASTM C1260 (now 2014) was observed in connection with the samples of quartzite, phyllite and schist indicating very small grain size, elongated grain shape and features reflecting deformation and recrystallization under low temperature regimes (*e.g.*, bulge nucleation and slow migration).

In conclusion, the following aggregates were found to be reactive either by field experience or after experimental testing in the Czech Republic:

- granodiorite quarried in Brno Massif in the SE of the Czech Republic, used in the highway D1, was found to show high alkali-silica reactivity according to extensive concrete deterioration and high amounts of alkali-silica gel in concrete (Lukschová *et al.*, 2010; Pertold *et al.*, 2010; Šachlová *et al.*, 2011);
- quartzite contained in quartz sands and gravels from sedimentary deposits in various places of the Czech Republic, presenting high alkali-silica reactivity

according to experimental laboratory testing (Lukschová, 2008; Pertold *et al.*, 2008; Lukschová *et al.*, 2009a);

- quartzite, phyllite and schist located in various parts of the Czech Republic (*e.g.*, Teplá-Barrandian Zone, Moldanubicum, and Silezicum) affected by different degrees of deformation and recrystallization, presenting high alkali-silica reactivity according to microscopical study combined with the standard acclerated mortar-bar test (Šachlová, 2013; Šachlova *et al.*, 2013);
- quartz sands and gravels originating from the Polabí area in the central part of the Czech Republic showed medium to high alkali-silica reactivity according to experimental testing of aggregates (Pertold *et al.*, 2008; Lukschová *et al.*, 2009a);
- quartz-and feldspar rich volcano-sedimentary rock types, quarried in the Barrandien area in the central part of the Czech Republic and used in highways D1 and D11, were found to show high alkali-silica reactivity according to extensive concrete deterioration as well as laboratory testing (Lukschová *et al.*, 2010; Pertold *et al.*, 2010; Šťastná *et al.*, 2013);
- acid volcanic rocks (melaphyre, rhyolite, porphyry) quarried in isolated deposits in the Barrandian Palaeozoic (SW of Czech Republic) and the Lugic Late Palaeozoic (N of Czech Republic) indicating medium to high alkali-silica reactivity according to experimental testing of aggregates (Seidlová *et al.*, 2012; Nekvasilová *et al.*, 2013);
- silica-rich limestone, quarried in the Barrandien area in the central part of the Czech Republic, used in some road constructions in Prague and indicating very high alkali-silica reactivity according to extensive concrete deterioration and extremely high amounts of alkali-silica gel in concrete (Lukschová *et al.*, 2009c).

8.12.3 Test methods and preventive measures

The Czech standards on determination of reactivity of aggregates were published in the 1960s and 1970s, although the application of these tests was not compulsory at that time. Traditionally the methods of assessment were the Chemical Method (ASTM C289, recently 2007, but now withdrawn) and expansion methods and an extensive study of the aggregates was performed.

The Czech Technical regulation TP 137 (2003) summarizes the modified test methods and an improved version of ASTM C1260 (now 2014) was also added. The methods summarized and/or modified are as follows:

- Modification of petrographic examination: Lukschová (2008) made some modifications to the RILEM AAR-1 petrographic method by using image analysis software [the RILEM method is updated in AAR-1.1 (in Nixon & Sims, 2016), but doesn't include image analysis].
- Microscopical investigation of mortar-bar specimens: after the ASTM C1260 (2014) test procedure, thin sections were prepared and analyzed by using polarizing microscopy and modal composition of mortar-bars is measured. This method was later accepted into an updated version of TP 137 (2003) (Lukschová *et al.*, 2009a).
- Modification of the gel-pat test: the gel-pat test method which was criticized, due to its insufficient testing period for slowly reactive aggregates, was modified in order to overcome this problem. Macroscopic identification of alkali-reactive aggregates was specified by microscopical investigation of gel-pat specimens (Lukschová, 2008).

8.13 FRANCE

8.13.1 Historical introduction and case studies

The research developed in France since the recognition of ASR in concrete strutures can be considered a case of success. Since the publication and the implementation of the document *Recommendations for the prevention of damages due to alkali-aggregate reactions* (LCPC, 1994), no single case of ASR has occurred in the country in new structures from that date.

The first cases of ASR in structures in France were diagnosed in the 1970s, namely Temple sur Lot, Castelnau, Maury, Bimont and Chambon sur la Romanche dams (Fasseu, 2001). In 1986, a bridge over highway A22 in the North of France presented signs of deteriorations just 10 years after construction and even before having been put into service. The diagnosis pointed to the coexistence of ASR with ettringite formation (Fasseu, 2001). A number of structures in the same area and built at the same time showed the same symptoms. The aggregate used in these structures was the siliceous limestone from Tournai region and the cements used were low in alkalis (0.60%). In 1987, three other bridges on the A4 motorway near Paris were diagnosed as having ASR. The aggregates used were also a siliceous limestone coming from the Avesnois region (south of the North Department), whose composition was similar to the Tournai limestone.

By 2008, more than 400 structures were affected by swelling internal reactions in concrete, including both ASR and ettringite formation (Godart & Le Roux, 2008). These structures included bridges, retaining walls and some dams, mainly built in the decades of 1970 and 1980. Some structures showed manifestations of damage from 3 to 10 years after construction, most of them just exhibited mild signs of deterioration and just a few were heavily damaged. About 10 structures were demolished before 2008. Most of the cases were recorded in the North of France and Brittany.

8.13.2 Aggregates, binders and admixtures

In many of the structures found to have ASR, two main types of reactive aggregates were identified: siliceous limestone with 10 to 15% SiO_2 and quartzite aggregate. Some Rhône river aggregates were also found to be reactive.

The use of tests for the assessment of aggregates started in the 1970s. By that time, the siliceous limestones were being studied, but the results obtained by the laboratory tests were contradictory: the aggregates were classified as non-reactive by the mortar-bar test ASTM C227 (now 2010) and potentially reactive according to the chemical test ASTM C289 (later 2007, but now withdrawn). These results triggered the research on the development of new laboratory tests for which the aggregates used in the Chambon dam concrete were considered as reference (Fasseu, 2001).

The regulation AFNOR NF P 18-594 (2004, revised in 2014) describes the test methods to be carried out in order to certify that an aggregate is not reactive. It establishes the reference methods and the alternative methods. These test methods aim at qualifying both the different fractions (sand and gravel) and the mixture of fractions according to the composition to be used in the manufacture of concrete. The aggregates are finally classified as: NR (not reactive), PR (potentially reactive) or PRP (potentially reactive at pessimum effect).

The recently published AFNOR FD P 18-464 (2014) aims at a better management of aggregates resources, as it allows the use of PR and PRP aggregates as long as preventative measures are adopted. In 2014, other important standards were also published, such as FD P 18-541 and FD P 18-543.

With respect to the the alkali content, the French standards determines the different alkali sources within the concrete (cement, aggregates, additions, admixtures and water). The content is expressed in kg/m^3 of concrete (Godart & Le Roux, 2008) according to a formula in which the alkalis from each component are weighted. The content in % of active alkalis of the cement is obtained by a formula which includes the alkalis measured according to NF EN 196-2 (2013) in the slags, the calcareous fines, the fly ashes, the pozzolanas, the clinker and the gypsum. The calculation of the alkali content depends on the type of the cement used and a complete set of chemical analyses of the cement components is needed.

The assessment of the alkalis in the aggregates follows the standard AFNOR FD P 18-544 (2014) measuring the soluble alkalis at 7 hours in boiling lime. The alkalis may be classified as the total alkalis obtained from the chemical analysis of the aggregate (expressed as alkali oxides, Na_2O and K_2O) and the active alkalis (those alkalis usually present in silicate minerals of the aggregate which are released and can contribute to the concrete pore solution).

8.13.3 Test methods and preventative measures

In 1981, the first two standards were prepared, regarding ASR, based on the ASTM C227 and on the ASTM C289 test methods.

In 1989, the Ministere de l'Equipement, du Logement, de l'Amenagement du Territoire et des Transports was faced with a significant number of structures affected by ASR. It was decided to take preventative measures in the construction of the structures for which it was responsible. The Laboratoire Central des Ponts et Chaussées (LCPC) was put in charge of a technical committee bringing together representatives of the various organizations involved in the construction process (suppliers of materials, contractors, operators, project supervisors, laboratories, engineering companies, etc.), to draft a document that would be applied as soon as it was finished. This document was published in January 1991 under the title *Provisional recommendations for the prevention of damage due to the alkali-aggregate reaction* (LCPC, 1991). It was later transformed into an official guidance (LCPC, 1994, *Recommendations for the prevention of damages due to alkali-aggregate reactions*). Following this guide, there were two milestones in the development of research and regulations regarding ASR: the publication of AFNOR NF P 18-594 (2004, 2014), which lists the laboratory tests to be performed for the characterization of the aggregates and AFNOR FD P 18-464 (2014), with new rules and improved test methods.

The assessment of aggregates is performed according to the following sequence:

1) Petrographic analysis (complemented by chemical analyis) covered by AFNOR FD P 18-543 (2014). This study aims at selecting the further tests to be performed according to AFNOR FD P 18-594 (2014). It includes the identification of the rock(s) and the component minerals, the description of the texture and the evaluation of the degree of alteration. The quantification of potentially reactive forms of silica is performed by point counting.

2)   Screening tests. The 'autoclave accelerated mortar-test' is the reference test and it is applied to sand, gravel and mixed aggregates. Bars of 40x40x160 mm are prepared for two different cement/aggregate ratios with the prisms being autoclave cured at 0.15 MPa and 127 °C for 5 hours. Another screening test is named the 'accelerated microbar test', which is performed on specimens of 10x10x40 mm cured in KOH 10% at 150 °C for 6 hours. As an alternative, the 'kinetic test' aims at making the interpretation of the other tests easier. It is based on the measurement of the dissoluted silica from the aggregates immersed in an alkaline solution at 80°C for 3 days and plotting the ratio SiO_2/Na_2O against time for 24, 48 and 72 hours. The test should be used for rocks in which Al_2O_3 is higher than 5%. If the aggregate contains chert, jasper or radiolaria, a further division has to be made: for more than 70%, the aggregate is considered as PRP; between 40 and 70%, the screening test will give PR or PRP; for less than 40%, the screening test indicates PR or NR.

3)   The long term test (concrete prism expansion test at 8 months) is performed on sand or gravel prepared in concrete prisms of 70x70x282 mm, cured at 38 °C and 100% RH. The aggregate is considered NR if expansion is lower than 0.04% at 8 months and has a flint content lower than 40%. It is PRP if expansion is lower than 0.04% at 8 months but the flint content is between 40 and 70%. The aggregate is PR if expansion is greater than 0.04%.

The standard AFNOR FD P18-464 (2014) states that the prevention of ASR is based on the assessment and classification of the aggregates as non-reactive (NR), potentially reactive (PR) or potentially reactive with a pessimum effect (PRP). This standard covers natural and recycled aggregates complying with AFNOR P-18-545 (2011) and is based on a new approach:

• The alkali-silica reaction is a problem of the cement-aggregate combination; the use of low alkali cement is not enough to avoid the ASR; limitation of alkalis is needed in the concrete and not just in the cement.

• It is focused on the optimization of the exploitation of the aggregate resources; PR aggregates are accepted as long as the concrete mixture formulation is first investigated.

When PR aggregates have to be used, it is mandatory that the concrete composition to be used in the structure is assessed by laboratory tests. In the performance criteria, in accordance with AFNOR NF P 18-454, concrete prisms of 70x70x282 mm are cured at 60 °C and 100% RH for 3 months (usually for aggregates with mineralogy considered not complex), 5 months (when complex aggregates are tested, mainly containing silicates) or 12 months (if additions are used). The concrete mixture includes the addition of NaOH in order to accomodate the variability of the cement alkali content. The interpretation of the results is based on AFNOR FD P 18-456.

   The French standards cover just the alkali-silica reaction, as the alkali-carbonate reaction is not considered to be a deleterious reaction in the country. It also specifies that the 'alkali-silicate reaction' occurs when altered minerals which can release silica and alkalis are used in concrete. Elsewhere, previously, the term 'alkali-silicate reaction' was used for a time to describe slow reactions involving certain silicate rock types, but today these reactions are regarded as particular types of ASR (see Chapter 10 on North America).

Although the cases of ASR in France are not as common as in other regions of the world, the regulation AFNOR FD P18-464 (2014), cited in the national annex of NF EN 206-1/CN (2014), aims at limiting the risks of damage due to the reaction. This document replaces and updates the recommendations published in 1994 by LCPC and it raises the concept of level of prevention based on the type of structure and the environmental conditions of exposure. It is assumed that the structures are built avoiding the accumulation of water and creating drainage systems for the quick evacuation of water. In addition, it is assumed that not only NR aggregates are used, but PR and PRP can be used as long as suitable preventative measures are applied. The regulation is applied to concrete and recommends an approach in four stages:

- Selection of the structure or part of the structure category (category I, II and III). based on the type of structure, objective, strategic importance and the effect of any damage on the safety of the structure (Table 8.7);
- Selection of the exposure class (XAR1, XAR2, XAR3, complementary to NF EN 206-1, 2004), considering the role of water, the environmental hygrometry, the risk of the use of de-icing substances, the marine environment and the input of alkalis from the environment;
- Determination of the level of prevention (A, B and C) (Table 8.8) in order to select the measures to avoid ASR, depending on the two previous factors;
- Choice of the type of precaution corresponding to the required level of prevention.

The prevention involves two stages: the definition of the level of prevention and the application of the possible solution depending on the level defined. It also needs the

Table 8.7 Examples of the type of structures or parts of the structures in each category (adapted from AFNOR FD P18-464, 2014).

Category I	Structures with concrete class ≤ C20/25, elements that are easy to replace, temporary structures, most of precast products
Category II	Most of buildings and civil engineering structures (e.g., small dams, roads, sewage plants, etc.)
Category III	Nuclear vessels, cooling towers, large dams, tunnels, exceptional bridges or viaducts, monuments or prestigious buildings

Table 8.8 Selection of the level of prevention regarding the type of structure and the environmental classes of exposure (adapted from AFNOR FD P18-464, 2014).

Environmental class Structure category	XAR1 dry or slightly humid environment (hygrometry ≤ 80%)	XAR2 hygrometry > 80%, or in contact with water (except marine environment)	XAR3 hygrometry > 80%, de-icing salts or marine environment
I Low or acceptable risk	A	A	A
II Low tolerance to risk	A	B	B
III Unacceptable risk	C	C	C

previous qualification of the aggregate as NR, PR or PRP or also as NQ (not qualified).

The standard defines the type of precaution, which depends on the level of prevention A, B and C. For level A, no special precautions with respect to the alkali-aggregate reaction are needed. The only requirements are the usual rules of construction. For level B, which is the most common, four possibilities exist: choice of NR aggregates; choice of PRP aggregates with two particular conditions on the concrete composition (the concrete composition contains only PRP aggregates, and either the whole mix of aggregates contains more than 70% flint or the concrete formulation satisfies the performance test on concrete at 60 °C); choice of PR aggregates with particular conditions of use (either limitation of the alkali contents in the concrete, with a threshold fixed at around 3 kg/m^3 of concrete or with respect to performance criteria for the expansion test at 60 °C, or use of a sufficient quantity of specific additions avoiding ASR, checked by performance criteria). Level C applies to exceptional structures.

The recommendations regarding ASR prevention imply that for structures of prevention levels B and C, an accurate knowledge of the aggregate is needed including homogeneity, petrography and the way of its occurrence in nature, as well as an evaluation of its performance in a concrete mix.

The principles of prevention in the recommendations published by LCPC in 1994 seem to have produced good results as there were no new cases of AAR diagnosed from the structures built thereafter according to the standards. These principles and procedures were integrated in the NF EN 206-1 standard.

8.13.4 Diagnosis and appraisal of structures

For the site investigation of manifestations of ASR, documents were prepared by LCPC in 1999 and in 2003 (LCPC, 1999, 2003) as guidelines to aid owners, managers and laboratories in coping with damaged structures. The methodology recommends a procedure in five steps, with specific investigation techniques, such as the monitoring of the cracking pattern, the monitoring of global expansion of the structure and the analysis in laboratory of samples taken from the structure, including residual expansion tests performed on concrete cores. In some cases all the investigation techniques must be applied simultaneously, but generally the methodology prioritises the monitoring of cracking and global deformations, allowing monitoring of the evolution of expansion with time. As there is no reliable and durable repair method to stop expansion of the structures, treatment methods to mitigate the effects of expansion are indicated.

This methodology allows the classification of the structures in two levels based on the localization, the intensity and the extent of damage: level 1 is attributed when damage affects vital elements of the structure and level 2 for less affected elements or those not vital to the serviceability of the structure. The frequency of measurements during monitoring depends on the cracking conditions. In cases in which the evolution of cracking and expansion exceeds the defined threshold, concrete cores have to be extracted from the structure and studied in the laboratory (LCPC, 2003) (Godart et al., 2004).

The quantification of cracks consists in the recording of all the cracks that intersect a frame of four axes plotted on the concrete surface (LCPC n° 47, 1997), with an area of one square metre. The width of the cracks needs to be measured with an accuracy of 0.05 mm using a fissurometer or a magnifying glass. Usually, one of the frames for measuring is plotted in each structural element, depending on the importance of the structure's degree of damage. A reference frame is also placed in an area free of cracking. The quantification is given in width/m and corresponds to the cracking index obtained by the ratio of the sum of the widths intercepted by the axis by unit length.

For the monitoring of deformations, different methods and equipments can be used such as distancemeter with Invar wire, infrared distancemeter, extensometer or Vernier caliper (Godart et al., 2004; Godart & Le Roux, 2008; Godart, 2009). The choice of the equipment depends on the size and accessibility of the structural element and the objective is to obtain the variation of the deformation with time.

For the diagnosis of ASR, the visual site inspection is not sufficient and supplementary methods are needed. For this, the uranyl ions fluorescence method is used (LCPC n° 36, 1993). It consists in spraying a solution of uranyl acetate on to split concrete cores and observing the surface with UV light. Any alkali-silica gel will fluoresce in yellowish-green colour. This first assessment has some limitations and the observation by SEM should always be used for confirmation of the results (Godart & Le Roux, 2008). SEM/EDS is considered in France as the preferential method for the identification of the ASR products. However, the presence of these products serves as a diagnostic indicator only if there is also significant cracking in the concrete. Supplementary methods include: concrete petrography (Le Roux et al., 1999), mineralogical composition of the concrete (Deloye & Divet, 1992) and quantification of the cracks (Salomon & Panetier, 1994).

In France, residual expansion tests are performed as a routine in damaged structures. They aim at evaluating if the structure will continue to expand and if its serviceability is likely to be affected. The tests are performed on concrete cores submitted to accelerated aging during one year in an environment of 38 °C and 100% RH. The concrete that shows deterioration due to ASR expands during the test and the expansion is larger at early ages and slows down with time. The LCPC method (LCPC n° 44, 1997) states that the evaluation of the residual expansion is based on the increase of deformation between weeks 8 and 52. However, Multon et al. (2007) recommend that the overall expansion curve is considered in order to evaluate the mechanical behaviour of the damaged structure (Godart & Le Roux, 2008).

The study of the damaged structures aims at providing managers and engineers with information about the residual strength of the structure and at evaluating the development of damage as well as at the simulation of the effects of the recommended treatment. The modelling must include the three following steps (LCPC, 2003):

- Mechanical modelling of the development of expansion with time at the material scale;
- Calibration of the model for the structural concrete based on the results of the residual expansion tests and on the monitoring of the global deformations of the structure or its cracking survey;
- Recalculation of the structure both in time and space, based on the model on stage 2 and integrating the history of thermo-hygrometric variations for the structure (Ulm et al., 2000; Li et al., 2004).

8.13.5 Protection, repair and rehabilitation

As regards treatment, there is no method accepted as reliable for repairing the structures. Concerning the various different solutions that are carried out, the following can be concluded (LCPC, 2003, 2010; Godart *et al.*, 2004):

–   Injection of cracks does not result in significant improvement of the concrete strength and does not prevent further ASR. The only advantage seems to be the avoidance of water ingress into the cracks;
–   Application of paint has a limited efficiency and it would be better to apply an impermeable cover with enough thickness to avoid the penetration of water into the structure. It is a temporary solution that allows an extension of the structure service life;
–   Application of a watertight coating with the thickness of a few millimetres contributes to the reduction of the penetration of water. It works as a provisional solution;
–   Strengthening with reinforcement or prestressing works as a mechanical treatment method but the long-term effectiveness is not proved;
–   Release of stress by sawing the structure has a temporary effect and does not prevent the development of the reaction. This solution is applied in specific structures, especially on dams (*e.g.*, Chambon Dam);
–   Demolition and replacement is sometimes inevitable, though expensive.

8.14 GERMANY

8.14.1 Historical introduction and case studies

Very first reports about a concrete phenomenon occurring in the USA, called ASR, were published in Germany in the Journal "Zement-Kalk-Gips" in the early 1950s (Zement-Kalk-Gips, 1950, 1951; Bödeker, 2003). Some years later, Bosschart (1958) and Forum (1965) reported in detail about the experiences with ASR in the USA and in Denmark. Based on geological considerations, however, it was assumed until 1965 that ASR could not occur in Germany. In 1965 the "Lachswehrbrücke" (Salmon Weir Bridge) in Lübeck (Northern Germany), built in 1964, had to be closed due to an alarming bending of the superstructure. In 1968, finally, the bridge had to be replaced. Subsequent investigations showed that the bridge was affected by ASR (in German: Alkali-Kieselsäure-Reaktion, AKR), caused by aggregates with opaline sandstone and flint from Northern Germany (Deutscher Beton-Verein, 1972).

After investigations in the following years, the very first guideline for preventing ASR-damages in concrete was published in 1974 (DAfStb, 1975, 2013). In former East Germany, the first examples of ASR damage were detected in the mid 1970s on precast concrete slabs for residential buildings near Rostock (Northern Germany) and a corresponding guideline was published in 1980 (StBA 4).

For a long time, ASR in Germany was mainly associated with aggregates containing opaline sandstone and flint from Northern Germany. The regulations and test methods in the guideline were designed for these types of aggregate. In coarse aggregates (gravels), only the amounts of opaline sandstone, flint and siliceous limestone were determined by petrographic examination. For the flint, only grains with a low density were assumed to be alkali-reactive, while the grains of opaline sandstone and siliceous limestone were tested for 60 minutes in a 10% NaOH solution. The mass loss as well as the amount of grains with signs of dissolution were considered for the evaluation. Fine aggregates were tested in a similar way for 60 minutes in 4% NaOH solution by determining again the mass loss afterwards.

But over the years, more cases of damage appeared and it was found that other, slowly reacting rock types were also alkali-reactive. In the mid 1980s, ASR damage was found on structures built with Precambrian greywacke from South-East Germany, after 10 to15 years in service. Thus, in 1997, an additional part was added to the guideline, containing a new test method, the German fog chamber test at 40 °C. The fog chamber test was intended for greywacke aggregates and other, slowly reacting aggregates. By regulation, however, other rock types than greywacke had to be tested only if ASR damage was evident in the field. Consequently, precautionary testing of many aggregates was not an obligation. At this time, the mortar-bar test was not yet part of the guideline, which further complicated the identification of reactive aggregates for some more years.

The situation got worse in the late 1990s. An increasing number of highway concrete pavements showed ASR damage after they had been in service for about 10 years and despite being built by following the regulations (Öttl, 2004). Highway pavements in Middle Germany, built in the early 1990s, contained gravels with greywacke, quartzite and siliceous schist as well as crushed rhyolite, granite and granodiorite aggregates (Stark *et al.*, 2010a, 2010b). All of them were assumed to be non-reactive, but none of them had been tested. Also, a 12 year old runway of an international airport, containing rhyolite aggregates, was severely damaged by ASR and had to be replaced. Due to the increase in the occurrence of damage to concrete pavements, an advisory circular was introduced in 2005, demanding an expert evaluation of concrete compositions for pavements regarding ASR (Bundesministerium, 2005).

In 2013, the latest edition of the advisory circular (ARS 04/2013) was released, containing a procedure for the evaluation of aggregates and concrete mixtures for highway pavements (FGSV, 2007; Bundesministerium, 2013). The key point of the evaluation procedure is ASR performance testing, based on the experiences made since 2004 (Müller *et al.*, 2009; Seyfarth *et al.*, 2009a; Müller *et al.*, 2013). On new concrete pavements, built after 2005 and based on the advisory circular (Bundesministerium, 2005), no signs of ASR had been detected so far.

8.14.2 Aggregates, binders and admixtures

The first cases of damage caused by ASR in Germany were related to aggregates containing opaline sandstone, siliceous limestone and flint from Northern Germany (Figure 8.11, Figure 8.12). Damage caused by such aggregates appeared very rapidly, usually after a couple of years. Certain slowly reactive aggregates were identified to cause damage after a much longer time of 10 to 15 years. The analysis of more than 50 ASR-damaged concrete structures revealed that slowly reactive aggregates were in fact mainly responsible for the damage (Freyburg & Schliffkowitz, 2006). Most cases (about 80%) were caused by gravels containing rhyolite, greywacke, radiolarite,

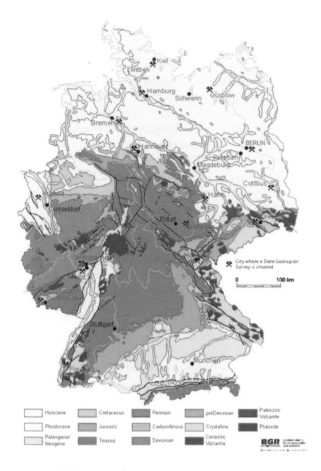

Figure 8.11 Geological map of Germany (© BGR, 2014).

siliceous limestone and often large amounts of strained quartz. The rest of the damage could be attributed to crushed rhyolites, granites, metamorphic rock and greywackes. Recently, also granodiorite, andesite and gneiss have been identified as alkali-reactive, especially when used in pavement concrete (Stark *et al.*, 2010a, 2010b). However, not all of these rock types are invariably reactive; comprehensive analysis and testing is necessary to evaluate their ASR potential in concrete (Stark & Wicht, 2013).

For a long time, the market in Germany was dominated by CEM I cements. But in the last decade there has been a clear shift towards CEM II and CEM III cements, mainly to lower CO_2 emissions by reducing the cement clinker production, but also because of an increasing demand for more effective and customized cements (Ludwig, 2012). While the market share of CEM I cements was about 60% in 2001, it dropped down to 30% by 2011 (Figure 8.13). This trend could have beneficial effects in mitigating ASR. However, for some applications, CEM I cements are used nearly exclusively, as for pavements.

Figure 8.12 Region of glacial deposit in Northern Germany containing alkali-reactive opaline sandstone, flint and siliceous limestone (DAfStb, 2013).

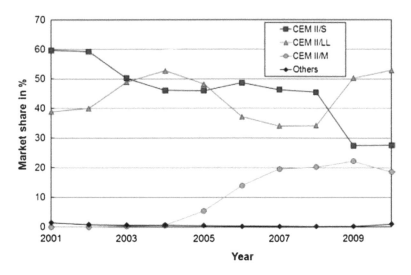

Figure 8.13 Market share of different cement types in Germany from 2001 to 2010 (Ludwig, 2012).

8.14.3 Test methods and preventive measures

The ASR test methods used in Germany today are petrography, an NaOH solution test for fast reacting aggregates (gravels and sands with opaline sandstone and flint), mortar-bar testing, the German fog chamber test at 40 °C, an alternative concrete prism test at 60 °C and ASR performance testing. Except the performance tests, all

Table 8.9 Concrete mix design for the German fog chamber test at 40 °C (DAfStb, 2013).

max. grain size:		16 mm	22 mm
CEM I 32.5 R (Na_2O_{eq} = 1.3 ± 0.1 wt.%)		400 kg/m³	
Aggregate proportions	Natural sand 0-2 mm (non-reactive)	30 vol.%	30 vol.%
	2-8 mm	40 vol.%	20 vol.%
	8-16 mm	30 vol.%	20 vol.%
	16-22 mm	—	30 vol.%
w/c ratio		0.45	

these test methods are part of the latest edition of the German '*Alkali-Guideline*' (DAfStb, 2013). Most of the test methods are similar to those used elsewhere in the world, *e.g.*, according to ASTM C1260 (2014) or RILEM AAR-3 (in Nixon & Sims, 2016). As an example, the fog chamber test runs for 9 months on concrete prisms (100×100×500 mm³) at 40.0 ± 2.0 °C and ≥ 99% RH with an standardized mixture (Table 8.9).

The ASR performance tests differ significantly from the regular test methods. The first ready-to-use ASR performance test for pavement concrete in Germany was developed at the Finger-Institute (Seyfarth & Giebson, 2005; Stark & Giebson, 2006; Stark & Seyfarth, 2006; Stark *et al.*, 2006a, 2006b; Stark *et al.*, 2007; Stark & Seyfarth, 2008; Seyfarth *et al.*, 2009a, 2009b; Stark *et al.*, 2010a, 2010b; Giebson, 2013; Stark & Wicht, 2013) and is now part of the latest advisory circular (ARS 04/2013) for concrete pavements (Bundesministerium, 2013). The development started back in the late 1980s with the first investigation into the durability of concrete regarding delayed ettringite formation (Seyfarth & Stark, 1991; Stark *et al.*, 1992; Stark & Seyfarth, 1995). Today, actual job concrete mixtures are tested in a special walk-in climate simulation chamber under a pre-defined cycle of alternating temperature and moisture conditions. One cycle lasts for 21 days and consists of 4 days of drying at 60 °C and ≤ 10% RH, 14 days of wetting at 45 °C and 100% RH and 3 days of freeze-thaw-cycling between +20 °C and –20 °C (Figure 8.14).

The concrete prisms (100×100×400 mm³) are cast with embedded stainless steel studs in each end for expansion measurements. After 24 hours, the prisms are demoulded, wrapped airtight in polyethylene foil and stored for 5 days at 20 °C. Subsequently, a flexible foam rubber tape is glued around the upper edges of the prisms to form a guard that will retain the applied test solution (Figure 8.15). Usually, the climate simulation concrete prism test (CS-CPT) starts 7 days after casting. At the end of the first drying phase, the initial length and mass are measured and 400 g of the test solution (de-icer or water, respectively) is applied to every prism for the first time and this solution remains on the prisms until the end of the cycle. After the cycle, the test solution is removed to measure length change and mass of the prisms and is replaced once the readings are taken. All measurements are done at 20 °C. During the second drying phase, the water of the test solution evaporates, leaving behind minor solid residues from the de-icer as well as leached substances from the concrete, *e.g.*, alkalis. At the end of the second drying phase, a new test solution is applied.

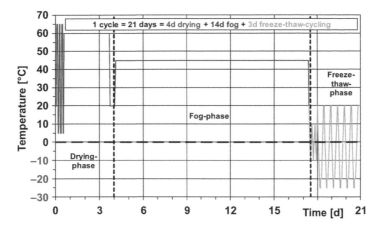

Figure 8.14 Temperature course (scheme) during one cycle of the climate simulation concrete prism test (CS-CPT) used for ASR performance testing.

Figure 8.15 Concrete prisms with NaCl solution in the climate simulation chamber.

For pavement concretes exposed to de-icers, it was found that 8 to 12 cycles (6 to 9 months) are sufficient to assess their ASR potential correctly compared with field performance (Seyfarth *et al.*, 2009a, 2009b). The expansion limits are defined as 0.5 mm/m for application of de-icer solutions (higher moisture impact) and as 0.4 mm/m for application of water only. In addition, the slope of the expansion curve between 6 and 8 cycles is evaluated and microscopical examinations are performed afterwards. The microscopical examinations comprise a thin section analysis (Figure 8.16) to detect the cause of a possible expansion and sometimes investigations by SEM/EDS to analyze certain phases.

The traditional ways to prevent or mitigate ASR in Germany had been the use of low alkali cements, the reduction of the total alkali loading by limiting the cement content and the use of non-reactive aggregates respectively (Table 8.10).

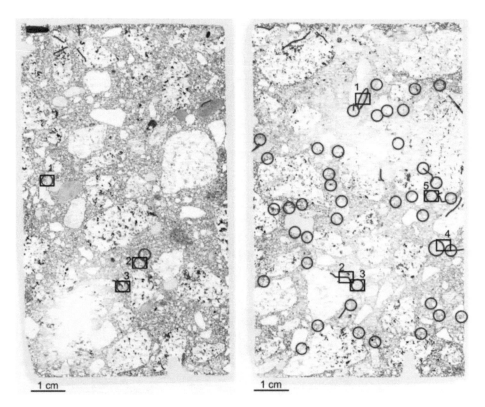

Figure 8.16 Two thin sections (60×100 mm) of a pavement concrete with markings of ASR-spots (left: exposed to water only, right: exposed to NaCl-de-icer solution).

The background of using low alkali cements and controlling the cement content is to keep the total alkali content in the concrete below 3.0 kg (Na_2O_{eq})/m³ and the OH⁻ concentration in the pore solution below 0.5 mol/l (pH = 13.7). Since most alkalis in slag are insoluble and hence not able to contribute to an ASR, a higher Na_2O_{eq} for cements with GGBS is allowed (Table 8.11). In certain cases, for example, in need for a higher cement content and the possibility of an external supply of alkalis, only aggregates classified as non-reactive are allowed to be used, otherwise they have to be replaced or the job mixture has to be evaluated by a performance test.

The use of SCMs (especially fly ash) is not an option within the existing regulations to mitigate ASR, but needs an expert recommendation, *e.g.*, based on performance testing results. Recent studies have confirmed international experiences by showing that fly ash (Schmidt, 2009; Dressler, 2013) and GGBS (Giebson *et al.*, 2010b) are effective in mitigating ASR in concrete, even when exposed to de-icers.

8.14.4 Diagnosis and appraisal of structures

ASR diagnosis starts with a visual inspection of the structure. Typical characteristics of ASR are map cracking, pop-outs and gel exudations (mainly with fast reacting

Table 8.10 Preventive measures for mitigating ASR acc. to the German "Alkali-Guideline" (DAfStb, 2013).

Aggregate reactivity	Cement content [kg/m³]	Preventive measure for exposure class		
		"WO" (dry environment)	"WF" (humid environment)	"WA" (humid environment and external alkalis, except concrete pavements)*
non-reactive (E I, E I-O, E I-OF, E I-S)	no restriction	—	—	—
moderate alkali-reactive regarding opaline sandstone and flint (E II-OF)	≤ 330	—	—	low-alkali cement
highly alkali-reactive regarding opaline sandstone (E III-O)	≤ 330	—	low-alkali cement	aggregate replacement
moderate alkali-reactive regarding opaline sandstone and flint (E II-OF)	> 300	—	low-alkali cement	low-alkali cement
highly alkali-reactive regarding opaline sandstone and flint (E III-OF)	> 300	—	low-alkali cement	aggregate replacement
other alkali-reactive aggregates, not containing opaline sandstone and flint, e.g. greywacke (E III-S)	≤ 300	—	—	—
other alkali-reactive aggregates, not containing opaline sandstone and flint, e.g. greywacke (E III-S)	≤ 350	—	—	low-alkali cement or expert evaluation
other alkali-reactive aggregates, not containing opaline sandstone and flint, e.g. greywacke (E III-S)	> 350	—	low-alkali cement or expert recommendation	aggregate replacement or expert recommendation

*for pavement concrete, the advisory circular ARS 04/2013 (Bundesministerium, 2013) has to be applied.

Table 8.11 Low-alkali cements according to DIN 1164 (2013) used in Germany ("Reproduced by permission of DIN Deutsches Institut für Normung e.V. The definitive version for the implementation of this standard is the edition bearing the most recent date of issue, obtainable from Beuth Verlag GmbH, Burggrafenstraße 6, 10787 Berlin, Germany").

Cement type	Requirements
CEM I – CEM V	$Na_2O_{eq} \leq 0.60\%$
CEM II/B-S	$\geq 21\%$ ground granulated blast furnace slag $Na_2O_{eq} \leq 0.70\%$
CEM III/A	$\leq 49\%$ ground granulated blast furnace slag $Na_2O_{eq} \leq 0.95\%$
	$\geq 50\%$ ground granulated blast furnace slag $Na_2O_{eq} \leq 1.10\%$
CEM III/B	Composition acc. to DIN EN 197-1:2011 (Table 1) $Na_2O_{eq} \leq 2.00\%$
CEM III/C	Composition acc. to DIN EN 197-1:2011 (Table 1) $Na_2O_{eq} \leq 2.00\%$

aggregates), discolorations along the joints (pavement slabs), displacements and deformations. However, laboratory tests are necessary to identify the existence of ASR clearly and to distinguish it from other deterioration processes. For this, the thin section analysis of concrete samples (Figure 8.16) supported by an inspection of fractured surfaces, SEM investigations and, where appropriate, the uranyl acetate treatment procedure are conducted. Residual expansion measurements, if necessary, are carried out on concrete cores by using the German fog chamber test at 40 °C or by using the climate simulation concrete prism test (CS-CPT) as well. Besides expansion, the static Young modulus has recently proved to be a reliable indicator to assess the degree of the microstructural damage caused by ASR (Mielich, 2009; Reinhardt & Mielich, 2012).

Early investigations on ASR-damaged concrete bridges showed that the usual repair methods, such as crack injection or applying outer reinforcement straps underneath the superstructure to control the bending, were useless because of the random cracking caused by the ASR (Bödeker, 2003). Hence, ASR-damaged bridges were often replaced. Later on it was recognized, based on laboratory studies, that avoiding a further ingress of water could be an effective way to stop a further reaction. Based on this concept, the main strategy to repair ASR-damaged structures has been the prevention of a further access of water. The method is based on (1) closing of existing cracks (0.3-1 mm) by treatment/injection with epoxy or polyurethane and (2) the application of a surface protection system, *e.g.*, based on acrylic resins, that limits the access of water but does not prevent water vapour leaving the concrete (DAfStb, 2002). At the moment, however, no data are available to evaluate how effective this strategy was to extend the service life of affected structures in the medium and long term respectively.

For pavements, the situation is more complicated, because of the large areas to be treated, plus grip and flatness requirements and the necessary joints which make it difficult to limit the ingress of water and de-icer solution reliably. Depending on the degree of the damage (Figure 8.17), the methods currently used in the field to slow down the deterioration or to repair severely damaged sections respectively are as follows (Vertreter des Bundes und der Länder, 2012):

• Thin overlay with cold mix asphalt;
• Overlay with rolled asphalt;

Figure 8.17 Severely ASR-damaged concrete slabs of a highway pavement.

Figure 8.18 Asphalt overlay of ASR-damaged concrete slabs after removing the damaged surface.

• Milling of the damaged concrete surface and applying an asphalt overlay (Figure 8.18);
• Complete replacement of damaged concrete slabs and casting new slabs with fast hardening concrete.

Several methods of current treatment are under test in the field since 2008 (Marquordt & Rother, 2011). After the first 4 years it became apparent that the application of a hydrophobic sealer could significantly slow down the ASR, while a lithium-based treatment was ineffective (Rother, 2013). Owing to the limited closing times for highways and especially for airfields, besides fast hardening concrete also

Figure 8.19 Installation of a precast concrete panel on a highway pavement (Rother, 2013).

the use of precast concrete panels is an option and first field site tests has been launched (Figure 8.19).

8.14.5 Cases

Two damaged German highway pavement concretes, built in 1991/93, were examined by thin section analysis in 2006 and it was found that ASR was the cause for the distress. In concrete 1, the gravel and the rhyolite aggregates were affected by ASR (Figures 8.20 & 8.21) and the damage occurred after 12 years in service. In concrete 2, the granodiorite aggregates were affected by ASR (Figures 8.22 & 8.23) and the

Figure 8.20 Highway pavement (concrete 1) after about 14 years in service (first damage after 12 years).

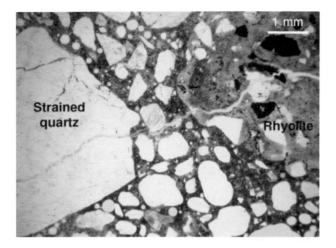

Figure 8.21 Thin section image of concrete 1 (14 years old) showing cracks in aggregate grains, partially filled with ASR gel.

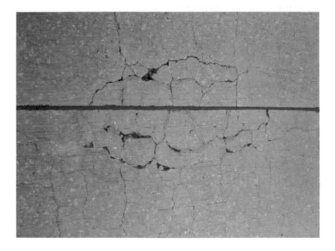

Figure 8.22 Highway pavement (concrete 2) after about 9 years in service (first damage after 8 years).

damage occurred after 8 years only. Both concretes contained 360 kg/m³ of a Portland cement (Na₂O_eq 0.90-0.95 wt.%) and had a w/c of 0.40-0.45. The concretes were air-entrained with 4.5-5.0% air. The objective of a subsequent study in Seyfarth *et al.* (2009a, 2009b) was to compare the field experience for those two concretes with ASR performance testing by using the climate simulation concrete prism test (CS-CPT). Therefore, concrete 1 was reproduced in the laboratory by using all the aggregates from the original deposits and a Portland cement (Na₂O_eq 0.90 wt.%) from the original plant. For concrete 2, the granodiorite aggregates from the original quarry were used in

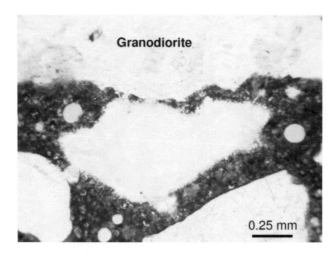

Figure 8.23 Thin section image of concrete 2 (9 years old) showing pores with ASR gel next to a granodiorite grain.

a typical mixture for pavement concrete that meets the latest requirements at that time for highway pavements.

To evaluate if the geological situation in the deposits had changed compared with the situation 15 years earlier, the coarse rhyolite aggregates from a core of concrete 1 were extracted for a mortar-bar test by means of shock-wave crushing. The mortar-bar test showed that the rhyolite from the core is highly alkali-reactive as well as the gravel and the rhyolite from the new batches (Figure 8.24). The regular German fog chamber test, however, identified the rhyolite as non-reactive (Figure 8.25). By

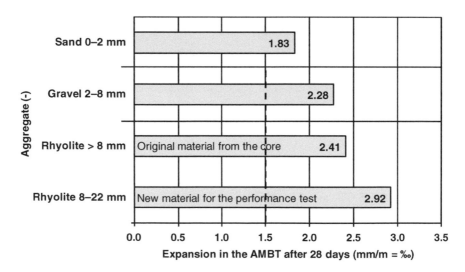

Figure 8.24 Mortar-bar test results for the aggregates used in concrete 1.

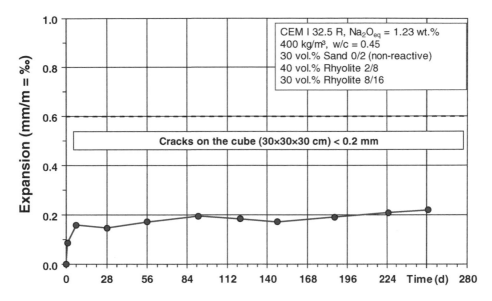

Figure 8.25 German fog chamber test result for the rhyolite used in concrete I.

contrast, ASR performance testing by using the climate simulation concrete prism test (CS-CPT) showed deleterious expansion after exposing the prisms for 7 cycles (21 weeks) to an NaCl solution (Figure 8.26) which finally corresponded with the field experience.

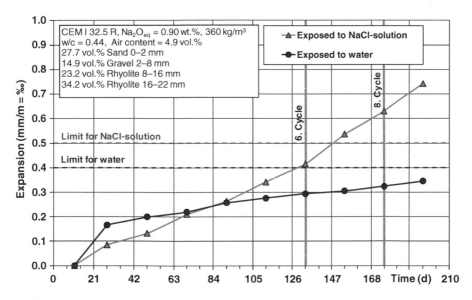

Figure 8.26 Expansions during climate simulation concrete prism test (CS-CPT) for the reproduced concrete I, exposed to NaCl solution and water (control).

Figure 8.27 Highway pavement (concrete 1) prior to the replacement in 2011 (Rother, 2013).

In the next couple of years, after the diagnosis and testing, the deterioration for concrete 1 in the field had further increased (Figure 8.27). After some repairs, the highway section had to be replaced in 2011.

For concrete 2, the granodiorite was assumed to be non-reactive according to the mortar-bar test and the German fog chamber test as well (Figures 8.28 & 8.29), but deleterious expansion occurred in the climate simulation concrete prism test (CS-CPT) after exposing the prisms for 8 cycles (24 weeks) to an NaCl solution (Figure 8.30). A subsequent thin section analysis provided clear evidence for an ASR (cracks, ASR gel), triggered by the granodiorite (Figure 8.31), which again clearly corresponded to the field experience.

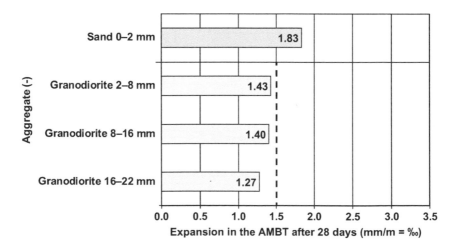

Figure 8.28 Mortar-bar test results for the aggregate used in concrete 2.

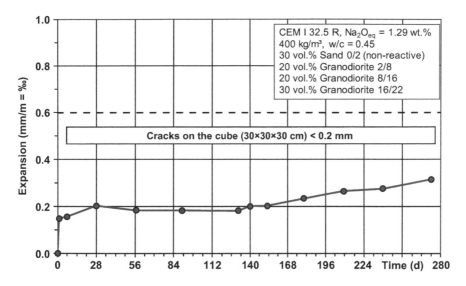

Figure 8.29 German fog chamber test result for the granodiorite used in concrete 2.

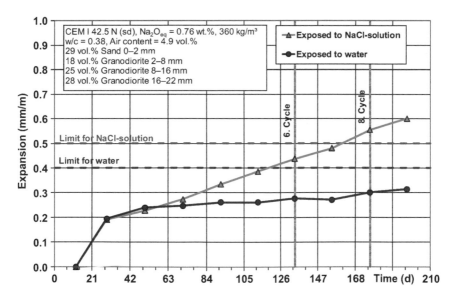

Figure 8.30 Expansions during the climate simulation concrete prism test (CS-CPT) for the reproduced concrete 2, exposed to NaCl solution and water (control).

Since 2004, more than 300 concrete mixtures comprised mainly of field representative concrete with different types of cement and mostly slow/late type alkali-reactive aggregates have been tested with the climate simulation concrete prism test (CS-CPT). The lessons learned are that ASR can be clearly triggered and accelerated considerably

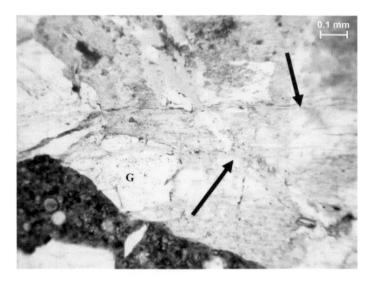

Figure 8.31 Thin section image of concrete 2 after the climate simulation concrete prism test (CS-CPT) (NaCl) shows a granodiorite grain (G) with microcracks.

in concrete with alkali-reactive aggregates and exposed to alkali-containing de-icers (Giebson, 2013). The alkali-reactivity of some aggregates is assessed incorrectly by mortar-bar tests and the German fog chamber test as well. Petrography provides powerful supplementary information about the aggregates, but is often not selective enough for a conclusive assessment. The climate simulation concrete prism test (CS-CPT) is an approved option for testing specific concrete pavement mixtures (*i.e.*, job mixtures), under widely realistic conditions, taking into account as many of the individual materials interactions as possible, for predicting a service life of 20-30 years.

8.15 HUNGARY

8.15.1 Historical introduction and case studies

Hungary is located in the Pannonian Basin, surrounded by the Alps, Carpathians and Dinarides. Most of Hungary's surface is lowland (about 68%), 30% is covered by hills between 200 and 400 metres and only 2% of the country rises above 400 metres. The largest part of the country is covered by soft Quaternary sediments, while hard rocks are located in hilly regions in Central Western, Northern and Southern Hungary. The resources of aggregates are composed mainly of volcanic and carbonate rocks (limestone and dolomite) (Török, 2015). Limited occurrences of granite and gabbro are not used as aggregates any more, especially due to environmental considerations. Crushed stones are sparsely used as aggregate in concrete since gravel deposits are more widespread throughout the country. Figure 8.32 presents the map of Hungary where the main quarries are marked.

Figure 8.32 Simplified map of Hungary with major quarries and gravel pits (updated from Török, 2007).

The aggregates for prefabricated and cast-in-situ concretes are obtained from alluvial deposits of sand and gravel. In addition, crushed basalt and andesite are used in some cases, such as for prestressed prefabricated bridge girders in order to increase the tensile strength of the concrete. The Hungarian experience with aggregates shows that the quartz gravel and sand are not potentially alkali-reactive. However, an andesite has shown to be reactive.

In Hungary there are only a few cases in which, among the causes of durability problems, the possibility of ASR was also listed, but finally not confirmed. Therefore, maintenance of buildings or constructions due to ASR is not happening in the country.

8.15.2 Aggregates, binders and admixtures

The standards followed regarding concrete and its components are those produced by CEN. The national MSZ 4798-1 (2004) is the addendum that gives guidances about the application of EN 206-1 (MSZ EN 206-1) in Hungary. The available types of cements are very wide. Within the five main cement types: ordinary Portland cements (CEM I), heterogeneous Portland cements (CEM II), blastfurnace slag cements (CEM III), pozzolanic cements (CEM IV) and ternary blended (composite) cements (CEM V), about 27 different cement types are produced. Due to the need of controlling the dust for environmental protection, the Na_2O_{eq} content of the cements made with high GGBS replacements (CEM III/A and CEM III/B) was increased, resulting in a larger possibility for alkali-silica reaction to occur. The threshold limit of alkali content is fulfilled by most of the cement types. Nowadays mainly the CEM I 42.5 and CEM II/A-S cement types are used.

8.15.3 Test methods and preventive measures

In Hungary just the accelerated test methods are used for the assessment of the potential reactivity of aggregates, namely:

• The Chemical Method in accordance with ASTM C289 (2007, but now withdrawn). In this method, crushed aggregate with grain size less than 0.3 mm is washed out and sieved to pass the 300 µm sieve and be retained on the 150 µm sieve. The aggregate crushed between the above mentioned grain size is subjected to 1 N NaOH solution during 24 h at 80 °C. The potential reactivity of an aggregate with alkalis in Portland cement concrete is indicated by the amount of reaction, which is determined by the decreasing concentration of the sodium hydroxide solution and the amount of dissolved silica. However, the gravels tested are plotted in a close but critical area.
• The petrographic examination following the DAfStb (2007, now 2013) method, made by sorting of the sensitive grains (macroscopic assessment). This method implies that an expert makes the sorting. The method usually classifies more aggregates reactive than the chemical method.

Furthermore, the grain sizes between 1 and 4 mm are subjected to alkali treatment. The reactivity is determined by the change in mass and decrease of grain sizes. For grain sizes above 4 mm the sensitive grains (flint, opal) are sorted by the petrographic method. The results are obtained by measuring the mass of flint as well as measuring the mass of the weakened (softened) grains of opal after treatment in alkali solution (treatment in NaOH solution for 60 min at 90 °C) related to the total mass of the sample.

8.16 ITALY

8.16.1 Introductory overview

The first structures with ASR were identified in the early 1980s, prevailingly located on the middle-southern Apennine mountains belts, and along the Adriatic coast. Several cases of alkali aggregate reaction in concrete were found in residential and industrial buildings (Baronio, 1983; Levi et al., 1985), in industrial pavements (Alunno Rossetti, 1981), in a chimney (Baronio & Berra, 1989) and in some dams (Bon et al., 2001; Giuseppetti et al., 2002; Saouma et al., 2005). Thus, the investigation of the suspicious aggregate types and affected concrete structures led towards the study on the reliability of testing procedures and the effectiveness of mitigation techniques (Berra, 1983; Baronio, 1984; Berra & Baronio, 1986). In 1983 Baronio examined 8 years old concrete of a building located in Lower Molise. Concrete cores were examined under the optical microscope in thin section and revealed the presence of abundant ASR gel around chert particles. Baronio and Berra (1989) examined a deteriorated chimney which was built at the early 1960s. The structure had shown map cracking, white exudations and detached pieces along its concrete columns and transverse beams. The chimney was also affected by chloride-laden winds coming from the sea, sea-water spray and high temperature combustion gases. Chemical and microstructural analysis on the deteriorated samples revealed that the aggressive environmental conditions might also have enhanced the alkali-silica reaction.

Gasparotto *et al.* (2011) studied a damaged pavement situated in Bologna (North Italy) for which concrete was manufactured with aggregates from a quarry in the Reno river valley composed of sand-gravel deposits of Quaternary age. Concrete petrography in thin section concluded that the siliceous limestone and chert/flint present in the aggregate did not show any reactivity and ASR had just developed from fine-grained silica rich marls.

Since the early 2000s, the Technical Committee "Cement & Concrete" of the National Institute for Standardization (UNI) has developed an upgrade programme of the standards on ASR and recommended criteria for prevention. The regulatory work was supported by experimental research funded by UNICEMENTO (Technical Organization of Italian Cement Producers).

8.15.2 Reactive Aggregates

The study of Italian aggregates has been quite intense, since the 1980s (Barisone, 1984; Barisone *et al.*, 1986, 1989; Barisone & Restivo, 1992a, 1992b, 2000). Barisone (2001) studied Quaternary alluvial sediments deposited by rivers rising from the Apennines range in Italy. In this study almost 150 rivers were sampled and more than 200 samples collected. The studies evidenced differences between the chert/flint of Marche and Molise districts and those from other districts. The alluvial deposits from the Apennine Mountains are composed of sedimentary rocks of Mesozoic and Cenozoic age: marls, limestones, shales, sandstones (all often siliceous with flint/chert, jasper and chalcedony in nodules and lenses) and clays. There are also areas with volcanic formations (sometimes with jasper, opal and chalcedony), granites and metamorphic rocks (gneisses, micaschists, granulites, etc.) with deformed quartz (Figure 8.33).

Regarding the characterization of Italian chert/flint, a detailed geological-petrographic study on the chert/flint presence covering over half of Italy (South of River Po, excluding Sicily and Sardinia isles) was carried out some years ago (Barisone *et al.*, 1986, 1989; Barisone, 2004). Barisone and Restivo (1996, 2000) stated that the Italian cherts/flints are less reactive than the ones from the North of Europe. The authors tested 30 samples from Great Britain, the Netherlands, Germany and Denmark, trying to find

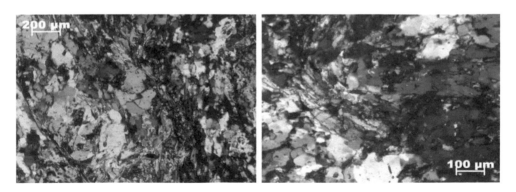

Figure 8.33 Examples of metamorphic rocks used as aggregates in Italy: samples from Partner Project (2006) expansion tests. Photomicrographs by Isabel Fernandes.

out the differences from the Italian examples and concluded that the higher reactivity of the north-European chert/flint might be due to a structural disorder at electron microscope scale or to a higher porosity mainly due to different weathering conditions during natural transportation. The chert/flint content of the natural alluvial Italian deposits was less than 7%. The opal from Quaternary deposits near Turin was found to be very reactive.

The classification of potential reactivity of Italian aggregates according to RILEM AAR-0, 2003 was presented by Vola *et al.*, 2011, who developed a study involving 60 Italian aggregates from 13 regions, mainly from the North of the country, especially from the Padana Valley (Figure 8.34 and Table 8.12). The authors confirmed that the

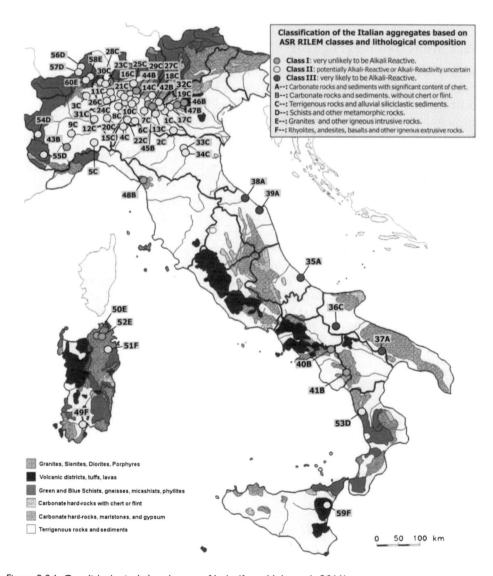

Figure 8.34 Geo-lithological sketch map of Italy (from Vola *et al.*, 2011).

Table 8.12 Classification of the Italian aggregates based on the geology and RILEM classification (Vola et al., 2011).

Aggregate class	Type of rock	ASR susceptible phases in the rock	RILEM ASR class	Reactivity assessment
A	Carbonate rocks and carbonate sediments with significant content of chert and flint	Chert or flint	II or III	Medium to strongly reactive
B	Carbonate rocks, marlstones, gypsum and carbonate sediments without chert or flint	Rarely strained or microcrystalline quartz	I	Not reactive
C	Terrigenous rocks and alluvial sediments (clays, silts, sands and gravels)	Chert or flint, microcrystalline and/or strained quartz	II or III	Medium to strongly reactive
D	Green and blue schists, gneisses, quartzites, phyllites and other metamorphic rocks	Microcrystalline and/or strained quartz	II	Slowly reactive
E	Granites, granodiorites, syenites, diorites, and other igneous intrusive rocks	Microcrystalline and/or strained quartz	II or I	Slowly or not reactive
F	Rhyolites, andesites, basalts, porphyries, and other igneous extrusive rocks	Microcrystalline and/or strained quartz; amorphous glassy phase, tridymite and cristobalite	II	Slowly to strongly reactive

reactivity of Italian aggregates is related to the presence of microcrystalline and crypto-crystalline quartz, generally defined as chert and/or flint, which are present in various sedimentary rocks, mostly carbonates (aggregate class A), but also in siliciclastic rocks (aggregate class C), and in alluvial deposits from the erosion of the Alpine and Apennine Mountain belts. The amorphous phase of volcanic rocks and strained quartz with undulatory extinction, mainly associated with metamorphic rocks (aggregate classes D, E and F), are also ASR susceptible, although to a much lesser extent (Barisone, 1984).

Aggregates with a significant content of chert/flint (aggregate class A) occur mainly in alluvial deposits of the Eastern Apennines on the Adriatic coast and in the central-eastern Padana Valley.

8.15.3 Standard Test Methods for alkali-silica reactivity assessment

The first Italian standard on the characterization of the alkali-reactivity of the aggregate for concrete was UNI 8520-22 *Aggregates for use in concrete – Determination of potential reactivity of alkali in aggregates*. This standard was published in 1999, and it included test methods and classification criteria for reactivity. UNI 8520-22 was revised to a limited extent in 2002 with the inclusion of the petrographic method according to UNI EN 932-3.

The implementation of this standard has been based on a background of previous studies and researches. Turriziani (1986) prepared an worldwide overview of the state of the art of investigations, test methods and prevention criteria for ASR, presented at the 8th International Congress on the Chemistry of Cement (Rio de Janeiro, 1986). In 1991 Collepardi *et al.* presented a summary on the tests performed in Italy for the identification of alkali-reactive aggregates. Berra *et al.* (1994, 1996) investigated the use of ASTM C1260 and ASTM C227 in assessing the effectiveness of fly ash and silica fume in reducing expansions caused by the alkali-silica reaction using fused quartz as a reactive aggregate.

According to UNI 8520-22 (1999), if opal and amorphous silica, volcanic glass, microcrystalline fibrous quartz such as chalcedony, or strained quartz, identified by petrographic analysis, are present, the aggregate must be tested using the accelerated mortar-bar test in 1 M NaOH solution at 80°C (25x25x285 mm size mortar-bars). If the 14 day expansion of such mortar-bars is less than 0.10%, the test aggregate is considered as non-reactive. Otherwise, long-term expansions using mortar-bars at 38° C and 100% R.H. should be considered. For this last test, 0.05% expansions at 3 months and 0.10% at 6 months are the suggested limits and, if they are exceeded, the aggregate is considered as potentially alkali-reactive.

Starting from 2010, a detailed revision of the Italian standards on the subject of ASR was carried out. This revision was addressed to the test methods and classification criteria. The standard methods for the evaluation of the alkali-silica reactivity currently in force in Italy are listed in Table 8.13.

UNI 11504 (2013) is relevant to mortar-bar expansion in 1 M NaOH solution at 80°C. The reactivity classification of the aggregate is based on expansion limits at 14 days of immersion: the aggregate is classed 'non-reactive' for expansion of less than 0.10%; the aggregate is 'reactive' for expansion of more than 0.20%; the test results are considered 'not conclusive' for expansion of between 0.10 and 0.20%. As proposed in an informative annex, in the case of a doubtful result, it could be useful to prolong the immersion time of mortar-bars until 28 or 56 days, with expansion limits of 0.33 and 0.48%, or correlate expansion data with the Avrami-Kolmogorov-Mehl-Jonhson kinetic model, allowing calculation of the kinetic

Table 8.13 List of standard methods recently published in Italy.

Number	Year of publication	Title
UNI 11504	2013	Alkali-aggregate reaction in concrete – Determination of the potential alkali- reactivity of aggregate for concrete – Test of accelarated expansion of mortar-bars
UNI 11530	2014	Alkali-aggregate reaction in concrete – Detection of the potential alkali- reactivity of aggregate for concrete – Detailed petrographic examination to detect potentially alkali-reactive constituents of aggregates
UNI 11604	2015	Alkali-aggregate reaction in concrete – Detection of the potential alkali- reactivity of aggregate for concrete- Concrete prisms accelerated expansion test

parameter ln K: the aggregate is considered reactive or not reactive if ln K is higher or lower than -6.0, respectively.

The petrographic method of UNI 11530 (2014) is focused on thin section analysis of aggregates, allowing a quantitative determination of the constituents and it is based on the recommendation RILEM AAR-1 (2003). If the microscopical examination does not reveal any reactive constituent, the aggregate is considered not reactive to alkali. If constituents, whose potential alkali reactivity in field concrete is well known, in doubt or unknown, are detected, the aggregate is classed as potentially reactive. This standard method represents an integration of UNI EN 932-3 (2004) and it is considered as the first stage of the aggregate characterization.

The concrete prism accelerated test method of UNI 11604 (2015) is based on a concrete mix having a specified composition (cement content 440 kg/m^3 and w/c ratio 0.5) and a defined aggregate grading. The concrete alkali content is prefixed to 5.5 kg Na$_2$O$_{eq}$/m^3. Test temperature is 38°C and the relative humidity inside the test prism storage chamber is 100%. The test duration is 1 year. The standard supplies a design of the container of the test prism that ensures limited alkali leaching from the concrete during the test. The aggregate is considerd as potentially alkali reactive if the 1-year expansion is higher than 0.04%.

The applications of the test methods and criteria for classification of aggregate reactivity are the subject of the new edition of UNI 8520-22 (2002). This standard is still being finalized with the adoption of the classification principles of the aggregate reactivity currently included in UNI 11417-2 (2014) *Durability of concrete works and precast concrete elements – Criteria for the avoidance of alkali-silica reaction*.

The classification principle is based on the parallel execution of the petrographic method and accelerated tests using mortar-bars. If the results of the petrographic and mortar-bar expansion tests agree in diagnosing the aggregate as a non-reactive, the aggregate is not subjected to the concrete prism expansion test and it is classified as non-reactive.

If the results of petrographic and accelerated mortar-bar tests are concordant in diagnosing the aggregate as reactive, it is classified as potentially reactive. Finally, if there is a discrepancy between the conclusions of petrographic examination and mortar-bar expansion testing, the aggregate is subjected to the concrete prism expansion test, the result of which being decisive for the classification of aggregate as reactive or non-reactive.

8.15.4 Mitigation measures

For the mitigation of ASR, Italian standard UNI 11417-2:2014 (2014) gives special focus to the following measures:

- Limiting the alkali content of concrete. Such a preventative measure can be attained by using low-alkali cements.
- Using composite cement type CEM III, CEM IV complying with EN 197-1 or enough amounts of active additions in concrete mix, such as pozzolanic materials (fly ash, natural pozzolana, silica fume) or granulated blastfurnace slag of proven effectiveness against ASR.
- Using chemical admixtures based on lithium compounds, especially lithium nitrate, which can change the properties of the silicate gel produced by ASR,

making it non-expansive. The effectiveness of these chemical admixture must be assessed by preliminary testing.

For normal building and civil engineering structures and in the absence of an external alkali supply, a limit of 3.0-3.5% Na_2O_{eq}/m^3 for the Total Effective Equivalent Sodium oxide has been proposed.

The use of blended cements as well as mineral additions (natural pozzolanas, fly ash, blastfurnace slags, silica fume) in concrete is a normal practice in Italy, particularly in those regions where there is a high risk of alkali-silica reaction.

However, UNICEMENTO promoted a pre-standard research activity with the aim of finding a suitable method to evaluate the effectiveness of blended cements manufactured with active mineral additions in preventing alkali-silica reaction in concrete (Costa *et al.*, 2011).

In a recent work that extended the above-mentioned research, Berra *et al.* (2013, 2015) proposed an innovative approach aimed at comparing the effectiveness of different types of ASR inhibitors, such as low-alkali Portland cements, lithium compounds and blended cements manufactured with active mineral additions.

The 'TAL' (Threshold Alkali Level) value (expressed in kg Na_2O_{eq}/m^3 concrete) is the basic parameter considered by this new approach. It is defined as the available alkali content of concrete above which deleterious ASR expansion will occur (Berra *et al.*, 2005). This parameter is specific for each aggregate: the lower the alkali-reactivity of aggregate, the higher its TAL value. Based on the TAL values, Berra *et al.* (2005) proposed the following classification of alkali-reactivity of silica/silicate aggregates:

- TAL $\leq$ 2.8 kg Na_2O_{eq}/m^3; quickly (highly) reactive aggregate
- 2.8 < TAL $\leq$ 5.5 kg Na_2O_{eq}/m^3; moderately reactive
- 5.5 < TAL $\leq$ 7.4 kg Na_2O_{eq}/m^3; slowly reactive
- TAL > 7.4 kg Na_2O_{eq}/m^3; non-reactive.

With reference to TAL, other parameters, such as the driving force of the deleterious expansion process in concrete (Δ_{ASR}) and the tolerable driving force by concrete (Δ_{tol}), have been defined and used as a more efficient approach for prevention of ASR (Berra *et al.*, 2013, 2015). The determination of TAL has been included in an informative Annex of the UNI 11604 (2015).

8.17 THE NETHERLANDS

8.17.1 Historical introduction and case studies

After the first appearance of alkali-silica reaction in Dutch literature (Bosschart, 1957), early research was focused on the reactivity of aggregates commonly used and quarried in the Netherlands (van de Fliert *et al.*, 1962). These authors concluded that sands and gravel from river deposits in the Netherlands were not alkali-silica reactive. After ASR was recognised in a bridge in Schoonhoven in 1991 (Heijnen & Van der Vliet, 1991; Heijnen, 1992; Heijnen *et al.*, 1996), three viaducts and one tunnel all in or over Dutch main road A59 were assessed in 1992 (Borsje *et al.*, 1992; Broekmans, 2002). The overall conclusion from concrete petrography analysis of cores extracted from two of the structures was that ASR did occur in the viaduct and that the tunnel concrete might contain some grains of alkali-reactive aggregate. The other structures were not included in the petrographic analysis. The reinvestigation of one of the viaducts, Wolput, by thin section petrography was justified by the presence of ASR induced cracking in the structure (Bubberman & Broekmans, 1994; Broekmans, 2002). During a quick survey along the A59 road including some structures, many more were found to show signs of deterioration potentially caused by ASR (Broekmans, 2002).

ASR has been diagnosed in 40 to 50 concrete structures in the Netherlands, mainly viaducts, tunnels, bridges and locks (Nijland & Siemes, 2002; Nijland *et al.*, 2003). The ASR-affected structures included approximately 20 bridges over the motorway A59. Most of the affected structures were built in the late 1960s to early 1970s. The affected concrete of the structures was usually manufactured with coarse grained ordinary Portland cement, with small numbers also containing either ground granulated blastfurnace cement with a maximum of 35-40% by volume of binder or up to 15% by volume of fly ash. Two structures affected by ASR with a blastfurnace slag > 65% by volume of total binder and three structures with an unknown slag content raised questions on the effectiveness of slag in inhibiting alkali-silica reactivity. However, the research has revealed that in all these cases, the concrete damage was found to be due to shrinkage and/or frost. Microscopical examinations revealed that, with a few exceptions, the water/cement ratio of the affected structures varied between 0.45 and 0.60 and the total alkali content of concretes generally exceeded 3.0 kg/m^3 (Nijland *et al.*, 2003).

In general, the ASR-induced deterioration of concrete was due to ordinary Portland cement and local river-dredged gravel and sand where the reactive constituent was porous chert, which was present in the coarse aggregate in varying amounts (1 to 6% by mass). Secondary ettringite formation often accompanied ASR (Heijnen, 1992). Most of the deteriorated concrete structures were more than 30 years old, revealing a slow reaction rate, mostly due to the nature of the reactive components of the aggregates (coarse porous chert grains, sericitic quartzite, sericitic sandstone, quartzite containing microcrystalline quartz or chalcedony) (Heijnen *et al.*, 1996).

8.17.2 Aggregates, binders and admixtures

Aggregates used in the Netherlands for concrete production may roughly be divided into four categories (Heijnen *et al.*, 1996; Nijland *et al.*, 2003):

- River gravel and sand, in particular those from the rivers Meuse, Rhone and Waal, with sand rich in quartz but containing subordinate amounts of porous and non-porous chert, quartzite, sandstones, limestones, glauconite, feldspars and opaque minerals;
- Marine gravel and sand, commonly dredged off-shore from the British North Sea coast;
- Secondary materials;
- Crushed natural rocks, such as granite from Glensanda, Scotland, UK.

A geological map of the Netherlands, showing provincial boundaries and aggregate extraction sites (active, abandoned and planned; circle size proportional to extraction dimension) is given in Figure 8.35.

Nearly all cases of ASR damage involve river derived aggregates. Marine aggregates, which have been used in the Netherlands for about a quarter of a century, have not been

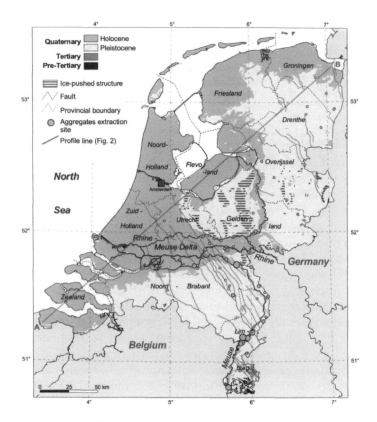

Figure 8.35 Geological map of The Netherlands (van der Meulen *et al.*, 2005, modified from Weerts *et al.*, 2004).

involved (Nijland & Siemes, 2002). Polarization and fluorescence studies of the concrete of ASR-affected structures show that the reactive aggregate grains are chalcedony, impure sandstones, greywacke, sericitic limestone, etc., in addition to porous chert. Microcrystalline quartz in quartzite and schist/slate has been identified to participate in the reaction. In one case, quartz porphyry was used as aggregate.

The types of cements that are produced in the Netherlands are blastfurnace slag (CEM III/A and CEM III/B according to EN 197-1) or Portland-fly ash cement (CEM II/B-V), respectively. The alkali limits of the cements are given in CUR-Recommendation 89 (2008) (see Table 8.14).

8.17.3 Test methods and preventative measures

The first national recommendation on the prevention of ASR, CUR Recommendation 38 (*Measures to prevent damage by the alkali-silica reaction (ASR) in concrete*) was published in 1994. In 2000, CUR regulations commission 62 was established, with the aim of specifying the procedures for the determination of deleterious ASR risk and the establishment of preventative measures and methods for assessing the reactivity of aggregates. The flow chart for preventing the ASR in new structures is given in CUR-

Table 8.14 Requirements for cement with preventative effect with regard to ASR (CUR Recommendation 89, 2008).

Type of cement	CEM II/B-V			CEM III/A	CEM III/B
Cement with a fly ash or slag content [%(m/m)] of	≥25		≥30	≥50	≥66
Na$_2$O$_{eq}$ of fly ash in cement [%(m/m)]	≤2.0	>2.0 & ≤3.0	>3.0 & ≤4.5	n/a	n/a
	Maximum alkali content of cement [%(m/m)]				
If alkali contribution from other components ≤ 0.6 kg/m^3	1.1	1.3	1.6	1.1	1.5
If alkali contribution from other components > 0.6 & ≤ 1.2 kg/m^3	0.9	1.1	1.5	0.9	1.3
If alkali contribution from other components > 1.2 & ≤ 1.6 kg/m^3	0.8	1.0	1.4	0.8	1.2

In this table "other components" is understood to include all constituents of the concrete, except cement and separate fly ash.

Recommendation 89 (2008), which was followed by CUR Recommendation 102 (2008) (*Inspection and assessment of concrete structures in which the presence of ASR is suspected or has been established*). In the Dutch recommendations, the reactivity of aggregates is first determined by measuring the amount of porous flint+chalcedony+opal present by means of polarising and fluorescence microscopy (PFM). This is followed by the accelerated mortar-bar expansion test, or in combination with the concrete prism expansion test (CPT).

The presence of moisture is a requirement for deleterious ASR expansions to take place. Thus, the Dutch Guidelines specify no precautions including the utilisation of additions for a dry environment with a further requirement that the structure thickness is less than 1 metre, while thicker elements are considered to have the possibility of being permanently humid during service. Regarding the amount of total alkali content of concrete, the commonly accepted limit of 3 kg/m^3 Na$_2$O$_{eq}$ is maintained in the Dutch guideline. It was also considered that external alkali infiltration may take place for the concretes subjected to environmental classes 3 (humid + de-icing salts), 4 (marine) and 5 (aggressive). Preventative measures for additive levels are suggested for concretes subjected to such environments. It was also considered that the amount of alkali available for ASR is lower for the concrete made with supplementary cementitious materials (additions) than the total alkali content of that concrete.

Petrographic examination is the major procedure used for identifying the susceptibility of aggregates to ASR. Petrographic examination, in combination with PFM and point-counting analysis, are used on a routine basis. Concrete aggregates are separated based upon their mineralogical composition (Heijnen *et al.*, 1996). According to CUR Recommendation 89 (2008), the assessment of aggregates starts with petrography (based on RILEM procedures: now see Nixon & Sims, 2016) as a screening method. The aggregates passing the petrographic examination should also be subjected to expansion testing. For 'non-reactive' aggregates, the accelerated mortar-bar test (AMBT) must show expansions less than 0.10% at 14 days. The AMBT is not reliable for the aggregates containing more than 2% by volume porous chert, chalcedony and opal (Larbi & Visser, 2002). For the concrete structures with alkali contents of less than

3 kg/m^3 having a desired service life of less than 50 years, the concrete prism test was considered as a reliable test method for evaluating the alkali-aggregate reactivity.

8.17.4 Diagnosis and appraisal of structures

The Civil Engineering Division of the Netherlands Ministry of Transport, Public Works and Water Management started a project to monitor the structural behaviour of two viaducts on highway A59 in 1998. In 2002, Borsje *et al.* summarized the results of this project, having two objectives: determining the changes in the behaviour of the structure with time and evaluating the effects of rehabilitation and maintenance practices on the structure. The project involved the monitoring of two viaducts, Wolput and Vlijmen-Oost. The former was strengthened in 1996 and the latter was renovated at the end of the 1980s. During the project, both structures were subjected to maintenance, preventing the ingress of moisture from the top of the bridge decks and application of a permeable surface layer on the undersides.

The monitoring system included the concrete moisture content, the deformations perpendicular to the reinforcement, the deformations within the plane of reinforcement, the changes in the height of the bridge decks (in order to understand if deflection was occurring) and finally the monitoring of crack propagation of concrete. The deformation measurements were taken once a season for a period of two years, in a total of eight measurements. The monitoring system has resulted in a number of findings about the bridge decks: relative humidity measurements between 85% and 95%, and the largest expansions observed at the locations where the highest relative humidity was measured. Monitoring of the concrete on viaduct Wolput indicated that the concrete was humid and continuous expansion was observed at those parts which were strengthened. Monitoring of Vlijmen-Oost indicated no deformations other than those derived from temperature changes. Effectiveness of monitoring techniques has also been discussed (Borsje *et al.*, 2002).

Investigations on the ASR-affected bridges by Siemes and Bakker (2000) on the motorway A59 have been used as basis for the Dutch guideline CUR Recommendation 102 (2008) on structural consequences of ASR. The basis of these investigations was the British guidance (ISE 1992) and included compressive, tensile splitting and uniaxial tensile strength determinations. During the investigation of 25 concrete bridges and other structures affected by ASR, it was found that the uniaxial tensile strength of concrete was extremely low in relation to both the compressive strength and tensile splitting strength. The reduction observed in some of the bridges was up to 82% with an expansion between 0.5% and 1% (Siemes & Visser, 2000; Siemes *et al.*, 2002).

The Dutch CUR Recommendation 102 (2008) provides the procedures used for the inspection and investigation of structures with medium or high risk (see Table 8.15). With the results obtained from the inspection and assessment of concrete structures, precautions are also recommended. However the propagation of the ASR mechanism and the consequences for the structure are not part of the evaluation with CUR Recommendation 102. Therefore, it lacks a precise recommendation on what the modification methods entail and possible consequences of modifications such as additives or SCMs.

A recent PhD research project, entitled *Development of a Performance Assessment Tool for Alkali Silica Reaction (PAT-ASR)*, is being carried out in TU Delft, in order to

Table 8.15 Selection table for control measures (Abbreviations: V: Sealing against moisture, targeted drying of concrete; M: Monitoring; VF: Measures for guaranteeing safety and functioning, e. g., limiting the load or applying a local reinforcement or support; C: Limiting the risk of reinforcement corrosion) (CUR Recommendation 102, 2008).*

Risk classification of the structure	Relatively low tensile strength indication *)	Level of crack formation (indicative)	Necessary measure(s)				Optional measures			
			V	M	VF	C	V	M	VF	C
Low	Yes	< 1 mm/m	1			2	4			
	Yes	> 1 mm/m	2	4		5	4	1	1	1
	No	< 1 mm/m				2	3			
	No	> 1 mm/m	1	4		3	4		1	1
Medium	Yes	< 1 mm/m	2	6	2		4		1	2
	Yes	> 1 mm/m	6	6	2	4			2	1
	No	< 1 mm/m	2	4	1		2	2	1	2
	No	> 1 mm/m	5	4	2	4	1	2		1
High	Yes	< 1 mm/m	5	4	4					2
	Yes	> 1 mm/m	6	4	5	4				2
	No	< 1 mm/m	6	5	2				2	2
	No	> 1 mm/m	6	6	3	4			1	2

*The numbers 1 to 6 indicate the importance of the measures. The number 6 indicates that the measures are highly recommended, the number 1 indicates that the measures are considered to be the least urgent. If a cell is empty this means that the measure will not be useful.

provide expansion data with well-known materials, microstructural modelling and structural modelling of both laboratory and site expansions. Main goals include analysis of reliability of test results from concrete prism testing for ASR (RILEM AAR-3 and AAR-4), damage quantification methods and fracture model implication possibilities. Previous studies by Çopuroğlu and Schlangen (2007) and Schangen *et al.* (2008) revealed that the Delft Lattice model is a promising tool for modelling ASR expansions.

8.18 POLAND

8.18.1 Historical introduction and case studies

Poland has been a significant aggregate producer in Europe since the 1990s, during a period of intense economic growth. Total (natural and manufactured) aggregate production in the country exceeded 230 million tonnes in 2012. Natural aggregates (sands and gravel) represent 70% of this production, which makes Poland (along with Germany) one of the biggest producers of this group of aggregates in Europe, still with 14.4 million tonnes of reserves. About 70% of all natural aggregate deposits are in Northern and Central Poland and the local exploitation represents about 60% of the total natural aggregate production. However the quality of these deposits is very variable. Deposits of low quality, genetically related with glacial activity, are dominant. The components considered potentially reactive include opal, silicified dolomites and, mainly, marl limestone with opal. The reactive components are most often concentrated in the 4-8 mm and 8-16 mm size fractions, and rarely in the finer fraction (below 4 mm).

Symptoms of concrete deterioration due to ASR were observed in a number of structures since the 1970s, coinciding with the production of cements by dry methods, but the quantification and recording of locations exhibiting symptoms of concrete deterioration are fragmentary (Góralczyk, 1984; Gościniak, 1989). Evidence comes from tests executed by research institutes dealing with this issue and certain building production organisations, including concrete element producers and building companies. Symptoms of concrete deterioration by ASR are often not recognised or are classified as being caused by other mechanisms (for example, aggressive water reactions, corrosion of concrete reinforcement, influence of atmospheric factors on concrete, etc.). Research and investigations performed show that the concrete deterioration takes place continuously in Northern, East and South-Eastern Poland. These effects are particularly observed in structures such as housing, public buildings and prefabricated concrete elements in the regions of Białystok, Olsztyn, Chełm, Gdańsk, Bydgoszcz and Toruń, and sporadically in the Rzeszów and Przemyśl, regions where crushed sandstone from the Karpaty Mountains is used.

Concrete spalling on roofs and wall elements, efflorescence, pop-outs and cracks are the most common symptoms of deterioration of concrete elements. The resultant necessary rehabilitation works add considerable costs to both new buildings and those that already exist. Research has confirmed that concrete damage arose from reactive aggregates and cements with an alkali content above 0.60%.

Góralczyk (2001) surveyed and made an inventory of concrete structures exhibiting symptoms of ASR and found that reactions occurred on the following concrete structures (Figure 8.36):

- Trasa Łazienkowska viaduct, near Torwar, Warsaw
- Ursynów viaduct, Warsaw
- Viaduct, near Grojec, Warsaw

Petrographic examination (Góralczyk, 2001) on concrete specimens from these structures confirmed the occurrence of characteristic alkali-aggregate reaction features: including dolomite conversion into calcite (dedolomitization), the occurrence of reaction rims and characteristic products of both alkali-carbonate reactions (calcite, brucite, natrite) and of alkali-silica reaction (alkali-silica gel).

8.18.2 Aggregates, binders and admixtures

About 20-30% of aggregates produced in Poland are potentially alkali-reactive. The producers are obliged to perform test methods in order to characterise the aggregates. The first studies performed concluded that Central and Northern Poland were considered to contain potentially reactive aggregates, mainly in the north-east, according to Polish requirements in the former PN-B-06712 (1986) (*Mineral aggregates for concrete*), replaced in 2004 by the PN-EN 12620. It now seems that the distribution of potentially reactive aggregates is very irregular. Most of these reactive aggregates are natural aggregates of the north-east and north-west regions, and crushed aggregates from dolomite rocks, porous limestones and weathered igneous rocks (melaphyres, porphyries, basalts, tuffs).

Figure 8.36 Trasa Łazienkowska viaduct, near Torwar. Deterioration of concrete elements. Characteristic cracks, spalling and exudation of ASR products.

The research undertaken in the country by different methods according to PN-B-06714-46 (1992) showed that natural aggregates containing opal and chalcedony, chert/flint and other silica-rich rocks, white marls and white marl limestone, clay rocks, dolomite sandstones with ferruginous-clayey binder, as well as weathered igneous rocks, could all be treated as reactive aggregates (Góralczyk & Kukielska, 2008). Aggregate deposits with the above-mentioned reactive components occur mainly in north-eastern Poland (regions of Białystok, Ciechanów, Olsztyn, Suwałki, and Toruń), where carbonate rocks (marls, dolomites), argillite and chert/flint containing amorphous silica outcrop. In north-western Poland (regions of Szczecin and Gorzów), silica rocks (chert/flint), carbonate rocks and clays (marls) are the main reactive components.

In Central Poland, potentially reactive natural aggregate deposits occur sporadically (regions of Łódź and Warszawa), as well as in Southern Poland (regions of Opole and Katowice) where the reactivity is associated with weathered igneous rocks (melaphyres, granites, volcanic rocks), which can be eliminated from the final aggregate product by processing (crushing, rinsing, sieving), although it implies increased costs of production.

Regarding the carbonate rocks, potentially reactive aggregates are mainly obtained from dolomite rocks that outcrop in the regions of Kielce, Częstochowa, Katowice and Sandomierz (Jazwica, Radkowice, Korzecko). Porous limestones with a considerable silica content (deposits in Karsy, Trawniki and Pińczów, also from Kielce, Tarnobrzeg and Lublin regions) also show to be potentially reactive. Silicified limestones and dolomites containing opal, chalcedony or cryptocrystalline quartz occur in the Piechcin – Bielawy and Pomorskie voivodeship deposits.

Crushed aggregates can also be produced from potentially reactive sandstones, namely (Figure 8.37):

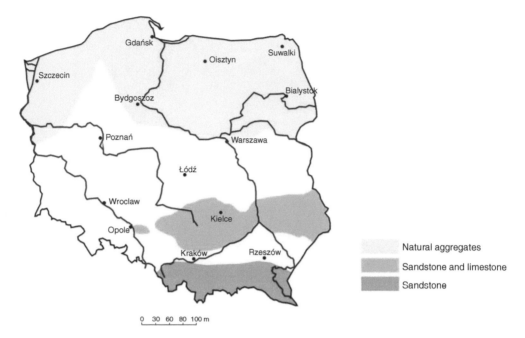

Figure 8.37 Areas of potentially reactive aggregates in Poland (Góralczyk, 2005).

a)   Sandstone from flysch in the Karpaty Mountains, containing carbonate rocks with silica and clay minerals in the rock cement (deposits of Bóbrka, Komańcza, Posiwieć, Osielec).

b)   Sandstone of Kielce region. Research showed that quartzites and quartzitic sandstones (Bieliny, Bukówka, Wiśniówka), ferruginous sandstones (Miniów, Kopulak) and sandstone with silica in the rock cement are reactive.

Other reactive aggregates are obtained from chalcedony (Inowłódz), diatomites (Leszczawka), radiolarites and opals (tamobrzeskie voivodeship). In these zones, the exploitation of aggregates for concrete is small and supplies only the local building industry.

In the South-Western region of Poland, crushed aggregates from igneous and metamorphic rocks showed to be less alkali-reactive with the exceptions of extrusive rocks (melaphyres, porphyries, tuffs, tuffites) and weathered rocks (granite, basalt) that may behave as potentially reactive. For this reason, tests are performed when lithological changes are found.

Regarding the alkali contents of cement, the standard PN-EN 197-1 (2002) is followed, but in practice the producers of concrete mixtures do not know the alkali content in the cement that is used. The determination of the alkalis in cement has been performed in recent decades by the Instytut Mineralnych Materiałów Budowlanych (Institute of Mineral Materials for Constructions) in Kraków, leading to the results presented in Table 8.16.

In order to reduce the alkali content in clinker below the admissible level of 0.60%, it is necessary to exclude 10-15% of exhaust gas. This requires the construction of

Table 8.16 Alkali content in selected Polish cements in common use.

Type of cement	Alkali content, % Na_2O_{eq}
CEM I	0.14-1.26; Dominant content 0.85
CEM II	0.41-1.55; Dominant content 0.80-0.90
CEM III/metallurgic	0.46-1.25
CEM IV	0.41-1.62

additional installations, and causes increases in energy consumption (*e.g.*, 65 kcal/kg of clinker and 8 kWh/Mg of clinker) and production costs. It can thus be stated that production of cement with reduced alkali content is technically possible. However, owing to the high costs involved, this is rarely undertaken in practice in Poland at present, although very sporadically there is a small production of special cements.

8.18.3 Test methods and preventative measures

The lack of standardised test methods has prevented the identification of reactive aggregates produced in Poland. Research on this issue at the Institute of Mechanised Construction & Rock Mining led to the development of test methods for quality assessment of the susceptibility of aggregates to alkalis, including quantitative criteria for the identification of reactivity (Filipczyk & Góralczyk, 2010; Góralczyk, 2011). Further research has been undertaken to study the reactivity of carbonate rocks (Góralczyk, 1993, 1994, 2000, 2003).

In 1978, an attempt was started in order to develop test methods and to evaluate the reactivity of Polish aggregates at the Centre of Investigation and Developments of Aggregate for Building Industry/Institute of Mechanized Construction and Rock Mining, which resulted in the publication of the first standards. Until that time, ASTM test methods were used by the researchers. For example, PN-B-06714-34 (1991) was based on the ASTM C227 (now 2010) method but took into consideration the results of research from 1985 to 1990. This Polish standard was recently withdrawn (2012). In these tests, an alkali content of 1.2% Na_2O_{eq} was defined and it was allowed to use alkali contents other than this limit for research purposes.

Further to these publications, two chemical methods were established:

*   The loss of mass resulting from the reaction with NaOH
*   The soluble silica in NaOH

PN-B-06714-46 (1992) (*Mineral aggregates: researches, designation of potential alkaline reactivity by rapid method*) is a chemical method which determines the reactivity of crushed aggregate particles by its mass loss after reacting with NaOH solution. The potential reactivity degree of the aggregates is classified by the loss of mass value in accordance with Table 8.17 and designated by one of the following procedures:

*   loss of fine and coarse aggregate mass by NaOH action,
*   content of flint with unit weight of 20-26 kN/m^3

Regarding the assessment of aggregates for their utilization in concrete elements, based on PN-B-0671-46 (1992) the following restrictions are stated in the Polish standards:

Table 8.17 Reactive aggregate classification according to PN-B-06714-46 (1992).

Feature	Degree of alkali-reactivity*		
	0	1	2
Loss of mass under NaOH action: #For given X1 or X4 aggregate fraction	≤ 0.5	> 0.5 ≤ 2.0	> 2.0
For reactive flint content X3	≤ 3.0	> 3.0 ≤ 10.0	> 10.0
For X4 + X3	≤ 4.0	> 4.0 ≤ 15.0	> 15.0

*Potential alkaline reactivity degrees:
  0 – non-reactive aggregate
  1 – potentially reactive aggregate
  2 – reactive aggregate
#Mass of:
  X1 – fine aggregate
  X3 – reactive flint
  X4 – coarse aggregate

- The aggregates tested and classified as 'degree 1' can be used in concrete if subjected to constant or occasional influence of moisture;
- The aggregates tested and classified as 'degree 2' can be used with low alkali cements ($Na_2O_{eq} < 0.60\%$) and the cement content should be lower than 500 kg/m^3. Mineral additions may be used for preventing the reactivity of aggregates. A recommendation defines the use of either 15% silica fume or 10% silica fume with 20% active fly ash replacing cement.

The standard PN-B-06714-47 (1988) (*Mineral aggregates: determination of silica soluble in sodium hydroxide content*) test method determines the soluble silica content of the aggregate kept in 80 °C NaOH solution for 24 hours with the qualitative criterion for reactive aggregates of 50 mmol/l of soluble silica.

At present, Poland is adopting the tests and procedures published by RILEM (Nixon & Sims, 2016).

8.19 PORTUGAL

8.19.1 Historical introduction and case studies

ASR was firstly recognized as a problem in Portuguese structures in the 1990s, thanks to the work undertaken at the National Laboratory for Civil Engineering (LNEC) by Braga Reis (1990), Silva (1992) and Silva and Rodrigues (1993). The late acknowledgement of ASR in the country was mainly due to the use of slow reactive aggregates, such as granites, which are quite abundant in the territory. Granitic rocks were used as aggregates in the construction boom that took place in the 1950 and 1960 decades, comprising the construction of several large dams and bridges (Fernandes, 2005; Fernandes & Noronha, 2010).

Since the 1990s, a lot of work has been carried out concerning the diagnosis of ASR damaged structures and also the evaluation of factors that can affect ASR in Portugal (*e.g.*, Braga Reis *et al.*, 1996; Silva *et al.*, 1996; Fernandes, 2005; Santos Silva, 2005; Fernandes & Noronha, 2010).

An updated inventory of the structural effects on the 16 Portuguese concrete dams affected by ASR was published in Lopes Batista and Piteira Gomes, 2012. These dams correspond to about 31% of the total number of large concrete dams in the country. The majority were built between 1940 and 1970, when there was not sufficient knowledge about ASR. The dams built with concrete using quartzite aggregates exhibited ASR much earlier, but the dams constructed with granite aggregate showed the effects of expansive reactions just about 20 to 30 years after construction. The affected dams are located in the North and in the interior South of the country, where granite and schist rock masses occur, and in the central part of the country, where there is a schist-greywacke massif, with several quartzite outcrops.

According to Camelo (2011), there are sixty-five concrete and masonry dams in Portugal. Among these, twenty-five are more than 50 years old. The same author states that in 2050, thirty-six dams will be more than 75 years old and nine will have been in service for more than 100 years. This clearly suggests that an increasing number of ASR-affected dams my be expected to exhibit ASR in the coming decades, as well as several of these structures being demolished, following the example of Alto Ceira dam, replaced in 2013.

Besides dams, other concrete structures, such as fourteen bridges, were also diagnosed as being affected by ASR (Figure 8.38). Some published work also refers to a number of structures suffering from ASR but for which names were not disclosed.

8.19.2 Aggregates, binders and admixtures

Portugal has a rich geological history, which is responsible for the abundance of distinct lithologies reflected in the variety of Portuguese aggregates used in concrete (Table 8.18).

The Iberian Variscan belt, also known as the Iberian Massif, is a large arcuate segment of the European Variscan Fold Belt, which extends over more than 3000 km from eastern Germany to the Iberian Peninsula, representing the largest continuous exposure of pre-Permian rocks within Europe (Ribeiro *et al.*, 1990). The Iberian Massif occupies most of the western and central part of the Iberian Peninsula. In Portugal, the margins of the Iberian Massif are covered by the Lusitanian and the Algarve sedimentary basins of Meso-Cenozoic age and also by the Lower-Tagus and Sado Basins. The Madeira and Azores archipelagos are part of the country. They are of volcanic origin and are located in the North Atlantic Ocean.

8.19.3 Test methods and preventive measures

The study of ASR-affected structures enabled the development of recommendations and specifications in order to prevent this deleterious mechanism. Therefore,

Table 8.18 Main materials used as aggregates in Portugal between the years 2000 and 2011 (from Ramos, 2013).

Sedimentary rocks	limestone; dolomitic limestone; sand; gravel; greywacke
Metamorphic rocks	schist; gneiss; marble; serpentinite; quartzite
Igneous rocks	granite; basalt (mainly Azores and Madeira Islands); trachyte (Azores Islands); andesite (Azores Islands); diorite; syenite; gabbro; dolerite

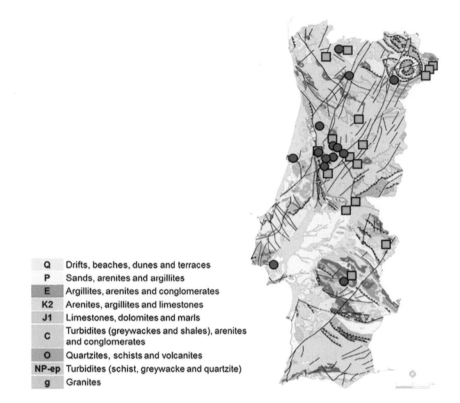

Q   Drifts, beaches, dunes and terraces
P   Sands, arenites and argillites
E   Argillites, arenites and conglomerates
K2   Arenites, argillites and limestones
J1   Limestones, dolomites and marls
C   Turbidites (greywackes and shales), arenites and conglomerates
O   Quartzites, schists and volcanites
NP-ep Turbidites (schist, greywacke and quartzite)
g   Granites

Figure 8.38 Location of damaged structures in Portugal (blue – bridges; red – dams) (IMPROVE Project, Santos Silva et al., 2014).

specification LNEC E 415 (1993), concerning the petrographic characterization of aggregates, was published. According to the same specification, the undulatory extinction of deformed quartz should be quantified in one of the three classes: weak (1-15°), moderate (15-25°) and strong (25-35°).

The studies performed by Santos Silva (2005) and Santos Silva and Gonçalves (2006) led to the conclusion that the test methods employed in Portugal up until 2004 were inappropriate for evaluating the alkali-reactivity of the aggregates. In addition, the use of the undulatory extinction angle in assessing the potential reactivity of aggregates was also questioned (RILEM AAR-1, 2003). Therefore, LNEC prepared the Specification LNEC E 461 (2007) for ASR inhibition in concrete, which is referenced by DNA 5.2.3.4 of the NP EN 206-1 Portuguese Concrete Standard and Annex G.3 of the Portuguese recommendation on aggregates for concrete NP EN 12620. Besides identifying potentially alkali-reactive rocks (jasper, chert, diatomite, siliceous schist, phyllite, greywacke, hornfels, quartzite, granitic rocks, rhyolite, dacite, andesite, basalt, limestone and dolomite), and also those that can behave as a source of alkalis (granitic rocks, syenite, trachyte, hornfels, arkose and greywacke), this specification defines the methodology for assessing the reactivity of aggregates and also evaluates the concrete mixtures. For this last purpose, the quantification of reactive silica is carried out by point counting under a polarizing microscope.

If necessary, petrographic characterization can be complemented by laboratory expansion tests: mortar-bar tests (ASTM C1260 or RILEM AAR-2) followed by concrete prism tests (RILEM AAR-4 and, eventually, RILEM AAR-3) (Figure 8.39). Aggregates are then classified in three categories as follows: Class I – non-reactive aggregates (reactive silica < 2.0 vol%); and Classes II and III – potentially reactive aggregates. The probability of occurrence of expansive reactions with Class III aggregates is higher than that with Class II aggregates. Some granitoids defined as Class I, after expanding less than 0.10% in the accelerated mortar-bar test, have demonstrated to be potentially reactive to alkalis in field structures. This type of aggregate is by default assumed to be of Class II. Also, if a carbonate rock contains a low SiO_2 content ($SiO_2 \leq 2.0\%$), it is classified as Class I.

Furthermore, LNEC E 461 (2007) establishes the measures to be adopted in order to avoid the occurrence of expansion reactions in concrete (Santos Silva & Gonçalves, 2006) for aggregates classified as Class II and III, which means that concrete prism tests of the concrete compositions have to be carried out. It should be noticed that the specification does not address the reaction involving carbonate aggregates (ACR), because dedolomitization of a carbonate aggregate has so far only been recognized in one Portuguese deteriorated structure.

The quantity of soluble alkalis allowed in concrete is given by the Portuguese specification LNEC E 461 (2007), as a function of the aggregate reactivity class

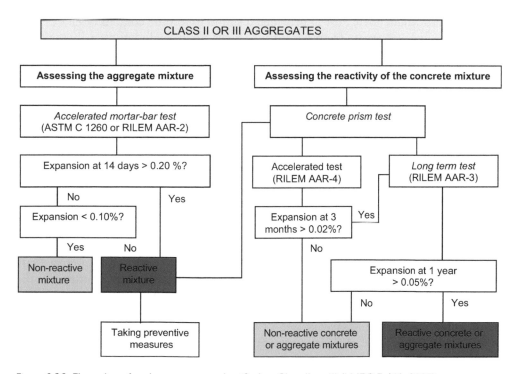

Figure 8.39 Flow chart for the aggregates classified as Class II or III (LNEC E 461, 2007).

Table 8.19 Maximum content of soluble alkalis in concrete (for reactivity classes: I – innocuous behaviour; II and III – potentially deleterious behaviour (LNEC E 461, 2007).

Aggregate reactivity	Alkali limit (kg/m³ of concrete)
Class I	no requirement
Class II	≤ 3.0
Class III	≤ 2.5

(Table 8.19). Notice that 3.0 kg/m³ Na_2O_{eq} is the highest permitted level of alkalis in concrete when potentially reactive aggregates are applied.

The Portuguese experience confirms that the monitoring systems and the visual inspection of large dams, designed in accordance with ICOLD and Portuguese guidelines, allow, in general, the early detection of the ASR phenomena, as well as the evaluation of their structural effects over time. The use of statistical models to make a preliminary interpretation of the observation results has been shown to be very useful in the identification of these ASR effects.

8.20 SLOVENIA

8.20.1 Historical introduction and case studies

Some scientific publications about aggregates used in Slovenia are available. Zatler-Zupančič and Mladenovič (1992) provided a summary on the content of chert in the sand and gravel aggregates from the Danube river basin, including its tributaries, in order to determine the concentration of reactive components in aggregates for which ASR expansion tests need to be performed. According to the results of petrographic analyses, four different types of mineral aggregates have been identified in Slovenia: crushed limestone, crushed dolomite, carbonate gravel (from the Sava river basin), and silicate gravel (from the Mura and Drava river basins). According to this study, the crushed limestone and dolomite that is used in Slovenia for the production of aggregates contains less than 1% silica, with traces of clay and organic material. However, the two above-mentioned river gravels contain up to 4% chert, which comprises more than 90% microcrystalline quartz including chalcedony and less than 10% opal. Although the alkali content of Slovenian cements was classified as 'low', it was recommended that expansion tests should be performed if the content of reactive components exceeded 3%.

Although the recognition of a few deteriorated concrete structures in Slovenia has been associated with ASR, this phenomenon has not been widely encountered. Only two well-documented cases of ASR are known. In 1999, damage was identified on the load-bearing reinforced concrete columns of a relay transmitter station at Kleče, near Ljubljana (Mladenovič et al., 2000). It was found that expansion of some cement mortar that was used, many years ago, to fill some parts of the hollow columns was the cause of the damage. ASR was confirmed as the cause of damage to the cement mortar which contained opal breccia. A second case of confirmed ASR was recorded in 2009, when pop-outs caused by this process were observed in the concrete beneath the finishing layer of an industrial epoxy-coated floor. The presence of anthropogenic pollution in the carbonate gravel from the Soča River, which was caused by crushed glass, was the reason for the ASR damage. It became visible two years after the floor had been completed (Mladenovič, 2010).

Between 2002 and 2006 a comprehensive investigation was carried out on silicate aggregates of different petrographic types from the Drava and Mura River Basins, which are commonly used for concrete (Mladenovič, 2006). The test methods that were selected for tests of the suceptibility of these aggregates to ASR were: ASTM C289 (2007, but now withdrawn) with variations in the time of dissolution and the grain size, ASTM C1260 (2014) and ASTM C1293 (2008). The types, quantities, chemical compositions, morphologies and microstructural properties of the reaction products were studied by means of scanning electron microscopy and energy dispersive X-ray spectroscopy.

The differences in the results between the samples from the Drava and Mura River Basins were found to be important with respect to the solubility of silica, but did not differ substantially with respect to expansion. In the case of all the samples from these locations, the expansion was high, and the mechanism of dissolution of the mineral grains, as well as the resultant reaction products and their composition, were the same.

8.20.2 Aggregates, binders and admixtures

Slovenia's geology is very complex and heterogeneous. Three major tectonic units meet here: the Alps, the Dinarides and the Pannonian basin. Both the Alps and the Dinarides have been strongly affected by tectonic processes, especially Alpidic tectogenesis. As a result, these rocks are highly brittle, although locally ductilely deformed. However, most of them are suitable for aggregate production: in the Dinaridic part, the major sources of aggregates are limestones and dolomites, whereas in the Alpidic part, igneous and metamorphic rocks dominate, such as Miocene tonalite/granodiorite and subvolcanic dacite (Trajanova et al., 2008) in the Pohorje area, Oligocene syenogranite and tonalite in the Eisenkappel igneous belt of the Karavanke Mountains, and andesitic tuff and dacite in the Smrekovec area. In the eastern part of the country, i.e., in the Pannonian basin, aggregates are produced from alluvial deposits.

Aggregates for the production of concrete are mostly obtained from clean and homogeneous limestones and dolomites, either in quarries or from gravel-pits (Figure 8.40). In the north-eastern part of the country, however, there are two sizeable river basins in one of which silicate rocks are dominant (the Mura river basin), whereas in the other both silicate and carbonate rocks occur (the Drava river basin). In both of these sediments, quartz is found as mono-mineralic grains in finer fractions, or as pebbles of metamorphic rocks (quartzite, gneiss) in the coarser fractions (gravel). In the western part of Slovenia, chert is a frequent component of the carbonate gravel in the Soča River basin; it originates from the cherty limestone which occurs in that area. In all cases these rocks and minerals may be mineralogically classified as slowly reactive.

In Slovenia, both ordinary Portland cement and Portland cement containing mineral additives (slag and natural pozzolanas) are produced. The Na_2O_{eq} content of these

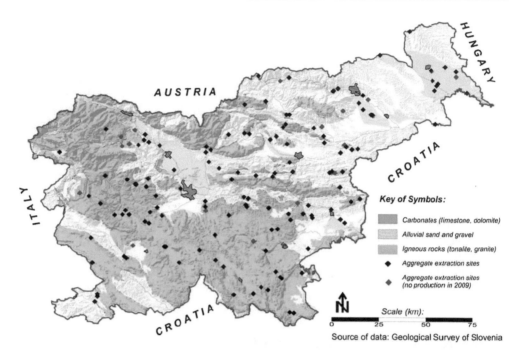

Figure 8.40 Geological formations in Slovenia suitable for aggregate extraction (for simplicity, the geological conditions given on the original geological map have been used). The sites of aggregate extraction, as of the year 2009, are shown (Žibret, SARMa project, GeoZS, 2009).

cements is between 0.7% and 0.9%. Since 2001, after the adoption of the European cement standard as SIST EN 197-1, cement production plants were able to produce limestone and fly ash cements. Portland cements incorporating calcareous fly ash contain more than 1% (approx. 1.25 %) of Na_2O_{eq}.

8.20.3 Test methods and preventative measures

In Slovenia, from 1969 onwards, aggregates for concrete were evaluated in accordance with ASTM C289 (2007, but now withdrawn). In 1980, other methods such as ASTM C295 (now 2012) and ASTM C227 (now 2010) were introduced in Slovenia. None of these tests were able to detect reactivity in local aggregates. During the period from 2003 to 2006, it was found that the use of ASTM C1260 (now 2014) could indicate the potential reactivity of some aggregates, even though information obtained from structures still showed no damage due to ASR. Up until now, despite the fact that, in all cases where pathology of concrete structures was identified, the possibility of ASR was taken into account, no new cases of ASR have been identified.

The Slovenian standard for concrete, SIST 1026 (2008), contains guidance on the application of SIST EN 206-1. In this standard, it is prescribed that, in the case of aggregates originating from the Mura or Drava river basins, it is necessary to verify their potential alkali-reactivity, although the methods to be used for such verification, as well as the criteria according to which the results should be evaluated, are not specified. Nowadays, the methods used for evaluations of this kind are ASTM C1260, or RILEM AAR-2, AAR-3 and AAR-4 (in Nixon & Sims, 2016).

During concrete production, different mineral additives are available for use: limestone, ground granulated blast furnace slag, fly ash (siliceous) and in special cases micro-silica. Taking into account that the use of cements with larger alkali contents appears to be increasing, as well as the fact that, apart from this increased quantity of cement alkalis, various additives are increasingly used in concrete and some can potentially release alkalis (*e.g.*, fly ash contains approximately 4% of alkalis as Na_2O_{eq}), it can be realistically expected that the risk of ASR will increase in Slovenia.

Slovenian carbonate aggregates have also been subjected to various investigations. Pinčič *et al.* (2013) studied the process of the dedolomitization of carbonate aggregate rocks in mortar-bars, and concluded that this process caused no expansion, but that temperature and the pH value of the solutions play a key role in its kinetics.

8.21 SPAIN

8.21.1 Historical introduction and case studies

Alkali-aggregate reactions were first identified in Spanish hydraulic structures in the 1990s. Although some cases in bridges are known (Carpintero & Bermudez, 2011; Carpintero *et al.*, 2014), most of the cases are associated with dams (Soriano, 1987; Martínez *et al.*, 1996; Soriano *et al.*, 2007). A recent study developed by Lanza (2012) states that 16 structures have been diagnosed in Spain with ASR. The reactivity in these structures is due to the presence of two different reactive components (Velasco *et al.*, 2010; Lanza & Alaejos, 2011; Alaejos *et al.*, 2014):

- Microcrystalline and cryptocrystalline quartz (rapid reaction), mainly related to quartzites and quartzarenites, but also identified in a granodiorite;
- Highly strained and microcracked quartz (slow reaction), identified in granite, monzonite or granodiorite.

Otherwise, the cases of ASR reported in the literature have been attributed to various aggregates, mainly granite, quartzite, granodiorite, dolerite, gneiss, schist and slate. The age at which AAR becomes evident is very variable and damage has been reported from 5 up to 45 years, with this wide range depending on the nature of the aggregate and the waterproofing surface treatments.

The best known cases of ASR in Spain are Las Salas and San Esteban dams (ICOLD, 1991; Alaejos & Bermúdez, 2003). Las Salas is a buttress dam of 50 m height and a gravity dam of 18 m height. It was built between 1969 and 1973 using granitic aggregates and 10% of fly ash. In 1975 the first horizontal cracks were detected on the downstream wall. In 1985 cracks were observed in the upstream wall. The damaged concrete, which was completed on August 1970, showed gel-like rims around the aggregate particles, when wet, and whitish-grey when dry. The repair work involved waterproofing of the structure and injection of cracks with epoxy resin.

San Esteban is an arch-gravity dam, which construction was finished in 1955. The aggregates included granite, gneiss, dolerite and schists with minor pyrite. Cracks were observed on the upper part of the dam as well as progressive deformation with the upstream displacement of the crest. The cement had a high content of free lime.

Some work has also been published without the identification of the structures, such as in Soriano *et al.* (2007), who published the study of the concrete of a buttresses-gravity dam built in the 1950s using granodiorite as aggregate. The rock was characterized by microscopical analysis and the crystals of quartz showed undulatory extinction angles of 30°. Portland cement was used, but no information was available about the composition of the concrete. The map cracking in the dam was observed mainly in the buttresses and the authors undertook a detailed study and identified the occurrence both of alkali-silica gel and ettringite, the latter deriving from the presence of pyrrhotite (an unstable variety of iron sulphide). The characterization of the reactivity of the aggregate was assessed by attack with chemical solutions (Soriano, 1987; Soriano & García Calleja, 1989) in order to observe the effect of dissolution, using X-ray diffraction and scanning electron microscopy. The products obtained were zeolites and alkali-silica gel similar to those found in the interior of the damaged concrete.

The occurrence of ASR with granitic aggregates is a basis of the research that is being developed regarding the contribution of some aggregates to the alkalis in concrete. The research about assessing alkali release by using high pH solutions has previously been based mainly on the French method (Menéndez *et al.*, 2012), but is now being developed by RILEM TC 258-AAA on the basis of extraction using a solution that models that of concrete pore solution (Menéndez *et al.*, 2016).

In Spain, research has also been developed regarding expansion in structures due to the presence of sulphides and some authors refer to the simultaneous occurrence of AAR and sulfate attack (Martínez *et al.*, 1996; Soriano *et al.*, 2007).

Gadea *et al.* (2010) studied the potential reactivity of different siliceous aggregates (slates, gneiss, hornfels, granites, quartzite and serpentine) by the chemical method for evaluating the potential reactivity of aggregates and by the accelerated mortar-bar method. The authors also used the surface reactivity method, by X-ray diffraction and scanning electron microscopy (SEM) to detect possible changes in the aggregate's surface before and after the attack, besides performing the mineralogical classification of the aggregates based on UNE-EN 932-3. By using the chemical method, significant differences in chemical reactivity were observed with all slates, two hornfels, the gneiss and the quartzite being found to be reactive (including expansion at 14 days of more than 0.10%).

8.21.2 Test Methods and Preventative Measures

In 1998, the Code on Structural Concrete (EHE-98) was published, expressing the methodology and aggregate tests that should be followed for ASR assessment: the UNE 146507-1 EX (chemical method), UNE 146508 EX (accelerated mortar-bar test), and UNE 146509 EX (accelerated concrete prism test). The chemical test was included in the Code but, as it proved not to be effective in detecting slow reactive aggregates, which correspond to most of the Spanish reactive aggregates, it has been abandoned (ASTM C289 has also been withdrawn by ASTM). Therefore, the revised version of the regulation, published in 2008 (EHE-08), establishes that the evaluation of the aggregates should start with the petrographic method to determine the possible ASR or ACR susceptibility. If the aggregate shows the possibility of ASR, then the accelerated mortar-bar test should follow. In the case of potential reactivity, the aggregate can only be used in favourable conditions or it must be assessed by the concrete prism test.

The methodology adopted in Spain is therefore similar to the one proposed by RILEM and used in many countries. The aggregates are first subjected to the simplified petrographic examination by UNE EN 932-3, but there is no specific provision therein for the petrographic identification of potentially reactive forms of silica and no limits indicated for any reactive components that are established. Nevertheless, the research developed by Lanza and Alaejos (2011) has proposed limits for reactive aggregates with micro-cryptocrystalline quartz, using the RILEM petrographic method (point counting method, AAR-1.1, see Nixon & Sims, 2016). If the aggregate is classified as potentially reactive, then two different laboratory tests can be performed:

- Accelerated mortar-bar test (UNE 146508 EX) using bars of 25x25x285mm, for alkali-silica reaction (or UNE 146507-2 EX for possible alkali-carbonate reaction). The aggregate is considered non-reactive for expansions less than 0.10%; when the expansion is higher than 0.20%, the aggregate is considered as potentially reactive. In case the expansion is between the described values, the measurements have to continue until 28 days of immersion. If during these 28 days, the value of the expansion is still between the previously described values, additional information would be required for determining the reactivity of the aggregate.
- Concrete prism test (UNE 146509 EX), using prisms of 75x75x275 mm, in case the aggregate is classified as potentially reactive by the preceeding tests. In this long term test the prisms are stored at 38° C for 52 weeks. The aggregate is classified as non-reactive for expansion less or equal to 0.04%.

As preventative measures, it is recommended that non-reactive aggregates are used or that the alkalis equivalent content of the cement is less than 0.60% by weight. In case these conditions cannot be accomplished, experiments should be carried out in order to establish the possibility of using cements with pozzolanic additions in accordance with UNE EN 197-1 and UNE 80307, such as: CEM III, CEM IV, CEM V, CEM II/A-D, CEM II/B-S and CEM IIB-V, as 'very adequate', or CEM II/B-P and CEM II/B-M, as 'adequate'.

For assessing the possible risk of alkali-carbonate reaction (ACR), the aggregate can be analysed for Al_2O_3, CaO and MgO, then the CaO/MgO ratio can be plotted against the percentage of Al_2O_3. This enables classification of the aggregates as non-reactive or potentially reactive aggregates in terms of ACR (this is based on a practice used in Canada: see Chapter 10). López-Buendía et al. (2006) studied several Spanish carbonate aggregates and determined that there are petrographic parameters to infer the alkali stability or reactivity risk in silicified, ferric, clayey carbonates with different proportions of dolomite and degrees of dedolomitisation. This additional evaluation test, that follows the premises of Katayama (2004) and the RILEM testing method (Sommer et al., 2001, also AAR-5 in Nixon & Sims, 2016), is nowadays in use in Spain as preliminary exercise to facilitate alkali-carbonate diagnosis. See Chapter 3 for an up-to-date international review of the 'so-called alkali-carbonate reaction'.

8.22 SWITZERLAND

8.22.1 Historical introduction and case studies

Since the first case of ASR in Switzerland was published in 1995 (Illsee dam, Regamey & Hammerschlag, 1995), several damaged structures have been reported (Leemann *et al.*, 2005), as revealed by the research undertaken on affected bridges in the scope of the project AGB2001/471 (Merz *et al.*, 2006). In Switzerland, the majority of aggregates used for concrete are potentially reactive (Merz & Leemann, 2011), but rapidly reactive aggregates are not common.

In Switzerland, several hundred structures are affected by ASR, ranging from supporting walls to bridges and dams (Figure 8.41). As the reactive mineral is quartz, ASR progresses at a relatively slow pace. An exception seems to be precast elements where cracks can appear after a few years. An overview of the affected structures is given in Merz *et al.* (2006). Since the publication of this document in 2006, more cases of ASR have been reported (TBA GR, 2007). The majority of cases are located in the Alpine region, although the highest density of structures is in the higher populated Swiss Middleland.

In a document of the Federal road authorities (Hunkeler *et al.*, 2007), recommendations about measures for affected structures are given. In general, a variety of methods has been used trying to reduce the effects of ASR and prolonging the service life of structures. These include, for example, hydrophobation, coating, strengthening with pre-stressed carbon lamella or concrete, as well as partial demolition and rebuilding. The repair of supporting walls in the Alps often poses problems as the accessibility of villages can be heavily impaired. In the majority of the affected bridges, the main

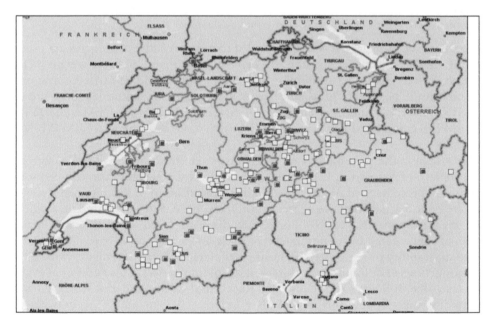

Figure 8.41 Map of Switzerland with affected structures marked (Merz *et al.*, 2006). Note that the survey didn't cover the south east area (Graubünden).

problem is the accelerated ingress of chlorides due to cracks caused by ASR. In two cases, slots had to be cut into dams to reduce stress and strain, ensuring structural safety. One small dam has been demolished and rebuilt as a result of the problems caused by ASR.

8.22.2 Aggregates, binders and admixtures

A good example of Swiss aggregates can be found in Leeman and Holzer (2005) and Leemann *et al.* (2005). In the latter, concrete and shotcrete in tunnels 19 to 82 years old were studied. The concrete petrography revealed aggregates containing deformed granite, gneiss and schist, sandstone, limestone, siliceous limestone and slate quarried from alluvial deposits. Deformed quartz with undulatory extinction and diffuse boundaries was observed in all samples. Cracks occurred in aggregates with diameter greater than 4 mm. Dissolution features were identified by ESEM investigation in quartz, feldspars and biotite. Schist, gneiss and granite showed a higher degree of reactivity than, for example, sandstones. This study was the basis of the use of preventative measures in the new Swiss tunnels, in which reactive aggregates were used in concrete containing pozzolanic admixtures. In these tunnels, liners were installed to prevent contact between the concrete and the ground water.

Aggregates used in Switzerland are quarried from fluvio-glacial deposits and mostly contain alpine rocks with various degrees of metamorphism: limestone, siliceous limestone, dolomite, sandstone, greywacke, quartzite, quartzarenite, rhyolite, basic rocks, weakly deformed granite and gneiss. Leemann and Holzer (2005) and Merz and Leemann (2011) studied Swiss aggregates from alluvial deposits containing the mentioned lithologies. They concluded that rapidly reactive aggregates with amorphous and cryptocrystalline quartz are the exception in Switzerland. The most reactive aggregates are siliceous sandstones, limestones and greywackes, as well as weakly deformed granites and gneisses (retrograde greenschist facies).

Petrographic analysis performed in accordance with SN 670 115 (2005) and SN 670 116 (2012) aimed at identifying the aggregate and at defining the capability for use in concrete. In principle, by this analysis, the reactive aggregates can be identified, leading to the possible application of the microbar test. This test is not mandatory, except to prove the non-reactivity of the aggregate.

The regional differences in the frequency of ASR damage in structures are caused by the differences in the petrographic composition of the aggregates. It was also concluded that there is a clear correlation between the south orientation of the constructions and the development of ASR damage, namely cracking. Although most of the structures that developed visible ASR damage are 30 to 50 years old (Merz *et al.*, 2006) and the time of latency is estimated to be about 20 years, Leeman & Griffa (2013) state that in several Swiss dams significant deformations have been measured without obvious visual signs of ASR.

Until the 1990s, only OPC was used in Switzerland. Since then Portland-limestone cement (CEM II / A-LL) has developed as the most widely used cement. In the last ten years, the use of other SCMs such as fly ash, silica fume, GGBS and burnt oil shale has increased.

8.22.3 Test methods and preventative measures

The first step in the assessment of aggregates with regard to ASR is the ultra-accelerated microbar test (AFNOR NF P 18-594, 2014) conducted on the grain sizes 0/4 mm and 8/16 mm. The results of the microbar test are valid for 5 years if the origin and composition of the aggregate do not change. The microbar test can give a faulty assessment of the potential reactivity for a few aggregates, namely weakly deformed granites. If an aggregate is potentially reactive according to the microbar test, or if no microbar test has been conducted, the second step comes into play. The concrete 'performance' test is based on the AFNOR NF 18-454 (2004). Job-site concrete is tested in concrete prisms at a temperature of 60 °C and a relative humidity of 100%. Test duration is 5 or 12 months, depending on recorded expansion. The experience has shown that the concrete technology measures have to be adapted for each region, meaning that different aggregates react differently. As such, it makes sense to verify the effect of SCM using the concrete performance test. The analysis of the expansions resulting in the microbar test shows that the aggregates in the Middleland exhibit higher values than the aggregates from the Alpine regions, contradicting the frequency of ASR observed in structures (Leemann *et al.*, 2008). However, the concrete performance test correlates well with the damage present in structures (Leemann & Merz, 2013). First efforts have been made to determine the residual expansion potential of concrete from affected structures (Merz & Leemann, 2013).

The guideline SIA Merkblatt 2042 (2012) establishes the principles for the prevention of ASR and it is applied in agreement with SN EN 206-1. It deals with the mitigation of ASR in new structures. Based on a risk class, taking into account the importance of the structure and the environmental class, a prevention class is defined for new structures. Depending on the classification, aggregates and concrete have to be tested.

In the standard three classes of ASR, prevention is specified (P1, P2, P3). These are derived from the combination of the environment classes based on SN EN 206-1 (U1, U2,U3,) with the risk classes (R1, R2, R3). The first takes into account the humidity, temperature variations and the contribution of alkalis from external sources, such as de-icers or underground water. The second is based on the importance of the structure or structural element, safety considerations, cost, intended service life (<50 years, 50-100 years, >100 years) and maintenance considerations. The standard contains a flow chart showing the required steps for the different prevention classes (Figure 8.42). First, a petrographic analysis of the aggregates in performed. Its result is only used for identification of the constituent rocks.

Prevention class P1 does not need any special measure, not even petrographic analysis of the aggregates. If P2 applies, measures related to the composition of the concrete (resistant to ASR) have to be taken. The sole exception is when the long-term experience with a specific aggregate has proven that it is non-reactive. In this case, an expert has to make this declaration based on the condition of structures older than 30 years containing this aggregate and on the documentation of the concrete mix design used for these structures.

8.22.4 Diagnosis and appraisal of structures

The diagnosis of ASR has been detailed in Leemann and Griffa (2013) regarding structures with signs of deformation and cracking such as dams. In this document the

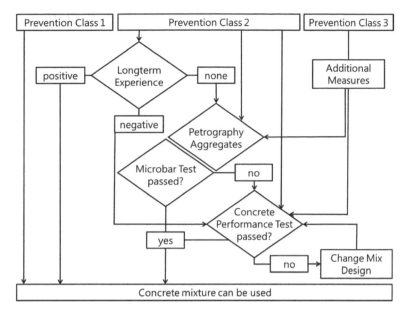

Figure 8.42 Actions to prevent ASR for the prevention classes P1, P2 and P3 (SIA Merkblatt 2042, 2012).

petrography methods are emphasised. The diagnosis should be developed in several steps:

- On-site inspection followed by concrete coring as the presence of ASR needs confirmation by concrete petrography;
- Once ASR is established, the analysis of different parts of the structure must be performed and tests to evaluate the mechanical properties of the concrete carried out;
- The prognosis can include the determination of the residual expansion potential;
- Modelling can provide information about the behaviour of the structure and be used in the prognosis of the performance of different parts of the structure. The repair or rehabilitation will depend on the type of structure or structural component.

The selection of the locations of the thin sections to be produced is usually made on the cores cut along the length axis (Figure 8.43).

The design of bridges related to roads follows the guideline of OFROU (ASTRA 2005a, 2005b) aiming at durable constructions. For a lifetime of 100 years, the standard generally assumes risk class R2 for such structures, but R3 for foundations.

For the protection and repair of existing structures, the guidelines in SN EN 1504-2 and SN EN 1504-3 are used. For protection in relation to water exposure, some measures can be adopted: water-resistant protection of the concrete surfaces exposed to aggressive agents in weakly damaged structures; adequate drainage of the buried elements and waterproof lining of the buried structures. Replacement of the cover concrete is adopted in more severe cases. Occasionally, mechanical reinforcement with

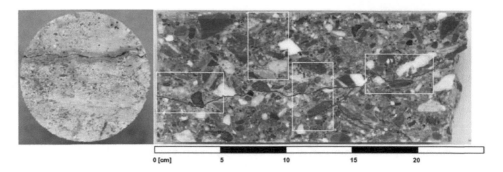

Figure 8.43 Core with a diameter of 100 mm cut along its length axis. The visually recognizable cracks are marked in red. Aggregates with a dark rim are indicated with a red star. The white rectangles indicate the position of areas intended for the preparation of thin sections. Core from the entrance area of an underpass (from Leemann & Griffa, 2013).

anchors or strengthening the construction, for example, with a new layer of reinforced concrete are carried out.

Preventive measures during construction seek to reduce the moisture and/or the alkali content or the pH in the concrete. In some cases, the aggregates can be replaced by non-reactive ones, depending on the prevention class. Usually just a certain grain size is replaced. The construction measures include the draining of superficial water and limited opening of the cracks according with national recommendations. It is also specified that recycled water and recycled aggregates cannot be used for ASR-resistant concrete.

The procedures to be taken when there are already signs of ASR in the structures, follow the guidance of SIA 269/2 (2011). After a site inspection (Figure 8.44), with crack characterization (crack index) (Figure 8.45), concrete petrography is performed on cores in order to identify the origin and extent of the damage. These elements allow

Figure 8.44 On site inspection of a tunnel entrance showing a dense cracking pattern (photograph by A. Leemann, EMPA).

Figure 8.45 Example of the crack assessment on a concrete wall (photograph by A. Leemann, EMPA).

the classification of the damage in four classes. The assessment of the static characteristics to verify the structural safety is performed as well as tests for the residual expansion of the concrete.

The protection and repair measures depend on the importance of the structure and on the intended service life, taking into account the environmental conditions and any moisture penetration. Eventually, replacement of the concrete may be required in the case of seriously damaged structures.

8.23 TURKEY

8.23.1 Historical introduction and case studies

The earliest documentation on Turkish aggregate sources from alkali-silica reaction viewpoint was published in 1975, reporting that 30% of the concrete aggregates used in some dam projects were susceptible to alkali-aggregate reactivity, according to the Chemical Method in ASTM C289 (2007, but now withdrawn) (Kocacitak, 1975). These reactive aggregates were used in concrete with low alkali cement or with the incorporation of fly ash. At that time, nearly 85% of the cement produced in Turkey could be classified as low alkali (Na_2O_{eq} <0.66%); however, after the conversion of wet system cement kilns into dry system kilns, the alkalinity of the cement produced in Turkey was significantly raised. Note that the alkali level of cement produced in the year 1996 was between 0.81% and 0.97% Na_2O_{eq} and today these levels are still in the same range. Still no case of ASR has been reported in dams.

First ASR damage in Turkey was observed during a routine inspection by highway engineers in 1995, when extensive cracks were discovered on the decks, piers and abutments of six highway bridges (Naldöken, Turgutlu, Buca, Hilal 2, Halkapınar and Turan), along with other structures which were all built in a period from the mid-1980s to the mid-1990s in the vicinity of Izmir. The deteriorated bridges were investigated as a part of the study on the maintenance and rehabilitation of highway bridges in the Republic of Turkey, in which Oriental Consultants Co. Ltd., Japan Overseas Consultants Co. Ltd. and the Turkish General Directorate of Highways (KGM) were involved, funded by the Japan International Cooperation Agency (JICA).

The findings of this study were reported by Tetsuya Katayama, who was chartered by JICA, for KGM engineers in Ankara, Istanbul and Antalya in May, 1996, and were later published as the first international documentation on the occurrence of ASR in Turkey (Katayama, 2000). It was stated that the local sand-gravel particles from the Gediz and Nif river systems, containing more than 3% reactive glassy rhyolite, were the cause of deleterious ASR expansions of several concrete bridges in the Izmir area. The sand used in the production of the affected structures was brought from Gediz and/or Nif river beds containing alkali-reactive minerals in 'pessimum' content. Crushed limestone aggregates used in these structures were found to be non-reactive by accelerated mortar-bar test (Katayama, 2000). Today, most of the ready-mixed concrete producers use the abundant sources of crushed fine and coarse non-reactive limestone aggregate near to Izmir, which in fact, is more economical than transporting the aggregate from the river basins around the city.

An interesting investigation on alkali-aggregate reaction was the rehabilitation studies of the aggregate used in the construction of Deriner Dam (Sağlık *et al.*, 2003). The dam is being constructed in the north-eastern part of Turkey on the Coruh River and will be the highest dam in Turkey and the fifth highest dam in the world, with its height being 252 m from the base when completed. The amount of concrete to be cast will be about 4 x106 m^3. Samples of two types of aggregate, having 20% and 30% of reactive silica, respectively, were tested in accordance with ASTM C1260 (now 2014) and their 14 day expansions were recorded as 0.115% and 0.157%, respectively. The expansions were reduced to safe levels with the use of 25% to 35% fly ash in the binder.

Recently a review of ASR in Turkey was published including two different case studies from Izmir (Andiç-Çakır *et al.*, 2012). In this paper the deteriorated apron and taxiway concrete pavements of Adnan Menderes Airport (ADB) located in Izmir, western Turkey, were investigated by concrete petrography (see Figure 8.46). Furthermore, Katayama's new findings on the core samples that were taken in 1996 has introduced new insights about the concrete that was being cast at that time. The

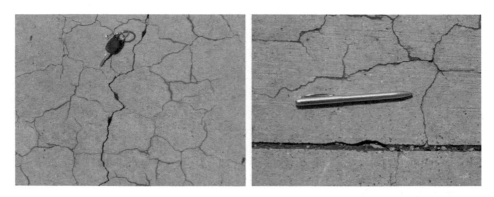

Figure 8.46 Adnan Menderes Airport concrete taxiway payment cracks (photographs by Ö. Andiç-Çakır, 2004).

deteriorated apron and taxiways were built in 1985 in addition to other existing airport structures. During this construction, it was stated by the authorities that the concrete was produced by using the same aggregate sources that were used in the deteriorated highway bridges. Approximately five years after the construction of above-mentioned taxiways, the operation engineers observed a closing of construction joints and cracking. Up to 2004, these signs of deterioration increased while the concrete continued expanding, thus, renovation became a necessity. Site observations revealed extensive map-cracking and moving joints, while the old apron and other airport buildings showed no sign of deterioration. In 2004, the cracked pavements were demolished and then they were rebuilt by using the local crushed limestone. Operation engineers state that they have observed no sign of deterioration of the newly build structures during the following seven years period.

The problem with ASR in the Izmir region was substantially decreased after the first studies in the late 1990s, by the crushed limestone aggregate sources replacing the reactive sources and by increasing utilization of blended cements. However, the maintenance of existing massive structures built in the 1980s is still under consideration. In order to investigate this case in more detail and to put forward the cause of deterioration, concrete petrography was applied. Polished thin sections were analysed by optical microscopy and by scanning electron microscopy (SEM). This revealed the presence of reacted chert and rhyolite sand particles, abundant ASR gel around the aggregates, as well as in cracks and in voids. Ettringite was found around the aggregates and in voids. Regarding the environmental conditions of the region, the possibility of an external sulphate attack and freeze-thaw attack (related to de-icing salt attack) was minimal. Petrographic methods were also used in this case to estimate mix proportions and/or the amount of reactive constituents of the deteriorated airport apron concrete. The results gave reasonable estimations for the deteriorated concrete: apparent water/cement ratio was between 0.48 and 0.50, cement content was about 250 kg/m^3 and the fine aggregate content was about 934 kg/m^3.

The alkali level of cements produced in 2011 (CEM I 42.5) had been unchanged at least since 1996, when the first ASR study was made (0.81-0.97%, PC 32.5), due to the raw materials used. Hence, the alkali contents of deteriorated bridges were estimated on the basis of EDS analysis of cement particles, water-soluble alkali of the bulk structure and unused fine aggregates, with the assumption that 300 kg/m^3 cement and 850 kg/m^3 fine aggregate was used in the concrete mix. As a result, the estimated total alkali content of the cement and concrete ranged from Na_2O_{eq} of 2.4 to 3.3kg/m^3 and from 2.5 to 4.0 kg/m^3, respectively, in which 0.3-0.7 kg/m^3 were attributable to the coarse aggregate and chemical admixture. Thus, the total alkali content of concrete was high enough to cause ASR in concrete, favouring ASR of highly reactive glassy rhyolite with 'pessimum' behaviour caused by cristobalite contained in the sand aggregates.

Typical ASR gel formation was observed in the deteriorated bridge and apron concretes. After long term storage, siliceous ASR gel was observed on the thin sections that were sampled from the affected bridge concrete. Such residual ASR gel in the reacted rhyolite and siliceous limestone (Hilal-2) presented an increase in the CaO content with decreasing alkali content. Regarding the environmental conditions of the region, the possibility of an external sulphate attack and freeze-thaw attack (related to de-icing salt attack) was minimal. During the microstructural examination of ADB apron concrete, ettringite was observed inside the air voids and the cracks around the aggregate presumably created by ASR (Andiç-Çakır *et al.*, 2012).

8.23.2 Aggregates, binders and admixtures

At the beginning of the 2000s, utilization of local mineral additives in concrete mixtures containing local reactive aggregates was studied by many researchers (Tosun, 2001; Andiç, 2002; Musal, 2003) and the documentation of deterioration has been continued (Baradan *et al.*, 2002). In addition to Gediz and Rif river sands from İzmir region (1) (given numbers in paranthesis represents the regions marked by numbers in stars, see Figure 8.47) many other aggregate sources in the country were also found to be reactive; including Bozdivlit basaltoids from Aliaga (2) (Çopuroğlu *et al.*, 2007), basalts from Niğde (3) (Korkanç & Tuğrul, 2004), sand and gravel from Koc River, Sivas (4) (Erik & Mutlutürk, 2004), sand from Firat River (5) (Aşık *et al.*, 2004), Deriner Çoruh River (6) (Sağlık *et al.*, 2003), cherts from Ankara (Bektaş *et al.*, 2004, 2008) and sand/gravel from Sakarya (Yıldırım *et al.*, 2011). It was stated that potential aggregate sources also exist in Gölcük (9) (Gümüş, 2009), Kırklareli (10) and Aksaray (11) regions (Mannasoğlu, 2010). Many of these publications refer to aggregate sources identified after routine aggregate reactivity tests, with a few of them showing concrete deterioration on site.

The investigated rocks from Niğde region can be classified as basalts, olivine basalts and basaltic andesites, in which reactivities are controlled by the presence of volcanic glass with SiO_2 content being more than 50% and alteration minerals. The basalts with volcanic glass having acidic-intermediate character show expansions higher than 0.10% in the accelerated mortar-bar test (AMBT) and ASR gel formation could be identified on the mortar-bars examined under SEM (Korkanç & Tuğrul, 2004).

The chert samples observed were taken from Soğukcam limestone (Cretaceous) near Ankara, where chert occurs as nodules or irregular layers within the medium to thin-bedded beige clayey limestone. The chert samples are largely composed of microcrystalline quartz with little chalcedony, both existing as small and irregular patches and

Figure 8.47 Map of alkali reactive aggregate sources, Turkey (THBB, 2013).

scattered crystals. The microcrystalline quartz displays a well-developed mosaic texture with a grain size of 3 to 12 μm. The chert examined by AMBT has shown 'pessimum' behaviour within 5–15% replacement level with non-reactive limestone (Bektaş et al., 2004). Bektaş et al. (2008) further examined 8 different chert samples from the same region, all of them showing a pessimum behaviour around 7.5–10% replaced by limestone, tested by AMBT.

Fırat river sand, which was found to be reactive (0.22% AMBT expansion at 14 days) was evaluated by thin section petrography. It was found that the aggregate was composed of sedimentary and igneous components. The sedimentary particles of the river sand contained recrystallized calcite and sandstone, and common components of the aggregate were granodioritic and basaltic rocks and volcanic glass with micro-/cryptocrystalline texture with both types showing reactivity potential (Aşık et al., 2004).

The basaltoid sample from Aliağa location is known as Bozdivlit Andesitic Basalts (Çopuroğlu et al., 2007). Upper Miocene rocks were previously designated as basalt, basaltoid and basaltic andesite by other researchers. However, the authors stated that these rocks are akin to andesitic rocks. The results have shown that glass composition, especially the silica content of the matrix, is critical on the reactivity of the andesite. Approximately 70% SiO_2 seems to be deleterious in terms of ASR reactivity. Dissolution of silica is overwhelming in the matrix of the andesite and it eventually generates four consequences: (1) Formation of ASR gel in the interfacial transition zone, (2) increased porosity and permeability of the andesite matrix, (3) reduction of mechanical properties of the andesite and (4) gel formation within the aggregate.

Regarding the exposed concrete structures that are built in Izmir, all the examined deteriorated concretes contained crushed limestone coarse aggregate, and the sources of sand-gravel aggregates were widespread. The local sand-gravel particles from Gediz and Nif river systems containing more than 3% reactive glassy rhyolite were the cause of deleterious ASR expansions of several concrete bridges in the Izmir area. Gravel-sized aggregate particles were sporadically found in the deteriorated bridges, while they were absent from the deteriorated airport pavements. Crushed limestone sand aggregate was also used in some of the deteriorated bridge concretes. It is interesting to note that chert particles in the sand had reacted in the airport concrete, however, this rock type had not reacted significantly in the bridge concretes at the time of examination in 1995 (Andiç-Çakır et al., 2012). Similarly, a study on the sand taken from the upstream of Gediz river have shown reactivity (Davraz & Gündüz, 2008).

Recently, Hasdemir et al. (2012) evaluated the reactivity of natural sands of different origins (terrestrial-river, dune and marine) sampled from Marmara Region. Among the seven samples examined, two of them were found to be reactive by the AMBT method, with the petrographic analysis supporting the mortar-bar expansion results. One of the potentially reactive sand samples had a high amount of SiO_2 content (85.67%), with appreciable amounts of quartz polymorphs, e.g., chert and chalcedony and lesser amounts of silicate based minerals. The other one contained the least amount of SiO_2, with a high percentage of silicate minerals and the presence of chert/chalcedony and volcanic rock grains.

8.23.3 Test methods and preventative measures

Until the discovery of the problem, TS 2517 (2010), which is similar to ASTM C289 (2007, but now withdrawn) was adopted as a screening test for reactive aggregates. However, this method was not suitable for detecting the reactivity of Turkish aggregates. Afterwards, the general approach has become to conduct accelerated mortar-bar tests (ASTM C1260 or similar) to detect the reactivity of the aggregates. If found to be deleterious, the suspicious aggregates are mixed with a known non-reactive aggregate, and/or mineral admixtures (generally fly ash or natural pozzolanas) are incorporated, and again tested by accelerated mortar-bar test.

Reported ASR expansions are investigated by some test methods, including the accelerated mortar-bar test, the concrete prism test and the concrete 'microbar' test (Andiç, 2007; Andiç-Çakır et al., 2009). The effect of gradation and the size of the particles on ASR expansions were also evaluated (Ramyar et al., 2005). These studies revealed that crushing the aggregate may alter the expansion properties of the aggregate by changing the microstructure. The 'microbar' test, with some modifications, is an accelerated method which permits investigation of alkali-silica reactivity of both fine and coarse aggregate separately or together in a mixture.

Among the test methods for determining the reactivity of aggregates, the chemical method (similar to ASTM C289, 2007, now withdrawn) had been used as a standard method in the 1980s and 90s, but was found to be unreliable for Turkish reactive aggregates. Today, both the mortar-bar and concrete prism expansion tests are used to evaluate the reactivity of aggregates. The complementary Turkish standard to EN 206-1, TS 13515 (2014), defines four different exposure classes of reinforcement corrosion in the presence of ASR. Annex M of the same standard defines a methodology of evaluating the aggregates, starting with the simplified petrographic analysis according to TS 10088 (1997) (EN 932-3), preceding the Chemical Test according to TS 2517 (2010), while evaluation of the supplementary cementitious materials are made according to mortar-bar and concrete prism tests. It was also stated that concrete in which the total alkali content exceeds 3 kg/m^3 necessitates special attention. A reliable performance test method is needed to be adapted for local specifications in Turkey.

ACKNOWLEDGEMENTS

The authors would like to express their gratitude to the colleagues who kindly contributed to this chapter both by providing information and/or revising the text: Ákos Török (Budapest University of Technology and Economics, Hungary), Ana Mladenovič (ZAG, Slovenian National Building and Civil Engineering Institute, Slovenia), Andreas Leemann (EMPA, Swiss Federal Laboratories for Materials Science and Technology, Switzerland), Bruno Godart (IFSTTAR, Institut Français des Sciences et Technologies des Transports, de l'Aménagement et des Réseaux, France), Caner Anaç (Delft University of Technology, the Netherlands), Christine Merz (Ungricht merz GmbH, Switzerland), Christodoulos Hadjigeorgiou (Geological Survey Department), Erika Csányi (Budapest University of Technology and Economics, Hungary), Gabriel Lorenzi (ISSeP, Institut Scientifique de Service Public, Belgium), Gabriele Vola (CİMPROGETTİ S.r.l. Lime Technologies, Italy), György L.

Balazs (Budapest University of Technology and Economics, Hungary), Jan Elsen (KU Leuven, Belgium), Jochen Stark (Bauhaus-University Weimar, Germany), Kambiz Ramyar (Ege University, Turkey), Katia Kontoghiorghe-Neocleou (State General Laboratory, Ministry of Health, Cyprus), Mario Berra (RSE, Ricerca Sul Sistema Energetico, Milan, Italy), Mário de Rooij (TNO, Netherlands Organisation for Applied Scientific Research, the Netherlands), members of the Technical Committee "Cement & Concrete" of the Italian Institute for Standardization UNI (Italy), Oğuzhan Çopuroğlu (Delft University of Technology, the Netherlands), Özgür Eren (Eastern Mediterranean University, TRNC), Philippe Demars & Pierre Gilles (Service Public de Wallonie, Direction Générale Opérationnelle Routes et Bâtiments, Belgium), Sárka Lukschová (Charles University in Prague, Czech Republic), Stefan Góralczyk (Institute of Mechanised Construction & Rock Mining, Poland), Stefan Krispel (Smart Minerals GmbH, Austria), Tetsuya Katayama (Taiheiyo Consulting Co. Ltd, Japan), Timo Nijland (TNO, Netherlands Organisation for Applied Scientific Research, the Netherlands), Umberto Costa (Italcement S.p.a, UNI/TC "Cement & Concrete", Italy) Victor Lanza (CEDEX, Spain).

REFERENCES

AFNOR FD P 18-456 (2004) Béton: Réactivité d'une formule de béton vis-à-vis de l'alcali réaction – Critères d'interprétation des résultats de l'essai de performance. (Concrete: Reactivity of a mix composition regarding the alkali-aggregate reaction. Criteria for the interpretation of performance test results). *Association Française de Normalisation*, Paris, France (in French).

AFNOR FD P 18-464 (2014) Béton: Dispositions pour prévenir les phénomènes d'alcali-réaction (Concrete: Guidelines for the prevention of the alkali-aggregate reaction). *Association Française de Normalisation*, Paris, France (in French).

AFNOR FD P 18-541 (2014) Granulats: Guide pour l'elaboration du dossier carrièrre dans le cadre de la prévention des désordres liés à l'alcali-réaction (Aggregates: Guidelines for the elaboration of the classification in the quarries for the prevention of damages due to alkali-aggregate reactions). *Association Française de Normalisation*, Paris, France (in French).

AFNOR FD P 18-542 (2004) Granulats: Critères de qualification des granulats naturels pour béton hydraulique vis-à-vis de l'alcali-réaction (Aggregates: Criteria for the classification of natural aggregates for hydraulic concrete regarding the alkali-aggregate reaction). *Association Française de Normalisation*, Paris, France (in French).

AFNOR FD P 18-543 (2014) Granulats: Etude pétrographiquedes granulats appliquée à l'alcali-réaction (Aggregates: Petrographic study of aggregates regarding the alkali-reaction). *Association Française de Normalisation*, Paris, France (in French).

AFNOR FD P 18-544 (2014) Granulats: Essai pour déterminer les alcalins solubles dans l'eau de chaux (Aggregates: Tests for the determination of the alkalis soluble in hot water). *Association Française de Normalisation*, Paris, France (in French) (to be published).

AFNOR NF P 18-454 (2004) Béton. Réactivité d'une formule de béton vis-à-vis de l'alcali-réaction – Essai de performance (Concrete: Reactivity of a mix composition regarding the alkali-aggregate reaction. Performance tests. *Association Française de Normalisation*, Paris, France (in French).

AFNOR NF P 18-545 (2011) Granulats: Éléments de définition, conformité et codification (Aggregates: Elements for definition, conformity and codification). *Association Française de Normalisation*, Paris, France.

AFNOR NF P 18-594 (2004, 2014) Granulats: Méthodes d'essai de réactivité aux alcalis (Aggregates: Test methods for the alkali reactivity). *Association Française de Normalisation*, Paris, France (in French).

Alaejos, P., & Bermúdez, M.A. (2003) Durabilidad y procesos de degradación del hormigón de presas. Estudio Bibliográfico. Monografía M-76. *Centro de Estudios y Experimentación de Obras Públicas*, Madrid. 168p.

Alaejos, P., Lanza, V., Bermúdez, M.A., & Velasco, A. (2014) Effectiveness of the accelerated mortar bar test to detect rapid reactive aggregates (including their pessimum content) and slowly reactive aggregates. *Cement and Concrete Research*, *58*, 13–19.

Alkaravlı, M. (2007) KKTC'de Madencilik ve Çevre, Jeoloji ve Maden Dairesi, Nisan, Lefkoşa (in Turkish).

Alunno Rossetti, V. (1981) Osservazioni di pop outs dovuti alla reazione alcali-aggregati su pavimentazioni in calcestruzzo in Italia. *La Prefabbricazione*, *6*, 263–265.

Andiç, Ö. (2002) Alkali-silis reaksiyonunun mineral ve kimyasal katkılar yardımı ile kontrol altına alınması, EÜ Fen Bilimleri Ensitüsü, *MSc dissertation* (in Turkish), 90p.

Andiç, Ö. (2007) Investigation of test methods on alkali aggregate reaction, *PhD Dissertation, Ege University*, Izmir, Turkey (in Turkish).

Andiç-Çakır, Ö., Çopuroğlu, O., & Katayama, T. (2012) A review of alkali-silica reactivity in Turkey: a case study from Izmir, West Anatolia, In: Fournier, B., & Ideker, JH. (eds.) *Proceedings of the 14th International Conference on Alkali-Aggregate Reaction*, 20–25 May, University of Texas, Austin, USA. 10p.

Andiç-Çakır, Ö., Çopuroğlu, O., & Ramyar, K. (2009) Evaluation of alkali silica reaction by concrete microbar test. *ACI Mater J.*, *106* (2), 184–191.

Asch, K. (2005) The 1:5 Million International Geological Map of Europe and Adjacent Areas (IGME5000). *BGR* (Hannover).

Aşık, İ., Şen, H., Ergintav, Y., Ünsal, A., Şentürk, E., & Bayrak, E. (2004) Alkali agrega reaksiyonu yönünden zararlı olan bir ocağın iyileştirilmesi, *Beton 2004*, Istanbul (in Turkish).

ASTM C227 (2010) Standard test method for potential alkali reactivity of cement-aggregate combinations (Mortar-bar method). *The American Society for Testing and Materials*, Philadelphia, USA.

ASTM C289 (2007) Standard test method for potential alkali–silica reactivity of aggregates (Chemical method). *The American Society for Testing and Materials*, Philadelphia, USA.

ASTM C295 (2012) Standard guide for petrographic examination of aggregates for concrete. *The American Society for Testing and Materials*, Philadelphia, USA.

ASTM C1260 (2014) Standard Test Method for the Potential Alkali Reactivity of Aggregates (Mortar Bar Method). *The American Society for Testing and Materials*, Philadelphia, USA.

ASTM C1293 (2008) Standard Test Method for Determination of Length Change of Concrete Due to Alkali-Silica Reaction. *The American Society for Testing and Materials*, Philadelphia, USA.

ASTRA (2005a) Richtlinie Projektierung und Ausführung von Kunstbauten der Nationalstrassen, Bundesamt für Strassen ASTRA, Abteilung Strassennetze, Standards, Forschung, Sicherheit, Bern.

ASTRA (2005b) Richtlinie Überwachung und Unterhalt der Kunstbauten der Nationalstrassen, Bundesamt für Strassen ASTRA, Abteilung Strassennetze, Standards, Forschung, Sicherheit, Bern.

Baradan, B., Yazıcı, H., & Ün, H. (2002) Betonarme Yapılarda Kalıcılık (Durabilite), *DEU Publications*, 2nd edition, Izmir (in Turkish), 318p.

Barisone, G. (1984) Petrographic analysis of aggregates related to alkali-silica reaction. *Bull Int Assoc Eng Geol.*, *30*, 177–181.

Barisone, G. (2001) Microscopical determination of flint content in the alluvial deposits of Italian Peninsula. *Proceedings of the 8th Euroseminar on Microscopy Applied To Building Materials (EMABM)*, Athens, Greece, pp. 1–6.

Barisone, G. (2004) Distribution of alkali-silica reactive minerals in Italian peninsula natural aggregates. *Proceedings of the 12th International Conference on Alkali-Aggregate Reaction in Concrete*, Beijing, China, Vol. I, pp. 236–242.

Barisone, G., Bottino, G., Cardu, G., & Paganelli, L. (1986) Italian peninsular aggregates and the Alkali-Silica Reaction. *Proceedings of 5th International I.A.E.G. Congress*, Buenos Aires, Argentina, pp. 1623–1632.

Barisone, G., Bottino, G., & Pavia, R. (1989) Alkali-Silica Reaction potentially bearing minerals in alluvial deposits of Calabria, Umbria, and Toscana regions (Italy). In: Okada, K, Nishibayashi, S., & Kawamura, M. (eds.) *Proceedings of the 8th International Conference on Alkali-Aggregate Reaction in Concrete (ICAAR)*, Kyoto, Japan, pp. 507–512.

Barisone, G., & Restivo, G. (1992a) Alkali silica potential reactivity of undulatory extinction quartz in the Western Alpine Arch. In: Poole, A. (ed.) *Proceedings of the 9th International Conference on Alkali-Aggregate Reaction in Concrete (ICAAR)*, London, UK, pp. 40–45.

Barisone, G., & Restivo, G. (1992b) Alkali-silica reactivity of alluvial deposits evaluated using chemical and psammographic methods. In: Poole, A. (ed.) *Proceedings of 9th International Conference on Alkali-Aggregate Reaction in Concrete (ICAAR)*, London, UK, pp. 46–52.

Barisone, G., & Restivo, G. (1996) Alkali-silica reactivity of some Italian and European flints. In: Shayan, A. (ed.) *Proceedings of the 10th International Conference on Alkali-Aggregate Reaction in Concrete (ICAAR)*, Melbourne, Australia, pp. 775–782.

Barisone, G., & Restivo, G. (2000) Alkali-silica reactivity of some Italian opal and flints tested using a modified mortar bar test. In: Bérubé, M.A., Fournier, B., & Durand, B. (eds.) *Proceedings of the 11th International Conference on Alkali-Aggregate Reaction in Concrete (ICAAR)*, Québec, Canada, pp. 239–245.

Baronio, G. (1983) Examples of deterioration by alkalis in Italian concrete structures. In: Idorn, G.M., & Rostam, S. (eds.) *Proceedings of 6th International Conference on Alkalis in Concrete*, Copenhagen, Denmark, pp. 503–510.

Baronio, G. (1984) Aggregates reactive to alkalis used in Italy. *Bull Int Assoc Eng Geol.*, 30, 183–186.

Baronio, G., & Berra, M. (1989) The occurrence of AAR in a concrete structure in Italy, In: Okada, K, Nishibayashi, S., & Kawamura, M. (eds.) *Proceedings of the 8th International Conference on Alkali-Aggregate Reaction in Concrete (ICAAR)*, Kyoto, Japan, pp. 71–76.

Bartuli, C., Cigna, R., Fumei, O., & Valente, T. (2012) A Critical Examination of the European Standard EN 1504 Products and Systems for Protection and Repair of Concrete Structures. *J Civil Eng Archit.*, 6 (2), 226–232.

BCA (1992) The diagnosis of alkali-silica reaction. Report of a working party, *British Cement Association*, Publication 45.042, 44p.

BCG (2013) The Cement Sector: A Strategic Contributor to Europe's Future. *Boston Consultancy Group Report*, 51p.

Bektaş, F., Topal, T., Goncuoglu, M.C., & Turanli L. (2008) Evaluation of the alkali reactivity of cherts from Turkey. *Constr Build Mater.*, 22, 1183–1190.

Bektaş, F., Turanli, L., Topal, T., & Goncuoğlu, M.C. (2004) Alkali reactivity of mortars containing chert and incorporating moderate-calcium fly as. *Cement Concrete Res.*, 34, 2209–2214.

Berra, M. (1983) Performances of some Italian cements tested for alkali-silica reaction. In: Idorn, G.M., & Steen Rostam (eds.) *Proceedings of the 6th International Conference on Alkalis-Silica Reactions in Concrete (ICAAR)*, Copenhagen, Denmark, pp. 372–382.

Berra, M., & Baronio, G. (1986) The Potential Alkali-Aggregate Reactivity in Italy: Comparison of some methods to test aggregates and different cement-aggregate combinations. In: Grattan-Bellew, P.E. (ed.) *Proceedings of the 7th International Conference on Alkalis-Silica Reactions in Concrete (ICAAR)*, Ottowa, Canada, pp. 231–236.

Berra, M., De Casa, G., & Mangialardi, T. (1996) Evolution of Chemical and Physical Parameters of Blended Cement Mortars Subjected to the NaOH Bath Test. In: Shayan, A. (ed.) *Proceedings of the 10th International Conference on Alkali-Aggregate Reaction in Concrete*, Melbourne, Australia, pp. 483–491.

Berra, M., Mangialardi, T., & Paolini, A.E. (1994) Application of the NaOH bath test method for assessing the effectiveness of mineral admixtures against reaction of alkali with artificial siliceous aggregate. *Cement Concrete Comp.*, 16, 207–218.

Berra, M., Mangialardi, T., & Paolini, A.E. (2005) Alkali–silica reactivity criteria for concrete aggregates. *Mater Struct.*, 38, 373–380.

Berra, M., Costa, U., Mangialardi, T., & Paolini, A.E. (2015) Application of an innovative methodology to assessing the alkali-silica reaction in concrete. *Mater Struct.*, 49 (9), 2727–2740.

Berra, M., Costa, U., Mangialardi, T., Paolini, A.E., & Turriziani, R. (2013) A new approach to assessing the performance of ASR inhibitors in concrete. *Mater Struct.*, 46 (6), 971–985.

BGR – Bundesanstalt für Geowissenschaften und Rohstoffe (2014) *Geologische Karte Deutschlands*, Hannover.

Bödeker, W. (2003) Alkalireaktion im Bauwerksbeton – Ein Erfahrungsbericht. Deutscher Ausschuss für Stahlbeton (DAfStb), Heft 539, Berlin.

Bon, E., Chillè, F., Masarati P., & Massaro, C. (2001) Analysis of the effects induced by alkali-aggregate reaction on the structural behavior of Piantelessio dam. *Proceedings of the 6th International Benchmark-Workshop on Numerical Analysis of Dams (Theme A)*. Salzburg, Austria.

Borsje, H., Peelen, W.H.A., Postema, F.J., & Bakker, J.D. (2002) Monitoring alkali-silica reaction in structures. *Heron*, 47 (2), 95–109.

Borsje, H., Spoon, A.C., Walthaus, W., & Bubberman, R.A.L. (1992) Onderzoek naar de conditie van kunstwerk 44G-117 "Heemraadsingel" in de Rijksweg 59 (Maasroute). Bijlage 8: Petrografisch onderzoek TNO. Excerpt Nebest-report PN813 (1), 11p.

Bosschart, R.A.J. (1957) Alkali-reactions of aggregate in concrete. *Cement* (in Dutch), 9, 494–500.

Bosschart, R.A.J. (1958) Alkali-Reaktion des Zuschlags im Beton. *Zement-Kalk-Gips*, Heft 3, S. 100–108.

Braga Reis, M.O. (1990) Implicação das reacções álcalis-agregado na durabilidade das estruturas de betão. *2as Jornadas Portuguesas de Engenharia de Estruturas*, LNEC, Lisboa, Portugal (in Portuguese).

Braga Reis, M.O., Silva, H.S., & Santos Silva, A. (1996) AAR in Portuguese structures. Some case histories. In: Shayan, A. (ed.) *Proceedings of the 10th International Conference on Concrete Alkali-Aggregate Reaction*, Melbourne, Australia. 10p.

Breitenbücher, R., & Sievering, Ch. (2012) Cracking in concrete pavements due to alkali-silica-reactions. *Proceedings of the 10th International Conference on Concrete Pavements*, Québec City, Canada, pp. 511–521.

Broekmans, M.A.T.M. (2002) The alkali-silica reaction: mineralogical and geochemical aspects of some Dutch concretes and Norwegian mylonites. *PhD-thesis University of Utrecht, Geologica Ultraiectina, 217*, 144p.

Broekmans, M.A.T.M., & Jansen, J.B.H. (1997) ASR in impure sandstone: mineralogy and chemistry. In: Sveinsdóttir, EL. (ed.) *Proceedings of the 6th Euroseminar on Microscopy Applied to Building Materials (EMABM)*, Reykjavik, Iceland, pp. 161–176.

Bubberman, R.A.L., & Broekmans, M.A.T.M. (1994) Petrografisch onderzoek aan viaduct 'Wolput', A59, hm 124,78. *Nebest-report* C1264 (3), 87p.

Bundesministerium für Verkehr, Bau- und Wohnungswesen (2005) Allgemeines Rundschreiben Straßenbau Nr. 15/2005, Sachgebiet 06.1: Straßenbaustoffe; Anforderungen, Eigenschaften / Sachgebiet 06.2: Straßenbaustoffe; Qualitätssicherung, Betreff: Vermeidung von Schäden an Fahrbahndecken aus Beton in Folge von Alkali-Kieselsäure-Reaktion (AKR).

Bundesministerium für Verkehr, Bau- und Stadtentwicklung (2013) Allgemeines Rundschreiben Straßenbau Nr. 04/2013, Sachgebiet 06.1: Straßenbaustoffe; Anforderungen, Eigenschaften / Sachgebiet 04.4: Straßenbefestigung; Bauweisen, Betreff: Vermeidung von Schäden an Fahrbahndecken aus Beton in Folge von Alkali-Kieselsäure-Reaktion (AKR).

Camelo, A. (2011) Durabilidade e vida útil das estruturas hidráulicas de betão e de betão armado. *Proceedings of 1ᵃˢ Jornadas de Materiais na Construção*, Porto, Portugal (in Portuguese), pp. 149–169.

Carpintero, I., & Bermudez, M.A. (2011) Evaluación estructural de un puente de hormigón pretensado afectado por una reacción álcali-sílice. *V congreso ACHE*, Barcelona.

Carpintero, I., Lanza, V., & Criado, J.E. (2014) Evaluación estructural de un tablero postesado de hormigón con reacción álcali-sílice. *VI congreso ACHE*, Madrid.

CEMBUREAU, Cements for a low carbon Europe, available from: http://www.cembureau.eu/sites/default/files/documents/Cement%20for%20low-carbon%20Europe%20through%20clinker%20substitution.pdf

CEN CR 1901 (1995) Regional specifications and recommendations for the avoidance of damaging alkali-silica reactions in concrete. *CEN Report, Comité Européen de Normalisation, Bruxells, Belgium*, 63p.

CEN TR 16349 (2012) Framework for a specification on the avoidance of a damaging Alkali-Silica Reaction (ASR) in concrete. *CEN Report, Comité Européen de Normalisation, Bruxells, Belgium.*

Chulliat, O., Grimal, E., Bourdarot, E., Boutet, J.M., & Taquet, B. (2012) Le gonflement des barrages in beton. Apports des recherches scientifiques: application on Barrage du Chambon et a son confortment. *XXIV Congrés des Grands Barrages*, Kyoto, Q.14–R.15.

Circulaire 42-03-06-05(01) (2006) Béton – Application de la NBN B15-001, *Ministère Wallon de l'Equipement et des Transports*.

Collepardi, M., Coppola, L., Moriconi, G., & Pauri, M. (1991) Diagnosi della reazione alcali-aggregato in calcestruzzi degradati. *L'industria Italiana del Cemento*, 10, 646–650.

Çopuroğlu O., Andiç-Çakır, Ö., Broekmans, M.A.T.M., & Kühnel, R. (2007) Mineralogy, geochemistry and expansion testing of an alkali-reactive basalt from western Anatolia, Turkey. In: Fernandes, I., et al. (eds.) *Proceedings of 11th Euroseminar on Microscopy Applied to Building Materials (EMABM)*, Porto, Portugal.

Çopuroğlu, O., & Schlangen, E. (2007) Modelling of effect of ASR on concrete microstructure. *Key Eng Mater.*, 348–349, 809–812.

Costa, U., Gallo, A., Ioni, G.P., Mangialardi, T., Migheli, A., Minoia, A., Paolini, A.E., Santinelli, G., Vola, G., Zanardi, G., & Zenone, F. (2011) A new approach to evaluating the effectiveness of pozzolanic and blastfurnace cements against alkali-silica reaction in concrete. *Proceedings of XIII International Congress on the Chemistry of Cement (ICCC)*, Madrid, Spain.

Cotelo Neiva, J.M., Lima, C., Plasencia, N., & Ferreira, F. (1995) Concrete deterioration of the Pracana dam: microscopy and rehabilitation. *Proceedings of the 5th Euroseminar on Microscopy Applied to Building Materials (EMABM)*, Leuven, Belgium, pp. 151–161.

ČSN 721153 (1984) Petrographic analysis of natural rocks. *Czech Standardization Institute*, Prague (in Czech).

ČSN 721179 (1967) Determination of alkali reactivity of aggregates. *Czech Standardization Institute*, Prague (in Czech).

ČSN 721180 (1968) Determination of distinguishable particles of aggregates. *Czech Standardization Institute*, Prague (in Czech).

CUR-Recommendation 38 (1994) Measures to prevent concrete damage due to alkali-silica reaction (ASR). Enclosure with Cement, *Centre for Civil Engineering Research and Codes*, Gouda, the Netherlands.

CUR Recommendation 72 (2008) Inspection and testing of concrete structures. Centre for Civil Engineering Research and Codes, *Centre for Civil Engineering Research and Codes*, Gouda, the Netherlands.

CUR Recommendation 89 (2008) Measures to prevent concrete damage by the alkali-silica reaction. (In Dutch.) 2nd revised edition. *Centre for Civil Engineering Research and Codes*, Gouda, the Netherlands.

CUR Recommendation 102 (2008) Inspection and assessment of concrete structures in which the presence of ASR is suspected or has been established. Engineering Research and Codes, Centre for Civil Engineering Research and Codes, Gouda, the Netherlands.

Cyprus Geological Survey (2013) Geology of Cyprus – Introduction. Cyprus Geological Survey, 1415 Lefkosia, Cyprus (also available via Wickapedia- Geology of Cyprus).

Cyprus Geological Survey (2016) Geological map of Cyprus Island.

DAfStb (Deutscher Ausschuss für Stahlbeton) (1975) Vorbeugende Maßnahmen gegen schädigende Alkalireaktion im Beton, Vorläufige Richtlinie – Fassung Februar 1974. In: Walz, K. (hrsg.): Betontechnische Berichte 1974, Beton-Verlag GmbH Düsseldorf, S. 71–89.

DAfStb (Deutscher Ausschuss für Stahlbeton, Hrsg.) (2002) Empfehlung für die Schadensdiagnose und Instandsetzung von Betonbauwerken, die infolge einer Alkali-Kieselsäure-Reaktion geschädigt sind.

DAfStb (Deutscher Ausschuss für Stahlbeton, Hrsg.) (2013) Vorbeugende Maßnahmen gegen schädigende Alkalireaktion im Beton (Alkali-Richtlinie). Beuth Verlag GmbH, Vertriebs-Nr. 65265.

Davraz, M., & Gündüz, L. (2008) Reduction of alkali silica reaction risk in concrete by natural (micronized) amorphous silica. *Constr Build Mater.*, 22, 1093–1099.

De Ceukelaire, L. (1988) Alkali-silicareactie nu ook in België. *Cement*, 10, 21–25.

Deloye, F.X., & Divet, L. (1992) The alkali-silica reaction: quantitative consideration. In: Poole, A. (ed.) *Proceedings of the 9th International Conference on Alkali-Aggregate Reaction in Concrete (ICAAR)*. London, pp. 27–31.

Demars, Ph. (2012) Les reactions alcalis-granulats dans les ponts géres para le MET. *Unpublished report.*

Demars, Ph., Gilles, P., Dondonne, E., Lefebvre, G., Darimont, A., Lorenzi, G., Henriet, G., & Marion, A.M. (2008) The degradation of the bridge deck slabs in Belgium mainly involves alkali-aggregate reactions. In: Broekmans, M.A.T.M. & Wigum, B.J. (eds.) *Proceedings of the 13th International Conference on Alkali-Aggregate Reactions in Concrete*, Trondheim, Norway, pp. 941–949.

Deutscher Beton-Verein e.V., Wiesbaden (1972) Rundschreiben Nr. 24 (1967), Nr. 30 (1968), Nr. 41 (1970), Nr. 50 (1972) und Nr. 53 (1972).

Deutscher Bundestag (2009) Antwort auf die kleine Anfrage der Fraktion BÜNDNIS 90/DIE GRÜNEN (Drucksache 16/12024): Zerstörung des Fahrbahnbelages durch die Alkali-Kieselsäure-Reaktion.

DIN 1164-10 (2013) Special cement – Part 10: Composition, requirements and conformity evaluation for cement with low effective alkali content. *Deutsches Institut für Normung.*

Dressler, A. (2013) Einfluss von Tausalz und puzzolanischen, aluminiumhaltigen Zusatzstoffen auf die Mechanismen einer schädigenden Alkali-Kieselsäure-Reaktion in Beton. *Dissertation, Technische Universität München.*

EHE-08 (2008) Instrucción de hormigón estructural. Ministerio de la Presidencia. *Real Decreto 1247/2008, de 18 de Julio, Boletín Oficial del Estado. Suplemento del Número 203.*

EHE-98 (1999) La Instrucción de Hormigón Estructural. *Boletines Oficiales del Estado de los días 13-1-99 (R.D. 2661/98) y 24-6-99 (R.D. 996/99). Real Decreto 2661/1998. REAL DECRETO 2661/1998, de 11-DIC, del Ministerio de Fomento B.O.E.:* 13-ENE-99.

Elsen, J., Desmyter, J., & Soers, E. (2003) Alkali-silica reaction in concrete in Belgium – a review. *Aardk. Mededel*, 13, 73–79.

EN 197-1 (2011) (DIN EN 197-1, NF EN 197-1, PN EN 197-1, SIST EN 197-1, UNE EN 197-1) Cement Part 1: Composition, specifications and conformity criteria for common cements. *CEN*, Brussels, Belgium.

EN 206-1 (2007) (NBN EN 206-1, MSZ EN 206-1, NF EN 206-1, NP EN 206-1, SIST EN 206, SN EN 206-1, UNI EN 206-1) Concrete – Part 1: specification, performance, production and conformity. *CEN*, Brussels, Belgium.

EN 450-1 (2012) (SN EN 450-1) Fly ash for concrete – Part 1: Definition, specifications and conformity criteria. *CEN*, Brussels, Belgium.

EN 932-1 (1996) Tests for general properties of aggregates – Part 1: Methods for sampling. *CEN*, Brussels, Belgium.

EN 932-2 (1999) Tests for general properties of aggregates – Part 2: Methods for reducing laboratory samples. *CEN*, Brussels, Belgium.

EN 932-3 (1996) +A1 (2003) (ÖNORM EN 932-3, UNE EN 932-3, UNI EN 932-3) Tests for general properties of aggregates – Part 3: Procedure and terminology for simplified petrographic description. *CEN*, Brussels, Belgium.

EN 932-5 (2012) Tests for general properties of aggregates – Part 5: Common equipment and calibration. *CEN*, Brussels, Belgium.

EN 932-6 (1999) Tests for general properties of aggregates – Part 6: Definitions of repeatability and reproducibility. *CEN*, Brussels, Belgium.

EN 934-2 (2009)+A1 (2012) (ÖNORM EN 934-2) Admixtures for concrete, mortar and grout – Part 2: Concrete admixtures – definitions and requirements. *CEN*, Brussels, Belgium.

EN 1504 Part 1 to Part 10 (2005 to 2008) (SN EN 1504 Parts 2,3) Products and systems for the protection and repair of concrete structures – Definitions, requirements, quality control and evaluation of conformity. *CEN*, Brussels, Belgium.

EN 12620 (2002)+A1 (2008) (NP EN 12620, ÖNORM EN 12620, PN EN 12620, UNI EN 12620) Aggregates for concrete. *CEN*, Brussels, Belgium.

EN 13263-1 (2005)+A1 (2009) Silica fume for concrete – Part 1: Definitions, requirements and conformity criteria. *CEN*, Brussels, Belgium.

EN 15167-1 (2006) Ground granulated blastfurnace slag for use in concrete, mortar and grout. Part 1: Definitions, specifications and conformity criteria. *CEN*, Brussels, Belgium.

Eren, Ö. (2017) Professor in The Department of Civil Engineering, Eastern Mediterranean University, personal communication.

Erik, D., & Mutlutürk, M. (2004) Alkali-silica reactivity features of gravel-sand aggregates in Koç River (Hafik-Sivas) *ROCKMEC'2004-VIIth Regional Rock Mechanics Symposium, Sivas, Turkey*.

Fasseu, P. (2001) Manifestation de l'alcali réaction en France. *Seminário Luso-Francês – Degradação de Estruturas por Reacções Expansivas de Origem Interna*, LNEC, Lisbon, Portugal (in French).

Fernandes, I. (2005) Caracterização petrográfica, química e física de agregados graníticos em betões. Estudo de casos de obra. *PhD thesis, University of Porto*, Portugal (in Portuguese). 334p.

Fernandes, I., & Noronha, F. (2010) A petrografia no diagnóstico de fenómenos de deterioração do betão em Portugal. In: Neiva, J.M.C., Ribeiro, A., Mendes Victor, L., Noronha, F., & Magalhães Ribeiro, M. (eds.) *Ciências Geológicas: ensino, investigação e sua história*, Vol. II, Geologia Aplicada, Associação Portuguesa de Geólogos and Sociedade Geológica de Portugal (in Portuguese), pp. 295–303.

Fernandes, I., Ribeiro, M.D.A., Broekmans, M.A.T.M., & Sims, I. (2016) *Petrographic Atlas: Characterisation of aggregates regarding potential reactivity to alkalis, RILEM TC 219-ACS Recommended Guidance AAR-1.2, for use with the RILEM AAR-1.1 Petrographic Examination Method*, Springer for RILEM.

FGSV (Forschungsgesellschaft für Straßen- und Verkehrswesen) (2007) TL Beton-StB 07 – Technische Lieferbedingungen für Baustoffe und Baustoffgemische für Tragschichten mit hydraulischen Bindemitteln und Fahrbahndecken aus Beton.

Filipczyk, M., & Góralczyk, S. (2010) Verification of new testing methods of sunburn basalt rocks. Prace Naukowe Instytutu Górnictwa Politechniki Wrocławskiej: Górnictwo i Geologia XIII, Wrocław, 2010.

Fischboeck, E.K., & Harmuth, H. (2009) An Austrian experience with identification and assessment of alkali-aggregate reaction in motorways. In: Alexander, M., Beushausen, H.-D., Dehn, F., & Moyo, P. (eds.) *Concrete Repair, Rehabilitation and Retrofitting II*, Taylor and Francis Group, London.

Forum, C.S. (1965) Alkali-Reaktion der Zuschlagstoffe im Beton. *Beton- und Stahlbeton, Heft 7*, S. 163–168.

Freyburg, E., & Schliffkowitz, D. (2006) Bewertung der Alkalireaktivität von Gesteinskörnungen nach petrografischen und mikrostrukturellen Kriterien. 16. Internationale Baustofftagung (ibausil), Weimar, Tagungsbericht, Band 2, S. 355–372.

Gadea, J., Soriano, J., Martín, A., Campos, P.L., Rodríguez, A., Junco, C., Adán, I., & Calderón, V. (2010) The alkali–aggregate reaction for various aggregates used in concrete. *Mater de Construcc.*, 60 (299), 69–78.

Gasparotto, G., Bargossi, G.M., Peddis, F., & Sammassimo, V. (2011) A case study of alkali-silica reactions: petrographic investigation of paving deterioration. *Periodico di Mineralogia*, 80 (2), 309–316.

Giebson, C. (2013) Die Alkali-Kieselsäure-Reaktion in Beton für Fahrbahndecken und Flugbetriebsflächen unter Einwirkung alkalihaltiger Enteisungsmittel. *Dissertation, Bauhaus-Universität Weimar*, F.A. Finger-Institut für Baustoffkunde.

Giebson, C., Seyfarth, K., & Stark, J. (2010a) Influence of acetate and formate-based de-icers on ASR in airfield concrete pavements. *Cement Concrete Res.*, 40, 537–545.

Giebson, C., Seyfarth, K., & Stark, J. (2010b) Effectiveness of ground granulated blast furnace slag in preventing deleterious ASR in concretes exposed to alkali-containing de-icer solutions. In: Brameshuber, W. (ed.) *Proceedings of the International RILEM Conference on Material Science (PRO 77)*, Vol. III, Aachen, Germany, pp. 221–230.

Giuseppetti, G., Donghi, G., & Marcello, A. (2002) Experimental investigations and numerical modeling for the analysis of AAR process related to Poglia dam: evolutive scenarios and design solutions. *Proceedings of the EPREM International Congress on Conservation and Rehabilitation of Dams*, Madrid, Spain.

Godart, B. (2009) Méthode de suivi dimensionnel et de suivi de la fissuration des structures avec application aux structures atteintes de réaction de gonflement interne du béton, *Guide technique des Laboratoires des Ponts et Chaussées*, Laboratoire Central des Ponts et Chaussées, 60p.

Godart, B., & Le Roux, A. (2008) Alcali-réaction dans les structures en béton: Mécanisme, pathologie et prévention. *Techniques de l'Ingenieur*, C2252, 19p.

Godart, B., de Rooij, M., & Wood, J.G.M. (2013) *Guide to diagnosis and appraisal of AAR damage to concrete in structures, Part 1 Diagnosis (AAR-6.1)*, RILEM State-of-the-Art Reports, Volume 12, Springer, Dordrecht, RILEM, Paris 89p.

Godart, B., Mahut, B., Fasseu, P., & Michel, M. (2004) A guide for aiding to the management of structures damaged by concrete expansion in France. In: Tang, M., & Deng, M. (eds.) Proceedings of the 12th International Conference on Alkali-Aggregate Reaction in Concrete (ICAAR), International Academic Publishers, Beijing World Publishing Corporation, Vol II, pp. 1219–1228.

Godart, B., Wood, J.G.M., & de Rooij, M. (2017) *Guide to diagnosis and appraisal of AAR damage to concrete in structures, Part 2 Appraisal and repair (AAR-6.2)*, RILEM State-of-the-Art Reports, Springer, Dordrecht, RILEM, Paris. *Expected completion for publication in 2017.*

Góralczyk, S. (1984) Foreign test methods of alkaline reactivity of aggregates. Building Materials no. 3/84.

Góralczyk, S. (1993) Researches of alkaline reactivity of carbonate aggregates. Institute of Mechanised Construction & Rock Mining, Warsaw.

Góralczyk, S. (1994) Alkaline reactivity of mineral aggregates. The scope of application and test methods. Article from a seminary. Institute of Mechanised Construction & Rock Mining, Warsaw.

Góralczyk, S. (2000) *Reactivity of carbonate aggregates*. Institute of Mechanised Construction & Rock Mining, Warsaw.

Góralczyk, S. (2001) Usefulness of carbonate aggregates to concrete, *PhD Dissertation, AGH University of Science and Technology*, Cracow.

Góralczyk, S. (2003) Occurence and assessment of reactive aggregates in Poland, Institute of Mechanised Construction & Rock Mining, Warsaw, Poland.

Góralczyk, S. (2005) Corrosion alkaline concrete. In: *Scientific and Technical Symposium Cement and Durability of Concrete*, Bydgoszcz, Poland (unpublished).

Góralczyk, S. (2011) New methods for testing and evaluation of alkaline reactivity of aggregates in the EU. Materiały Budowlane, No. 11/2011.

Góralczyk, S., & Kukielska, D. (2008) Analysis of the quality of national aggregates. *Prace Naukowe Instytutu Górnictwa Politechniki Wrocławskiej*, 121, 50.

Gościniak, H. (1989) Qualitative-quantitative survey of minerals reacting with alkalis from cement. Institute of Mechanised Construction & Rock Mining, Warsaw.

Gregerová, M., & Všianský, D. (2008) Geovisualization of aggregates for AAR prediction and its importance for risk management. In: Broekmans, M.A.T.M., & Wigum, B. (eds.) *Proceedings of the 13th International Conference on Alkali-Aggregate Reaction in Concrete (ICAAR)* Trondheim, Norway, pp. 1059–1073.

Guédon-Dubied, J.-S., Cadoret, G., Durieux, V., Martineau, F., Fasseu, P., & Van Overbecke, V. (2000) Study on Tournai limestone in Antoing Cimescaut Quarry. Petrological, chemical and alkali reactivity approach. In: Bérubé, M.A., Fournier, B., & Durand, B. (eds.) *Proceedings of the 11th International Conference on Alkali-Aggregate Reaction (ICAAR)*, Québec, Canada, pp. 335–344.

Gümüş, S. (2009). Kocaeli Bölgesindeki Agregaların Alkali Silika Reaksiyonu Bakımından İncelenmesi, Masters Thesis, Sakarya Üniversitesi Fen Bilimleri Enstitüsü (in Turkish), 85s.

Haas, J. (2001) Geology of Hungary. Eötvös Kiadó, Budapest, 316p.

Hasdemir, S., Tuğrul, A., & Yılmaz, M. (2012) Evaluation of alkali reactivity of natural sands. *Constr Build Mater.*, 29, 378–385.

Heijnen, W.M.M. (1992) Alkali-aggregate reactions in the Netherlands. In: Poole, A. (ed.) *Proceedings of the 9th International Conference on Alkali-Aggregate Reaction in Concrete (ICAAR)*. London, UK, pp. 432–439.

Heijnen, W.M.M., & Van der Vliet, J. (1991) Alkali-silicareactie ook in Nederland. *Cement, 43* (718), 6–11.

Heijnen, W.M.M., Larbi, J.A., & Siemes, A.J.M. (1996) Alkali-silica reaction in The Netherlands. In: Shayan, A. (ed.) *Proceedings of the 10th International Conference on Alkali-Aggregate Reaction (ICAAR)*, Melbourne, Australia, pp. 109–116.

Hunkeler F., Merz, C., & Kronenberg, P. (2007) Alkali-Aggregat-Reaktion (AAR). Grundlagen und Massnahmen bei neuen und bestehenden Kunstbauten. ASTRA Dokumentation 8213, 132p.

ICOLD (1991) Alkali-aggregate reaction in concrete dams. International Commission On Large Dams, Paris. Bulletin 79, 1991.

ISE (1992) Structural effects of alkali-silica reaction. The Institution of Structural Engineers, SETO, London, 45p.

JICA/KGM (1996) The study on the maintenance and rehabilitation of highway bridges in the Republic of Turkey, Oriental Consultants Co. Ltd and Japan Overseas Consultants Co. Ltd.

Katayama, T. (2000) Alkali-aggregate reaction in the vicinity of Izmir, western Turkey. In: Bérubé, M.A., Fournier, B., & Durand, B. (eds.) *Proceedings of the 11th International Conference on Alkali-Aggregate Reaction (ICAAR)*, Québec, Canada, pp. 365–374.

Katayama, T. (2004) How to identify carbonate rock reactions in concrete. *Mater Charact.*, *53* (2–4), 85–104.

Kocacitak, S. (1975) A note of information regarding to alkali-aggregate reactions in Turkey. *Symposium on Alkali-Aggregate Reaction*, Reykjavik, Iceland, pp. 259–262.

Korkanç, M., & Tuğrul, A. (2004) Evaluation of selected basalts from Niğde, Turkey, as source of concrete aggregate. *Eng Geol.*, *75*, 291–307.

Krispel, S. (2008) Portland-slag cements – reduction of the residual risk of aggregates containing reactive components. In: Broekmans, M.A.T.M., & Wigum, B.J. (eds.) *Proceedings of the 13th International Conference on Alkali-Aggregate Reaction in Concrete (ICAAR)*, Trondheim, Norway, pp. 873–882.

Lanza, V. (2012) Estudo de la reactividade álcali-sílice originada por componentes reactivos minoritarios. *PhD thesis. Escuela Técnica Superior de Enginieros de Caminos, Canales y Puertos, Universidade Politécnica de Madrid* (in Spanish), 426p.

Lanza, V., & Alaejos, P. (2011) Optimized Gl Pat Test for detection of alkali-reactive aggregates. *ACI Mater J.*, *109*, 403–412.

Larbi, J.A., & Visser, J.H.M. (2002) ASR of aggregate with high chert content. *Heron, 47* (2), 141–159.

LCPC (1991) Recommandations provisoires pour la prévention des désordres dus à l'alcali-réaction (Provisional recommendations for the prevention of damage due to the alkali-aggregate reaction) *Laboratoire Central des Ponts et Chaussées* (in French).

LCPC (1994) Recommandations pour la prévention des désordres dus à l'alcali-réaction (Recommendations for the prevention of damage due to alkali-aggregate reactions). *Laboratoire Central des Ponts et Chaussées* (in French).

LCPC (1999) Manuel d'identification des réactions de dégradation interne du beton dans les ouvrages d'art. (Handbook for identification of the reactions of internal degradation of the concrete in the structures) *Laboratoire Central des Ponts et Chaussées* (in French).

LCPC (2003) Aide à la gestion des ouvrages atteints de réaction de gonflement interne, (Aid for the management of structures damaged by an internal swelling reaction). *Guide technique du Laboratoire Central des Ponts et Chaussées* (in French).

LCPC (2010) Protection et réparation des ouvrages atteints de réactions de gonflement interne du béton: Recommandations provisoires (Protection and repair of the structures affected by expansive internal reactiosn in concrete). *Guide technique du Laboratoire Central des Ponts et Chaussées* (in French).

LCPC n° 36 (1993) Essai de mise en évidence du gel d'alcali-réaction par fluorescence des ions uranyl. Projet de méthode d'essai LCPC n° 36. *Laboratoire Central des Ponts et Chaussées* (in French).

LCPC n° 44 (1997) Alcali-réaction du béton – essai d'expansion résiduelle sur béton durci. Techniques et méthodes des LPC. Projet de méthode d'essai LCPC n° 44. *Laboratoire Central des Ponts et Chaussées* (in French).

LCPC n° 47 (1997) Détermination de l'indice de fissuration d'un parement en béton, Projet de méthode d'essai LPC, *Laboratoire Central des Ponts et Chaussées*. (in French).

Le Roux, A., Thiebaut, J. Guedon, J.S., & Wackenheim, C. (1999) Pétrographie appliquée à l'alcali-réaction – *Etudes et Recherches des LPC, série Ouvrages d'Art*, OA 26, LCPC (in French).

Leemann, A., & Griffa, M. (2013) *Diagnosis of alkali-aggregate reaction in dams*. SFOE-Project SI/500863-01. EPMA, Swiss Federal Laboratories for Materials Science and Technology, Duberdorf, 68p.

Leemann, A., & Holzer, L. (2005) Alkali-aggregate reaction – identifying reactive silicates in complex aggregates by ESEM observation of dissolution features. *Cement and Concrete Composites, 27*, 796–801.

Leemann, A., & Merz, C. (2013) An attempt to validate the ultra-accelerated microbar and the concrete performance test with the degree of AAR-induced damage observed in concrete structures. *Cement Concrete Res.*, *49*, 29–37.

Leemann, A., Hammerschlag, J.G., & Thalmann, C. (2008) Inconsistencies between different accelerated test methods used to assess alkali-aggregate reactivity. In: M.A.T.M. Broekmans, B.J., & Wigum, B. (eds.) *Proceedings of the 13th International Conference on Alkali-Aggregate Reaction in Concrete (ICAAR)*, Trondheim, Norway, pp. 944–953.

Leemann, A., Thalmann, C., & Studer, W. (2005) Alkali-aggregate reaction in Swiss tunnels. *Mater Struct.*, *38*, 381–386.

Leroy, R., Boldea, L.I., Seignol, J-F., & Godart, B. (2011) Re-assessment and treatment-design of an ASR-affected gravity da. In: Schleiss & Boes (eds.) *Dams and Reservoirs under Changing Challenges*. Taylor & Francis Group. London, ISBN 978-0-415-68267-1.

Levi, F., Mancini, G., & Napoli, P. (1985) Structural damages due to the alkali silica reaction (ASR). *L'iIndustria Italiana delle Costruzioni*, *169*, 59–73.

Li, K., Coussy, O., & Larive, C. (2004) Modélisation chimico-mécanique du comportement des bétons affectés par la réaction d'alcali-silice. Expertise numérique des ouvrages d'art dégradés, *Etudes et recherches des Laboratoires des Ponts et Chaussées, OA 43, LCPC* (in French), 202p.

LNEC E 415 (1993) Inertes para argamassas e betões – determinação da reactividade potencial com os álcalis. Análise petrográfica. *Especificação LNEC*, Lisboa (in Portuguese), 6p.

LNEC E 461 (2007) Betões. Metodologias para prevenir reacções expansivas internas. Laboratório Nacional de Engenharia Civil, Lisboa (in Portuguese), 6p.

Lopes Batista, A., & Piteira Gomes, J.M. (2012) Practical assessment of the structural effects of swelling processes and updated inventory of the affected Portuguese concrete dams. *Proceedings of the 54° Congresso Brasileiro do Concreto*, 30p.

López-Buendía, A.M., Climent, V., & Verdú, P. (2006) Lithological influence of aggregate in the álcali-carbonate reaction. *Cement Concrete Res.*, *36*, 1490–1500.

Lorenzi, G., Jensen, J., Wigum, B., Sibbick, R., Haugen, M., Guédon S., & Åkesson, U. (2006) Petrographic atlas of the potentially alkali-reactive rocks in Europe. Professional paper 2006/1 – N.302. Geological Survey of Belgium.

Ludwig, H.-M. (2012) Future cements and their properties. *Cement International*, *10*, 80–89.

Lukschová, Š. (2008) Alkali-silica reaction of aggregates in real concrete and mortar specimens, *PhD Thesis, Institute of Geochemistry, Mineralogy and Mineral Resources, Faculty of Science, Charles University in Prague, Prague, Czech Republic* (+ 5 app.), 70p.

Lukschová, Š., Burdová, A., Pertold, Z., & Přikry, R. (2010) Alkali-silica reactivity of aggregates in concrete pavements. *Proceedings of the 32nd International Conference on Cement Microscopy (ICCM)*, New Orleans, USA, 14p.

Lukschová, Š., Burdová, A., Pertold, Z., & Přikryl, R. (2011) Macro- and micro-indicators of ASR in concrete pavement. *Mag Concrete Rese.*, *63* (8), 553–571.

Lukschová, Š., Pertold, Z., & Hromádko, J. (2009a) Factors affecting DEF and ASR in concrete. *Proceedings of the 4th Concrete Future – Twin Coimbra International Conferences, Coimbra*, Portugal, CF189.

Lukschová, Š., Přikryl, R., & Pertold, Z. (2009b) Petrographic identification of alkali-silica reactive aggregates in concrete from 20th century bridges. *Constr Build Mater.*, *23* (2), 734–741.

Lukschová, Š., Přikryl, R., & Pertold, Z. (2009c) Evaluation of the alkali-silica reactivity potential of sands. *Mag Concrete Res.*, *61*(8), 645–654.

Mannasoğlu, N. (2010) Granitik Kayaçların Alkali-Agrega Reaktivitelerinin Karşılaştırılması, Masters Thesis, İstanbul Üniversitesi Fen Bilimleri Enstitüsü (in Turkish), 169s.

Marquordt, D., & Rother, K.H. (2011) Erfahrungen aus Erhaltungsmaßnahmen an AKR-geschädigten Betonfahrbahndecken. Griffig – Broschüre der Gütegemeinschaft Verkehrsflächen aus Beton e.V., Nr. 2, S. 2-5.

Martínez, M.C.A., Orosas, I.F., López, A.L., Ramos, S.L., Sanz, B.M., & Rubiera, N.P. (1996) Patologías del hormigón. Interacción entre el ataque por sulfatos y la reacción álcali-árido. *Geogaceta, 20* (3), 723–724.

Menegaki, M.E., & Kaliampakos, D.C. (2010) European aggregates production: Drivers,correlations and trends. *Resour Policy, 35,* 235–244.

Menéndez, E., Garcia-Rovés, R., & Ruiz, S. (2016) Alkali release from aggregates: contribution to ASR. *Proceedings of the Institution of Civil Engineers: Construction Materials*, Themed issue on Alkali-Aggregate Reactivity in Concrete, 169 (CM4) August, Institution of Civil Engineers, London, 9p.

Menendez, E., Préndez, N., Márquez, C., & Aldea, B. (2012) Evaluation of granitic aggregates behavior in relation with the alkaline extraction and compositional change in their phases, In: Fournier, B., & Ideker, J.H. (eds.) *Proceedings of the 14th International Conference on Alkali-aggregate Reaction (ICAAR)*, University of Texas, Austin, USA, 10p.

Merz, C., & Leemann, A. (2011) Validierung der AAR-Prüfungen für Neubau und Instandsetzung, Betonbauten in der Schweiz. UVEK, Forschungsaufträge AGB 2005/023 und AGB 2006/003, Bericht Nr. 648, Bern.

Merz, C., & Leemann, A. (2013) Assessment of the residual expansion potential of concrete from structures damaged by AAR. *Cement Concrete Res., 52,* 182–189.

Merz, C., Hunkeler, F., & Griesser, A. (2006) Damages due to alkali-aggregate reaction in concrete structures in Switzerland (Schäden durch Alkali-Aggregat-Reaktion an Betonbauten in der Schweiz). UWEK, Forschungsauftrag AGB2001/471, Bericht Nr. 599, Bern.

Mielich, O. (2009) Beitrag zu den Schädigungsmechanismen in Betonen mit langsam reagierender alkaliempfindlicher Gesteinskörnung. Dissertation, Otto-Graf-Institut, Materialprüfungsanstalt Universität Stuttgart.

Mladenovič, A. (2006) Alkali-silica reaction of Slovenian aggregates for concretes and mortars. *PhD Thesis.*

Mladenovič, A. (2010) Expert opinion No. P 585/05 – 420 – 114 – int. K 459/09 about the reasons for damage to the floor at AET Tolmin, Slovenian National Building and Civil Engineering Institute (in Slovenian).

Mladenovič, A., Strupi-Šuput, J., & Capuder, F. (2000) Re-occurring damage to hollow PC columns caused by ASR in the hardened grout mixture used in previous repair works. In: Bérubé, M-A. (ed.) *Proceedings of the 11th International Conference on Alkali-Aggregate Reaction in Concrete*, Québec, Canada, pp. 879–887.

Modrý, S., Dohnálek, J., Gemrich, J., & Hörbe, M. (2003) Elimination of alkali reaction of aggregates in the concrete motorways. *Project n. 803/120/114. Czech Technical University, Prague.* Unpublished report (in Czech).

MSZ 4798-1 (2004) *Conditions of use of MSZ EN 206-1 in Hungary.* Hungarian Standards Institution. Budapest.

Müller, Ch., Borchers, I., & Eickschen, E. (2013) Experience with ASR test methods: advice on obtaining practical evaluation criteria for performance testing and aggregate testing. *Cement International, Jg. 11* (3), 86–93.

Müller, Ch., Borchers, I., Stark, J., Seyfarth, K., & Giebson, C. (2009) Beurteilung der Alkaliempfindlichkeit von Betonzusammensetzungen – Vergleich von Performance-Prüfverfahren. *17. Internationale Baustofftagung (ibausil)*, Weimar, Tagungsbericht, Band 2, S. 261–266.

Multon, S., Barin, F.X., Godart, B., & Toutlemonde, F. (2007) Estimation of the residual expansion of concrete affected by alkali-silica reaction. *Journal of Matererials in Civil Engineering, 20, SPECIAL ISSUE: Durability and Service Life of Concrete Structures: Recent Advances*, pp. 54–62.

Musal, B. (2003) The effect of zeolite and glass fibres on alkali-silica reaction, Dokuz Eylül University, *MSc dissertation*, 141p.

NBN B 12-109 (2015) Cement – low alkali content cement. IBN. *Bureau for Standardisation*. Brussels.

NBN B 15-001 (2012) Concrete – Specification, performance, production and conformity – National supplement to NBN EN 206-1 (2001) IBN. *Bureau for Standardisation*. Brussels.

Nekvasilová, Z., Šachlová, Š., & Přikryl, R. (2013) Experimental testing of ASR potential of volcanic rocks. *Proceedings of the 12th SGA Conference*. Mineral deposit research for a high-tech world, Uppsala, Sweden, Elanders Sverige AB, pp. 1818–1821.

NEN 8005 (2008)/A1 (2011) Dutch supplement to NEN-EN 206-1: Concrete – Part 1: Specification, performance, production and conformity. *Stichting Nederlands Normalisatie-instituut*, Delft.

Neubauer, F. (2003) Geology of Europe. In: De Vivo, B., Grasemann, B., & Stuwe, K. (eds.): Volume Geology. Encyclopedia of Life Supporting Systems. (17 printed pages + 18 figures), *UNESCO Publishing-Eolss Publishers*, Oxford, UK (Online version: www.eolss.net).

NF EN 196-2 (2013) Method of testing cement. Chemical analysis of cement. *Association Française de Normalisation*, Paris, France.

Nijland, T.G., & Siemes, A.J.M. (2002) Alkali-silica reaction in the Netherlands: Experiences and current research. *Heron*, 47 (2), 81–85.

Nijland, T.G., Larbi, J.A., & Siemes, A.J.M. (2003) Experience of ASR in the Netherlands, European marine sand and gravel – shaping the future, *EMSAGG Conference*, Delft University, The Netherlands.

Nixon, P.J., & Sims, I. (1992). RILEM TC106 alkali aggregate reaction – accelerated tests. In: *Proceedings of the 9th International Conference on Alkali-Aggregate Reaction in Concrete*. London, UK: Reunion Internationale des Laboratoires et Experts des Materiaux, pp. 70–100.

Nixon, P.J., & Sims, I. (Eds.) (2016) *RILEM Recommendations for the prevention of damage by alkali-aggregate reactions in new concrete structures, State-of-the-Art report of the RILEM Technical Committee 219-ACS*, RILEM State-of-the-Art Reports, Volume 17, Springer pp RILEM, Paris, 184p.

ÖNORM B 3100 (2008) Beurteilung der Alkali-Reaktion im Beton (Assessment of the alkali silica reactivity in concrete). Österreichisches Normungsinstitut, Viena, 19p.

ÖNORM B 3309-1 (2010) Aufbereitete hydraulisch wirksame Zusatzstoffe für die Betonherstellung (AHWZ) (Processed hydraulic additions for concrete production – Part 1: Combination products (GC/GC-HS)). Österreichisches Normungsinstitut, Vienna, 26p.

ÖNORM B 3309-2 (2010) Aufbereitete hydraulisch wirksame Zusatzstoffe für die Betonherstellung (AHWZ) (Processed hydraulic additions for concrete production – Part 2: Ground granulated blast furnace slag for use in concrete, mortar and grout – National specifications to ÖNORM EN 15167-1 (GS bzw. GS-HS)). Österreichisches Normungsinstitut, Vienna, 11p.

ÖNORM B 3309-3 (2010) Aufbereitete hydraulisch wirksame Zusatzstoffe für die Betonherstellung (AHWZ) (Processed hydraulic additions for concrete production – Part 3: Fly ash for concrete – National specifications to ÖNORM EN 450-1 (GF bzw. GF-HS). Österreichisches Normungsinstitut, Vienna, 12p.

ÖNORM B 3327-1 (2005) Zemente gemäß ÖNORM EN 197-1 für besondere Verwendung. Teil 1: Zusätzliche Anforderungen (Cements according to ÖNORM EN 197-1 for special use – Part 1: Additional requirements). Österreichisches Normungsinstitut, Vienna, 14p.

ÖNORM B 4710-1 (2007) Beton Teil1: Festlegung, Herstellung, Verwendung und Konformitätsnachweis (Regeln zur Umsetzung der ÖNORM EN 206-1) (Concrete – Part 1: Specification, production, use and verification of conformity (Rules for the implementation of ÖNORM EN 206-1 for normal and heavy concrete)). Österreichisches Normungsinstitut, Vienna, 160p.

Öttl, C. (2004) Die schädigende Alkalireaktion von gebrochener Oberrhein-Gesteinskörnung im Beton. Dissertation, Otto-Graf-Institut, Universität Stuttgart.

Partner Project (2006) European Standard Tests to prevent Alkali Reactions in Aggregates, *GROWTH Project GRD1-2001-40103*

Pertold, Z., & Lukschová, Š. (2007) Quantitative microscopic description of concrete samples and identification of ASR employing SEM/EDS method. *Unpublished technical report* (in Czech), Faculty of Science, Charles University in Prague, 21p.

Pertold, Z., & Lukschová, Š. (2008) Quantitative microscopic description of concrete samples from Prague and identification of ASR. *Unpublished technical report* (in Czech), Faculty of Science, Charles University in Prague, 14p.

Pertold, Z., Chvátal, M., Pertoldová, J., Zachariáš, J., & Hromádko, J. (2002) AAR damages of concrete pavement in highway D11 caused by alkali reaction of aggregates, *Beton* (in Czech), 2 (2), 21–24.

Pertold, Z., Lukschová, Š., & Přikryl, R. (2008) Evaluation of alkali-reactivity of aggregates. *Final report. Unpublished technical report, Faculty of Science*, Charles University in Prague (+ 7 app.) (in Czech), 68p.

Pertold, Z., Lukschová, Š., Přikryl, R., Burdová, A., Seidlová, Z., & Šťastná, A. (2010) Degradation of concrete pavements caused by AAR. *Final report. Unpublished technical report, Faculty of Science, Charles University in Prague* (+ 6 app.) (in Czech), 74p.

Piteira Gomes, J.M. (2008) Modelação do comportamento estrutural de barragens de betão sujeitas a reacções expansivas. *PhD thesis, Universidade Nova de Lisboa, Portugal* (in Portuguese).

PN-B-06712 (1986) Mineral Aggregates for concrete. Polish Committee for Standardization, PKN (replaced in 2012 by PN EN 12620).

PN-B-06714-34 (1991) Mineral aggregates. Testing. Determination of alkaline reactivity. Polish Committee for Standardization, PKN (withdrawn in 2012).

PN-B-06714-46 (1992) Mineral aggregates. Determination of silica soluble in sodium hydroxide content. Polish Committee for Standardization, PKN.

PN-B-06714-47 (1988) Mineral aggregates. Determination of silica soluble in sodium hydroxide content. Polish Committee for Standardization, PKN.

Poole, A.B. (1975) Alkali-silica reactivity in concrete from Dhekelia, Cyprus. *Symposium on alkali-aggregate reaction, preventative measures, Reykjavik.* Iceland Building Research Institute, pp.113–130.

Prinčič, T., Štukovnik, P., Pejovnik, S., De Schutter, G., & Bosiljkov, V.B. (2013) Observations on dedolomitization of carbonate concrete aggregates, implications for ACR and expansion. *Cement and Concrete Res.*, 54, 151–160.

Ramos, J.M., Batista, A.L., Oliveira, S.B., Castro, A.T., Silva, H.S., & Pinho, J.S. (1996) Reliability of arch dams subjected to concrete swelling. Three case histories. Memória No. 808, *Laboratório Nacional de Engenharia Civil, Lisboa*, Portugal.

Ramos, V. (2013) Characterization of the potential reactivity to alkalis of Portuguese aggregates for concrete. *PhD thesis. University of Porto/University of Aveiro.*

Ramyar, K., Topal, A., & Andiç-Çakır, Ö. (2005) Effects of aggregate size and angularity on alkali-silica reaction. *Cement and Concrete Res.*, 35, 2165–2169.

Regamey, J.M., & Hammerschlag, J.G. (1995) Barrage d'Illsee – assaintment. *Research and Development in the Field of Dams. Proceedings of the ICOLD-Symposium* (Crans-Montana, Switzerland).

Reinhardt, H.W., & Mielich, O. (2012) Mechanical properties of concretes with slowly reacting alkali sensitive aggregates. In: Drimalas, T., Ideker, J.H., & Fournier, B. (eds.) *Proceedings of the 14th ICAAR*, Austin, Texas, USA, 7p.

Rezaieh, A.Z. (2014) *The preliminary evaluation of the susceptibility of cyprus aggregates to alkali aggregate reaction.* MSc thesis. The graduate School of Applied Sciences, Near East University, Nicosia. 79p.

Ribeiro, A., Pereira, E., & Dias, R. (1990) Structure in the northwest of the Iberian Peninsula. In: Dallmeyer, R.D., & Martínez-Garcia, E. (eds.) *Pre-Mesozoic geology of Iberia*, Springer-Verlag, pp. 220–236.

RILEM AAR-0 (2003) Recommended Test Method: Detection of potential alkali-reactivity of aggregates – Outline guide to the use of RILEM methods in assessments of aggregates for potential alkali-reactivity. *Mater Struct.*, 36, 472–479.

RILEM AAR-1 (2003) Detection of potential alkali-reactivity of aggregates – petrographic method. TC 191-ARP, Alkali-reactivity and prevention – assessment, specification and diagnosis of alkali-reactivity. Prepared by Sims, I and Nixon, P. *Mater Struct.*, 36, 472–479.

RILEM AAR-2, TC 191-ARP (2000) AAR-2: detection of potential alkali-reactivity of aggregates-the ultra-accelerated mortar-bar test. *Reunion Internationale des Laboratoires et Experts des Materiaux*, 5, 1–32.

RILEM AAR-5, TC 191-ARP (2005) AAR-5: rapid preliminary screening test for carbonate aggregates. *Reunion Internationale des Laboratoires et Experts des Materiaux*, 8, 1–18.

Rother, K.H. (2013) 5 Jahre Erfahrungen mit baulichen Erhaltungsmaßnahmen an AKR-geschädigten Fahrbahndecken aus Beton. Präsentation, VSVI Halle.

Šachlová, Š. (2013) Microstructure parameters affecting alkali–silica reactivity of aggregates. *Constr Build Mater.*, 49, 604–610.

Šachlová, Š., Burdová, A., Pertold, Z., & Přikryl, R. (2011) Macro- and micro-indicators of ASR in concrete pavement. *Mag of Concrete Res.*, 63 (8), 553–571.

Šachlová, Š., Šťastná, A., Přikryl, R., Pertold, Z., & Nekvasilová, Z. (2013) Factors affecting ASR potential of quartzite from a single quarry (Bohemian Massif, Czech Republic). *Proceedings of the 12th SGA Conference*. Mineral deposit research for a high-tech world, Uppsala, Sweden, Elanders Sverige AB, pp. 1833–1836.

Sağlık, A., Kocabeyler, M.F., Orkun, Y., Halıcı, M., & Tunç, E. (2003) Deriner barajı ve HES inşaatı kütle betonunda kullanılması planlanan agregalarda alkali-silika reaksiyonu riski ve önlenmesine yönelik yürütülen çalışmalar. 5. *Ulusal Beton Kongresi* (in Turkish), pp. 205–224.

Salomon, M., & Panetier, J.L. (1994) Quantification du degré d'avancement de l'alcali-réaction dans les bétons et de la néofissuration associée. *Third Canmet/ACI international Conference on durability of concrete*, Nice, France, Supplementary papers, pp. 383–402.

Santos Silva, A. (2005) Degradação do betão por reacções álcalis sílica. Utilização de cinzas volantes e metacaulino para a sua prevenção. *PhD thesis, Laboratório Nacional de Engenharia Civil e escola Superior de Engenharia da Universidade do Minho*, Portugal (in Portuguese), 340p.

Santos Silva, A., & Gonçalves, A. (2006) Appendix A – Portugal. In: Wigum, B.J,, Pedersen, L.T., Grelk, B., & Lindgard, J. (eds.) *PARTNER Report 2.1, State-of-the-art report: key parameters influencing the alkali aggregate reaction*, SINTEF, Trondheim, Norway, pp. 57–61.

Santos Silva, A., Soares, D., Fernandes, I., Custódio, J., & Bettencourt Ribeiro, A. (2014) Prevenção das reações álcalis-agregado (RAA) no concreto – melhoria do monitoramento da reatividade aos álcalis de agregados. *In*: Véras Ribeiro, D. (coord.). *Proceedings do 1° Encontro Luso-Brasileiro de Degradação de Estruturas em Concreto Armado (DEGRADA)*, Bahia, Brasil, 293–306.

Saouma, V, Perotti L., & Uchita, Y. (2005) AAR analysis of Poglia dam with Merlin numerical model. *Proceedings of the 8th ICOLD Benchmark Workshop on Numerical Analysis of Dams*, Wuhan, China.

Schlangen, H.E.J.G., Çopuroğlu, O., Andiç-Çakir, O., & Garcia-Diaz, E. (2008) Modelling ASR expansions based on measurements of local properties of expanding gel. In: *Proceedings of the 13th ICAAR Alkali-Aggregate Reaction in Concrete*, Trondheim, Norway, 10p.

Schmidt, K. (2009) Verwendung von Steinkohlenflugasche zur Vermeidung einer schädigenden Alkali-Kieselsäure Reaktion im Beton. Dissertation, Technische Universität München.

Seidlová, Z., Pġikryl, R., Pertold, Z., & Šachlová, Š. (2012) Alkali-Silica Reaction of Volcanic Rocks, In: Fournier, B., & Ideker J.H. (eds.) *Proceedings of the International Conference on Alkali-aggregate Reaction*, University of Texas, Austin, USA, 10p.

Seyfarth, K., & Giebson, C. (2005) Beurteilung des AKR-Schädigungspotentials von Betonen mittels Klimawechsellagerung. Beitrag zum 45. *Forschungskolloquium des DAfStb (6./7. Oktober 2005, Wien), Beton- und Stahlbetonbau 100*, 189–192.

Seyfarth, K., & Stark, J. (1991) Zur Dauerhaftigkeit wärmebehandelter Betone. 11. Internationale Baustofftagung (ibausil), Weimar, Tagungsbericht, Band 1, S. 179–189.

Seyfarth, K., Giebson, C., & Stark, J. (2009a) AKR-Performance-Prüfung für Fahrbahndecken aus Beton: Erfahrungen aus Labor und Praxis im Vergleich. 17. *Internationale Baustofftagung (ibausil), Weimar, Tagungsbericht Band 2*, 255–260.

Seyfarth, K., Giebson, C., & Stark, J. (2009b) Prevention of deleterious ASR by assessing aggregates and specific concrete mixtures. *Proceedings of the 3rd International Conference on Concrete and Development*, Tehran, Iran, pp. 159–169.

SIA 505 269/2 (2011) Erhaltung von Tragwerken – Betonbau. SIA, Zürich.

SIA Merkblatt 2042 (2012) Prévention des désordres dus à la réaction alcalis-granulats (RAG) dans les ouvrages en bétonVorbeugung von Schäden durch die Alkali-Aggregat-Reaktion (AAR) bei Betonbauten, SIA, Zürich.

Siemes, A.J.M., & Bakker, J.D. (2000) Evaluation of the Institution of Structural Engineers procedure on concrete structures with alkali-silica reaction in the Netherlands. In: Bérubé, M. A., Fournier, B., & Durand, B. (eds.) *Proceedings of the 11th International Conference on Alkali-Aggregate Reaction in Concrete*, Québec City, pp. 1195–1204.

Siemes, A.J.M., & Visser, J. (2000) Low tensile strength in older concrete structures with alkali-silica reaction. In: Bérubé, M.A., Fournier, B., & Durand, B. (eds.) *Proceedings of the 11th International Conference on Alkali-Aggregate Reaction in Concrete*, Québec, Canada, pp. 1029–1038.

Siemes, A.J.M., Han, N., & Visser, J.H.M. (2002) Unexpectedly low tensile strength in concrete structures. *Heron, 47* (2), 111–124.

Silva, H.S. (1992) Estudo do envelhecimento das barragens de betão e de alvenaria – alteração físico-química dos materiais, Tese apresentada a concurso para acesso à categoria de Investigador Auxiliar e para obtenção do grau de Especialista do LNEC, Lisboa, (in Portuguese) 385p.

Silva, H.S., & Rodrigues, J.D. (1993) Importância, ocorrência em Portugal e métodos de diagnóstico das reacções álcali-agregado em betões. Contribuição dos conhecimentos geológicos. *Geotecnia* (67) (in Portuguese), pp. 65–76.

Silva, H.S., Braga Reis, M.O., & Santos Silva, A. (1996) Geological conditioning of ASR development. A brief evaluation of Portuguese mainland. In: Shayan, A. (ed.) *Proceedings of the 10th International Conference on Alkali-Aggregate Reaction in Concrete*, Melbourne, Australia, pp. 142–149.

SIST 1026 (2008) Concrete – Part 1: Specification, performance, production and conformity – Rules for the implementation of SIST EN 206-1, *Slovenian* Institute for standardization. Slovenia.

Škarková, J. (2005) Report no.1-2/2005/DSP/BET Identification of the possible reason of concrete deterioration in the airport Ostrava-Mošnov. Identification of ASR products. *Unpublished technical report (in Czech) Highway constructions, Ltd.*, pp. 115–117.

Škarková, J. (2006) Concrete deterioration in the airport Mošnov – a summary (in Czech). *Proceedings of International Conference Concrete Roads, Karlova Koruna, Chlumec nad Cidlinou*, November 9, 2006, pp. 93–100.

SN 670 115 (2005) Aggregates. Qualitative and quantitative mineralogy and petrography (Gesteinskörnungen – Qualitative und quantitative Mineralogie und Petrographie), *Schweizerischer Verband der Strassen- und Verkehrsfachleute, VSS, Zürich, Switzerland.*

SN 670 116 (2012) Fillers. Qualitative and quantitative mineralogy and petrography (Fillers. Minéralogie et pétrographie qualitative et quantitative). *Schweizerischer Verband der Strassen- und Verkehrsfachleute, VSS, Zürich, Switzerland.*

SN EN 1504-2 (2004) Products and systems for the protection and repair of concrete structures – Definitions, requirements, quality control and evaluation of conformity – Part 2: Surface protection systems for concrete (Produkte und Systeme für den Schutz und die Instandsetzung von Betontragwerken – Definitionen, Anforderungen, Qualitätsüberwachung und Beurteilung der Konformität – Teil 2: Oberflächenschutzsysteme für Beton). *Schweizerische Normen-Vereinigung, SNV, Winterthur., Switzerland.*

SN EN 1504-3 (2005) Products and systems for the protection and repair of concrete structures – Definitions, requirements, quality control and evaluation of conformity – Part 3: Structural and non-structural repair (Produkte und Systeme für den Schutz und die Instandsetzung von Betontragwerken – Definitionen, Anforderungen, Qualitätsüberwachung und Beurteilung der Konformität – Teil 3: Statisch und nicht statisch relevante Instandsetzung). *Schweizerische Normen-Vereinigung, SNV, Winterthur., Switzerland.*

Soers, E., (1988) Preliminary results on a petrographical examination of alkali-silica reaction damage in Belgium. In: *Proceedings of the 10th International Conference on Cement Microscopy, Texas,* pp. 298–305.

Soers, E. (1995) ASR-reactions in concrete: Belgian experience. In: Elsen, J. (ed.) *Proceedings of the 5th Euroseminar on Microscopy Applied to Building Materials (EMABM),* Leuven, Belgium, pp. 121–127.

Soers, E., & Meyskens, M. (1990) Alkali-aggregaat reactie. Uitbreiding kwaliteitscontrole noodzakelijk. *Cement, 42* (11), 20–27.

Sommer, H., Steigenberger, J., & Zückert, U. (2001) Vermeiden von Schäden durch Alkali-Zuschlag-Reaktion. Schriftenreihe Straßenforschung Bundesministerium für Verkehr, Innovation und Technologie Vienna: Heft 504.

Soriano, J. (1987) Reactions d' interaction entre certains granulats et la phase interstitielle du beton. Pore Structure and Materials Properties. *Proceedings of the International Symposium RILEM/IUPAC, Vol. 2. Chapman & Hall Edition,* London, pp. 25–32.

Soriano, J., & García Calleja, M.A. (1989) Áridos reactivos. Acción del hidróxido cálcico sobre áridos silicatados. *III Congreso Geoquímica España, 1,* pp. 9–15.

Soriano, J., Alaejos, P., Bermúdez, M.A., García Calleja, M.A., & Lanza, V. (2007) Estudio del hormigón de una presa afectada por una reacción álcali-árido. *Ingeniería Civil, 146,* 49–54.

St. John, D.A. (1992) Alkali-aggregate reaction – New Zealand experience. In: Swamy, R.N. (ed.) *The alkali-silica reaction in concrete, Blackie and Son Ltd,* London, UK, 249–269.

Stanton, T.E. (1940) The expansion of concrete through reaction between cement and aggregate. *Proc AmSoc Civil Eng., 66,* 1781–1811.

Stark, J., & Giebson, C. (2006) Assessing the Durability of Concrete Regarding ASR. In: Malhotra V.M. (ed.) *Proceedings of the 7th CANMET/ACI International Conference on Durability of Concrete,* Montreal, Canada, pp. 225–238.

Stark, J., & Giebson, C. (2008) Influence of acetate and formate based de-icers on ASR in airfield concrete pavements. In: Broekmans, M.A.T.M., & Wigum, B.J. (eds.) *Proceedings of the 13th International Conference on Alkali-Aggregate Reactions in Concrete,* Trondheim, Norway, pp. 686–695.

Stark, J., & Seyfarth, K. (1995) Schädigende Ettringitbildung im erhärteten Beton. Wissenschaftliche Zeitschrift der HAB Weimar 41, *Heft 6/7,* S. 37–64.

Stark, J., & Seyfarth, K. (2006) Performance Testing Method for Durability of Concrete Using Climate Simulation. In: Malhotra V.M. (ed.) *Proceedings of the 7th CANMET/ACI International Conference on Durability of Concrete,* Montreal, Canada, pp. 305–326.

Stark, J., & Seyfarth, K. (2008) Assessment of specific pavement concrete mixtures by using an ASR performance-test. In: Broekmans, M.A.T.M., & Wigum, B.J. (eds.) *Proceedings of the 13th International Conference on Alkali-Aggregate Reactions in Concrete*, Trondheim, Norway, pp. 320–329.

Stark, J., & Wicht, B. (2013) *Dauerhaftigkeit von Beton*. 2. Auflage, Springer Verlag.

Stark, J., Bollmann, K., & Seyfarth, K. (1992) Investigations into delayed ettringite formation in concrete. In: Mullick, A.K. (ed.) *Proceedings of the 9th International Congress on the Chemistry of Cement*, New Dehli, India, pp. 348–354.

Stark, J., Seyfarth, K., & Giebson, C. (2006a) Beurteilung der Alkali-Reaktivität von Gesteinskörnungen und AKR-Performance-Prüfung Beton. 16. Internationale Baustofftagung 20.-23.09.2006, Weimar, Tagungsbericht Band 2, pp. 399–426.

Stark, J., Freyburg, E., Seyfarth, K., & Giebson, C. (2006b) AKR-Prüfverfahren zur Beurteilung von Gesteinskörnungen und projektspezifischen Betonen. Beton – Die Fachzeitschrift für Bau +Technik, Verlag Bau+Technik GmbH, Nr. 12/2006 (56. Jahrgang), pp. 574–581.

Stark, J., Freyburg, E., Seyfarth, K., Giebson, C., & Erfurt, D. (2007) *Bewertung der Alkalireaktivität von Gesteinskörnungen*. Beton- und Stahlbetonbau (102), Heft 8, Ernst & Sohn Verlag für Architektur und technische Wissenschaften GmbH & Co. KG, Berlin, pp. 500–510.

Stark, J., Freyburg, E., Seyfarth, K., Giebson, C., & Erfurt, D. (2010a) 70 Years of ASR with No End in Sight? Part 1, *ZKG International*, 4, 86–95.

Stark, J., Freyburg, E., Seyfarth, K., Giebson, C., & Erfurt, D. (2010b) 70 Years of ASR with No End in Sight? Part 2, *ZKG International*, 5, 55–70.

Šťastná, A., Nekvasilová, Z., Přikryl, R., & Šachlová, Š. (2013) Microscopic examination of alkali-reactive volcanic rocks from the Bohemian Massif (Czech Republic). *Proceedings of the Third International Conference on Sustainable Construction Materials and Technologies*, Kyoto Research Park, Kyoto, Japan, 548p.

Šťastná, A., Šachlová, Š., Pertold, Z., Přikry, R., & Leichmann, J. (2012) Cathodoluminescence microscopy and petrographic image analysis of aggregates in concrete pavements affected by alkali-silica reaction. *Mater Charact.*, 65, 115–125.

StBA 4 (1980) Vorschrift 96/80, Vermeidung von betonschädigenden Alkali-Kieselsäure-Reaktionen. Ministerrat der DDR (Hrsg.) 11, S. 81–91.

TBA GR (2007) AAR-Schäden an Kunstbauten des Tiefbauamtes GR. Tiefbauamt Graubünden, Bericht 070101-2.

THBB (2013) Hazır Beton Dergisi, SAyı 120, Kasım-Aralık, 70–82 (in Turkish)

Török, Á. (2007) *Geológia mérnököknek (Geology for Engineers)*. Műegyetemi Kiadó, Budapest, ISBN 978-963-420-934-8 (In Hungarian with English summary), 384 p.

Török, Á. (2015) Los Angeles and Micro-Deval values of volcanic rocks and their use as aggregates, examples from Hungary. In: Lollino, G., Manconi, A., Guzzetti, F., Culshaw, M., Bobrowsky, P., & Luino, F. (eds.) *IAEG XII Congress 2014, Engineering Geology for Society and Territory. Volume 5, Urban Geology, Sustainable Planning and Landscape Exploitation*, Springer, pp. 115–118.

Tosun, K. (2001) Uçucu kül ve silika tozunun alkali silika reaksiyonuna etkisi, DEU Fen Bilimleri Enstitüsü, MSc dissertation, (in Turkish) 122p.

TP 137 (2003) Previous technical conditions – Ministry of Transport CZ and Road and Motorway Directorate of the Czech Republic. Elimination of alkali reaction of aggregates in concrete. Road and Motorway Directorate of the Czech Republic, Prague, (in Czech).

Trajanova, M., Pecskay, Z., & Itaya, T. (2008) K-Ar geochronology and petrography of the Miocene Pohorje Mountains batholith, Slovenia. *Geol Carpathica*, 59 (3), 247–260.

TS 21 (2015) Cement – Composition, specifications and conformity criteria for white Portland cement, Turkish Standardisation Institute (TSE), Ankara, Turkey, 10p, (in Turkish)

TS 2517 (2010) Determination of potential alkali silica reactivity of aggregates – Chemical method, Turkish Standards Institute (TSE), in Turkish.

TS 13515 (2014) Complementary Turkish Standard for the Implementation of TS EN 206-1, Turkish Standards Institute (TSE), in Turkish.

TS 10088 (EN 932-3) (1997) Tests for general properties of aggregates-Part 3: Procedure and terminology for simplified petrographic description.

Turriziani, R. (1986) Internal degradation of concrete: alkali-aggregate reaction, reinforcement steel corrosion. *Proceedings of the 8th International Congress on the Chemistry of Cement*, Rio do Janeiro, Brazil, Vol. I, 388–437.

Ulm, F.J., Coussy, O., Li, K., & Larive, C. (2000) Thermo-Chemo-Mechanics of ASR-expansion in concrete structures. *J Eng Mech., ASCE, 126* (3), 233–242.

UNE 80307 (2001) Cements for special uses. *AENOR*, Spain, 10p.

UNE 146507-1 EX (1999) Ensayos de áridos. Determinación de la reactividad potencial de los áridos. Método químico. Parte 1: Determinación de la reactividad álcali-sílice y álcali-silicato. *AENOR*, Spain, 10p.

UNE 146507-2 EX (1999) Ensayos de áridos. Determinación de la reactividad potencial de los áridos. Método químico. Parte 2: Determinación de la reactividad álcali-carbonato. *AENOR*, Spain, 6p.

UNE 146508 EX (1999) Ensayos de áridos. Determinación de la reactividad potencial álcali-sílice y álcali-silicato de los áridos. Método acelerado en probetas de mortero. *AENOR*, Spain, 8p.

UNE 146509 EX (1999) Determinación de la reactividad potencial de los áridos con los alcalinos. Método de los prismas de hormigón. *AENOR*, Spain, 8p.

UNI 11417-2 (2014) Durabilità delle opere di calcestruzzo e degli elementi prefabbricati di calcestruzzo – Parte 2: Istruzioni per prevenire la reazione alcali-silice (Durability of concrete works and precast concrete elements – Part 2: Criteria for the avoidance of alkali-silica reaction). *UNI – Ente Nazionale Italiano di Unificazione*, Italy.

UNI 11504 (2013) Reazione alcali-aggregato in calcestruzzo – Determinazione della potenziale reattività agli alcali degli aggregati per calcestruzzo – Prova di espansione accelerata di barre di malta. *UNI – Ente Nazionale Italiano di Unificazione*, Italy.

UNI 11530 (2014) Reazione alcali-aggregato in calcestruzzo – Determinazione della potenziale reattività agli alcali degli aggregati per calcestruzzo – Esame petrografico di dettaglio dell'aggregato per la determinazione dei costituenti potenzialmente reattivi agli alcali. *UNI – Ente Nazionale Italiano di Unificazione*, Italy.

UNI 11604 (2015) Reazione alcali-aggregato in calcestruzzo – Determinazione della potenziale reattività agli alcali degli aggregati per calcestruzzo – Prova di espansione accelerata in calcestruzzo. *UNI – Ente Nazionale Italiano di Unificazione*, Italy.

UNI 8520-2 (2005) Aggregati per calcestruzzo – Istruzioni complementari per l applicazione della EN 12620 – Requisiti. *UNI – Ente Nazionale Italiano di Unificazione*, Italy.

UNI 8520-22 (1999, 2002) Aggregati per la confezione di calcestruzzi – Determinazione della potenziale reattività degli aggregati in presenza di alcali, *UNI – Ente Nazionale Italiano di Unificazione*, Italy.

van de Fliert, C., Hove, J.F., & Schrap, L.W. (1962) Alkali-aggregate reaction in concrete. *Cement* (in Dutch), *14*, 20–28.

van der Meulen, M.J., van Gessel, S.F., & Veldkamp, J.G. (2005) Aggregate resources in the Netherlands. *Netherlands J Geosci., 84* (3), 379–387.

Velasco, A., Alaejos, P., & Soriano, J. (2010) Comparative study of the alkali-silica reaction (ASR) in granitic aggregates. *Estudios Geológicos, 66*, 105–114.

Vertreter des Bundes und der Länder sowie weitere Fachleute (2012) *Empfehlungen für die Schadensdiagnose und die Bauliche Erhaltung von AKR-geschädigten Fahrbahndecken aus Beton*. Fortschreibung.

Vola, G., Berra, M., & Rondena, E. (2011) Petrographic Quantitative Analysis of ASR susceptible aggregates for concrete. *Proceedings of the 13th Euroseminar on Microscopy Applied to
Building Materials (EMABM)*, Ljubljana, Slovenia.

Weerts, H.J.T., Schokker, J., Rijsdijk, K.F., & Laban, C. (2004) Geologische overzichtskaart van
Netherland/Geological map of the Netherlands, *Netherlands Institute of Applied Geoscience
TNO*, Utrecht, NL.

West, G. (1996) Alkali-aggregate reaction in concrete roads and bridges. Thomas Telford
Publications, London.

WGPNCOLD, Work Group of the Portuguese National ICOLD (2003a) Ageing process and
rehabilitation of Pracana dam. *Vingt et unieme Congres des Grands Barrages*, Montreal,
Q.82, 18p.

WGPNCOLD, Work Group of the Portuguese National ICOLD (2003b) Observed behaviour
and deterioration assessment of Pracana dam. *Vingt et unieme Congres des Grands Barrages*,
Montreal, Q.82, 21p.

Wigum, B.J., Pederson, L.T., Grelk, B., & Lindgard, J. (2006) Partner Report 2.1 – State-of-the-
art report: Key parameters influencing the alkali aggregate reaction. PARTNER-project-
GRD1-CT-2001-40103. *SINTEF Building and Infrastructure. SBF52 A06018.* 131p.

Yıldırım, K., Sümer, M., & Uysal, M. (2011) Uçucu külün alkali silis reaksiyonuna etkisinin
araştırılması, 8. Ulusal Beton Kongresi, TMMOB İnşaat Mühendisleri Odası,5-7 Ekim 2011,
Izmir, pp. 99–107.

Zatler-Zupančič, B., & Madlenovič, A. (1992) Alkali reactive components in the sand and gravel
of the river Danube and its tributaries. In: Poole, A. (ed.) *Proceedings of the 9th International
Conference on Alkali-Aggregate Reaction in Concrete (ICCAR)*, London, UK, pp. 1121–
1128.

Zement-Kalk-Gips, 3. Jahrgang (1950) Untersuchungen über alkaliempfindliche Zuschlagstoffe.
Heft, 11, 277–278.

Zement-Kalk-Gips, 4. Jahrgang (1951) Zur Frage der alkaliempfindlichen Zuschlagstoffe und
ihrer Untersuchung. *Heft, 8*, 216–221.

Žibret, G. (2009) Baseline study reports (BSR) for national SSM (Slovenia), *SARMa, Sustainable
Aggregates Resource Management*. Geological Survey of Slovenia, 29p.

Chapter 9

Russian Federation

Vyatcheslav R Falikman & Nikolai K Rozentahl

9.1 GEOLOGICAL STRUCTURE AND TECTONIC ZONING OF THE RUSSIAN FEDERATION

Russia occupies 17125407 km^2 (2015) and borders 18 countries, including two partially recognized states. Russia has borders overland with Norway, Finland, Estonia, Latvia, Lithuania, Poland, Belarus (Byelorussia), Ukraine, Abkhazia, Georgia, South Ossetia, Azerbaijan, Kazakhstan, China, North Korea, Mongolia and, offshore, with Japan and the USA. Such a very large area is difficult to represent with clarity on a single geological map, but one useful two-sheet overview is given in Text-Box 9.1 following this Section 9.1.

Within Russia, both platforms and orogens can be distinguished. Russia's European region is located on the east European platform. The central part of the platform is composed of Pre-Cambrian igneous and metamorphic rock. The territory between the Urals and the Yenisei River is occupied by the younger West Siberian platform. To the east of the Yenisei River there is the ancient Siberian Platform, extending up to the Lena River and corresponding primarily to the Mid-Siberian upland. The Russian orogens are represented by the Baltic shield, the Urals, the Altai, the Urals-Mongolian epi-Palaeozoic folded zone, the north-western part of the Pacific folded zone and a minor segment of the outer part of the Mediterranean folded zone. The highest mountains are the Caucasus and are confined to the younger folded zones.

9.1.1 East European platform

The East European platform foundation is represented by Lower and Upper Archaean metamorphic rock and locally by Lower Proterozoic rock ruptured by granitoid intrusives. The mantle is formed by Riphean, Vendian and Phanerozoic deposits. The platform's basic structures include the Baltic shield (its eastern part), and the Russian Plate that comprises the Voronezh and the Volga-Urals anticlines, and the Moscow and the Mezen anticlines. The platform foundation is dissected by a series of Riphean aulacogens (sediment filled grabens), including Pachelmsky, Sernovodsko-Abdulinsky, Kazansko-Sergiyevsky, Kirovsky, Mid-Russian, Moskovsky, Kandalaksha, Keretzko-Leshukonsky.

The foundation internal structure is characterized by availability of major Archean rock blocks and narrow belts consisting of Lower Proterozoic rock masses. At the turn of the Lower Proterozoic and Riphean, in the Russian platform's western regions a rapakiwi-granite batholith was intruded. The platform mantle is divided into two parts: the lower one formed by Riphean and Lower Vendian rock; the upper one formed by the Upper Vendian of the Precambrian, then into synclines and anticlines during the Cainozoic. Trappean magmatism developed on the Russian plate in Riphean, Vendian and Devonian times. Alkali intrusions of the Middle Palaeozoic era occur on the Kola Peninsula.

9.1.2 Siberian platform

The Siberian platform is of Epi-Archean age. Within the platform, the Aldan shield and the Lena-Yenisei plate may be identified among the basic structural elements, of which there are: the Aldan and the Anabar anticlines; the Tunguska and the Vilyui synclines; the Angara-Lena, Lena-Anabar, Angara-Vilyui and Yenisei-Khatanga troughs; the Olenyok, Turukhan-Norilsk and Peledui upheavals; and the Nyuisk, Brezovsky, Irkutsk, Kansky, Lindensky, Ust-Aldan, Chulman and Tokkinsky depressions.

The platform foundation is dissected by Piphean aulacogens – the Irkinivsky, Urinsky, Udzhinsky, Kyutyungdinsky, Kotuikansky and Markhinsky as well as the Devonian Patomsko-Vilyui aulacogen, along the axis of the Vilyui syncline. The platform foundation is composed mainly from the Archean deeply metamorphosed rocks that are overlapped by the Lower Proterozoic terrigenous deposits of the Udokan series (protoplatform mantle). The upper layer is divided into a number of complexes differing by the rock composition and the geotectonic plan. The Siberian platform manifests ultrabasic alkali, granitoid alkali and trappean Riphean magmatism - the Earlier Cambrian, Middle Palaeozoic, Late Palaeozoic – the Earlier and the Late Mesozoic. A special place in the Siberian platform structure is occupied by the Tunguska trappean syncline.

9.1.3 The Urals-Mongolian folded zone

In the structure of the Urals-Mongolian Epi-Palaeozoic orogeny, which divides two ancient platforms, it is possible to distinguish regions of Ripheic, Baikal, Salair, Caledonian and Hercynian folding. In the Yenisei-Sayano-Baikal region, the Riphean and Baikal folding enframes the Siberian platform. It includes the Yenisei ridge, the major (north-eastern) part of the eastern Sayany, Kamar-Daban and all of the Western Trans-Baikal Region up to the Nichatsky fault in the east and the Main Mongolo-Okhotsky fault in the south. The Timano-Pechora plate encompasses the East European platform from the north-east. It is composed of rock of the Timano-Kaninsky upheaval and the Pechora syncline, which is split by the Pechora-Kozhvinsky, Kolvinsky and Sorokina swells into the Izhma-Pechora, Denisovsky and Khoreiversky depressions. Major oil and gas fields are related to the Palaeozoic deposits.

The East Sayan – Kuznetsk folded system consists of the Kizir zone, Kuznetsky Alatau and Highland Shoria divided by the Pre-Cambrian Khakass rock mass overlaid by the Minusinsk depression formed in the Devonian era. In the system's south-east, there is the Tuva rock mass of the Riphean consolidation with the superimposed Salair Kharalsky trough. In West Sayany, the Highland Altai Caledonian folded system is composed of both igneous and sedimentary series of the Upper Riphean-Vendian and Cambrian. The Zaisan-Gobi Hercynian folded region occupies the axial position in the

Urals-Mongolian zone and consists of the Tom-Kolyvansky, Salairsky, Anuisko-Chuisky, Rudnoaltaisky and West Kalbinsky systems. The geosynclinal complex is primarily represented by Devonian and Lower Carbonaceous formations. The Urals Hercynian folded system extends in the meridional direction for 2500 km. Along the border with the East European platform there is the Pre-Urals fore deep filled with rocks of Permian age.

The West Siberian plate has a heterogenic foundation composed of Hercynian, Caledonian, Salair, Baikal and pre-Baikal rock complexes. Oil deposits in the positive structures are related to the sandstones of the Jurassic and Lower Cretaceous, which overlie the basement rocks, while gas deposits are concentrated in formations of the Cenomanian and Campanian substages. Manganese deposits are confined to the Trans-Urals palaeogene rock. To the south-east of the Siberian platform, there is the Mongolo-Okhotsk folded zone, separated from the more ancient northern regions by the great tectonic suture – the Main Monglo-Okhotsk deep fault. The region comprises three sectors: the East-Trans-Baikal, Upper Amur and Okhotsk regions. To the south there is the Bureya rock mass within the Zeya-Bureya depression and the Bureya trough, which is filled with continental Jurassic, Cretaceous and Palaeogene sediments.

9.1.4 Mediterranean folded zone

The Mediterranean folded zone occurs within Russian Federation territory in its outer part (Scythian plate, the northern flank and the western part of the Greater Caucasus). The Front Range and the Pre-Caucasus contain recognisable fore deep facies.

9.1.5 Pacific folded zone

The Pacific folded zone in Russian Federation territory is represented in the extreme north-west, within which there are ancient Doriphean rock masses, areas of Mesozoic and Cainozoic orogeny and contemporary tectonically active areas. In the north-east there is the Verkhoyano-Chukotka orogen with the Okhotsk, Omolonsky, Chukotka and Kolyma ancient median masses. Within this area, the Verkhoyano-Kolyma system can be identified occurring mainly on the Archean continental crust, and in the Novosibirsk-Chukotka system. These systems are separated by the Svyatonosko-Oloyisky Cretaceous volcanic belt. The Verkhoyano-Chukotka zone contains Jurassic and Lower Cretaceous granite intrusions. Extensive Molass facies are widely distributed in the Pre-Verkhoyansk trough and the Zyryanskaya depression.

The Sikhote-Alinsksya folded system is confined by the Bureya and Khankai masses and consists of several submeridional zones, of which the western members are super-imposed on the Pre-Cambrian continental base, while the eastern ones lie on the oceanic crust of the Pre-Upper Permian period. The shoal-like Cambrian limestones are known in the western zone, along the eastern edge of which the Devonian volcanic belt was formed. The Carboniferous and Permian periods are represented by limestones and extrusive rock. The eastern zones are formed of a thick series of terrigenous-tufaceous-siliceous geosynclinal Triassic and Jurassic deposits.

The Koryak folded zone is divided into orogenous systems of very complex imbricate overlapping and overburden structures. The transverse section in the western zones is represented by a series of geosynclinal siliceous-igneous and carbonate-terrigenic

TBI 9.1 A Geological Map of the Russian Federation (formerly the main part of the USSR), in two sheets (western and eastern), from inserts in Nalivkin (1960).

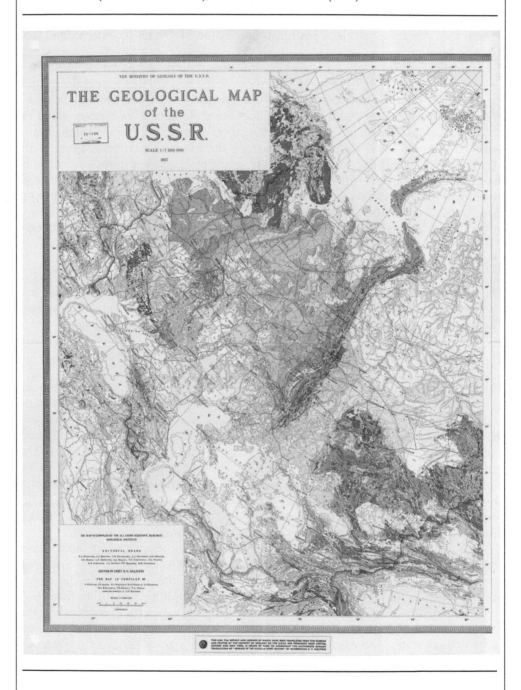

TB1 9.1 (Cont.)

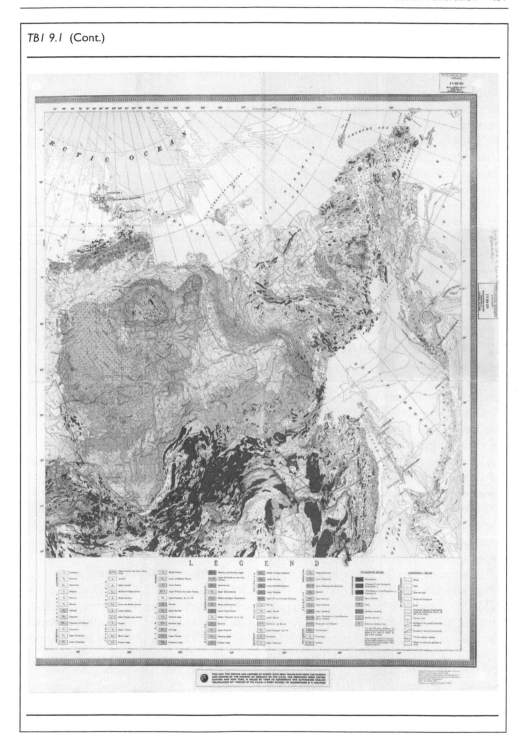

(Ordovician period – Cretaceous Aptian stage) rock, overlapped with beds of a molass complex of marine and continental sediments. All Palaeozoic and Mesozoic troughs were laid on oceanic type crustal material represented by ophiolites.

The West-Kamchatka folded system is an Upper Cretaceous terrigenous geosynclinal complex that was superimposed on a granite-gneiss and shale basic foundation and, after folding, was overlapped with palaeogene-neogene rock. In the Central and East Kamchatka-Olyutorskaya systems, the Upper Cretaceous complex was built up with the igneous and sedimentary Paleogenic series. In the late Pliocene and early Pleistocene, major shield basalt volcanoes were formed within the Central zone. The Eastern zone is characterized by the recent superimposed volcanism (28 active volcanoes) confined to the young keystone faults.

The Kurils island arc, consisting of the Great and Small Ranges, contains 39 active volcanoes. It is formed of Cretaceous and Quarternary volcanogenic-sedimentary and volcanogenic bodies. The arc is split by a system of young transverse fault troughs, while to its front and fronting east Kamchatka there is a deep water trough.

The Sakhalin Cainozoic folded zone is divided into the Eastern and Western areas separated by the Central Sakhalin fault trough. The North Sakhalin depression is formed by Middle Miocene rocks.

9.2 NON-ORE MINERALS: PRODUCTION AND CONSUMPTION OF AGGREGATES

Production of non-metallic building materials (primarily, crushed stone obtained from natural rock quarrying, gravel and sand) in the Russian Federation is spread over than 1080 major, medium and minor companies. Sand and gravel represents over 20 % of the total output of these materials.

The market for non-metallic building materials is highly concentrated. The greatest number of producers are in the Southern Federal District of Russia, where about 210 of the market participants are in this industry. In the North-Western Federal District, over 190 market participants deal with non-metallic materials mining and marketing. The greatest volumes of sand, crushed stone and gravel quarrying and marketing are undertaken in St. Petersburg, the Leningrad Region, Moscow, the Moscow Region, the Archangelsk Region and the Republic of Karelia.

The share of the Central Federal District (CFD) is about 30 %, involving at least 170 companies. In 2014 the CFD was the leader in non-metallic building materials (NMBM) production in Russia. The North-Western Federal District occupies a leading position in the hard-rock crushed stone market in Russia's European region. The basic reserves of magmatic and metamorphic rock building stone are located in the Republic of Karelia (approximately 2 billion m^3), the Murmansk area (approximately 450,000 tonnes of stone and 10,000 m^3 sand-gravel mixtures), Leningrad, Archangelsk Regions (approximately 170,000m^3 stone and over 20,000 m^3 of sand for concrete) and the Komi Republic (over 20 sources of building stone and 350 sources of sand and gravel) (Russian Academy of Sciences, 2003).

In Russia, reserves that have industrial value are divided into explored categories 'A', 'B' or 'C' depending on the degree of knowledge; category 'C_1' is a reserve or deposit proven by drilling, whereas 'C_2' is a revealed deposit with possibly usable material.

According to Bekrenev (2013), the resource potential of sand-gravel material (SGM) in the Russian Federation is represented by over 2400 deposits with total reserves of A+B+C$_1$ categories of 10.9 billion m^3. Of them, designated material is estimated as – 45.9% of deposits with 5.3 billion m^3 of reserves, while in the non-designated stock is estimated at – 54.1% with 5.5 billion m^3 of reserves. Sand deposits for building works number over 2 thousand with industrial grade reserves of 7.1 billion m^3. The proportions of these in the designated stock are 53.7% with 3.4 billion m^3 of reserves, while those in the non-designated stock are 46.3% with 3.7 billion m^3 of reserves.

The main reserves of SGM and sands for building works (SBW) of A+B+C$_1$ categories are located in Russia's European region, comprising 65%, or 82% including the Urals. This represents 74% of the country's SBW output, while Siberia and the Far East produces the remaining 26%. SGM quarrying in the European region produces 87% of the whole of Russia's output.

The greatest quantities of explored SBW reserves of A+B+C$_1$ categories are divided into Siberian (25.4%), Central (21%) and the Volga (11%) Federal Districts (FD). The SBW reserves are estimated for the Central (21%) and Volga (11%) FDs.

In general the SBW deposit developments and their output are affected by the availability of other building stone materials sources that produce crushed stone, rough stone, manufactured sand, etc. For example, in the Urals FD, where SBW and SGM quarrying is traditionally smaller than in other regions (about 0.7 million m^3 of SBW and 1.5 million m^3 of SGM, respectively), while 27 million m^3 building stone is quarried. In other FDs, the total SBW and SGM quarrying production exceeds 2–3 times that of the building stone quarrying. In the Siberian and Far East FDs, these figures are comparable.

According to the data of Lopatnikov and Levkova (2008), the raw material resources base for crushed stone and sand extraction amounts to 195 separate deposits with 128 of them in operation. Average silica-bearing contents of the sand-gravel deposits may be divided into three groups:

- boulder and gravel materials in deposits of the Kaluga and Vladimir Regions comprise flint/chert stone up to and above 30%;
- in the Smolensk, Tver and Moscow Regions, 15 to 30% fint/chert;
- in the Ivanovo, Yaroslavl and Kostroma Regions, less than 15% fint/chert.

Zolotykh (1990) reports that there are 105 igneous (effusive) rock deposits, 199 metamorphic rock deposits and 82 sand and gravel deposits on Russia's territory.

According to Ivanov (1992), in Russia rocks containing minerals capable of interacting with alkalis are widespread. These include magmatic rocks: granites and granodiorites with deformed lattice quartz and weathered feldspar inclusions; rhyolites, dacites, andesites, trachyandesites and basalts, containing silicate and basalt glass devitrificated to various degrees as well as some tridymite, cristobalite and opal; obsidian, 'cinerites' (volcanic ash), retinitis (water-rich rhyolite glass) and silica-rich glass, often with microcracks. They also include gneisses and schists that are characterized by open grain contacts and contents of deformed lattice quartz, microquartz, feldspar and micaceous minerals.

The hazardous metamorphic rock varieties are quartzites, meta-sandstone and amphibolites containing opaline silica and secondary microcrystalline quartz, as well as quartz schists, meta-greywackes, phyllites and related rocks containing opaline silica and microcrystalline quartz. The potentially reaction-hazardous sedimentary rocks include limestones, dolomitic limestones and dolomites containing chalcedony and opal in intermediate layer, micrograin or finely disseminated forms.

In Russia, rocks containing potentially reactive silica are the sandstones in the middle Volga region, and flint comprising gravel and sand deposits (Urals, East Siberia, North-West of the country's European part). In the Far East, in the Maritime Territory, Sakhalin Island and Kamchatka, some acid rock (andesites, scorias, etc.) are considered to be potentially reactive.

The large igneous rock reserves are located in the Leningrad Region, Karelia and the Urals Region. The gravel-sand and sand deposits are located mainly in the European region of the Russian Federation and are unavailable in most Siberian areas (Buyanov, 2005). Among the producing deposits, particularly in the central and north-western zones of Russia, considerable volumes of aggregate are obtained in deposits formed in glacial transfer of rock masses. Morainic (fluvioglacial) sediments in the Moscow, Tver, Smolensk Regions and the Klin-Dmitrov ridge are composed of these deposits. The stone mineral composition in these sediments is rather diverse. The constituents include large proportions of flint, and there are opals and other non-crystalline and microcrystalline minerals mainly comprising amorphous silicon dioxide.

The raw materials for crushed rock aggregates are igneous, sedimentary and metamorphic rocks as well as over-sized gravel from sand-gravel mixtures. In other regions (Tatarstan, the Krasnodar Territory, the Moscow and Nizhny Novgorod Regions), only medium- and low-strength rocks are available. In Bashkortostan, the Perm Territory, the Rostov, Sverdlovsk and Chelyabinsk Regions, there are deposits that are suitable for any grade of crushed stone production. In Russia's European region, igneous rocks represent only 14 of the 46 sources. In some other regions, only sedimentary rock and sand-gravel mixtures are available (Lopatnikov & Tedeyev, 2007).

In some regions, such as the Khanty-Mansi Autonomous District, no sources of stone are available (Senatorov, 2008). The volcanic rock in Russia's Far East regions has been studied in some detail. Among these rocks, a considerable number exhibit reactivity with cement alkalis (Buyanov, 1999).

The majority of active companies engaged in non-metallic materials production were set up 30 and more years ago. The present processing technologies correspond to the 1970s level of production oriented for production of 2 or 3 fractions of crushed rock and gravel and one fraction of sand. Just a few domestic companies produce a greater number of product types. According to the CMPro analytical agency (2015), production of non-metallic materials in 2014 was 407.1 million m^3, (Figures 9.1 and 9.2). Consumption of non-metallic materials in the same year was 425.6 million m^3. Import of non-metallic materials in 2014 was 18.8 million m^3, while the import of the same materials from Ukraine was 14.5 million m^3. In its turn, the export of non-metallic materials in 2014 was 758000 m^3. The leading importer of Russian non-metallic materials is Kazakhstan. Over 50% of the non-metallic building materials (NMBM) output relates to pebble stone, gravel and crushed stone, 37% is sand, and 8% comprises other NMBM. In recent years, the NMBM production structure has been essentially constant.

Similar data are presented by Peshkova (2014). The output of non-metallic building materials in Russia in 2013 was 411.5 million m^3, including 226.6 million m^3 of pebble stone, gravel and crushed stone and 152.6 million m^3 of sand. The total reserves of stone identified in the European region were 15.7 billion m^3 (Figure 9.3 and Table 9.1).

Figure 9.1 Non-metallic materials outputs in the Russian Federation by District, in 2014, million m³ (CMPro, 2015).

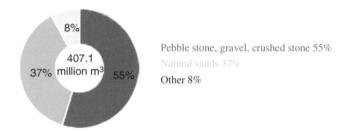

Figure 9.2 Output of non-metallic building materials in the Russian Federation in 2014 (CMPro, 2015).

The expected growth in long-term requirements for non-metallic building materials will encourage quarry development in all Federal Districts. The largest developments are expected to be in the North-Western (Leningrad Region), Central (Moscow and the Moscow Region), Southern (Krasnodar Territory) and Far Eastern (Vladivostok) Federal districts, with somewhat less growth in the Volga, Urals and Siberian Districts. In this connection, in the Volga, Urals and Far East districts, the reserves of $A+B+C_1$ categories currently identified will be worked out in the next 15–20 years.

In Moscow and the Moscow Region, a part of the demand is met currently by imported raw materials from other parts of the Central district and neighbouring regions. This is a result of the urbanisation rate in this region and the industrial reserves will be worked out much faster. This necessitates additional exploration and transfer of the reserves from C_2 category to industrial categories. Analysis shows that the reserves

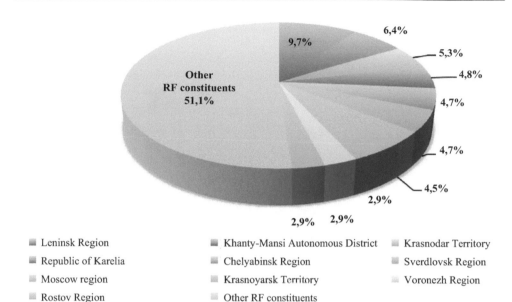

Figure 9.3 Proportions of the total output of non-metallic building materials from 10 regions of the Russian Federation in 2011 (from Peshkova, 2014).

Table 9.1 The major suppliers of non-metallic materials in 2014 (after Peshkova, 2014).

Supplier	Thousand m^3
Orsk quarry management	3717
Pavlovsk Nerud	3529
Lenstroykomlektatsiya	2996
Uralasbest	2315
LSR-Bazovyiye	2167
Karelprirodresurs	1934
Novosibirsk quarry management	1580
Biyankovsky crushed stone plant	1348
Nerud Invest PFK	1300
Kamnerechensk crushed stone plant	1228
Berdyaush non-metallic company	1143
Pervouralsk mine group	1115
Sangalyk diorite quarry	1084

of unappropriated (not allocated) stock fit for development in the Russian Federation will be about 3.9 billion m^3 of the sand-gravel material and 2.6 billion m^3 of sand for building works. This can potentially meet the demands up to 2040 or 2050. To create reserves for the later periods it will be necessary to increment reserves in the majority of the Federation and, firstly, in individual regions of the Central, Volga, Southern and North Caucasus districts.

At present the concrete deterioration processes caused by the alkali-aggregate reaction and the appropriate protection methods have been studied in detail, which has allowed for the active use of reactive rock for manufacture and use of aggregates in concrete.

9.3 TECHNICAL REGULATION, SPECIFICATIONS AND TEST METHODS

In Russia, test requirements for aggregates are regulated by the standards: GOST 8267 (1993) for coarse aggregate, and GOST 8736 (1993) for sand. The test methods for aggregate alkali-reactivity are specified in GOST 8269.0 (1997). When developing the last-named standard, in addition to recognising the domestic findings, the RILEM test developments (now in Nixon & Sims, 2016), with ASTM C289 (recently 2007, but now withdrawn) and ASTM C1260 (now 2014) were also adopted. According to GOST 8269.0, rock or crushed stone (gravel) are assessed as being potentially reactive if the petrographic analysis finds the presence of one or more varieties of minerals that contain reactive silica in the amounts equal to or exceeding the values given in Table 9.2. According to GOST 8269.0 (1997), the aggregate reactivity is determined at the exploration stage by the rock mineralogical and petrographic evaluation procedure. In case of finding a potentially reactive rock, the chemical method is used to estimate the amount of alkali-soluble silica.

Russian standards permit potential use of overburden and enclosing rock, as well as ore- and non-metallic mineral processing wastes.

All standards are of a prescriptive nature. If the soluble silica quantity exceeds the specified value from the ASTM C289 chemical method (not more than 50 mmol/l of alkali-soluble silica), then the accelerated mortar-bar test method is used (with an expansion criterion of not more than 0.1 %). The test series ends with testing concrete specimens for expansion over one year (with an expansion criterion of not more than 0.04 %). The test methods set out in the standard are generally co-ordinated with the methods developed by RILEM (now in Nixon & Sims, 2016). The only difference is

Table 9.2 Potentially reactive rock types and specified limit of reactive component (from GOST 8269: 1997).

Mineral and silica type	Types of potentially reactive rock	Maximum content of mineral % by mass
Opal	Basalts and other lavas. Limestones, hornfels, opaline shale	0.25
Crystalline cristobalite, tridymite	Silica-comprising melts (materials obtained by melting)	1.0
Deformed or strained quartz	Quartz vitrophyres, quartzites, sandstones, volcanic and metamorphic acid rock	3.0
Amorphous acid glass	Vitreous-based obsidians, perlites, liparites, andesite-dacites, andesites, tuffs and these rock analogues	3.0
Crypto-/micro-crystalline chalcedony	Flints, limestones, dolomites, sandstones with opal-chalcedony and chalcedony-quartz rock cement, jaspers, hornfels	5.0

that the accelerated method additionally accounts for variation dynamics of expansion at the test end. The value obtained in the last measurement shall not exceed the values in three preceding measurements by more than 15%.

It should be noted that the tests performed have shown that the criterion that is assumed in the chemical method is not a reliable indicator. In a number of cases, specimen expansion was observed when the soluble silica content had been less than 50 mmol/l; at the same time, some aggregates with essentially higher than 50 mmol/l quantity of soluble silica showed no excessive expansion in tests. Based on these results, the conclusion was reached was that the expansion tests are mandatory.

Taking account of RILEM developments of accelerated tests, expansion tests were conducted at 80°C in different environments: distilled water, sodium chloride solution and alkali solution. The tests in water and in 1M sodium chloride solution have not proved to be as useful as the tests in the alkali solution [as standardised in C1260 (2014) and RILEM AAR-2 (in Nixon & Sims, 2016)].

The basic requirements to provide concrete protection against destructive effects of the alkali-silica reaction (ASR) are specified in the Code 28.13330 (2012). The Code envisages the following ASR-caused damage prevention procedures:

- concrete proportioning with minimum cement content to reduce the total alkali content in the concrete;
- use of Portland cement with alkali content (as Na_2Oeq) not more than 0.6% of cement mass;
- manufacture of concrete using Portland cements with mineral additions, Portland pozzolana cement and Portland slag cement;
- introduction of water-repelling and air-entraining admixtures into the concrete composition.

When reactive silica is present in aggregates, it is not allowed to introduce sodium and potassium salts as admixtures into concrete. In practice, in many cases, recommendations such as cement content reduction, use of cements with low alkali contents and non-use of alkali-silica reactive aggregates, are hardly feasible. In a number of regions the basic commercial cement tends to be high in alkali. In the same regions, a number of aggregates contain substantial quantities of potentially reactive silicon dioxide (for example, quarries using fluvioglacial sediments in the north-west of Russia's European region, sand-gravel sediments from the Kama River, and the fluvial sediments and volcanic rocks of Siberia). The concrete protection methods for the prevention of ASR are presented in greater detail in STO 36554501-022 (2010).

9.4 AGGREGATES PRESENTING AN AAR HAZARD IN CONCRETE

The above-mentioned regulatory documents have served as a basis for numerous tests conducted by different researchers. It was established that crushed stone from Sychevsky and Terelesovsky quarries, Salsk quarry, Kamennogorsk integrated works and Novokievsk crushed stone plant all contained alkali-soluble silica in concentrations of not more than 50 mmol/l. Crushed stone from Selizharovsky, Khromtsovsky, Oreshkinsky and Akademichesky quarries contained alkali-soluble silica in a quantity exceeding 50 mmol/l. In Tatarstan, the main volume of commercial aggregates are represented by the sand and gravel mixtures extracted at the Kama and Vyatka Rivers and contain large quantities of amorphous silica (Morozova, 2005). The content of soluble silica in these aggregates can be 112 to 254 mmol/l. The aggregates comprise flints, quartzites, limestones, sandstones and slates.

Sand-gravel mixtures are also used in the Republic of Bashkortostan. These mixtures comprise sandstones, siliceous materials, metamorphic quartzites and silicified rock inclusions. The reactive silica content exceeds by 1.7 to 12 times the maximum content of 50 mmol/l permitted by GOST 8269.0-97 (Oratovskaya, 2005). The use of cement with mineral additives, as a rule, precludes development of hazardous damage to concrete with the above-specified aggregates; this has been established by long-term (up to 30 months) tests of concrete (Yakovlev, 1999). It should be specifically noted that, in the last century in the USSR, there were relatively few examples of concrete damage caused by alkali-silica reactions. This was the result of the use of cements containing high percentages of mineral additives at that time.

The coarse aggregate from the Kuchugurovsky deposit is composed of quartzitic sandstone with a chalcedonic cement. The sand-gravel mixture from the Kama River basin comprises 35% flints and siliceous rock, 30%, quartzite-like sandstones , 20%, weathered sandstones, 5%, volcanic rock, 5%, carbonaceous flinty shales, and 5% carbonaceous rock with silicified grain inclusions. The sand comprised 60% quartz, 30% flint, and 10% sandstones. The sandstone, in turn, exhibited 764 mmol/l of soluble silicon dioxide. In the sand-gravel mixture, the overall quantity of soluble silicon dioxide was 99.8 mmol/l in the gravel and 41.7 mmol/l in the sand. Zolotykh (1990) provides a list of rocks and minerals (Table 9.3), which was included in the recommendations (Guidelines, 1984).

9.5 ASR STUDIES IN RUSSIA

In the former USSR, the study of this type of damage was studied for the first time by Moskvin & Royak (1962). This summarizes the research results obtained in the USSR up to 1962. Further research on this issue was discussed in a number of other domestic studies into concrete and reinforced concrete durability. Belogurova *et. al.*, (1990), Moskvin and Royak (1979), Moskvin *et al.* (1980), Royak (2001, 2002), Ivanov (1992, 1995), Viktorov and Lozhkin (1992) and others carried out investigations of individual aggregates and the interaction processes of cement alkalis and additives with reactive aggregates, and identified the list of reactive rocks and estimated the raw material base.

A considerable volume of experimental research was undertaken in the period from 1963 to 1969 by NIS Gidroproyekt Laboratory of stone materials, where 179 prism specimens 70 x 70 x 220 mm in size were prepared from concrete based on 9-20 mm grain-size aggregates with varying reactivities (Viktorov, 1982, 1986). The samples were selected from industrial aggregate sources in a variety of domestic and foreign quarries. Portland cement for concrete mixes contained 0.4 to 0.8% Na_2O eq. One half of specimens was made using mix water with an NaOH addition of 1% of cement by mass, the other half of the prisms used pure water. Three concrete prisms (aggregates: crushed granite and quartz sand) and two mortar prisms served as reference standards.

Aggregates for concrete (flint, sandstone, andesites, diabases etc.) were selected from gravel or crushed stone, crushed and sieved to 5-20 mm size providing rock types with

Table 9.3 Types of rock containing minerals capable of reaction with alkalis.

Rock	Minerals capable of reaction with alkalis
Igneous	
Granites	Quartz with undulating extinction deformed lattice, feldspar
Granodiorites	weathered minerals, open contact grains
Rhyolites	Silicate or basalt glass with various degrees of devitrification, tridymite,
Dacites	cristobalite, opal
Andesites	
Trachyandesites	
Basalts	
Volcanic obsidians:	More or less devitrified glass with the great silica content, often with
Tuff, slag	microcracks. Retinite is a water-rich rhyolite glass that can expand.
'Retinites'	
Quartz porphyry	
Perlites	
Metamorphic	
Layered mica gneiss	Undulatory quartz, secondary micro-quartz, open contact grains,
	feldspar and micaceous minerals
Quartzites	Quartz or opal-consolidated quartz, secondary micro-quartz.
Amphibolites	Phyllosilicates. Undulatory quartz or micro-fissured quartz.
Sedimentary	
Sandstones	Undercrystallized silica cement, grain expansion. Phyllite minerals.
Quartzites	Microcrystalline opal.
Greywackes	
Quartz	
schists	
'Silicites'	
Siliceous	Chalcedony, opal
Limestones	Opaline silica in micrograins or disseminated, linked together or not
Dolomitic limestones	with sedimentary sulfur compounds and phyllosilicates
Dolomites	

a range of potentially reactive components, including opal, chalcedony, volcanic glass. A Quartz sand was used to make prisms.

To determine the soluble silica content, all aggregate samples prior to use in concrete were subjected to petrographic studies and chemical analyses. Concrete prisms were stored in a moist air cabinet, and every 6 months the dynamic modulus of elasticity was determined using an IChM K-2 modulus meter. For the convenience of comparison, the absolute value of the modulus was not used, but rather the elastic modulus coefficient K_{ynp}; the latter is derived as the arithmetic mean of all prism measurements of this programme for the ordinary and alkali-enhanced concretes.

The concrete prisms are prepared into 4 groups depending on the petrographic composition and the aggregate shape:

1)   Crushed stone from potentially reactive (PR) igneous rock – andesite, diabase, basalt and ignimbrite as well as from a non-reactive granite aggregate for concrete as an elasticity reference (Figure 9.4a);

2)   Crushed stone from PR sedimentary and metamorphic rock – flint and argillite as well as crushed sand (for test prisms) from flint (Figure 9.4b);
3)   Gravel from a PR rock – tuffaceous siltstone, sandstone and andesites (Figure 9.4c);
4)   Gravel from a PR rock – flint and sandstone with chalcedony cement (Figure 9.4d).

During this testing, the Leningrad Institute of Railway Engineers and the Gidroproyekt Central Asia Division in Tashkent prepared test beams (according to the procedure recommended by ASTM for US laboratories with five types of gravel aggregates). After 1 year their measurements fully coincided with the results of the above-described prism tests. The test procedure was confirmed taking account of the following considerations.

Gravel and crushed stone behaviour differs in concrete. The gravel particle surfaces are coated with a weathering "crust" that inhibits interaction with alkalis in concrete, while crushed stone has fresh and active surfaces. Sand particles in concrete typically amount to less than about 25% of its volume, and even in cases of interaction with alkalis the sum of stresses in the envelope around 0.5 mm grains will be less than the paste matrix cohesion. In mortar samples this stress can be lower, and so degradation can occur in such samples.

Portland cement-based prisms with gravel aggregate containing 0.4-0.8% alkalis, after 3 to 7 years of hardening, were characterized by K_{ynp} over 1.1, thereby indicating the absence of explicit aggregate reactivity. Specimens based on gravel and cement, with the introduction of 1% alkali into the mix water, exhibited the same elastic modulus coefficient (Fig. 9.4c, 9.4d). Partial reduction of K_{ynp} down to 1.02 was established only for the prisms based on gravel containing tuffaceous sandstone and andesites from the Yenisei River valley, or for concrete with gravel from the Prut River containing flint.

Thus, only when the content of potentially reactive silica is increased in Portland cement concrete, is it possible for concrete damage to occur and only when crushed flint aggregate with water absorption less than 0.7% was used.

In recent decades the Research Institute of Concrete and Reinforced Concrete (NIIZhB) has studied about 50 different aggregates: crushed stone, gravel, sand with contents of alkali-soluble SiO_2 from 20 to 1100 mmol/l (Rozentahl et al., 2014).

Considering the data on the amounts of aggregates supplied to the Moscow precast concrete plants, 4 sources have been studied in detail: Abramovo, Akademicheskoye, Vyazemskoye and Oreshkinskoye. At the plants using aggregates from the above deposits, representative samples of crushed stone from gravel and sand were taken to determine chemically the alkali-soluble silicon dioxide. The majority of the aggregates were included in Table 9.3 and have shown high contents of alkali-soluble silicon dioxide: over 50 mmol/l. For a number of the aggregates, the tests have been repeated many times in different years (Rozentahl et al., 2014). Table 9.4 presents the test results for the rock aggregates from the four deposits. For a number of aggregates, the tests were repeated multiple times in different years.

Taking into account the chemical analysis results, it was considered for a long time that the crushed stone from Vyazemsky quarry would have reactivity with cement alkalis. Reactivity of this crushed stone was studied at NIIZhB in different years for 12 batches of crushed stone. The soluble SiO_2 content ranged from 141.9 to 682.4 mmol/l. The twelfth sample was prepared specifically by the Vyazemsky quarry administration

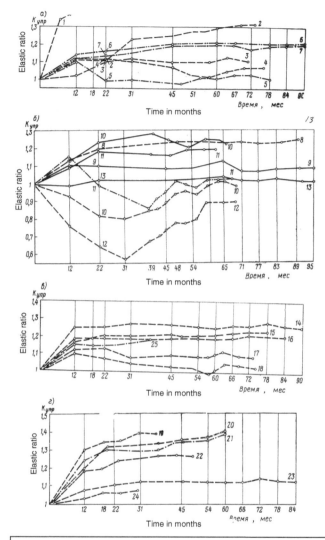

Figure 9.4 Elastic ratio variation of concrete with reactive aggregates in time (Viktorov, 1972).

Table 9.4 Results of alkali-soluble silicon dioxide quantity determination.

Deposit and aggregate	Grain size, mm	Soluble SiO$_2$ content, mmol/l
Abramovo (crushed stone from gravel)	5–10	294; 250
	10–20	274
Akademicheskoye (crushed stone from gravel)	5–10	230
	10–20	377; 465
Vyazemskoye (crushed stone from gravel)	5–10	339; 361; 364
	10–20	142; 302; 406; 437; 508; 533; 594; 682
Oreshkinskoye (crushed stone from gravel)	5–10	120; 144: 137; 338;
	10–20	201; 241; 255; 261; 684
Vyazemskoye (sand)	0–5	21; 52

for flint stone grain sampling using a microscope. The soluble SiO$_2$ content in this sample was 782.0 mmol/l. The accelerated tests of all samples have shown expansion values 0.03 to 0.069%, which is less than the 0.1% value specified by GOST 8269.0. The year-long tests were conducted with two samples of crushed stone with the soluble SiO$_2$ content of 363.9 and 437.2 mmol/l, respectively. Expansion values after 12 months were 0.024 and 0.016%. Again this is less than the 0.04% specified by GOST 8269.0. Thus, the crushed stone from Vyazemskoye deposit was recognized as being non-reactive with cement alkalis. These laboratory findings have also been corroborated by the long-term experience of using the Vyazemskoye crushed stone in construction.

Integrated studies of a wide range of aggregates by three methods were conducted according to GOST 8269.0. In concrete testing, the Portland cements from three cement plants were used with the alkali contents ranging from 0.6 to 1.1%. The mineralogical composition of Portland cement clinkers is given in Table 9.5. The test results by three methods according to GOST 8269.0 are given in Table 9.6.

The accelerated test results identified 13 aggregates containing over 50 mmol/l of alkali-soluble SiO$_2$, but these aggregates proved to be non-reactive. While out of five aggregates comprising less than 50 mmol/l of soluble SiO$_2$, three aggregates were reactive.

In a joint RF – UK research project, similar procedures were used to test the gravel from Listvennichnoye deposit (SiO$_2$: 20.5 mmol/l) and the sand from Okhotskoye deposit (SiO$_2$: 46.7 mmol/l). By the expansion test measurements both in the RF and the UK the aggregates were recognized as reactive, though the alkali-soluble SiO$_2$ content in samples was less than 50 mmol/l.

Table 9.5 Mineralogical composition of cement clinkers.

Cement manufacturer	Mineral composition, %				Alkali content, R$_2$O %
	C$_3$S	C$_2$S	C$_3$A	C$_4$AF	
Belgorodsky	64	17	7	15	0.62
Voskresensky	62	12	8	13	1.05
Maltsovsky	63	15	6.5	13	0.56

Table 9.6 Results of aggregate tests by chemical, accelerated and long-term methods (in approximate order of increasing quantity of soluble silica).

Ser. No.	Aggregate Material	Aggregate reactivity evaluation by three methods according to GOST 8269.0 (1997)		
		Chemical method. Soluble SiO_2 quantity, mmol/l and evaluation	Accelerated method*	Long-term tests. Deformations, %, and evaluation
1	Kamchatka shale	19.9 Non-reactive	0.145 – Reactive	0.063 Reactive
2	Quartzite sandstone	19.2 Non-reactive	0.061 14 Non-reactive	–
3	Crushed stone from gravel, Listvennichnoye deposit	20.5 Non-reactive	0.132 – Reactive	–
4	Crushed stone from Ryboretskoye deposit. Karelia. Quartzite sandstone	38.6 Non-reactive	0.072 16 Reactive	0.019 Non-reactive
5	Sand from Okhotskoye deposit	46.7 Non-reactive	0.192 – Reactive	–
6	Sand from Vyazemskoye deposit	52.0 Reactive	0.070 17 Reactive	0.163 Reactive
7	Sand from Vyatskoye deposit	72.5 Reactive	0.012 5 Non-reactive	–
8	Kamchatka perlite	134.0 Reactive	0.106 – Reactive	0.071 Reactive
9	Kamchatka pumice	134.0 Reactive	0.047 13 Non-reactive	0.027 Non-reactive

No.	Material			
10	Crushed stone from gravel, Oreshkinskoye deposit	201.5 Reactive	0.125 / – / Reactive	0.015 Non-reactive
11	Crushed stone from gravel, Abramovskoye deposit	214.3 Reactive	0.041 / – / Non-reactive	0.020 Non-reactive
12	Crushed stone from gravel TPI-1	345.0 Reactive	0.051 / – / Non-reactive	0.022 Non-reactive
13	Crushed stone from gravel, Vyazemskoye deposit	437.2 Reactive	0.034 / 7.0 / Non-reactive	0.016 Non-reactive
14	Crushed stone from gravel, Vyazemskoye deposit	682.4 Reactive	0.033 / – / Non-reactive	–
15	Crushed stone from gravel, Vyazemskoye deposit	533.2 Reactive	0.064 / 12 / Non-reactive	–
16	Crushed stone from Akademicheskoye deposit	464.7 Reactive	0.077 / 14 / Non-reactive	–
17	Gravel TPI-2. Shale and quartzite mixture	1014 Reactive	0.071 / Cracks / Reactive	Destruct. in 3 months Reactive
18	Sakhalin jasper	1082 Reactive	0.022 / 4 / Non-reactive	0.028 Non-reactive
19	Pyrex glass	4768 Reactive	Cracks in 7 days / Reactive	Destruct. in 3 months Reactive

*In this column, first line: % expansion, second line: % difference between the latest result and the three preceding ones; third & later lines: aggregate evaluation.

The tests of 12 aggregates by the accelerated and long-term methods have shown in 10 cases the same evaluation of their reactivity. By the accelerated test results, the crushed stone from Oreshkinskoe deposit gravel gave an expansion value of 0.125% and was evaluated as reactive, though the long-term test results indicated that it is non-reactive. The crushed stone from Ryboretskoye deposit has shown expansion of 0.07% (less than specified 0.1%) in accelerated tests, but the increment of the latter result was 16%, *i.e.*, higher than the specified value of 15%. However, since the key method is the year-long test, when the concrete test for this crushed stone gave an expansion of 0.019% (less than the 0.04% criterion), the crushed stone from the Ryboretskoye deposit was recognized as non-reactive.

In the course of numerous studies, reactive sands from different deposits have been identified. These sands are characterized by a crystalline structure, the weathering degree, availability of amorphous silica and the presence of various impurities. The sand comprised metamorphic rock containing quartz, veined quartz, acid volcanic rock and quartz with a disturbed crystalline structure. It has been found that reactivity is inherent for sands from deposits at Okhotskoye, Vyazemskoye and Malkinskoye.

Of particular interest is the granite reactivity evaluation. The tests of crushed granite from deposits in Karelia and Ukraine have shown the alkali-soluble SiO_2 content 5 to 140 mmol/l. However the accelerated and long-term tests of granite aggregate-based concrete have shown the absence of any significant expansions (Table 9.7).

NIIZhB has also conducted tests of crushed aggregate from dolomitic limestone of from the Dankovskoye and Zubtsovskoye deposits. The carbonate content in the Dankovskoye aggregate gave values of 41.07% $CaCO_3$ and 35.91% $MgCO_3$. The content of alkali-soluble SiO_2 was 0.85 mmol/l. The Zubtsovskoye aggregate gave 38.98% $CaCO_3$ and 42.44% $MgCO_3$, with an alkali-soluble SiO_2 content of 28.36 mmol/l. The accelerated tests were performed by the AAR-5 method developed by RILEM for carbonate rock studies (see AAR-5 in Nixon & Sims, 2016), with the specimen expansion measurements in the course of storing in 1M NaOH solution at 80°C. Expansions were measured over a one month period at 7 day intervals and the results are shown in Table 9.8.

In conformance with AAR-5 criteria, the aggregate from the dolomitic limestone is considered reactive provided specimen expansion on completion of the test (28 days) is equal to 0.1% or more. The accelerated test results of aggregates from dolomitic limestone from the Dankovskoye and Zubtsovskoye aggregates are consequently recognized as non-reactive with cement alkalis.

In addition, long-term tests of crushed aggregate from the dolomitic limestone of Dankovskoye and Zubtsovskoye deposits were performed according to GOST 8269.0

Table 9.7 Accelerated and long-term test results for granite aggregate-based concrete.

Sample No.	Soluble SiO_2 quantity, mmol/l	Concrete expansion, %	
		in accelerated tests	*in long-term tests*
1	5.5	−0.009	−0.0040
2	13.9	0.006	−0.035
3	91.8	0.016	−0.037
4	140.6	0.005	−0.042

Table 9.8 Expansion of dolomitic limestone-based concrete specimens in tests in 1 M NaOH solution at 80°C, using RILEM AAR-5 (in Nixon & Sims, 2014).

Deposit	% Expansion, after tests 3 to 30 days				
	3	10	17	24	30
Dankovskoye	0.011	0.015	0.019	0.026	0.030
Zubtsovskoye	0.017	0.020	0.032	0.037	0.046

Table 9.9 Expansion of dolomitic limestone-based concrete specimens in tests at 38°C

Deposit	Cement kg/m³	W/C	Slump cm	Bulk weight, kg/m³	Expansion in %, after tests up to 12 months					
					2	4	6	8	10	12
Dankovskoye	341	0.50	3.0	1875	0.027	0.023	0.023	0.022	0.019	0.019
Zubtsovskoye	325	0.62	1.5	1828	0.028	0.037	0.036	0.040	0.040	0.040

with the concrete specimens at 38°C and with the alkali content 1.5% as Na_2Oeq of cement by mass. The concrete mix proportions were (cement:sand:coarse aggregate) was: 1:1.4:2.6. The crushed aggregate grading was from 5 to 20 mm. The results are presented in Table 9.9.

As can be seen from the table, the decisive results of the long-term tests (12 months) indicate that the Zubtsovskoye dolomitic limestone is evaluated as reactive with cement alkalis. However, by both the accelerated and long-term test results, the Dankovskoye dolomitic limestone aggregate is evaluated as non-reactive with cement alkalis.

It should be noted that one more process (Alekseyev, 1990) is well known and characterized by the reaction between the pore-water alkalis and dolomitic limestone aggregates; this is the so called process of 'dedolomitization', when the solid phase expansion and the paste matrix structure impairment may occur. This process may be superimposed on the above interaction with activated silica alkalis that may be contained within the carbonate. Though carbonate rock is generally alkali-resistant, however, some varieties of dolomitic limestones can be affected by deterioration. The dedolomitization reaction in the general form is described by equation:

$$CaCO_3.MgCO_3 + NaOH = CaCO_3 + Mg(OH)_2 + Na_2CO_3$$

Taking into account that calcium hydroxide is always available in the concrete paste matrix, a similar effect also may be caused by some sodium and potassium salts. Sodium carbonate formed as a result of interaction with calcium hydroxide will maintain the NaOH content in the mortar. Such regeneration of sodium hydroxide, participating in the further reaction with dolomite crystals, may lead to hydrocarbonate formation and complex compounds of the magnesium hydrocarbonate type. These reactions result in considerable concrete expansion deformations because of the formation of higher volumes of reaction product (particularly, brucite) compared with the initial components (Fedosov & Bazanov, 2003). See Chapter 3 for an up-to-date account of the so-called alkali-carbonate reaction.

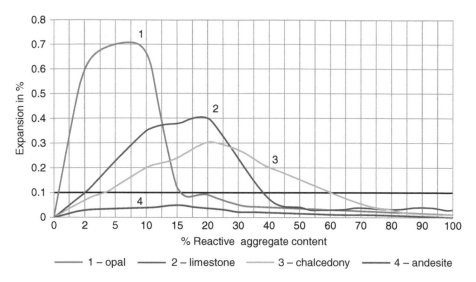

Figure 9.5 Effect of various mineral and rock contents on fine grained concrete specimen expansion at the age of 6 months: 1 – opal; 2 – limestone; 3 – chalcedony; 4 – andesite.

Petrova and Sorvacheva (2012a) have studied the dependence of fine grained concrete sample expansion on the content of potentially reactive component minerals in the aggregate at the age of 6 months (Figure 9.5).

According to research data and Zhukov (1972), opal is most reactive with alkalis, and its presence in the aggregate in amounts of about 1% may result in the concrete deterioration. Moskvin (1962) points out that the severity of the deterioration process is affected by the opal content in the rock (Figure 9.6) as well as by the mineral particle size. The most reactive size range is 0.3–2.5 mm. With a more finely dispersed material, the alkali compounds are bonded by the chemically active silica, while with coarser grains the ratio of concrete surface and the volume decreases, and the force turns to be insufficient for concrete structure disturbance.

By using the accelerated method at 80°C the relative expansion of concretes prepared with various aggregates including one that resulted in structural failures, Petrova and Sorvacheva (2012a have shown that, in a number of cases, the expansions noticeably exceeded the specified value of 0.1% (Figure 9.7). Specimen composition:

1 – Portland cement, alkali content 0.89%; aggregate 1 (initial materials used to manufacture deteriorated structures);
2 – same cement, aggregate 2;
3 – same cement, aggregate 3;
4 – same cement, aggregate 4.

Morozova (2005) analyzed the content of reactive silica in gravels from different deposits near the Kama River. The material was used by concrete and mortar mixing plants and precast concrete plants in Kazan during the period of 2001–2005. The content of soluble active silica in the gravel was determined by the chemical method according to GOST 8269.0-97 (1997). The data obtained show that the gravel

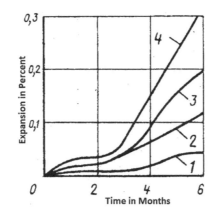

Figure 9.6 Effect of the opal content in the aggregate on concrete expansion deformation. Curves 1 – 4 represent increasing opal proportions in the aggregate (from Moskvin, 1962).

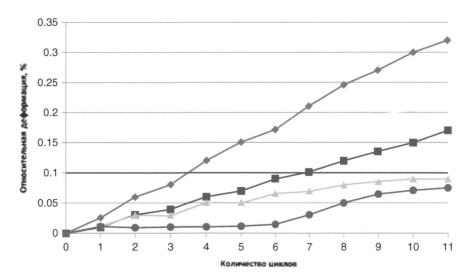

Figure 9.7 Concrete relative expansion versus curing time in 1M solution of NaOH at *t* = 80 °C (from Petrova & Sorvacheva, 2012a).

from the Kama deposits contained alkali-soluble silica in the range of 112 to 254 mmol/l, this being 2.2 to 5.1 times higher than the allowable values according to GOST 8267 (1993). This directly indicates high potential reactivity of these aggregates in cement concretes.

Generally all cements used in Kazan and the Republic of Tatarstan contain increased (over 0.6%) quantities of alkali oxides above the requirement specification of Code 28.13330.2012 (updated SNiP 2.03.11-85). In combination with potentially reactive aggregates, it predetermines a possibility of alkali-aggregate damage development. Since the quantity of alkali-soluble silica exceeds the specified values as well, the authors have estimated a possibility of alkali damage development using accelerated methods with expansion measurements (GOST 8269.0-97, item 4.22) in fine grain concrete/mortar with these aggregates and cements supplied in the Republic of Tatarstan. It follows from the data obtained that the cements in use do not cause expansions under alkali exceeding the permissible limit (0.1%). However, in the researchers' opinion, they may develop in longer term in concrete tests (or in service use). To estimate kinetics of their development the authors recommend conducting more additional long-term testing (1 year).

Thus, from the results of numerous studies, it is possible to state that the soluble silicon dioxide content limit in the aggregate (50 mmol/l) cannot be a satisfactory criterion for assessing the risk of concrete deterioration due to AAR. In using this criterion some aggregates can be invalidly assessed as reactive. At the same time with the soluble silicon dioxide content below the specified value, some aggregates do cause significant concrete deformations resulting in crack formation.

The use of active mineral additives, as a rule, prevents development of concrete deformations (Royak et al., 1986; Buyanov, 1999). In his paper, Izotov (2003) presents the results of studies on the effect of mixed binder composition on the alkali-aggregate development in the concrete, which was prepared using a 5–20 mm fraction of gravel aggregate from the Kama deposit. The mixed binder contained thermal power plant fly ash or a zeolite-containing rock (ZCR) that has high hydraulic activity. The Kama deposit gravel used in the test was designated as potentially reactive. The soluble silica content determined by the GOST 8269-76 method was 279 mmol/l. As is seen from the data obtained, the relative expansions of specimens on the basis of the mixed binder with the mineral additive content over 10% were less than the critical value. Therefore, in the author's opinion, this binder can be recommended to make concretes using potentially reactive aggregates.

Using solid-state NMR spectroscopy with ^{29}Si nuclei, Voronkov (2013) has studied transformations of highly dispersed silica-comprising additives (silica fume, metakaolin, precipitated silica, low-calc fly ash) within the paste matrix of cement-sand mortars under accelerated and long-term alkali expansion test procedures for Portland cement mortars and concretes. It was established that in the conditions of the accelerated test method, the highly active mineral additives (silica fume, precipitated SiO_2, metakaolin) fully enter into the pozzolanic reaction and lose the phase identity prior to the start of expansion (prior to immersion into NaOH solution). Consequently, the ratio of the additive interaction rate and the cement hydration rate is higher than at normal temperatures. The C-S-H gel formed by the joint Portland cement hydration and the additive pozzolanic reaction under the accelerated procedure is characterized by the long-chain structure (about 10 units) and the increased Al/Si ratio (~0.13-0.25). At the same time it was shown that the inhibitory effect of highly active mineral additives exercised under the accelerated procedure is primarily governed by the low basic capacity of the pozzolanic product and its ability to absorb considerable amounts of alkali compounds directly during formation.

The study has found that the long-term test procedure conditions do not affect the ratio of cement hydration and pozzolanic reaction rates and the hydration product compositions significantly. The mineral additives, depending on their activity, enter the pozzolanic reaction over a period of several days (in the case of precipitated SiO_2) to several months (with fly ash). In conditions of the accelerated test method, the differences in the inhibiting efficiency of mineral additives, including fly ash, are equalised. Compared with the long-term tests, the accelerated method overestimates the inhibitory activity of the additives. The tests of mineral additives within cement-sand mortars under long-term test conditions allow the differentiation of the mineral additives by their capacity to inhibit alkali expansion. The results obtained with this long-term method that do not modify the nature of transformations within cement systems with active additives as compared to usual conditions, should be considered as the most reliable.

Among the highly active additives studied, metakaolin appears to have the highest efficiency in that activity can be controlled by the participation of aluminum ions to provide additional passivation of the reactive aggregate particles. Similar results were obtained by Rozental *et al.* (2014), which have shown that introduction of the mineral additives, fly ash, silica fume, metakaolin or MB 10-50 modifier (50% of silica fume and 50% of fly ash) in the amount of 20%, reduced expansions of concrete with potentially reactive aggregates down to the safe level. Substitution of PC D0 cement for PC-D20 cement with a slag additive has reduced expansion of specimens with the aggregates noted above by 19-45%. Royak (2001) also showed that reduction of concrete expansion could be attained by the introduction of active mineral additives of a sedimentary and volcanic origin, or granulated slag.

Petrova and Sorvacheva (2012a) have attempted to reduce potential reactivity of cement alkalis with the aggregate silica by introducing a polycarboxilate-based super-plasticizer into the concrete mix composition, dependent on the basis of the initial materials, in the proportions of 0.3 to 0.7% of cement by weight. Such superplasticizers are used in reinforced concrete element manufacture. Judging by the results obtained, introduction of the superplasticizer in this amount does not prevent development of alkali-silica reaction, but an increase in the admixture content leads to an increase in concrete expansion. In the authors' opinion, to reduce cement consumption and to increase the material density, it is possible to use an additive in the amount of 0.3–0.4%, but in combination either with the aggregate comprising a lesser quantity of amorphous silica, or with a low-alkali cement.

Table 9.10 Expansion of concrete samples containing pozzolanic additions, after testing for 1 year at 38°C (from Royak, 2001).

Additive	Amount of additive, %	Soluble SiO$_2$, mmol/l	Expansion, %
Spongolite	–	625	0.990
	17,0		0.380
	30,0		0.070
Gaize 1	17,0	606	0.960
	30,0		0.090
Gaize 2	17,0	533	0.620
	30,0		0.067
Tuff	17,0	373	0.610
	40,0		0.110
Perlite	17,0	202	0.400
	40,0		0.065
Pumice	17,0	132	0.590
	40,0		0.090
Granulated slag	17,0	5	0.530
	60,0		0.090

Brykov *et al.* (2012) reports studies concerning features of the conversion processes that appear to take place during the mortar-bar tests (GOST 8269.0: 1997), where high-dispersion silica-containing additives (silica fume, metakaolin, and precipitated silica) in the cement are used for rock aggregate and sand-cement mortar-bars stored at 20 and 80°C. According to solid-state NMR spectroscopy on the bars stored at 80°C, the additives rapidly lose their phase individuality by reacting with $Ca(OH)_2$ to form calcium silica hydrogel (C-S-H). The Portland cement hydration in the presence of these mineral additives proceeds more slowly than in mortar-bars without these additions. Compared with Portland cement gel, the C-S-H product formed by the additives is characterized by lower Ca/Si ratio, longer aluminum-silicon-oxygen structural chains, and by higher aluminium content.

In later studies by Brykov *et al.* (2014), additions of silica fume, precipitated silica, metakaolin and siliceous fly ash behaviour as constituents of mortars were studied. The mortar samples were tested for long-term alkali-silica reaction expansion in accordance with the GOST 8269.0 specification. Solid-state ^{29}Si-MAS NMR spectroscopy and thermo-gravimetric analysis were used to describe Portland cement hydration. Investigation of the effect of pozzolanic reactions with these supplementary cementitious materials was undertaken in order to establish a structure of products resulting from these processes. It was found that the long-term test conditions, in contrast to the accelerated test results, do not affect the composition of products formed significantly, as compared with normal conditions. This suggests that the results obtained with long-term testing will be more relevant in predicting the inhibiting effect of supplementary cementitious materials used in concretes.

In Rosentahl *et al.* (2014), some methods of concrete protection against alkali-reaction effects were presented. Test results of concrete with reactive aggregates after treatment with a solution of lithium compounds (nitrate, formate and lithium hydroxide) at a temperature of 20 °C were presented. The initiation process of the concrete treatment was begun when the expansion reached the value of 0.1%. It was shown that, in principle, this treatment can stop the ongoing expansion of concrete.

In summary these recent results suggest:

1. The accelerated mortar-bar test does not differentiate between different ultrafine silica additives in their ability to control the concrete expansion due to ASR. This is thought to be because all these ultrafine materials become quickly bound into $Ca(OH)_2$.
2. The long-term concrete prism test procedure allows the various additives to react more slowly from several days to months. This makes it easier to separate them in terms of their ASR inhibiting ability.
3. It is found that C-S-H gel formed in the presence of metakaolin has the highest Al - Si ratio (0.15 – 0.19) and has the longest chains' length (approximately 6x Si, Al nuclei).

9.6 PRACTICAL EXAMPLES OF STRUCTURAL DAMAGE CAUSED BY AAR

In Russia and the former USSR (now The Russian Federation), the well-known cases of damage are of house footing structures, railway sleepers, concrete elements in a number of public and industrial buildings and structures, and port facilities. In some cases in a few months or even years after manufacture of structures, there appeared visible signs of concrete degradation, including crack networks and white jelly-like exudations. The chemical analysis showed the presence of great quantities of SiO_2 and the alkali metals sodium and potassium in the exudations. This damage in humid conditions proved to be most favourable for concrete strength retention, and even gain over long-term periods, where strength gradually increased while the structural concrete was deteriorating. Such deterioration processes occurred within concrete without special environmental components, except for the necessary presence of moisture.

The first examples of deterioration of this type became apparent in the early 1960s (Moskvin, 1962). Then the Gidroproyekt Institute and the Research Institute of Transport Construction registered and described in detail the individual cases associated with the alkali-silica reaction development in the vicinity of Simbirsk (Ulyanovsk) and at hydraulic engineering projects in the Volga Region (Viktorov & Osipov, 1982).

This required detailed field inspections of hydraulic structures on the Kama, Dniester and Belaya Rivers, where the gravel with a considerable content of potentially reactive silicon was being used in construction (Viktorov & Osipov, 1982). Sand used as a fine aggregate also comprised up to 30% of silica grains (according to later measurements). Since, to prepare concrete, the Portland cement had been used with additives of tripolith (diatomite), gaize or blastfurnace granulated slag, with a cement content of 250-320 kg/m^3, none of the inspected structures had specific cracks or spalling of the concrete surface that indicated cement-alkali interaction with aggregates under damp conditions.

The accumulated experience and proven approaches, connected with limitation of the concrete alkali content up to 3 kg of alkali per m^3, have allowed for the development of concrete compositions resistant to alkali-reactivity using any Portland cement with this limited alkali content. These were used to construct hydro-power plants on the Euphrates River in 1980 using reactive gravel. For structures of the Nizhnekamsk and the Cheboksary Hydro schemes in 1978–1979, the concrete composition was proportioned using potentially reactive gravel and sand comprising up to 25% of siliceous additives. The experience of dealing with similar aggregates 10-30 years ago was taken into account by reference to construction of the Votkinsk and the Kama Hydroes on the Kama River, the Dubossary Hydro on the Dniester River as well as in concrete slope protection in the Belaya River valley in the vicinity of Ufa in 1965–1980.

The detailed field inspections of concrete pile foundations of high-rise residential buildings in the city of Novgorod, conducted by NIIZhB in 1998–2002, disclosed indicators of alkali-aggregate reactivity, and have resulted in development of similar recommendations providing for the secondary protection of structures using special impregnating compounds.

In 2004 in Russia there were cases of mass-scale deterioration of precast concrete railway sleepers manufactured in 2001 (Petrova & Sorvacheva, 2012b, 2013a). The deterioration was accompanied by the appearance of cracks with various opening widths (Figure 9.8). Photographs of core samples drilled out from these sleeper structures show coarse aggregate grains that deteriorated with gel release, being one of the characteristic features of the alkali-silica reaction (Figure 9.9). Of similar nature were failures of transmission tower footings of the USSR railway overhead system after 3 years of service (Petrova & Sorvacheva, 2012b).

Рис. 5. Повреждения шпал

Figure 9.8 (top) General view of deterioration of railway sleepers, (bottom) Close up view of cracks in railway sleepers.

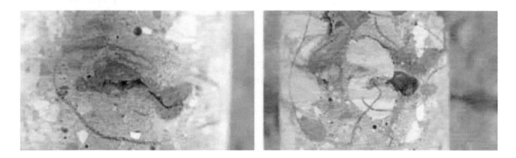

Figure 9.9 Two examples of gel exudation from coarse aggregate in concrete.

Sorvacheva and Chistyakov (2014) inspected a highway pavement (Figure 9.10), from which a core sample was cut out and cured for 30 minutes in the special solution, presumably uranyl acetate (Figure 9.11a). In the highway construction, the Portland cement used comprising 0.8% alkalis in terms of Na_2Oeq, fine and coarse aggregates comprising amorphous silica 8.2 mmol/l and 35 mmol/l. According to the present codes, these materials are not potentially reactive, since the total content of amorphous silica is less than the allowable value of 50 mmol/l. However, the photograph under UV-lighting (Figure 9.11b) clearly shows the alkali-silicate gel formed both around and inside the concrete aggregate. Development of internal degradation in the highway pavement is also demonstrated by microscopical studies (Figure 9.12).

Figure 9.10 Cracking near joints in a concrete highway pavement

(a)

(b)

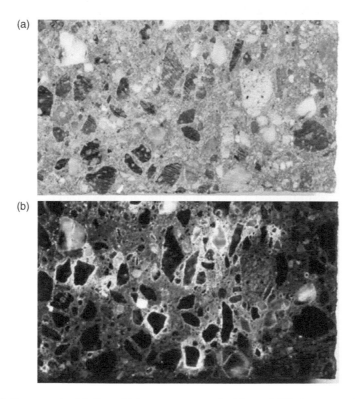

Figure 9.11 (a) Cut sample slab from highway pavement (in Figure 9.10), in conventional light. (b) Sample slab as in (a) but seen in UV light.

Quite a few cases are known, when the internal deterioration was explained by other causes, including low frost-resistance, technological defects or concrete composition. The fact that concrete damage was a result of internal alkali-aggregate reaction can only be established by a thorough analysis when a sufficient alkali content in the concrete and reactive silica presence in the aggregates are determined. Correct detection

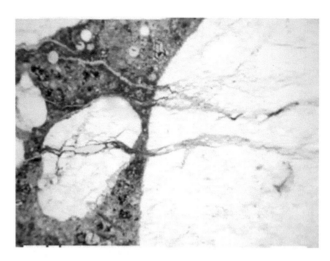

Figure 9.12. Alkali-silica gel in concrete cracks and pores: view in thin-section under a polarizing microscope.

and diagnosis of internal deterioration is very important, since once initiated, the process of concrete deterioration by AAR can be stopped only by creating dry operating conditions, which is not practically feasible in most cases.

Attention to internal alkali-aggregate deterioration in Russia is increasing every year, because its consequences are rather serious, and the causes of structural damage are intensifying. First of all, the alkali content is increasing in cements as a result of dust collection and recovery improvements (the dust contains increased quantities of alkali metal compounds sublimated during cement clinker calcinations). Simultaneously, the alkali content rises in the raw materials for cement production, while in some cases the alkalis are components of raw materials and alkali-comprising wastes such as aluminum plant, nepheline tailings and some fly ash types from thermal power plants.

REFERENCES

Alekseyev, S.N. (1990) *Reinforced concrete durability in aggressive environments.* Alekseyev, S.N., Ivanov, F.M., Modry, S., Shissl, P. – M.: Stroyizdat. 316p.

ASTM C289 (2007) (now withdrawn). *Standard test method for potential alkali-silica reactivity (chemical method).* West Conshohocken, USA, American Society for Testing and Materials.

ASTM C1260 (2014) *Standard test method for potential alkali reactivity of aggregates (Mortar-Bar method).* West Conshohocken, USA, American Society for Testing and Materials.

Bekrenev, I.V. (2013) Investigation of specific development of the non-metallic building materials market in Russia. *Mining Information Analysis Bulletin (GIAB)*, 6, 340–348.

Belogurova T. P., Krasheninnikov O. N., Royak G. S., Traktirnikova T. L. (1990) On the reactivity of the overburden of the PE "Apatit". In the book *"Physico-chemical fundamentals of processing and use of mineral raw materials"*, Academy of Science of the USSR, Apatity, 94 p. pp.32–35.

Brykov, A.S., Vasil'ev, A.S. & Mokeev, M.V. (2012) Hydration of Portland cement in the presence of high activity aluminum hydroxides. *Russ J Appl Chem+*, *85* (12), 1793–1799.

Brykov, A.S., Voronkov, M. & Mokeev, M.V. (2014) Ultrafine silica additives behaviour during alkali-silica reaction long-term expansion test. *Materials Sciences and Applications*, ISSN Print: 2153-117X, ISSN Online: 2153-1188, Website: http://www.scirp.org/journal/msa, 66–72.

Butkevich, G.R. (2003) Nonmetallic building materials industry: Achieved status and prospects *Building materials*. No. 11. pp. 2–5.

Buyanov, Yu. D. (1999) Effect of mineral aggregate quality on durability and corrosion resistance of structures. In: Buyanov, Yu. D., & Levkova, N.S. (eds.) *Proceedings of the International Conference "Durability and corrosion protection of structures. Construction and rehabilitation"*, Moscow. pp. 255–260.

Buyanov, Yu. D. (2005) Prospects of improving concrete aggregate qualities. In: Buyanov, Yu. D., Kharo, O.Ye., Butkevich, G.R. & Levkova, N.S. (eds.) *Proceedings of the 2nd All-Russia (International) Conference on Concrete and Reinforced Concrete. 5–9 September 2005*, Moscow; in 5 volumes. Volume 3. pp. 236–241.

Buyanov, Yu. D. & Lopatnikov, M.I. (2004) Mineral resources of the building materials industry and the main issues of their development. In: *Proceedings of the XI International Conference "Technologies, equipment and raw materials base of mining plants in the building materials industry"* September 2004. Collection of reports – St. Petersburg. pp. 7–9.

CM Pro Analytical Agency (2015) *Annual Report: Non-metallic building materials – 2014*, CMPro, 120p.

Fedosov, S.V. & Bazanov, S.M. (2003) *Sulfate corrosion of concrete*. "ASV" Publishing house, 192p.

GOST 8267-93 (1993) *Crushed stone and gravel from dense rock for construction works. Specifications*. Euro-Asian Council for Standardization, Metrology and Certification (EASC), Commonwealth of Independent States (CIS).

GOST 8269.0-97 (1997) *Crushed stone and gravel from dense rock and industrial wastes for construction works. Physical and mechanical test methods*. Euro-Asian Council for Standardization, Metrology and Certification (EASC), Commonwealth of Independent States (CIS).

GOST 8736-93 (1993) *Sand for construction works. Specifications*. Euro-Asian Council for Standardization, Metrology and Certification (EASC), Commonwealth of Independent States (CIS).

Guidelines (1984) *Guidelines for the prevention of internal corrosion of concrete during the chemical interaction of alkali cement with reactive species of aggregates*, VNII Trans Stroy, Moscow, 54p.

Guidelines for determining of the concrete aggregates reactivity with alkalis of cement. (1972) NIIZhB, Gosstroy of the USSR. – M., p. 24 p. 1–4.

Gusev, B.V. (2005) Corrosion processes and methods of dealing with them. *Abstracts of International scientific-practical conference "Protection against corrosion in construction and municipal economy."* In the framework of the 3rd International specialized exhibition "ANTICOR-Galvanic Service". Official catalogue. pp. 59–60.

Ivanov, F.M. (1992) Internal corrosion of concrete. In: *Beton i zhelezobeton (Concrete and reinforced concrete). Stroyizdat*; No. 8, 8–10.

Ivanov, F.M. (1995) Interaction of concrete aggregates with cement alkalis and admixtures. In: Ivanov, F.M., Lyubarskaya, G.V., & Rosental, N.K. (eds.) *Beton i zhelezobeton (Concrete and reinforced concrete)*, No.1. pp. 15–18.

Izotov, V.S. (2003) Peculiarities of alkaline corrosion and efflorescence in concrete based on blended binders. *Proceedings of KhASA*, No. 1. pp. 68–69.

Izotov, V.S. and Gizzatullin, A. R. (1988) Influence of complex admixtures on alkaline corrosion of concrete. In: *The performance of construction materials under the influence of various operational factors: Intercollegiate Proceedings*: – Kazan': Kazan' State Technological University. pp. 22–26.

Kramar, L.Ya. (1994) State and prospects of scientific and technical potential development in the South-Urals Region: The report run-down. Magnitogorsk. pp. 55–57.

Lopatnikov, M.I. (2007) Sand-gravel deposits as possible sources of local durable raw materials. In: Lopatnikov, M.I. & Tedeyev, T.R. (eds.) *Building materials*, No. 9. pp. 18–19.

Lopatnikov, M.I. (2008) Raw material base of crushed stone and sand extraction in the Moscow Region. In: *Building materials, equipment and technologies of the 21st century*, No.8. pp. 40–41.

Morozova, N.N. (2005) Problem of alkali corrosion in concretes in the Republic of Tatarstan and the ways of its handling. In: Morozova, N.N., Khozin, V.G., Mateyunas, A.I., Zakharova, N.A., & Akimova, E.P. (eds.) *KGASU News*, No. 2. pp. 58–63.

Moskvin, V.M. (1962) Concrete corrosion under interaction of cement alkalis and the aggregate active silica. In: Moskvin, V.M., & Royak, G.S. (eds.) *Gosstroyizdat, 1962.* 164p.

Moskvin, V.M. & Royak, G.S. (1979) Corrosion processes in concrete and their prevention methods. In: *Transport construction*, 1931. *Transstroyizdat*, No. 9, pp. 47–48.

Moskvin, V.M., Ivanov, F.M., Alekseev, S.N. & Guzeev, E.A. (1980) *Corrosion of concrete and reinforced concrete, their protection methods* (Ed.) Stroyizdat, 536p.

Nalivkin, D.V. (1960) *The geology of the USSR, a short outline (including a 1:7,500,000 scale geological map of the USSR in full colour)* (translation by Tomkeieff, S.I. & Richey, J.E.), International Series of Monographs on Earth Sciences, Volume 8, Oxford, UK, Pergamon Press. 182p. & 2 maps.

Nixon, P.J. & Sims, I. (eds.) (2016) RILEM recommendations for the prevention of damage by alkali-aggregate reactions in new concrete structures. *RILEM State-of-the-Art Reports*, Volume 17, Springer, Dordrecht, pp RILEM, Paris, 168p.

Oratovskaya, A.A. (2005) The use of reactive aggregates in cement concretes. In: Oratovskaya, A.A. & Yakovlev, V.V. (eds.) *Proceedings of the 2nd All-Russia (International) Conference on Concrete and Reinforced Concrete*. 5–9 September 2005, Moscow; In 5 volumes. Volume 4. pp. 383–388.

Peshkova, G. Yu. (2014) Analysis of nonmetallic building materials market: basic trends and prospects of development. *Corporate Governance and Innovative economic development of the North: Bulletin of Research Center of Corporate Law, Management and Venture Investment of Syktyvkar State University.* – Syktyvkar, SSU. No. 4, 206 p. pp, 53–64.

Petrova, T.M. & Sorvacheva, Yu. A. (2012a.) Internal corrosion of concrete as a factor of the durability decreasing of the objects of transport construction. *Science and Transport (series Transport Construction)*. No. 4. pp. 56–60.

Petrova, T.M. & Sorvacheva, Yu. A. (2012b) The reasons for development of internal corrosion and fall of concrete sleepers' durability. *Bulletin of PGUPS*. No. 2. pp. 87–92.

Petrova, T.M. & Sorvacheva, Yu. A. (2013a.) Revisiting the durability of concrete structures. *"Scientific World"*, March 2013, pp. 19–30.

Petrova, T.M. & Sorvacheva, Yu. A. (2013b) On the issue of durability of reinforced concrete structures. In: *Current trends in theoretical and applied sciences*, 43. p. 48.

Rosental, N.K. (2002) On the causes of early deterioration of concrete and reinforced concrete structures. In: Rosental, N.K. , Cekhny, G.V. & Lyubarskaya, G.V. (eds.) *Industrial and civil construction*, No. 9. pp. 41–43.

Rosental, N.K., Chekhny, G.V., Lyubarskaya, G.V., & Rosental, A.N. (2009) Internal corrosion protection of concrete with reactive aggregate. *Building materials*, No. 3. pp. 68–71.

Rozental N.K., Lyubarskaya G.V. & Rosental A.N. (2014). Reactivity of aggregates and corrosion resistance of concrete. Concrete and Reinforced Concrete – Glance at Future. In: ed. B. Gusev, A. Zvezdov, A. Tamrazyan & V. Falikman Proceedings of III Russian (II International) conference on concrete and reinforced concrete. 2014, Moscow; M., MSUCE, In 7 volumes. Volume 3. 463p. pp. 377–387.

Rosental, N.K., Lyubarskaya, GT.V., & Rosental A.N. (2014b) Test of concrete with reactive aggregates. In: *Beton i zhelezobeton (Concrete and reinforced concrete)*, No. 5, pp. 24–29.

Royak, G.S. (2001) Prevention of internal corrosion in concrete. In: *Proceedings of the 1st all-Russian conference on concrete and reinforced concrete "Concrete at the turn of the third Millennium"*. Volume 3, pp. 1431–1434.

Royak, G.S. (2002) Internal corrosion of concrete. TsNIIS Proc., 2002, No. 210, 156p.

Royak, G.S. (2003) *Internal corrosion of the concrete*. Autoabstract. Dis. on competition of Doct. of Sci. academic grade– M.: TsNIIS, 78p.

Royak G. S., Granovskaya I. V., Traktirnikova T. L. (1986) Prevention of alkaline corrosion of concrete by active mineral additives. *Concrete and reinforced concrete*. No. 7. pp. 16–17.

Russian Academy of Sciences (2003) *Proceedings of the International Scientific Conference*. Fundamental problems of complex use of natural and technologic raw materials of the Barents region in the technology of building materials – Apatity, April 2003. pp. 1–4.

Sal'nikov, N.S. & Ivanov, F.M. (1971) Corrosion deterioration of concrete, containing large dosage of potash. In: *Beton i zhelezobeton (Concrete and reinforced concrete)*, No. 10. pp. 17–19.

Senatorov, P.P. (2008) The state of mineral resources and building stone extraction in the Russian Federation's European part. In: Senatorov, P.P., Vlasova, R.G., Sadykov, R.K., Sibgatullina, E.A., & Senatorov, S.Z. (eds.) *XIII International Conference "Technologies, Equipment and Raw Material Base of Mining Companies in the Building Materials Industry"*. Collection of Conference Proceedings (4–6 June 2008), Moscow. pp. 8–31.

Sorvacheva, Yu. A. & Chistyakov, E. Yu. (2014) Demonstration of internal corrosion of concrete in existing structures. *XX international scientific and practical conference "Modern equipment and technologies"* Section 6: Materials Science. pp. 93–94.

SP 28.13330 (2012) Corrosion protection of structural elements. *Updated living edition of SNiP*. 2.03. pp. 11–85.

STO 36554501-002-2010 (2010) Concrete protection against corrosion caused by the reaction of aggregate SiO_2 with cement alkalis. NITs "*Stroitel'stvo*" JSC.

Viktorov, A.M. (1972) Predictions of alkali corrosion of the concrete. *Beton i zhelezobeton*, No.7. pp. 18–19.

Viktorov, A.M., & Lozhkin, A.N. (1992) Problems of alkali corrosion of concrete dams. *Hydraulic Engineering, 116*, No. 11. pp. 52–53.

Viktorov, A.M., & Osipov, A.D. (1982) Alkaline concrete corrosion control measures. *Beton i zhelezobeton (Concrete and reinforced concrete)*, No. 1. pp. 46–47.

Viktorov, A.M. (1986) Prevention of alkaline corrosion of moist concrete. *Beton i zhelezobeton (Concrete and reinforced concrete)*, No. 8. pp. 38–39.

Voronkov, M.E. (2013) Inhibiting activity and transformations of mineral additives under the alkali expansion testing of cement compositions. In: Voronkov, M.E. & Brykov, A.S. (eds.) *Proceedings of the third scientific-technical conference of young scientists "Science Week – 2013"*. SpbSTI(TU), 96p.

Voronkov, M.E. (2013) *The interactions of silica-containing additives in the cement compositions under the conditions of the alkali expansion*. Autoabstract of the Cand. Diss. SpbSTI (TU).

Yakovlev, V.V. (1999) On the concrete alkali-silica corrosion studies. In: Yakovlev, V.V., Oratovskaya, A.A., & Smirnova, N.F. (eds.) *Proceedings of the International Conference "Durability and corrosion protection of structures. Construction and rehabilitation"*, Moscow. pp. 427–434.

Zhukov, Yu. A. (1972) *Effect of calcium hydroxide on the development of destructive processes in the concrete exposed to alkaline corrosion.* Autoabstract … Candidate Tech. Sciences. Leningrad: LIIZhT. p. 19.

Zolotykh, H.B. (ed.) (1990) *Typification of potentially reactive minerals from nonmetallic building material deposits".* *State-of-the Art Report.* Scientific-research and design-prospecting Institute for mining, transportation and processing of mineral raw materials in the building materials industry - FSUE "VNIIPIstromsyrye". 90p.

Chapter 10

North America (USA and Canada)

Michael D.A. Thomas, Kevin J. Folliard & Jason H. Ideker

10.1 INTRODUCTION

This chapter deals with alkali-silica reaction (ASR) in the North American countries of the United States of America (USA) and Canada. Note that ASR in other parts of this continent (*i.e.*, Central America) is discussed in Chapter 11. Also note that although cases of alkali-carbonate reaction (ACR) have been found in the USA and Canada, this form of alkali-aggregate reaction is discussed in Chapter 4.

The phenomenon of ASR was first discovered in the 1930s by Thomas Stanton of the California State Division of Highways (Stanton, 1940a), who correctly diagnosed a chemical reaction between the alkali hydroxides of the cement and certain silica minerals within the aggregate as the cause of cracking in numerous concrete highway structures in Monterrey County, California. Stanton is widely regarded as a pioneer with regards to ASR, as he not only discovered the mechanism but also made numerous other contributions in his landmark paper (Stanton, 1940b). These included the development of a test method to detect reactive aggregates [the method evolved into ASTM C227 (2010)], the importance of aggregate mineralogy, the existence of a 'pessimum' proportion for some aggregates, the possibility of preventing damage using low-alkali cement, the potential for pozzolanas to control expansion, the importance of the particle size of reactive aggregates (including the observation that reactive aggregate did not produce damaging expansion when ground to cement fineness) and that, although expansion may not lead to failure of the structure, the resulting cracking would shorten the life of the structure as it would promote other deterioration mechanisms such as the penetration of seawater and corrosion of embedded steel reinforcement.

Following Stanton's discovery, ASR was diagnosed as the cause of abnormal cracking in a number of dams operated by the U.S. Bureau of Reclamation, such as the Parker Dam in Arizona (Meissner, 1941). During the 1940s a number of agencies initiated studies on ASR in the USA (Army Corps of Engineers, Bureau of Public Roads, Portland Cement Association) and for a number of decades the USA led the field in ASR research. Some key studies from this period were focused on clarifying the mechanisms including Hansen's (1944) proposal that osmotic pressure generated within the gel reaction product was responsible for expansion and Powers and Steinour's (1955a, 1955b) later attempt to refine this 'model'. A map showing some of the cases of ASR identified in the USA is given in Figure 10.1.

It was more than a decade after its discovery in the USA before ASR was first confirmed as the cause of some concrete deterioration in Canada. Swenson (1957a) reported his

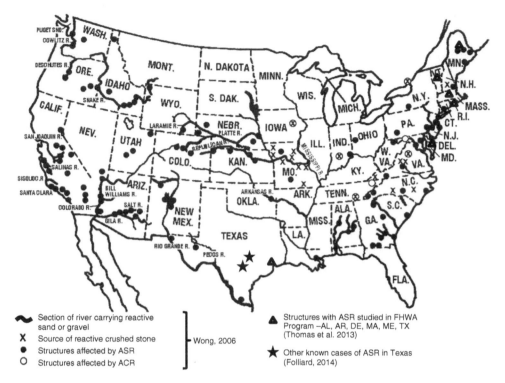

Figure 10.1 Map Showing Location of some ASR-Affected Structures in the USA (modified from Wong, 2006).

findings for an investigation on a 25-year-old bridge in Montreal, Quebec, that exhibited signs of *"continuing growth"* in 1953. It is interesting that some of the aggregates implicated as being the source of the reaction were not identified by the mortar-bar test (ASTM C 227: 2010) which had become one of the standard tests for detecting alkali-silica reactive aggregates at that time. It is further noteworthy that in the same year, Swenson (1957b) also discovered a different form of alkali-aggregate reaction, which became known as the alkali-carbonate reaction (ACR, see Chapter 4). Today cases of ASR have been found in every Canadian Province, as shown in Figure 10.2.

North America has played host to the International Conference on Alkali-Aggregate Reaction (ICAAR) in Concrete four times, twice each in the USA (Purdue, IN, in 1978 and Austin, TX, in 2012) and Canada (Ottawa, ON, in 1986 and Quebec City, QC, in 2000).

10.2 DISTRIBUTION OF ASR CASES AND REACTIVE AGGREGATES

10.2.1 USA

According to the map (Figure 10.1) presented by Wong (2006), there are only a few states in the U.S.A. that do not have well-known cases of ASR and recent studies conducted by the U.S. Federal Highways Administration (Thomas *et al.*, 2013a, 2013b) have

Figure 10.2 Map Showing Location of ASR-Affected Structures in Canada (modified from CSA A864: 2000).

confirmed cases of ASR in a number of those states. In addition, the authors are aware of cases of ASR in most of the other states that are shown to be apparently free of ASR in Figure 10.1. It is probably reasonable to state that cases of ASR have occurred in most, if not all, of the contiguous states of the U.S.A. Indeed, Stark (2006) reported that *"potentially deleteriously reactive rock types probably exist in most every state in the United States"*. However, to the authors' knowledge there are no confirmed cases of ASR in structures in the non-contiguous states of Hawaii and Alaska. While ASR stemming from aggregate reactivity has not been implicated in Hawaii, a NAVSTA Pearl Harbor bridge is known to have suffered from alkali-silica reaction as a result of "poor silica fume dispersion" (Malvar *et al.*, 2001). The cases of large silica fume agglomerates resulting in ASR-related damage are incredibly rare and generally the addition of silica fume can effectively reduce the potential for damaging ASR.

The authors are not aware of any publications that provide a comprehensive description of the distribution of reactive aggregates in the U.S.A.; however, the following is a brief summary (U.S. Department of the Air Force, 2006):

- Atlantic Seaboard (Maine to Georgia): primarily metamorphic rocks such as gneiss, granite-gneiss, schist, quartzite, metagreywacke, metavolcanics, chert.
- Southern States (Florida to Texas): primarily chert and quartzite with some opaline and chalcedonic carbonates and shales.
- Midwest (Ohio to Minnesota and Missouri): opaline to chalcedonic carbonates, shales, and sandstones.

- Great Plains (North Dakota to Oklahoma and Colorado): opaline to chalcedonic carbonates, shales, and sandstones.
- Basin and Range (Montana to Arizona): glassy to cryptocrystalline rhyolite to andesite volcanics and chert.
- Pacific Coast (Washington to California): glassy to cryptocrystalline rhyolite to andesite volcanics, chert, opaline sedimentary rocks.

This list is not intended to be exhaustive and there are doubtless reactive rocks other than those listed in many of these regions.

10.2.2 Canada

In 2000, in conjunction with the 11th ICAAR (International conference), which was held in Quebec City, the Canadian Journal of Civil Engineering published a special issue, which included seven papers that provide detailed information on the distribution of alkali-silica (and alkali-carbonate) reactive aggregates and AAR-affected structures in Canada on a regional basis. These are: British Columbia (Shrimer, 2000), Prairie Provinces of Alberta, Saskatchewan and Manitoba (Roy & Morrison, 2000), Ontario (Rogers *et al.*, 2000), Quebec (Bérubé *et al.*, 2000), New Brunswick (DeMerchant *et al.*, 2000), Nova Scotia (Langley, 2000) and Newfoundland (Bragg, 2000). Table 10.1 provides a summary of this information and the reader is referred to the individual papers for more comprehensive data.

Table 10.1 Regional Variations in ASR in Canada (modified from CSA A23.1 (2009, now 2014) Appendix B and other sources listed in the table).

Region (Reference)	Observations
British Columbia (Shrimer, 2000; Shrimer *et al.*, 2008)	Most reactive aggregates derived from sand and gravel deposits (quarried sources are less common in B.C.) containing variable amounts of sandstone, quartzite, chert, opal, volcanic rock, granitic and metamorphic rock.
	Local cements historically low in alkali (0.30 to 0.55% Na_2Oeq) resulting in a low incidence of ASR in the province.
	Approximately 90% of B.C. aggregates are identified as being reactive by the AMBT[1] but only 45% by the CPT[2].
Alberta (Scott and Morrison, 2000)	Reactive aggregates include chert, arenite, sandstone, greywacke, cherty sandstone and quartzite.
	Well-documented cases are confined to Southern Alberta but CPT indicates potentially reactive aggregates exist across the province.
Saskatchewan and Manitoba (Scott and Morrison, 2000)	Siliceous (opaline) shale and aggregates derived from the Canadian Shield granitic rocks
	Use of relatively low-alkali cements (0.50 to 0.80% Na_2Oeq) combined with the generally low relative humidity of the region has resulted in relatively low incidence of ASR.
	Gardiner Dam is a notable case where ASR has occurred in concrete containing fly ash of moderate calcium and very high alkali content (see Section 10.7.2)

Table 10.1 (Cont.)

Region (Reference)	Observations
Ontario (Rogers *et al.*, 2000)	Northern Ontario: Palaeozoic cherts in gravels, quarried rhyolitic porphyry, quartz/mica schist and silicified volcanic rocks in gravels. Sandstones, argillites, quartz arenites, quartzites and greywackes of the Huronian Supergroup (includes 'Sudbury gravel' widely used in ASR studies in Canada and worldwide).
	Southern Ontario: quarried (Precambrian) granite and limestones containing chert and finely disseminated silica (includes 'Spratt's limestone' widely used in ASR studies in Canada and worldwide).
	ASR is relatively widespread across the province where moderate to high-alkali cements have been used historically.
	Lower Notch Dam is a notable structure where low-calcium fly ash was used to control expansion with a reactive aggregate and high-alkali cement (see Section 10.7.2).
Quebec (Bérubé *et al.*, 2000)	St. Lawrence Lowlands: siliceous limestones, Potsdam sandstone and greywacke.
	Appalachian Region: rhyolitic tuffs, chloritic schists, phyllites and gravels containing various quantities of greywackes, quartzites, volcanics and metavolcanic rock fragments.
	Laurentian Shield: granitic gneisses, metagreywackes and biotite schists.
	Large number of dams and other hydraulic structures have been affected by ASR. Also damage to highway structures in Montreal and Quebec City is widespread.
New Brunswick (DeMerchant *et al.*, 2000)	Argillites, greywackes, basalts, granites, schists and gneisses.
	ASR-damaged structures located throughout province.
	Large proportion of aggregates are identified as reactive by the AMBT but not by the CPT.
	Mactaquac Dam is a notable ASR-affected structure that has drawn attention from around the world because of extensive remediation work (e.g., slot-cutting) that has been conducted (see Section 10.7.2)
Nova Scotia (Langley, 2000)	Main reactive rock types consist of metamorphosed greywackes, argillites and phyllites and some quartzites, schists and rhyolites.
	Damage due to ASR widespread across the province except in Cape Breton.
	Fly ash (Class F) has been widely used to mitigate ASR expansion with reactive aggregates in the Halifax-Dartmouth area.
Newfoundland (Bragg, 2000) and Labrador	Newfoundland: limestones containing chert nodules and 'quartz eyes', siliceous 'cherty' siltstones and sandstones, argillites, greywackes, arkose, phyllites, gneisses, schists, granites, felsic volcanics, psammites and pelites.
	Labrador: greywackes, metasediments and gneisses

[1] AMBT refers to the accelerated mortar-bar test (CSA A23.2-25A (2014) or ASTM C 1260 (2014)
[2] CPT refers to the concrete prism test (CSA A23.2-14A (2014) or ASTM C 1293 (2008)

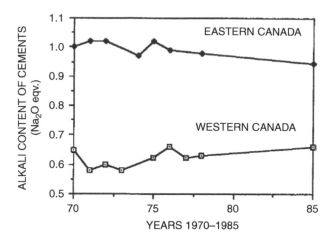

Figure 10.3 Variations in Cement Alkali Levels in Western and Eastern Canada from 1970 to 1985 (from Grattan-Bellew, 1992, using data from the Cement Association of Canada).

Table 10.1 and Figure 10.2 indicate a much greater prevalence of ASR-affected concrete structures in Eastern Canada compared with the B.C. and the Prairie Provinces. This is largely due to the differences in the level of alkali in the cements that have been historically used in these regions. as can be seen from Figure 10.3 (from Grattan-Bellew, 1992). Damage due to ASR is widespread in Eastern Canada and has occurred to a great many dams, other hydraulic structures and highway structures. Together with corrosion of embedded reinforcement and damage caused by cyclic freezing and thawing (especially surface scaling in the presence of deicing salts), ASR is one of the main causes of premature deterioration of concrete in Canada.

10.3 TEST METHODS AND PROTOCOLS FOR IDENTIFYING REACTIVE AGGREGATES

A wide range of test methods has been developed and used for evaluating aggregate reactivity in North America. These include expansion tests performed on mortars or concretes, chemical tests for determining the potential for both ASR (based on the amount of soluble silica and reduction in alkalinity when a ground aggregate sample is exposed to NaOH) and ACR (based on the lime, magnesia and alumina content of the aggregate) and petrographic examination of the aggregate; service history can also be used in lieu of laboratory testing where appropriate. A general discussion of the various test methods that have been used in North America is provided in ACI 201.2R-08 *Guide to Durable Concrete* and ACI 221.1R-98 *State-of-the-Art Report on Alkali-Aggregate Reactivity* (ACI, 1998, 2008). Briefly, the standard tests that have been most widely used at one time or another are described as follows [note that standard tests developed for alkali-carbonate reactivity are not included here (see Chapter 4)]:

Table 10.TB1

Test method	Description
Mortar-bar method (ASTM C227, 2010)	Expansion test with mortar-bars (25 x 25 x 250-mm gauge length) stored over water in sealed containers at 38°C.
	First published by ASTM in 1950 based on the test protocol established in Stanton's landmark paper (Stanton, 1940b)
	The test fails to identify the potential reactivity of many slowly reacting aggregates (such as greywackes and argillites) due to alkali leaching during test.
'Quick chemical' test (ASTM C 289, 2007, but now withdrawn)	Crushed aggregate sample immersed in 1 Molar NaOH solution at 80°C. Solution analysed to determine hydroxide and silica contents. Results plotted on a nomogram of 'reduction in alkalinity' versus 'dissolved silica' to determine potential reactivity.
	First published by ASTM in 1952 based on test developed by U.S. Bureau of Reclamation (Mielenz et al. 1948). The test is generally considered to have poor reliability (ACI 201.2R-08) and has recently been withdrawn.
Petrographic examination (ASTM C 295, 2012, CSA A23.2-15A, 2014)	Potentially reactive components identified and quantified by an experience petrographer.
	First published by ASTM in 1954.
	Although an essential part of any aggregate evaluation, it may be risky to rely solely on a petrographic examination as some reactive phases (e.g. finely disseminated silica in some siliceous limestones) may not be detected during a routine evaluation using optical microscopy.
'Pyrex' mortar-bar test (ASTM C 441, 2011)	ASTM C227 mortar-bar test (see above) using crushed and graded borosilicate glass (e.g. Pyrex) as a standard reactive aggregate.
	Used to evaluate the effectiveness of pozzolans or slag in terms of controlling damaging ASR.
	First approved by ASTM in 1959 based on work by Moran and Gilliland (1950).
	The test does not predict how a pozzolan or slag will perform with different reactive aggregates.
Concrete-prism test (ASTM C 1293, 2008, CSA A23.2-14A, 2014)	Expansion test with concrete prisms (75 x 75 x 250-mm gauge length) stored over water in sealed containers at 38°C.
	First published by CSA in 1986 and by ASTM in 1995.
	Test method considered by many to be the most reliable for identifying aggregate reactivity based on generally good correlation between test and field performance.
Accelerated mortar-bar test (ASTM C 1260, 2014, AASHTO T303, 2008, CSA A23.2-25A, 2014)	Expansion test with mortar-bars (25 x 25 x 250-mm gauge length) stored in 1 Molar NaOH at 80°C.
	First approved by ASTM in 1989 and by CSA in 1990. It is based on the NBRI Accelerated Test Method from South Africa (Oberholster and Davies, 1986).
	Rapid test notorious for producing a large number of "false positives" (i.e. it identifies as deleteriously reactive many aggregates that have good field performance and do not cause expansion in concrete-prism tests). Conversely it is also now documented that this test provides false negatives as well (i.e. it identifies as non-deleteriously-reactive certain aggregates that show poor field performance and/or show reactivity in the more reliable concrete prism test).

In Canada, specifications for minimizing the risk of ASR (including test methods) are covered by national standards (CSA A23.1 and A23.2) and the general framework is provided in CSA A23.2-27A (2014). In the USA specifications have tended to vary regionally with many state highways agencies (and other state or federal agencies) developing different specifications. However, recently the Association of American State Highway Transportation Officials (AASHTO) developed a national specification (or standard practice) for identifying reactive aggregates (and evaluating preventive measures, which is discussed in the next section); this has the designation AASHTO PP065-11 (2011). The development of this standard practice including the rationale behind the selection of tests and performance limits has been described in detail elsewhere (Thomas *et al.*, 2008, 2012a, 2012b).

The AASHTO PP65 practice is based on and evolved from the CSA approach (CSA A23.2-27A) and will be discussed further in this section (and the next section dealing with preventive measures). In both cases, in terms of identifying aggregate reactivity, the procedures consider information from one or more of the following: (i) field performance or service history of the aggregate, (ii) petrographic examination of the aggregate, (iii) laboratory testing of the aggregate using the accelerated mortar-bar test (AMBT) or the concrete prism test (CPT). If the aggregate is a quarried carbonate the potential for alkali-carbonate reactivity is first evaluated using the chemical composition (CaO, MgO and Al_2O_3) of the aggregate. If the chemical composition indicates that there is potential for ACR, the rock must be tested using the concrete prism test (see below).

The general approach of PP65 for evaluating aggregates is laid out in the flow chart presented in Figure 10.4. The aggregate may be accepted based solely on satisfactory service history (field performance) and/or petrographic examination; however, a certain risk is assumed in such cases and it is recommended that aggregate is tested in the accelerated mortar-bar test (AMBT) and/or preferably the concrete prism test (CPT).

The AMBT has been standardized in the USA as AASHTO T303 (2008) and ASTM C1260 (2014), and in Canada as CSA A23.2-25A (2014). In this test mortar-bars ($25 \times 25 \times 250$ mm gauge length) with stainless steel pins for length-change measurements are stored in 1 Molar NaOH solution at 80°C. Typically an expansion limit of 0.10% at 14 days is used as the pass-fail criterion to determine aggregate reactivity, although some agencies use much more stringent criteria such as an expansion limit of 0.08% after 28 days in the hot caustic solution.

The CPT has been standardized as ASTM C1293 (2008) in the USA and as CSA A23.2-14A (2014) in Canada. In this test concrete prisms ($75 \times 75 \times 250$mm gauge length) with stainless steel pins are stored over water in sealed containers at a temperature of 38°C. An expansion limit of 0.040% at 1 year is used to determine aggregate reactivity. There has been some interest in an accelerated version of this test using a storage temperature of 60°C and a concomitantly reduced test duration, but the test has not yet been standardized in North America due to concerns regarding excessive leaching of alkalis and increased silica solubility in certain aggregates (among other

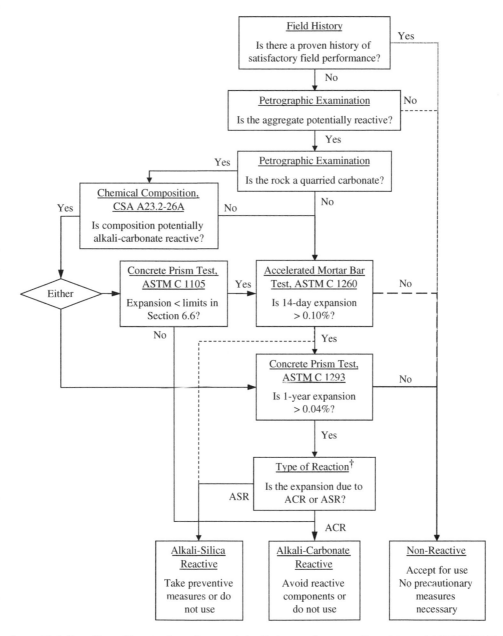

Figure 10.4 Flow Chart Showing Basic Approach for Evaluating Aggregate Reactivity in AASHTO PP65 (2011) and CSA A23.2-27A (2014).

challenges due to the increased temperature) leading to reduced ultimate expansion at the elevated temperature (Ideker *et al.*, 2013).

Note that there are three possible outcomes of the PP65 practice for evaluating aggregates; these are:

Outcome of tests	Appropriate action(s)
Aggregate is **non**-reactive:	Accept for use; no other precautions are necessary
Aggregate is alkali-**silica** reactive:	Reject for use **or** use with suitable preventive measures
Aggregate is alkali-**carbonate** reactive:	Reject for use

The selection of suitable preventive measures for ASR is covered by the second part of the AASHTO PP65 practice (discussed in the next section of this chapter).

In the case of quarried carbonate aggregate, consideration is given to the potential for alkali-carbonate reactivity (ACR) based on the chemical composition of the rock (tested according to CSA A23.2-26A: 2014). If the composition indicates a potential for ACR, the rock must be tested in accordance with the CPT. If the rock results in expansion (> 0.040%) in the CPT, the prism must be examined after test to determine if ACR contributed to expansion; if this is the case the rock must be rejected for use in concrete. If the rock results in expansion in the CPT but no evidence of ACR is detected in a post-test forensic evaluation, the rock is deemed to be alkali-silica reactive and may be used with suitable preventive measures. If the chemical composition indicates that there is no potential for ACR, the rock may be tested using the AMBT and/or CPT and the results need only be evaluated in terms of the risk of ASR.

10.4 PREVENTIVE MEASURES USED IN NORTH AMERICA

Options for preventing or minimizing the risk of alkali-silica reaction that have been used in North America include:

- Use of non-reactive aggregates
- Limiting the alkali content of the concrete (*e.g.*, use of low-alkali cement)
- Use of supplementary cementing materials (SCMs), a.k.a. pozzolans or slag, or blended cements
- Use of lithium-based admixtures

A full discussion on these preventive measures is given in Chapter 4 of this book.

10.5 AASHTO PP65 STANDARD PRACTICE FOR SELECTING PREVENTIVE MEASURES

The recently-developed AASHTO PP65 (2011) 'Standard Practice for Determining the Reactivity of Concrete Aggregates and Selecting Appropriate Measures for Preventing Deleterious Expansion in New Concrete Construction' evolved from CSA A23.2-27A (now 2014) 'Standard practice to identify degree of alkali-reactivity of aggregates and to identify measures to avoid deleterious expansion in concrete' which was first published in 2004. Only the PP65 practice is described here as both protocols are very similar. Note that the PP65 methodology for determining the reactivity of aggregates was covered in Section 10.3. This present section only describes the process for evaluating and selecting the preventive measure once it has been established that the aggregate is alkali-silica reactive.

The PP65 practice includes both (i) a performance-based approach using either the AMBT or the CPT to evaluate the effectiveness of the preventive measure and

(ii) a prescriptive approach for selecting an appropriate level of prevention. The performance-based approach can be used to determine the level of lithium or SCM required with a particular reactive aggregate, whereas the prescriptive approach can be used to select the appropriate maximum alkali content for the concrete or the level of SCM required. In other words, either approach can be used to determine the level of SCM required, but the lithium dose must be determined by performance testing and the maximum alkali content by prescription.

10.5.1 AASHTO PP65 Performance Approach

For the performance approach the AMBT or CPT would be conducted using a range of lithium doses or SCM replacement levels to determine the minimum amount required to control expansion to an acceptable level (*e.g.*, maximum expansion of 0.10% at 14 days in the AMBT or 0.040% at 2 years in the CPT). Although PP65 does include a modified version of the AMBT for evaluating lithium-reactive aggregate combinations, the authors strongly recommend that only the CPT is used for this purpose.

10.5.2 AASHTO PP65 Prescriptive Approach

The prescriptive approach of AASHTO PP65 can be summarized in the following steps (Thomas *et al.*, 2012a):

Step 1. Determine aggregate reactivity class: the aggregate is tested in either the accelerated mortar-bar test (AMBT), ASTM C1260, 2014 (or equivalent AASHTO T303, 2008), or, preferably, the concrete prism test (CPT), ASTM C1293 (2008). The criteria in Table 10.2 are used to classify the aggregate reactivity, which can range from 'R0 – non-reactive' through to 'R3 – very highly reactive'.

Step 2. Determine level of ASR risk: based on the aggregate-reactivity class determined in Step 1 (Table 10.2) and the size and exposure conditions of the concrete under construction, the level of ASR risk is determined using the criteria in Table 10.3. The risk may range from 'Level 1' (lowest or negligible risk) through to 'Level 6' (highest risk).

Step 3. Determine level of prevention: based on the level of ASR risk determined in Step 2 (Table 10.3) and the classification of the structure (see Table 10.5), the level of prevention required is determined using Table 10.4. The level of prevention required may range from 'Level V' (no measures necessary) to 'Level ZZ' (extreme preventive measures necessary).

Step 4. Identification of preventive measures: based on the level of prevention required that was determined in Step 3 (Table 10.4), a number of options are presented as acceptable measures for preventing ASR; these are:
Option 1 – limiting the alkali content of the concrete (Table 10.6)
Option 2 – using supplementary cementing materials, SCMs[1] (Tables 10.7 and 10.8)
Option 3 – limiting the alkali content of the concrete and using SCM (Table 10.9).

---

[1] Note that for Option 2 the minimum amount of SCM determined from Table 10.7 may be adjusted based on the alkali level of the Portland cement using Table 10.8.

Table 10.2 Classification of Aggregate Reactivity, after AASHTO PP65 (2011).

Aggregate-Reactivity Class	Description of Aggregate Reactivity	One-Year Expansion in CPT (%)	14-Day Expansion in AMBT (%)
R0	Non-reactive	≤ 0.04	≤ 0.10
R1	Moderately reactive	> 0.04, ≤ 0.12	> 0.10, ≤ 0.30
R2	Highly reactive	> 0.12, ≤ 0.24	> 0.30, ≤ 0.45
R3	Very highly reactive	> 0.24	> 0.45

Table 10.3 Determining the Level of ASR Risk, after AASHTO PP65 (2011).

Size and exposure conditions	Aggregate-Reactivity Class (Table 10.2)			
	R0	R1	R2	R3
Non-massive[1] concrete in a dry[2] environment	Level 1	Level 1	Level 2	Level 3
Massive[1] elements in a dry[2] environment	Level 1	Level 2	Level 3	Level 4
Concrete exposed to humid air, buried or immersed	Level 1	Level 3	Level 4	Level 5
All concrete exposed to alkalis in service[3]	Level 1	Level 4	Level 5	Level 6

[1] A massive element has a least dimension > 3 ft (0.9 m).
[2] Corresponds to an average ambient relative humidity lower than 60%, normally only found in buildings.
[3] Examples include marine structures exposed to seawater and highway structures exposed to deicing salts (e.g., NaCl) or anti-icing salts (e.g., potassium acetate, potassium formate, sodium acetate, sodium formate, etc.).

Table 10.4 Determining the Level of Prevention, after AASHTO PP65 (2011).

Level of ASR Risk (from Table 10.3)	Classification of Structure (see Table 10.5)			
	S1	S2	S3	S4
Risk Level 1	V	V	V	V
Risk Level 2	V	V	W	X
Risk Level 3	V	W	X	Y
Risk Level 4	W	X	Y	Z
Risk Level 5	X	Y	Z	ZZ
Risk Level 6	Y	Z	ZZ	††

†† It is not permitted to construct a Class S4 structure (see Table 10.5) when the risk of ASR is Level 6. Measures must be taken to reduce the level of risk in these circumstances.
The level of prevention V, W, X, Y, Z or ZZ is used in Tables 10.6 to 10.9.

Although the performance approach allows the option for using lithium compounds as a preventive measure, the prescriptive approach does not. Research has shown that the efficacy of lithium compounds in controlling expansion due to ASR is highly influenced by the nature of the reactive aggregate (Tremblay *et al.*, 2007). Currently, it is not possible to prescribe the required lithium dose based on aggregate reactivity or mineralogy and, consequently, lithium compounds must be tested using the prescriptive approach to determine the minimum dose required with a specific aggregate.

Table 10.5 Structures Classified on the Basis of the Severity of the Consequences Should ASR1 Occur, after AASHTO PP65 (2011) [modified for highway structures from RILEM AAR-7.1, now in Nixon & Sims, 2016].

Class	Consequences of ASR	Acceptability of ASR	Examples[2]
S1	Safety, economic or environmental consequences small or negligible	Some deterioration from ASR may be tolerated	• Non-load-bearing elements inside buildings • Temporary structures (e.g. < 5 years)
S2	Some safety, economic or environmental consequences if major deterioration	Moderate risk of ASR is acceptable	• Sidewalks, kerbs and gutters • Service-life < 40 years
S3	Significant safety, economic or environmental consequences if minor damage	Minor risk of ASR acceptable	• Pavements • Culverts • Highway barriers • Rural, low-volume bridges • Large numbers of precast elements where economic costs of replacement are severe • Service life normally 40 to 75 years
S4	Serious safety, economic or environmental consequences if minor damage	ASR cannot be tolerated	• Major bridges • Tunnels • Critical elements that are very difficult to inspect or repair • Service life normally > 75 years

1 This table does not consider the consequences of damage due to ACR. This practice does not permit the use of alkali-carbonate aggregates.
2 The types of structures listed under each Class are meant to serve as examples. Some owners may decide to use their own classification system. For example, sidewalks or kerbs and gutters may be placed in the Class S3.

Table 10.6 Maximum Alkali Contents in Portland Cement Concrete for Various Levels of Prevention, after AASHTO PP65 (2011).

Prevention Level	Maximum Alkali Content of Concrete (Na_2Oeq)	
	lb/yd^3	kg/m^3
V	No limit	
W	5.0	3.0
X	4.0	2.4
Y	3.0	1.8
Z[1]	Table 10.9	
ZZ[1]		

[1] SCMs must be used in Prevention levels Z and ZZ.

Table 10.7 Minimum Levels of SCM to Provide Various Levels of Prevention, after AASHTO PP65 (2011).

Type of SCM[1]	Alkali level of SCM (% Na$_2$Oeq)	Minimum Replacement Level[3] (% by mass of cementitious material)				
		Level W	Level X	Level Y	Level Z	Level ZZ
Fly ash (CaO ≤ 18%)	≤ 3.0	15	20	25	35	Table 10.9
	> 3.0, ≤ 4.5	20	25	30	40	
Slag	≤ 1.0	25	35	50	65	
Silica Fume[2] (SiO$_2$ ≥ 85%)	≤ 1.0	1.2 x LBA or 2.0 x KGA	1.5 x LBA or 2.5 x KGA	1.8 x LBA or 3.0 x KGA	2.4 x LBA or 4.0 x KGA	

[1] The SCM may added directly to the concrete mixer or it may be a component of a blended cement.
[2] The minimum level of silica fume (as a percentage of cementitious material) is calculated on the basis of the alkali (Na$_2$Oeq) content of the concrete contributed by the Portland cement and expressed in either units of lb/yd^3 (LBA in Table 10.7) or kg/m^3 (KGA in Table 10.7). For example, for a concrete containing 300 kg/m^3 of cement with an alkali content of 0.91% Na$_2$Oeq the value of KGA = 300 x 0.91/100 = 2.73 kg/m^3. For this concrete, the minimum replacement level of silica fume for Level X is 2.5 x 2.73 = 6.8%. Regardless of the calculated value, the minimum level of silica fume shall not be less than 7% when it is the only method of prevention.
[3] The use of high levels of SCM in concrete may increase the risk of problems due to deicer salt scaling if the concrete is not properly proportioned, finished and cured.

Table 10.8 Adjusting the Minimum Level of SCM Based on the Alkali Content in the Portland Cement, after AASHTO PP65 (2011).

Cement Alkalis (% Na$_2$Oeq)	Level of SCM
≤ 0.70	Reduce the minimum amount of SCM given in Table 10.7 by one prevention level[1]
> 0.70, ≤ 1.00	Use the minimum levels of SCM given in Table 10.7
> 1.00, ≤1.25	Increase the minimum amount of SCM given in Table 10.7 by one prevention level
> 1.25	No guidance is given

[1] The replacement levels should not be below those given in Table 10.7 for prevention level W, regardless of the alkali content of the Portland cement.

Table 10.9 Using SCM and Limiting the Alkali Content of the Concrete for Exceptional Levels of Prevention, after AASHTO PP65 (2011).

Prevention Level	SCM as Sole Prevention	Limiting Concrete Alkali Content Plus SCM	
	Minimum SCM Level	Maximum Alkali Content, lb/yd^3 (kg/m^3)	Minimum SCM Level
Z	SCM level shown for Level Z in Table 12.7	3.0 (1.8)	SCM level shown for Level Y in Table 10.7
ZZ	Not permitted	3.0 (1.8)	SCM level shown for Level Z in Table 10.7

10.6 OTHER GUIDELINES FOR PREVENTING ASR

In addition to the national standard practices for identifying reactive aggregates and selecting preventive measures that exist in the USA (AASHTO PP65: 2011) and Canada (CSA A23.2-27A: 2014) guidance on ASR (testing and prevention) can be found in numerous other locations.

For example, Appendix X1 of ASTM C33 (2016) provides some limited guidance on interpretation of the results from ASTM test methods for ASR and measures for mitigating alkali-silica reaction. If the aggregate produces little or no expansion in C1260 (2014) or C1293 (2008), or has a satisfactory service history (with similar cementitious materials), no mitigation is necessary. On the other hand, if the aggregates are considered to be deleteriously alkali-silica reactive, Appendix X1 (section 4.3) recommends one of the following preventive measures:

- Use of ASTM C150 Portland cement meeting the low-alkali option ($\leq 0.60\%$ Na_2Oeq)
- Use of ASTM C595 (2016) blended cement meeting the optional mortar-bar expansion requirement
- Use of hydraulic cement meeting the ASTM C1157 (2011) performance specification, including 'Option R' – *Low Reactivity with Alkali-Reactive Aggregates.*
- Use of pozzolans or slag meeting the optional requirements of the relevant material specifications [C618 (2015) for fly ash and natural pozzolans, C1240 (2015) for silica fume and C989 (2014) for slag] for preventing excessive expansion due to ASR.

The optional requirements for blended cements, C1157 (2011) cements, pozzolans and slag all make use of the C441 (2011) 'Pyrex' mortar-bar method for demonstrating the effectiveness in controlling expansion due to ASR. However, each material specification has different performance requirements and these may be summarized as follows:

Table 10.10

ASTM Specification	Expansion Limit of Mortars with Pyrex Glass
C595 (2016) Blended cements	Maximum expansion of 0.020% at 14 days **and** 0.060% at 56 days
C1157 (2011) Hydraulic cements (performance-based specification)	Maximum expansion of 0.020% at 14 days
C618 (2015) Fly ash and natural pozzolans	Expansion of fly ash mortars not greater than expansion of control mortars with low-alkali cement ($\leq 0.60\%$ Na_2Oeq) at 14 days
C989 (2014) Slag	Expansion of the job cement plus slag should not expand by more than 0.02% at 14 days or, if the job cement is not known, the slag should reduce the 14-day expansion of a mixture with high-alkali cement by at least 75% when compared with a mix with high-alkali cement on its own.
C1240 (2015) Silica fume	Blend of high-alkali cement plus silica fume must reduce expansion by at least 80% compared with high-alkali cement alone

A joint C09/C01 Task Group recently recognized that numerous ASTM specifications provide requirements and guidance for avoiding deleterious ASR expansion for individual concrete materials (*e.g.*, cementitious materials, aggregates and SCMs), whilst, together, these specifications do not provide coherent guidance for preventing deleterious expansion in concrete; also, the specifications utilize standard test methods that are in many cases considered to be unreliable. The Task Group thus recommended that clear and consistent guidance/specification for the prevention of ASR be developed that addresses performance at the concrete level and includes requirements for aggregates. A joint ASTM technical subcommittee, C01/C09-50, was then formed to develop new global requirements for the Risk Management of Alkali-Aggregate Reactions. This new subcommittee started its activities in 2010 and decided to adopt and modify the AASHTO PP65 (2011) Standard Practice. This was published as ASTM C1778 'Standard Guide for Reducing the Risk of Deleterious Alkali-Aggregate Reaction in Concrete' in 2014 (ASTM, 2014).

ACI 301 (2010) *'Specifications for Structural Concrete'* states that potentially reactive aggregates may be used either with low-alkali cement ($\leq 0.60\%$ Na_2Oeq) or supplementary cementing materials *"in an amount shown to be effective in preventing harmful expansion due to alkali-aggregate reaction in accordance with ASTM C441 (2011)"* and defers to ASTM C33 (2016). ACI 318 (2008) *'Building Code Requirements for Structural Concrete'* requires aggregates to meet the requirements of ASTM C33 (2016) but makes no reference to alkali-aggregate reactions (ASR or ACR) or the use of deleteriously reactive aggregates in concrete. Guidance for preventing damage due to AAR is provided in ACI 201.2R (2008) *'Guide to Durable Concrete'* and ACI 221.1R (1998) *'Report on Alkali-Aggregate Reactivity'*.

10.7 SOME NOTABLE CASES OF ASR IN NORTH AMERICA

As discussed earlier in this chapter, cases of ASR have occurred in all 48 contiguous states of the USA and all 10 Provinces of Canada and there are literally thousands of ASR-affected structures including dams, other hydraulic structures, bridges, tunnels, airfield and highway pavements, sidewalks (pedestrian pavements), kerbs and gutters, marine structures, retaining walls, foundations and other types of elements. This section only illustrates a few cases of particular interest.

10.7.1 Cases in the USA

Parker Dam

Shortly after Stanton's discovery of ASR (Stanton, 1940), the United States Bureau of Reclamation (USBR) diagnosed the mechanism as the cause of expansion and cracking in the Parker Dam (Meissner, 1941). The Parker Dam spans the Colorado River between Arizona and California and was constructed between 1934 and 1938. What makes this structure particularly noteworthy is that not only was it one of the first major structures to be diagnosed with ASR worldwide, but it is still in use today after 80 years (Figure 10.5). *"Growing cracks"* were observed just 2 to 3 years after the dam was complete and an early investigation showed that the growth

Figure 10.5 Parker Dam in 2013 showing map cracking (right).

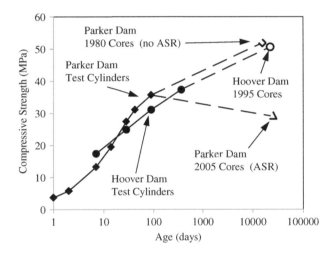

Figure 10.6 Effect of ASR on Strength of Cores in Parker Dam (produced using data from Dolen, 2005).

was due to the newly-discovered cement-aggregate interaction probably involving andesitic aggregate; in addition concrete cores taken from the dam showed inferior compressive strength and low modulus of elasticity (Meissner, 1941).

Figure 10.6 shows data from the USBR (Dolen, 2005) for the Hoover and Parker Dams comparing strength data from cylinders cast during construction and cores taken at 60 years or later. Cores taken from parts of the Parker Dam that were constructed with low-alkali cement and have no ASR show similar strength development as the Hoover Dam. However, cores taken from ASR-affected concrete that was constructed with high-alkali cement show a much reduced strength. Fortunately, continued monitoring of the dam indicated that the expansive ASR process terminated at some point in time and currently there are no progressive movements occurring. This is contrary to some other cases where ASR expansion has not been observed to slow down after numerous decades (see case history for Mactaquac Dam below).

Figure 10.7 Davis Dam in 2013 (no evidence of ASR).

Interestingly, the same materials (cement and aggregate) were used to build the nearby Gene Wash and Copper Basin arch dams as part of the Colorado River Aqueduct System and these structures also show ASR, but the Intake and Gene Pumping Plants of the same system which were built with the same aggregate but low-alkali cement (0.38% Na$_2$Oeq) do not (Tuthill, 1982).

Davis Dam

The Davis Dam spans the Colorado River between Arizona and Nevada, and is approximately 75 miles upstream of the Parker Dam. The concrete structures which include spillway, gravity structure, intake structure and powerhouse were built between 1946 and 1950 using aggregate that *"is similar to that used in the Parker Dam"* (Anon, 1949). Petrographic examination showed 45% of the coarse aggregate to be deleteriously reactive, which was confirmed by expansion testing of mortar-bars (Gilliland & Moran, 1949). To prevent deleterious reactions from occurring in the concrete structures, a *"siliceous admixture"*, namely a calcined shale from the Puente formation in the vicinity of Pomona, California, was used at a cement replacement level of approximately 20% (by volume). Three activity tests were used to evaluate the calcined material; these were: (i) a lime-pozzolan strength test (minimum mortar strength of 600 psi, 4 MPa, after 1 day at 70°F and 6 days at 130°F), (ii) a reduction-in-alkalinity test (24 hours in NaOH at 80°C), and (iii) a reduction-in-expansion test (minimum 75% reduction in the 14-day expansion of mortar-bars produced with Pyrex aggregate and high-alkali cement). A recent visit to the structure by one of the authors in 2013 (see Figure 10.7) confirmed the long-term effectiveness of this preventive measure as there was no evidence of damage due to ASR.

Bibb Graves Bridge

The Bibb Graves Bridge, built in 1931 across the River Coosa in the town of Wetumka, AL, is a reinforced concrete parabolic arch structure with a suspended roadway, with a total of seven arches supporting the roadway as shown in

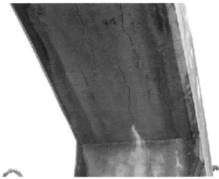

Figure 10.8 Bibb Graves Bridge in 2010; cracking in the underside of one of the arches (right).

Figure 10.8. ASR was detected in some of the concrete components of the bridge as early as 1956 and the occurrence of the reaction is surprising considering that the bridge was reportedly (Hester and Smith, 1956) constructed using cement with a very low-alkali cement (range 0.21 to 0.28% with an average of 0.25% Na_2Oeq). Detailed inspections of the structure in 2005 and 2009 (Thomas *et al.*, 2013a, 2013b) revealed that significant cracking was limited to a single concrete arch and only those parts of the arch above the bridge deck; the remaining components of the structure showed very little or no signs of ASR, with only very minor cracking after more than 70 years of service.

This raised the hypothesis that a few concrete loads might have been supplied with a cement with a higher alkali content; however, chemical analysis of the concrete revealed a similar and low water-soluble alkali content in both the ASR-affected an non-damaged concrete. The explanation for the occurrence of ASR in just a few isolated components and its absence in the remaining concrete, which apparently used the same materials and is exposed to the same conditions, remains a mystery at this time.

One of the alarming features of this case is that the expansion rate for the ASR-affected concrete between 2005 and 2012 was estimated to be as high as 200 micro-strain/year (0.02%/y). Efforts were made to retard the rate of expansion of the ASR-affected arch by treating the concrete in an attempt to reduce the internal relative humidity (RH); the treatment consisted of sealing all open cracks with a flexible waterproof caulk, sealing the upper surfaces of the arch with a low-viscosity epoxy "flood-coat" and "sealing" the remaining three sides of the arch (vertical surfaces and underside) with a "breathable" silane (Thomas *et al.*, 2013). Early post-treatment monitoring (up to 30 months) indicates that the treatment has had little impact on the internal RH or the expansion rate; longer-term monitoring is ongoing.

Manette Bridge

The Manette Bridge, located in Kitsap County, Washington (west of Seattle), is a concrete pier and steel truss bridge constructed in 1930. In 1949 the original wooden deck and timber trusses were replaced with concrete and steel (Washington DOT,

2008). In 2011 this bridge was removed from service and subsequently deconstructed. A new bridge was built immediately adjacent and completed in February of 2012 (Washington DOT, 2008, update accessed 2014).

The main concerns were structural deficiencies in the concrete piers and structural steel trusses. These were elements built in 1930. From 1976 through 2003 concrete samples were extracted from the foundations of the bridge (main piers) and it was determined that alkali-silica reaction was the cause for deterioration. The foundations generally exhibited signs of cracking, leaching from cracks, rust stains (indicative of reinforcing steel corrosion), delaminations, soft concrete and formwork holes. Some exposed reinforcement bars were also present. In 1949 (Piers 4 and 6), 1991 (Pier 5) and 1996 (Piers 4 and 6) were repaired in an attempt to 'encase' the concrete. However, these repairs were deemed ineffective since the core of the concrete continued to suffer from ASR. An assessment in 1993 by Washington DOT showed that there was no practical means to restore or prevent ASR in the concrete column and footing concrete. As a result the bridge was scheduled for replacement which was completed in 2012.

10.7.2 Cases in Canada

Mactaquac Dam

Mactaquac Generating Station is located on the St. John River approximately 20 km west of the city of Fredericton in the province of New Brunswick (Hayman *et al.*, 2010). The concrete structures, which consist of an intake structure, a powerhouse and two spillways, started to exhibit distress, which was subsequently attributed to ASR, approximately 10 years after construction (Figure 10.9). This case is of particular interest because of the substantial growth that has been observed, the extent of the remediation work necessitated by ASR, and the severe reduction to the expected life of the structure. To date, the height of the intake structure is estimated to have increased by more than 175 mm since construction and an estimated 450 mm of concrete has been removed by repeated slot cutting perpendicular to the longitudinal axis of the same structure.

The unrestrained expansion of the concrete is currently estimated to be between 120 and 150 microstrain per annum ($\mu\varepsilon/y$); there is no indication that the expansion rate is decreasing as the structure ages. Because of continued operational problems and escalating maintenance costs, current projections are that the concrete structures (approx. 500,000 m^3) will have to be replaced by 2030 (to coincide with the projected

Figure 10.9 Mactaquac Dam in 2008 showing map cracking (right).

Figure 10.10 Lower Notch Dam in 2011 (no evidence of ASR).

replacement of the turbines and generators) after a service life of just over 60 years. This replacement is also of interest and economy dictates that the same aggregate (*i.e.*, the greywacke/argillite rock excavated during construction of the intake channel) will have to be used again if at all possible. Currently extensive testing is being conducted to evaluate various measures to prevent deleterious reaction with this rock aggregate (Hayman *et al.*, 2010).

Lower Notch Dam

Construction of the Lower Notch Dam (Figure 10.10), located on the Montreal River just before it flows into Lake Timiskaming south of New Liskeard, Ontario, was completed in 1969. This case is of interest as the reactivity of the aggregate was well known at the time of construction and a decision was made to prevent expansion by using Class F fly ash (20 and 30% for structural and mass concrete, respectively) in combination with a high-alkali cement (as high as 1.08% Na_2Oeq). This case has been well documented (Thomas, 1996; Thomas *et al.*, 2012c) and a recent visit by one of the authors confirmed that there is no damage due to ASR in evidence after 40 years.

Gardiner Dam

The Gardiner Dam was constructed between 1958 and 1968, and is situated on the South Saskatchewan River in Saskatchewan. The concrete spillway piers are exhibiting

Figure 10.11 Gardiner Dam Spillway in 2013 showing map cracking (right).

minor to moderate map cracking due to ASR (see Figure 10.11). The concrete contains 25% fly ash (Roy, 2000); however, the fly ash used was of moderate calcium content (13% CaO) and very high alkali content (7–8 % Na_2Oeq). This case serves to demonstrate that the incorporation of fly ash in the concrete may not always be foolproof; it is necessary to use a sufficient quantity of a fly ash with the appropriate composition (low to moderate calcium and alkali content) for effective control.

REFERENCES

AASHTO PP65 (2011) *Standard practice for determining the reactivity of concrete aggregates and selecting appropriate measures for preventing deleterious expansion in new concrete construction.* Washington DC, USA, American Association of State and Highway Transportation Officials. 20p.

AASHTO T303 (2008) *Standard method of test for accelerated detection of potentially deleterious expansion of mortar-bars due to alkali-silica reaction.* Washington DC, USA, American Association of State and Highway Transportation Officials. 6p.

American Concrete Institute. ACI 221.1R-08. (2008) *Report on Alkali-Aggregate Reactivity.* Farmington Hills, MI, USA, ACI.

American Concrete Institute. ACI 201.2R-08. (2008) *Guide to durable concrete.* Farmington Hills, MI, USA, ACI.

American Concrete Institute. ACI 301-10. (2010) *Specifications for structural concrete.* Farmington Hills, MI, USA, ACI.

American Concrete Institute. ACI 318-08 (2008) *Building Code Requirements for Structural Concrete.* Farmington Hills, MI, USA, ACI.

Anon. (1949) Admixture combats alkali reaction in Davis Dam concrete. *Eng News-Rec., 20* (January), 83–85.

ASTM C33 (2016) *Standard specification for concrete aggregates.* West Conshohocken, USA, American Society for Testing and Materials.

ASTM C227 (2010) *Standatd test method for potential alkali-silica reactivity of aggregates (mortar-bar method).* West Conshohocken, USA, American Society for Testing and Materials.

ASTM C289 (2007) (now withdrawn). *Standard test method for potential alkali-silica reactivity of aggregates (chemical method).* West Conshohocken, USA, American Society for Testing and Materials.

ASTM C295 (2012) *Standard guide for petrographic examination of aggregates for concrete.* West Conshohocken, USA, American Society for Testing and Materials.

ASTM C441 (2011) *Standard test method for effectiveness of pozzolans or ground blast-furnace slag in preventing excessive expansion of concrete due to the alkali-silica reaction.* West Conshohocken, USA, American Society for Testing and Materials.

ASTM C595 (2016) *Standard specification for blended hydraulic cements.* West Conshohocken, USA, American Society for Testing and Materials.

ASTM C618 (2015) *Standard specification for coal fly ash and raw or calcined natural pozzolan for use in concrete.* West Conshohocken, USA, American Society for Testing and Materials.

ASTM C989 (2014) *Standard specification for slag cement for use in concrete and mortars.* West Conshohocken, USA, American Society for Testing and Materials.

ASTM C1157 (2011) *Standard performance specification for hydraulic cement.* West Conshohocken, USA, American Society for Testing and Materials.

ASTM C1240 (2015) *Standard specification for silica fume used in cementitious mixtures.* West Conshohocken, USA, American Society for Testing and Materials.

ASTM C1260 (2014) *Standard test method for potential alkali-reactivity of aggregates (mortar-bar method).* West Conshohocken, USA, American Society for Testing and Materials.

ASTM C1293 (2008) *Standard test method for determination of length change of concrete due to alkali-silica reaction.* West Conshohocken, USA, American Society for Testing and Materials.

ASTM C1778 (2014) *Standard Guide for Reducing the Risk of Deleterious Alkali-Aggregate Reaction in Concrete.* West Conshohocken, USA, American Society for Testing and Materials.

Bérubé, M-A., Durand, B., Vézina, D., & Fournier, B. (2000) Alkali–aggregate reactivity in Québec (Canada). *Can J Civil Eng.,* 27, 226–245.

Bragg, D. (2000) Alkali–aggregate reactivity in Newfoundland, Canada. *Can J Civil Eng.,* 27, 192–203.

CSA A23.1 (2014) *Concrete materials and methods of concrete construction.* Mississauga, Ontario, Canada, Canadian Standards Association.

CSA A23.2 (2014) *Test methods and standard practices for concrete.* Mississauga, Ontario, Canada, Canadian Standards Association.

CSA A23.2-14A (2014) *Potential expansivity of aggregates (procedure for length change due to alkali-aggregate reactions in concrete prisms at 38°C).* Mississauga, Ontario, Canada, Canadian Standards Association. pp. 350–362.

CSA A23.2-15A (2014) *Petrographic examination of aggregates.* Mississauga, Ontario, Canada, Canadian Standards Association.

CSA A23.2-25A (2014) *Test method for detection of alkali-silica reactive aggregate by accelerated expansion of mortar bars.* Mississauga, Ontario, Canada, Canadian Standards Association. pp. 425–433.

CSA A23.2-27A (2014) *Standard practice to identify degree of alkali-reactivity of aggregates and to identify measures to avoid deleterious expansion in concrete.* Mississauga, Ontario, Canada, Canadian Standards Association. pp. 439–451.

CSA A864-00 (2000) *Guide to the evaluation and management of concrete structures affected by alkali-aggregate reaction.* Mississauga, Ontario, Canada, Canadian Standards Association.

DeMerchant, D.P., Fournier, B., & Strang, F. (2000) Alkali–aggregate research in New Brunswick. *Can J Civil Eng,* 27, 212–225.

Dolen, T.P. (2005) *Materials Properties Model of Aging Concrete.* Report DSO-05-05, Bureau of Reclamation, Denver, Colorado.

Gilliland, J.L. & Moran, W.T. (1949) Siliceous admixture for the Davis Dam. *Eng News-Rec.*, 3 (February), 62–64.

Grattan-Bellew, P.E. (1992) Alkali-silica reaction – Canadian experience. In: Swamy, R.N. (ed.) *The alkali-silica reaction in concrete.* Blackie, Glasgow, chapter 8. pp. 223–248.

Hansen, W.C. (1944) Studies relating to the mechanism by which the alkali-aggregate reaction proceeds in concrete. *J Am Concrete I.*, 15 (3), 213–227.

Hayman, S., Thomas, M., Beaman, N., & Gilks, P. (2010) Selection of an effective ASR-prevention strategy for use with a highly reactive aggregate for the reconstruction of concrete structures at Mactaquac generating station. *Cement Concrete Res.*, 40, 605–610.

Hester, J.A. & Smith, O.F. (1956) *The alkali aggregate phase of chemical reactivity in concrete: Part II Deleterious reactions observed in field concrete structures.* Bureau of Materials and Tests of the Alabama Highway Department Research Report, Montgomery, July 1956.

Ideker, J.H., East, B.L., Folliard, K.J., Fournier, B., & Thomas, M.D.A. (2010) The current state of the accelerated concrete prism test. *Cement Concrete Res.*, 40 [4], April, 550–555.

Langley, W.S. (2000) Alkali–aggregate reactivity in Nova Scotia. *Can J Civil Eng*, 27, 204–211.

Malvar, L.J., Cline, G.D., Burke, D.F., Rollings, R., Sherman, T., & Greene, J. (2001) *Alkali-Silica Reaction Mitigation: State-of-the-Art*, Naval Facilities Engineering Service Center TR-2195-SHR, 40p.

Meissner, H.S. (1941) Cracking in concrete due to expansive reaction between aggregate and high-alkali cement as evidenced in Parker Dam. *Proceedings, Am. Concrete Inst.*, 57, pp. 549–568.

Mielenz, R.C., Greene, K.T., & Benton, E.J. (1948) Chemical test for reactivity of aggregates with cement alkalies: Chemical processes in cement-aggregate reaction. *Proceedings, Am. Concrete Inst.*, 44, 193.

Moran, W.T. & Gilliland, J.L. (1950) *Summary of Methods for Determining Pozzolanic Activity. Symposium on Use of Pozzolanic Materials in Mortars and Concretes*, ASTM STP 99, ASTM International, p. 109.

Nixon, P.J. & Sims, I. (eds.) (2016) RILEM recommendations for the prevention of damage by alkali-aggregate reactions in new concrete structures. *RILEM State-of-the-Art Reports*, Volume 17, Springer, Dordrecht, ppRILEM, Paris, 168p.

Oberholster, R.E. & Davies, G. (1986) An accelerated method for testing the potential alkali reactivity of siliceous aggregates. *Cement Concrete Res.*, 16, 181–189.

Powers, T.C. & Steinour, H.H. (1955a) An investigation of some published researches on alkali-aggregate reaction. I. The chemical reactions and mechanism of expansion. *J Am Concrete I.*, 26 (6), 497–516.

Powers, T.C. & Steinour, H.H. (1955b) An interpretation of some published researches on the alkali-aggregate reaction. Part 2: a hypothesis concerning safe and unsafe reactions with reactive silica in concrete. *J Am Concrete I.*, 26 (8), 785–811.

Rogers, C.A., Grattan-Bellew, P.E., Hooton, R.D., Ryell, J. & Thomas, M.D.A. (2000) Alkali–aggregate reactions in Ontario. *Can J Civil Eng.*, 27, 246–260.

Roy, S.T.R. (2000) The identification of alkali–aggregate reaction at the Gardiner Dam, Outlook, Saskatchewan. In: *Proceedings of the 11th International Conference on Alkali–Aggregate Reaction in Concrete*, Quebéc, Qué.

Roy, S.T.R. & Morrison, J.A. (2000) Experience with alkali–aggregate reaction in the Canadian prairie region. *Can J Civil Eng.*, 27, 261–276.

Shrimer, F.H. (2000) Experience with alkali–aggregate reaction in British Columbia. *Can J Civil Eng.*, 27, 277–293.

Shrimer, F.H., Briggs, A., & Hudson, B. (2008) Alkali-Aggregate Reaction in Western Canada: Review of Current Trends. In: *Proceedings 12th International Conference on Alkali-Aggregate Reactions (ICAAR)*, Trondheim, Norway. pp. 32–41.

Stanton, T.E. (1940a) Influence of cement and aggregate on concrete expansion. *Eng News-Rec*, *1*, February, 59–61.

Stanton, T.E. (1940b) Expansion of concrete through reaction between cement and aggregate. *Pro. ASCE*, *66* (10), 1781–1811.

Stark, D. (2006) Alkali-silica reactions in concrete. In: Lamond, J.F., & Pielert, J.H. (eds.) *Significance of Tests and Properties of Concrete and Concrete-Making Materials (Chapter 34)*, ASTM STP-169D. West Conshohocken, PA, American Society of Testing and Materials. pp. 401–409.

Swenson, E.G. (1957a) Cement aggregate reaction in concrete of a Canadian bridge. *ASTM Pro*, *57*, 1043–1056.

Swenson, E.G. (1957b). A reactive aggregate undetected by ASTM tests. *ASTM Proc*, *57*, 48–51.

Thomas, M.D.A. (1996) Field Studies of Fly Ash Concrete Structures Containing Reactive Aggregates. *Mag Concrete Res.*, *48* (177), December, 265–279.

Thomas, M.D.A., Fournier, B., & Folliard, K.J. (2008) *Report on determining the reactivity of concrete aggregates and selecting appropriate measures for preventing deleterious expansion in new concrete construction*. Federal Highways Administration, Report FHWA-HIF-09-001, National Research Council, Washington D.C.

Thomas, M.D.A., Folliard, K.J., Fournier, B. & Ahlstrom, G. (2012a) AASHTO Standard Practice for prevention of AAR. In: *Proceedings of the 14th International Conference on Alkali-Aggregate Reactions (ICAAR)*, Austin, TX, May 2012.

Thomas, M.D.A., Fournier, B., & Folliard, K.J. (2012b) Selecting Measures to Prevent Deleterious Alkali-Silica Reaction in Concrete: Rationale for the AASHTO PP65 Prescriptive Approach. FHWA-HIF-13-002, Federal Highway Administration, Washington, DC.

Thomas, M., Hooton, R., Rogers, C., & Fournier, B. (2012c) 50 years old and still going strong: Fly ash puts paid to ASR. *Concr. Int.*, January, 35–40.

Thomas, M.D.A., Folliard, K.J., Fournier, B., Rivard, P., & Drimalas, T. (2013a) *Methods for Evaluating and Treating ASR-Affected Structures: Results of Field Application and Demonstration Projects: Volume I: Summary of Findings and Recommendations*. FHWA-HIF-14-0002, Federal Highway Administration, Washington, DC.

Thomas, M.D.A., Folliard, K.J., Fournier, B., Rivard, P., Drimalas, T., & Garber, S.I. (2013b) *Methods for Evaluating and Treating ASR-Affected Structures: Results of Field Application and Demonstration Projects: Volume II: Details of Field Applications and Analysis*. FHWA-HIF-14-0003, Federal Highway Administration, Washington, DC.

Tremblay, C., Berube, M-A., Fournier, B., Thomas, M.D.A., & Folliard, K.F. (2007) Effectiveness of lithium-based products in concrete made with Canadian reactive aggregates. *ACI Mater J*, *104* (2), 195–205.

U.S. Department of the Air Force (2006) Alkali-Aggregate Reaction in Portland Cement Concrete (PCC) Airfield Pavements, Engineering Technical Letter (ETL) 06-2. Available from: http://www.wbdg.org/ccb/AF/AFETL/etl_06_2.pdf [Accessed 14th March 2014].

Wong, G.S. (2006) Petrographic evaluation of concrete aggregates. In *Significance of Tests and Properties of Concrete and Concrete-Making Materials* (chapter 33), ASTM STP-169D. West Conshohocken, PA, American Society of Testing and Materials. pp. 377–400.

Washington Department of Transportation (2008) Manette Bridge – Condition Summary, March 2008 Update.doc. Available from: http://www.wsdot.wa.gov/NR/rdonlyres/7DC4C58C-E45E-444D-8267-4D44A791FC70/0/Mane. [Accessed 8th April 2014].

Chapter 11

South and Central America

Eduardo M. R. Fairbairn

11.1 INTRODUCTION

This chapter presents an overview of Alkali-Aggregate Reaction in South and Central America. To accomplish this task, a thorough literature search was carried out. Several sources (in English, French, Portuguese and Spanish) have been consulted, including journal papers, international, regional and local conferences, workshops and symposia, theses, bulletins of professional associations and engineering institutes. This in-depth research indicated that, in the region covered by this Chapter, no incidences of AAR have so far been reported outside Argentina, Brazil, Mexico and Uruguay.

For some countries there is explicit reference to the absence of constructions affected by AAR. This is the case of Chile (Gonzales, 2013), Bolivia (Cenzano *et al.*, 2011) and Peru (Cotera, 1991). For other countries, such as Ecuador (García, 2013), Colombia (Bolivar, 2003) and Paraguay (Acosta *et al.*, 2007), although there is no case of AAR reported, academic researches have detected the potential to develop the reaction by means of laboratory tests. Potential reactive aggregates have also been reported in the Dominican Republic (hydropower plant Palomino) and Panama (hydropower plant Proyecto dos Mares), where silica fume has been used to inhibit the development of AAR (Tecnosil, 2009).

As the lack of reported AAR cases may appear to be paradoxical, some of the most expressive Latin American consultants and researchers have been contacted: Francisco Andriolo, Nicole P. Hasparyk, Roberto Torrent, Selmo C. Kuperman, Silvina Marfil, Francisco G. Holanda and Walton Pacelli de Andrade. They confirmed the lack of reported cases of AAR for several countries in the continent. However, there is a consensus view that AAR-affected structures are likely to exist in several of the countries although these cases have not been identified or reported.

Therefore, in the following sections, the overview of AAR cases and studies has been confined to Argentina, Brazil, Mexico and Uruguay and these will be described in greater detail.

11.2 ARGENTINA

Research on the alkali-silica reaction (ASR) in Argentina started in the 1950s by a study on the pavement of a route in the Province of Buenos Aires and since then the subject has been addressed by a number of researchers (Marfil & Maiza, 2008). Furthermore,

Maiza *et al.* (1999) reported cases of four deteriorated pavements in the region near Cordoba in north-eastern Argentina.

More recently, the most significant cases reported in the specialized literature occurred in concrete pavements (Marfil & Maiza, 2001; Maiza *et al.*, 2011); there is also the occurrence of AAR in a building (Marfil & Maiza, 2008) and in an overpass (Priano *et al.*, 2012). All of these cases occurred in the provinces of Cordoba, Corrientes and Buenos Aires, in north-eastern Argentina.

The pavements are: (a) a highway from the Province of Córdoba, built in 1998; (b) a road in the north of the Province of Buenos Aires built in 1983; (c) an urban road surface in the city of Bahía Blanca (in the south of the Province of Buenos Aires) built in 1986; (d) three concrete pavements situated in the Province of Corrientes, all built before 1999. The building, from 1935, is located in the city of Buenos Aires, and the overpass in Bahia Blanca.

In the example of the highway of Cordoba (a): the coarse aggregate consists mainly of quartz rocks which contain microcrystalline quartz, while the fine aggregate contains unweathered volcanic glass as deleterious material. The road in Buenos Aires (b) has a coarse aggregate composed of granitic migmatite, quartzites, and schists and the fine aggregate contains quartz with undulatory extinction. The city pavement (c) had quartzites and granitic rocks predominant in the coarse aggregate and sand with a high content of glassy volcanics. For the cases (a), (b) and (c), microscopical studies of thin sections showed widespread microcracking that affected mainly the mortar matrix and some reactive aggregate particles. For these three pavements the thin-sections show widespread microcracking. Both AAR gel and crystalline reaction products were identified at the aggregate–mortar interface and inside air voids. Ettringite was also identified in the concrete samples studied, but the authors of the report (Marfil & Maiza, 2001) stressed that its significance with respect to the observed damage was unknown.

The north-eastern region called 'Mesopotamia Argentina' is a humid and verdant land between the rivers Paraná and Uruguay. Its basalts are one of the largest sources for the production of coarse aggregates in the region, mainly in the provinces of Corrientes and Misiones. Maiza *et al.* (2011) reported degradation by AAR in a road, an urban street and an access taxiway that connects the parking of aircraft to the runaway at the airport. The assessment of the reaction comprised petrographic analysis, determination of concrete composition, plus physical and mechanical properties such as density, absorption, porosity and Young´s modulus.

It was shown that the basaltic coarse aggregates from Mesopotamia Argentina, with petrographic characteristics similar to those employed in the analysed pavements, are capable of reacting with alkalis originated from cement, producing deleterious phenomena of sufficient magnitude seriously to affect the structural integrity and durability of the concrete pavements.

It should be noted that a comprehensive study of potential reactive aggregates from the Provinces of Corrientes and Entre Rios was published by Marfil *et al.* (2010). The petrographic mineralogical study showed that all the samples of basaltic rock contain materials capable of reacting with alkalis, such as volcanic glass, microcrystalline silica, and clay from the montmorillonite group. The laboratory tests displayed expansions largely beyond the prescribed limits, indicating the reactive potential of the aggregates.

Besides the AAR in the pavements in Argentina, there was one case of a building affected by the alkali-silica reaction in the city of Buenos Aires (Marfil & Maiza, 2008).

The reaction was identified because the columns of the building presented significant cracking, putting the structure at risk. Reaction products were identified mainly in the reactive aggregate-cement paste interfaces. Ettringite was common inside entrapped air voids and in microcracks, sometimes associated with aluminosilicates. It was found that the coarse aggregate was mostly reactive due to the presence of oal, sandstones with cryptocrystalline and amorphous silica cement, chalcedony, tridymite and strained quartz and that the fine aggregate also contained the same mineralogical species.

A comprehensive study encompassing several structures in the city of Bahia Blanca and surroundings has been carried out by Priano et al. (2012). Eleven structures have been studied. The materials used as aggregate had generally similar lithological compositions. Coarse aggregates were 50% granitic crushed stone and 50% polymictic rolled pebbles. The composition of the latter was dominated by volcanic rocks (andesites, rhyolites and tuffs), most with vitreous matrices that were generally altered (devitrified), and to a lesser extent quartz, granitic rocks and metamorphic rocks. Fine aggregate in all cases were natural sand, though the source varies (wind, river and marine). The lithological composition of these sands is similar and in turn similar to the rolled pebbles coarse aggregate. Although the aggregates can be considered as potentially reactive from the petrographic point of view, only three of the structures were found to be affected by alkali-silica reaction: two pavements and an overpass. This is due to the fact that the development of the reaction depends also on two other factors: high contents of alkalis and moisture.

10.3 BRAZIL

Brazil is the largest country in South America (8,500,000 km^2), corresponding to half of the continent's area. Most of the electrical energy in the country comes from hydropower. The installed generating capacity is 128,500 MW, being 86,700 MW (67%) from hydropower (ANEEL, 2014). As 29% of the system corresponds to thermal electric plants, whose operation depends on the level of the hydroelectric reservoirs, the hydroelectric power accounts, on average, for about 85% of the electrical energy supply, depending on the season.

The Brazilian Chapter of ICOLD lists about 730 dams higher than 15 m (CBDB, 2014), and Andriolo (2000) reported that about 63,000,000 m^3 of concrete had been used to build hydroelectric structures in the 20th century. These figures show the importance that the durability of concrete used in HPP has for the economy of the country. This also explains why alkali-aggregate reaction had been reported over a long period only in hydraulic structures in Brazil.

A summary review of Brazilian dams higher than approximately 15 m affected by AAR is given in Table 11.1. The occurrence of the reaction is reported for more than 30 hydro-electric plants. These structures are extremely important and strategic for the infrastructure of the country. Besides flood control and water supply for metropolitan São Paulo (20,000,000 inhabitants) and for north-eastern regions subject to cyclical droughts, these plants are responsible for the generation of 9,304 MW of electrical energy, more than 10% of all the installed generating capacity in the country.

Table. 11.1 Cases of major dams affected by AAR in Brazil – Main sources: Andriolo (2000), Sabbag (2003), Battagin *et al.* (2009), Kuperman (2013) and CBDB (2014).

Construction decade	Dam	Height (m)	Reservoir capacity $(x10^3\ m^3)$	Functions[1]	Power (MW)	State
1920	Ilha dos Pombos	19	9,000	H	187[5]	RJ/MG
	Rio das Pedras	35	49,000	H, WS	Aux.	SP
1930	Billings-Pedras	31	997	H, WS	889[6]	SP
	Jurupará	27	42,040[2]	H	7[2]	SP
	Pedreira pumping plant	-	-	FC	-	SP
	Pedro Beicht	22	14,100	FC	-	SP
	Salto do Meio (Chaminé)	12	NA	H	18[14]	PR
1940	Peti	46	43,600[3]	H	9	MG
	Traição - pumping plant	-	-	FC[4]	-	SP
	Vossoroca	21	35,700	FC		PR
1950	Guaricana	30	6,840	H	36[14]	PR
	Mascarenhas de Moraes (former Peixoto)	72	4,040,000	H	476[8]	MG
	Paulo Afonso I	20	26,000[9]	H	180[9]	BA
	Pirapora	40	59,000	H	22[10]	SP
	Ribeirão do Campo	26	13,900	WS	-	SP
	Sá Carvalho	15	50	H	78[7]	MG
1960	Barra Bonita	33	3,160,000	H, N	140[13]	SP
	Furnas	127	22,950,000	H	1,216[8]	MG
	Paulo Afonso II	37[9]	26,000[9]	H	443[9]	BA
	Pedra	58	1,640,000	H, I, WS	20[9]	BA
	Santa Branca	54	434,000	H	56[5]	SP
1970	Atibainha (Cantareira System)	46[11]	289,000	WS	-	SP
	Cascata (Cantareira System)	12[11]	NA	WS	-	SP
	Jaguara	55	420,000	H	424[7]	SP/MG
	Jaguari (Cantareira System)	62[11]	NA	WS	-	SP
	Jaguari (CESP)	77	793,000	FC, H	28[12]	SP
	Joanes II	15	128,000	WS	-	BA
	Moxotó (Apolônio Sales)	61[9]	1,150,000	H	400[9]	BA/AL
	Paiva Castro (Cantareira System)	22[11]	NA	WS	-	SP
	Paraibuna	84	2,636,000	FC, H	85[12]	SP
	Paulo Afonso III	47[9]	26,000	H	794[9]	BA
	Paulo Afonso IV	35	127,500	H	2462[9]	BA
	Porto Colombia	49	1,525,000	H	320[8]	MG/SP
	Sobradinho	43	34,116,000	H, I, N, WS	1050[9]	BA
	Tapacura	46	94,200	WS	-	PE
	Tunnels 2, 6 and 7 (Cantareira System)[11]	-	-	WS	-	SP

[1] functions from main to secondary: H – Hydroelectric; FC – Flood Control; WS – Water Supply; I – Irrigation; N – Navigation;
[2] source: CBA – Companhia Brasileira de Aluminio;
[3] source: Wikimapia;
[4] source: EMAE – Empresa Metropolitana de Águas e Energia – SP;
[5] source: Light Energia;
[6] integrated to the Henry Borden hydroelectric compound (Gramulia Jr., 2009);
[7] source: CEMIG;
[8] source: FURNAS;
[9] source: CHESF;
[10] generated by the hydroelectric Rasgão;
[11] see reference Tung et al. (2006);
[12] source: CESP;
[13] source: AES Tietê;
[14] source: COPEL.

The occurrence of alkali-aggregate reaction in hydraulic structures has been widely documented in journal papers, conferences and technical reports for most of the structures. A summarized list of these references is given in Table 11.2.

Quartzites, which are essentially quartz rocks, were used in a small number of dams. However, in the great majority of the dams the aggregates were granitic in type, which are widely used to produce concrete in Brazil. They are composed of feldspar-quartz rock, like granite and gneiss. A great part of these granitic rocks were affected by tectonic forces that made the aggregates potentially reactive through recrystallisation (Andriolo, 2000). The nature of aggregates as well as literature references that give details of the AAR cases in dams are included in Table 11.2.

Table 11.2 Bibliographic references and nature of aggregates for Brazilian dams affected by AAR.

Construction decade	Dam	Reference	Nature of aggregate
1920	Ilha dos Pombos	Corrêa (1997)	Gneiss, Biotite
	Rio das Pedras	Battagin et al. (2009)[2]	Gneiss
1930	Billings-Pedras	Guerra et al. (1997), Braun (2006), Kuperman (2013)	Granite
	Jurupará	Paes Filho and Paulon (1997)	Gneiss/Biotite-Granite
	Pedreira pumping plant	Battagin et al. (2009)[2]	Gneissic Granite
	Pedro Beicht	Tung et al. (2006)	Granite-Gneiss
	Salto do Meio (Chaminé)	Pires (2009)	Basalt
1940	Peti	Magalhães and Moura (1997)	Gneiss
	Traição – pumping plant	Piasentim et al. (2004), Braun (2006)	Mylonite
	Vossoroca	Battagin et al. (2009)[2]	Gneiss
1950	Guaricana	Portela et al. (2012)	Mylonite, Basalt, Granite
	Mascarenhas de Moraes (former Peixoto)	Galletti et al. (1997)	Granite, Gneiss and Migmatite
	Paulo Afonso I	Silva et al. (2008)	Granite, Gneiss and Migmatite
	Pirapora	Braun (2006), Santos et al (2016)	-
	Ribeirão do Campo	Tung et al. (2006)	Cataclastic biotite gneiss
	Sá Carvalho	Andriolo (2000),	Gneiss
1960	Barra Bonita	Strong et al. (1999)[1], Battagin et al. (2009)[2]	Basalt
	Furnas	Galletti et al. (1997), Hasparyk et al. (2008), Kuperman (2013)	Quartzite
	Paulo Afonso II	Silva et al (2008)	Granite, Gneiss and Migmatite
	Pedra	Cavalcanti et al. (2008), Kuperman (2013), Cavalcanti et al. (2016)	Gneiss (with strained quartz)
	Santa Branca	Andriolo (2000)	Gneiss
1970	Atibainha (Cantareira System)	Tung et al. (2006)	Mylonite
	Cascata (Cantareira System)	Tung et al. (2006)	Granite, Gneiss
	Jaguara	Carvalho et al. (1997)	Quartzite
	Jaguari (Cantareira System)	Tung et al. (2006)	Mylonitic Gneiss
	Jaguari (CESP)	Kuperman (2013), Carneiro et al (2016), Bernardes et al. (2016)	Mylonitic Gneiss

Table 11.2 (Cont.)

Construction decade	Dam	Reference	Nature of aggregate
	Joanes II	Hasparyk et al. (2006)	Gneiss
	Moxotó (Apolônio Sales)	Juliani et al. (2008)	Granite, Gneiss
	Paiva Castro (Cantareira System)	Tung et al. (2006)	Gneissic Granite
	Paraibuna	Kuperman et al. (2003)[3]; Battagin et al. (2009)[2]	Mylonite
	Paulo Afonso III	Silva et al. (2008)	Granite, Gneiss and Migmatite
	Paulo Afonso IV	Silva et al. (2008)	Granite, Gneiss and Migmatite
	Porto Colombia	Galletti et al. (1997)	Basalt
	Sobradinho	Battagin et al. (2009)[2]	Quartzite
	Tapacura	Hasparyk et al. (2006), Silva (2007)	Cataclastic Gneiss and Granite
	Tunnels 2, 6 and 7 (Cantareira System)	Kuperman et al. (2000), Tung et al. (2006)	Granite, Gneiss

[1] indicates very small magnitude.
[2] indicates alkali-silicate reaction.
[3] alkali-silicate reaction not present.

The structures affected by AAR were mainly water intakes, power houses, spillways and dams. The evidence of AAR was in most cases, stressed quartz, gel exudation, cracks, open construction joints, displacements between blocks, crest movement, turbine movement and turbine base deformation. The measures that have been taken to manage the effects of AAR are, mainly: monitoring and analysis; joint cutting (seldom); grouting of leaking cracks; waterproofing of some parts that are out of water (Kuperman, 2013).

One of the most important Brazilian contributions to the prevention of AAR has been the use of pozzolanic materials in major works. Although this is noted here, engineering applications have already been described in other references (e.g., Andriolo, 2000). Some further details will be given in subsection 11.3.1, as well as an overview of the research, which is concerned with the use of new sustainable materials.

Besides the occurrences of AAR in hydraulic works, a new feature in the Brazil has been the reported appearance, in the mid-decade of 2000, of alkali-aggregate reaction in buildings of the metropolitan area of the city of Recife in north-eastern Brazil. This was a turning point in the Brazilian concrete technology and is described in subsection 11.3.1.

As a consequence of the resumption of Brazilian economic growth in the 1990s, the construction of large dams has been increased to meet the increasing energy demand in the country. Examples of these major hydropower plants in construction are the HPPs of Jirau [3,750 MW, 2,700,000 m^3 of concrete, 2010] (Zanoti, 2011), Santo Antonio [3,150 MW, 2,500,000 m^3 of concrete, 2010] (Guimarães et al., 2011) and Belo Monte [11,200 MW, 3,900,000 m^3 of concrete, 2011] (Rufato et al., 2011). Development projections into 2020 plan to install a new hydroelectric capacity of the order of 19,000 MW in HPPs, with capacities varying from 45 to 6,100 MW (Ferreira & Amaro, 2011).

The prospects concerning the growing development of the hydropower plants, together with the cases of alkali-aggregate reaction recently identified in residential and commercial buildings, imply that the studies concerning the detection and prevention of AAR will be in the forefront of the Brazilian technical and scientific research in future years.

11.3.1 Pozzolans, new materials and sustainability

AAR was first identified in Brazil as an issue of concern in the beginning of the 1960s for Jupiá (1550 MW) hydropower plant.

Jupiá HPP (owned by CESP) used a volume of concrete of the order of 1,600,000 m^3. The aggregates came from the basaltic rock of the foundations and rolled gravel from a sedimentary deposit near the construction site. These aggregates (agate, quartzite) contained highly reactive chalcedony as indicated by ASTM C295 (now 2012) petrographic analysis (Andriolo & Sgarboza, 1986) and were considered potentially alkali-aggregate reactive. This was confirmed by chemical analysis (ASTM C289: 2007, now withdrawn) and by tests carried out on mortar-bars (ASTM C227, now 2010). An attempt was made to adopt preventative measures and, initially, in the beginning of the construction, cement with an alkali content around 0.2% was used (Andriolo, 2000). Since the supply of large volumes of cement with varying characteristics was required, the best solution was to use a common cement with artificial pozzolana. During an initial period, fly ash from thermoelectric plants in the South of Brazil was used. Meanwhile, a processing plant at Jupiá dam was installed to produce pozzolana from the calcination of kaolinitic clays existent in the vicinity of the construction site. The effectiveness of this metakaolin in combating AAR could be evaluated by ASTM C441 (now 2011) tests (Andriolo, 1986).

Similar solutions, using calcined kaolinitic clay pozzolana (metakaolin), were used in the construction of other CESP dams (Ilha Solteira -3450 MW, Capivara – 620 MW and Água Vermelha – 1400 MW) during the period 1963 to 1979, corresponding to a total volume of 5,990,000 m^3 of concrete and 225,000 tons of pozzolana (Andriolo & Sgarboza, 1986).

The solution for these dams was applied and used for most of the dams and other infrastructure constructions in Brazil since the 1970s. These do not show symptoms of AAR (Sanchez et al., 2010). It should be stressed that the use of pozzolana or ground granulated blastfurnace slag (GGBS) is generally suitable for massive concrete construction, because it also reduces the heat of hydration of the material. The ideal content of the mineral addition varies according to the type of aggregate, necessitating experimental tests to define the optimum content. Some contents that have been investigated, or used in practical applications, are presented by Hasparyk (2005): silica fume from 10% to 15%; metakaolin from 10% to 25%; natural pozzolana from 20% to 30%; GGBS from 40% to 65%.

It should be stressed that some agro-industrial pozzolanas widely available in Brazil have been studied for mitigation of AAR in the presence of reactive aggregates. Besides the benefit of the use of these residuals in the construction of massive structures (Fairbairn et al., 2010a), it was demonstrated that they can substantially reduce the CO$_2$ emissions to the atmosphere (Fairbairn et al., 2010b). Hasparyk et al. (2000) demonstrated the benefits of using rice husk ash in contents above 12%. Águas (2014) studied the effects of sugar-cane bagasse ashes in AAR. The results indicate that contents greater than 12% should be used (Águas et al., 2016).

The use of powdered rock (<75 μm) is also mentioned by Andriolo (2000). It was obtained from crushed sand and used in the construction of Rio Jordão (PR, 1996, 7MW), Bertarello (RS, 1996, mainly water supply), Salto Caxias (PR, 1999, 1240 MW) and Val de Serra (RS, 1972, mainly water supply).

Besides the use of pozzolana, Brazilian researchers have been actively studying the effect of fibres on the development of AAR. This was motivated by a recent comprehensive study, which indicated that up to 50% of the reinforcement used in the construction of spillways and power houses of HPPs could be substituted by fibres (Fairbairn et al., 2008), significantly reducing the time of construction and consequently its costs. Consequently, it also allowed study of the effect of fibres on the development of alkali-aggregate reaction. An early study indicated that the fibres can reduce the AAR expansion (Carvalho et al., 2010). This research is now being continued with complementary tests to verify the beneficial effect of fibres on alkali-silica reaction.

11.3.2 AAR in buildings in Recife – PE and the Brazilian standard

Recife, the capital of the State of Pernambuco, is one of the most important Brazilian cities. Known as the 'Brazilian Venice' because of its many rivers, small islands and several bridges, Recife is the largest city in north-eastern Brazil, with a metropolitan population of approximately 3,800,000 inhabitants.

Until 2005 there were only two reported cases of structures affected by AAR. The Tapacurá dam, built in 1973, had the first symptoms detected in 1990 (Silva, 2007). This case did not attract the attention of the professional community, because it was one more of the many cases of AAR detected in Brazilian dams (see Tables 11.1 and 11.2). Another isolated appearance of AAR was the Paulo Guerra bridge (Ávila & Fonte, 2002). The pile caps of this bridge presented map cracking with crack openings reaching 5 mm. Even taking into account that, up to that date, there were very rare cases of AAR in structures other than the dams, this case also did not call the attention of Brazilian engineering to a possibly significant occurrence of AAR in Recife.

The existence of alkali-aggregate reaction in a great number of commercial and residential buildings in the metropolitan region of Recife has been observed due to the interest generated by the inspection of the foundations of several buildings, after the collapse of the Areia Branca building in October 2004. Paradoxically, this building did not present evidence of the reaction, but among 30 buildings that had its foundations examined by the ABCP (Brazilian Association of Portland Cement), almost 20 proved to have AAR in their foundation blocks and base plates (Pechio et al., 2006; Battagin et al., 2009; Sanchez et al., 2010).

The main characteristics shown by the petrographic evaluation of drilled cores from twelve buildings (Pechio et al., 2006) were: map-cracking pattern; crushed mylonitic granite as coarse aggregates; presence of a dark reaction rim around the aggregate particles; presence of expansive alkali-silicate potassium-rich gel; characteristic fibre-radial crystal, potassium-rich products. The main reaction products were the expansive massive cracked gel, found close to the aggregate boundaries, and the crystallized spear-shaped products located in the inner part of the aggregates. Neither the rosette-shaped crystals found in some national dam concretes, nor the sodium-rich forms have been observed in these concretes. The analyses also indicate the presence of high alkali contents in the mortar matrix. The structures are aged from 3 to 25 years and all of their foundations are placed in regions of shallow groundwater, exhibiting direct correlation between the intensity of exposure to moisture and the degree of damage (Pechio, 2006).

A detailed study and overview of the alkali-aggregate reaction in the Recife metropolitan is presented by Andrade *et al.* (2008). Besides visual inspection of seven affected buildings, aged from 9 to 21 years, the analyses were based on concrete cores extracted from the structures, and aggregates both fine and coarse collected from different regions and quarries that traditionally supply aggregates for the Recife metropolitan area. With these samples, petrographic analyses and potential reactivity tests were carried out, in addition to scanning electron microscopy (SEM) of mortar cast with artificial sand from some coarse aggregates.

Rock types were identified as cataclastic gneiss, mylonite, cataclasite, porphyritic granite and gneiss mylonite. The coarse aggregates always contained strained quartz minerals with undulating extinction in their composition, most of the time associated to microcrystalline and recrystallized quartz in a matrix structure. It was verified that all the concrete samples taken from the buildings presented evidence of AAR, either from petrographic analysis, or by optical and SEM microscopy. Additionally, the analyses carried out with the fine and coarse aggregates indicated their potential reactivity. Andrade *et al.* (2008) concluded that *"alkali-aggregate reaction is now an additional likely cause of cracking in the reinforced concrete structures of Recife, especially for those in permanent contact with water, such as the concrete elements of buildings foundations and concrete monuments"*.

The events in Recife provoked the mobilization of the technical community in Brazil to take measures for the evaluation and prevention of AAR. That was routine for the engineering of dams but was really a novelty for buildings located in urban environments. In May 2008, the Brazilian standard ABNT NBR 15577 (ABNT, 2008) was published, for the prevention of alkali-aggregate reaction, comprising six parts:

(1) Guide for the evaluation of potential reactivity of aggregates and preventative measures for its use in concrete;
(2) Sampling, test sample preparation and testing periodicity of aggregates for use in concrete;
(3) Petrographic analysis for evaluation of the potential reactivity of aggregates with alkali compounds from concrete;
(4) Determination of expansion of mortar-bars by the accelerated mortar-bar method;
(5) Determination of mitigation of expansion by accelerated mortar-bar method;
(6) Determination of expansion of concrete prisms.

This standard is mainly based on the Canadian standard CSA A.23.1/A.23.02, and the test methodology was mainly based on ASTM methods that were known and used in the country.

The AAR cases in Recife were a turning point in the way that the Brazilian engineering community dealt with the prevention and mitigation of this expansive reaction. In fact, until 2005, the technology for the assessment and prevention of alkali-aggregate reaction was mainly dominated by a few research centres, especially those related to electricity generating companies, public institutions and some universities. As a matter of fact, after 2005, the number of AAR tests and petrographic analyses has grown exponentially in the country, as was demonstrated by Battagin *et al.* (2009) indicating that AAR in urban environments is a new challenge for Brazilian engineering.

11.4 MEXICO

In Mexico, potential alkali-reactive zones, which have potentially reactive concrete aggregate sources, were identified by the Central and South Mexico Chapter of ACI in 1996 (Hernández-Castañeda & Mendoza-Escobedo, 2006). A comprehensive study of the reactive potential of aggregates for concrete pavements in the State of Chihuauha was developed by Olague *et al.* (2002, 2011). A survey of the Chihuauha province using ASTM C289 testing concluded that 13% of aggregate sources were potentially alkali-reactive in concrete. Potentials for alkali-silica and alkali-carbonate reactions in some physiographic regions have also been suggested by Reyes (2009).

In the technical literature, one can find one occurrence of AAR in Mexico (Olague *et al.*, 2003). These are cases of 4 concrete pavements in the streets of Chihuauha city.

Petrographic analysis of the aggregates used (crushed limestone and river sand) indicated the presence of chalcedony, andesite, quartz, lithic rhyolites and lithic andesites. What appeared paradoxical in these cases was that, although the cement had a low alkali content, the deterioration was indeed shown to be due to AAR. This was demonstrated by SEM analyses of the reaction products in cracks, identification of fluorescent gel and petrographic examination of the concrete. It was then demonstrated that the alkalis deriving from the aggregates had an important influence on the development of the reaction.

11.5 URUGUAY

The most significant case of AAR in Uruguay occurs in the Baygorria hydroelectric power plant built from 1956 to 1960. It has 3 Kaplan turbines and installed capacity of 108 MW. The main characteristics of the evolution of the reaction and its structural consequences have been described by Patrone (2008, 2013).

According to the original design of the turbine, the blades had mounting clearances with respect to the upper and lower distributor rings of 0.30 mm. During routine inspections in 1962 and 1966, engineers reported total loss of the mounting clearances. Initially the problem was mitigated by grinding roughing the top ring, in order to recover the original mechanical clearance. Such measures became incorporated into scheduled and preventative maintenance, performed periodically until 2001, because the origin of the problem was unknown. After a consulting project with Éléctricité de France (EDF) in 1978, an instrumentation system was installed to help to understand the origin of the expansion. The results of the instrumentation, along with other complementary investigations showed that the problem was caused by a process of irreversible volumetric deformation of the concrete cone that supports the fixed blades.

A set of laboratory tests carried out at LCPC/France (two phases, in 1978 and 1987) and at the Universidad Nacional de La Plata/Argentina (in 1990) indicated that the concretes at different stages of casting were not reactive, with the exception of the last, or fourth phase. Clear signs of ASR were observed. The aggregates used had been basaltic and the sand was constituted almost exclusively of silica in various forms. Alkalis corresponded essentially to cement, additives and to additional alkali released by the basaltic aggregates.

The samples showed a significant expansion of the order of 0.08% after a year. Extrapolating this result to concrete 4 metres high in the affected cone area, would represent a residual expansion of about 3 mm, in addition to the 7.5 mm produced since 1960.

In early 1990, it was determined that the grinding of the upper ring could not continue indefinitely. It was also assessed that the accumulated stresses surrounding metal parts, caused by the deformation of the concrete, could be very close to the limit of failure of the respective materials. This was particularly worrying in the case of the lower ring, made of cast iron. Several definitive solutions have been studied, among which may be mentioned: lithium injection followed by the application of an electrical field, and the introduction of joints within the concrete structure.

The adopted solution was limited to the replacement of the lower ring, together with the removal of the concrete cone immediately below the bottom ring and not all the concrete affected by ASR. This solution (see details in Patrone, 2008, 2013) was implemented in the three turbines in 2007, 2008 and 2009, respectively. The field monitoring carried out since the execution of the mitigation measures has indicated a drastic reduction of the vertical displacement axes. As a consequence, it has not been necessary to remove the adjustment plates incorporated as a regulatory mechanism between the blades and the bottom ring. Therefore, the owners expect that the hydro-electric power plant will operate, without any further remedial work, for at least the next 20 years.

REFERENCES

ABNT – Associação Brasileira de Normas Técnicas (Brazilian Association of Standards) (2008) NBR 15577-1:2008 corrected version:2008, parts 1 to 6. (in Portuguese).

Acosta, A., Villalba, J., Rojas, R., & Cabrera, R. (2007) Determination of alkali-aggregate reactivity potential by the accelerated mortar bar method, *Primer Congreso Nacional de Ingenieria Civil Asunción*, November 2006, Assunción (in Spanish).

Águas, M.F.F. (2014) Effects of sugar-cane bagasse ash on the alkali-aggregate reaction, Ph. D. thesis, COPPE/Universidade Federal do Rio de Janeiro (in Portuguese).

Águas, M.F.F., Fairbairn, E.M.R., Toledo-Filho, R.D., Hasparyk, N.P., & Cordeiro, G.C. (2016) Influence of sugarcane bagasse ash in the expansions of mortars affected by alkali-silica reaction, *Proceedings of the 15th International Conference on Alkali Aggregate Reaction (ICAAR)*. São Paulo. Digital Proceedings.

Andrade, T., Silva, J.J.R., Silva, C.M., & Hasparyk, N.P. (2008) History of some AAR cases in the Recife region of Brazil, *Proceedings of the 13th International Conference on Alkali Aggregate Reaction (ICAAR)*. Trondheim. Digital Proceedings.

Andriolo, F.R. & Sgarboza, B.C. (1986) The use of pozzolan from calcined clays in preventing excessive expansion due to the alkali-aggregate reaction in some Brazilian dams, *Proceedings of the 7th International Conference on Alkali-Aggregate Reaction (ICAAR)*, Ottawa, August.

Andriolo, F.R. (2000) AAR dams affected in Brazil – report on the current situation, *Proceedings of the 11th International Conference on Alkali-Aggregate Reaction ICAAR*, Québec. pp. 1243–1252.

ANEEL – Agencia Nacional de Energia Elétrica (2014) BIG – Banco de Informações de Geração [Online] Available from: http://www.aneel.gov.br/aplicacoes/capacidadebrasil/capacidadeb rasil.cfm [Accessed 16th May 2014] (in Portuguese).

ASTM C227 (2010) *Standard test method for potential alkali-silica reactivity of aggregates (mortar-bar method)*. West Conshohocken, USA, American Society for Testing and Materials.

ASTM C289 (2007, now withdrawn), *Standard test method foir potential alkali-silica reactivity of aggregates (chemical method)*. West Conshohocken, USA, American Society for Testing and Materials.

ASTM C295 (2012) *Standard guide for petrographic examination of aggregates for concrete.* West Conshohocken, USA, American Society for Testing and Materials.

ASTM C441 (2011) *Standard test method for effectiveness of pozzolans or ground blast-furnace slag in preventing excessive expansion of concrete due to the alkali-silica reaction.* West Conshohocken, USA, American Society for Testing and Materials.

Ávila, J.I.S.L. & Fonte, A.O.C. (2002) Alkali-aggregate reaction in the pile caps of a bridge, *IBRACOM 44° Congresso Brasileiro do Concreto*, belo Horizonte (in Portuguese).

Battagin, I.L.S., Battagin, A.F., & Sbrighi Neto, C. (2009) The alkali-aggregate reaction Brazilian technical standard makes its first birthday, *IBRACON Concreto e Construções* [Online] 54, 34–47. Available from: http://www.ibracon.org.br/publicacoes/revistas_ibracon/rev_constru cao/revistas.asp [Accessed 16th May 2014] (in Portuguese).

Bernardes, H.M., Rodrigues, R.O., & Bertolino, Jr.,R. (2016) Structural effects of AAR on the Jaguari hydropower plant water intake, *Proceedings of the 15th International Conference on Alkali Aggregate Reaction (ICAAR)*. São Paulo. Digital Proceedings.

Bolivar, I.C. O.G. (2003) *Manual of aggregates for concrete*, Universidad Nacional de Colombia, Facultad de Minas, Escuela de Ingenieria Civil, Medellin (in Spanish).

Braun, P.V.C.B. (2006) Living with expansive reactions in dams- the EMAE's experience, *RAA 2006 IBRACON II Simpósio sobre reação álcali-agregado em estruturas de concreto*, Rio de Janeiro (in Portuguese).

Carneiro, E.F., Pinfari, J.C., Cappi, T.P.A., & Covre, M.H.L. (2016) Maintenance and repairs on the water intake of Jaguari hydropower plant affected by AAR, *Proceedings of the 15th International Conference on Alkali Aggregate Reaction (ICAAR)*. São Paulo. Digital Proceedings.

Carvalho, C.J., Carim, A.L.C., & Silveira, J.F.A. (1997) Research of alkali-aggregate reactivity in concrete structures of the HPP Jaguara, *Simpósio sobre Reatividade Álcali-Agregado em Estruturas de Concreto*, Comitê Brasileiro de Barragens (CBDB), Goiânia. [Online] Available from: http://www.engipapers.com.br/ [Accessed 21st May 2014] (in Portuguese).

Carvalho, M.R.P., Fairbairn, E.M.R., Filho, Toledo-Filho, R.D., Cordeiro, G.C. & Hasparyk, N.P. (2010) Influence of steel fibers on the development of alkali-aggregate reaction, *Cement Concrete Res., 40*, 598–604.

CBDB – Comitê Brasileiro de Barragens – (2014) *Register of dams*. [Online] Available from: https://cadastrodebarragens.pti.org.br/#/home [Accessed 14th May 2014].

Cavalcanti, A.J.C.T., Juliani, M.A.C., Tristão, G., & Silveira, J.F.A. (2008) Evaluation of alkali-aggregate reaction expansion on Pedra dam by mathematical model, *Proceedings of the 13th International Conference on Alkali Aggregate Reaction (ICAAR)*. Trondheim. Digital Proceedings.

Cavalcanti, A.J.C.T., Silva, P.N., Silva, T.A.A. & Soares, R.C. (2016) Expansion slot cutting to counteract alkali aggregate reaction at Pedra Dam, *Proceedings of the 15th International Conference on Alkali Aggregate Reaction (ICAAR)*. São Paulo. Digital Proceedings.

Cenzano, M.C., Maita, A.R. & Quispe, C.R.U. (2011) Inhibition of alkali-aggregate reaction using natural pozzolan, *Rev. Inv. Des.* 6 (6), 20–26. (in Spanish).

Cotera, M.G. (1991) *Ataque quimico al concreto*, [Presentation], ACI, Peruvian Chapter, Corrosion in concrete structures (in Spanish).

Corrêa, W.G. (1997) HPP Ilha dos Pombos – Studies of the álcali-aggregate reactivity of the concrete of HPP Ilha dos Pombos and protective measures taken, *Simpósio sobre Reatividade Álcali-Agregado em Estruturas de Concreto*, Comitê Brasileiro de Barragens (CBDB), Goiânia. [Online] Available from: http://www.engipapers.com.br/ [Accessed 21st May 2014] (in Portuguese).

CSA A23.1 (2014) *Concrete materials and methods of concrete construction*, Canadian Standards Association, Mississauga, Ontario, Canada.

CSA A23.2 (2014) *Test methods and standard practices for concrete*, Canadian Standards Association, Mississauga, Ontario, Canada.

Fairbairn, E.M.R., Toledo-Filho, Velasco, R.V., & Araújo, D.L. (2008) Steel fiber reinforced concrete for the construction of hydropower plants, *Journée technique BFM*, Paris, Fondation École Française du Béton, 1, 97–110 (in French).

Fairbairn, E.M.R., Ferreira, I.A., Cordeiro, G.C.C., Silvoso, M.M., Toledo-Filho, R.D. & Ribeiro, F.L.B. (2010a), Numerical simulation of dam construction using low-CO2-emission concrete, *Mater Struct.*, 43, 1061–1074.

Fairbairn, E.M.R., Americano, B.B., Cordeiro, G.C.C., Paula, T.P., Toledo-Filho, R.D. & Silvoso, M. M. (2010b), Cement replacement by sugar cane bagasse ash: CO2 emissions reduction and potential for carbon credits, *J Environ Manage.*, 91, 1864–1871.

Ferreira, T.V.B. & Amaro, P.R. (2011) Prospects for expansion of hydropower park, *IBRACON Concreto e Construções* [Online], 63, 44–47, Available from: http://www.ibracon.org.br/pub licacoes/revistas_ibracon/rev_construcao/revistas.asp [Accessed 23rd May 2014] (in Portuguese).

Galletti, A.A.B., Silveira, J.F.A., Andrade, M.A.S., & Peres, R.G. (1997) Inspections undertaken in Furnas hydropower plants to detect alkali-aggregate reaction, *Simpósio sobre Reatividade Álcali-Agregado em Estruturas de Concreto*, Comitê Brasileiro de Barragens (CBDB), Goiânia [Online] Available from: http://www.engipapers.com.br/ [Accessed 21st May 2014] (in Portuguese).

García, F.D.E. (2013) *Alkali silica reactivity potential of combinations of cementitious materials and aggregates – Method of mortar bar*. Escuela Politécnica Nacional, Facultad de Ingeniería Civil y Ambiental, Quito, Ecuador, Proyecto previo a la obtención del título de inginiero civil (in Spanish).

Gonzáles, T. M. A. (2013) *Alkali silica reaction: microscopic characterization tools and discussion about their treatment in the Chilean standard – in consultation – NCh170*, [Presentation], 19ª Jornadas Chilenas Hormigón Sustentable, 5th November (in Spanish).

Guerra, M.O., Frbella, C.A.C., Guedes Sobrinho, E.F., Carone, R., & Silveira, J.F.A. (1997) Plan of research and auscultation of alkali-aggregate reactivity in Billings-Pedras Regulatory dam, *Simpósio sobre Reatividade Álcali-Agregado em Estruturas de Concreto*, Comitê Brasileiro de Barragens (CBDB), Goiânia [Online] Available from: http://www.engipapers.com.br/ [Accessed 21st May 2014] (in Portuguese).

Guimarães, A.P.B., Nascimento, J.F.F. & Barbin, A.S. (2011) Techniques and concrete technology in the UHE Santo Antônio, *IBRACON Concreto e Construções* [Online], 63, 97–105, Available from: http://www.ibracon.org.br/publicacoes/revistas_ibracon/rev_construcao/revistas.asp [Accessed 23rd May 2014] (in Portuguese).

Gramulia, Jr., J. (2009) *Contribution of the Henry Borden hydroelectric plant for the operation planning of hydrothermal power systems*, M. Sc. Thesis, Universidade Federal do ABC (in Portuguese).

Hasparyk, N.P., Monteiro, P.J., & Carasek, H. (2000) Effect of silica fume and rice husk ash on alkali-silica reaction, *ACI Mater J.*, 97, 486–492.

Hasparyk, N.P. (2005) *Investigation of concretes affected by álcali-aggregate reaction and advanced characterization of exuded gel*, Ph.D. Thesis, Universidade Federal do Rio Grande do Sul [Online] Available from: http://www.lume.ufrgs.br/bitstream/handle/10183/6350/000528715.pdf?sequence=1 [Accessed 23rd May 2014] (in Portuguese).

Hasparyk, N.P., Monteiro, P.J.M., & Dal Molin, D.C.C. (2008) AAR in Furnas dam, Brazil, residual expansion and the effect of lithium, *Proceedings of the 13th International Conference on Alkali Aggregate Reaction (ICAAR)*. Trondheim. Digital Proceedings.

Hasparyk, N.P., Cavalcanti, A.J.C.T., & Andrade, W.P. (2006) Alkali-aggregate reaction in dams, *IBRACON Concreto e Construções* [Online], 42, 38–43, Available from: http://www.ibracon.org.br/publicacoes/revistas_ibracon/rev_construcao/revistas.asp [Accessed 21st May 2014] (in Portuguese).

Hernández-Castañeda, O. & Mendoza-Escobedo, C.J. (2006) Durability and infrastructure: Challenges and socio-economic impact, *Ingeniería Investigación y Tecnología*, 7(1), 57–70. (in Spanish).

Juliani, M., Cavalcanti, A.J.C.T., Carrazedo, R., & Gaspare, J.C. (2008) Analysis of the effects caused by alkali-aggregate reaction on the structures of Moxotó power plant, *Proceedings of the 13th International Conference on Alkali Aggregate Reaction (ICAAR)*. Trondheim. Digital Proceedings.

Kuperman, S.C., Fabbro, J.C., Cifú, S., Kako, H., Tavares, F., Ferreira, W.v.F., Werneck, A.C.A., & Sardinha, V.L.A. (2000) Management of a water intake affected by alkali-aggregate reaction, *Proceedings of the 11th International Conference on Alkali Aggregate Reaction (ICAAR)*. Québec. pp. 1323–1332.

Kuperman, S.C., Cifú, S., Moretti, M.R., & Re, G. (2003) Review of the auscultation instrumentation installed in dams owned by CESP, *ICOLD-CBDB XXV Seminário Nacional de Grandes Barragens*, Salvador, October 2003 (in Portuguese).

Kuperman, S.C. (2013) *Brazilian experience with AAR, International Workshop on Managing AAR in Dams*, [Presentation] ICOLD 81st Annual meeting, Seatle.

Maiza, P., Marfil, S., & Milanesi, C. (1999) Minerals developed in concrete damaged by alkali-silica reaction (Córdoba Province). *VII Jornadas Pampeanas de Ciencias Naturales*, COPROCNA. pp. 193–200. Santa Rosa (in Spanish).

Maiza, P., Marfil, S. Rocco, C., Fava, C., & J. Tobes, J. (2010). Concrete pavements made with basalt aggregates affected by alkali silica reaction (ASR). Case studies. *I Congreso Hormigón Premezclado de las Américas*, Mar del Plata. 12p. (in Spanish).

Magalhães, R.A. & Moura, C.N. (1997) Characterization, history and assessment of alkali-aggregate of reaction parameters for tha mathematical modelling of Peti dam, *Simpósio sobre Reatividade Álcali-Agregado em Estruturas de Concreto*, Comitê Brasileiro de Barragens (CBDB), Goiânia. [Online] Available from: http://www.engipapers.com.br/ [Accessed 21st May 2014] (in Portuguese).

Marfil, S., Batic, O., Maiza, P., Grecco, L. & Falcone, D. (2010) Behavior of basaltic rocks of the Province of Corrientes and Entre Rios vis-à-vis the alkali-silica reaction. *VI Congreso Uruguayo de Geología*. Parque UTE Lavalleja [Online] Available from: http://www.sugeologia.org/documentos/ACTAS%20VI%20CONGRESO%20URUGUAYO/ [Accessed 12th May 2014] (in Spanish).

Marfil, S. & Maiza, P. (2008) Petrographic study of a building deteriorated due to alkali-silica reaction, in Buenos Aires city (Argentina), *Proceedings of the 13th International Conference on Alkali Aggregate Reaction (ICAAR)*. Trondheim. Proceedings. pp. 983–993.

Marfil, S.A. & Maiza, P.J. (2001) Deteriorated pavements due to the alkali–silica reaction, a petrographic study of three cases in Argentina, *Cement Concrete Res.*, *31*, 1017–1021.

Olague, C., Castro, P., & Lopez, W. (2002) Alkali-silica reaction of aggregates for concrete pavements in Chihuahua´s State, Mexico, *Materiales de Construcción* [Online], 52 (268), 19–31 Available from: http://materconstrucc.revistas.csic.es [Accessed 7th May 2014].

Olague, C., Wenglas, G. & Castro, P. (2003) Influence of alkalis from different sources than cement in the evolution of alkali-silica reaction, *Materiales de Construcción* [Online], 53 (271–272), 189–198, Available from: http://materconstrucc.revistas.csic.es [Accessed 11th May 2014].

Olague-Caballero, C., Wenglas-Lara, G., & Astorga-Bustillos, F. (2011) Mineral admixtures as mitigating alkali silica reaction in hydraulic concrete structures, *Tecnociencia Chihuahua*, 6(1) (in Spanish).

Paes Filho, J.R. & Paulon, V.A. (1997) *Project for the rehabilitation of Jurupará dam,*. 28th Jornadas Sul-Americanas de Engenharia Estrutural, São Carlos (in Portuguese).

Patrone, J.C. (2008) Effects and Remedy of the expansive concrete of the "Baygorria" dam, *Memorias* [Online], No. 6, Available from: http://www.um.edu.uy/_upload/_investigacion/web_investigacion_51_Memoria_3_ExpansionHormigon.pdf [Accessed 11th May 2014] (in Spanish).

Patrone, J.C. (2013) Intervention in the turbines of the Baygorria dam affected by alkali-aggregate reaction, *Congreso Argentino de Presas y Aprovechamientos Hidroeléctricos – CAPyAH 2013*, Comité Argentino de Presas, San Juan (in Spanish).

Pechio, M., Kihara, Y., Battagin, A. F., & Andrade, T. (2006) Alkali-Aggregate reaction products from Recife buildings, *RAA 2006 IBRACON II Simpósio sobre reação álcali-agregado em estruturas de concreto*, Rio de Janeiro (in Portuguese).

Piasentin, C., Juliani, M., Becocci, L., & Barreto, C. R. N. G. (2004) Traição dam – the evolution of the alkali-aggregate reaction (AAR) in the last 20 years, *Proceedings of the 12th International Conference on Alkali Aggregate Reaction (ICAAR)*. Beijing, 1323–1332, 1111–1119.

Pires, K. O. (2009) *Investigation of the residual potential alkali-aggregate reaction -a case study*, M. Sc. Thesis, Universidade Federal do Paraná [Online] Available from http://www.prppg.ufpr.br/ppgcc/ [Accessed 21st May 2014] (in Portuguese).

Portella, K. F., Joukoski, A., Swinka Filho, V., Soares, M. A. & Ferreira, E. S. (2012) Physical chemistry research of a concrete dam with over 50 years of operation, *Cerâmica.*, *58*, 374–380.

Priano, C., Marfil, S., Maiza, P., & Señas, L. (2012) Petrography of concrete constructions in the south of the Province of Buenos Aires, *Congreso internacional sobre patología y recuperación de estructuras (CINPAR 2012)*, La Plata (Buenos Aires). 20pp. (in Spanish).

Reyes, A. (2009) *Reacción Álcali-Sílice*, [Presentation], ACI, Colombian Chapter, Seminario reacción Álcali-Agregado, 20th March (in Spanish).

Rufato, L.F., Franco, H.C.B., & Bandeira, O.M. (2011) The Belo Monte hydropower compound, *IBRACON Concreto e Construções* [Online], 63, 48–56, Available from: http://www.ibracon.org.br/publicacoes/revistas_ibracon/rev_construcao/revistas.asp [Accessed 23rd May 2014] (in Portuguese).

Sabbag, A. F. (2003) *Determination of alkali-aggregate reaction in the concrete of the Mascarenhas hydropower plant*, M. Sc. Thesis [Online] Available from: http://www.prppg.ufpr.br/ppgcc/ [Accessed 14th May 2014] Universidade Federal do Paraná (in Portuguese).

Sanchez, L., Kuperman, S. C., & Helene, P. (2010) A brief description of alkali-aggregate reaction occurrence and prevention in Brazil, *Concrete under Severe Conditions Environment and Loading, Proceedings of the 6th International Conference*, Mérida, June, pp. 305–312.

Santos, R.T., Tristão, G.A., Baima, L.Q.G., Bizarro, L.E., & Sollero, M.B.S. (2016) Study of the structural behavior of Pirapora dam affected by alkali-aggregate reaction, *Proceedings of the 15th International Conference on Alkali Aggregate Reaction (ICAAR)*. São Paulo. Digital Proceedings.

Silva, P.N., Cavalcanti, A.J.C. T., Kuperman, S.C., Helene, P., & Hasparyk, N.P. (2008) AAR at Paulo Afonso hydroelectric complex, Part I: influence on the mechanical and elastic properties of the concrete, *Proceedings of the 13th International Conference on Alkali Aggregate Reaction (ICAAR)*. Trondheim. Digital Proceedings.

Silva, G. A. (2007) Recovery of pile caps affected by alkali-aggregate reaction, M. Sc. Thesis [Online] Available from: http://www.unicap.br/tede//tde_busca/arquivo.php?codArquivo=161 [Accessed 14th May 2014] Universidade Católica de Pernambuco (in Portuguese).

Strong, H.D., Astolphi, J.C., Lindquist, L.N., & Oliveita, P.J.R. (1997) Verification of the existence of alkali-aggregate reaction in the concrete of the HPP Barra Bonita, *ICOLD-CBDB XXII Seminário Nacional de Grandes Barragens*, São Paulo, April 1997 (in Portuguese).

Tecnosil (2009) *Soluções* [Online], no. 4, August/September, Available from: http://www.tecno silbr.com.br/wp-content/uploads/Tecnosil4_site.pdf [Accessed 8th May 2014] (in Portuguese).

Tung, W.S., Kuperman, S.C., Melo, I.A., Dardis, C.R., Augustinis, J.A., & Ohara, M.T. (2006) SABESP and the Alkali-Aggregate Reaction, *RAA 2006 IBRACON II Simpósio sobre reação álcali-agregado em estruturas de concreto*, Rio de Janeiro (in Portuguese).

Zanoti, J.A.C. (2011) Hydroelectric power plant of Jirau, *IBRACON Concreto e Construções* [Online], 63, 26–30, Available from: http://www.ibracon.org.br/publicacoes/revistas_ibra con/rev_construcao/revistas.asp [Accessed 23rd May 2014] (in Portuguese).

Chapter 12

Southern and Central Africa

Mark Alexander & Geoff Blight

12.1 INTRODUCTION

Much of this chapter necessarily concentrates on AAR in South Africa, where a lot of work and research on the subject has been carried out. However, available information on other southern and central African countries is collated in sections 12.7 and 12.8.

The first edition of South Africa's best known and most widely used handbook on concrete appeared in 1957. Originally called 'Concrete Technology' with F.S. (Sandy) Fulton as the editor and major author, it contained the following statement: *"Cement-alkali reaction would only be suspected with cements containing more than 0.6 per cent alkali, (calculated as Na_2O). In general, South African cements have alkali contents well below this value, so this type of reaction is very improbable in this country. On the other hand, T.E. Stanton has stated: 'Under certain circumstances a low alkali cement will not prevent abnormal expansion', and this has been substantiated by other experts. It would appear in all cases of this type of expansion that 'the reactive constituent usually occurs in the form of a silica readily soluble in sodium or potassium hydroxide' "*.

Twenty four years later, in April, 1981, South Africa hosted the Fifth International Conference on AAR, in Cape Town. In his opening address, Professor T.L. (Pluto) Webb had the following to relate: *"My first personal contact with it [AAR] was 20 years ago, when as a result of suspecting that it might be of some importance in South Africa, I made enquiries in a number of countries while on an overseas study visit. I recall with mixed feelings and claustrophobic terror a day in Denmark when Dr Idorn put me into a small boat and took me on a conducted tour of exploration of almost submarine concrete structures. We meandered through kilometres of pitch dark cavernous passages, with concrete sheet piling to the left and right of us, with a concrete roof sometimes only 50 centimetres above the gunwales of the boat and with Dr Idorn waving his flash-light to only slightly break the Stygian gloom and saying, more or less alternatively: 'Isn't this a beautiful example of alkali-aggregate reaction' and 'mind your head' – Thank you Dr Idorn for teaching me something about alkali-expansion, for demonstrating that all Scandinavians still have the spirit of the Vikings, and most importantly, for saving me from concussion."*

And later in his address: *"The South African research on alkali-aggregate reaction really commenced in 1977 at the National Building Research Institute. Once the nature and extent of the problem became apparent the agencies concerned in both the public and building sectors gave the whole project their support"*.

By 1994, in the seventh edition of what is now called 'Fulton's Concrete Technology', the brief statement on AAR that appeared in the first edition had grown to a 23 page

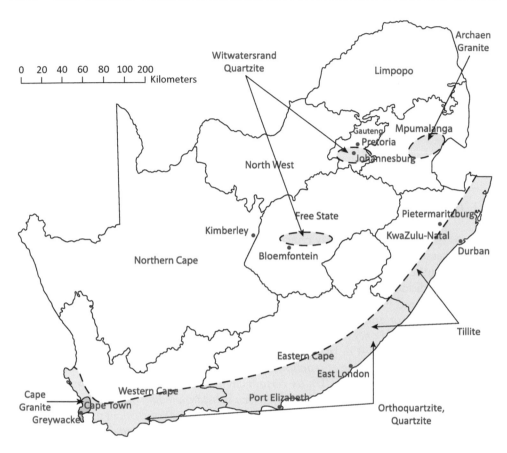

Figure 12.1 South Africa and its provinces.

chapter, and it is now recognized that AAR can occur in the South Western Cape Province, the Eastern Cape Province, Gauteng and Free State Provinces, as well as Kwa-Zulu Natal and Mpumalanga Provinces. (See the affected areas marked on the map of South Africa, Figure 12.1.)

12.2 SUSCEPTIBLE ROCK TYPES

The rocks that have proved susceptible to AAR in South Africa are:

Gauteng and Free State: quartzites and shales associated with gold-bearing reefs (common name: Witwatersrand quartzite, or Wits quartzite)
Western Cape: metasediments (various hornfels, meta-greywacke)
   orthoquartzites
   arkoses
   tertiary and quaternary quartzite pebbles
   granites
Eastern Cape: orthoquartzites

Kwa-Zulu Natal: quartzite, sandstone, tillite (Dwyka Formation - no deleterious service record)

Mpumalanga: Archaean granite and gneiss

12.3 OVERVIEW OF INCIDENCES OF AAR IN SOUTH AFRICA

The important incidences of AAR are noted below (from Fulton, 2009), region by region. In all the cases observed, the type of AAR observed is alkali-silica reaction (ASR).

12.3.1 Gauteng and Free State

In this region, extensive ASR has been noted in structures using Witwatersrand Supergroup quartzites and shales. However, the reaction generally does not proceed unless additional moisture sources are available due to causes such as poor drainage, ponding, and poor detailing allowing moisture to accumulate. Affected structures include reservoirs, bridges including portal frames of a major urban highway, and an airport runway.

12.3.2 Western Cape

ASR is very prevalent in the Cape Peninsula and surrounding region. As indicated above, the main source is the greywacke rocks of the Malmesbury Group, which are very commonly used as coarse aggregates. All types of structures are affected, including *inter alia* bridges, various buildings, culverts, dams, a hydro-electric power station, and sports stadia. ASR has been detected further north in the prestressed concrete railway sleepers of the Sishen-Saldanha railway line, where the offending aggregate is granite.

12.3.3 Eastern Cape

ASR occurs on certain exposed structures in the Eastern Cape, but at a lesser scale than in the Western Cape. According to Fulton (2009), the affected structures are irrigation and water storage dams, an irrigation water concrete pipe line, bridges, airport structures (apron and building), pile caps, retaining walls, and lighting mast foundations.

12.3.4 Mpumalanga

In this north-eastern region of South Africa, ASR has been noted in structures with coarse granite aggregate, such as an airport apron and mast footings.

12.4 DIAGNOSIS, EVALUATION & TEST METHODS USED IN SOUTH AFRICA FOR AAR

In South Arica, tests for assessing susceptibility of aggregates to alkali reaction are often considered in three main categories (from Blight & Alexander, 2011).

12.4.1 Initial non-quantitative screening tests (used to make a provisional assessment)

In most cases, this involves the use of aggregate petrography. Petrography is used as an essential tool to confirm whether cracking in a structure is AAR-related or not, as well as giving an initial indication of the likelihood of AAR occurring with a given aggregate source, by the process of identifying possibly alkali-reactive minerals in the aggregate. This pre-supposes the presence of competent qualified concrete and aggregate petrographers – currently in short supply in the region.

12.4.2 Indicator tests (to differentiate between potentially reactive and innocuous aggregates)

These tests are used in South Africa to differentiate between potentially reactive and innocuous aggregates. By far the most popular test locally is the SA version of the accelerated mortar-bar test, ASTM C1260 (2014), which has the standard test designation of SANS 6245 (2006). The method involves monitoring expansion of mortar-bars or prisms containing the test aggregate and immersed in a 1M sodium hydroxide solution at 80°C. The monitoring period is usually 12 to 14 days. This test was in fact originally devised at the erstwhile NBRI of the CSIR in South Africa in the 1980s, under the direction of Dr (Bertie) Oberholster and colleagues, Oberholster (1983).

Other tests in this category need to be carefully judged as to whether the results they provide are applicable in any given situation, since in some cases aggregates with a known poor service record may show up in the tests as innocuous. Some tests can take a very long time (up to 1 year) and are therefore not particularly useful, besides being more expensive. All the tests require a high degree of skill to perform. See Chapter 2 for more details on AAR testing.

12.4.3 Performance tests (to obtain alkali content limits to avoid damaging expansion)

Laboratories may use a selection of these tests to assess the reactivity performance of particular concrete mixes [*e.g.*, ASTM C1293 (2008): 38°C concrete prism test; RILEM AAR-4.1 60°C concrete prism test (in Nixon & Sims, 2016)]. They can also be used to evaluate the effect of supplementary cementitious materials (SCMs) on potential AAR. A drawback is that in general at least three months is required in order to obtain meaningful results.

Regarding the above test methods used in SA, those aimed at assessing the potential of aggregates for AAR may not be particularly useful in predicting the rate and degree of damage that is likely to occur in a real structure made with such aggregates. Such tests may screen out aggregate sources in terms of AAR susceptibility, but these aggregates could possibly still be used successfully in concrete structures provided the proper precautions are taken. Therefore, it is often stressed that it is necessary to consider not only the potential of an aggregate for AAR, but also an assessment of the conditions under which the aggregate might still be used in an actual structure.

Particular mention should be made of the benefits of long-term structural monitoring of AAR-affected structures. This represents the ultimate in performance tests and is dealt with later in a series of illustrative case histories.

Further detailed information concerning general and current methods of diagnosis, appraisal and testing concrete and its constituent materials may be found in Chapters 2 and 3 of this book.

12.5 REGIONAL APPROACHES TO MINIMISE RISKS OF AAR

AAR, or more specifically ASR, can occur only if three necessary conditions are met: an alkali source, reactive aggregate, and sufficient moisture to sustain the reaction and induce expansion. Thus, preventative measures must address one or more of these factors. In South Africa, the commonly accepted methods to avoid or minimise the risk of AAR are used, and they are briefly summarised here.

12.5.1 Reducing the effect of alkalis, including use of cement extenders

Where possible or practical, the total alkalis in a concrete mix are limited to a certain lower limit, below which the risk of ASR is minimal (depending on the environment of the structure). South Africa adopted the EN 197-1 (now 2011) 'common cements' specification in the late 1990s, with the fairly common use at that time of plain Portland cements (*i.e.*, CEM I types). However, these have largely been replaced in recent times by blended and composite cements. A positive consequence is that current cements tend to be less liable to initiate ASR in concretes containing susceptible aggregates, due to their lower clinker contents.

While CEM I cements are still available in the 52.5N grade, the great majority of current cements fall into the EN 197 categories of CEM II (A and B classes with slag, fly ash, and ground limestone being the main supplementary materials), and CEM IIIA. A certain proportion of CEM IV and CEM V cements are also produced commercially. The most common class of cement typically used in South Africa is a CEM IIA or IIB 42.5, which include supplementary materials depending on regional variations.

According to Fulton (2009), the limits of total alkali content (Na_2Oeq), in order to avoid ASR with the various rock types discussed earlier are given in Table 12.1.

Table 12.1 indicates that most South African aggregates will not cause ASR problems as long as the concrete alkali content is less than about 2.8 kg/m^3, although two of the most alkali-reactive aggregates, the Western Cape greywacke and the Witwatersrand

Table 12.1 Limits of total alkali content from clinker component of cement, per m^3 of concrete (after Chapter 10 of Fulton, 2009). Reproduced from Owens, G. (Ed.) Fulton's Concrete technology, 2009. Midrand, Cement and concrete Institute, with permission from The Concrete Institute

Rock type	Total Na$_2$Oeq kg/m^3 (of concrete)
Witwatersrand Supergroup quartzite, shale	2.0
Dolomite Group chert	2.8
Malmesbury Group metasediments	2.1
Table Mountain Group orthoquartzite	2.8
Bokkeveld Group arkose	2.8
Natal Group quartzite, sandstone	2.8
Dwyka Formation tillite	2.8
Enon Formation quartzite pebbles	2.8
Quarternary quartzite gravels	2.8
Archaean granite	Not determined
Cape granite	4.0
Salem granite	Not determined

quartzite, require alkali contents to be less than about 2.0 kg/m^3 in order to avoid long-term problems.

In the event that it is not possible or practical to limit the total alkalis in the concrete, particularly in mixes with high cement contents, the common practice in South Africa is to use cement extenders such as ground granulated blastfurnace slag (GGBS), fly ash, or silica fume. However, cement extenders can themselves contribute some alkalis to the mix. The advantage of extenders is that they have considerably lower active alkali contents than Portland cement, because their alkalis are generally tied up in their glassy phases and are released at a much slower rate than for Portland cement. Typical values for active alkali contents are given in the literature (Alexander & Mindess, 2005); for example in South Africa, such values are typically 42-50 % for GGBS, 17-40 % for FA, and 30 % for CSF (condensed silica fume).

The commonly accepted minimum proportions of cement extenders in concrete to alleviate ASR problems in South Africa may be listed as:

Ground granulated blastfurnace slag (GGBS)	40–50 %
Fly ash (FA)	20–30 %
Silica fume (CSF)	10–15 %

Further comments concerning this approach can be found in Alexander & Mindess (2005).

12.5.2 Avoiding the use of alkali-reactive aggregates

The obvious way to prevent ASR is to avoid the use of alkali-reactive aggregates. Nevertheless, there are occasions when it is not possible to avoid an alkali-reactive aggregate. In these cases, blending with a non-reactive aggregate may reduce the problem. Another measure is to exclude ASR-susceptible constituents by beneficiation (selective quarrying and crushing, heavy media separation, etc.). Alternatively, attention must be paid to reducing or minimising the alkalis in the concrete mix or modifying the environment to eliminate moisture.

12.5.3 Modifying the environment to reduce the moisture content of the concrete

If it is impossible to avoid or minimise ASR by the measures given above, then the only other possible alternative is to attempt to prevent moisture penetrating the structural fabric, or reduce its moisture content to less than 85% RH, by, variously, shrouding or cladding the structure with a protective cover; providing sufficient fall in flat elements to ensure rapid drainage; paying attention to details of drainage such as expansion joints, embedded gutters and drain pipes, etc. and use of hydrophobic coatings to shed external water while allowing the concrete to 'breathe'.

However, as indicated in the illustrative case histories later in this chapter, it is extremely difficult to dry out concrete structures effectively, particularly if they are of large dimension and if any surface or environmental moisture is allowed to enter the structure. Thus, in many cases, sufficient moisture will be present in the concrete to allow AAR to continue.

12.5.4 Use of admixtures

Admixture usage – mainly water-reducers, including plasticisers and superplasticisers – are in common use in the South African ready-mix industry and by larger and more experienced concrete contractors. These tend to be beneficial in that they permit a reduction in water content, and consequently often in cement content, of concrete. To the authors' knowledge, the use of lithium-based admixtures specifically to control ASR has not been reported in South Africa. Information relating to the use of lithium salts for controlling ASR is provided in Chapter 5, section 5.5 of this book.

12.6 ILLUSTRATIVE CASE HISTORIES: SOUTH AFRICA

This section will describe three illustrative case histories, one each from the Provinces of Gauteng, Free State and the Western Cape.

All three case histories are well documented and the structures concerned were observed and/or subjected to repeated load testing over periods of several years. Several additional case histories from South Africa are included in Blight & Alexander (2011).

12.6.1 Motorway portal frame, Johannesburg, Gauteng

Loading tests

Two full-scale loading tests were carried out, six years apart, on a reinforced concrete portal frame that was showing severe deterioration as a result of AAR. The structure, one of a series supporting a double-decker stretch of urban freeway in Johannesburg, had been designed in 1963 when the occurrence of AAR was unknown in South Africa. The dimensions and layout of the frame and the locations of the measurements are shown in Figure 12.2.

Prior to the loading tests, an elastic finite element analysis of the frame was done using a reduced value of elastic modulus for the concrete that had been established by means of laboratory measurements on cores taken from the structure. Measured deflections, rotations and strains were then compared with the predicted quantities. Figure 12.3 gives the results, with experimental points shown for the 1988 test, and the 1982 test results shown as chain-dotted lines. In every case, close agreement was found between prediction and measurement. Over the short time duration of each of the loading tests, the structure behaved almost completely elastically and deformations were nearly fully recovered on removal of the load. Figure 12.3a shows the predicted and measured load-deflection curves for mid-span of the upper beam of the asymmetrical portal and Figure 12.3b shows the corresponding predicted and measured strains in the mid-span tensile reinforcing.

Overall, the agreement obtained between strains, rotations and displacements predicted by analysis on the one hand and observation on the other was excellent, although measured quantities were consistently less than predicted. The results of the two tests, made six years apart in 1982 and 1988, are almost indistinguishable. Note from Figure 12.3a that even though the knee J of the portal was apparently badly damaged by AAR, it still behaved as if fully continuous. Note from Figure 12.3b that the beam of the portal was still behaving as if the concrete was uncracked. Figure 12.3c shows the

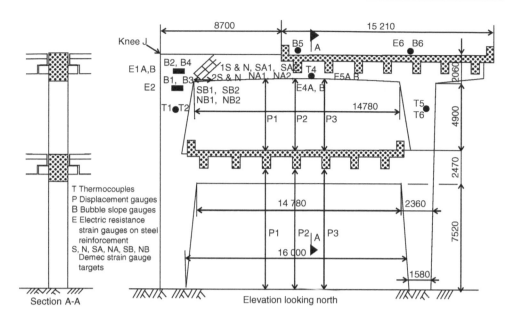

Figure 12.2 Full-scale loading tests and subsequent repair of a motorway portal frame, Johannesburg, Gauteng. Elevations of portal frame showing positions and types of instruments used in load tests in 1982 and 1988. (Dimensions are in mm).

compressive surface strain of the concrete at knee J, which proved to be considerably less than predicted and also indicated that J was behaving as if fully continuous.

Surface diagonal strains on the concrete were measured with gauges 1N to 3N and 1S to 3S, mounted on each face of the beam as shown in Figure 12.3d. The structure was loaded in five approximately equal load increments, and the portal is orientated east-west. Since load increments 1 to 3 were placed on the northern deck span, the torsional strain would have reached a maximum with the application of the third load increment. It would be anticipated, therefore, that diagonal strains due to torsion would increase up to load increment 3 and that thereafter as loads were applied to the southern span, the effect of torsion would reduce to zero at full load and that strains due to pure shear would dominate. This pattern is evident in Figure 12.3d, where the measured diagonal strains at gauge 2N have been compared with the predicted behaviour. Considering the limited resolution of the Demec strain gauge (5×10^{-6}), the measured strains follow the calculated trend very well and were also less than predicted.

The predicted movements and strains shown in Figure 12.3 were based on a value of the elastic modulus derived from measurements made on cores drilled from the structure. This value was 18 GPa. With two exceptions, measured movements and strains, both in 1982 and 1988, were less than predicted by between 21% and 33% of the measured value. The two exceptions were the in-plane rotation of joint J, where the measured values averaged the predicted value for full continuity of the joint, and the strain in the reinforcing at midspan of the beam, where the measured strain was close to the lower limit of the predicted strain, assuming that the concrete took full tension, but 6.3% less than the prediction of no tension in the concrete. This means that the actual elastic modulus of the concrete was not 18 GPa, but between 22 and 24 GPa, *i.e.*, 22% to 33%

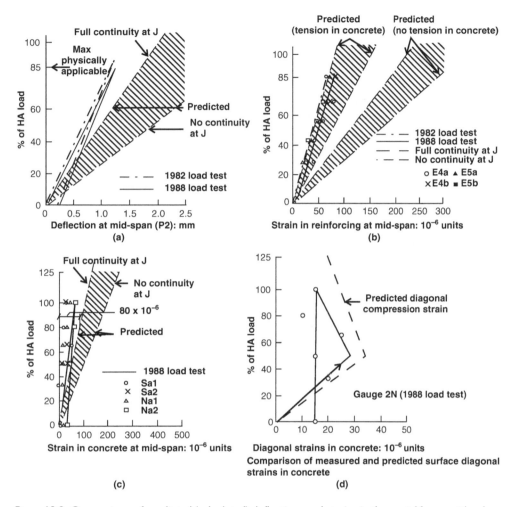

Figure 12.3 Comparison of predicted (calculated) deflections and strains in the portal frame with values measured in the two full-scale loading tests of 1982 and 1988, under 100% NA loading (= 81% of HA loading).

higher than the value from measurements on cores. In the measurements on cores, for which 18 GPa had been the mean, individual measurements, all on concrete damaged by AAR, had varied from16 to 22 GPa (–11% to +22% from the mean.)

The overall conclusion from the 1982 and 1988 tests was that the strength margin of the portal was adequate, and that it could continue in service with only cosmetic repairs.

Repair of portal frame

Although the structural response of the portal frame appeared to be quite adequate as discussed above, it seemed to some observers that the structure had deteriorated unacceptably, based on the visual outward appearance. New cracks had appeared and existing cracks had continued to widen. It was therefore decided by the owners that the structure must be rehabilitated, and investigations were started to find the most effective form of rehabilitation.

A number of possible solutions to the repair and rehabilitation of the apparently badly cracked section of the upper portal beam were considered. These included:

(a)   filling the major surface cracks with an elastomeric sealer and sealing the surface of the exposed concrete to exclude moisture from the surface and from minor surface cracks;

(b)   encasing the beam in a water-excluding ventilated metal or glass fibre reinforced resin sheath to exclude incident rain, but allow the concrete to dry out gradually to equilibrium with the surrounding atmosphere;

(c)   demolishing the damaged length of the upper beam and reconstructing it in reinforced concrete using aggregate that is not susceptible to AAR; and

(d)   different variations of (c) above, including augmenting the strength of the upper beam with bolt-on steel members, and replacing the damaged length with a bolt-on steel beam.

If measure (a) was to be adopted, it would be important to seal the cracks and concrete surface at a time of the year when the concrete was at its driest. Johannesburg has well-defined wet and dry seasons, and the obvious time appeared to be August/September, at the end of the dry season. However, it was not known to what extent the moisture in the concrete varied seasonally, or if the concrete dried out to any significant extent during the dry season. To provide this information, a series of *in situ* thermocouple psychrometer measurements described below was undertaken.

The width of the portal beam and columns is 1250 mm. To ensure that information would be available for most of the thickness of the concrete, the psychrometers were installed in pairs in holes drilled to depths of 150 and 400 mm from the concrete surface. The probes were attached to the ends of wooden dowel sticks that had been heavily varnished with polyurethane varnish, and inserted into the holes. The varnish prevented absorption of water by the dowels, which were push-fitted in the holes. Soft rubber discs attached to the outer ends of each dowel, through which the psychrometer leads were passed, further helped to seal each cavity at its outer end. Since the psychrometers were installed at a height of 16 m above ground level, they had to be installed and accessed for reading via a truck-mounted hydraulic boom used for repairing overhead power lines.

Measurements were taken at intervals for a period of 20 months and the results are summarized in Figures 12.4 (period from April to December 1989) and 12.5 (period from January to November 1990). Readings were discontinued after it was decided, on the basis of the measurements, to rehabilitate the portal by demolishing and reconstructing the upper portal beam.

The upper portion of each of Figures 12.4 and 12.5 show the variations of moisture suction with time, while the lower portion shows the corresponding rainfall, plotted on a daily basis, and the 30-year mean monthly atmospheric relative humidity (plotted for only one year). The measurements are identified by the depth of installation and the direction faced by the side of the portal, *e.g.*, 150 N is installed at 150 mm depth on the north-facing side of the beam. The measurements were started in April 1989, towards the end of the wet season, and it was found that suctions were negligible, indicating that

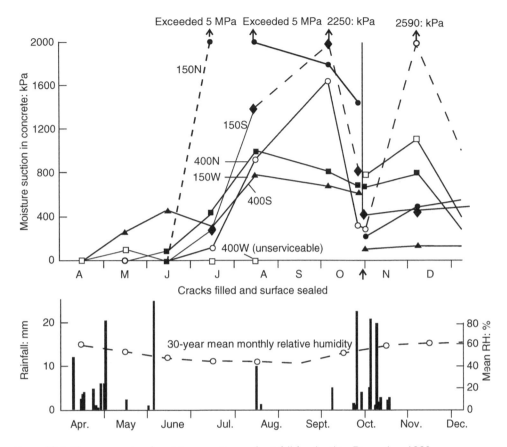

Figure 12.4 Measurements of moisture suction and rainfall for April to December 1989.

the concrete was extremely wet. However, it is possible that at this stage the psychrometers were not yet in moisture equilibrium with the concrete.

During the 1989 dry season (roughly May–September) the 150 mm-deep psychrometers recorded large suctions, with 150 N (which faces the sun in winter) going out of the range of measurement and 150 S (permanently in shade in winter) recording 2250 kPa. 150 W showed a disappointingly low maximum suction of 1600 kPa. Only two of the 400 mm-deep psychrometers were in working order, and they too showed very disappointing maximum suctions of less than 1000 kPa. On the evidence presented in Figures 12.4 and 12.5, concrete that has cracked as a result of AAR will not dry out to below the limiting relative humidity of 97% (a suction of 4100 kPa) in a single dry season. Once the 1989 dry season was ended by rain in October, suctions plummeted once again to low values.

At the end of October (against the advice of the second author) the major surface cracks were caulked with a stiff cement mortar and all surfaces of the beam were treated with a cement-slurry based polymer waterproofing coating. This was the worst possible time to carry out such a treatment, as the concrete was moist and the waterproofing treatment served to seal the moisture into the concrete; subsequent

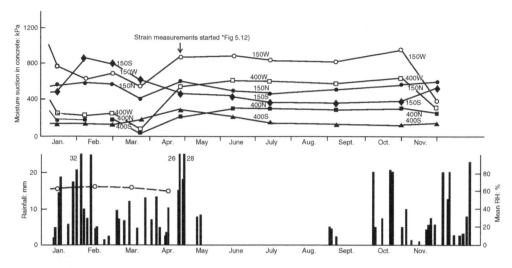

Figure 12.5 Measurements of moisture suction and rainfall for January to November 1990.

measurements (Figure 12.5) were to confirm this. After some wide fluctuations in suction between November 1989 and January 1990, conditions in the concrete stabilized and suctions remained virtually constant for the ensuing 11 months. Very little change in suction occurred during the 1990 dry season, even though there was no rain for 14 weeks. During these months, suctions varied from 100 to 900 kPa, which proved to be far too low to have any influence on arresting the progress of AAR. A very stable moisture regime had been established as a result of sealing the surface of the structure, and also sealing in the moisture that had gained entry via the cracked surface. While sealing the cracks might prove to be a good solution in other circumstances, in this case it was applied at the worst possible time, and exacerbated the problem, rather than ameliorating it.

It was eventually decided by the owner to adopt solution (c), and the repair was carried out in 1991 (Blight & Alexander, 2011). The replacement beam comprised dolomite aggregate, which not only avoids ASR, but also imparted the desired stiffness to the concrete. In 2003 the opportunity arose to inspect the portal from close quarters from a hydraulic lift. It was found that cracking had changed very little since the replacement beam was de-propped 12 years earlier and that the repair was in excellent condition.

12.6.2 Loading Tests on an Underground Mass Concrete Plug (Free State)

Pneumatic tools and machinery are extensively used in underground mining, usually in conjunction with a reservoir or receiver for the compressed air supply that often takes the form of a chamber or blind drive excavated in rock and sealed by shotcrete-lining the rock walls and closing the entrance with a pressure-retaining concrete plug. In this case the reservoir or receiver was located at a depth of 2000 m in a gold mine in the Free State province, measuring 3.5 m wide by 3.3 m high in vertical section, and was closed by a 1.5 m thick unreinforced concrete plug, as shown in Figure 12.6a. The compressor

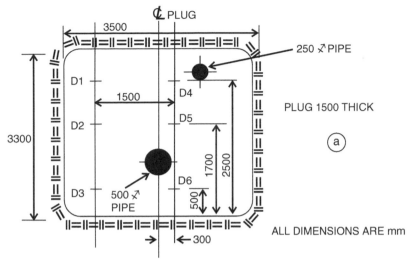

Positions of dial gauges (D1, D2, D3, D4, D5, D6 and
LVDTs (D3 and D5)

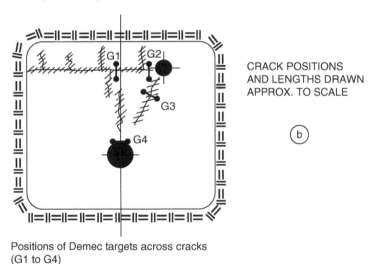

Positions of Demec targets across cracks
(G1 to G4)

Figure 12.6 Layout of measuring points on plug face.

fed the receiver through a 250 mm diameter pipe and the air was drawn off through
a 500 mm diameter pipe. The receiver operated at pressures of between 1400 kPa
(compressor cut-out pressure) and 600 kPa (compressor cut-in pressure).

The concrete plug was of 50 MPa concrete and had been designed and constructed
without any thought of the possibility of AAR, although the aggregate used consisted of
Witwatersrand quartzite. However, in 1981 when the plug was designed, it was not
generally known that concrete containing Witwatersrand quartzite could be subject to
attack by AAR. The air on both sides of the plug was at 100% relative humidity and the
temperature a constant 40 °C on the outside of the plug and up to 55 °C on the inside.

Pressure vessel safety regulations required that the receiver be pressure-tested to 130% of its working pressure (*i.e.*, 1800 kPa) in the following stages:

0 – 70% of working pressure	Hold 1 hour	Measure deformation and leakage
70 – 100% of working pressure	Hold 1 hour	ditto
100 – 130% of working pressure	Hold 10 hours	ditto
130% – 0		Measure recovery of deformation

A request was received from the mine management to plan and carry out the proof test. A system of dial gauges and LVDTs was designed, supported on a rigid framework constructed of welded steel scaffold poles that stood free of the plug and were wedged against the rock walls, floor and roof. Both mechanical dial gauges and LVDTs were used as it was not certain if electrical gauges would work satisfactorily in the 100% humidity underground. The exposed face of the plug was white-washed so as to show up any cracks that might form during testing.

Because of the explosive nature of compressed air, the test pressurization was carried out using water as the pressurized fluid. The results of the test are shown in Figure 12.7. Average outward movement of the plug face, measured by 0.01 mm dial gauges and LVDTs calibrated to 0.01 mm (see Figure 12.6a for location), was a very small 0.4 mm, no creep was observed during the 10 hour pressure hold, and the loss of pressure was negligible. The plug movement was completely reversed on unloading. Hence the plug passed the test and was put into service as a compressed air receiver in February 1983.

Late in 1984 it was noticed that some cracking of the plug had occurred, and in January 1985 the mine management asked for movement of the plug under pressure

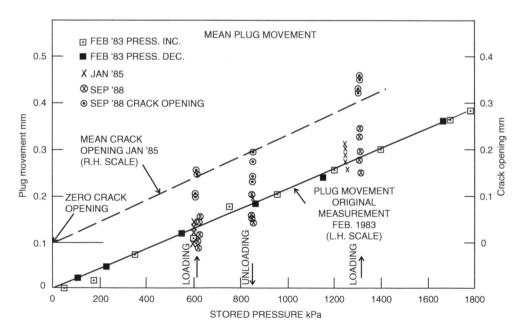

Figure 12.7 Measurements of plug movement and crack opening in 1983, 1985 and 1988.

to be re-measured. The location and extent of the cracks are shown in Figure 12.6b. All the cracks and their limits were marked on the concrete to check on possible further extension. Outward movement of the face of the plug was measured, as well as opening of the cracks, as the air pressure in the receiver fluctuated between the limits of the working pressure of 600 to 1400 kPa. The same system of measurement (dial gauges and LVDTs in the same locations and supported on the same frame) was used to measure the movement of the face, while crack-openings were measured by a 200 mm gauge length Demec gauge. The results (also given in Figure 12.7) showed close to the same relationship between air pressure and plug movement found in the original proof-test, although the measurements were somewhat more randomly variable. There was no indication that the plug was moving progressively. The cracks opened and closed in sympathy with the face movement, following a line parallel to that for the face movement (see Figure 12.7) and pressure losses were negligible. The air receiver was allowed to continue in service without any attempt to repair the cracks, but it was decided to repeat the measurements in 1986.

As it happened, these measurements were not made until September 1988 when, fortunately, they showed no deterioration of the condition of the plug. As shown in Figure 12.7, the September 1988 measurements were almost identical with those made in January 1985, 3.7 years earlier and the cracks had not extended. (The experimental points for crack-opening in 1985 have been omitted from Figure 12.7 to avoid graphical congestion.)

In 1990, the air receiver was decommissioned because the sector of the mine it served had been worked out. The mine was asked to provide a sample of the concrete for examination and, eventually, a piece with a mass of about 5 kg was provided that had been broken out of the plug using a jack-hammer. This showed all the visual signs of AAR attack, including reaction halos around the large particles of aggregate, cracks through the aggregate particles, etc. This was the first time that it was confirmed that the plug had been subject to attack by AAR, a suspicion that had been growing since the cracking was noticed in 1984.

A number of points are illustrated by this case history:

- All of the necessary conditions for AAR to develop were present, *i.e.*, an AAR-susceptible aggregate, a constant supply of free moisture, a high and constant temperature and, with a required strength of 50 MPa in difficult mixing and placing conditions, (probably) a cement-rich concrete mix. In retrospect, the plug acted as a giant accelerated AAR-susceptibility specimen.
- The initial cracking may have resulted from shrinkage, but any shrinkage must rapidly have been reversed by expansion caused by AAR as well as compressive creep in the highly stressed rock walls (overburden stress of 56 MPa) confining the sides of the plug. The cracks obviously did not open into the void of the receiver because leakage of air remained negligible. The directions of the major principal stresses on the plug were probably vertical and horizontal, resulting from closure of the walls of the rock excavation as well as restrained swelling of the concrete caused by AAR. If tension cracks parallel to the free faces of the plug had occurred, they did not lead to any significant pressure losses.
- The measurements show that the strength and elastic modulus of the concrete did not deteriorate to any significant degree. Although it had been attacked by AAR,

the expansion caused by AAR had been restrained by the confining rock walls, resulting in compressive confining stresses on the concrete.

12.6.3 Repair of a Sports Grandstand (Western Cape)

The reinforced and prestressed concrete grandstand was completed in 1978. Figure 12.8 shows the main dimensions in plan and a side elevation of a typical cantilever frame. The main cantilever beams, supporting a light galvanized steel roof, were prestressed back to the columns. Cracking, mainly parallel to the lengths of the beams and heights of the columns, became apparent in 1987 and a full condition survey was carried out in 1989 to assess the severity of the cracking. At this time, some of the cracks were 3.5 mm wide and penetrated the 60 mm cover to the reinforcing steel. Wide cracks ran the full length of some of the cantilever beams. It was decided to repair the structure by applying a hydrophobic surface treatment and injecting the cracks with an epoxy resin. The repair work was carried out in July/ August 1992, which was, incidentally, towards the end of the wet season. The hydrophobic surface treatment and resin injection were followed by application of a water-proofing elastomeric coating in 1994.

Starting in 1992, precise survey measurements were made of the deflections of the beams relative to the tops of the columns at positions F, M and B, marked on the elevation and plan in Figure 12.8. Figure 12.9 shows the variation with time of the measured deflections of positions M and F relative to position B for the 10 years from July 1992 to July 2002. It will be noted that each beam from No. 1 to No. 8 had deflected by a different amount relative to position B. It is also evident that from 1992 onwards, the downward deflection of every beam was decreasing. These upward movements are plotted in Figure 12.10 for April 1996 and July 2002, when decreases at position F of from 3.5 to 14 mm were recorded. The most likely explanation for the continuing upward movement is that, during the period of observation, the concrete continued to swell against the prestress forces, thus pushing the beams upwards. A second possibility is that, simultaneously, the compression modulus of the concrete in the upper part of the beam section had been slowly deteriorating, allowing the prestress force to compress the top of the beam and add to the upward movement. Whatever the explanation, the net effect is on the side of safety, the deflections have occurred very slowly and also at an approximately constant rate – see Figure 12.10.

Figure 12.9 shows the deflected shapes of the eight beams in 1992 ($1'$, $2'$, etc.) and in 2002 (1, 2, etc.). It is interesting that the beam with the greatest deflection (8) has also recovered the most (Figure 12.10), with beam 3 which initially deflected least, showing the second largest recovery of deflection.

The calculated deflection profiles shown in Figure 12.9 were calculated under the estimated dead load, using the second moment of area of the gross beam cross-section. Fitting the calculated line to the measured deflections gives a concrete elastic modulus E of 20 GPa for beams 2 and 3, 9 GPa for beam 4, 5 GPa for beams 1, 5 and 7, and 3 GPa for beam 8. All of these are very possible values: ranging from 20 GPa for almost undamaged concrete to 3 GPa for concrete severely damaged by AAR.

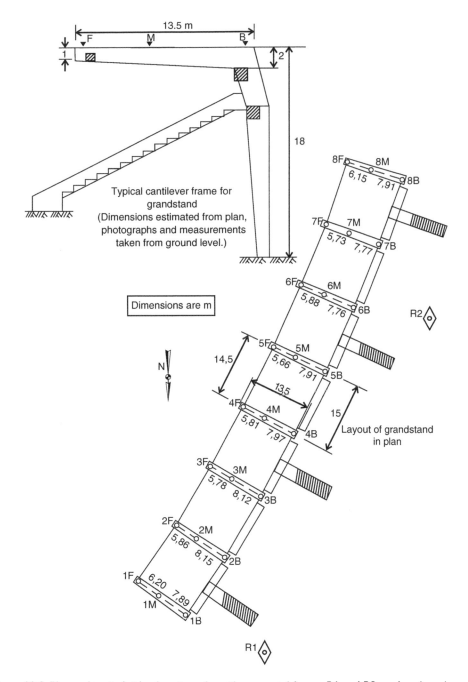

13.5 m

F   M   B

1

2

18

Typical cantilever frame for
grandstand
(Dimensions estimated from plan,
photographs and measurements
taken from ground level.)

Dimensions are m

N

14,5

13,5

15

Layout of grandstand
in plan

8F   8M
6,15 7,91 8B

7F   7M
5,73 7,77 7B

6F   6M
5,88 7,76 6B

R2

5F   5M
5,66 7,91 5B

4F   4M
5,81 7,97 4B

3F   3M
5,78 8,12 3B

2F   2M
5,86 8,15 2B

1F 6,20 7,89
1M   1B

R1

Figure 12.8 Plan and typical side elevation of cantilever portal frame. R1 and R2 are benchmarks.

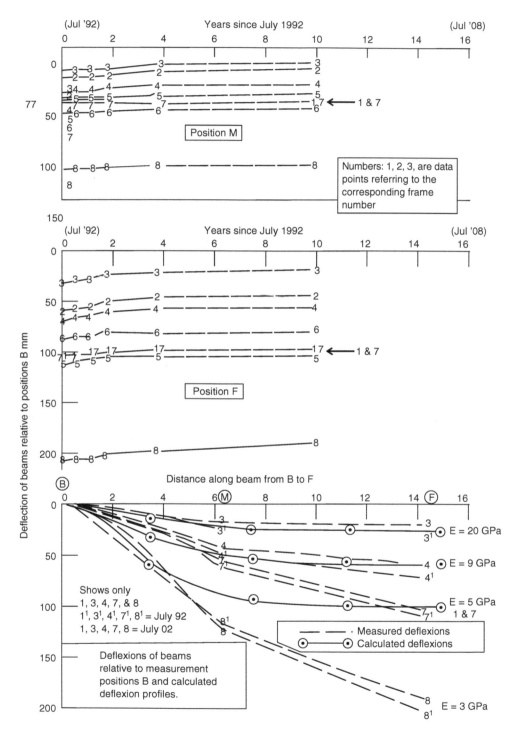

Figure 12.9 Deflections of beams relative to measurement positions B and calculated deflection profiles.

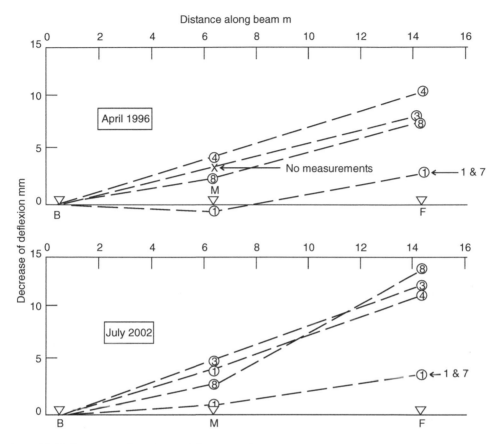

Figure 12.10 Decrease of cantilever deflection with time.

In 2013 the structure showed no signs of new or re-opened old cracks, although this may be due to recent (possibly unrecorded) crack filling and the application of an elastomeric paint. Both visually and in terms of the beam deflection measurements, the 1992 repairs appear to have been completely successful. This case is a powerful reason for believing that, in many cases, even if AAR swelling cannot easily be stopped, it need not inevitably be a continuing threat to the integrity of the structure.

12.7 ILLUSTRATIVE CASE HISTORIES: OTHER AFRICAN COUNTRIES

Very few documented case histories are available for AAR problems in Southern and Central African countries other than South Africa. Those that are available will briefly be reported, specifically from Namibia, Kenya, Zambia, Mozambique, Zimbabwe, Uganda and Zaire. The geographical locations are illustrated in Figure 12.11. Based

Figure 12.11 Geographical locations of countries in Southern and Central Africa, other than South Africa, for which information on AAR is given.

on a fundamental understanding of the causes of AAR, it is to be expected that this form of deterioration might be observed widely throughout the sub-continent, and this is indeed the case as illustrated. Unfortunately, in some cases, early observations of damage caused by AAR and remedial measures are described, but no follow-up concerning the success (or otherwise) of the repairs is available.

12.7.1 Namibia

Alkali-susceptible rocks in Namibia are Archaean granite and gneiss. According to Fulton (2009), mast foundations at Swakopmund have been affected by ASR, where the coarse aggregate was from a granite quarry at Walvis Bay.

12.7.2 Kenya

Kamburu spillway (Sims & Evans, 1988)

The Kamburu 94 MW hydroelectric scheme is situated on the Tana River, 160 km north of Nairobi, fed by water impounded behind a 56 m high asphaltic concrete–faced rockfill dam. The scheme was built between 1970 and 1974. The spillway is set in a rock–cut channel 30 m deep and 50 m wide. The spillway control structure, shown in upstream elevation and section in Figure 12.12, consists of a mass concrete rollway supporting 4 concrete piers. Between these are three radial gates, 13 m square in elevation.

In 1982, eight years after construction, movement was observed between pier 1 and the contiguous stop-log store. By 1985 it was found that pier 1 had deformed to the extent that it was no longer possible to open gate 1. Examination of pier 1 showed the

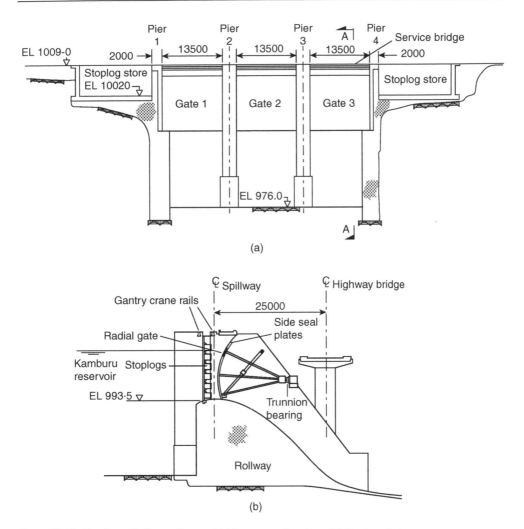

Figure 12.12 Kamburu Spillway, Kenya. (a) Upstream elevation (b) Section A-A.

cracking in elevation illustrated by Figure 12.13, and the outward movement contours (of up to 40 mm) shown by Figure 12.14.

The coarse aggregate for the concrete was crushed rock from the spillway excavation and the underground power station. It was almost entirely gneiss with leucocratic white and grey varieties predominating and some pink granitic gneiss. The rock contained strained quartz and veins of opal. The fine aggregate consisted of a mixture of crusher sand and river sand, the latter consisting mainly of single crystal grains of quartz. The cement was ordinary Portland cement, with a mean alkali content varying between 0.67 and 0.79% Na_2O equivalent.

Cores taken from the affected concrete showed the following evidence of the development of ASR:

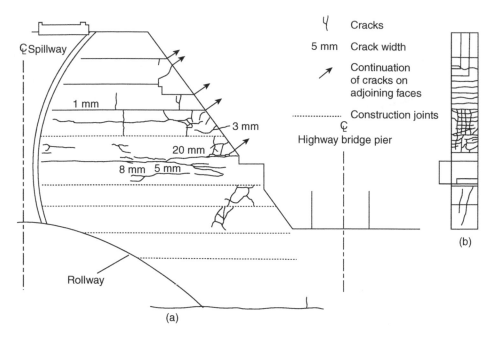

Figure 12.13 Kamburu Spillway, Kenya. Part elevation on Pier I showing cracking: (a) Side (b) End.

- gel within microcracks both within and remote from aggregate particles
- microcracking due to gel pressure
- internally cracked reacted particles of opal and quartz.

Quoting Sims & Evans (1988), *"Experience at Kamburu spillway is that when opal is present as a contaminant in the concrete, there is no safe level of alkali content"*.

Core samples exhibited a wide range of compressive strengths, but there was no evidence of weakening due to ASR. Within grade C40 concrete, for example, the strength was in the range 35.0 to 56.5 MPa with a mean of 44.0 MPa and a standard deviation of 8.7 MPa. The tensile splitting strength was 2.7 – 4.3 MPa with a mean of 3.5 MPa and a standard deviation of 0.4 MPa, the ratio of tensile to compressive strength being 0.05 – 0.10.

The following steps were taken as remedial measures:

- gate 1 was released by trimming 140 mm off the side of the gate adjacent to pier 1,
- grouting and drainage holes were used to keep water from accumulating in the stop-log store where it could possibly seep into the concrete of pier 1,
- pier 1 was strengthened by installing rock anchors into the rock behind it,
- the larger cracks in the faces of the piers were sealed by injecting epoxide resin, after removal of loose or otherwise deteriorated concrete.

Borehole extensometers and inclinometers were installed to monitor possible future movements of the piers. These are shown diagrammatically in Figure 12.15 a) and b). It would be most interesting to follow these up, as it is now 24 years since the paper describing the remedial measures appeared.

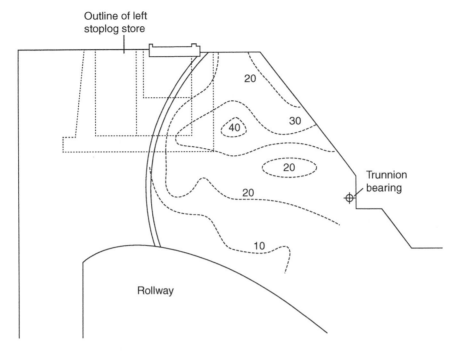

Figure 12.14 Kamburu Spillway, Kenya. Deflection contours for Pier 1 (in mm).

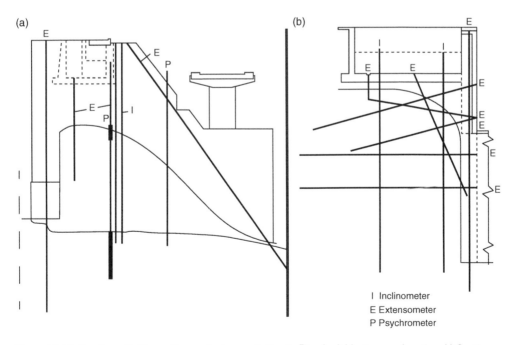

Figure 12.15 Kamburu Spillway, Kenya. Instrumentation in Pier 1. a) Upstream elevation, b) Section.

12.7.3 Zambia

Itezhitezhi Dam Project, Zambia. (Thaulow, 1983)

The Itezhitezhi dam on the Kafue River in Zambia is an earthfill embankment (8.5 million m^3 of earthfill) that provides storage for the Kafue Gorge Power Plant, 300 km downstream. The dam was constructed between 1973 and 1976 and the 5000 million m^3 capacity reservoir was full by the middle of 1978.

In 1980, substantial swelling of the concrete in the intake towers to the diversion tunnels became apparent. Swelling of about 40 mm over the tower height of 30 m, as well as horizontal bending had occurred. The aggregate had been tested according to ASTM C227 (now 2010), before use, and was found by that test to be non-expansive. The alkali content of the cement varied between 0.65 and 0.85% with an average of 0.72% Na_2O equivalent.

The investigation of the problem by Thaulow (1983) comprised a visual inspection of the concrete, strength determinations by CAPO test on the intake tower and regulation gate shaft, and measurement of the strain of a reinforcement bar in the intake tower. Visual inspection showed clear signs of AAR in the form of coarse map cracking (typical crack width 0.5 mm) of the intake towers. To quote Thaulow: *"Measured strain on one vertical 16 mm diam. rebar was 0.18%"*. He does not elaborate on how the measurement was done, but the most likely way would have been to expose a length of bar, glue Demec targets or an ERS gauge to the bar, and then cut the bar to measure the elastic rebound. $0.18\% = 1.8 \times 10^{-3}$ in strain units, which is equivalent to an elastic stress release in the steel of 360 MPa. Assuming a modular ratio of 15, this would indicate a compressive stress in the concrete of 6 MPa, which is high, but not impossible, especially locally. Also, the average compressive strength of the concrete was determined to be 31.6 MPa, and hence a 6 MPa swelling stress, in addition to the gravity stresses carried by the tower, probably of the order of 2 to 3 MPa at most, would not be excessive.

Figure 12.16 shows levelling measurements of the gantry rails on the intake towers, starting in early 1980, at which time the expansion of the towers had been measured at about 40 mm.

The cores of concrete taken from the intake tower showed that the concrete was made with red granite coarse aggregate. The cores contained silica gel that was filling voids in the concrete. Examination of thin sections of concrete showed that the coarse aggregate contained slightly weathered feldspar grains. Veins in the granite were filled with a uniform brown mineral believed to be opal. Some aggregate particles contained cracks radiating into the cement paste, which were filled with silica gel. The sand fraction of the aggregate contained mainly quartz and feldspar grains and showed no signs of reactivity.

As Figure 12.16 shows, the expansion of the towers seemed to have ceased by mid–1982, and all things considered, no repair measures were deemed necessary at that time.

12.7.4 Mozambique

Cahora Bassa Dam (Ramos *et al.*, 1995; Batista *et al.*, 2012)

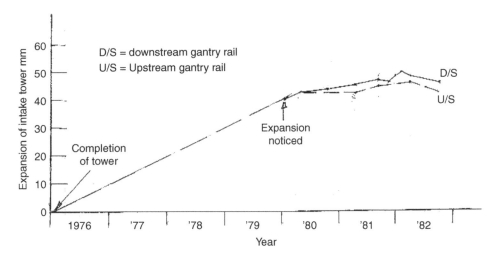

Figure 12.16 Itezhitezhi Dam Project, Zambia. Expansion of intake towers, 1976–1982.

Cahora–Bassa Dam is a 170 m high arch dam with a crest length of 300 m. The dam is situated on the Zambezi river in Mozambique and was constructed between 1971 and 1974. A summary of pertinent information on the dam is given below, based on the reports by Ramos *et al.* (1995) and Batista *et al.* (2012).

Measurements from no-stress strain meters are available from 1975 and precise levelling data are available from 1977. The no-stress strain meter data indicate that the effects of concrete swelling became measureable in 1979. The measured expansions correspond to rates of expansion of about 13 to 26.10^{-6} per year. The results from the no-stress strain meters were used to establish zones of swelling strain and evolution of swelling in the dam.

The precise levelling measurements indicate a height increase of 11 mm from 1977 to 1994, and 24 mm for the 30 years from 1977 to 2007, over a total concrete height of about 125 m. The profile of measured vertical displacement changes from the precise levelling is symmetric about the centreline of the dam. The 1994 measurements indicate a vertical strain of 88 microstrain or 5.2 microstrain per year, while the 2007 measurements correspond to a vertical strain of 190 microstrain, or an average strain of 6.3 microstrain per year. Hence vertical strains appear to have been almost constant over time, but have remained small. Over the same 30 year period, upstream radial displacements have corresponded to changes of reservoir water level and have reached a maximum at mid-crest of 85 mm (280×10^{-6} of the 300 m crest length).

Compressive stresses, caused by swelling alone, amount to about 12 MPa at the abutments of the upper parts of the arch, whereas the highest tensions amount to 3 MPa at mid-span of the arch. The tensile stresses are reduced by the dead weight and water pressure to about 8 MPa compression. Compression stresses at the abutments remain at 12 MPa. Hence, despite the action of AAR, in 2012, 35 years after completion, serviceability and safety conditions for Cahora Bassa remain good.

12.7.5 Zimbabwe

Kariba dam (Charlwood *et al.*, 2012).

It is reported that Kariba dam, also a major arch dam on the Zambezi river upstream of Cahora Bassa dam and completed in 1959, has developed problems with stop-log and vertical gate guides. The *"approximate accumulated free expansion strain as a proportion of the estimated service limit state"* is only 0.04% over a life of 53 years as compared with 0.08% over 35 years for Cahora Bassa. Hence AAR-related problems at Kariba appear minor.

Several accounts of ASR and scour problems related to the Kariba Dam have been published recently (Parker, 2014 and in The Times of 17 May 2014). These highlight an 80 mm expansion of the dam due to ASR attributed to the gneiss coarse aggregate. The expansion is causing difficulties in operating the floodgates.

12.7.6 Uganda

The Owen Falls (now Nalubaale) power plant (Figure 12.17), on the Victoria (White) Nile near Jinja, was completed between 1951 and 1954 and is now known to exhibit ASR, associated with schistose aggregates excavated from the site, and has been subjected to some mitigation action and careful management. Investigation in the 1990s (Mason & Molyneux, 1998) found that a second stage of concreting was probably exerting the primary expansive force, causing marked movement in the upper gallery, in contrast to the lower gallery (see Figure 12.18). This has given rise to difficulties in alignment of equipment. The initial solution was to insert removable/adjustable spacers under the turbines; in the worst case, a 90 mm spacer was used. ASR in the power house concrete is expected to continue for the life of the plant, but further remedial action is now being considered.

12.7.7 Zaire

The N'Zilo (formerly Delcommune) Dam in Zaire was completed in 1952. It is a thin, double curvature arch dam about 246 feet (80.7 m) high. It has suffered an expansion of approximately 1000 microstrain in 30 years and this has resulted in extensive cracking. This damage has not been positively ascribed to ASR in the concrete, but investigations of the causes of expansion continue.

12.7.8 Summary

Incidences of AAR have been reported from a number of structures in southern and central Africa, mostly large water-retaining structures such as dams. Despite the occurrence of AAR, all these structures remain serviceable to the present time, although in some cases remedial works were required to ensure continued operation of ancillary equipment. Thus, it seems that in many cases, depending on the specific conditions, AAR-affected structures can continue to fulfil their functions over many years. Further case studies, and follow-up studies on structures known to exhibit AAR, are however required.

Figure 12.17 Downstream (north) views of Owen Falls (Nalubaale): a) the main dam, b) the tailrace and power station (photographs courtesy of Dr Rene Brueckner of Mott MacDonald Limited).

12.8 CENTRAL AFRICAN COUNTRIES WHERE NO INFORMATION IS AVAILABLE

The usual definition of 'Central Africa' is that it consists of Angola, Zambia, Zimbabwe, Mozambique, Congo (DRC), Tanzania, Uganda, and Kenya, and was the definition used here (although note that some of these countries can equally be regarded as part of southern Africa). Countries that lie to the north of these are defined as West or North

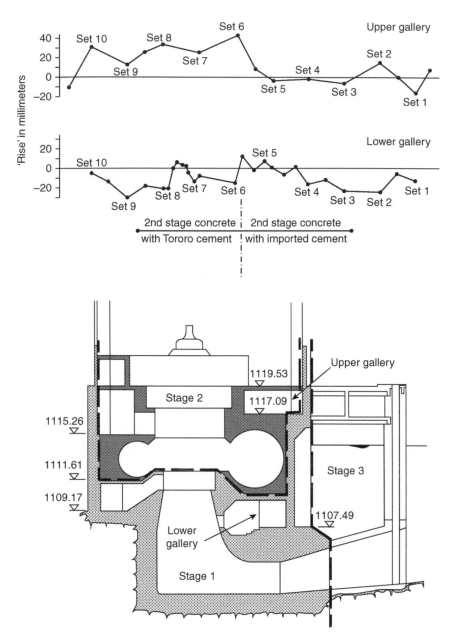

Figure 12.18 Cross-section through one of the generator installations in the power station of Owen Falls (Nalubaale), where the three concreting stages were found to exhibit different degrees of movement (from Fig. 9, Mason & Molyneux, 1998) Republished with permission of ICE Publishing, from The effect of concrete expansion at Owen Falls Power Station, Uganda in Proceedings of the Institution of Civil Engineers. Water, maritime and energy, Mason and Molyneux, 130, 1998; permission conveyed through Copyright Clearance Center, Inc.

Africa, comprising inter alia Senegal, Guinea, Mauritania, Mali, Niger, Chad, Sudan, Ethiopia, Gabon, Cameroon, Congo (Brazzaville), and Somalia. Many of these countries have been, or are involved in long drawn out and bloody civil or proxy wars that started when the previous colonial borders crumbled some 50 years ago. The golden age for large construction in many regions of Africa ended in about 1970. However, with the inevitable and needed development in Africa beginning to show signs of occurring, doubtless much construction, including of large dams, is likely to take place in coming decades.

The current authors have been unable to obtain information about AAR in concrete in all of these regions at the present time. However, concerning Ethiopia, a review by Dinku and Bogale (2004) identified the general rock types present as basalts, ignimbrites, tuffs, sandstones, limestones and granite. They identified the granites as being the most probable sources for any potentially alkali-reactive aggregates. They found no examples or accounts of ASR in concretes in Ethiopia and ascribed this to the very low alkali contents of the cements used in construction.

REFERENCES

Alexander, M.G. & Mindess, S. (2005). *Aggregates in concrete*. London: Taylor & Francis. 435p.

ASTM C227 (2010). *Standard test method for potential alkali-silica reactivity of aggregates (mortar-bar method)*. Conshohocken, USA, American Association for Testing and Materials.

ASTM C1260 (2014). *Standard test method for potential alkali-reactivity of aggregates (mortar-bar method)*. Conshohocken, USA, American Association for Testing and Materials.

ASTM C1293 (2008). Standard method for determination of length change of concrete due to alkali-silica reactivity. Conshohocken, USA, American Association for Testing and Materials.

Batista, A.L., Gomes, J.P., Carvalho, E.F., & Tembe (2012) I.M. Analysis and interpretation of the structural behaviour of Cahora Bassa dam (Mozambique). In: *Proceedings of the 54th Congress of the Brazilian Concrete Institute* (Ibracon). Maceio, October 2012, 14p.

Blight, G.E. & Alexander, M.G. (2011). *Alkali-Aggregate Reaction and Structural Damage to Concrete – Engineering assessment, repair and management*. London: CRC Press (Taylor and Francis Group), 246p.

Charlwood, R., Scrivener, K., & Sims, I. (2012). Recent developments in the management of chemical expansion of concrete in dams and hydro projects – Part 1: Existing structures. Hydro 2012, Bilbao, Spain.

Dinku, A. & Bogale, B. (2004) Alkali-Aggregate reaction in concrete: A review of the ethiopian situation. *J EEA.*, 21, 47–58.

EN 197-1 (2011). Cement. Composition, specifications and conformity criteria for common cements. *Committee for European Normalisation* (Standardisation), Brussels.

Fulton's Concrete Technology (2009) (ed.) G. Owens. 9th edn. *Cement & Concrete Institute*, Midrand, SA.

Mason, P.J. & Molyneux, J.D. (1998) The effect of concrete expansion at Owen Falls power station, Uganda. In: *Proceedings of the Institution of Civil Engineers: Water, Maritime & Energy*, 130, December, pp. 226–237 (Paper 11726).

Nixon, P.J., Sims, I. (eds.) (2016) RILEM recommendations for the prevention of damage by alkali-aggregate reactions in new concrete structures. *RILEM State-of-the-Art Reports*, Volume 17, Springer, Dordrecht, RILEM, Paris, 168p.

Oberholster, R.O. (1983). Alkali reactivity of siliceous rock aggregates: Diagnosis of reactivity, using cement and aggregate and description of preventive measures. In: *Proceedings. of the 6th International Conference on AAR in Concrete*, Copenhagen, Denmark. pp. 419–433.

Parker, D. (2014). Kariba dam in scour threat. Alkali silica reaction unknown when the dam was built. *New Civil Engineer*. ICE, London, 01-08.05.2014, 6–7.

Ramos, J.M., Batista, A.L., Oliveira, S.B., de Castro, A.T., Silva, H.M., & de Pinho, J.S., (1995). Reliability of Arch Dams Subject to Concrete Swelling – Three Case Histories. In: *Proceedings. of the Second International Conference on Alkali-Aggregate Reaction in Hydroelectric Plants and Dams*, USCOLD, Chattanooga, Tennessee. pp. 259–274.

SANS 6245 (2006). *Potential reactivity of aggregates with alkalis (accelerated mortar prism method)*. South African Bureau of Standards, Groenkloof, Pretoria, Republic of South Africa.

Sims, G.P. & Evans, D.E. (1988). Alkali-silica reaction: Kamburu spillway, Kenya, case history. *Instn Civ Engrs.*, Part 1, *84*, 1213–1235.

Thaulow, N. (1983). Alkali-silica reaction in the Itezhitezhi dam project, Zambia. In: G.M. Idorn, & S. Rostam, (eds.) *Proceedings of the 6th International Conference Alkalis in Concrete* Copenhagen, Denmark. pp. 471–477.

Chapter 13

Japan, China and South-East Asia

Kazuo Yamada & Toyoaki Miyagawa

13.1 INTRODUCTION

The original edition of this book included a chapter on the experience of AAR in Japan. Since then, the recognition and understanding of AAR have spread and grown around the world (the principal reason for preparing this new edition), so that this chapter now embraces China and other south-east Asian countries, as well as updating the situation in Japan.

13.1.1 Advances since the 1st edition

In the first edition of this book, 'The Alkali-Silica Reaction in Concrete' (Swamy, 1992), Japanese researchers reported the most advanced knowledge at that time, with examples of damage, and countermeasures for suppressing ASR in Japan. At that time, andesite, (Kawamura *et al.*, 1983) and chert, (Tatematsu *et al.*, 1986; Marino *et al.*, 1987) were considered as important rock types and many of the basic mechanisms were well studied excepting petrographic considerations. As is stated in the 1st edition, ASR was 'confirmed' in Japan in 1983. After intensive studies for several years, Japan established a methodology for suppressing ASR in 1986 as JIS standards based on:

- JIS A 1145 chemical method (modified ASTM C289),
- JIS A 1146 mortar-bar test (modified ASTM C227),
- JIS A 1804 rapid method using an autoclave,
- A total concrete alkali limit less than 3.0 kg/m^3,
- An application of effective supplementary cementitious materials (SCMs)
- JASS 5N "Japanese Architectural Standard Specification for Reinforced Concrete Work relating Nuclear Power Generation Facilities" by Architectural Institute of Japan, AIJ with JASS 5N T603 "Test Method for Reactivity of Concrete" at varying alkali content at 40 °C for half year including wet cloth wrapping of 10x10x40cm concrete prism,
- Additional supplementary standards of Japan Concrete Institute, JCI for water soluble alkali measurement, petrographic identification and quantification methods of harmful minerals in aggregate, expansion measurement of core sample from concrete structures.

These methods are still considered effective in Japan with slight modifications. There have been misunderstandings and too much simplification, such as an ignorance of proportional 'pessimum' effects of highly reactive aggregate or of slow expansion by cryptocrystalline quartz.

After the establishment of countermeasures, the general activity of researchers on ASR had become limited. Then, after reinforcement breaks were reported on TV and in newspapers in 2003, major concerns were raised about the mechanical performance of ASR damaged structures and the mechanism of reinforcement breakage. The Japanese Society of Civil Engineers (JSCE) published a report of the technical committee chaired by Miyagawa (2005). The effect of reinforcement break is thought to be limited and the major reason of reinforcement break was sharp bending processing (Miyagawa *et al.*, 2006). However, the countermeasures against ASR were not themselves considered incomplete. Small numbers of researchers including Torii and Morino have been continuing basic studies on reaction mechanisms. Torii reported severe damage in the Hokuriku-district and various ways of repairing and reinforcing methods were tried (Torii *et al.*, 2008).

Recently, a technical committee in JCI on ASR, chaired by K. Torii, re-discussed the working mechanism (JCI – TC062A: 2008) and many researchers and engineers in Japan now recognize that the Japanese methods were not perfect and required reconsideration. Another technical committee in JCI, chaired by K. Yamada, has worked on preparing a procedure for ASR diagnosis based on petrographic observation, JCI – TC 115FS (2014). During this period, ASR studies in Japan have involved intensive and continuous studies undertaken by T. Katayama characteristically from a petrographic viewpoint, but his work has been published in English only. JCI – TC115FS proposed TC standards for petrographic observation of concrete, concrete prism testing with alkali wrapping, total alkali content estimation by SEM/EDS, and core expansion test in 1N-NaOH solution and saturated NaCl solution.

Now another technical committee in JCI chaired by K. Yamada on ASR controlling design and management scenario based on performance specification, JCI -TC152A (2015–2017) is active and mechanical performance evaluation not only expansion in long term is now under discussion.

13.1.2 Extension from Japan to China and South-East Asia

Reported examples of AAR damage in East and South-East Asia are limited except for Japan, China and Taiwan. Only a brief summary is available, Katayama (1997). Even in Japan, the majority of examples are limited to concrete containing highly reactive andesite and chert, but with a few exceptions there is little information concerning late expansive aggregate (Hamada *et al.*, 2004; Katayama *et al.*, 2004; Katayama, 2012), although late expansion effects are not rare, but are general in certain areas of Japan. "Late expansion" was defined for slow reactive aggregate not showing pessimum effects by Katayama (1997) separated from rapid expansion aggregates showing pessimum effects. There are many cryptocrystalline quartz-bearing metamorphic rocks resulting from orogenic movements and contact metamorphism of granite has developed hornfels containing small quartz. The potential for AAR is dependent on the petrographic and mineralogical characteristics of aggregates as is recognised worldwide. Therefore, when the recognition of AAR in a particular country is discussed, it is very important to understand the local geology. In this chapter, in order to cover the limited number of reported examples of AAR, a general geology in the region is explained and the possibilities of AAR relating to specific aggregate rock types are explained.

Here, one point should be remembered; there is a significant discrepancy between required information for the studies of pure geology and for recognition of AAR. With AAR, major concerns are related to the occurrence of reactive silica mineral or acidic glass as components of rock aggregate, although from the viewpoint of geology they are only a small component of various rock forming minerals and are considered much less important than the main rock forming minerals.

Nowadays, reactive minerals have been clearly identified and include opal of various origins, cristobalite, tridymite, chalcedony, cryptocrystalline or microcrystalline quartz occurring as agglomerates, single grains or grain boundary fillings, and acid volcanic glass. With a basic knowledge of geology, the typical formation processes of these phases are also well-known. The procedure to observe and to describe aggregate is already documented in ASTM C295 or RILEM AAR-1 (in Nixon & Sims, 2016), and it is possible to point out where the rocks bearing these minerals can be found if one is familiar with local geology. Typical rock types which may include reactive minerals are summarized in Table 13.1. Difficulties arise if there is little experience or knowledge of the local geology in specific countries, although every country will have a geological map.

Table 13.1 Typical rock types and reactive minerals in some Japanese rocks. The effect of geological age depends on the geothermal gradient.

Rock type	Reactive minerals	Remarks
Neogene and Quaternary acidic volcanic rocks such as andesite, dacite and some types of basic volcanic rocks such as basalt and tuff	Cristobalite/ tridymite, acidic volcanic glass	Typically rapidly expansive, one example showing a proportional pessimum effect.
Chert, (Flint,) Siliceous shale or schist, Silicified rocks	Opal, chalcedony, crypto/ microcrystalline quartz	Wider range of reactivity depending on crystallite size.
Altered igneous rocks. Hydrothermal activity affected rocks Partially weathered rocks in hot and humid climates	Secondary opal, cristobalite	Rapidly expansive. Especially damaging when used in pessimum proportion. Includes igneous rocks and also various altered rocks.
Sandstones, mudstones, greywacke, quartzite	Crypto/ microcrystalline quartz	Generally show late expansive behaviour in concrete
Argillaceous dolostone/ limestone	Crypto/ microcrystalline quartz	Wide range of potential reactivity.
Sheared rocks such as Pre-Cambrian rocks, metamorphic rocks, cataclasite, mylonite, hornfels	Crypto/ microcrystalline quartz	Typically late expansive types in concrete.
Natural gravel or sand	Possibility of mixing of a range of reactive types	Wider range from rapid and severe when mixed in pessimum composition to late expansive.

This chapter is based on the concept that instead of introducing numerous examples in specific countries and limited number of examples in developing countries, an outline of geology in each area and the possibilities of ASR are explained. The first, important examples reported are from Japan and the general geology of Japan is explained from the viewpoint of ASR, *i.e.*, the distribution of typical alkali reactive rock types which may be used as aggregates. Then, some examples from other Asian countries are explained with attention to the geology and aggregate sources.

13.2 TYPICAL EXAMPLES OF ASR IN JAPAN

13.2.1 Andesite, coarse and fine aggregates and reinforcement breakage

Steel reinforcement can be broken by ASR, as shown in Figure 13.1, and are detected in many places in Japan. Okinawa, the Hanshin highway near Osaka and Hokuriku district are typical examples [Figure 13.2 (Torii, 2008)]. The distribution of reported ASR damage and reinforcement breakages occur when the major reactive minerals in the aggregate are cristobalite, tridymite and acid volcanic glass. According to some studies [Nomura and Torii (2008), Katayama *et al.* (2004)] the total alkali in concrete that suffered serious ASR damage in Hokuriku district was estimated as less than 3.0 kg/m^3, but additional alkali release from the aggregate is suggested, although it is not clear whether released alkali increased pH of pore solution. The way of estimating alkali amount from cement by Katayama is unique. This calculates total alkalis from cement in two forms as the ratio of soluble alkali sulfate and insoluble alkalis in clinker, by measuring alkali concentration in major clinker minerals by using electron probe microanalysis (EPMA) or secondary electron microscope with energy dispersive

Figure 13.1 Reinforcement breakage in concrete containing reactive andesite aggregate, Hokuriku District, Japan.

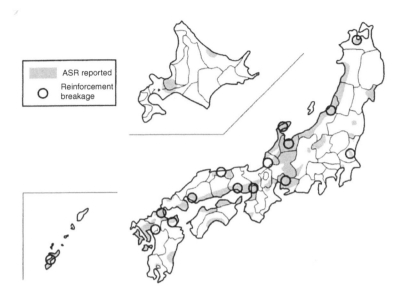

Figure 13.2 Distribution of reported ASR (grey shading) and reinforcement breakage (circles), Torii (2008).

spectrograph (SEM/EDS), then the alkali concentration in the original cement can be estimated. Some of the severely damaged concretes containing this kind of highly reactive andesite aggregate, as both coarse and fine aggregates, in piers of a bridge that have recently been replaced showed extraordinary deterioration of decreased rigidity of concrete, as reported by Daidai and Torii (2008).

However, andesite does not always cause ASR damage. The oldest example of ASR in Japan is in breakwaters of Otaru port in western Hokkaido. The oldest concrete example examined was made in 1899 and andesite aggregate was used. Alkali silica gel (ASG) was detected but no serious damage was observed. Volcanic ash was also used in this concrete and this might have suppressed the ASR damage (Katayama & Sakai, 1998).

13.2.2 Pessimum effect of opal in tuff particles included in a sand with innocuous limestone coarse aggregate

Examples explained above are from structures 30 years or older, which were constructed before the establishment of ASR countermeasures in Japan in 1986. However, after the introduction of ASR countermeasures, there are still examples of ASR damage. In the Tokyo bay area, in order to avoid drying shrinkage and ASR, limestone aggregate is preferentially used as coarse aggregate, taking advantages of sea transport from distant big limestone quarries. Fine aggregate marks use of several kinds of transported sand blended to adjust the poor particle size distribution of local sand. Some specific kinds of sand include particles with highly reactive minerals, such as opal and in some cases these have been partially combined with the Japanese pure limestone coarse aggregate which has limited ability to absorb alkali; this opal may result in a proportional 'pessimum' effect and this is thought to be the reason for the damage.

Cases of ASR found in the Tokyo bay area, constructed during a particular period (Yamada *et al.*, 2007), involved an innocuous limestone combined with a reactive type of sand, giving the possibility that ASR expansion might become serious. ASR damage of this type is shown in Figure 13.3 (Yamada *et al.*, 2007). Many cracks were observed, but only in the parts of buildings or bridges affected by rain. The sand used in this concrete was estimated as the mixture from several sources. One source of sands contained relatively soft tuff particles containing opal. Only a few % of reactive particles caused severe ASR expansion after several years.

The pessimum phenomenon depends on the reactivity of minerals in the aggregate. Typical pessimum compositions of various minerals are summarised in Table 13.2 (Katayama, 1997). Proportional pessimum effects of the sand used in the structure mentioned above was confirmed by mortar-bar test, with various alkali contents, and the results are shown in Figure 13.4 (Inoue *et al.*, 2010). One important thing to remember is the suppression of ASR by fly ash, which decreases at the pessimum composition, as shown in Figure 13.5 (Kawabata *et al.*, 2012). When highly reactive aggregate is used in the mix, the required amount of fly ash needed to be increased significantly.

13.2.3 Opal veins in andesite aggregate

In a railway bridge concrete, andesite aggregate was used and caused serious expansion several years after construction. Petrographic investigation revealed the

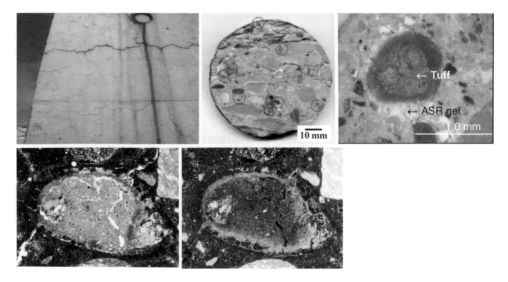

Figure 13.3 Example of early age cracking resulting from the pessimum effect of the combination of reactive opal bearing tuff particles in the sand and limestone coarse aggregate, after Yamada *et al.* (2007).

Table 13.2 Minerals causing alkali-silica reaction and pessimum compositions (Katayama, 1997).

Silica mineral	Pessimum	Volcanic glass	Pessimum
Opal	< 5 %	Rhyolite	100%
Cristobalite	< 10 %	Dacite	100%
Tridymite	< 10 %	Andesite	Boundary of innocuous/ harmful
Chalcedony	20 %		
Cryptocrystalline quartz[*]	50 %	Basalt	Innocuous
Microcrystalline quartz	> 80%		

* Grain size is less than 4 μm.

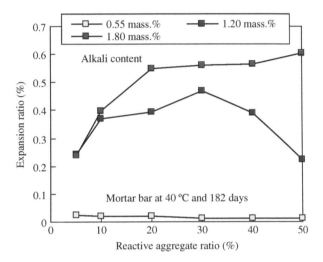

Figure 13.4 Compositional pessimum behaviour of land sand containing opal bearing tuff (after Inoue *et al.*, 2010).

existence of opaline veins accompanying cristobalite in the andesite aggregate, as shown in Figure 13.6 (Hayashi *et al.*, 2009). The coarse aggregate was composed of highly reactive andesite and innocuous dolerite. The aggregate used was assessed as 'innocuous' in a test before construction, but it is now difficult to imagine why this kind of andesite was originally considered innocuous. There might have been some contamination during construction. The sand used was a land sand found to be marginally innocuous by a mortar-bar test. The basic problem is in the ASR expansion with limited total alkali content of only approximately 2.6 kg/m³, less than the 3.0 kg/m³ in the Japanese standard specification. The best way to avoid this ASR risk is to eliminate this kind of highly reactive aggregate by petrographic observation. As noted previously, use of fly ash as a preventative measure is increasing, but it is difficult to estimate how much fly ash has to be added for this kind of sensitive pessimum condition.

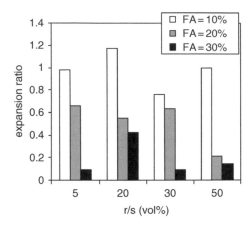

Figure 13.5 Decreased suppressing effect of ASR by fly ash (vol%) at the pessimum composition (Kawabata *et al.*, 2012).

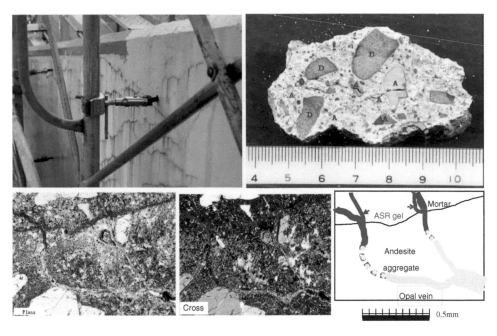

Figure 13.6 Serious ASR damage of a prestressed concrete member by extremely reactive opal bearing andesite (A) in coarse aggregate. D is innocuous dolerite, Hayashi *et al.* (2009:1).

13.2.4 Chert aggregate

Chert aggregate has also caused serious ASR damage in many concretes in Japan, but especially in central Japan. However, the reported number of cases of ASR damage by chert is smaller than those due to andesite. Reactivity of chert strongly depends on the crystal size. Most reactive chert including chalcedony is too reactive to be detected by ASTM C1260 (2014) because silica is dissolved; according to Iwasuki and Marino (2002), alkaline solution reacts so quickly no expansion is detected. This may be a form of size pessimum effect. Chert composed of recrystallized larger quartz can be non-reactive.

13.2.5 Cryptocrystalline quartz in metamorphic cataclasite and hornfels

The first report of ASR damage by cryptocrystalline quartz in Japan was by Katayama (2004). Even after this report, the number of reports on cryptocrystalline quartz remains very limited. Cryptocrystalline quartz is observed in various cataclasites. In an example shown in Figure 13.7 (Yamada *et al.*, 2007), greenschist is changed to cataclasite. In the area of Fukuoka, there are varieties of schist that contains crypto-crystalline quartz and is used widely, and in some structures ASR damage is observed. It is interesting that, although this kind of schist is widely used and the appearance of concretes is similar, ASR damage is only observed in some limited examples. Other types of late expansion by cryptocrystalline quartz is caused by various rock types affect by contact metamorphism. One example, of hornfels from a sandstone, is shown in Figure 13.8 (Yamada *et al.*, 2011). This kind of hornfels is common all over Japan because the intrusion of plutonic rock such as granite is widespread. Therefore, it is

Figure 13.7 Cryptocrystalline quartz in cataclasite of greenschist causing ASR (from Yamada *et al.*, 2007).

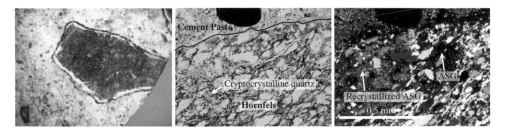

Figure 13.8 Cryptocrystalline quartz and partially recrystallized ASG in hornfels (from Yamada *et al.*, 2011).

important to be aware of the possibility of this type of ASR and a wide survey is required, including petrographic observation based on geological knowledge. Without a complete survey, it is difficult to establish reasonable countermeasures for suppressing ASR.

13.2.6 Challenges in Japanese Construction

13.2.6.1 The history of combating ASR in Japan

After 1986, the Japanese cement industry uniquely started to switch the alumina source for making Portland cement from natural clay to other sources, such as mainly coal ash from a pulverised coal fired power plant or ashes from incineration of municipal solid wastes. Those alternative recycled materials contain much lower amounts of alkalis than clay and this resulted in the level of total alkali in cement being reduced to less than 0.6 % by mass. The timing coincided with the introduction of countermeasures for ASR. Obviously after the introduction of the regulations, the incidence of damage from ASR became drastically reduced. However, it is difficult to conclude whether it was the regulations, or the reduction of alkali level in cements, that was the most effective reason. After the success of the methodology for suppressing ASR, a majority of researchers and engineers had been convinced that ASR problems were finished.

However, based on the continuation of some basic researches, a technical committee on ASR in JCI (JCI-TC062A) met during 2006–2008. This pointed to certain misunderstandings in Japan. When the Japanese methodology was first established, the regulation was designed to be simple, making it possible for an easier acceptance by users. Surprisingly, the importance of geological observation relating to the chemical method, as stated in the original ASTM C289 (later 2007, but now withdrawn) were ignored so that the distinction between the 'potentially harmful' and 'harmful' lines was erased, making it difficult to estimate the proportional 'pessimum' effect. Even in such situations, the introduction of new methodology and its strict execution had significant meaning for engineers working with concrete in opening their eyes to the risk of ASR at that time. A total concrete alkali limit of 3.0 kg/m^3 was determined, based on the results of 210 concrete prism tests using various kinds of aggregates and having alkali contents from 1 to 9 kg/m^3 (Public Works Research Institute, 1989). The results showed that there was a clear difference below 3.0 kg/m^3. However, the duration of test period was only one year and the behaviour after that was not monitored.

It takes a long time to improve standards, but concrete users cannot wait. Japan Railway East introduced their own standard (Koga et al., 2010). In this method, the traditional mortar-bar test and chemical method are used, but the criteria for assessing reactivity and necessary countermeasures were modified to be more on the safe side. Of course, the risk was reduced, but the limitations of the mortar-bar and chemical methods still remained and some problems may cause difficulties in the future.

Recently, several important reports have been published. The Japanese Public Works Research Institute disclosed the expansive behaviour of various Japanese rock types measured over 23 years, although unfortunately the important chert or pessimum compositions of andesite were not included (Koga et al., 2013). In some cases, the innocuous results from the chemical method and mortar-bar test, and a total alkali limit of less than 3.0 kg/m^3, have failed to suppress ASR.

In order to improve existing ASR countermeasures in Japan, it was firstly necessary to recognize that ASR damage still occurred and problems remain, even with new concrete when the current Japanese methodology for suppressing ASR had been applied. It was not long before the mechanisms of serious damage were recognized and the damaged structures in Hokuriku district were repaired in appropriate ways. Very recently in Hokkaido, the existence of ASR began to be recognized, at least as one of the mechanisms of the combined degradation and cracking having been previously considered as simple freeze-thaw damage. It then becomes important to re-examine whether ASR damage had occurred before the present Japanese mitigation methods were introduced and consider whether the cases might be considered to be at risk from the present mitigation methods. According to these present criteria, aggregates were only judged reactive from expansion in mortar-bar tests and then only in cases where the concrete contained more than 3.0 kg/m^3 alkalis, without pozzolanic materials.

13.2.6.2 The new ASR avoidance countermeasures

In order to overcome this unsatisfactory present situation, a detailed diagnosis including petrographic observation is essential. With this more detailed procedure, the reasons why there are examples of alkali-reactive concrete in spite of application of the present criteria may be understood. The anomalies might be seen to have been caused by a pessimum effect, the presence of cryptocrystalline or microcrystalline quartz in the aggregate which passed a mortar-bar test, or additional alkali release from aggregate itself. Some researchers have recently started to prepare new diagnosis procedures in JCI-TC 115FS (2014).

Simultaneously, appropriate new countermeasures must be studied. How much is the total alkali limitation effective in suppressing ASR? How much replacement by ground granulated blast furnace slag (GGBS) or fly ash is required in order to suppress the pessimum effect? How can we eliminate extremely reactive aggregate including very small amounts of opal? How long is it possible to suppress ASR?

One method to be used will be an accelerated concrete prism test such as RILEM AAR-4.1 (in Nixon & Sims, 2016), with estimation of expansive behaviour by ASR based on a concept of accelerated ratio (Kawabata et al., 2014). Based on many examples of expansive behaviour, these expansion curves over time are fitted to an equation described by three parameters, relating ultimate expansion amount, time to starting expansion and expansion rate. In the concrete prism test, expansion is accelerated by higher temperature and higher alkali concentrations. Therefore, once dependence of three parameters on these two accelerating factors is determined, it will be possible to convert the expansion curve from accelerated expansion results to the expansion behaviour of real concrete in real exposure conditions. In this procedure, the concrete prism test is applied for an actual concrete mix, but not for a specific aggregate.

The suppression of ASR expansion by fly ash is mainly due to the lowering of the pH of the pore solution phase in concrete (Kawabata et al., 2012) and the effect of addition of pozzolanic materials can also produce reductions in total alkali concentration (Kawabata et al., 2013; Kawabata & Yamada, 2015). The use of very highly reactive aggregate that has caused rapid and serious expansion is shown in Figure 13.6. A concrete prism test, using 10x10x40 cm size prisms containing 5.5 kg/m^3 of total alkalis, wrapped in wet paper exposed at 60 °C showed varying acceleration ratios, as indicated in Figure 13.9 (Kawabata et al., 2014) depending on temperature and alkali amount. The alkali leaching was limited to a few percent in this experiment. As can be understood, acceleration of expansion varies significantly depending on total alkali content and environmental temperature.

Recently, Yamada and Kawabata et al., published a series of papers pointing the limitation of conventional concrete prism test having problems in alkali leaching and drying depending on alkali content and temperature during curing. Both difficulties can be avoided by wrapping concrete prisms by wet cloth with alkaline solution having the

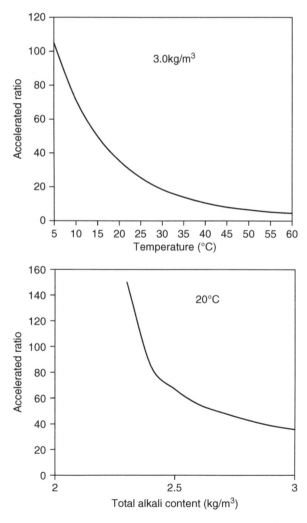

Figure 13.9 Accelerated ratio of concrete prism test (total alkali = 5.5 kg/m^3, 60 °C) against temperature and total alkali content (after Kawabata et al., 2014).

same pH of pore solution (Yamada *et al.*, 2016a). Yamada suggested that acceleration condition of concrete prism test should be optimized depending on the reactivity of aggregate. For rapid reactive aggregates, mild conditions should be applied and for slowly reactive varieties, severe conditions should be applied in order to evaluate the reactivity quantitatively. By this alkali-wrapping, it became possible to evaluate the effects of alkali contents and temperature separately (Yamada *et al.*, 2016b). Based on these expansion data, by taking into account temperature and moisture especially precipitation condition in exposure site, it is possible to reproduce the expansion behaviour of exposed concrete blocks in field. One important reason of disagreement between laboratory concrete test and field exposure relating ASR expansion seems in the difference in the moisture supply not only in the difference in the temperature.

Compared with countermeasure specification for new construction, an effective way to suppress further ASR in already deteriorated concrete is limited. As is well known, by the pressurized injection method, lithium can act as a suppressing agent. Miyagawa *et al.* (2012) described the distribution of lithium by time of flight secondary ion mass spectrograph (TOF-SIMS) and concluded that ingress of lithium into ASG can suppress further ASR.

During the period from the first edition of this book, there are some other important developments relating to use of recycled by-products or wastes as aggregate in Japan. One of these is ASR prevention criteria for recycled aggregates, for which three standards have been prepared: JIS A 5021, 5022, and 5023, for high H, medium M and lower L grades, respectively. A major difference is the residue of paste surrounding aggregates. Every grade of aggregate is regarded as reactive, excepting when its origin is able to be identified and it passes an alkali-reactivity test. However, there are some areas of uncertainty in this standard, which concern the test procedures such as acid dissolution of paste portion before chemical analysis giving a possible limitation to chemical method (Iwatsuki & Morino, 2007).

Another type of recycling concerns types of slag from various industrial processes, such as blast furnace slag, electric furnace slag, copper slag, ferro-nickel slag, and slag of molten ash from solid municipal waste incineration. Although, in general, these slags are innocuous, depending on the chemical composition and cooling procedure, there is a slight possibility of an excess silica-rich phase being produced that might show alkali-reactivity (Kayatama, 1992). Also it is necessary to pay attention to possible expansion by residual free lime.

13.3 GEOLOGY OF EAST AND SOUTH-EAST ASIA – CONSIDERING ASR AND POTENTIAL ASR

13.3.1 General geological setting

Asia is the eastern part of the Eurasian continent. The eastern end is characterized by a subduction zone of an ocean plate under a continental plate. This is the north-western part of the inter pacific volcanic belt and many active volcanos occur here, continuing from the Kuril Islands through Japan, the Philippines, eastern Papua New Guinea, to the Indonesian islands of Sumatra, Java and others. In this area, there are risks of rapidly expansive aggregate, such as andesite. The western part is characterized by very old strata on the continental plate from Precambrian to Palaeozoic in age, composed of various metamorphic rocks. The south-western part relates to the Himalayan orogenic movement, continuing from Indonesia and passing through western Thailand, eastern Myanmar (formerly Burma), Arunachal Pradesh, Bhutan, Nepal and northern India, having a wide range of metamorphic and sheared rocks such as mylonite or cataclasite (see Chapter 15 for an account of AAR in India).

In general, in the area having relatively new volcanos after the Neogene, there is a possibility of rapidly expansive aggregates, relating to cristobalite, tridymite, and volcanic glass. In other areas, comprising older geology and orogenic belts, there is a possibility of slowly expansive aggregates relating to micro/cryptocrystalline quartz and rapid expansion by chalcedony.

Of course, the characters of aggregate used locally in concrete depend on local detailed geology and it is impossible to describe everything here. In this chapter, some examples of the relation between domestic geology and ASR are introduced. General geology of countries is based on two review reports published by Limestone Association of Japan (1996, 2000).

13.3.2 Geology of Japan and reported ASR damage

Japan is located in an active volcanic belt of circum Pacific Rim and composed of complex geological structure from Paleozoic era to Quaternary period. Typical alkali reactive rocks are considered as two types; relatively new Quaternary and Neogene andesite containing cristobalite, tridymite, and acid volcanic glass; reactive chert included in Jurassic accretionary complex. Siliceous sediments in the Pacific Ocean moved to attach Japan in this geological age. Excepting some highly reactive one including chalcedony, Japanese chert is usually composed of recrystallized cryptocrystalline quartz after opal-CT typically existing in Cenozoic European flint. Alteration by hydrothermal activities in Miocene and later ages can make various rock types more reactive. Besides, there are many other cryptocrystalline or microcrystalline quartz bearing rocks such as cataclasite of various rock types along median tectonic line or hornfels caused by contact metamorphism accompanied by granite intrusion although these are not usually recognized as reactive in Japan.

Figure 13.10 (Yamada *et al.*, 2011) shows general geological structure of Japan composed of four main islands, Hokkaido, Honshu, Shikoku, and Kyushu and many small islands such as Ryukyu Islands including Okinawa continuing to Taiwan in Geological sense. The detailed geological structure is complicated and is not easy to describe in a limited space. However, from the viewpoint of alkali reactive rock causing ASR, required information is relatively limited and there is no need to understand the genetics of geology of Japan perfectly. By simplifying the geology of Japan, important geological characters for ASR is summarized in order to give general information even for concrete engineers who are not familiar with geology.

Paleozoic to Paleogene sedimentary and metamorphic rocks make a basic structure of Japan. This structure is divided to Northeast and Southwest Japan by a big fault system, Tanakura tectonic line. Volcanic activities in all geological ages overlapped these older structures. Among them, important volcanic activities are limited in "green tuff" of Neogene and in active volcanoes of Quaternary. Chert in Japan is mainly accompanied by Jurassic accretionary complexes as explained before. Another important geological character is the median tectonic line, a quite big and long fault

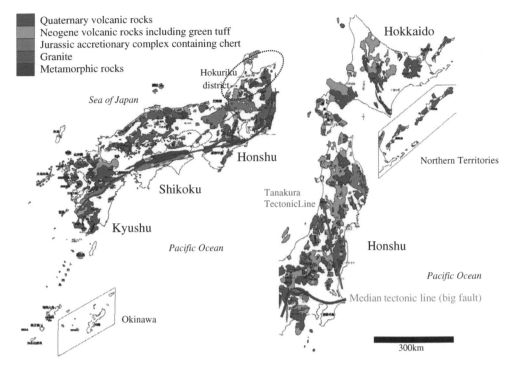

Figure 13.10 General geological map of Japan categorized from the viewpoint of ASR (Yamada *et al.*, 2011).

accompanying metamorphic rocks. Besides, there are numerous faults causing cataclasite accompanying cryptocrystalline/ microcrystalline quartz.

From the standpoint of ASR, major reactive rock types in Japan can be classified in several categories.

1) Volcanic rocks: Majority is andesite and less rhyolite containing cristobalite/ tridymite and acid glass accompanied by volcanic actives after Neogene.

2) Opal and cristobalite bearing rocks affected by green tuff activities in Miocene or originated from altered volcanic glasses. By the effect of alteration, various non-reactive rock types can be modified into reactive.

3) Reactive chert composed of micro/ cryptocrystalline quartz and occasional chalcedony mainly accompanied by Jurassic accretionary complexes.

4) Cryptocrystalline quartz bearing rocks: Sedimentary rocks affected by contact metamorphism, pelitic/ siliceous schist, various rock types affected by faults such as cataclasite or mylonite.

5) River or sea gravels and sands having their hinterland in the area bearing reactive rock types. Hokuriku district is a typical area. The reactivity depends on the river system even in the same district (Daidai *et al.*, 2012).

6) Various sediments on land: Although these were formed by old river systems or sea, the source area is not easy to identify always. Therefore, there are possibilities that reactive rock types can be included. There has been a myth that "good"

natural gravels in one-time were exposed nature for long time and having no reactivity. Of course completely wrong.

7)   Proportional pessimum effect regarding natural gravels and sands is another important point of Japanese ASR. If reactive rocks containing highly reactive minerals are included in a pessimum composition, the ASR expansion can be much bigger than totally reactive aggregate.

8)   Japanese limestone is usually pure and does not contain cryptocrystalline quartz in harmful range that causes so-called "Alkali Carbonate Rock Reaction" like argillaceous dolomite. However, because of the limited ability of alkali binding, at the pessimum composition, limestone might make the situation worse.

9)   Transportation of aggregate can have completely different characters from the local geology where aggregate is used. Okinawa is an example. Imported from Hualian in Taiwan caused serious ASR damages in Ryukyu Islands (Katayama *et al.*, 2008).

10)   Mixing of various aggregate in metropolitan areas may cause pessimum effects by highly reactive aggregate containing opal (Yamada *et al.*, 2007).

11)   Effects of de-icing salt or sea salt enhance the reactivity of aggregates (Katayama *et al.*, 2004; Habuchi & Torii, 2012). Therefore, if the same aggregate is used, the reactivity can be different in cold and marine environment from warm inland.

Apart from geological structure, as is shown in the example above, aggregate transport is a characteristic point of Japanese aggregate supply. There are several areas of high population density such as Tokyo, Osaka, Nagoya, Fukuoka, or so. Especially to Tokyo and Osaka, significant amounts of aggregate are transported by sea from far away such as Hokkaido, Kyushu, or even from China. In Ryukyu Islands including Okinawa, there are limited amount of appropriate aggregate source and significant amount of aggregate have been transported from Hualian in Taiwan because the distance from Taiwan is less. Highly reactive aggregate had been imported from without recognition of the reactivity and this resulted in serious ASR damages including reinforcement break. In other places sea side such as Setouchi area around inland sea between Honshu and Shikoku, aggregate transport is not so special in small scales. A problem of aggregate transport is in the blending of two innocuous aggregate in origin that can make proportional pessimum effect. These situations are same for some cities such as Hong Kong or Singapore. Aggregates are usually imported from other places and the reactivity depends on the sources and combination.

It is interesting that there are significant discrepancies between the distributions of reported ASR damages in Figure 13.2 and risk map in Figure 13.10. The discrepancies suggest that the imperfect diagnosis of damaged structures or the existences of hidden or unpublished damages.

Geological age affects the crystalline forms of reactive minerals strongly as shown in Table 13.3. Many of siliceous rocks are biogenic origin such as diatoms, radiolarians and siliceous sponges. When these deposited under marine, crystalline form was opal-A. By passing geological ages, opal-A changes to opal-CT, chalcedony, cryptocrystalline quartz, microcrystalline quartz, and granular quartz finally. Similarly for volcanic rocks, there is an important difference in the reactivity of silica minerals between after Neogene and before Palaeogene (Katayama & Kaneshige, 1986). Highly reactive rapid expansive minerals such as opal, cristobalite, tridymite presented after Neogene change mineral form to late expansive cryptocrystalline quartz found before Palaeogene. These

Table 13.3 Distribution of silica minerals in siliceous sedimentary rocks in Japan (Katayama & Futagawa, 1989).

AGE	CENOZOIC		MESOZOIC			PALEOZOIC				QUARTZ	
	MIOCENE	PRE-MIO	CRETA	JURA	TRIAS	PERM	CARB	DEV	SIL	C.I.	G. SIZE
OPAL-A ZONE	▬										
OPAL-CT ZONE	▬										
CHALCEDONY*	▬▬									1–2	<2–5?m
CHALCEDONY/ CRYPTOCRYSTALLINE QUARTZ ZONE	─{ }─────				...					3–6	<2–5?m
MICROCRYSTALLINE QUARTZ ZONE			... ─────					{ }─		6–8	<10?m 20–40?
GRANULAR QUARTZ ZONE			·········· METAMORPHISM							9–10	>20– 50?m

*HYDROTHERMAL PRODUCTS IN THE GREEN TUFF REGIONS

C.I.: CRYSTALLINITY INDEX

are the examples in Japan. Geothermal and climate condition such as temperature and precipitation are different in the sense of geological time scale, the transformation of reactive minerals will be completely different.

13.3.3 China

China is very large and the geology is diverse, including the existence of highly reactive volcanic rocks and it is not easy to show general geology. However, Deng summarized typical examples of ASR and so-called ACR and the locations of alkali-reactive rocks in China (Fig. 13.11). There are many reports. Deng *et al.* (2004) described the reactive minerals in a variety of rock types obtained from various parts of China. Examined rock types are volcanic rocks (tuff, rhyolite, porphyrite, basalt), sedimentary rocks (chert, sandstones, cataclastic quartzite, river sand) and carbonate rocks (siliceous dolostones [dolomite rocks] and limestone, dolomitic limestones, calcitic dolostones, argillaceous dolostones, dolostones with middle-size crystals). Detected alkali reactive minerals are micro-/crypto-crystalline quartz and chalcedony. Unfortunately, there is no information on geology.

In their following paper (Deng *et al.*, 2008), more detailed observation results were described for many aggregates form the southern part of China. Sandstones and slates from Sichuan, Yunnan and Jiangxi provinces included micro-/crypto-crystalline quartz. Tuffs from Chongqing city and Zhejiang province included micro-/crypto-crystalline quartz and chalcedony. Because the geological age is old, the original glass phase in tuff is thought to be changed into fine quartz grains and chalcedony as Katayama and Kaneshige (1986) suggested. Gneisses from Lanping, Yunnan province, include micro-/crypto-crystalline quartz.

Yang *et al.* (2004) reported river sand of the Yichang channel segment of the longest Yangtze river, for which the drainage area covers 19 % of China. The aggregates are differentiated by mineralogy:

1) Upper part before Wan country composed of: basalt, granite, rhyolite, tuff, breccia, andesite, porphyrite, syenite, trachytes, quartzite and gneiss.
2) From Wan country to Fengjie composed of: Jurassic gritstone, siltstone, shale and Triassic limestone and marl.

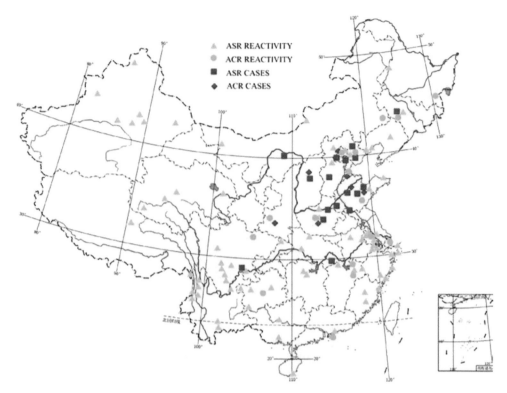

Figure 13.11 Examples of ASR cases and alkali silica reactive rocks (Deng, 2014).

3)   From Fengjie to Xiangxi composed of: Jurassic limestone, Triassic gritstone, shale and limestone.
4)   From Xiangxi to Yichang composed of: Huangling anticline complex rock, granite, quartz dolerite and post-Sinian limestone.

As reactive rock types, flint and rhyolite are listed. Flint includes chalcedony and cryptocrystalline quartz. Rhyolite includes cryptocrystalline quartz. However, reactivity was examined by the chemical method and mortar-bar test, so there may be overlooked reactive aggregate containing cryptocrystalline quartz.

An example of reactive basalt is reported by Deng *et al.* (2008). Permian Emeishan (one stratigraphic time scale of China) group of basaltic rocks located in Sichuan, Guizhou and Yunnan provinces of China are 3000 m thick and cover more than 260 km^2. Another comparable size of basalt is the Deccan plateau in India (see Chapter 15). This basalt included chalcedony and cryptocrystalline quartz as reactive minerals. These minerals are estimated to be formed by hydrothermal activities following the magmatic stage. Judging from other examples, initially basalt might contain cristobalite, but during a long geological time the crystalline form has changed.

In the northern part of China, there are examples of ASR damage reported. Gravel aggregates form the Yongding River produced ASR in concrete bridges (Tang *et al.*, 1996, 1997; Fu, 1997; Fu *et al.*, 2004) and contained chalcedony, micro-/cryptocrystalline quartz in siliceous limestone, siliceous dolostone, altered basalt and altered andesite. The cases of concrete damage spread over Beijing, Hebei, Shandong, Shaanzi and Riangxi Provinces. Hao (2004) reported that the concrete alkali content in pre-stressed girders of a railway bridge was 6-10 kg/m^3, caused by both high alkali cement and externally-derived alkalis.

Li (2004) reported the ASR diagnosis of dam concrete, such as Tuoxi hydropower plant in the middle reaches of the Zijiange river in Anhua Country, Hunan Province, Fengman hydropower station on the second Songhuajian river, Jilin Province in north-east China, and Daheiting reservoir in Qian'an Country, Hebei Province on the Luanhe river. Reactive aggregates, including rhyolite, andesite and tuff were reported for the Tuoxi hydropower plant. From the Luanhe river, flint was identified as reactive, whilst tuff, rhyolite and 'rusticated' rock were assessed as potentially reactive; andesite, mixed aggregates and sands were found to be non-reactive by the chemical method.

Xie *et al.* (2004) reported microcrystalline quartz and chalcedony in most aggregates along the Qinghai-Tibet railway. The Tibet plateau, occupying one fourth of China, forms the south-western area of China and has been called 'the Roof of the World'. The Tibet plateau was formed by the collision of the Indian and Eurasian plates, with the Indian plate being subducted under the Eurasian plate. So, the Tibet plateau is formed by various deformed metamorphic rocks. In some coarse aggregates, the following rock types were observed: biotite-andesite, biotite-quartz-schist, fluidal structure tuff, welded tuff, metamorphic rocks, siltstone, diabase, rhyolite and biotite-quartz-diorite. In some sand, the following rock types are described: perlite, gneiss, siliceous rock, sandstone, tuff, rhyolite, phyllite, clay rock, pyroxenite, granite, diabase, quartzite, carbonatite, reddish marble, indicating a diverse variety of geological origins. From the viewpoint of ASR, in these old rocks, cryptocrystalline quartz and chalcedony are always suspicious as reactive minerals.

In China, it is important to mention the 'so-called alkali-carbonate reaction' (ACR). There are many reports on this issue (Fu, 1997; Tang *et al.*, 1997, 2000; Deng *et al.*, 2012). Recently Katayama (2010) concluded that the fundamental mechanism of ACR is the same as ASR, caused by cryptocrystalline quartz within a largely carbonate rock (see Chapter 3).

In Hong Kong, the main aggregate is local and imported granitic rocks, but some ASR damage has been reported (Leung *et al.*, 2000). Tse *et al.* (1996) reported damage to sewage tanks by late expansion by devitrified rhyolitic tuff coarse aggregate from south China (Figure 13.12).

13.3.4 Korea

Damage to a concrete pavement at Seohae Expressway, on the western coast of South Korea, has been reported (Yun *et al.*, 2008), although the detailed analysis was not carried out to identify the reactive rock type. However, judging from the location, the major local rock type seems Pre-Cambrian, with some Palaeozoic or Mesozoic rocks accompanying intrusive rocks; potential alkali-reactive rocks will probably be late-expansive type by cryptocrystalline quartz, but perhaps more rapidly expansive when chalcedony is present. In Cheju Island and the north-eastern area of North Korea, Neogene and Quaternary volcanic activities are observed and there will be rapidly

Figure 13.12 Sewage works in Hong Kong affected by ASR: a) General view, b) Map cracking of concrete tank wall, c) cracking along the top concrete wall surface. Photographs courtesy of Dr Ian Sims of RSK (of Sandberg when taken).

expansive aggregate. Actually, there is a report of ASR damage in Cheju Island caused by cristobalite in silica-rich tholeiite basalt (Joon, 1997).

13.3.5 Taiwan

Hualian is a known source of reactive aggregate in Taiwan (Katayama *et al.*, 2008).

Taiwan is located at a collision type orogenic zone between the Philippine sea plate and the Eurasia plate. Taiwanese geology shows zoning, having an orientation NNE-SSW, and a majority is composed of sedimentary rocks relating to formations after the late Palaeozoic. The eastern side of the central mountain range is composed of metamorphosed rocks and the western part is of folded and solidified rocks. In eastern marginal and western foothill areas, there are relatively new strata, later than Neogene. Therefore, reactive aggregate in Taiwan is expected as late-expansive type, caused by micro-/crypto-crystalline quartz in metamorphosed or deformed rocks, accompanying rapidly expansive chalcedony.

ASR damage of Hwa-Lian and Keelung harbours in eastern and northern areas, respectively, is reported (Lee *et al.*, 2000). One aggregate from the eastern area was sandstone and dissolved silica in the chemical method was extremely high (1204 and 1489), with rapid expansion in ten days by mortar-bar test. Detailed petrographic information is not available, but the high reactivity suggests the existence of chalcedony in chert. Lee *et al.* (2000) reported a little bit more detailed rock types from four other locations, in eastern, northern, north-western and eastern areas. Identified rocks were sandstone, quartzite, metamorphic sandstone, slate, vein-quartz, siliceous sandstone, slate, quartz-schist, marble, schist, serpentine, andesite, gneiss and limestone. Combinations are obviously different from place to place. There is no information on which kinds of these rocks has high reactivity.

13.3.6 Thailand

Thailand is located in the middle of the Alps-Himalayan orogenic belt and is divided into three parts; the western Shan Thai Para-platform, eastern Khorat Kontum Platform, and Yunnann Malay mobile belt. In Thailand, important characteristics are in Mesozoic and Palaeozoic strata and deformation by various orogenic movements. This means every rock in Thailand has a possibility for slow expansion by crypto- or micro-crystalline quartz formed by metamorphism.

Recently, in Bangkok, severe ASR damage was found in housing and piers of highway bridge structures after a few years of construction. Petrographic observation revealed that the ASR was mainly caused by cryptocrystalline quartz in granite mylonite (Yamada *et al.*, 2013) but there are many alkali silica gel (ASG) formations in various rock types such as limestone, pelitic hornfels and chert. In this structure, varieties of aggregate sources are used and in other reports, sericite rock, quartzite are considered as reactive rock types (Jensen & Sujjavanich, 2016a, 2016b).

The textures of replacement of ASG by ettringite are observed in almost every part in cement paste, especially surrounding parts of coarse aggregate suggesting delayed ettringite formation, DEF ((Baingam *et al.*, 2012; Jensen & Sujjawanich, 2016: 2). Judging from the texture of ettringite at the aggregate-paste boundaries, there is a possibility of DEF in the early stage. In a hot climate, concrete temperature in thick housing or foundation can rise enough for DEF and humid condition may have generated ettringite in cement paste to expand and made gaps between aggregate and paste filled with secondary ettringite later. In general, cryptocrystalline quartz has been considered slow-reactive rock type in North America and Europe. However, in Thailand, cracks were observed in early stages such as three or four years and this is another reason of the suspicion of DEF. However, major cracks are originated from aggregate filled with ASG being judged from the texture of cracks suggesting major reason of damages are ASR. In hot and humid climates, every cryptocrystalline quartz may react fast enough to generate cracks in a few years although it is difficult to deny the possibility of DEF. In various parts of the world, examples of DEF are reported recently. However, it should be careful for the diagnosis of combination of ASR and DEF. Secondary ettringite in cracks or voids are not the evidence of DEF. Detailed analysis of invisible ettringite formation in cement paste is required but it is not sure how long ettringite in cement paste can be retained in hot and humid conditions in the original position when it generated cracks.

13.3.7 Common possible risks in South-East Asia

Figure 13.13 (Yamada *et al.*, 2013) shows the sawn surface of deteriorated concrete from a highway bridge in Bangkok. There are typical textures of ASG surrounding

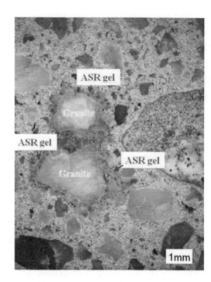

Figure 13.13 Cross section of concrete showing ASG bleeding from granite particles (Yamada *et al.*, 2013).

reactive aggregate that looks like wet by being of darker colour than other parts. This darker part is formed by ASG bleeding from core aggregate. The crack in granite is filled with typical ASG, as shown in Figure 13.14.

There are interesting textures in other parts of this specimen. One position at a grain boundary in granite shows possible opal and allophane (hydrated aluminosilicate) materials (Figure 13.15, after Yamada *et al.*, 2013). Mica is generally present in granite and thought to be altered to form these opal and allophane minerals. This reaction can be expected to be common in hot and humid climates like those in South-east Asia. The relationship between annual rainfall and change of clay minerals is shown in Figure 13.16 (Yamada *et al.*, 2013 after Barshad, 1966). With more rain, aluminosilicate releases silica and alumina-rich minerals remain. This means silica will move and there is a possibility to deposit as amorphous silica, such as opal. Simultaneously there may be a possibility of deposition of amorphous silica from ground water. Therefore, it is important to expect the existence of opal in any aggregate in South-east Asian countries. Opal is an amorphous material and not necessarily detected by X-ray diffraction. The only reliable way of detecting opaline silica is through the examination of thin sections by polarising optical microscope.

13.3.8 Other countries and general points

In other countries, Katayama (1997) reported ASR damage in various countries such as India, Pakistan, Iran, Iraq, Turkey, Israel, Bahrain, Saudi Arabia, and Yemen (see Chapters 8, 15 & 16 for information on these countries). Of course, the examples reported are limited, but it is likely that some ASR damage could be found in all countries. Here, from the viewpoint of geological structure, a general estimation of ASR possibility is suggested.

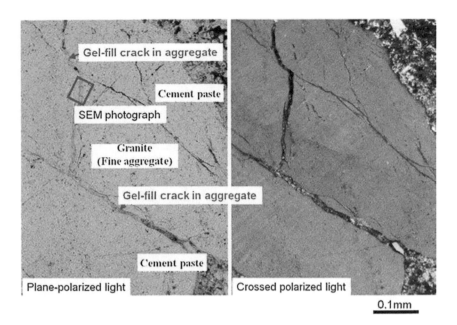

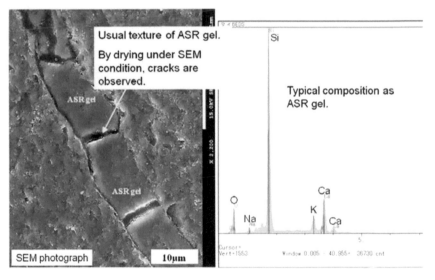

Figure 13.14 Polarizing optical microscope photographs of cracks in a granite particle filled with ASG, and the analysis by SEM/EDS (Yamada *et al.*, 2013).

(1)   Volcanos after the Neogene

From north to south, the following countries are located in the subduction zone of an ocean plate with active volcanos: East end part of Russia such as the Aleutian Islands, north eastern part of North Korea, some north eastern parts of China, Cheju Island, Japan, Philippines, southern islands of Indonesia continuing from Sumatra Island, Java Island, and to Timor-Leste, eastern end

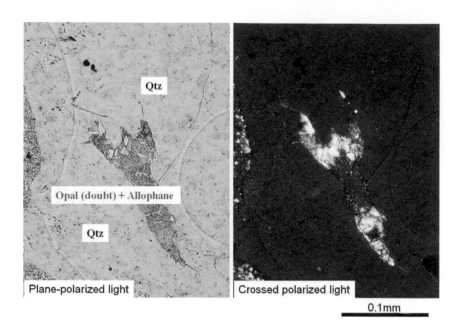

Figure 13.15 Alteration products found in granite (Yamada *et al.*, 2013).

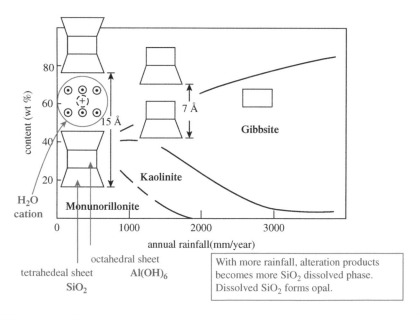

Figure 13.16 Relationship between annual rainfall and clay mineral composition changes (Yamada *et al.*, 2013 after Barshad, 1966).

of Papua-New-Guinea, then continuing to the eastern Islands of Melanesia such as Solomon islands, Vanuatu and ends at the North Island of New Zealand. This is the western part of the circum-Pacific volcanic belt and in the eastern part is the western parts of north and south American Continents. Not all parts of these countries have volcanos, but further detailed geology relating volcanos has to be considered. However, in these countries rapidly expansive aggregate such as andesite or rhyolite, and some kind of basalt have to be paid special attention. Of course, there are possibilities of cryptocrystalline quartz, but the damage from rapid expansion will be conspicuous. An important point is the geological age of the rocks. By diagenesis, highly reactive minerals such as cristobalite and tridymite or volcanic glass change their crystalline forms to more stable quartz that is less reactive (Katayama & Kaneshige, 1986). Also in sedimentary rocks, similar changes are observed, such as reactive opal changing into quartz (Katayama & Futagawa, 1989).

(2)   Continental plateau and old orogenic belt

There are stable geologies forming parts of the continental crust and old orogenic belts that are now stable: such as Mongolia, the eastern part of China, the main part of the Korean Peninsula, Vietnam, Laos, Cambodia, the eastern part of Thailand, India and Sri Lanka. In these countries, there are various types of metamorphic rock, sedimentary rock and igneous rocks, but relatively old from the sense of ASR. Every rock has to be checked from the viewpoint of late-expansion by cryptocrystalline quartz and rapid expansion by chalcedony. Sometimes, rapid expansion may be caused by the 'so-called alkali-carbonate rocks'.

(3)   Orogenic zone

There are many plates in East and South-east Asia such as the North American plate, Pacific Ocean Plate, Eurasian Plate, Philippine Sea Plate, Austrian Plate and Indian Plate. By the collision of plates, strong shearing stress takes place. Following are in this region: Taiwan, western part of China such as Szechwan, Tibet, Borneo Island, Indonesia except southern island with volcanos, Papua New Guinea except its eastern end, western part of Thailand, Myanmar, Bangladesh, Arunachal Pradesh, Bhutan, Nepal, northern end of India and Pakistan. Strong deformation affects all the geology in this area and various metamorphic rocks and deformed rocks, such as cataclasite or mylonite, will be found including cryptocrystalline quartz. Late-expansion has to be considered and again chalcedony may cause rapid expansion.

(4)   Innocuous aggregate

From the nature of the rock type, there are some kinds of rock considered as innocuous: pure limestone or dolostone (dolomite rock) without harmful cryptocrystalline quartz, basic plutonic rocks such as gabbro or peridotite without significant alteration, and acid plutonic rocks such as granitoids without alteration and mylonitization.

ACKNOWLEDGEMENT

The authors of this chapter present their sincere gratitude for suggestive and fruitful discussion on geological aspects with Dr Tetsuya Katayama.

REFERENCES

ASTM C289 (2007, now withdrawn). *Standard test method for potential alkali-silica reactivity of aggregates (chemical method).* West Conshohocken USA, American Society for Testing and Materials.

ASTM C295 (2012) *Standard guide for petrographic examination of aggregates for concrete.* West Conshohocken, USA, American Society for Testing and Materials.

ASTM C1260 (2014) *Standard test method for potential alkali-reactivity of aggregates (mortar-bar method).* West Conshohocken, USA, American Society for Testing and Materials.

Baingam, L., Waengsoy, W., Choktaweekarn, P., & Tangtermsirikul, S. (2012) Diagnosis of a Combined Alkali Silica Reaction and Delayed Ettringite Formation. *Thammasat Int. J of Sci. and Tech.*, *17*, 4, 22–35.

Barshad, I. (1966) The effect of variation in precipitation on the nature of clay mineral formation in soils from acid and basic igneous rocks. In: *Proceedings of International Clay Conference.* pp. 167–173.

Daidai, T., Andrade, O., & Torii, K. (2012) The maintenance and rehabilitation techniques for ASR affected bridge piers with fracture of steel bars. In: *Proceedings of the 14th International. Conference on Alkali-Aggregate Reaction (AAR) in Concrete*, No. 021411-DAID.

Daidai, T. & Torii, K. (2008) A Proposal for Rehabilitation of ASR-affected Bridge Piers with Fractured Steel Bars, In: *Proceedings of the 13th International Conference on Alkali-Aggregate Reaction (AAR) in Concrete*, pp. 42–49.

Deng., M. (2014.) original figure provided personally.

Deng, M., Lan, X., Xu, Z., & Tang, M. (2004) Petrographic characteristics and distributions of reactive aggregates in China, In: *Proceedings of the 12th International Conference on Alkali-Aggregate Reaction in Concrete.* pp. 87–98.

Deng., M., Song, X., Lan, X., Huang, X., & Tnag, M. (2012) Expandability of alkali-dolomite reaction in dolomitic limestone. In: *Proceedings of the 14th Internationl Conference. on Alkali-Aggregate Reaction in Concrete*, 031212-MIN.

Deng, M., Xu, L., Lan, X., & Tang, M. (2008) Microstructures and alkali-reactivity of Permian Emeishan group basaltic rocks. In: *Proceedings of the 13th International Conference on Alkali-Aggregate Reaction in Concrete*, No. 030.

Deng, M., Xu, Z. & Tang, M. (2008) Suitability of test methods for alkali-silica reactivity of aggregates to Chinese aggregates. In: *Proceedings of the 13th Internationl Conference on Alkali-Aggregate Reaction in Concrete*, No. 032.

Fu, P. (1997) Alkali-aggregate reaction in concrete in Beijing and Tianjing areas, East Asia Alkali-Aggregate Reaction Seminar, Tottori University, Japan, 73–80, 1997.

Fu, P.X., Liu, Y., & Wang, J.M. (2004) Alkali reactivity of aggregate and AAR-affected concrete structures in Beijin. In: *Proceedings of the 12th International Conference on Alkali-Aggregate Reaction in Concrete.* pp. 1055–1061.

Habuchi, T. & Torii, K. (2012) Corrosion characteristics of reinforcement in concrete structures subject to ASR and seawater attack marine environment. In: *Proceedings of the 14th International Conference on Alkali-Aggregate Reaction in Concrete*, No. 021411-HABU.

Hamada, H., Sagawa, Y., Inoue, Y., & Hayashi, K. (2011) An example of deteriorated concrete structure being used sedimentary rock as coarse aggregate by ASR. *JCI Proceedings*, *33* (1), 1073–1078.

Hao, T.Y. (2004) Diagnosis of a railway bridge damaged by alkali-silica reaction. In: *Proceedings of the 12th International Conference on Alkali-Aggregate Reaction in Concrete.* pp. 882–887.

Hayashi, K., Kono, K., Yamada, K., & Hara, K. (2009) Diagnosis of ASR damages of concrete structures containing limestone aggregate and sea sand. In: *Proceedings of the JCI, 31* (1), 1249–1254.

Hayashi, K., Yamada, K., Kono, K., & Oba, M. (2009) Diagnosis of ASR degradation observed in a prestressed concrete bridge. In: *Proceedings of the 63rd JSCE Annual Meeting, 64* (5) (in Japanese), p. 99.

Inoue, Y., Hamada, H., Kawabata, Y., & Yamada, K. (2010) Effectiveness of ASR suppression by fly ash of mortar containing aggregate showing pessimum effect. *Proceedings of JCI, 32* (1), 953–958 (in Japanese).

Iwatsuki, E. & Morino, K. (2002) Expansive behaviours and microstructures of ASR mortar bar according to ASTM C1260 and JIS A5308. In: *Proceedings of JCI, 24* (14), 687–692 (in Japanese).

Iwatsuki, E. & Morino, K. (2007) Results of chemical method of aggregate immersed in nitric and choric acids solution. In: *Proceedings of JSCE Annual Meeting, 62* (5), 913–914.

Japan Concrete Institute, JCI-TC062A (2008) Technical Committee on Mitigation and Diagnosis of Alkali Silica Reaction Considering the Action Mechanisms, Technical Committee Report.

Japan Concrete Institute, JCI-TC115FS (2014) Technical Committee on Diagnosis of ASR-affected Structures 2012.4-2014.3.

Japan Society of Civil Engineers (2005) State-of-the -Art Report on the Countermeasures for the Damage Due to Alkali-Silica Reaction, Concrete Library, No.124, 2005.8.

JASS 5N T-603-2013 Test method for alkali aggregate reaction, Japanese Architectural Standard Specification for reinforcing concrete work in nuclear power generation plant.

Jensen, V. & Sujjavanich, S. (2016a) ASR and DEF in concrete foundations in Thailand. In: *Proceedings of 15th International Conference on Alkali-Aggregate Reaction.* 193p.

Jensen, V. & Sujjavanich, S. (2016b) Alkali Silica Reaction in concrete foundations in Thailand. In: *Proceedings of 15th International Conference on Alkali-Aggregate Reaction.* 201p.

JIS A 1145-2007 Method of test for alkali-silica reactivity of aggregates by chemical method, Japanese Industrial Standard, published by the Japanese Standards Association, Tokyo.

JIS A 1146-2007 Method of test for alkali-silica reactivity of aggregates by mortar-bar method, Japanese Industrial Standard, published by the Japanese Standards Association, Tokyo.

JIS A 1804-2009 Method of test for production control of concrete – Method of rapid test for identification of alkali-silica reactivity of aggregate, Japanese Industrial Standard, published by the Japanese Standards Association, Tokyo.

JIS A5021-2011 Recycled aggregate for concrete-class H, Japanese Industrial Standard, published by the Japanese Standards Association, Tokyo.

JIS A5022-2012 Recycled aggregate for concrete-class M, Japanese Industrial Standard, published by the Japanese Standards Association, Tokyo.

JIS A5033-2012 Recycled aggregate for concrete-class L, Japanese Industrial Standard, published by the Japanese Standards Association, Tokyo.

Joon, J.H. (1997) Study of alkali-aggregate reaction in Korea, East Asia Alkali-Aggregate Reaction seminar, Supplementary papers, A61-72.

Katayama, T. (1992) Petrographic study on the potential alkali-reactivity of ferro-nickel slags for concrete aggregates. In: *Proceedings of the 9th International Conference on Alkali-Aggregate Reaction in Concrete.* pp. 497–507.

Katayama, T. (1997) Petrography of alkali-aggregate reactions in concrete – Reactive minerals and reactive products, East Asia Alkali-Aggregate Reaction seminar, Supplementary papers, A43-59.

Katayama, T. (2010) The so-called alkali-carbonate reaction (ACR) – Its mineralogical and geochemical details, with special reference to ASR. *Cement Concrete Res., 40,* 643–675.

Katayama, T. (2012) Late-expansive ASR in a 30-year old PC structure in eastern Japan. In: *Proceedings of the 14th International Conference on Alkali-Aggregate Reaction in Concrete,* No. 030411-KATA-05.

Katayama, T. & Futagawa, T. (1989) Diagenetic changes in potential alkali-aggregate reactivity of siliceous sedimentary rocks in Japan – A geological interpretation. In: *Proceedings of the 8th International Conference on Alkali-Aggregate Reaction in Concrete*. 525p.

Katayama, T. & Kaneshige, Y. (1986) Diagenetic changes in potential alkali-aggregate reactivity of volcanic rocks in Japan –A geological interpretation. In: *Proceedings of the 7th International Conference on Alkali-Aggregate Reaction in Concrete*. p. 489.

Katayama, T. & Sakai, K. (1998) Petrography of 100-year-old concrete from Otaru port, Japan. In: *Proceedings of the 2nd International Conference, on Concrete under severe Conditions*. pp. 250–261.

Katayama, T., Sarai, Y., Higashi, Y., & Honma, A. (2004) Late-expansive alkali-silica reaction in the Ohnyu and Furikusa headwor structures, central Japan. In: *Proceedings of the 12th International Conference on Alkali-Aggregate Reaction in Concrete*. pp. 1086–1094.

Katayama, T., Tagami, M., Sirai, Y, Izumi, S., & Hira, T. (2004) Alkali-aggreate reaction uder the influence of deicing salts in the Hokuriku district, Japan. *Mater Charact*, 53, 105–122.

Katayama, T., Oshiro, T., Sarai, Y., Zaha, K., & Yamato, T. (2008) Late-expansive ASR due to imported sand and local aggregates in Okinawa Island, southwestern Japan. In: *Proceedings of 13th International Conference on Alkali-aggregate Reaction in Concrete*. pp. 862–873.

Kawabata, Y., Ikeda, T., Yamada, K., & Sagawa, Y. (2012) Suppression effect of fly ash on ASR expansion of mortar/ concrete at the pessimum proportion. In: *Proceedings of 14th International Conference on Alkali-Aggregate Reaction in Concrete*. No. 031711-KAWA-01.

Kawabata, Y. & Yamada, K. (2015) Evaluation of alkalinity of pore solution based on the phase composition of cement hydrates with supplementary cementitious materials and its relation to suppressing ASR expansion. *J Advanced Concrete Tech.*, 13, 538–553.

Kawabata, Y., Yamada, K., & Matsushita, H. (2012) The effect of composition of cement hydrates with supplementary cementitious materials on ASR expansion. In: *Proceedings of 14th International Conference on Alkali-aggregate Reaction in Concrete*, No. 031711-KAWA.

Kawabata, Y., Yamada, K., & Matsushita, H. (2013) Relation of phase composition of cement hydrates with supplementary cementitious materials to the suppressing effect on ASR expansion. *Journal of Japan Society of Civil Engineers, Ser. E2 (Mater. Concrete Struct.)*, 69 (4), 402–420 (in Japanese).

Kawabata, Y., Yamada, K., Ogawa, S., Martin, R.P., Sagawa, Y., Seignol, J.F., & Toutlemonde, F. (2016) Correlation between laboratory expansion and field expansion of concrete -Prediction based on modified concrete expansion test-. In: *Proceedings of 15th International Conference on Alkali-Aggregate Reaction*. p. 34.

Kawabata, Y., Yamada, K., Ogawa, S., & Sagawa, Y. (2014) Simplified prediction of ASR expansion of concrete based on accelerated concrete prism test. *Cement Concrete Tech.*, 67, 449–455 (in Japanese).

Kawamura, M., Takemoto, K., & Hasaba, S. (1983) Case studies of concrete structures damaged by the alkali-silica reaction in Japan, Review of the 37th General Meeting, The Cement Association of Japan, pp. 88–89.

Koga, H., Hyakutake, T., Watanabe, H., Wakizaka, Y., Nishizaki, I., & Moriya, S. (2013) Alkali-silica reactivity of aggregate in Japan verified by 23-years exposure test. *J of Japan Society of Civil Engineers, Ser. E2 (Mater. Concrete Struct.)*, 69 (4), 361–376 (in Japanese).

Koga, M., Kino, J., & Matsuda, Y. (2010) About countermeasures against alkali silica reaction, SED, Structure Technical Center, Japan Railway East, pp.184–187.

Lee, C., Sheu, S.W, Chen, K.C., Ko, J.L, & Rau, C.C. (2000) Field AAR inspection for the four harbors in Taiwan. In: *Proceedings of 11th International Conference on Alkali-aggregate Reaction in Concrete*, pp. 869–878.

Lee, C., Shien, W.K., & Lou, I.J. (2000) Evaluation of the effectiveness of slag and fly ash in preventing expansion due to AAR in Taiwan. In: *Proceedings of 11th International Conference on Alkali-aggregate Reaction in Concrete*, pp. 703–712.

Leung, W.C., Shen, J.M., Lau, W.C., & Chan, C.Y. (2000) Testing aggregates for alkali aggregate reactions in Hong Kong. In: *Proceedings of 11th International Conference on Alkali-aggregate Reaction in Concrete*, pp. 395–404.

Li, J. (2004) Problems on alkali-aggregate reactions of dam concrete in China. In: *Proceedings of 12th International Conference on Alkali-aggregate Reaction in Concrete*, pp. 1078–1085.

Miyagawa, T., Seto, K., Sasaki, K., Mikata, Y., Kuzume, K., & Minami, K. (2006) Fracture of reinforcing steel in concrete structures damaged by alkali-silica reaction -Field survey, mechanism and maintenance. *J Adv Concr Technol., 4* (3), 339–345.

Miyagawa, T., Yamamoto, T., Mihara, T., & Era, K. (2012) Controlling ASR expansion by lithium ion pressurized injection method. In: *Proceedings of 14th International Conference on Alkali-aggregate Reaction in Concrete*, No. 022111-MIYA.

Morino, K., Shibata, K., & Iwatsuki, E. (1987) Characterization of alkali aggregate reactivity of cherty rock. *Clay sci., 27*, 199–210 (in Japanese).

Nixon, P.J. & Sims, I. (eds.) (2016) RILEM recommendations for the prevention of damage by alkali-aggregate reactions in new concrete structures. *RILEM StateArt Rep., 17*, Springer, Dordrecht, pp RILEM, Paris, 168p.

Nomura, M. & Torii, K. (n.d) The alkali-reaching property of sands and inspection on alkali-leaching from aggregate in structures. In: *Proceedings of 13th International Conference on Alkali-aggregate Reaction in Concrete*.

Public Works Research Institute, Japan Cement Association (1989) Collaboration report on controlling of alkali aggregate reaction by limiting alkali amount in cement. *Collaboration Report* No. 25.3 (in Japanese).

Swamy, R.N. (ed.) (1992) *The alkali-silica reaction in concrete*. Glasgow & London, Blackie. 352pp.

Tang, M., Deng, M., & Xu, Z. (2000) Comparison between alkali-silica reaction and alkali-carbonate reaction. In: *Proceedings of 11th International Conference on Alkali-aggregate Reaction in Concrete*, pp. 109–118.

Tang, M., Deng, M., Xu, Z., Lan, X., & Han, S. (1996) Alkali-aggregate reaction in China. In: *Proceedings of 10th International Conference on Alkali-aggregate Reaction in Concrete*, pp. 195–201.

Tang, M., Xu, Z., Deng, M., Lu, Y., Han, S., & Lan, X. (1997) Alkali-aggregate reaction in China. In: *Proceedings of East Asia Alkali-Aggregate Reaction Seminar*, Tottori University, Japan. pp. 1–12.

Tatematsu, H., Takada, J., & Tanigawa, S. (1986) Characterization of alkali aggregate reaction products. *Clay Sci., 26*, 143–150 (in Japanese).

Torii, K. (2008) Suggestion for prolonging life of concrete. *Bridge and Foundation, 42* (8), 82–84 (in Japanese).

Torii, K., Wasada, S., Sasatani, T., & Minato, T. (2008) A Survey on ASR-affected Bridge Piers with Fracture of Steel Bars on Noto Expressway. In: *Proceedings of 13th International Conference on Alkali-aggregate Reaction in Concrete*. pp. 1304–1311.

Tse, W.L. & Gilbert, S.T. (1996) A case study of the investigation of AAR in Hong Kong. *Proceedings of 10th International Conference on Alkali-aggregate Reaction in Concrete*, pp. 158–165.

World limestone resources (East Asia part), Limestone Assoc. of Japan, 1996 (in Japanese).

World limestone resources (Asia-Oceania part), Limestone Assoc. of Japan, 2000 (in Japanese).

Xie, Y.J., Jia, Y.D., Yang, F.M., Zhong, X.H., Zhang, Y., & Zhu C.H. (2004) Alkali-aggregate reaction in the concrete of Qinghai-Tibet railway. In: *Proceedings of 12th International Conference on Alkali-aggregate Reaction in Concrete*. pp. 458–465.

Yamada, K., Hirono, S., & Ando, Y. (2013) ASR problems in Japan and a message for ASR problems in Thailand. *J Thailand Concrete Assoc., 1* (2), 1–18.

Yamada, K., Hirono, S., & Miyagawa, T. (2011) New Findings of ASR Degradation in Japan. In: *Proceedings of 13th International Congress on the Chemistry of Cement*, No. 589.

Yamada, K., Kawabata, Y., Kawano, K., Hayashi, K., & Hirono, S. (2007) Actual ASR diagnosis including petrologic considerations and its importance. In: *Proceedings of the Concrete Structure Scenarios*, JSMS, Vol. 7, pp. 21–28 (in Japanese).

Yamada, K., Sagawa, Y., Nagase, T., Ogawa, S., Kawabata, Y., & Tanaka, A. (2016a) Importance of alkali-wrapping in concrete prism tests. In: *Proceedings 15th International Conference on Alkali-Aggregate Reaction*. p. 84.

Yamada, K., Tanaka, A., Oda, S., Sagawa, Y., & Ogawa, S. (2016b) Exact effects of temperature increase and alkali boosting in concrete prism tests with alkali wrapping. In: *Proceedings 15th International Conference on Alkali-Aggregate Reaction*. p. 203.

Yang, H.Q., Li, P.X., & Dong, Y. (2004) Research on the alkali-activity of the natural aggregates from Yichang channel segment of Yangtze river. In: *Proceedings of 12th International Conference on Alkali-aggregate Reaction in Concrete*. pp. 466–4472.

Yun, K.K., Hong, S.H., & Han, S.H. (2008) Expansion behaviour of aggregate of Korea due to alkali-silica reaction by ASTM C 1260 method. *J Korea Concrete I., 20* (4), 431–437 (in Korean).

Chapter 14

Australia and New Zealand

Ahmad Shayan (Australia) & Sue Freitag (New Zealand)

This chapter describes alkali aggregate reactions in Australia and New Zealand. Although Australia and New Zealand are geographically close, issues relating to alkali aggregate reaction and the nature of the research carried out to address them in each country are different. Therefore the chapter is divided into two independent parts: sections 14.1 to 14.10 by Ahmad Shayan covering alkali-aggregate reaction in Australia, and sections 14.11 to 14.17 by Sue Freitag covering alkali-aggregate reactions in New Zealand.

PART ONE – AUSTRALIA (AHMAD SHAYAN)

14.1 INTRODUCTION

The first part of this chapter (Sections 14.1 to 14.10) summarises the early history of alkali-aggregate reaction (AAR) in Australia in the mid-20th Century, and covers more recent advances made in its study since the early 1980s. Reported cases of AAR in Australia and the types of reactive aggregates used in them are outlined. Important aspects of the improvements achieved in this field include development of new test methods for detecting reactive aggregates, clarification of the interactions between AAR and some other deleterious chemical mechanisms, such as delayed ettringite formation (DEF) and impressed current cathodic protection (ICCP) of reinforcing steel. It also covers the AAR mitigation approaches used in Australia, as well as aspects of rehabilitation of AAR-affected structures.

14.2 EARLY HISTORY OF AAR IN AUSTRALIA

AAR work in Australia consists of two distinct periods. The first period relates to the 1940s and 1950s, soon after Stanton (1940) attributed the cause of damage to some bridge structures in California to this phenomenon. The initial AAR research, which was very extensive and lasted nearly two decades, was initiated in 1942 by the Commonwealth Scientific and Industrial Research Organisation (CSIRO), under a research programme entitled, 'Studies in Cement –Aggregate Reaction' (see Table 14.1).

This research programme explored several aspects of the AAR mechanism and factors that affect it. Most of the work was conducted by Harold Vivian, who was honoured for his work at the 10th International Conference on AAR, held in Melbourne in August 1996. This excellent effort is well recognised worldwide. In a paper presented at that conference, Idorn (1996) elaborated on the significant contributions made by Vivian in those years.

Table 14.1 List of publication by CSIRO under the research programme, 'Studies in Cement-Aggregate Reaction'.

Part	Published in	Year	Authors	Title of paper
—	Bulletin No. 161, 19 pages	1943	Alderman, A.R.	A review of evidence concerning expansive reaction between aggregate and cement in concrete
I	Bulletin No. 229, pp. 7–46	1947	Alderman, AR, Gaskin, AJ, Jones, RH, & Vivian, HE	Australian aggregates and cements
II	Bulletin No. 229, pp. 47–54		Vivian, H.E.	The effect of alkali movement in hardened mortar
III	Bulletin No. 229, pp. 55–66			The effect of void space on mortar expansion
IV	Bulletin No. 229, pp. 67–73			The effect of expansion on tensile strength of mortar
V	Bulletin No. 229, pp. 74–77			The effect of void space on the tensile strength changes of mortar
VI	Bulletin No. 229, pp. 78–84		Gaskin, A.J.	The effect of carbon dioxide
VII	J. CSIR, Vol. 20, No. 4, pp 585–594	1947	Vivian, H. E.	The effect of storage conditions on expansion and tensile strength changes of mortar
VIII	J. CSIR, Vol. 21, No. 2, pp 148–159	1948		The expansion of composite mortar bars
IX	CSIR Bulletin No. 256, PP. 7–12	1950	Jones, R.H. and Vivian, H.E.	Some observations on mortar containing reactive aggregate
X	CSIR Bulletin No. 256, PP. 13–20		Vivian, H.E.	The effect on mortar expansion of amount of reactive aggregate component in the aggregate
XI	CSIR Bulletin No. 256, PP. 21–30			The effect on mortar expansion of amount of available water in mortar
XII	CSIR Bulletin No. 256, PP. 31–47			The effect of amount of added alkali on mortar expansion
XIII	CSIR Bulletin No. 256, PP. 48–52			The effect of added sodium hydroxide on the tensile strength of mortar
XIV	CSIR Bulletin No. 256, PP. 53–59			The effect of small amounts of reactive component in the aggregate on the tensile strength of mortar

Table 14.1 (Cont.)

Part	Published in	Year	Authors	Title of paper
XV	CSIR Bulletin No. 256, PP. 60–78			The reaction product of alkalis and opal
XVI	Aust. J. Appl. Sci. Vol. 2, No. 1, pp. 108–113.	1951	Vivian, H.E.	The effect of hydroxyl ions on the reaction of opal
XVII	Aust. J. Appl. Sci. Vol. 2, No. 1, pp 114–122			Some effects of temperature on mortar expansion
XVIII	Aust. J. Appl. Sci. Vol. 2, No. 1, pp 123–131			The effect of soda content and of cooling rate of Portland cement clinker on its reaction with opal in mortar
XIX	Aust. J. Appl. Science, vol 2, No.4, pp 488–494.			The effect on mortar expansion of the particle size of the reactive component in the aggregates
XX	Aust. J. Appl. Science, Vol. 3, No. 3, pp. 228–232.	1952	McGowan J. K. and Vivian, H.E.	The correlation between crack development and expansion of mortar.
XXI	Aust. J. Appl. Science, Vol. 6, No. 1, pp. 78–87.	1955	Gaskin, A. J., Jones. R. H. and Vivian, H. E.	The reactivity of various forms of silica in relation to the expansion of mortar bars
XXII	Aust. J. Appl. Science, Vol. 6, No 1, pp 88–93.		Bennett, I. C. and Vivian, H. E.	The effect of fine-ground opaline material on mortar expansion
XXIII	Aust. J. Appl. Science, Vol. 6, No 1, pp 94–99		McGowan J. K. and Vivian, H.E.	The effects of superincumbent load on mortar bar expansion
XXIV	Aust. J. Appl. Science, Vol. 6, No 1, pp 100–104		Roberts, J. A. and Vivian, H. E.	The restoration of cracked mortar which has deteriorated through alkali-aggregate reaction.
XXV	Aust. J. Appl. Sci. Vol. 8, No. 3, pp 222–234	1957	Davis, C.E.S.	Comparison of the expansion of mortar and concrete
XXVI	Aust. J. Appl. Sci. Vol. 9, No. 1, pp 52–62.	1958		Comparison of the effect of soda and potash on expansion

Before the commencement of the early AAR research in Australia, Alderman (1943) conducted a literature review on the reported cases of damage to concrete structures overseas, including those reported by Stanton to that date (particularly Bradley Highway concrete pavement and King City Bridge in California), Parker Dam on Colorado River, Buck Hydroelectric Plant in Virginia, as well as laboratory test results obtained by others. This review highlighted the fact that the presence of both high alkali cement and reactive aggregate are necessary for AAR to develop and cause cracking.

The outcome of the early AAR research programme in Australia is chronologically summarised in the following paragraphs.

At the start of this research programme, Alderman *et al.* (1947) tested 68 aggregates from the various States in Australia in combination with cements from the same States, and found that only a few of them appeared to be reactive, when tested in mortar-bars (water: cement: aggregate ratios of 0.4–0.5: 1: 2, with cement alkali levels in the range of 0.12%–1.10% and storage conditions of sealed moist air at 20°C. It was recognised that the presence of high cement alkali content and opaline material or cryptocrystalline quartz were necessary for the AAR to cause deleterious expansion. Petrographic examination was found to be a useful tool for detecting reactive aggregates.

In 1947, in Parts II to V of the series 'Studies in Cement-Aggregate Reaction', Vivian used opal as reactive fine aggregate in combination with cements of various alkali contents and found that alkali can move from high alkali regions through to low alkali regions in the mortar-bar to react with reactive particles; that void spaces in the mortar reduces the level of expansion; that increasing levels of expansion results in progressive reduction in tensile strength of the mortar; and that void space also reduces the tensile strength.

Vivian (1947) also investigated the effect of storage conditions on expansion and tensile strength of mortars containing reactive fine opal aggregate, and found that storage in water caused significant leaching of alkali from the mortar and loss of expansion. Storage in sealed humid air increased the expansion, whereas storage in dry air reduced it. Despite the difference in expansion, cracking was always associated with reacted opal particles, which significantly reduced the tensile strength of mortar by more than 50%.

Gaskin (1947) noted that exposure to carbon dioxide reduced the AAR-induced expansion of mortar by converting the alkali hydroxide to alkali carbonate, which is inactive with respect to AAR.

Vivian (1948) compared the expansion of mortar comprising a uniformly reactive matrix of high alkali cement and reactive aggregate (opal) with that of composite mortar-bars comprising alternate sections of reactive and non-reactive mortar. The latter was made by combining non-reactive aggregate with low and high alkali cement in different segments of the mortar-bar. He showed that water availability and amount of alkali were important factors in the reaction. Water and alkali were shown to move towards reactive particles. Excessive amounts of water and alkali led to formation of sol rather that gel, and consequently to reduction in expansion. The water movement was said to occur through vapour phase rather than liquid phase, and that reactive aggregate particles generated low water vapour pressure around them causing water to move towards them.

Jones and Vivian (1950) found, through microscopical examination of reacted mortar, that membranes did not form across cracks in aggregate particles, induced by AAR swelling. They interpreted this observation to indicate that the expansion mechanism concerning the formation of an osmotic pressure cell was not valid, as no membrane was noted. They showed that a dark rim (gel) had developed around the reacted particles, which may have had the effect of an osmotic membrane. However, Vivian (1950, Part XV) states that this may be merely a precipitate that doesn't act like a true semi-permeable membrane.

In Part X of the study, Vivian (1950) found that, for the highly reactive opal, when the amount of reactive component in the mortar is small, the expansion depends on the amount of reactive component, but when the amount is large, the amount of alkali and water available per reactive particle decreases, causing reduced expansion. The maximum expansion was achieved when 5% opal was present in the fine aggregate (*i. e.*, the 'pessimum' proportion at the alkali content used).

In Part XI, Vivian (1950) showed that mortar expansion was highly influenced by the amount of available water in the mortar and that the expansion of the mortar-bar was correlated to the water content of the mortar and that of the reacted particles. It was also shown in Part XII (Vivian, 1950) that the expansion of mortar is determined by an interrelationship between the amounts of reactive component, alkali and available water. Usually small amounts of alkali (around 0.50–1.0%) were needed for large mortar expansion. Increased alkali levels shifted the pessimum proportion to higher values. However, the addition of alkali reduces the tensile strength of the mortar (Vivian, 1950, Part XIII). The reduced tensile strength was also related to the formation of cracks around the reacted particles (Vivian, 1950, Part XIV). The reaction product of opal and alkali consists of only silica, alkali and water, and the volume increase of the product, *i.e.*, the swelling, is a function of the water content of the product (Vivian, 1950, Part XV).

Vivian (1951), using NaOH and [(CH$_3$)$_4$N] OH, which are strongly dissociated in water and generate large OH concentrations that can attack silica, established that it was the OH ions which are the important component in the reaction with opal, not the alkali metal ions associated with them. The nature of alkali metal affects the properties of the reaction product and whether it causes expansion. By contrast, Ca(OH)$_2$ cannot dissociate strongly in water and does not react strongly with opal.

Vivian (1951, Part XVII) found that higher curing temperatures had a considerable effect on the reaction between alkali and aggregate, and on the amount of water held by the reaction product, and on the rigidity of the product, all of which influence the magnitude of the expansion. The initial rate of expansion was higher at 110 °F than at room temperature, but total expansion was lower. This was said to arise because the particles reacted initially fast and became liquid and could not cause further expansion, whereas at room temperature the initial rate was lower but particles continued to remain more rigid and reacted over longer periods of time. It should be noted that these arguments may only apply to very reactive aggregates – not necessarily to less reactive aggregates used in construction.

Davis (1951) studied 22 cements of varying alkali content (0.03% to 2.0%) in mortars containing 5% opal, and noted that both the alkali content and the cooling rate of the clinker had a large influence on the AAR-induced expansion. He suggested that the boundary for low alkali cement should be reduced from 0.6% to 0.35%. However, it should be noted that this observation could apply only to highly reactive materials, not necessarily to normal commercial reactive aggregates.

Vivian (1951) used a mortar composition of 0.5 water/cement ratio and 1 part of cement to 2 parts of aggregate, with a cement alkali level of about 0.90% Na$_2$Oeq and storage in a sealed container over water at room temperature. He found that the largest expansion was achieved at different percentages of opal content, depending on the mean particle size of opal, as shown in Figure 14.1.

Particles finer than 300 mesh (about 40 µm particle size) caused no expansion, and acted like pozzolanic materials. Maximum expansion occurred at 5–10% for various particle sizes, and mortars containing more the 40% opaline material exhibited very small expansions. It should be noted that the amounts that cause largest expansion would vary at different alkali levels.

McGowan and Vivian (1952) found a good correlation between expansion of mortar-bar measured by comparator and that calculated by summing the widths of

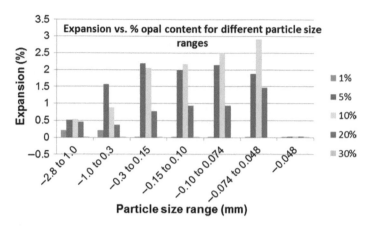

Figure 14.1 Effect on the expansion of particle size and amount of opal in mortar (after Vivian, 1951).

observed cracks in the mortar-bar, and concluded that the expansion is related to the widening of cracks. The slope of the straight line of the plot of calculated expansion against measured expansion was not 1.0 but 0.65, *i.e.*, the calculated values were lower than the corresponding measured values. The authors attributed this to the difficulty in measuring the crack width. However, the reasons could also include the fact the some expansion takes place before cracking occurs, and that microcracks inside the mortar-bars cannot be seen or measured.

Bennett and Vivian (1955) confirmed that opal ground to pass a 300 mesh sieve (about 40 μm) did not cause mortar-bar expansion. Moreover, when 20% of this material was used in mortar containing reactive aggregate, it caused a considerable reduction in mortar-bar expansion (from 1.18% at 12 months to 0.02% at 12 months). Therefore, sufficient amounts of finely ground opal acted as a pozzolana in suppressing AAR expansion.

Gaskin *et al.* (1955) established that all forms of silica, except macro-crystalline quartz (such as Pyrex glass, tridymite, cristobalite, chalcedony, opaline material, fused silica), react in concrete or mortar to produce expansion. Cristobalite and fused silica had not been studied before, but this study showed that they are reactive and cause mortar-bar expansion.

McGowan and Vivian (1955) found that imposing sufficiently large loads in the vertical direction, along the length of mortar-bars that contain reactive aggregate and high alkali, reduced the expansion, and the expansion resumed on removal of the load. Loads greater than 10 lbs /in^2 (0.07 MPa) caused the cracks to become oriented in the direction of bar length, otherwise they were random. There were indications that vertical load caused lateral expansion of bars. The largest expansion occurred at 0.75% alkali content. At 1% alkali, larger loads showed lower expansion than at 0.75% alkali, possibly because a more liquid gel formed at higher alkali and was squeezed out by a larger load of 50 lbs/in^2 (0.35 MPa). At 0.5% alkali, unloaded bars expanded 0.12% at 1 year.

Roberts and Vivian (1955) reported that mortar specimens cracked by AAR suffered a loss in tensile strength of about 87% (70 lbs/in^2) compared with similar specimens made with non-reactive aggregate (560 lbs/in^2) after 28 days to 1 year of reaction. Grouting with sodium silicate solution (specific gravity of 1.41) was reported to restore over 50% of original strength after 7 days and 63% after 28 days. The repaired specimens became highly impermeable to water.

Davis (1957) found that if mortar and concrete have the same fine aggregate and high alkali cement, then they will both expand to similar amounts, despite the presence of non-reactive coarse aggregate. However, with lower alkali cement, the expansion of concrete is delayed and is smaller in magnitude.

Davies (1958) found that KOH initially caused more expansion than NaOH, but the total expansion was the same. He stated that the total amount of alkali was the important factor rather than the type of alkali.

Lawrence and Williams (1961) studied the dissolution rate of opal in lithium hydroxide (LiOH), which was found to be slower than those in NaOH or KOH in the initial 16 hours, but later the rate was the same for all the three hydroxides. LiOH was also found to inhibit the reaction of the latter hydroxides with silica. LiOH reacted differently with silica and did not produce a gel. When LiOH was combined with NaOH, it inhibited the expansion. Both $Li_2O.SiO_2$ and $2Li_2O.SiO_2$ are insoluble in cold water and could have formed at the surface of silica particles, inhibiting further reaction of silica. This study showed that lithium hydroxide had the potential to prevent AAR under certain conditions.

The work of Vivian and others outlined above showed that many factors influence the AAR-induced expansion of mortar or concrete, including the amount and particle size of reactive particles, aggregate to cement ratio, alkali content, water/cement ratio, compaction (void content), storage temperature and humidity, and rate of drying. It also showed that alkali can move within the specimen from areas of high concentration to low concentration to react with reactive particles, and that LiOH behaved differently from NaOH and KOH, and had the potential to prevent AAR. Imposed mechanical loading was also found to suppress the AAR-induced expansion of mortar. Moreover, it was concluded that the mechanical properties of AAR-affected specimens were adversely influenced by AAR.

Work on AAR does not appear to have continued in the 1960s and 1970s. The only other published work was that of Vivian (1983) who reviewed the use of the mortar-bar test (ASTM C227, now 2010) and the 'quick chemical test' (ASTM C289, recently 2007, but now withdrawn) in assessing aggregate reactivity. During this period, and up to 1994, the then current Australian Standards for testing aggregates for AAR were the mortar-bar test AS1141-38 and the quick chemical test (AS1141-39), which were essentially copies of ASTM C 227 and ASTM C289, respectively. Vivian (1983) concluded that the chemical test was a reliable test and proposed that the limits of the mortar-bar test, which was current in Australia at the time, be changed from 0.05% at 3 months to 0.05% during the test period (one year). He also stated that the value of 0.60% as the boundary of low alkali cement was too high and that it should not exceed 0.40% under any circumstance. However, the proposed limits were not implemented in the standards current at the time.

As a result of the early AAR work in Australia, a number of reactive aggregates were found in different areas in Australia. Table 14.2 presents the list of the aggregates identified as reactive at the time, which is complemented by reactive aggregates more recently found in both AAR-affected structures and in laboratory testing using new test methods. The locations of AAR cases are shown in Figure14.2.

Table 14.2 List of Australian reactive aggregates from the 1950s work.

State	Source
Australian Capital Territory	Deformed granitic (gneissic) rock.
New South Wales	River sand, river gravel, quartzitic rocks, deformed granitic (gneissic) rock, dacite, meta-greywacke, meta-argillite, hornfels, quartzite, quartz gravel of gneissic origin, rhyodacite tuff.
Northern Territory	Tennant's Creek gravel (sandstone), Katherine river gravel (quartzite), Matarauka gravel (quartzite), Rhyolite.
Queensland	Brisbane river sand, quartz-feldspar porphyry, Beerburrum trachyte, ignimbrite/rhyolite, greywacke, quartzite, siliceous limestone, river gravel, glass-bearing basalt.
South Australia	Oodnadatta surface gravel, Cooper Pedy quartzite, meta-sedimentary rock (phyllite/slate).
Tasmania	Derwent river sand and other siliceous coarse and fine aggregates, meta-quartzite and deformed granitic rock, basalt.
Victoria	Opal from Gippsland, Scoria from Mount Noorat, several river gravels, quartz gravel of gneissic origin, dacite, rhyodacite, sandstone, phyllite, hornfels, quartz from gold mine tailings, deformed granitic (gneissic) rock.
Western Australia	Various deformed granitic rocks, metadolerite, quartzite, sandstone, various gravels, chert.

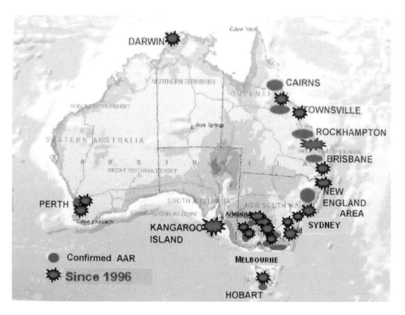

Figure 14.2 Geographical distribution of identified AAR cases in Australia. Note that each location could include more than one structure.

The reactive components in the aggregate types listed in Table 14.2 include amorphous/ or glassy silica; deformed/strained quartz with stress lamellae and patchy extinction; microcrystalline quartz and perhaps some silicate minerals such as various feldspars.

Despite the extensive history of the AAR work outlined above, no field cases of AAR were detected in the period 1940s to 1970s. It was only from the 1980s that such cases have been discovered.

14.3 DIAGNOSIS OF AAR AND CHARACTERISATION OF CONCRETE

Several approaches have been used in Australia for the diagnosis of AAR and other deterioration mechanisms in concrete, and for characterisation of the affected elements for prognostic and maintenance purposes; the most relevant methodologies are:

- Visual and stereoscopic observation/ assessment of the interior of the affected concrete through examination of drill cores. This is useful as it gives a broad appreciation of the main features or defects in the concrete and the extent of the reaction in the concrete (*e.g.*, cracking in the matrix, extent of aggregate rimming and internal cracking of aggregate pieces).
- Petrographic examination of concrete thin/polished sections, which is used to determine the severity of microcracking in the concrete, and identification of the nature of ingredients, distribution of phases, etc.
- Microstructural examination of phases formed in the concrete by scanning electron microscopy (SEM), combined with energy-dispersive X-ray (EDX) analysis. This is a very powerful technique for determining the chemical and morphological nature of phases present in the concrete and their spatial relationships, which aid the understanding of mechanisms of reaction and consequent damage.
- Measurement of the residual alkali content of concrete, to assess whether sufficient amounts of alkali remain in the concrete to fuel the reaction in the future.
- Measurement of the residual expansion potential of concrete, to understand whether further cracking would occur in the future, which is important in planning rehabilitation actions.
- Determination of strength properties of concrete, including compressive, splitting tensile and flexural strength as well as elastic modulus, which are important in addressing loss in structural capacity and need for strengthening. This would also assist in planning strategies for rehabilitation.
- Supplementary mineralogical analyses of concrete and reaction products, using X-ray diffraction (XRD), infrared spectroscopy (IRS), differential thermogravimetric analyses (DTA/DTG) could be useful in the characterisation of the reaction products.
- Supplementary chemical analyses to determine cement content, as well as chloride and sulphate contents, are used in cases where the reaction products is to be characterised in detail.

Examples of various Australian AAR-affected structures are given in Table 14.3, with the aggregate type and relevant reference. These structures were identified by the application of the above methodologies in the diagnosis, assessment and prognosis of AAR. A review of the varied visual manifestations of AAR in Australian concrete structures is given in Section 14.3.1.

Table 14.3 Reported cases of AAR-affected structures in Australia.

Name or type of structure	Aggregate	Location/Year identified	Reference
Upper Yarra Dam	sandstone/phyllite	Victoria – 1980	Cole *et al.* (1981); Shayan (1989a)
Canning Dam	deformed granitic	Western Australia – 1993	Shayan (2000)
Dam	Dacite	Victoria – 1987	Shayan (1988)
Water tank	Hornfels/Schist	North Victoria – 1992	Leaman and Shayan (1996)
Dam	deformed quartz gravel	Victoria – 1996	Shayan (1999)
Dam	phyllite/slate	N.A. – 1997	Shayan (1998, Unpublished report)
Dam	deformed granitic	Victoria. – 1998	Shayan (1998, Unpublished report)
Two Dams	deformed granitic	N.A. – 1998	Shayan (1998, Unpublished report)
Dam	deformed granitic	N.A. – 2002	Shayan (2002, Unpublished report)
Gordon Dam Intake Tower	Quartzite	Tasmania 1992	Blaikie *et al.* (1996)
Cooling Tower, Tarong	Quartzite	Queensland – 1992	Carse (1993)
Bridges (95 No.)	Greywacke, ignimbrite/ chert	Queensland- 1980s	Carse (1988)
Load-out Jetty	Not reported	Western Australia – 1992	Davies *et al.* (1996)
Causeway Bridge	deformed granite, metadolerite	Western Australia – 1983	Shayan and Lancucki (1986) Ross and Shayan (1996)
Bridges (4 No.)	metadolerite, deformed granite	Western Australia – 1997	Shayan (1997, Unpublished report)
Bridges (7 No.)	quartz gravel	Victoria 1991–1994	Shayan (1994)
Bridges (2 No.)	hornfels/schist	Victoria – 1992	Shayan *et al.* (2003)
Culvert	Hornfels	Victoria – 1999	CIA Paper Shayan *et al.* (2003)
Building	sandstone	Victoria – 1999	Shayan *et al.* (2003)
Railway Bridge	quartz gravel	Victoria– 1999	Shayan (1999, Unpublished report)
Precast planks in Jetty approach	deformed granitic	Queensland – 1999	Shayan (1999, Unpublished report)
Railway Bridges (2 No.)	quartz gravel	Victoria – 1996	Shayan (1996, Unpublished work)
Bridge	hornfels	North New South Wales – 1998	Shayan *et al.* (1998)
Bridge	hornfels	North New South Wales– 1998	Shayan (1996, Unpublished report)

Table 14.3 (Cont.)

Name or type of structure	Aggregate	Location/Year identified	Reference
Deep Creek Bridge	quartz gravel	North New South Wales– 2000	Shayan and Morris (2003)
Bridge	quartz gravel	South New South Wales – 2000	Shayan and Morris (2002)
Bridges (4 No.)	hornfels/greywacke	South New South Wales – 2000	Shayan and Morris (2002)
Bridges (4 No.)	deformed granitic/ sand-stone/acid igneous/hornfels	New South Wales – (2003)	Shayan et al. (2003) Unpublished
Railway sleepers (1,000,000)	deformed granitic rock	N.A.– (1990)	Shayan and Quick (1992)
Bridge (4 No.)	Quartz gravel	Victoria – 2005	Shayan and Andrews-Phaedonos (2005)
Bridge (2 No.)	River gravel	NSW – 2005	Shayan and Morris (2005)
Industrial Tank foundation	Hornfels & deformed acid igneous rock	NSW – 2005	Shayan (2006)- Unpublished
Dam	Deformed granite	NA – 2005	Shayan and Grimstad (2006)
Arch Bridge (2 No.) and footbridge	Acid igneous rock	NA – 2008	Shayan (2009) - unpublished report
Weir	Acid igneous	NA – 2008	Shayan (2009) - unpublished report
Bridge	River gravel	Como Bridge, Sydney - 2008	Shayan et al. (2008)
Dam	Acid porphyry	NA – 2011	Shayan (2012) unpublished report
Bridge-Prestressed Planks	Meta-sediments	QLD – 2010	Shayan (2010) unpublished report
Dam	Acid igneous rock & meta-sediment	N/A –2010	Shayan (2010) unpublished report
Bridge	Quartzite	N/A – 2012	Shayan and Thomas (2014)
Dam	Quartzite	N/A – 2012	Shayan and Cribbin (2015)
Bridge	Hornfels, mica-rich	NSW, 2013	Shayan (2014) unpublished report
Bridge	Gneissic quartz gravel	VIC, 2014	Shayan et al. (2015)
Bridge	Mixed Schistose and gneissic crushed rock	VIC, 2014	Shayan et al. (2015)
Dam	Gneissic granite	ACT	Shayan (2015, unpublished work)
Wharf structures and bridges (several)	Various	Miscellaneous work for consultants	Shayan (2003–2015)

14.3.1 Visual manifestation of AAR in various Australian structures

It is generally known that AAR often causes random cracking or map-cracking in affected elements. However, this pattern mainly occurs in mass concrete or lightly reinforced elements. Examples of this type of cracking are shown in Figure 14.3, for retaining walls in a dam, end of crosshead beam in a bridge, large pylon in another bridge, and columns in a spillway bridge. These structures are located in South Australia, Victoria and Western Australia, although similar situations exist elsewhere in Australia.

Other than map-cracking, illustrated above, columns in a number of bridges have developed a single longitudinal crack, an example of which is shown in Figure 14.4, for

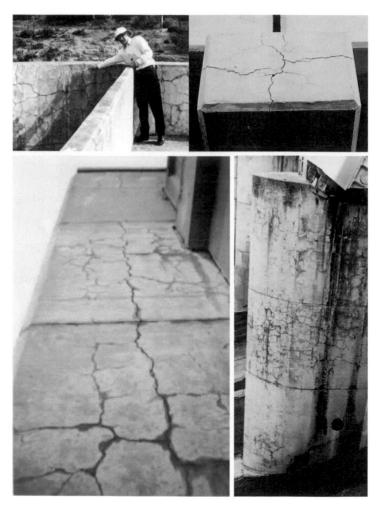

Figure 14.3 Random map-cracking caused by AAR in: retaining wall of a dam (top left); end of crosshead beam in a bridge (top right); bridge pylon (bottom left); bridge pier in dam spillway bottom right).

Figure 14.4 Vertical AAR-induced cracking (3–5 mm wide) in a bridge column built in 1930.

a bridge constructed in 1930, which shows significant internal signs of AAR, indicated by strong rimming of the gneissic quartz gravel aggregate particles used in this concrete. This bridge crosses a water reservoir and piles in the water show more widespread cracking.

Reactive to slowly reactive gneissic quartz gravels are relatively common in different regions in Victoria. The slowly reactive type was identified in some bridges and a dam in the western parts of Victorian, whereas more reactive varieties are present in a number of AAR-affected bridges in the eastern parts. These structures are included in the list AAR-affected structures presented in Table 14.3.

Shayan and Ferguson (1996) found that many of the gneissic quartz gravels in the eastern regions of Victoria originated from the weathering of schists and gneissic rocks in the central Victorian highlands, and were spread out to both the northern and southern sides of the highlands through river flows. These sources are currently being mined and used in concrete, but the present standard of practice would require implementation of proven precautions to ensure mitigation of AAR in the structures concerned.

Heavily reinforced concrete elements have greater resistance to cracking, and the cracks are often oriented in the direction of the greatest restraint, i.e., greatest reinforcement or prestress. Examples of this type of cracking are given in Figure 14.5 for precast, prestressed bridge deck plank, deteriorated submerged bridge pile, railway sleeper, and cast-in-situ crosshead beam and dam wall in NSW, Queensland, Western Australia and Victoria. Such directional cracks are common in AAR-affected, prestressed, precast sleepers.

A case (Shayan & Morris, 2002), where both the prestressed deck planks and the concrete deck overlay had developed severe cracking, and were subsequently replaced

Figure 14.5 Parallel and longitudinal AAR-induced cracking in various precast prestressed elements (except the crosshead beam, lower right, which is cast-in-situ), all of which exhibit signs of AAR, and most exhibit signs of combined DEF (except the crosshead beam) due to excessive heat curing. The pile shown in the third row is severely disintegrated.

due to losses in strength properties, is shown in Figure 14.6. This is the first reported bridge in which the deck was removed as a result of AAR-induced damage. Subsequent tests on the removed planks (Al-Mahaidi, *et al.*, 2008; Shayan, *et al.*, 2008; Shayan & Xu 2012) showed that, compared with the design values, they had significantly lower strength properties, particularly elastic modulus (reduction from 37.5 GPa to 15.5 GPa). This was due to the AAR-induced microcracking in the concrete and was sufficient justification for their replacement.

Another visual manifestation of AAR in some Australian dam structures has been in the form of distortion, misalignment and rotation of the affected elements in relation to the neighbouring elements. For example, Figure 14.7 shows the displacement and rotation of a massive concrete block in a dam wall, relative to its neighbouring block, which was caused by significant expansion and cracking in one block (Figure 14.7, right) compared to the other block which had no sign of reaction in it.

Similar rotational movements and cracking were identified in the retaining wall of a dam which contained phyllite and slate as coarse aggregate (Table 14.2, reported in

Figure 14.6 Longitudinal parallel cracking in pre-stressed concrete deck planks (top left) concrete overlay (top right), and side of a removed plank (bottom left) showing three longitudinal cracks highlighted by moisture. Both concretes contained the same reactive aggregate and exhibited strong AAR.

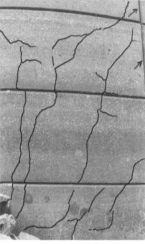

Figure 14.7 Rotational displacement of massive concrete blocks in a dam (left), and extensive cracking in the displaced block. Arrows indicate the joint which is showing considerable misalignment with the next block. The adjoining block shows very little, if any, cracking.

1997), in buttresses of one of the two dams which contained slowly reactive deformed granite as coarse aggregate (Table 14.2, reported in 1998), and in the main body of Canning dam (Shayan, et al., 2000). Some of these movements have structural and stability consequences, which need to be addressed by specialist structural engineers.

In the case of the latter dam, the culprit aggregate was of the slowly reactive nature. It appears that, given sufficient reaction time and adequate amounts of alkali, the slowly reactive aggregates could cause significant damage.

A further example of type of structures affected by AAR is an arch bridge (Figure 14.8), in which precast arch units, as well as the main body of the arches exhibited cracking. The aggregate was an acid igneous rhyodacite rock.

Figure 14.8 AAR-induced cracking in an arch bridge. Cracking is parallel to the length of arch in the crest of the precast arches, but random in the body of the bridge.

The external cracking patterns, if present, provide an easy way of identification of AAR-affected structures, although they may not be present at early ages of the affected structures. After a preliminary visual identification of suspect structure has been made, the concrete would then need to be subjected to a number of diagnostic techniques in order to determine its present condition and assess its prognosis and rehabilitation requirements.

14.3.2 Visual and stereoscopic examination of concrete samples

Visual and low magnification observations of concrete samples taken from suspect structures can reveal useful information. For example, Figure 14.9 shows definite strong reaction rims around the aggregate particles and confirms that AAR exists in the concretes concerned, and that it is probably the cause of cracking. Visual observations can also indicate the severity of reaction and its consequent damage, based on the distribution and number of reacted particles. Figure 14.10 shows examples of white AAR gel filling voids next to some reacted aggregate particles, and aggregate internal cracking which indicates the AAR cases concerned to be strong. In addition to the AAR symptoms, visual examination has revealed features such as penetration of cracks to the reinforcement and delamination of cover concrete (Figure 14.10), which assist in establishing the status of deterioration of concrete and likely repair needs.

14.3.3 Petrographic examination of aggregates and concrete samples

In Australia, all commercially available aggregates (for both concrete and flexible pavement) are subjected to petrographic examination in the process of assessment of source rock, and at various production intervals. Identification of reactive components is one of the reasons for the examination of concrete aggregates. Australian Standard AS 1141.65, *"Alkali-aggregate reactivity - Qualitative petrological screening for alkali-*

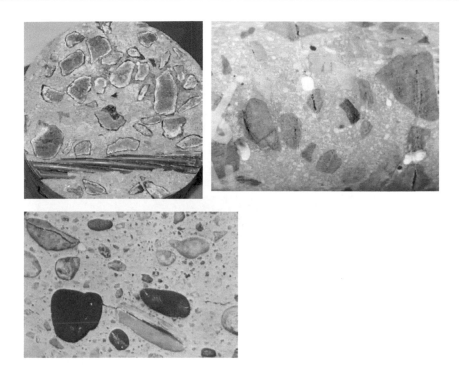

Figure 14.9 Visual appearance of concrete samples taken from cracked concrete elements showing: strong reaction rims on reacted aggregate particles (top left); aggregate internal cracking and AAR gel filling void (top right and bottom).

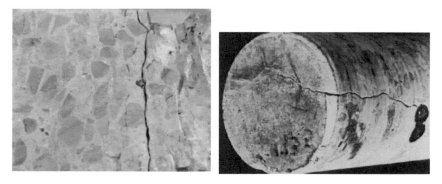

Figure 14.10 Longitudinal section through a core from an AAR-affected prestressed pile showing delamination (left) and penetration of AAR-induced crack to the level of reinforcement.

silica reaction" outlines the procedures for the petrographic examination of aggregates in Australia.

The most common reactive components found in commercial aggregates are largely of microcrystalline quartz and strained quartz, examples of which are shown in Figure 14.11, although cryptocrystalline quartz and other unstable forms of silica, such as chalcedony, cristobalite and tridymite have also been seen in some reactive rocks.

Figure 14.11 Photomicrographs of microcrystalline quartz and strained quartz (top) fine siliceous matrix of dacite (bottom left) and cryptocrystalline texture of hornfels (bottom right) found in some reacted concretes.

Rocks containing these minerals are generally of metamorphosed sedimentary or igneous origin, although some acid igneous rocks, tuffs, greywacke and chert-bearing gravels also contain reactive microcrystalline quartz and other unstable forms of silica. Glassy/opaline components in some Australian basaltic rocks have also been associated with alkali reactivity (Shayan & Quick, 1988; Shayan, 2004).

Detection of considerable amounts of reactive components in an aggregate can be taken as evidence that it could be susceptible to AAR, and classed as reactive or potentially reactive, without further testing. For example, this can be done when about 5% of opaline silica is detected in the aggregate phase. However, many old Australian rocks, particularly those in Western Australia, are very finely textured, and petrographic identification of reactive phases in them is not possible, yet they are reactive. For this reason, in Australia, aggregates cannot be classed as non-reactive based on petrographic examination alone. This decision is possible only after further testing, based on other methods, as discussed later in this Chapter.

Petrographic examination is also performed on concrete samples taken from suspect structures, in order to verify whether they contain reactive components and the distribution of reaction sites, AAR gel filling cracks and the extent of micro-cracking in the concrete. The geological nature of the aggregate and nature of the cement phase could also be established by the petrographic examination. Figure 14.12 shows some features mentioned above, where microcracking in the concrete matrix and AAR gel formation are evident.

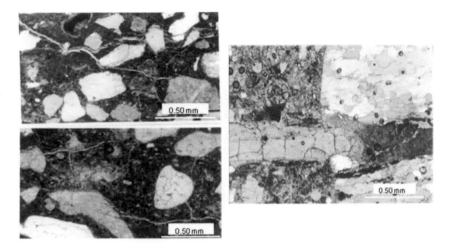

Figure 14.12 Micro-cracking pattern and presence of AAR gel observed in petrographic section (centre, lower left) and AAR gel flowing from aggregate into crack (right).

Not only AAR products, but other phases such as ettringite and its distribution in concrete can be observed, which help in establishing whether DEF has occurred in the concrete. The presence in concrete of mineral additives, such as ground granulated blastfurnace slag and fly ash, etc., can also be verified using this technique.

14.3.4 Scanning electron microscopy of concrete

Visual and petrographic examinations are very useful, but they cannot provide information on the chemical composition of the phases of interest in concrete, nor detailed microstructural and morphological information on such phases. Scanning electron microscope (SEM) examination, accompanied by energy-dispersive X-ray (EDX) spectroscopy of phases observed in concrete, are much more powerful techniques and can also be used to provide semi-quantitative chemical composition of the phases concerned.

No specific standard is followed in Australia for conducting SEM/EDX. It is often conducted in the secondary electron mode on fracture surfaces of concrete to provide morphological details under high magnifications, up to 10,000 times, when some individual crystals or regions of interest could be chemically analysed. For more accurate chemical compositions, polished slabs or slides can be analysed, and in this case the back-scattered electron mode is utilised to provide better phase contrast. Wavelength-dispersive spectroscopy (WDS) can provide even more reliable compositions. Due to the high magnifications achievable by this method, it can detect even small amounts of reaction products, which may remain undetected by petrographic examination. Knowledge of the composition of the reaction products is very important in relation to the interpretation of the reaction mechanisms involved. Therefore, SEM/EDX provides the most unambiguous evidence of whether AAR is present in a concrete sample, and on the nature of the AAR products.

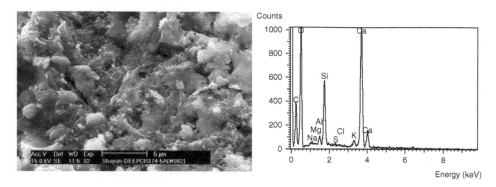

Figure 14.13 SEM view of hydrated cement paste (left) and its Ca-rich composition, as shown by the EDX spectrum (right).

The author routinely conducts SEM/EDX in the diagnosis of deterioration problems in various concrete structures, and this has enabled very important information to be obtained in relation to the mechanisms involved. As the reactions take place in the cement paste, and SEM/EDX work determines the compositions of AAR products, it is relevant to know of the composition of the cement paste in the absence of AAR.

Figure 14.13 shows an area of the cement paste which is relatively free of aggregate phases and comprises largely the calcium silicate hydrate (CSH) phase with small amounts of Aluminium (Al), derived from calcium aluminate. The EDX spectrum indicates Ca to be by far the dominant ingredient with much smaller amounts of silicon (Si). Very small amounts of alkali metals, sodium (Na) and potassium (K) are also indicated. As gypsum is added to Portland cement, another element that could be present is sulphur (S), which is normally associated with the sulpho-aluminate phases, but the area analysed was free of this phase.

In contrast with the features of the cement paste, AAR products show very different morphological features and chemical compositions. Typical examples of the morphological and compositional ranges of reaction products are presented in Figures 14.14 to 14.17.

The compositions of both the AAR gel and the crystalline products can vary depending on the alkali content of the cement and the dominant alkali species, as well as the mineral assemblage in the reacted aggregate. In the case of Figure 14.14, potassium (K) dominates the composition with much smaller amounts of sodium (Na), but the AAR gel contains similar amounts of both alkali species. The calcium (Ca) content of the products is derived from the hydrated cement due to their close proximity to this source of calcium. Prolonged exposure of AAR gel to the cement paste, particularly under wet conditions, can lead to replacement of Na and K by Ca, which increases the proportion of Ca in the AAR gel. Figure 14.15 shows corresponding AAR products in a different structure, and in this case Na dominates the composition of both products. However, these differences do not imply different mechanisms of reaction.

Figure 14.14 SEM views of AAR gel in a crack (top) and crystalline AAR rosettes (bottom) formed at the reacted aggregate rim, in concrete cores taken from the AAR-affected block shown in. The compositions of the gel and rosettes are given by the EDX spectra.

Mineralogical investigations of crystalline AAR phases have not yielded definitive mineral species. Cole *et al.* (1981) determined that the AAR rosettes found in the rims around reacted aggregate particles or in voids of an old concrete dam in Victoria, were similar to Okenite, but the basal spacings of the AAR products did not exactly match those of the natural mineral Okenite (the largest spacing of about 21Å was missing, but the other reflections matched reasonably well). They argued that this could have occurred due to dehydration of the AAR product. Shayan and Lancucki (1987) isolated similar AAR products from an AAR-affected bridge in Western Australia, and found it to be very similar to that reported for the Victorian dam. The X-ray powder diffraction pattern of the material isolated from the bridge is presented in Figure 14.16, and is very close to that for the former product.

The 12.04 Å reflection shown in Figure 14.16 is characteristic of several products examined by the author. Partial dehydration can produce peaks around 9.5-10 Å in the XRD pattern, as seen in Figure 14.16. The XRD pattern of the AAR product also indicated small amounts of calcite and quartz impurities which are labelled C and Q on the respective peaks. In addition, the chemical compositions of the AAR products from

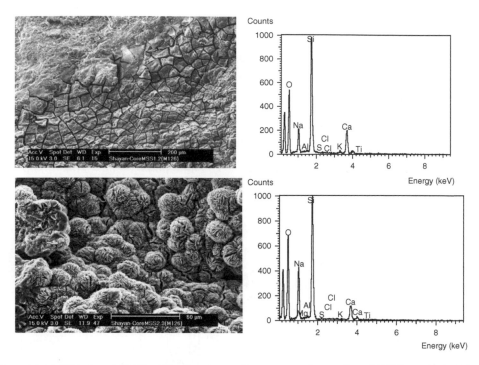

Figure 14.15 SEM views of AAR gel (top) and crystalline rosettes (bottom), with EDX spectra (right), in an AAR-affected element from a bridge in northern New South Wales, Australia.

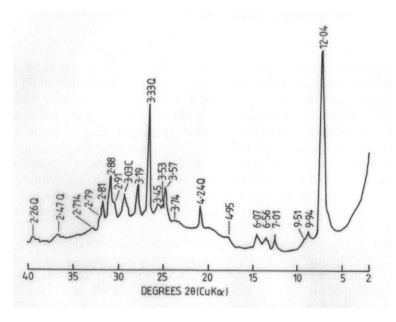

Figure 14.16 XRD pattern of the crystalline AAR product found in AAR-affected concrete.

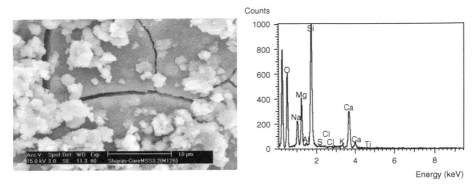

Figure 14.17 SEM view of AAR gel in a pile submerged in seawater (left). The EDX spectrum (right) indicates the AAR gel to be covered or intermixed with brucite Mg(OH)$_2$.

the two regions were also similar. In fact, these materials appear to be globally similar and within the same broad compositional range.

It has been noted that exposure environment, such as marine conditions, can significantly alter the composition of the AAR gel and that this may reduce the residual expansion potential of the concrete (Shayan, 2006). As an example, the AAR gel found in bridge piles that are exposed to seawater, was found to incorporate significant amounts of magnesium (Mg) from seawater, as shown in Figure 14.17. Very fine crystallites of brucite, Mg(OH)$_2$, were sometimes found to be associated with the AAR gel. This change was noted in the centre of large AAR-affected piles which have been submerged for some 30 years.

The interpretation of SEM/EDX data in relation to AAR as a cause of damage to concrete is not always straightforward, particularly when other mechanisms are also implicated. A prominent example of this situation concerns the combined presence of AAR and the so-called 'delayed ettringite formation' (DEF), and a number of important cases have been reported in Australia. The investigations conducted in Australia on the interaction of AAR and DEF are briefly covered later in this Chapter (see Section 14.4.1).

14.3.5 Determination of residual alkali content of concrete

In Australia, residual alkali content is determined on most structures suspected of AAR damage, which usually contain reactive aggregates. The residual alkali content is an important parameter that determines the potential of concrete for further reaction and expansion. The method used is the hot water extraction procedure applied to pulverised concrete samples, using a water to solid ratio of 5. In this method the solid-liquid equilibrium in hardened concrete is disturbed due to the large dilution factor, so that the alkali concentrations achieved may not reflect those prevailing in the concrete pore solution. Nevertheless, the aim of the method is to determine the reserve of the alkali hydroxides available in the concrete, which the method achieves.

Table 14.4 Residual alkali content of concrete samples from a bridge and a dam.

Structure	Location	Core No.	Soluble Alkali content: kg/m^3		
			Na$_2$O	K$_2$O	Na$_2$Oeq
Bridge	Abutment	1	0.43	1.99	1.74
	Wing wall	5	0.37	1.14	1.12
Dam	SL 203	1	2.70	1.65	3.80
Dam	SL 277	1	3.33	1.60	4.38
		2	1.83	0.92	2.44
		3	2.69	1.26	3.52

Expression of concrete pore solution has also been used to assess the potential of concrete for further reaction, this method would extract the alkali which is already in the solution, not the alkali associated with the solid (*e.g.*, adsorbed), which may become available when the liquid phase is depleted of alkali due to reaction with silica. On this basis, the hot water extraction method is considered satisfactory, particularly when it is used for comparative purposes. Results of Berube and Tremblay (2004) confirmed that more alkali is released by hot water extraction than expected from pore solution analysis.

The interpretation of the quantity of water extractable alkali with respect to promoting further AAR is not clear. For new concrete, generally, a limit of 3.0 kg/m^3 of equivalent alkali content applies in order to limit AAR development, although for very reactive aggregate this limit is lowered to 2.5 kg/m^3. However, these values are not appropriate for concrete elements which have already undergone deleterious expansion and cracking, and a limit below which no further reaction would occur is not established. However, when the residual alkali content is found to be greater than these limits, then it could be expected that further reaction and cracking would take place under favourable humidity conditions.

An example of determination of residual alkali content is given in Table 14.4, which includes specimens from a bridge and a dam; the latter incorporating a high alkali fly ash.

In the case of the bridge elements, significant AAR has already taken place, which has consumed a large proportion of the original alkali content. What remains is not likely to promote further strong AAR, but may maintain additional mild reaction. Monitoring of the structure would clarify whether further expansion and cracking takes place.

The higher alkali contents in the samples from the dam arose from the high alkali content of the fly ash, which was used at around 30% replacement for Portland cement. The concrete in the sampled areas has not developed any cracking some 40 years after construction, as fly ash concrete has larger tolerance for alkali.

14.3.6 Determination of residual expansion of concrete

Concrete elements which have undergone some degree of AAR-induced expansion usually exhibit further potential for cracking, which is termed the residual expansion potential. This is usually measured on core samples drilled from the structure, which

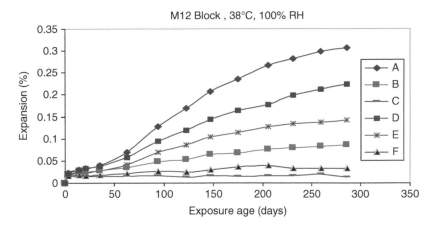

Figure 14.18 Residual expansion measured on different block portions of deck plank M12.

means that the measured expansion represents the free expansion of concrete, *i.e.*, without the influence of any confinement from surrounding concrete. On this basis, the measured value of this free expansion could overestimate the expansion of the structural element concerned. However, due to size effect and leaching of alkali from the relatively small mass of the core, the expansion could also be underestimated.

An example of residual expansion measurement conducted on large concrete blocks from a prestressed deck plank is shown in Figure 14.18. Measurements A, B, D and E were made across in the transverse direction (*i.e.*, across the width of the plank), but C and F were along the length (*i.e.*, parallel to the direction of prestress). It is clear that significant residual expansion exists in the plank.

Figure 14.19 shows the residual expansion results for core samples drilled from plank M12. In comparison with results obtained on the blocks, the core expansion is more rapid and larger.

In fact, comparison of these results with measurements conducted on the same plank in outdoor conditions (Shayan & Xu, 2012) showed that the expansion of planks under field conditions was around 40% of core expansion at 38°C, 100% RH. This is in agreement with results obtained overseas (Tomita *et al.*, 1989), which concluded that the field structure may only achieve around 50% of the residual expansion of cores drilled from the same structure and then stored and measured in the laboratory.

The results of residual expansion and alkali content determinations are complementary, and are used to assess the future behaviour of the affected element. However, for some of the elements in the structures listed in Table 14.2, there have been occasions when the results of these two determinations have appeared to be at odds, *i.e.*, the amount of available alkali is large but the residual expansion is low, despite the fact that reactive components always exist in affected concretes. This conflict was caused by the presence of excessive numbers of microcracks, when any new AAR gel formed was being used up to fill the cracks, rather than causing additional stress and expansion.

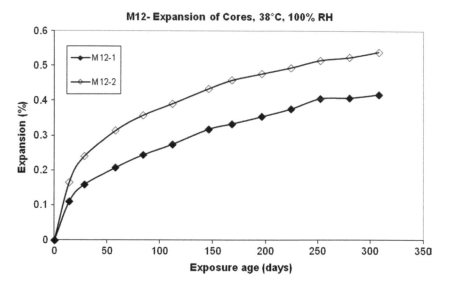

Figure 14.19 Residual expansion measured on cores drilled from plank M12.

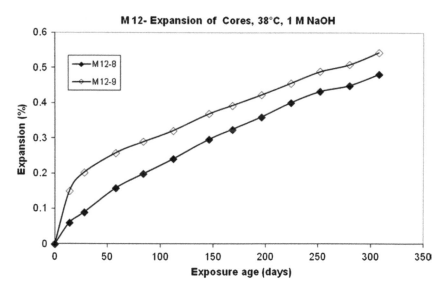

Figure 14.20 Core expansion measured on cores drilled from plank M12 and stored in 1M NaOH at 38°C.

Another expansion test is sometimes conducted on cores stored in 1M NaOH solution at 38°C to assess whether reactive components still remain in the concrete. Figure 14.20 shows the results of this test for cores from plank M12, and clearly indicates that the concrete contains reactive components, as the expansion keeps increasing, compared with the results in Figure 14.19.

14.3.7 Mechanical properties of AAR-affected concrete

The adverse effects of AAR on the mechanical properties of concrete has been known since the early days of AAR studies; Vivian (1947 – Part IV of the studies in cement-aggregate reaction) showed that the tensile strength of AAR-affected mortar specimens was considerably reduced compared with control mortar specimens. There is a significant volume of worldwide literature showing the adverse effect of AAR on strength properties of concrete. It would be expected that the magnitude of such adverse effects would depend on the reactivity of the culprit aggregate and the extent of the reaction, which would be driven by the alkali content of concrete.

Investigation of Australian concrete bridges that have been damaged by AAR has shown, in both the published work (Shayan & Morris, 2000; Al-Mahaidi *et al.*, 2008; Shayan *et al.*, 2008; Shayan *et al.*, 2015) and unpublished reports, that the losses in compressive strength and elastic modulus have been up to 35% and 50%, respectively. It is well known that the elastic modulus is more sensitive than compressive strength to AAR-induced expansion.

The compressive strength of concrete is determined on drill cores using the Australian standard AS1012-9-2014, and the determination of the static chord modulus of concrete is conducted in accordance with the Australian Standard AS1012.17-1997- Appendix A (Methods of testing concrete. Method 17: Determination of the static chord modulus of elasticity and Poisson's ratio of concrete specimens). Compressive strain measurements are recorded by two bonded strain gauges attached at 180° along the central 120 mm of the cores.

The compressive strength of concrete is first determined on companion cores. In the modulus test, three loading/unloading cycles are applied up to 40% of compressive strength, and at this point the load is released before repeating the loading cycle two more times. The fourth loading cycle is then applied up to the peak compressive stress. A final test to failure is then carried out.

Examples of the application of these tests to concrete cores taken from prestressed deck planks, which have undergone different extents of reaction and expansion, have been reported elsewhere (Shayan & Morris, 2000; Shayan *et al.*, 2008; Al-Mahaidi *et al.*, 2008; Shayan *et al.*, 2015). The results for different types of elements in different bridges are presented in Table 14.5.

It should be noted that the cast-in-situ concretes in Bridges A, B and C, which are located in Victoria, were probably of VicRoads grade VR400/40, with specified 28-day compressive strength of 40 MPa and elastic modulus of 35 GPa. The strength and modulus values at the current ages of the bridges would be expected to be considerably higher than the specified 28-day values. The precast, prestressed planks use higher strength concrete and required transfer strength of 35 MPa and elastic modulus at transfer of 36.7 GPa at the release of prestress (*e.g.*, 24 hours).

It is clear that the strength and particularly the elastic moduli have significantly dropped as a result of AAR, and that the elastic modulus is more sensitive than compressive strength to the effects of AAR. It should also be noted that cores are often taken from sound portions of the AAR-affected elements, and that the strength values determined on them may not represent the deteriorated properties of the whole element.

Table 14.5 Strength properties of AAR-affected concrete in four bridges.

Bridge	Location	Elements tested	Compressive strength (MPa)	Modulus of elasticity (GPa)
A	Victoria	Columns	33.6	29
		Abutment crosshead	41.1	29
B		Pier wall	39.2	27
		Abutment crosshead	44.8	19
C		Pile cap	32.4	17.3
		Columns	41.5	30.3
		Pier crosshead	36.6	25.3
D	New South Wales	Three Prestressed deck planks	38	10
			41.8	16
			34.7	13
E	Queensland	Prestressed deck plank	47.6	31.7

Examples of the stress-strain curves for the elements tested are presented in Figure 14.21 for the cast-in-situ elements in bridges A, B and C, which shows the significant loss of stiffness in the abutment crosshead of Bridge B and the pile cap of bridge C. The prestressed planks from bridge D behaved similarly to the latter elements, whereas that from bridge E was similar to the columns of Bridge A. These elements show much greater compressive strain than the other elements at the same level of compressive stress. The larger reduction in the stiffness arises from the greater extent of AAR-induced microcracking in these elements.

Shayan *et al.* (2003) investigated the effect of AAR expansion on the compressive and flexural strength of concrete. Various aggregates, exhibiting different levels of AAR-induced expansion, were used to provide a range of expansion values at the age of one year after being used in concrete of high alkali content ($5.8 \text{ kg Na}_2\text{O/m}^3$). The strength properties of the concretes were also determined. No specific pattern was noted in the plot of compressive strength against expansion. The relationship between flexural strength and expansion is presented in Figure 14.22, which shows that it is sensitive to AAR-induced expansion and deteriorates even at low levels of expansion, but particularly beyond 0.010%.

These effects were also confirmed in the laboratory by Abdullah *et al.* (2013) on specimens that were designed to generate different levels of expansion, using a highly reactive fused silica in the fine aggregate in combination with non-reactive coarse aggregate. Fused silica appeared to have a 'pessimum' proportion around 7.5%, as shown in Figure 14.23. Note that the level of expansion achieved by using fused silica was an order of magnitude greater than that caused by most natural aggregates (*cf.*, Figure 14.22).

Such large levels of expansion significantly reduce the mechanical strength of concrete, including compressive strength, modulus of rupture and elastic modulus (Figure 14.24).

Kubat *et al.* (2015) also obtained similar results, using fused silica as reactive aggregate, and showed the significant reduction in stiffness caused by AAR-induced expansion, as demonstrated by the stress-strain behaviour of the concretes exhibiting varying levels of expansion (Figure 14.25).

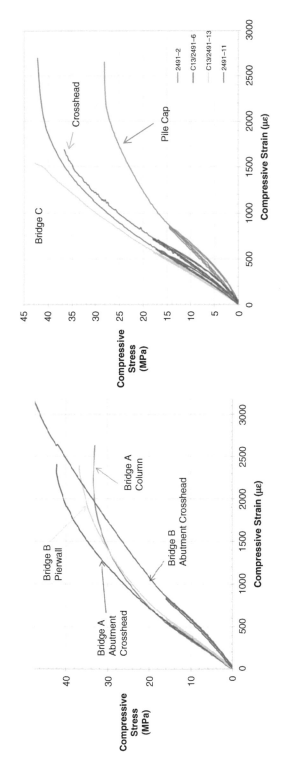

Figure 14.21 Stress–strain curves for various AAR-affected elements in bridges A, B and C.

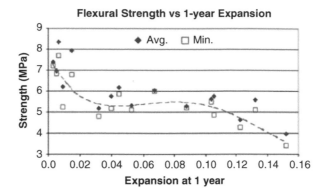

Figure 14.22 Effect of AAR expansion on flexural strength of concrete (after Shayan et al., 2003).

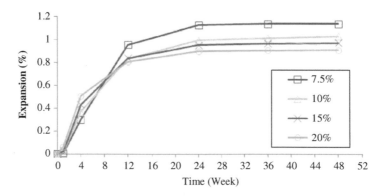

Figure 14.23 Expansion behaviour of concrete prisms containing various amounts of fused silica (after Abdullah et al., 2013).

The results reported here are in agreement with results observed elsewhere (e.g., Rigden et al., 1995; Jones & Clarke, 1988; Ahmad et al., 2003) regarding the effects of AAR on the mechanical properties of concrete.

Despite the losses observed in the strength properties of concrete specimens, the load-bearing capacity of actual reinforced concrete elements in structures may not suffer to the same extent. In fact, despite significant losses in the compressive strength and elastic modulus of prestressed bridge deck planks, which had developed multiple longitudinal AAR cracking, full scale loading of the planks after removal from the deck showed that the effect of AAR on the load capacity of the planks was small, and that after 25 years in service they were still able to carry the design load (Shayan et al., 2008; Al-Mahaidi et al., 2008). As mentioned earlier, the bridge deck was replaced mainly because of the large residual expansion potential of the concrete, which could have affected the bond strength of the prestressing cables in an unpredictable manner.

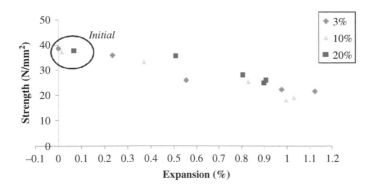

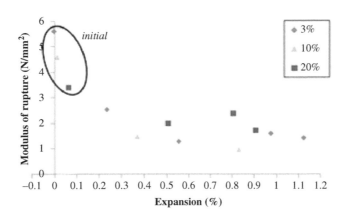

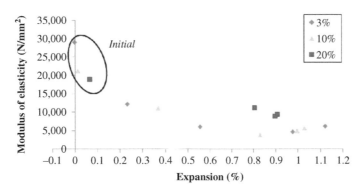

Figure 14.24 Effect of expansion on compressive strength (top), modulus of rupture (middle) and elastic modulus (bottom) of concrete incorporating different amounts of fused silica.

This experience shows that determining the strength properties of concrete cores taken from an element can only be used as a guide, and the effect of the reinforcement should also be considered in the assessment of the load capacity of the element concerned.

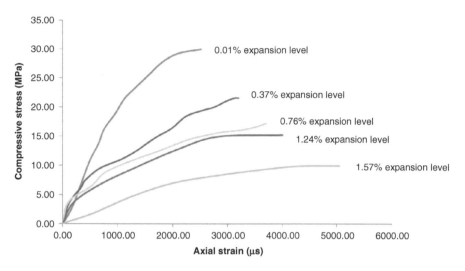

Figure 14.25 Stress-strain curves showing loss in concrete stiffness with expansion level (after Kubat *et al.*, 2015).

14.4 INTERACTION BETWEEN AAR AND OTHER MECHANISMS

AAR is a chemical reaction which can be influenced by factors that influence the quantity and availability of the reactants involved, and which can be complex within the hydrated cement and various exposure environments, such as seawater. The manufacturing processes of concrete, *i.e.*, whether cast-in-situ or heat-cured, and the presence of other materials such as various additives and steel reinforcement, and their interactions with the hydrated cement and the exposure environment, can also have an effect.

However, two main mechanisms are well known and have been investigated in some detail and will be discussed in this chapter, including DEF and impressed cathodic protection currents applied to steel reinforcement to mitigate the corrosion process.

14.4.1 Interaction between AAR and DEF

The combined occurrence of AAR and DEF arises largely in heat-cured concrete elements that include reactive aggregates and are subjected to high temperatures of curing, above 70°C, or in large concrete pours in which the hydration heat reaches similar or even higher temperatures. Commercial interest in some of the larger cases of combined AAR and DEF, and litigations involved in such cases, has made this topic very controversial, leading to blame being attributed to AAR by some and to DEF by some other investigators of the same elements or structures. Since the late 1970s, a large volume of publications has accumulated on this topic, but it is not the intention of this section to present a critical review of this literature.

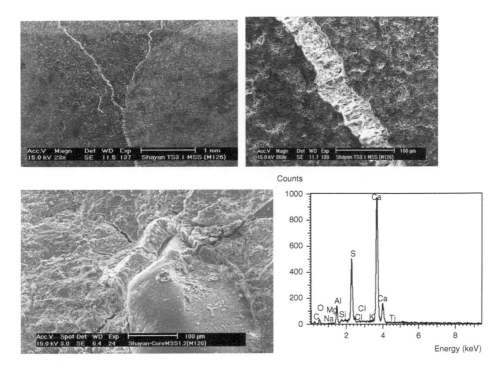

Figure 14.26 SEM observations of DEF signs in concrete in the form of ettringite-filled cracks around coarse aggregate, branching into cement paste (top left, magnified in top right), and around a sand grain (bottom left). The EDX spectrum (bottom right) indicates the composition of ettringite.

Symptoms and manifestation of AAR in affected elements have already been discussed in earlier sections. DEF symptoms occur in the cement paste as pockets of ettringite, formed from *in situ* reaction of sulphate ions with calcium aluminate grains in the cement, as well as in the form of dense ettringite layers formed in cracks around aggregate particles and in the cement paste. Ettringite-filled voids are also observed in concretes subjected to DEF.

Examples of these features are presented in Figures 14.26 and 14.27. The morphological features of DEF observed by SEM depend on the orientation of the ettringite, *i.e.*, whether the ettringite is viewed in the plane of the crack, or at right angles to the crack. The two different orientations of ettringite are shown in both Figures 14.26 and 14.27. The formation of this type of dense, compact ettringite found in AAR-DEF cases is different from the more common, secondary formation of loose ettringite crystals in open spaces (Figure 14.28), which is considered to be harmless.

Although there are some overseas reports in which the cause of deterioration has been attributed to DEF alone, no such cases have been observed in Australia, where all field cases, in which DEF has been detected, have also suffered from AAR. In most cases the co-existence of AAR and DEF is easily recognisable, such as those illustrated in Figure 14.29.

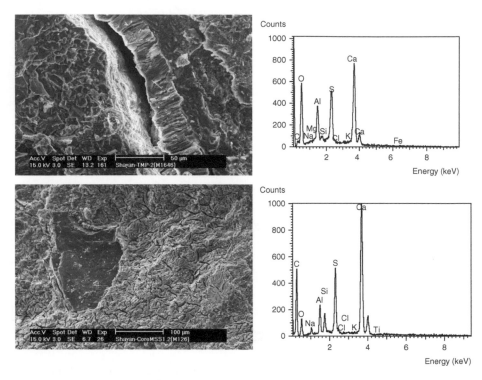

Figure 14.27 SEM observations of DEF signs in concrete: cross section of ettringite layer filling a crack at the aggregate interface, with ettringite crystals formed at right angles to aggregate surface (top), and ettringite layer viewed in the plane of a crack (bottom). The presence of Si and Na in the EDX spectrum of the latter (bottom, right) may indicate that some AAR gel could be mixed with ettringite.

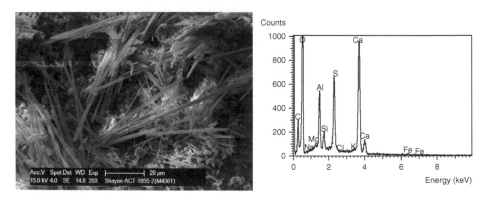

Figure 14.28 SEM micrograph (left) showing harmless occurrence of loose secondary ettringite formation in void space (EDX spectrum, right).

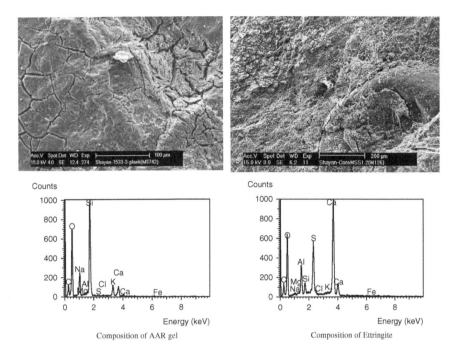

Figure 14.29 SEM views of association of AAR and DEF in two different structures. In both images, the AAR gel (left) has formed on a reacted aggregate particle in near vicinity of ettringite (right). The ettringite mat seen in the right-hand image is on and around a sand grain.

On some other occasions, the AAR gel and ettringite appear to be mixed (Figure 14.30), and the EDX composition clearly reflects the mixed nature of the two materials. The coexistence of the two products has been observed in many structures in the cases of both precast, prestressed elements, and cast-in-situ components of large dimensions, in which the hydration heat of cement could reach high temperatures exceeding 70°C. Examples of such cases are given in Table 14.6.

The close association of AAR products and ettringite complicates the interpretation of cause(s) of damage to concrete structures. The first such cases studied in Australia were those of the concrete railway sleepers, listed in Table 14.4 (Shayan & quick, 1992). Based on the field evidence and laboratory experimentations, Shayan (1993a) demonstrated that the deformed granitic aggregate used in the sleepers was reactive and caused significant AAR-induced expansion and cracking in concrete blocks of the same cross sectional area as the concrete sleepers (300 mm cubes), without the steam-curing process. Similar aggregates in the general area were also shown to be of similar reactive nature (Shayan, 1993b).

The steam-curing of the concrete would have enhanced the rate of AAR-induced cracking and enabled the effects of DEF to be manifested in the form of ettringite-filled cracks. The reduced alkalinity of the pore solution arising from AAR promotes precipitation of ettringite, and this may be the reason why ettringite forms at or near the

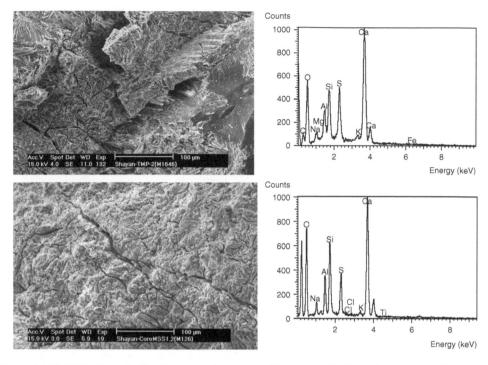

Figure 14.30 SEM views (left) of samples from two different AAR-affected structures, showing a mixture of AAR gel and ettringite at the aggregate-cement paste interfacial zones (EDX spectra, right).

Table 14.6 Examples of Australian cases of the association of AAR and DEF.

Structure	Concrete element	Type of concrete	Reference
Railway	Sleepers	steam cured	Shayan and Quick (1992)
Railway (Finland)	Sleepers	steam cured	Shayan and Quick (1994)
Railway (USA)	Sleepers	steam cured	Shayan (1993, unpublished)
Bridge	Column & pier wall	cast in-situ [#]	Ross and Shayan (1996)
Dam	spillway	mass concrete	Shayan (1999)
Dam	Spillway	mass concrete	Shayan (1998/99, unpublished)
Bridge	Piles	precast, prestressed	Shayan et al. (1998)
Bridge (2 No.) [*]	Deck planks	precast, prestressed	Shayan and Morris (2002)
Bridge	Column	cast in-situ [#]	Shayan and Morris (2003)
Bridges (5 No.)	Piles	precast, prestressed	Shayan (2004, unpublished)
Bridge (2 No.)	Deck planks	precast, prestressed	Shayan and Morris (2005)
Storage tank	Foundation	Cast-in-situ	Shayan (2006, unpublished)
Bridge	Piles	Prestressed	Shayan et al. (2008)
Bridge	Arch Units	Precast	Shayan (2010, unpublished)
Loading Jetty	Piles	Precast	Shayan (2017, unpublished)

[*] Several other bridges have this type of element which have exhibited similar cracking.
[#] Large size of concrete element may have caused elevated temperatures exceeding 75°C

AAR gel formation sites. The author concluded that AAR was the main cause of cracking of the sleepers, and that the DEF was either a secondary contributor to the cracking, or merely a consequence of the AAR.

To verify the above conclusions, laboratory investigations were undertaken (Shayan & Ivanusec, 1996) in which 7.5% opaline material was added to one batch of non-reactive sand and both used in mortar-bars to generate innocuous behaviour as well as AAR-induced expansion under enhanced alkalinity.

Some of the mortar mixtures also incorporated 2.5% additional gypsum by cement mass. The specimens were initially cured at either 23°C or 75°C for 8 hours and then stored at 38°C, 100% RH for expansion measurements. These conditions enabled the laboratory specimens to be innocuous (control), or have potential for either AAR or DEF, or both. Results of expansion measurements showed that only those with AAR potential developed deleterious expansion. Evidence of additional list-induced expansion was noted only in the specimens that had developed AAR expansion.

In further extensive investigations (Shayan et al., 2004), several variables, including type of aggregate, aluminate content (5–8%), alkali content (0.37–1.4%), type of alkali (KOH, NAOH), sulphate content (2.6–5.0%), and curing temperature (23, 65 and 85°C) were incorporated in mortar and concrete tests to understand factors that contribute to expansion and deterioration of concrete. Reactive and non-reactive sand were used for mortar testing, whereas reactive and non-reactive coarse aggregate plus non-reactive sand were used in concrete testing. Based on expansion curves obtained on the various mortar-bars and concrete prisms, it was concluded that the main cause of expansion in most cases was AAR and that DEF contributed to the AAR expansion when elevated amounts of sulphate and aluminate were present and the curing temperatures was 85 °C. In the absence of reactive aggregate, DEF could occur only when all the factors including high alkali, high sulfate, high aluminate and high temperature of curing (around 85 °C) were present concurrently. Incorporation of good quality class F fly ash in the concrete was able completely to suppress expansion caused by both AAR and DEF.

Considering that the cements available for the manufacture of Australian concrete railway sleepers were probably of about 2.6% SO_3 and 5% aluminate contents, the likelihood of DEF alone as a cause of expansion and cracking would be minimal, but it could have contributed to the AAR-induced deterioration.

Diamond (1994) suggested that growth of ettringite in pre-existing microcracks could cause expansion, whereas Johansen et al. (1994), Scrivener and Taylor (1993), and Glasser et al. (1995) suggested that it did not. However, Deng and Tang (1994), and Xie and Beaudoin (1992) state that if micro-cracks are filled with a solution which is supersaturated with respect to ettringite, then crystallisation of ettringite could cause expansive forces through hydrostatic pressure in the confined areas. This supports the experimental observations made by Diamond (1994) and Shayan and Ivanusec (1996) of additional expansion in AAR-affected specimens, which they suggest was due to ettringite formation in microcracks. However, the extent of damage to concrete does not seem to be related to the amount of ettringite detected in the element (Stark & Seyfarth, 1999), as ettringite tends to grow in existing spaces.

In another case of significant deterioration of bridge piles submerged in tidal waters, a portion of the pile was completely destroyed, which caused concern over the load capacity of the bridge and led to closure of the bridge and emergency replacement of critical piles. This case also involved both AAR and DEF. An example of this serious delamination was given earlier in Figure 14.4. The asset owner, based on two erroneous consultants' reports, originally believed that this was a clear case of DEF. However, further investigations (Shayan & Morris, 2005) showed that the main cause of expansion and cracking was a strong case of AAR (as evidenced by strong reaction rims, and the fact that coarse aggregate particles extracted from the piles still exhibited significant expansion in mortar-bar and concrete prism tests), which was exacerbated by heat-curing of the piles. In this case DEF also contributed to the deterioration but was probably not the main cause.

In subsequent investigations (Shayan *et al.*, 2006, 2008) to confirm these conclusions, sixteen large scale, reinforced laboratory model piles were manufactured in which the following parameters were incorporated: three levels of alkali reactivity (non-reactive, slowly reactive, reactive), two levels of sulphate (2.6% and 5.0 %), one cement alkali level (1.4%), two levels of initial curing (20 °C and 85 °C), one level of calcium aluminate content (5%) and two levels of exposure (partial immersion in tap water or saltwater). The cement content of the concrete mix used in all the piles was 450 kg/m^3. The piles, half-immersed in large containers, were wrapped inside plastic sheeting to allow high humidity to develop in the non-immersed portion, and kept at 38 °C for expansion measurements. Embedded concrete strain gauges were place in both the submerged and drier portions of each pile. These piles had the potential for AAR alone, DEF alone, combined AAR and DEF and no deterioration.

Results of this work showed that:

- The reactivity of the aggregate played the major role in the expansion and deterioration. As expected, expansion was greater in the submerged wet zone of the piles than the aerial zone. Steam-curing increased the reactivity of the slowly reactive aggregate.
- Steam-curing at 85°C, *per se*, did not cause any deterioration in the concrete containing non-reactive aggregate, at the native SO_3 and C_3A contents of the cement.
- Steam cured piles containing reactive aggregate expanded more than their corresponding ambient-cured piles, indicating the influence of elevated temperature on AAR expansion, as well as possibility of DEF in these specimens. Additional sulphate increased the list-induced expansion.
- With non-reactive aggregate, all the other parameters (sulphate and alkali contents, and curing temperature) need to be simultaneously high for deleterious expansion to occur. This may not often be encountered in practice, whereas AAR damage can occur more often.
- It was recommended that the allowable cement sulphate content of 5% in
- AS 3600 be reduced to around 3.0% – 3.5%, and the steam curing temperature be limited to below 65°C.

These results paralleled those obtained earlier for small laboratory samples (Shayan *et al.*, 2004) and reached the same conclusions in relation to the AAR-DEF phenomena, *i.e.*, that damage to Australian concrete structures has been mainly due to AAR, and that DEF may have contributed to the deterioration in some of these structures.

In relation to prediction of DEF susceptibility of concrete, there is no specific test in Australia, but the test used by Scott and Duggan (1987) for assessing alkali-reactivity of aggregate, is now used overseas for assessing the DEF potential of concrete mixes. A modification of this test was more recently proposed by Tagnit-Hamou and Petrov (2003). This test involves an initial heat curing and 7 days of storage in lime saturated water, followed by thermal cycling (50–10 °C). This test can be used to determine the maximum temperature at which DEF would not occur in the given concrete.

Prevention of DEF damage in new structures to be built with Portland cement uses different parameters than those for AAR damage, and involves control of cement composition, such as sulphate, aluminate and alkali contents, in addition to controlling the curing temperature to below 70 °C. However, the use of sufficient amounts of suitable 'supplementary cementitious materials' (SCMs), as cement replacement in concrete, would eliminate the risk of both these deleterious mechanisms (Shayan *et al.*, 1993, 2004).

14.4.2 Interaction of AAR and impressed current Cathodic Protection

Investigations conducted by Natesaiyer and Hover (1986), Sergi and Page (1992), Page *et al.* (1992), Ali and Rasheeduzzafar (1993), Page and Yu (1995), Kuroda *et al.* (1996), Alkadhimi and Banfill (1996), Torii *et al.* (1996) and Shayan (2000) demonstrated that application of an impressed current cathodic protection (CP) system, to prevent the corrosion of reinforcement, could aggravate AAR in the presence of reactive aggregates. Shayan and Xu (2001) carefully inspected the specimens (Shayan, 2000) that were subjected to current levels of 25–75 mA/m^2 for 2–3 years and found that considerable longitudinal cracking (parallel to the reinforcement bar) was present at the bottom and side surfaces of the specimens, and that the degree of cracking and crack width increased at higher current levels.

In addition to the adverse effect of the impressed current on AAR, Rasheeduzzafar *et al.* (1993) found that high levels of current could also damage the concrete-steel bond strength. Ali and Rasheeduzzafar (1993) observed that very high current densities of 215 and 1076 mA/m^2 applied to alkali-enriched, reinforced mortar specimens that contained reactive glass, caused enhanced AAR and reduced the compressive strength and hardness of mortar in the vicinity of the cathodically-protected reinforcement.

Results of Kuroda *et al.* (1996), using CP current densities of 25 – 200 mA/m^2 on reinforced concrete prisms that contained reactive aggregate, found that maximum expansion occurred at 50 mA/m^2 when the alkali content was 1.5% (NaOH), and that the expansion significantly decreased at a current level of 200 mA/m^2 as a result of formation of a more fluid, high-alkali gel at high currents, which was unable to create high expansive pressures in the concrete.

In most of the previous studies, relatively large CP currents were applied to small reinforced specimens of concrete containing reactive aggregates, and significant increases in expansion (*e.g.*, 20%, Shayan, 2000) were noted under certain conditions. Carse and Dux (2004) used large model pile specimens and subjected them to immersion in salt water and a CP current of 30 mA/cm^2 for seven years in the laboratory (which is a large current for the long-term CP application), and verified that the impressed CP current significantly increased the expansion of concrete that contained a very reactive Australian aggregate.

To study the influence of a more practical range of current levels, Shayan *et al.* (2008) applied three CP currents (5, 10 and 20 mA/m^2) to six reinforced model columns (300 × 300 × 2100 mm), three of which contained slowly reactive and three reactive aggregate, some with and some without chloride contamination. These specimens, which included embedded concrete strain gauges in four locations in each column, two in the upper half and two in the bottom half, were submerged in either water or 3.5% NaCl solution.

After nearly 4.5 years of expansion monitoring, the CP currents were applied to the specimens, using mixed titanium metal oxide anodes, which were installed separately to the top and bottom halves of the columns, with a gap of 100 mm between them, and then covered in a layer of cement-based mortar. Figure 14.31 shows the arrangement of anodes and the constant current supply to the columns. The top half of each column was wrapped with a cloth which extended into the solution to maintain high humidity. The half-immersed columns in their containers were wrapped in plastic sheeting and stored in a room at 38 °C for expansion monitoring, which continued for some 800 days.

Figure 14.32 shows the expansion results for the three columns obtained over some 1600 days prior to application of CP (marked as a vertical line on the graphs), as well as data obtained over some additional 800 days after the CP application.

Results of this work were progressively reported by Shayan *et al.* (2011, 2012), and examples are given in Figure 14.32 for two columns with slowly reactive aggregate and

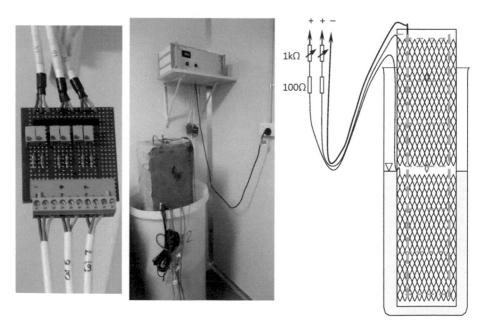

Figure 14.31 Example of finished column (centre), arrangements for application of CP (right), and current regulators for top and bottom parts of columns (left). The negative terminal is connected to the reinforcement cage in the column.

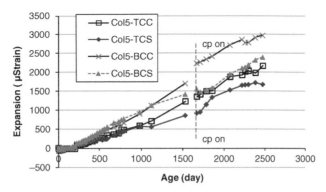

Column with slowly reactive aggregate and
CP current of 5 mA/ m^2

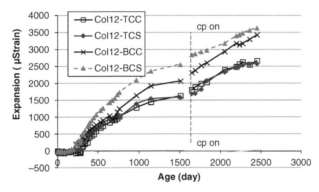

Column with reactive aggregate and CP
current of 5 mA/m^2

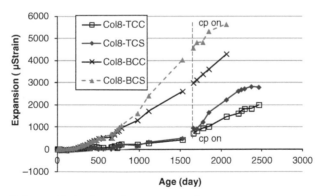

Column with slowly reactive aggregate and
CP current of 20 mA/m^2

Figure 14.32 Contribution of different CP currents to the AAR-induced expansion of columns containing slowly reactive and reactive aggregates.

reactive aggregate, which were subjected to the lowest CP current of 5 mA/cm^2 and a higher current density.

Concrete expansion was measured in four locations in each column, as designated below, which are also shown on the graphs in Figure 14.32:

- TCC: Top part and centre of column; strain gauge being installed perpendicular to column length.
- TCS: Top part and surface of column; strain gauge being installed parallel to column length.
- BCC: Bottom part and centre of column; strain gauge was installed perpendicular to column.
- BCS: Bottom part and surface of column; strain gauge was installed parallel to column length.

The bottom part of the columns, which was immersed in water or salt solution, showed more expansion than the top part, which reflects the higher availability of moisture for the reaction.

Generally, the expansion of the columns, except Column 5 and Column 6, slowed down after two to three years of storage at 38°C. The expansion rate resumed after the CP current was applied, and appeared to be the same or higher than that before the CP application, especially for the upper part of the columns. The extent of expansion depended on both the CP current density and the level of aggregate reactivity.

This work showed that application of CP current to AAR-affected reinforced concrete elements, to suppress steel corrosion, can increase the AAR expansion. Higher current densities caused larger expansions. It was also noted that:

- The CP effect was more pronounced for the relatively dry top part of columns, probably due to a better supply of oxygen to the cathodic area.
- The expansion rate of concrete depended on aggregate reactivity, and CP current level. Additional expansion of 1000-2000 µS were observed for both the slowly reactive aggregate subjected to 20 mA/m^2 and the highly reactive aggregate subjected to 10 mA/m^2 CP. The 5 mA/m^2 caused less expansion for the aggregates tested.

The results have highlighted the need for caution to determine the AAR status of concrete, when application of impressed current CP is contemplated for protection of concrete structures at risk of corrosion damage. Consequently, in Australia, petrographic examination of concrete is conducted before CP application to identify the nature of the aggregate involved, so that the minimum possible level of current could be applied.

14.5 TESTING AGGREGATES FOR ALKALI-REACTIVITY

Incremental developments in AAR testing in Australia have been documented by Shayan (1995, 2003, 2015). As mentioned earlier, the AAR test methods adopted in Australia in the 1960s and up to 1994 were the Australian Standards AS1141-38 (mortar-bar test) and AS1141-39 (quick chemical test), which had been adopted from ASTM C 227 and C289, respectively, without the applicability of these tests to the Australian aggregates having been addressed.

The paper by Vivian (1983) indicated that acceptance criteria of the mortar-bar test (ASTM C227, or AS 1141-38) were unsatisfactory for some of the Australian aggregates.

The Department of Main Roads in Western Australia had conducted the chemical test (AS 1141-39) on numerous aggregate sources, including those which were later found to have reacted in structures, and none of them had been classed as reactive.

In the period between 1983 and 1985, research conducted by the author on a major AAR-affected structure and testing of the culprit aggregates, as well as testing of other aggregate sources intended for the construction of bridges for the Bicentennial Roads Project in Western Australia, led to the development of an accelerated mortar-bar test procedure (AMBT), which was later published (Shayan et al., 1988). This test included acceptance criteria for the detection of both reactive and slowly reactive, as well as non-reactive Australian aggregates, which were based on the service performance of the aggregates. The AMBT involved immersion of mortar-bars in 1M NaOH solution at 80°C, similar to the test described by Oberholster and Davies (1986) and much later as ASTM C1260 (now 2014) [see sub-section 14.5.3 for comparison of the Australian and ASTM methods].

In the same period, a concrete prism test (CPT) and acceptance criteria were also developed, which were later described by Shayan et al. (1987). The CPT employed a concrete mix design which was used in Western Australia for bridge construction, with a cement content of 410 to 420 kg/m^3. However, for the CPT a cement alkali level of 1.38% Na_2Oeq was adopted (i.e., concrete alkali content of 5.66- 5.80 kg/m^3), with storage conditions of 38°C and 100% RH. Note that the equivalent Canadian CPT (now CSA, 2014) at the time used a cement content of 310 kg/m^3 and a lower cement alkali level. The author believes that these low cement and alkali contents were the reason why some reactive Canadian aggregates of greywacke rock types could not be detected by their CPT method in that period. Later, in the mid-1990s, the ASTM C1293 (now 2008) version of CPT was introduced which employed a cement content of 420 kg/m^3 and cement alkali level of 1.25%, i.e., concrete alkali content of 5.25 kg/m^3, which is lower than the alkali content used in the Australian method (Shayan et al., 1987).

Other AMBT parameters, such as mixing of the 1M NaOH into the mortar mix and storage of mortar bars over water, rather than storage in the 1M NaOH solution, and variation in the period of pre-curing before soaking in 1M NaOH solution were explored (Shayan, 1989) in order to seek to optimise the test method, but it was concluded that storage in 1M NaOH solution was the best option. Moreover, these studies also showed that storage of concrete prisms in 1 M NaOH at 50 °C or 80 °C did not yield consistent or satisfactory results, which was attributed to the denser nature of the paste in the CPT, compared with the mortar in the AMBT. The denser paste inhibited penetration of the alkali into the concrete specimen and its reaction with the coarse aggregate.

Incorporation of alkali in the mixture of concrete prisms produced more consistent expansion results, but the strength of the concrete was significantly reduced, a phenomenon which would not occur in field concrete. The loss in strength as a result of alkali addition to mortar or concrete mixes was reported by Shayan and Ivanusec (1989) and Shayan (1995). An example of the results is presented in Figure 14.33, in which the compressive strength of concrete made with a non-reactive aggregate (top line) is compared with those of five different reactive aggregates. It is evident that the alkali addition significantly reduced the strength of the non-reactive concrete. Therefore, the

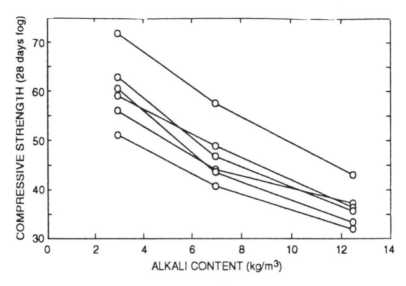

Figure 14.33 Effect of alkali addition to concrete on concrete strength for concretes made with a non-reactive aggregate (top curve) compared with five different reactive aggregates [after Shayan and Ivanusec (1989) and Shayan (1995)].

reduction in strength of reactive concretes, observed in laboratory testing, is partly due to the addition of alkali to the concrete mix and partly due to the effects of AAR.

In later work (Shayan, 1988), it was also found that storage of concrete prisms or mortar-bars in saturated NaCl solution at 50 °C, as used in Denmark, was unsatisfactory in detecting reactive aggregates. Moreover, Shayan *et al.* (1992) showed that the Australian AMBT method was superior to a Japanese autoclave test method in identifying a number of reactive/ slowly reactive Australian aggregates.

It was further established (Shayan & Quick, 1989) that the mechanism of reaction and nature of AAR products formed in specimens subjected to the new AMBT and the older methods were the same, and only differed in the reaction rate.

This observation provided more confidence in the use of the AMBT method. As a result of testing aggregates that had caused damage to various structures by the two test methods AS1141-38 and AS1141-39, it was recognised in the mid 1980s (Shayan, 1987) that these tests were inappropriate for the slowly reactive Australian aggregates. This finding and other work eventually led to the withdrawal of these test methods from the Australian Standards in 1994. After this date there was no formal Australian Standard for AAR testing of aggregates.

The new AMBT and CPT tests were first adopted in 1992 by the then Roads & Traffic Authority (RTA) of New South Wales (NSW) under the designations RTA T363 and RTA T364, respectively, and were included in their structural concrete specification B80. The name of this organisation has now changed to the Roads & Maritime Services (RMS) and, therefore, the designations of the tests have also changed to RMS T363 and RMS T364 test methods (Roads & Maritime Services, 2012).

Some other Road Authorities later adopted the same tests under their own designations, such as VicRoads test methods RC376.03 and RC376.04 in Victoria, for the

Table 14.7 Aggregate reactivity classification based on AMBT method, RMS T363.

Expansion under storage conditions of 1M NaOH solution at 80°C		Classification
Measured at Immersion Age:		
10 days	21 days	
E < 0.10%[*]	E < 0.10%[*]	Non-reactive
E ≥ 0.10%[*]	E >> 0.10%[*]	Reactive
E < 0.10%[*]	E ≥ 0.10%[*]	Slowly reactive

* The expansion limit for natural fine aggregates is 0.15% at 21 days

AMBT and CPT methods, respectively. The Main Roads Department in Western Australia later adopted the AMBT test method as WA 624.11, although Queensland DTMR has instead adopted a concrete prism test, using 50 °C steam curing temperature (Carse & Dux, 1990). The latter has had limited application and only in Queensland.

14.5.1 Expansion limits for Australian AMBT and CPT

The expansion limits established for the Australian AMBT, which was later adopted by the road agencies under their own designations, are given in Table 14.7. These test limits were established based on test results (Figure 14.34) obtained on reactive and slowly reactive aggregates which have known field records, as described by Shayan *et al.* (1988). The limits were chosen such that the aggregate designated 'UY' in Figure 14.34, which has caused significant damage to a dam structure, is on or above the lower boundary of the chosen limit, and is classed as slowly reactive. This lower limit has been confirmed for a number of other similar slowly reactive aggregates which have also caused AAR-induced cracking in dam and bridge structures.

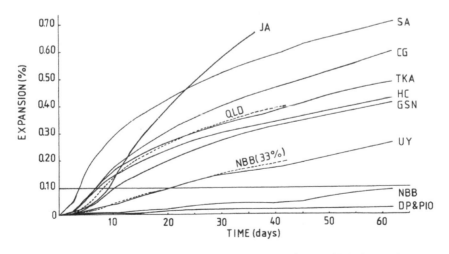

Figure 14.34 AMBT results (Shayan et al., 1988) which were used to establish the test limits.

Table 14.8 Differences between RMS T363 and ASTM C1260 AMBT methods.

Test method	Mortar Preparation	Moist curing before immersion	Hours in water at 80°C for zero reading	Total age & immersion period in 1M NaOH	Test expansion limits
ASTM C1260	Fixed W/C Ratio, 0.47	None, Place in cold water for heating to 80°C	24	16 days: 2 days pre-curing; 14 days in 1M NaOH at 80°C	Single point @ 14 days: Non-reactive: <0.10%; Uncertain: 0.10–0.20%; Reactive: >0.20%
RMS T363	Flow Table	48 hours, then Place in cold water for heating to 80°C	4	24 days: 3 days pre-curing; 21 days in 1M NaOH at 80°C	Non-reactive: <0.10% in 21 days, Slowly Reactive: <0.10% in 10 days but >0.10% in 21 days; Reactive: >0.10% in 10 days, (For natural sand the limit is 0.15%)

(1996). The ASTM method uses the same size mortar-bars, specimen demoulding time of 24 hours and storage solution of 1M NaOH solution at 80 °C, but differs from RMS T363 in some details of test procedure and acceptance limits, as presented in Table 14.8.

To compare the results of the two AMBT methods, RMS (then RTA) and ARRB conducted a programme of testing on 18 aggregates with known service records, and the results were reported by Shayan and Morris (2001). The conclusion of this comparative testing are summarised as follows:

- Specimens may be prepared and tested by either RMS T363 or ASTM C1260 methods.
- Both methods correctly classify moderately to highly reactive aggregates.
- Both methods correctly classify non-reactive aggregates.
- However, for aggregates which are classed as slowly reactive by RMS T363, and which have caused serious AAR-induced cracking in concrete structures in Australia, the ASTM C1260 method gave non-reactive or uncertain classifications.

The first three points are illustrated in Figure 14.37 for non-reactive and reactive aggregates. However, the erroneous classification by ASTM C1260 in the fourth point is unacceptable, as it would lead to damage to structures which would use such aggregates. Examples of such outcomes are illustrated in Figure 14.38.

Shayan (2007) presented several major cases of damage to Australian structures for which the RMS T363 was able to detect the aggregates as reactive, but the ASTM C1260 method would class the culprit aggregates as non-reactive or at best uncertain. Therefore, the ASTM C1260 limits cannot be accepted for slowly reactive aggregates in Australia.

Based on the results of these and other studies, Shayan and Morris (2001) and Shayan (2007) recommended that the test limits of RMS T363 be also applied to the ASTM C1260 method, or that the 14-day limit of the latter test be lowered to ≥0.08% (rather than 0.10%) for reactive aggregate and <0.08% for non-reactive aggregate.

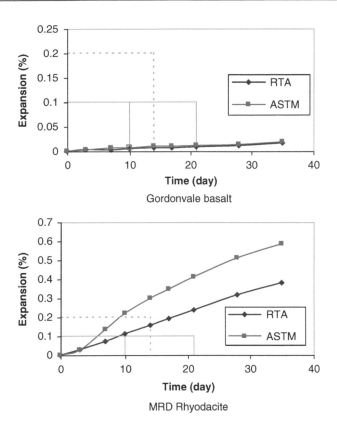

Figure 14.37 Results of AMBT by the two methods for non-reactive aggregate (left) and reactive aggregate (right), for which they give the same reactivity classification. Test limits for the two methods are shown with dashed lines for the ASTM C1260 method.

14.5.4 Aggregates requiring further testing

Some aggregates, particularly those of slowly reactive nature, do not follow the simple pass-fail test, and their classification is more difficult. Shayan (2007) detailed some examples of such slowly reactive aggregate; two of which relate to the Australian railway sleepers (Shayan & Quick, 1992) and a dam structure (Shayan, 1999). Both cases describe significant AAR-induced damage to the elements or structures concerned.

In the case of railway sleepers, Figure 14.39 shows AMBT results for four aggregate varieties, all of deformed granitic nature, taken from the dedicated quarry that produced the aggregate required for manufacture of one million sleepers. Except for one variety, which falls on the borderline for slowly reactive aggregate, those for the other three varieties fall below the reactivity limits. This is despite the fact that many thousands of sleepers have developed AAR-induced cracking. These results show that the AMBT does not unambiguously identify these aggregates as slowly reactive.

Table 14.9 shows the expansion results for concrete prisms, made in accordance with the RMS T364 test method, which incorporated the different aggregate varieties at two alkali levels. The higher alkali level was used for comparison with the standard level.

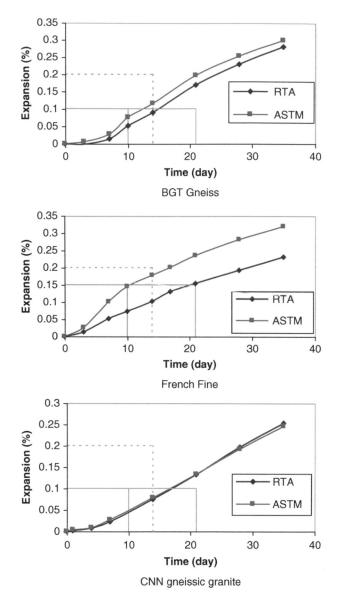

Figure 14.38 Results of AMBT by the two methods for three aggregates which are classed as slowly
reactive by RTA T363 (in agreement with service performance) but are classed as
uncertain (top row) or non-reactive (bottom) by ASTM C 1260. Test limits for the two
methods are shown with dashed lines for the ASTM C1260 method.

The results showed that the order of expansion in the CPT method (1.38% alkali)
was different from that observed for AMBT, but again only one variety ('Coarse-grain
Quarry') gave an expansion value which is close to the lower boundary of reactive
aggregate (0.03%). The other varieties are classed as non-reactive. At 2% alkali level,
two more varieties exceeded the 0.03% limit of the test. Therefore, it was clear that the

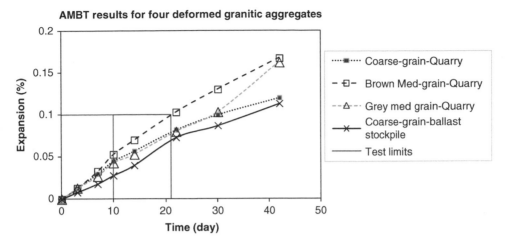

Figure 14.39 AMBT (RMS T363) results for the four aggregate varieties.

Table 14.9 One-year expansions in concrete prism tests (%).

Cement alkali level:	1.38%	2%
Aggregate variety	% expansion @ 1 year (38 °C, 100% RH)	
Coarse grain - Quarry	0.032	-
Brown Mid-grain - Quarry	<0.03	0.035
Grey Mid-grain - Quarry	<0.03	0.035
Coarse grain - Ballast stockpile	0.018	0.024

CPT method, as formulated, would not be suitable for unambiguous detection of such aggregates, *i.e.*, neither AMBT nor CPT could be used with confidence for this type of aggregate.

Shayan (1992a, 1993a, 2007) reported that concrete prisms and concrete blocks (300 mm cubes) made with such aggregates in combination with elevated alkali levels (similar to that used in the CPT RMS T364, or higher), and stored under wet conditions (1–2 mm layer of water surrounding the block) at 50°C, generated far greater expansion than concrete prisms stored at 38°C, 100% RH. The expansion was associated with cracking, which was very fine at the beginning but became wider with time. Figure 14.40 shows the cracking patterns of two such blocks, which strongly resemble AAR-induced cracking.

Shayan (2007, 2011) documented a similar case for slowly reactive quartz gravel, which has caused significant AAR-induced damage to a dam structure (Shayan, 1999), but could not confidently be identified by AMBT or CPT to be slowly reactive. However, concrete blocks made at several alkali contents, and stored in water at 40 °C, developed considerable expansion and cracking at the age of 18 months, when the alkali content was above 1.5% Na_2Oeq. Blocks which contained cement alkali levels of up to 1.2% did not expand significantly even after 3 years.

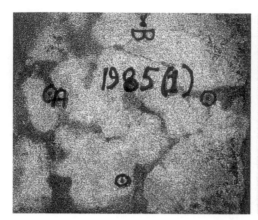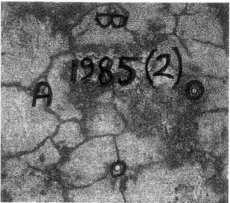

Figure 14.40 View of 3-year old concrete blocks (300 mm cubes) made with the same aggregate as used in the railway sleepers, incorporating 400 kg/m³ cement at alkali levels of 1.4% (left) and 1.9% (right). The blocks were cast at normal temperature and placed in plastic boxes which allowed 1–2 mm layer of water around the blocks, and stored in a room kept at 50 °C.

It appears that under these test conditions slowly reactive aggregates generate AAR-induced cracking similar to, but much more rapidly than, that in field structures under natural conditions.

These case studies indicate that AMBT results which fall just on the borderline, or even just below it, could be indicative of a very slowly reactive (nonetheless dangerous) aggregate and should not be accepted as non-reactive. They also indicate that some aggregates require additional testing such as that described above for concrete blocks. However, although this test is effective as a research tool to demonstrate the reactive nature of very slowly reactive aggregates, it is very slow and impractical as a routine aggregate assessment test. Shayan *et al.* (2008) applied the concrete prism test, under the storage conditions of 60 °C, 100% RH, to a wide range of aggregates and found it to be superior to other tests for slowly reactive aggregates, as described below.

14.5.5 Improved detection of slowly reactive Australian aggregates

It is noted from the previous section that storage of concrete prisms and blocks at 50 °C enhanced the level of expansion. Gogte (1973) appears to be the first researcher who used concrete prisms cured at 60°C to assess the AAR-susceptibility of aggregates containing microcrystalline quartz, and found the test to be superior to tests which used a lower curing temperature. Therefore, the traditional CPT method (38°C, 100% RH), designated here as 'CPT38', which takes one year to complete, can be made faster and probably more reliable by increasing the curing temperature from 38°C to 60°C (designated here as 'CPT60'), as reported for some 50 Australian aggregates by Shayan *et al.* (2008). These authors noted that the CPT60 test can shorten the testing time to 3–4 months, as

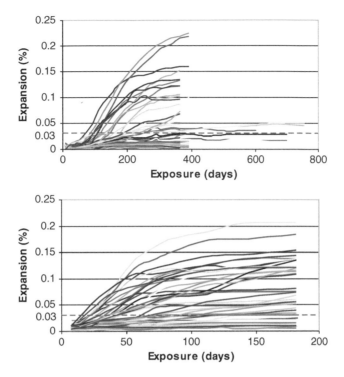

Figure 14.41 Results of CPT38 (top) and CPT60 (bottom) obtained on 50 Australian aggregates for exposure to test conditions of 2 years and 6 months, respectively.

compared to at least one year for CPT38. Based on the performance of these known aggregates, they proposed an expansion limit of 0.03% at 4 months for this test.

Figure 14.41 shows the trends of expansion obtained by Shayan *et al.* (2008) on 50 aggregates using CPT38 and CPT60. They cover a broad range of expansion in both test methods. The reactive aggregates are easy to detect due to the large and relatively rapid expansion of concrete prisms incorporating them. The slowly reactive aggregates cause more gradual and lower levels of expansion, and are more difficult to identify.

Figure 14.42 shows examples of the assessment of two slowly reactive aggregates by three test methods, which illustrate the much better performance of CPT60 compared with CPT38, for these aggregates. These aggregates have caused significant AAR damage to major concrete structures in Australia. It is evident that CPT60 clearly identified the two aggregates as reactive, whereas CPT38 failed to detect their reactivity potential. The AMBT results were reasonable.

Given these encouraging results, Shayan (2011) proposed that the Australian Road Authorities, and ultimately the Australian Standards, adopt the CPT conducted at 60° C (CPT60) rather than the conventional RMA T364, which has a curing temperature of 38°C (CPT38). This recommendation is supported by the results of Gogte (1973), Mullick *et al.* (1986) and Rao and Sinha (1989), who show that increasing the curing

expansion trend, but still below the limit of 0.10% at 28 days. On this basis, the aggregate can be classed as non-reactive, but this test has not been standardized; its primary purpose is to distinguish the differing expansive behaviour of some carbonate rocks from that of alkali-silica reactive aggregates, on the basis that the coarser particle size of AAR-5 (4-8mm) favours any carbonate reactivity in contrast to the finer particle size used in the AAR-2 version of the AMBT that favours any ASR.

Microstructural examination of concrete specimen taken from the AAR-5 prism showed sporadic occurrences of ASR gel, which indicates that the continued expansion of the prisms in the RILEM AAR-5 test could probably be due to mild reaction of the flint particles present in the dolomitic aggregate, but the amount of gel formation was probably insufficient to cause significant expansion. It can be assumed that this aggregate is not susceptible to ACR. A summary of this work was presented by Shayan & Katayama (2016). An up-to-date review of the 'so-called alkali-carbonate reaction' is given in Chapter 3.

14.7 NEW AUSTRALIAN STANDARDS AND GUIDELINES

Even before the old Australian Standards for detecting reactive aggregates were withdrawn in 1994, the then newly developed AMBT method (Shayan et al., 1988) was being used by Road agencies (e.g., Shayan, 1992a; Shayan & Carse, 1994). In 1994, RMS T363 and RMS T364 (and other designations used by other road authorities) were adopted by road agencies as their internal test methods and were specified in the jurisdictions' Specification for Structural Concrete, e.g., 'Specification B80' for RMS and 'Specification Section 610' for VicRoads.

As mentioned earlier, detailed study of ASTM C1260 and RMS T363 showed that they can lead to different classification outcomes for slowly reactive aggregates, and indicated that that the former cannot be adopted in Australian.

However, as a result of several years of consolidation of data, for example, Shayan and Morris (2001) and Shayan (2007), it was recognised that the procedure of testing can use either of the RMS T363 or ASTM C1260 methods, but particularly that the acceptance limits of RMS T363 must be used as the criteria for assessing the alkali-reactivity potential of aggregates. The same also applied to the comparison between the outcomes of RMS T364 and ASTM C1293, but to a lesser extent.

After several years of discussion, consensus was reached amongst the stakeholders in the form of two proposed Australian Standards for alkali-aggregate reactivity (AMBT and CPT), which incorporated a modified version of the respective ASTM test procedures, but using the limits of the corresponding RMS test methods.

The Australian Standards for AMBT (AS 1141.60-1) and CPT (AS 1141.60-2) were finally published in 2014. These new standards are now referenced in the Standards Australia Specification for concrete aggregate (AS2758.1), and consequently in AS 3600 (Concrete Structures) and AS 5100 (Bridge design code). However, due to the differences that exist between the test procedures, road agencies still use their internal methods (on which they have accumulated data for more than two decades) until such time that sufficient amounts of data have accumulated on the new standard tests.

Also, the guidelines on minimising the risk of AAR damage to concrete structure (HB79- 1996) was substantially reviewed and updated, including incorporation of the standard test methods and specifications mentioned above, and was recently published (SA Handbook HB79- 2015).

In addition to these standard tests, petrographic examination of aggregates (AS1141.65) was adopted in 2008 and is also used as a screening test for assessing the potential of aggregates for AAR. However, it is not always definitive for fine-grained aggregates as it cannot detect very fine-grained reactive components, such as fine silica minerals in aggregates derived from the very old Western Australian metamorphic rocks. Therefore, AS1141.65 is not a pass/fail test, but a useful tool in determining the nature of mineral phases in the aggregate. However, should sufficient amounts of known reactive components be detected in the aggregate, then AS1141.65 can be sued to reject the aggregate as potentially reactive. Nevertheless, quantitative measurement of expansion caused by the aggregate would be needed for its classification.

14.8 MITIGATION APPROACHES FOR AAR IN AUSTRALIA

It is obvious that the best approach to avoid AAR in concrete structures is to select non-reactive aggregates when alternative aggregates exist in the vicinity of construction sites, and when this approach is possible. However, often the practicality and economic aspects of construction necessitates the use of aggregates available at the site, particularly remote sites, which is often the situation for dam construction.

In such cases, the best approach is the use of adequate amounts of appropriate supplementary cementitious materials (SCM) as partial replacement for Portland cement. In Australia, such materials commonly include low calcium fly ash, ground granulated blast furnace slag and silica fume.

The beneficial effects of pozzolanic materials, such as fly ash, in improving the properties of hardened concrete, e.g., resistance to AAR, sulfate attack, and frost attack, and reduction in permeability of concrete has been known in Australia for many decades (Alexander, 1954), although specific published reports on the use of SCMs for AAR mitigation are rare before the 1980s. This may be related to the fact that the test methods used for detection of reactive aggregates in that period were inadequate in detecting reactive aggregates, and there was probably no serious concern over the risk of AAR-induced damage to concrete structures.

Based on the work conducted in the late 1980s on the alkali-reactivity of some Queensland aggregates, a summary of which was reported by Shayan (1992a), some general guidelines for minimising AAR risks in concrete structures were published by Shayan and Carse (1992). The guidelines recommended the amounts of SCMs in the binder to be in the range of 8–10% for silica fume, 20–30% for fly ash and 60–65% for slag. The newly published Australian Guidelines HB79 also recommends similar ranges.

Table 14.10 gives the compositions of some Australian SCMs which are commonly used in concrete, and which are required to comply with the requirements of the Australian Standards on SCMs; AS 3582.1 (fly ash), AS 3582.2 (slag) and AS 3582.3 (amorphous silica). These standards are currently under revision.

The effectiveness of the fly ash listed in Table 14.10 in suppressing AAR expansion was demonstrated by Shayan (1990), using two well-known reactive Canadian aggregates, Spratt and Sudbury. Two cements were used, one with 0.90% native alkali

Table 14.10 Examples of chemical compositions of commonly used Australian SCMs

Oxides	GP Cement	Slag[†]	Fly ash	Silica Fume
SiO_2	19.3	32.0	51.5	91.40
TiO_2	0.30	0.51	1.5	< 0.10
Al_2O_3	5.5	13.0	27.6	0.20
Fe_2O_3	2.9	0.40	11.8	–
Mn_3O_4	0.20	0.30	0.15	–
MgO	1.4	4.9	1.3	0.40
CaO	62	41.5	2.2	0.30
Na_2O	0.06	0.20	0.4	< 0.10
K_2O	0.4	0.33	0.6	0.37
P_2O_5	0.11	0.02	0.73	0.10
SO_3	2.46	2.1	0.2	< 0.10
LOI	5.3	4.9	1.8	6.90
Total	99.93	100.16	99.78	100
Na_2O_{equiv}[#]	0.32	0.42	0.80	0.33

† LOI in slag includes 3.7% sulfur in the form of SO_3
# Na_2O_{equiv} = % Na_2O+0.658 (%K_2O)

content (Na_2O equiv.) and the other 0.55%. In each case 0.49% alkali was added to boost the alkali level. The results of concrete prism tests conducted on these aggregates at different alkali levels and with or without fly ash (Figure 14.44) clearly showed that the fly ash suppressed the reaction even when the concrete contained high alkali contents. The expansion was even below that achieved by the cement which had no added alkali.

Comparison of Gladstone and Tarong fly ashes used in this work showed that the former, which contained larger amount of glassy phases had a larger pozzolanicity index and was a more effective fly ash.

Shayan (1992a) used both the AMBT and CPT methods to assess the efficiency of fly ash and slag in suppressing AAR, and found both to be effective. It was confirmed that, at a given dosage of cement replacement, the efficiency of the SCMs in suppressing AAR depended on the level of aggregate reactivity, and the alkali content of concrete. A very reactive aggregate from Queensland (Australia) needed 30% fly ash for AAR mitigation, whereas most other aggregates responded positively to 25% addition of the same fly ash.

In a practical application described by Shayan (1995b), use of silica fume for manufacture of new concrete railway sleepers was shown to be very effective (Figure 14.45 in eliminating the deleterious AAR-induced expansion caused by two types of reactive granitic aggregates, which had damaged numerous concrete railway sleepers in service. Figure 14.45 (a) shows that the slowly reactive granite did not cause exopansion at cement alkali level of 1.3%, but deleterious expansion occurred at 1.8% alkali level. The more reactive Granite 2 caused deleterious expansion at both alkali levels. However, Figure 14.45 (b) shows that incorporation of 10% silica fume in the concrete is highly effective in suppressing the AAR-induced expansion that the reactive aggregates would have otherwise caused.

Shayan *et al.* (1993) showed that the use of silica fume in steam-cured concrete not only suppressed AAR, but also prevented DEF, because of its effects on reducing the

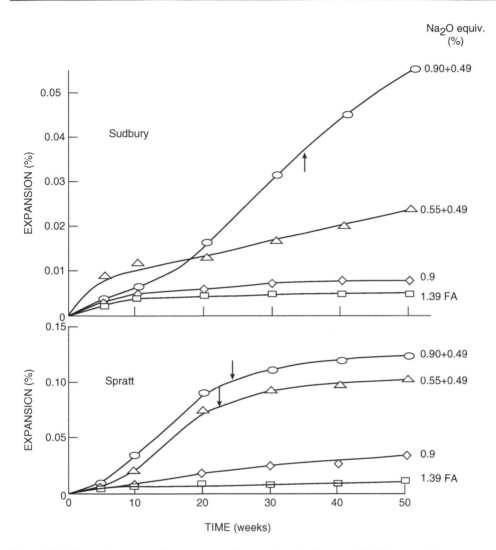

Figure 14.44 Expansion curves for concrete prisms made using Spratt and Sudbury reactive aggregates at different alkali levels, with or without fly ash. Arrows refer to time of cracking.

alkalinity of the concrete pore solution which, in turn, encourages ettringite precipitation at early ages. This reduces the concentration of sulfate in the pore solution, which removes the risk of DEF.

Shayan *et al.* (1996) using several reactive aggregates and curing conditions of 23 °C and 40 °C (both at 100% RH), demonstrated that a fly ash which was used to control AAR remained effective in the long term even under the accelerated conditions (Figure 14.46).

In another practical application, the reactivity of another slowly reactive granitic aggregate, which has caused significant damage to a dam structure could be controlled by using a high volume fly ash mix (HVFA) and even more efficiently by a triple blend

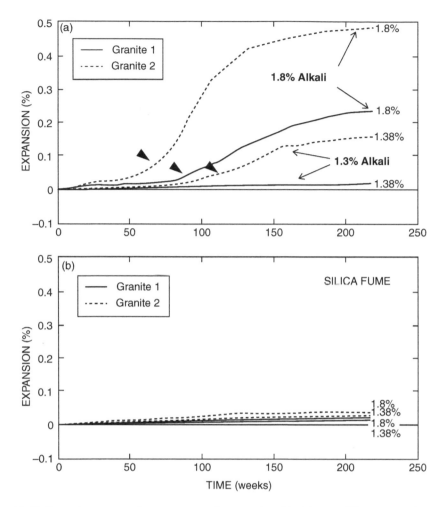

Figure 14.45 Expansion curves for concrete prisms made with (a) two different reactive granitic aggregates at two alkali levels, and (b) similar concrete mixtures which incorporated 10% silica fume as cement replacement.

(TB) concrete mix, comprising Portland cement, silica fume and fly ash (Shayan *et al.*, 2000), as indicated by the results given in Figure 14.47.

Several parameters influence the amount of SCM which would successfully suppress the AAR expansion. These include the degree of reactivity of the aggregate, the composition of SCM, the amount of SCM in the binder, the total alkali content of the concrete, and the curing conditions, particularly the curing temperature. Therefore, an amount of a SCM which would sufficiently suppress the reactivity on one aggregate may not be as efficient with an aggregate which is much more reactive.

Figure 14.48 shows examples of how the effectiveness of fly ash in suppressing AAR-induced expansion depends on the nature of the aggregate concerned and the alkali

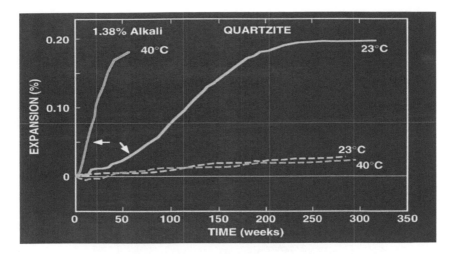

Figure 14.46 CPT results (solid lines) for a reactive quartzite aggregate under different curing conditions, and companion results for specimens that contained 25% fly ash to control AAR (dashed lines). Arrows refer to time of cracking of specimens.

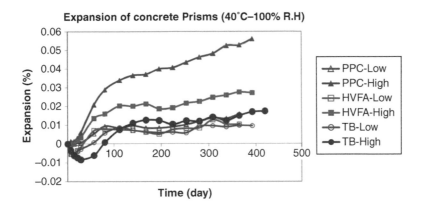

Figure 14.47 Expansion curves for concrete prisms made with slowly reactive granitic aggregate and either of plain Portland cement (PPC), High volume fly ash (HVFA) or triple blend of cement- silica fume and fly ash (TB), at low and high alkali contents.

content of the concrete (Shayan *et al.*, 1996). Concrete prisms containing the quartzite developed cracking at the lowest alkali content in the absence of fly ash. Incorporation of 25% fly ash by cement mass suppressed the AAR-induced expansion at the lower alkali content of 1.38%, but not at 2.5%, where cracking occurred at around 110 days of storage. The effect of fly ash was similar for the less reactive greywacke aggregate, buy cracking occurred later at around 170 days. The implication of these results is that if the fly ash contains large amounts of available alkali, then it may not suppress the AAR expansion at commonly used dosage rates.

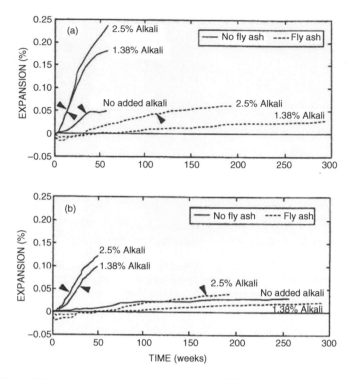

Figure 14.48 Effect of fly ash on expansion of concrete prisms (38°C, 100% RH) containing different alkali contents and aggregates with different degrees of reactivity; (a) very reactive quartzite; (b) reactive greywacke. No added alkali = 0.90% Na$_2$O equiv. Arrows indicate age of cracking.

Long-term results (up to 300 weeks) presented in this work have shown that the two Australian fly ashes studied have been effective in preventing deleterious AAR damage in concretes with alkali contents as high as 7.0 kg Na$_2$O/m^3, but they produced only a delaying effect in concretes containing 12.5 kg Na$_2$O/m^3. The delay was between two and six years, depending on the type of aggregate. A measureable chemical shrinkage occurred in the first few months in the presence of fly ash in the latter concretes, although some of them later expanded and cracked due to AAR.

For highly reactive aggregates and at high alkali contents, the fly ashes were more effective in preventing AAR expansion under the storage conditions of 23°C (fog room) than at 40°C 100% RH, due to the much faster rate of AAR under the latter conditions. As field concretes usually contain less than 7.0 kg Na$_2$O equiv/m3, the two fly ashes should be effective in suppressing AAR in practical applications.

14.8.1 Influence of SCMs on ionic concentrations in pore solution

The alkali hydroxides present in the pore solution of concrete facilitate the reaction of the SCMs and are consumed in the process. It has been shown in several studies that the alkalinity of the concrete pore solution is significantly reduced by incorporation of SCMs in the concrete mix.

The suppression of AAR-induced expansion by the Australian SCMs was also shown to result from the reduction in the alkalinity of the concrete pore solution (Shayan, 1995), as shown in Figure 14.49. Compared to the relatively high hydroxyl concentration in the pore solution of the plain cement mortar (Figure 14.49 (a)), incorporation of fly ash and highly reactive silica mineral opal in the mortar reduced the concentration to

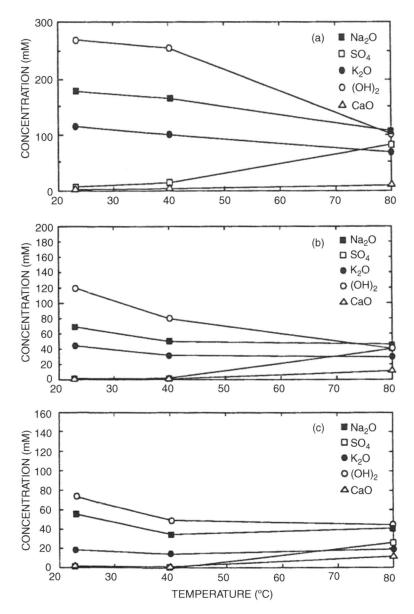

Figure 14.49 Influence of incorporation of fly ash and highly reactive opal particles in mortar and of curing temperature on the ionic composition of pore solution extracted from mortar cylinders after 28 days: (a) Control; (b) 25% fly ash; (c) 10% opal.

one half and one third, respectively, as shown in Figure 14.49 (b) and Figure 14.49 (c). The drop in concentration due to these additions was sharper at higher temperature, which would be expected from the effect of temperature on the reaction rate.

It is noted that the drop in hydroxyl concentration at 80°C was associated with corresponding increases in sulfate concentration, particularly in the pore solution of plain cement mortar. This is why delayed ettringite formation can occur at high curing temperature. However, the concentration is much lower in the presence of the fly ash and opal, indicating that the incorporation of these materials in concrete would also suppress the risk of DEF.

The reduction in hydroxyl concentration of concrete pore solution as a result of SCM addition occurs, in turn, because they modify the nature of the hydrated paste by increasing the silica content (lower Ca/Si ratio) which enables the paste to retain more alkali to maintain its charge balance. This is illustrated in Figure 14.50, which shows the SEM views of hydrated pastes of plain Portland cement and blended fly ash cement. Partially assimilated fly ash particles are evident in the blended cement paste. The EDX spectra clearly indicate larger amounts of silicon and aluminium, derived from fly ash, in the paste from blended cement.

Shayan and Thomas (2014) determined the extractable soluble alkali from a concrete which contained plain Portland cement and had undergone AAR, caused by a reactive aggregate, as well as another concrete which contained the same aggregate but also contained a blended fly ash cement of high alkali content, and which had not developed AAR. The fly ash-bearing concrete was capable of retaining more alkali than the plain

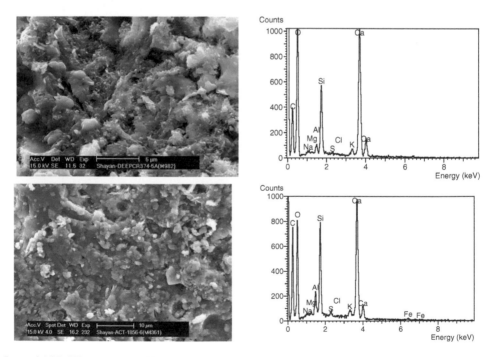

Figure 14.50 SEM views and EDX compositions of Portland cement paste (top) and blended fly ash cement paste (bottom), indicating lower Ca/Si ratio in the latter.

concrete, *i.e.*, despite the much higher alkali content of the fly ash concrete, the amount of alkali that could be released from the solid phase into the pore solution of concrete was only about 50% of that released from plain Portland cement concrete. This would indicate that the alkali in the fly ash concrete was not available for deleterious AAR in this concrete, despite the reactive nature of the aggregate phase.

14.8.2 Mitigation of AAR by glass powder

In Australia most of the post consumer container glass is recycled for manufacture of new containers, but still a considerable amount of contaminated glass is not suitable for this purpose, and used to end up in landfill. However, in the past 10–15 years the latter material has also been separated and used for various purposes including manufactured sand for concrete footpaths.

More importantly, fine glass powder of less than 10 µm particle size, made from the waste glass, was shown to be effective as a cement replacement material in suppressing AAR expansion (Shayan & Xu, 2004). Extensive laboratory research was conducted on the use of glass powder as a pozzolanic material in concrete, leading to a field trial (Shayan & Xu, 2006), which produced very positive and promising results.

In the above studies it was shown (Figure 14.51) that glass powder was effective in suppressing AAR expansion in long-term laboratory expansion tests (3.2 years), *i.e.*, the high alkali content of the glass powder (about 13%) did not appear to cause expansion of concrete prisms that contained reactive aggregate in combination with low alkali cement (no added alkali) and glass powder. Therefore, the large alkali content of the glass powder did not seem to contribute sufficiently to the soluble alkali content of concrete to cause deleterious AAR expansion in the presence of reactive aggregate. It was only when the alkali content of the cement was raised to 1.4% that

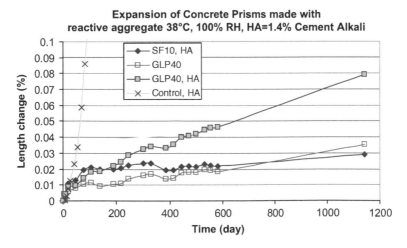

Figure 14.51 Expansion of concrete prisms containing a very reactive coarse aggregate in combination with the materials indicated. (HA denotes 1.4% cement alkali). GLP 40 denotes mixture with low alkali cement but containing 40% glass powder.

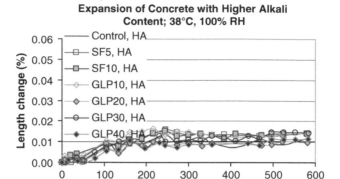

Figure 14.52 Expansion of concrete prisms containing non-reactive coarse aggregate and various amounts of GLP and silica fume, at high alkali content (HA) in concrete of 5.8 kg Na_2O equivalent/m^3

concrete developed mild expansion of about 0.045% expansion after some 580 days the expansion increased to about 0.08% after some 1150 days (3.2 years), indicating that in high alkali concrete the glass powder did not entirely remove the expansion caused by a very reactive aggregate. Silica fume, which had much lower alkali content than glass powder was more effective in suppressing the expansion.

Nevertheless, comparison with the expansion of the control prisms (HA), shows the 40% GLP significantly reduced the expansion rate of the concrete as well as the magnitude of the expansion. This reduction of the expansion occurred despite the high level of alkali in the concrete. The results indicate that the pozzolanic effects of the GLP were more important than its possible contribution to the alkali level of concrete.

Moreover, the glass powder itself did not cause deleterious expansion even in the presence of high alkali contents (Figure 14.52). The expansion behaviour of concrete prisms with 40% GLP was similar to that of concrete prisms containing 10% silica fume at a cement alkali level of 1.4%Na_2Oeq. This means that the GLP and silica fume behaved similarly and both acted as pozzolanic materials. It was also found that for a 40 MPa mortar mixture, the strength gain of mortar containing silica fume and glass powder were comparable on the basis of equivalent cement content.

Utilisation of chemical compounds such as lithium salts or lithium-bearing glass (*e.g.*, Stark, 1992) are not currently favoured in Australia for suppressing AAR expansion, although original work on lithium salts dates back several decades (McCoy & Caldwell, 1951; Lawrence & Vivian, 1961) was conducted in Australia. This is because more certain and cheaper alternatives, such as Class F fly ash, are available.

14.8.3 Examples of AAR mitigation in Australian concrete structures

The data presented above show that the Australian SCMs, when used at the recommended dosage rates, are effective in mitigating against AAR in laboratory investigations. Such materials, particularly fly ashes, have been used in a few concrete structures in Australia, not necessarily purposefully to control AAR, but mostly for other reasons (*e.g.*, reducing heat of hydaration, long-term strength considerations and also reducing project costs). However, in these cases the aggregates used were found later to be reactive, and the use of fly ash was very fortuitous as potentially dagerous AAR-induced damage was avoided.

Such cases include the use of fly ash in Tallowa dam in NSW, which was mainly to reduce the cement content and hydration heat, but it was shown that without it the aggregate employed would have caused AAR-induced damage (Shayan, *et al.*, 1997). Unpublished work by the present writer in 2004, showed that some large concrete anchor blocks ($8m^3$) which support tall power transmission towers, had developed significant AAR-induced cracking, caused by a meta-quartzite aggregate, and that some other blocks were without cracking. The AAR-affected cracked blocks were found not to contain fly ash, whereas those without cracking were shown to contain fly ash and were free of deleterious AAR.

The same reactive meta-quartzite aggregate was reported to have caused damage, a few years after construction, to some bridge components (unpublished report, 1991) and to the intake tower of a dam (Blaikie *et al.*, 1996). However, the same aggregate was used in a dam structure in combination with a fly ash, which has remained essentially free of deleterious AAR over the past 40 years (Shayan & Thomas, 2014). Very small and limited areas of the same dam structure, which for some reason do not include fly ash, exhibit mild AAR-induced craze-cracking (Shayan & Cribbin, 2015), which remain confined by the bulk of the structure.

The long-term efficacy of fly ash in suppressing AAR has also been documented by Thomas *et al.* (2012) for dam structures in Wales (UK) and Ontario (Canada), 50 years after their construction.

As mentioned earlier, AAR was mitigated in the manufacture of some 100,000 concrete railway sleepers by incorporating 8–10% silica fume in the concrete mix (Shayan, 1995b), which use the same slowly reactive aggregate that had caused significant damage to earlier concreye sleepers (Shayn & Quick, 1992). To this date, the writer in not aware of any problem with these sleepers, which have been in service for over 20 years.

It should be noted that even the same type of SCMs show compositional and mineralogical differences amongst various sources, and not all such materials show the same degree of effectiveness in suppressing AAR. For example, low calcium fly ashes (Class F) are much more effective than those with high CaO contents (Class C). These variations, together with different degrees of reactivity of aggregates need to be taken into account in formulating the concrete mix to mitigate AAR. For example a very reactive Australian aggregate needed 30% class F fly ash (Shayan *et al.*, 1996), whereas 25% was sufficient for most other Australian aggregates.

14.8.4 Use of AMBT methods for rapid assessment of the efficacy of SCMs

It would be of great advantage if the AMBT method could be validated for assessing the efficacy of the SCM in suppressing AAR expansion. The results of such tests should ideally correlate with the results of longer term tests and field performance.

Davies and Oberholster (1987) were the first to use the AMBT method to assess the effectiveness of various proportions of different SCMs in preventing AAR expansion. They also used laboratory and field concretes and showed that the AMBT overestimated the minimum amount of SCM that was required to suppress AAR expansion.

Shayan (1990, 1992), using a non-reactive and a number of reactive aggregates, as well as various cements and SCMs, found that the Australian AMBT could be used to assess the effectiveness of the SCMs tested, and that the results agreed with those of concrete tests. However, these tests were conducted only at the recommended dosage rates of the SCMs.

Subsequently several other studies have been conducted on the application of the AMBT method to assess the effectiveness of SCMs in suppressing AAR, as well as determining the minimum amount of SCM which would achieve this, including Berra (1994), Barringer (1999), Carpenter and Cramer (1999), Fournier and Malhotra (1999) and Fournier et al. (2000), Thomas and Innis (1999), Thomas et al. (2006, 2007, 2012).

Carpenter and Cramer (1999), also using ASTM C1260 AMBT method, stated the acceptance limit of the AMBT for blended binders, incorporating silica fume, fly ash or glass powder, as 0.20% expansion at 14 days, which is far too liberal compared to the Australian limit of 0.10% in 21 days. The modified AMBT method which allows the binder to contain SCMs and the soaking solution to vary from 1.0 M NaOH, is designated as ASTM C1567-08, 'Standard test method for determining the potential alkali-silica reactivity of combinations of cementitious materials and aggregate (Accelerated mortar bar method).

Thomas and co-workers, using the ASTM C1260 test method showed that an AMBT limit of 0.10% expansion at 14 days generally correlated well with the CPT limit of 0.04% at two years. They stated that an amount of SCM incorporated in the binder of mortar mix, in combination with reactive aggregate, which would pass the AMBT limit of 0.1% in 14 days, would be likely to be successful in suppressing AAR in field concrete.

This statement is in agreement with the Australian situation, but with the applicable limit of the Australian AMBT of 0.10% in 21 days. This is shown in Figure 14.53, which presents the results of the assessment of efficacy of various SCMs in suppressing AAR-induced expansion of a reactive rhyodacite aggregate. In this example, silica fume was not entirely effective, whereas incorporation of 20% and 25% fly ash and 65% slag in the binder appeared to be effective, yielding 21-day expansion values below 0.10%. A similar set of data for a slowly reactive gneissic granite aggregate (which is less reactive than the rhyodacite aggregate), indicated that 10% silica fume and as low as 15% fly ash were also effective in suppressing the expansion below 0.10% in 21 days. However, as Figure 14.53 shows expansion continues to increase beyond the age of 21 days, which is due to the continual supply of alkali by the 1M NaOH solution. Currently work is being conducted by the writer to verify the results of the AMBT and their correlation with long-term CPT results. Results indicate that a given dosage of SCMs exhibits more effectiveness in the CPT than AMBT, but this appears to vary with the degree of reactivity of the aggregate.

The US Bureau of Reclamation recommendation for AAR mitigation uses the hierarchy of ASTM C1260, petrography and ASTM C1567, with greatest emphasis on petrography for acceptance/ rejection of aggregate or aggregate binder combinations. This approach heavily relies on the petrographic examination for determining the aggregate reactivity, and would not be considered appropriate in Australia.

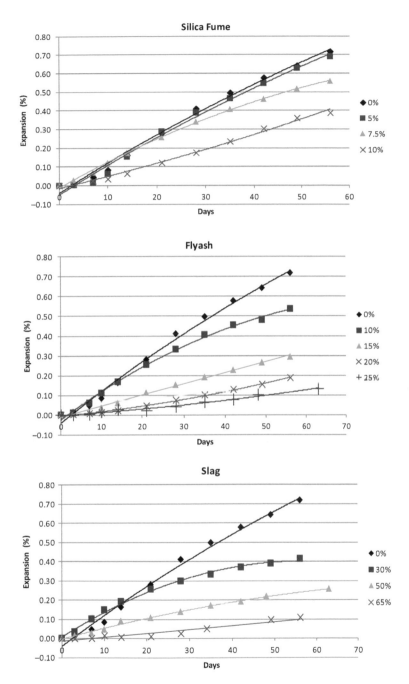

Figure 14.53 Performance of different dosage rates of various Australian SCMs in suppressing the reactivity of a reactive rhyodacite aggregate

It is clear that several parameters influence the amount of SCM which would successfully suppress AAR, including the degree of reactivity of the aggregate, the composition of SCM, the amount of SCM in the binder, the total alkali content of the concrete, and the curing conditions, particularly the temperature of curing. Therefore, an amount of a SCM which would suppress the reactivity on one aggregate to safe levels may not be as efficient with an aggregate which is much more reactive.

14.9 REHABILITATION OF STRUCTURES AFFECTED BY AAR

The rehabilitation needs of a deteriorating AAR-affected structure depend on the state of deterioration and its current condition with respect to serviceability and long-term durability, which are determined mainly by the residual strength properties and residual expansion potential of the affected elements. On this basis, a wide range of rehabilitation options may be applicable, varying from minor repairs such as crack injection and surface coating to significant mechanical strengthening (*e.g.*, post-tensioning, application of fibre composites), slot cutting to relieve AAR expansion pressure, and finally replacement. Examples of major rehabilitation projects for AAR-affected structures can be found in the proceedings of the international conferences on AAR since the early 1990s.

It is important to determine the residual strength and AAR potential of the concrete prior to any repair action, so the current mechanical properties of the concrete are known and can be taken into account in the design of the rehabilitation. It is also important to monitor the behaviour of the repaired structure through embedded sensors, so that the effectiveness of the repair can be assessed. Torii *et al.* (2000) applied a new concrete prestressing technique involving confinement of AAR-affected columns in a large scale laboratory model as well as in a bridge. Embedded sensors have shown the effectiveness of the technique in confining the AAR expansion and cracking, and preventing strength reduction. However, this technique was very expensive.

Mohamed *et al.* (2006) confined small non-reinforced cylinders of reactive and non-reactive concretes and found that confinement by CFRP (uni-dimensional carbon fibre of HEXCEL of 200 g/m^2) reduced the longitudinal strain by 21% and transverse stain by 75%. The difference may have arisen because of the fibre direction, being more effective along the fibre direction (*i.e.*, transverse direction on the cylinders).

Unlike North America, the UK and Japan, research and practice on the rehabilitation of AAR-affected structures is rather limited in Australia. The Australian cases of repair have included silane treatment (Shayan, 1995b), concrete jacketing (Carse, 1996), patching and surface coating (Davies *et al.*, 1996), crack injection and surface coating (unpublished), and post-tensioning of Canning Dam which was investigated by Shayan *et al.* (2000). The surface coatings are often found to be ineffective in the long term.

The dam reported by Shayan (1988) was later post-tensioned vertically through cables embedded into bedrock. Badly deteriorated sections of retaining walls of a dam in Victoria were recently replaced (unpublished). The AAR–affected prestressed deck planks in a five span bridge, which was investigated by Shayan and Morris (2002), were also replaced and subsequently subjected to full scale testing as reported by

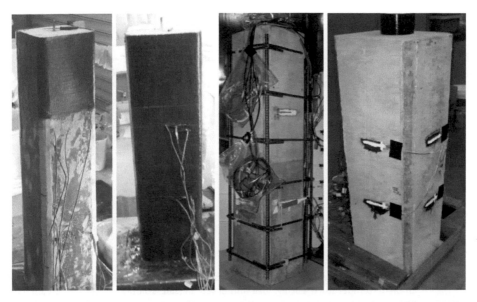

Figure 14.54 AAR-affected columns with square section wrapped with one layer of CFRP of 240 GPa modulus (two left columns); steel reinforcement installed for jacketing and after casting concrete to complete the jacketing (two right columns).

Shayan *et al.* (2008) and Al-Mahaidi *et al.* (2008). Another bridge of the same construction was also investigated by the author in 2013 and the authority concerned has decided to replace the deck, due to a strong case of AAR in the prestressed deck planks.

In the past two decades, the use of fibre-reinforced polymer (FRP) materials, such as carbon-fibre reinforced polymer (CFRP) products, has been promoted for the repair of AAR-affected structures. Shayan *et al.* (2008) used AAR-affected columns with square cross sections to investigate the effectiveness of confinement by epoxy-bonded CFRP wrapping, as compared with the conventional jacketing with reinforced concrete. Figure 14.54 shows the elements used for wrapping and concrete jacketing.

The columns included embedded strain gauges which were placed in the concrete and on the steel bars, as well as on the CFRP wrapping (and within the concrete jacket) to monitor the development of AAR-induced expansion (Figure 14.55). In the case of square section columns, the CFRP used was one layer of 240 GPa modulus. The concrete jacketing used a 40 MPa grade concrete with 7 mm maximum aggregate size.

Figure 14.55 indicates that both types of jacketing reduced the rate of expansion, but did not suppress it to safe levels. The strain developed in the CFRP was significant but did not exceed its elongation capacity (Shayan *et al.*, 2009). Given the highly reactive nature of the aggregate used, it is possible that the tensile strain capacity of the CFRP could be exceeded over a few years.

For the jacketed column, the reinforced concrete jacket, which contained non-reactive aggregate, was noted to have developed cracking after the completion of the measurements, as shown in Figure 14.56. Therefore, the cracking must have been generated due to the expansion of the original column.

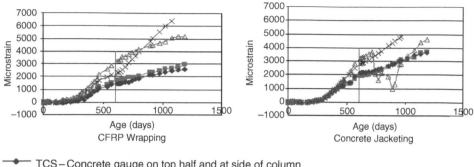

Age (days)
CFRP Wrapping

Age (days)
Concrete Jacketing

—◆— TCS – Concrete gauge on top half and at side of column

—■— TCC – Concrete gauge on top half and at centre of column

—△— BCS – Concrete gauge on bottom half and at side of column

—✕— BCC – Concrete gauge on bottom half and at centre of column

Figure 14.55 Concrete strain development in columns with reactive aggregate: (left) – cast under ambient temperature curing, half-immersed in salt solution, (right) – Steam cured, half-immersed in salt solution. Strain gauges place in the immersed portion (BCS and BCC) recorded larger expansion than those in the upper part (TCS and TCC). Vertical line indicates time when CFRP wrapping was applied.

Figure 14.56 Crack developed in the concrete jacket of a column containing reactive aggregate.

Table 14.11 Properties of CFRP Wrapping Materials.

Name	Specific gravity	Modulus (GPa)	Thickness (mm)	Tensile strength (MPa)	Ultimate elongation (%)
MBRACE CF240/4400	1.7	240	0.235	3800	1.55
MBRACE CF640/2650	2.1	640	0.19	2650	0.4

In the case of circular section columns up to 3 layers of CFRP of both 240 GPa and 640 GPa elastic moduli were applied and the columns were cured at 38 °C and 100% RH. The properties of the wrapping materials used are presented in Table 14.11.

The results of this study were presented by Shayan *et al.* (2012), and some of the results are shown in Figure 14.57. Again, the wrapping reduced the expansion rate and the higher modulus CFRP was more effective. Early age application of the CFRP was also more effective in confining the expansion than later application. However, in practice, often application of CFRP is contemplated after significant expansion has already taken place and caused cracking. Nevertheless, the epoxy-bonding of CFRP would still be effective in strengthening the cracked elements, as it is often used to repair even seismic damage.

With respect to expansion control, it was found that application of one CFRP layer significantly reduced the expansion and two layers further increased the effectiveness, but wrapping with three layers did not improve the confining effect of CFRP compared to two layers (Figure 14.58). These results related to the effect of the number of CFRP layers (both normal modulus and high modulus) on confining the expansion of columns wrapped at the age of one month, and the age at final expansion measurement was

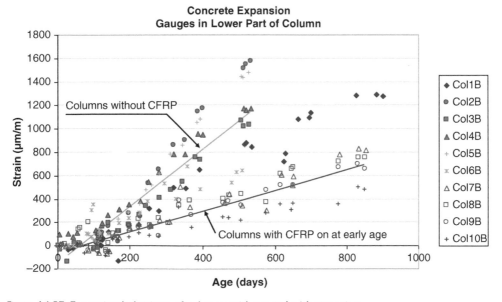

Figure 14.57 Expansion behaviour of columns without and with wrapping.

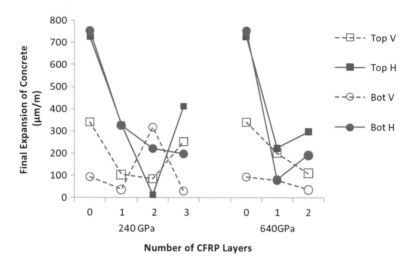

Figure 14.58 Effect of number of CFRP layers on confining the expansion of columns.

513–540 days. Measurements were made through strain gauges placed in the columns in the horizontal (H) and vertical (V) directions. Lettering 'Top' and 'Bot' refer to the top (dry) and bottom (wet) halves of the half-submerged columns, respectively.

The lower expansion measured in the vertical direction resulted from the confining effects of the main steel bars (8x12mm diameter, which gives a confinement ratio of 1.27%). The lateral confinement by stirrups (4x10mm diameter) was much smaller. This is in agreement with the results of Hobbs (1988) and Jones and Clark (1996), which showed that a reinforcement ratio of 1% reduced the expansion by 30–50% compared with free expansion of the same concrete.

It is believed that the reduced efficiency of a higher number of CFRP layers, compared with the first layer, arises from the fact that the bond strength of the first epoxy layer is greater due to the interlocking effects of the concrete interface, which is impregnated by the epoxy, whereas subsequent layers increase the thickness of the epoxy layer which becomes softer under the curing conditions of 38 °C and 100% RH.

Similar results were obtained in a parallel study (Abdullah *et al.*, 2012a) with respect to strain development in wrapped columns. These authors found that strain development in wrapped columns varied with the number of CFRP layers and shape of the columns. They also found that the time of CFRP wrapping is most important in confining the expansion. Columns wrapped earlier with sufficient layers of CFRP exhibited better active confinement of the expansive concrete, whereas later application of CFRP provides only passive confinement to the concrete (*i.e.*, the CFRP just acted as a protective layer).

Another phase of the latter study (Abdullah *et al.*, 2012b) also showed that the CFRP wrapping significantly enhanced the mechanical properties of the affected columns. More recently, the results of Kubat *et al.* (2014) showed that confinement by one layer of CFRP increased the compressive strength of reactive and non-reactive concrete by 466% and 179%, respectively, whereas two layers of CFRP increased the strength by 683% and 300%, respectively. In the latter study, the stress-strain relationship of the reactive concrete, confined with even two layers of CFRP, showed that it was still more sensitive than non-reactive concrete to strain development, *i.e.*, exhibited considerably larger strains under a given stress, compared to non-reactive concrete.

As a result of confining the AAR-induced expansion of concrete, application of CFRP also provides effective confinement of strains in the steel reinforcement, preventing it from approaching yield strains. Findings by Shayan *et al.* (2012) clearly demonstrated this effect, as seen in Figure 14.59.

Lack of information on the long-term durability of the FRP systems under aggressive environmental conditions is considered a barrier to the use of these products under

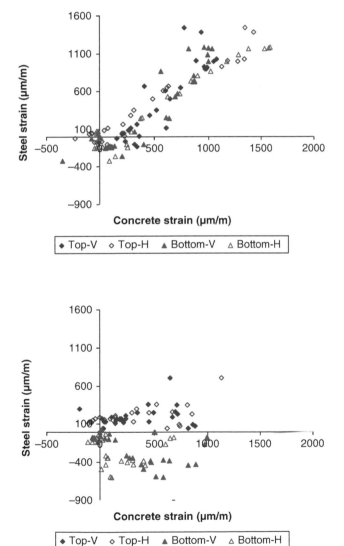

Figure 14.59 Correlation between concrete expansion and strain development in vertical steel bars. Left: columns without wrapping; Right: columns wrapped with 3-layers of 240GPa CFRP at early age.

moist and warm conditions. Nevertheless, the use of such composite systems is expected to increase in the future due to its flexibility and ease of application.

It is obvious that the best method of mitigation against AAR is to use non-reactive aggregates by using appropriate tests in selecting the aggregate before it is used, or employing appropriate concrete mix formulations which include sufficient amounts of SCMs to suppress the aggregate reactivity. In any case, repair/rehabilitation methodologies should be provided as part of the durability planning for the structures concerned with respect to AAR issues.

14.10 SUMMARY

The history of AAR studies in Australia goes back some 70 years, and a great deal of knowledge has accumulated in this period. However, the first field cases of AAR were discovered as late as in the early 1980s and new cases have emerged continually to the present time. A variety of AAR-affected concrete structures and elements have been identified, including bridges, buildings, dams, large water storage tanks, railway sleepers, retaining walls and transmission tower anchorage blocks. Precast elements and large cast-in-situ sections, in which cement hydration temperatures exceed 75 °C, have developed cracking, induced by both AAR and DEF, but the main cause of damage is attributed to AAR.

As a result of studies conducted on structures damaged by AAR, the methodologies for diagnosis of AAR and DEF are well-established in Australia. Studies on test methods for detecting reactive aggregates, led to the development of a new accelerated mortar-bar test (1M NaOH, 80 °C) in the mid-1980s, which was much more reliable than the old test methods, which were discarded in the mid 1990s. This test method was adopted by most Australian Road Authorities.

Consequently, many aggregates covering a variety of geological origins have been found to be reactive. Recently, Australian Standards have also been developed and are designated as AS1141.60.1 and AS1141.60.2 for the accelerated mortar-bar test (1M NaOH, 80 °C) and the concrete prism test (38 °C, at 100 % RH), respectively.

Methodologies for mitigation of AAR in concrete are also well-established, and commercially used Australian SCMs have been found to be very efficient in suppressing the AAR-induced expansion of these aggregates, when used in sufficient amounts in concrete. Moreover, finely ground glass powder was also found to be effective.

New Australian guidelines (HB 79) have been developed for the mitigation of AAR in new concrete. Various repair methods have been tried in Australia, including the application of carbon fibre-reinforced polymers for confinement of AAR-induced expansion, and these aspects are still under further development.

PART TWO – NEW ZEALAND (SUE FREITAG)

14.11 INTRODUCTION

New Zealand lies along the south-west boundary of the Indo-Australian and Pacific plates to the south-east of Australia. Its geology is characterised by the circumpacific 'ring of fire' belt of volcanic rocks and seismic activity. Subduction of the Pacific Plate under the Indo-Australian Plate causes volcanic activity in the northern and central North Island and has produced fault lines extending the length of the country, with frequent earthquakes and uplift of axial mountain ranges.

Rock derived from recent volcanic activity is found over a significant area in the North Island (see Figure14.60). As the northern half of the North Island includes up to half the population of New Zealand, the aggregates derived from these rocks are a valuable resource for construction and could potentially be used in a substantial proportion of the concrete manufactured. Many of these volcanic materials are acidic or intermediate in composition and are therefore potentially alkali reactive. Consequently, appropriate controls are required to prevent alkali-silica reaction (ASR) causing premature deterioration of concrete structures in these and other areas where potentially reactive aggregates may be used.

Sections 14.12 to 14.17 review the approaches taken in New Zealand since the 1940s to minimise the risk of ASR damage to concrete structures. Much of the information presented about the reactivity of specific materials is presented in CCANZ TR3 (2012) and the references cited therein. Readers are referred to TR3 for further information. Other information herein was gleaned from investigations related to the proposed use of specific aggregates or suspected ASR damage on particular structures. The findings from such investigations were targeted to the needs of the client and are not published.

Limestone is not widely used in New Zealand as concrete aggregate. No cases of alkali-aggregate reaction associated with limestone aggregate have been observed, and no precautions to manage such reactions have been developed for New Zealand concrete. See Chapter 3 for an up-to-date review of the 'so-called alkali-carbonate reaction'.

14.12 THE EFFECT OF MATERIALS ON THE RISK OF ASR DAMAGE IN NEW ZEALAND

14.12.1 The geology of New Zealand aggregates

Most concrete aggregate used in New Zealand comprises greywacke. The term 'greywacke' covers a broad group of indurated sand/silt/claystones available in many parts of the country. Providing it is not heavily weathered and the silt/claystone fractions are excluded, greywacke makes an excellent concrete aggregate. To date no cases of ASR associated with this material have been recorded.

Most of the volcanic aggregates used in concrete are geologically recent and fresh. Many of the acidic/intermediate rock types have glassy matrices, and the more siliceous types may contain cristobalite and tridymite. Rhyolitic varieties include obsidian, pitchstone, pumice, ignimbrite and lithoidal rhyolite. Dacites are less common. A wide range of andesites occur. For example, the Egmont andesites from Mt Taranaki (formerly Mt Egmont) have about 30% glassy matrix, while in the Tongariro and Eastern andesites the glassy matrix varies from 10% to 75%, as does the degree of devitrification. The glassy matrix has been considered the main reactive constituent in these volcanic rocks.

The large area of North Island acid-intermediate volcanic rocks shown in Figure 14.60 contains some areas of andesite, particularly in the Coromandel Peninsula and Bay of Plenty. Most of the area between Lake Taupo and the Bay of Plenty is underlain by acid rocks, including widespread ignimbrite sheets.

The alluvial aggregates from these areas comprise varying combinations of rhyolite, dacite, andesite and basalt, sometimes with greywacke. For example, sand from the lower

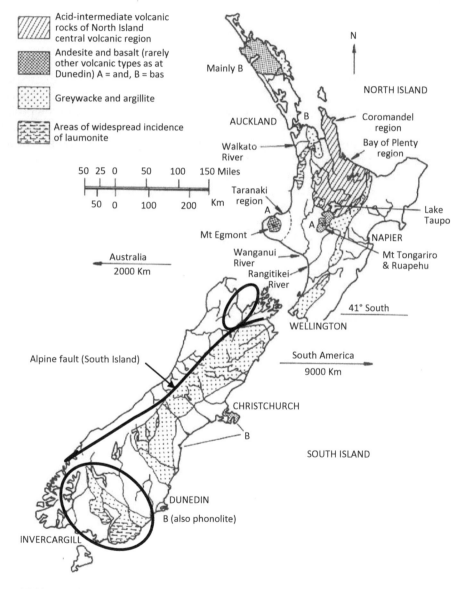

Figure 14.60 Rock/aggregate types and geographical features discussed in the text. South Island regions where mixed alluvial aggregates have been found to be potentially reactive are circled (original map from Watters, 1969).

reaches of the Waikato River consists of 30–40% rhyolite, dacite and andesite grains mixed with non-reactive materials. It has formed extensive alluvial terrace deposits and has been widely used in concrete in the greater Auckland and Waikato regions. Most if not all cases of ASR in these regions are associated with the use of Waikato River sand.

The acid/intermediate volcanic materials found in New Zealand are similar to those found in Japan. Findings from Japanese research have therefore been relevant to understanding the alkali-reactivity of similar New Zealand materials. In particular, the cristobalite and tridymite found in rhyolites and dacites have been found to exhibit 'pessimum' proportions. In contrast, volcanic glass shows little pessimum proportion, consequently New Zealand andesites with glassy matrices show little or no such pessimum behaviour unless they also contain cristobalite or tridymite.

A range of basaltic rocks from the Auckland isthmus area has also been used widely as concrete aggregate. Most do not have a glassy texture and are therefore considered non-reactive. Basalt from one Auckland quarry, however, was found to be associated with many cases of ASR. Elevated alkali contents were measured in some of the affected concretes containing this aggregate and Waikato River sand, and it was concluded that the basalt had released alkali into the concrete pore solution. Quarrying of this aggregate eventually ceased for other reasons. A slightly different Auckland basalt, slightly glassy and doleritic in texture, is suspected to have reacted in one structure investigated for ASR, but most of the ASR observed in this concrete was related to Waikato River sand.

Major deposits of fresh volcanic rock in the South Island are limited to basalt and phonolite. They have generally been assumed non-reactive. Phonolite from Dunedin has been used in concrete. To date no cases of ASR associated with these materials have been recorded.

Basalts with glassy textures or silica contents higher than 50% are recognised as potentially alkali-reactive and therefore not recommended for use as concrete aggregate. Localised alkali- and silica-rich phonolites and trachytes are present in eastern/central Otago, but it is not known whether they are potentially reactive or alkali-releasing, or how much is used as concrete aggregate.

Alluvial aggregates in Southland/southeast Otago and Nelson contain a wide range of lithologies, including quartz, quartzite, schist, granite and gneissic materials. Some also contain minor quantities of acid to intermediate volcanics, some of which are partly metamorphosed. Isolated cases of ASR associated with these aggregates have been observed, and attributed to the presence of strained or microcrystalline quartz and/or quartzite. Chalcedony, glass and devitrified glass were also observed in these aggregates.

A review of the distribution of potentially alkali-reactive rock types in the South Island concluded that potentially alkali-reactive aggregates are limited to the above materials, and that the granites, schists and gneisses present in alluvial gravels, including those on the West Coast, are unlikely to be reactive (Freitag et al., 2011).

ASR associated with amorphous silica minerals, such as opal and chalcedony, has not been recorded in New Zealand. Also limestone is not widely used as concrete aggregate in New Zealand, therefore the alkali-reactivity potential of local limestones has not been investigated or recorded.

Increasing interest has been shown in use of recycled material as aggregate in concrete, though in practice the ability to utilise waste material cost-effectively is constrained by the relatively small, widely-spread population base that limits the supply of appropriate material.

If the recycled aggregate is manufactured by crushing hardened concrete (left over from ready mix concrete production, or recycled from demolition sites) or recovering aggregate from returned surplus fresh ready mixed concrete, then its potential alkali-reactivity is determined by the reactivity of the parent aggregate. If the original aggregate source is unknown, or the recycled material consists of a mix of masonry, concrete and asphalt, then the recycled material is considered potentially reactive (CCANZ TR14: 2011).

Individual concrete manufacturers have assessed the use of recycled glass as aggregate for general use and for specific projects. Although concrete with acceptable properties has been produced and used, the actual use of recycled glass has been limited by the need to manage ASR as well as the economics of supply and of managing the quality of an extra mix component of variable composition.

Apart from the basalt mentioned above, no cases of ASR damage associated with alkali- release from other aggregates have been identified.

14.12.2 Cement

Since 1945, when records of alkali analyses became available, most cement manufactured in New Zealand has contained less than 0.9% alkalis as Na_2Oeq (St John, 1988). The main exceptions are ordinary and rapid hardening cements produced from 1958 until the late 1960s at one national supplier's plant (see Figure 14.61) and at two

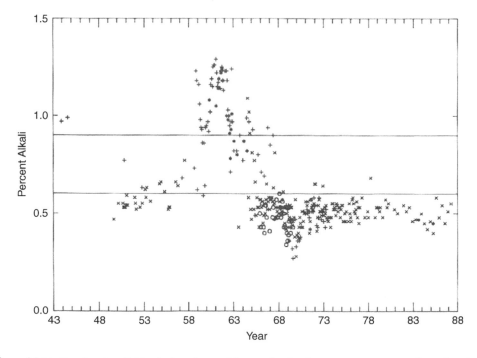

Figure 14.61   Results of available alkali analyses of the production from one cement manufacturing plant for the period 1943 to 1988, showing the marked production of high alkali cements between 1958 and 1968 (St John, 1988).

key: + ordinary; x low alkali; * rapid hardening; o special.

smaller plants supplying local markets during approximately the same period. Cement with alkali equivalent between 0.6% and 0.9% was produced for short times at five plants. Cement with alkali equivalent between 0.6% and 0.8% was imported during major periods of infrastructure construction between 1947 and 1956. Cement with alkali equivalent less than 0.6% has been available from at least one plant since 1945. Since the mid 1970s, the alkali equivalent of all cement produced in New Zealand has been less than 0.7%.

Where potentially reactive aggregate is available, low alkali cement has been specified for all major public works since the risk of ASR was first recognised in the 1940s.

Initially, low alkali cement was produced largely by chance rather than by regulation, resulting from low alkali contents of raw materials, 'hard burning' to meet strength requirements, or individual kiln features. The requirement for low alkali cement to be used in public works where reactive aggregates were available would have given a market advantage to producers of low alkali cement in these areas. In 1974, an optional limit of 0.6% alkali equivalent for cement to be used with aggregates that may be potentially alkali-reactive was introduced to the New Zealand specification for Portland cement (NZS 3122). The combination of this clause plus passive protection provided by the default use of locally-produced low alkali cement over many decades resulted in a widespread misconception that all cement complying with NZS 3122 was by definition low alkali, and that this would be sufficient protection against ASR damage. Consequently, in 2009 the requirement was removed from NZS 3122 to focus attention on concrete alkali content instead.

Changes in the economics of local cement production resulted in increasing amounts of cement and clinker being imported into New Zealand from about 2010. Concrete producers expressed concern that higher alkali cement could be used in concrete containing potentially reactive aggregates if cement alkali contents were not regulated. Thus in 2014, a maximum limit of 0.6% alkali equivalent was introduced into NZS 3122 as a specified requirement, with an option that cement exceeding this limit may be supplied with prior agreement of the purchaser or cement user.

The ready availability and widespread use of low alkali cement has been the single most important factor in limiting ASR damage in New Zealand structures, resulting in relatively low concrete alkali contents being achieved in most concrete by default. Indeed, most cases of ASR damage are in structures built between the 1950s and early 1970s and are likely to be related to the use of cements with alkali contents exceeding 0.6%, in particular where other sources of alkali were present. Adherence to the cement alkali limit introduced in 2014 will further help to minimise the incidence and severity of ASR damage in structures built in future.

14.12.3 Supplementary cementitious materials

Supplementary cementitious materials (SCMs) sourced from industrial waste streams, such as silica fume, type F fly ash, and ground granulated blastfurnace slag, are not produced in New Zealand and must be imported if they to be used in concrete. A type C fly ash has been produced in limited quantities by the one remaining coal-fired thermal power station, which burns coal from a variety of sources. This plant has partly been converted to natural gas and may be fully converted by 2020 unless the coal and energy markets change. The extra logistics and costs associated with the use of fly ash, ground granulated blastfurnace slag and silica fume mean they are not routinely added to concrete except in particular circumstances, such as to reduce heat evolution in mass concrete, improve the durability of concrete exposed to marine/coastal environments or mild aggressive chemical exposure, or minimise ASR damage if other solutions are not economic or practical.

Natural pozzolanas sourced from volcanic and diatomaceous deposits have been used from time to time in concrete for major infrastructure projects since the 1950s, to improve the workability of low-cement concrete mixtures, reduce heat evolution and minimise ASR expansion and cracking (Smith, 1977). A national Standard for the production of Portland pozzolana cement was developed in the 1970s (NZS 3123: 1974). With the eventual replacement of Portland pozzolana cements by other types of blended cement, which were included in NZS 3122 from 2009, NZS 3123 was revised in 2009 to cover natural pozzolanas only, which were not covered by Australian and New Zealand Standards for SCMs. At the time of writing (2016), only one natural pozzolana, an amorphous silica of volcanic origin, has been fully commercialised on a long term basis for use in concrete. This material was developed in the 1990s to compete with the more expensive imports of silica fume.

Use of natural pozzolana in the Waikato River hydro-electric power development scheme prevented significant ASR damage while allowing a potentially reactive local sand to be used.

Interest has been shown in using waste glass as an aggregate, but the cost of producing powdered glass for use as an SCM has not been considered economic.

Use of SCM in new concrete is one of the methods recommended by CCANZ TR3 for reducing the risk of ASR damage. The additional costs incurred in using SCM, however, means it is not the primary means of managing ASR damage in NZ.

14.13 THE EFFECT OF ENVIRONMENT ON THE RISK OF ASR DAMAGE IN NEW ZEALAND

14.13.1 Climate

The climate ranges from sub-tropical/warm temperate in the north to cool temperate in the south, with alpine conditions in mountainous areas, Mackintosh (2015). Mountain chains extending NE-SW along much of the length of New Zealand provide a barrier to the prevailing westerly winds and divide the country into different climate regions. The west coast of the South Island is the wettest area, and the area to the east of the mountains is the driest. Most of the country is within 150 km of the coast, which means temperatures are generally mild. Mean annual temperatures range from 10 °C in the south to 16 °C in the north, with relatively small variations between summer and winter, though inland and to the east of the main mountain ranges the seasonal variation in average temperature is up to 14 °C. Temperatures rarely exceed 40 °C or fall below freezing for prolonged periods.

Annual rainfall exceeds 750 mm per year in most of the North Island, and 500 mm in most of the South Island. These areas include areas where potentially reactive aggregates have been used.

Average atmospheric relative humidity is generally 70–90%. Diurnal variations may be up to 20% RH. Diurnal temperature ranges may reach 20 °C in some parts of the country, but are not often higher. Therefore condensation is unlikely to provide sufficient moisture for ASR to proceed where exposure to other sources of moisture is low.

Most snow falls in the mountain areas, which are largely uninhabited but host infrastructure related to hydro-electric electricity generation. Snow is rare in the coastal areas and west coast of the South Island, though the east and south of the South Island do experience some snow in winter. Frost occurs throughout the country, but freeze-thaw attack on concrete is uncommon except in some elevated and inland areas, and severe damage resulting from the expansive freezing of water in cracks caused by ASR has not been recorded.

Overall, New Zealand climate generally lacks the extremes and variations in temperature and humidity that have been reported to exacerbate ASR in other countries. Where potentially reactive aggregates are widely used, the climate is cool temperate to subtropical, with rainfall exceeding 1000 mm per year. Thus the environmental exposure conditions most likely to exacerbate ASR damage are direct exposure to rain and run-off.

14.13.2 Exposure to external alkali sources

A large part of New Zealand is subject to deposition of salt spray carried by prevailing and onshore winds. Reinforcement corrosion associated with the ingress of seawater and sea-spray is the most common and severe concrete durability problem encountered, and exposure to marine chlorides is taken into account in durability design to manage the risk of reinforcement corrosion (NZS 3101). In contrast, observed incidences of ASR are generally related to direct exposure to moisture rather than exposure to sea-spray or seawater. The most severe ASR-related damage observed in the South Island was in marine/estuarine structures, but is thought to have been caused by a combination of ASR and DEF related to the combination of raw materials and curing temperatures, and therefore not directly attributable to ingress of additional alkali from seawater (Freitag *et al.*, 2011).

Chloride-based de-icing salt is rarely used in New Zealand, therefore ASR is unlikely to be exacerbated by the ingress of alkalis from this source. Apart from in geothermal areas, ground water and soil are unlikely to contribute significant alkalis.

14.13.3 Effect of structural design

Most significant structures are heavily reinforced for seismic reasons. Therefore ASR cracking tends to reflect the position of reinforcement rather than exhibiting the predominantly random map cracking patterns often considered characteristic of ASR. Nevertheless, random cracks do occur on less heavily reinforced elements and in more severe cases.

Reinforcement and prestressing restrain ASR expansion, and therefore generally reduce the risk of ASR causing structural damage in buildings and civil structures. However, beam and deck soffits on one critical motorway bridge have been strengthened because of concern about ASR damage. ASR cracking is considered a serviceability risk because it may increase the ingress of agents that can promote reinforcement corrosion.

14.14 THE EFFECT OF INFRASTRUCTURE DEVELOPMENT ON ASR RESEARCH IN NEW ZEALAND

New Zealand's population in 2014 was approximately 4.5 million, in a land area slightly larger than that of the UK. This relatively small population supports well developed national, regional and local infrastructure networks, providing transportation, electricity, water, telecommunications and other essential services.

In 1940, New Zealand had a population of 1.5 million people. By 1970 this population had doubled. During this period of considerable growth, which called for major road-building, extensive construction of hydro and geothermal power projects and other major public works, the Public Works Department (PWD) and its successor the Ministry of Works and Development (MWD), led both the design work and construction.

As early as 1943, the PWD recognised that a potential alkali-aggregate problem existed in New Zealand. Since volcanic aggregates were to be used in many of the hydro power stations to be built on the Waikato River after World War II, PWD initiated investigations into ASR by the Department of Scientific and Industrial Research (DSIR) (Hutton, 1945). It also took active steps to prevent ASR by using low alkali cements and pozzolana. At times, PWD/MWD was purchasing up to 50% of all the cement produced in New Zealand and was able to exert pressure through its specifications and purchasing contracts to keep the alkali contents of most cements below 0.6% alkali equivalent. Low alkali cements were thus specified in all public works projects, although in practice they were not always available and were not necessarily used in smaller projects. It is this early application of Stanton's work (Stanton, 1940, 1943), learnt by New Zealand engineers training at the US Bureau of Reclamation, that has been an important contributing factor to minimising ASR in New Zealand.

Two divisions of DSIR were involved in investigating ASR. The New Zealand Geological Survey investigated prospective aggregate sources, while the Chemistry Division investigated aspects of the chemistry and tested cements, pozzolanas and concrete. Until the 1960s, the Chemistry Division focussed on materials for the Waikato River hydro power development; this work still forms the main body of fundamental knowledge about ASR in New Zealand.

The long term collaboration between MWD and DSIR, plus participation in local aggregates and concrete industry forums, and international forums such as RILEM and the series of international conferences on ASR, meant that by the early 2000s, the causes and extent of ASR in New Zealand were generally well understood, and protocols for minimising damage had been established and accepted by industry. The maturing of understanding about ASR coincided with changes in procurement practices and governance of infrastructure assets (see also section 14.16.2), as well as changes in government research funding mechanisms, and resulted in the end of government-funded general ASR research.

Assessment of the potential reactivity of new and existing aggregate resources was then largely based on data generated in the 1950s-60s for similar or related materials. Specialists in cement and concrete chemistry retired, and capabilities in assessing aggregates or investigating suspected cases of ASR devolved largely to consulting geologists and engineers. With the loss of local specialist chemistry expertise, increasing reliance was placed on services of offshore ASR specialists. This has in turn increased the direct costs of ASR-related testing and investigation, resulting in relatively little testing or research related to ASR being carried out. Even on major infrastructure development projects, the preference has been to use aggregates known from historic testing and experience to be non-reactive, importing them from outside the region if necessary, rather than investigating how local aggregates may be utilised.

Consequently, ASR investigations since the early 2000s have focussed on answering questions about specific structures or materials. Investigations related to suspected cases of ASR have been commissioned by the owners of individual structures for the purposes of identifying remedial work that may be necessary. One such investigation led to government-funded research focussing on a wider population of structures for general interest (Freitag et al., 2011). Other investigations have been commissioned by suppliers of specific materials.

With limited ASR expertise now available within New Zealand to support decisions about materials selection for specific applications, precautions to minimise the risk of ASR damage remain conservative. Although this approach has been successful in managing concrete durability to date, it is a barrier to the optimisation of increasingly scarce aggregate resources and uptake of new materials.

14.15 METHODS USED IN NEW ZEALAND TO DEVELOP UNDERSTANDING ABOUT ASR

Lack of consistent demand for measuring potential alkali-reactivity and concrete expansion means that test methods used internationally are not available as routine commercial testing services within New Zealand. Similarly, local standardised testing protocols with associated acceptance criteria specific to New Zealand materials have not been developed. Instead, investigations since the 1940s have utilised a combination of approaches depending on the question that needs to be answered, the time available, and the economics of the specific project: a bespoke 'problem solving' approach rather than compliance testing.

The following sections describe how various experimental approaches for investigating ASR have contributed to the body of knowledge about ASR in New Zealand.

14.15.1 Petrographic examination

Petrographic examinations by the New Zealand Geological Survey during the 1950s and 1960s characterised aggregates from many sources and potential sources. Findings were recorded by St John (1988a). In the 1980s, DSIR documented all aggregate resources in NZ, recording the locations and geological and petrographic descriptions in a database (St John et al., 1988b). Resources were not available to maintain the database, however, and public-funded petrographic examination of aggregate resources has not been carried out since then.

Instead, industry tends to utilise existing published data about aggregate reactivity as much as possible. Individual aggregate producers are increasingly commissioning petrographic descriptions of their materials in response to specific requests from their customers. Most such testing is related to supply for projects or resources large enough to justify the cost.

Large-area thin sectioning techniques for the petrographic examination of concrete were developed by DSIR (Abbott & St John, 1981). Examining a large area of sample made it possible to identify more reliably the presence of ASR, the rock types under-going reaction, and other features relating to the quality and durability of the concrete. This enabled cases of ASR to be identified where external signs were subtle or insig-nificant and thus added significant value to the information collected from visual observations during field inspections (see section 14.15.8). Much of this pioneering expertise was recorded by St John *et al.* (1998), subsequently updated by Poole & Sims (2016).

Petrographic examination of a commercial concrete sand containing a small amount of potentially reactive materials gave very similar results to those obtained by a different analyst from a sample taken from the same source some 25 years earlier. Thus although seemingly a subjective technique, petrographic examination by suitably experienced analysts can be very reliable.

14.15.2 ASTM C289 – Chemical method

Results from testing to the method described by ASTM C289 (recently 2007, but now withdrawn), *'Test Method for Potential Reactivity of Aggregates (Chemical Method)'*, generally reflect the behaviour expected from petrography and the observed reactivity of New Zealand volcanic and greywacke aggregate in concrete structures. For example, the method clearly distinguished between non-reactive basalt and material of basaltic appearance but with silica contents tending to that of andesite, which is more reactive. Today, however, such materials would be classi-fied more easily by X-ray fluorescence (XRF) than by ASTM C289. In addition, the test was found to be highly sensitive to sample preparation, and required experience to obtain repeatable and accurate results. Thus the curves plotted to delineate the three categories of reactivity (based on aggregates originally tested in the USA) imply greater test accuracy than the inherent variabilities in rock composition and testing allow in practice, and an individual result from an aggregate source may not be reliable, particularly if it falls near one of the division lines. Neither does the test detect the 'pessimum' proportion behaviour exhibited by some andesite and rhyolite aggregates.

Overall, the test is potentially adequate as a screening tool for New Zealand rocks when supported by data from geological and petrographic data and under-standing of reactivity demonstrated by other tests and *in situ* behaviour. However, ASTM C289 testing has not been available as a routine commercial testing service in New Zealand since the mid-1990s. In addition, it was not adopted by RILEM (Nixon & Sims, 2016) and was withdrawn by ASTM in 2016. This limits its on-going use for compliance testing. Nevertheless, the collation of data in CCANZ TR3 (2012), summarised in Figure 14.62, remains a valuable generic record of potential aggregate reactivity and the test may still have a role in wider investigations.

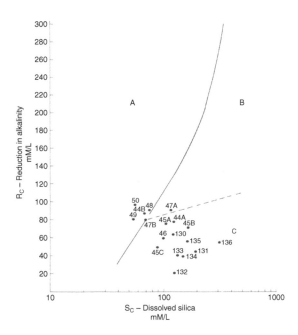

Figure 14.62 (a to f) Results of ASTM C289 testing applied to some New Zealand aggregates (after CCANZ TR3: 2012), divisions A (innocuous), B (potentially deleterious) and C (deleterious): 14.62a Egmont andesite from Taranaki. Re-testing of samples 49 and 50 showed them to be deleterious. Mortar and concrete testing showed that most of these andesites do not exhibit pessimum proportion.

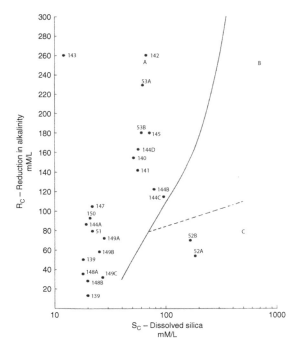

Figure 14_62b Basalt. The samples plotting as deleterious (division C) had a silica content of 51% and are doleritic in texture.

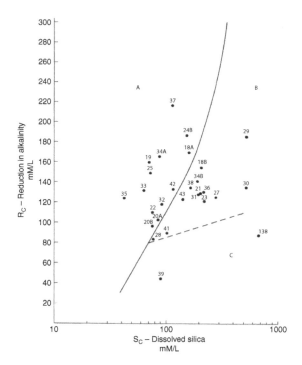

Figure 14_62c Central North Island andesites. The materials are from multiple sources, some alluvial, some hard rock, and some of which show pessimum proportion behaviour in mortar tests (Figure 14.63b). Some of the apparently non-reactive material is gravel containing grey-wacke, while some also contains rhyolite, presumably in amounts away from the pessimum proportion.

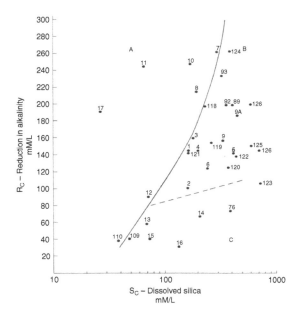

Figure 14.62d Eastern North Island andesites. These are older and more complex than the aggregates represented in Figures 14.62a and 14.62c. Some of the apparently innocuous samples are from altered rocks.

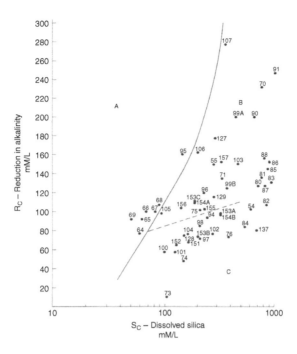

Figure 14.62e Rhyolite and dacite, including alluvial materials containing these rock types. The lower group of apparently innocuous aggregates are alluvial greywackes containing rhyolite. Results from another sample from the same source as the uppermost 'innocuous' sample (95) plotted well within the lower part of division B (96).

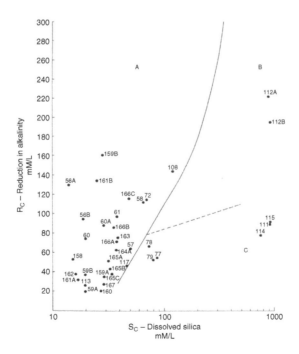

Figure 14.62f Greywackes, miscellaneous, foreign and synthetic materials. Greywackes all plot as unreactive, reflecting *in situ* behaviour. The group at the left side of division C are quartzite; the other samples in divisions B and C are known highly reactive imported limestone (div B) and pyrex and cristobalite (div C, right side) .

14.15.3 ASTM C227 – Mortar-bar test (38°C)

The more important results from testing to the principles of ASTM C227 (2010), *'Potential Alkali Reactivity Cement—Aggregate Combinations (Mortar-Bar-Method)'*, are given in Figure 14.63 a to f (from St John, 1988a). These illustrate the need to test volcanic aggregates for *'pessimum proportions'*, which vary not only with composition of the aggregate but also to some extent with concrete alkali content.

The reactivity of natural gravels and sands depends on the amount of non-reactive material present, which varies with time and with source location. Thus systematic laboratory testing of all materials is the only way to establish unequivocally the potential reactivity of a source.

Generally the use of ASTM C227 for testing New Zealand materials has been satisfactory. The problems encountered elsewhere when the test has been used with sedimentary and some metamorphic materials (Duncan *et al.*, 1973) did not significantly affect DSIR results from New Zealand volcanics. Significant leaching was observed in other tests (Freitag, 1990). Lack of ongoing testing to this method reflects lack of resources to carry out the testing and a preference for faster or more reliable tests such as petrography, accelerated mortar-bar testing, or concrete prism testing.

14.15.4 Mortar-bar tests (80 °C)

A limited investigation was carried out to determine the value of the rapid approach ('AMBT-80') described in methodologies such as ASTM C1260 (2014) for assessing New Zealand aggregates (Freitag, 1998). It examined several concrete sands in the gradings and blends used commercially. The results indicated that an andesite sand, a rhyolite sand blended with basalt, and a greywacke sand contaminated with rhyolite were all potentially reactive. These materials had been observed to be reactive in earlier mortar-bar tests and/or in concrete structures.

A greywacke concrete sand tested in its commercially-supplied grading also tested as potentially reactive. Subsequent investigation of concrete from structures likely to contain this greywacke sand and cement with moderately high alkali content revealed no clear evidence that alkali-silica reaction had occurred in situ (Freitag *et al.*, 2000) or that future *in situ* expansion was likely (Freitag, 2002). In contrast, unpublished findings from an investigation into the effectiveness of SCMs in mitigating ASR expansion found that a control mix consisting of a different greywacke ('as supplied') had tested as non-reactive. The reason for the difference between the findings of the two sets of mortar-bar test results is not known.

The ASTM C1260 approach has not found widespread use for evaluating individual aggregate sources because it gives conservative results and, for New Zealand aggregates, does not seem to add value to information from petrographic assessment or experience unless used in conjunction with appropriate concrete testing procedures. Consequently there has been no demand to establish associated acceptance criteria for New Zealand aggregates. This means that when the test is used, results should be compared with results from known reactive and non-reactive control aggregates or concrete testing, which increases the cost of testing and complexity of interpreting results. Overall, petrography is currently preferred by industry as a screening test because it is generally more convenient and cheaper.

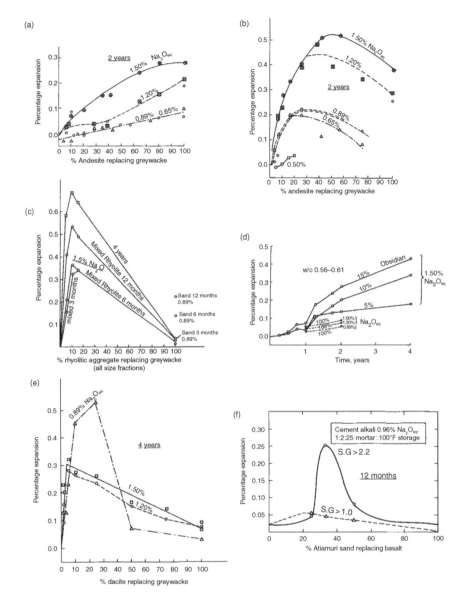

Figure 14.63 a to f Results from ASTM C227 mortar-bar testing (from St John, 1988a).
(a) Egmont andesite, no pessimum proportion; (b) Tongariro andesite, showing pessimum proportion 15–50% depending on alkali content; (c) Whakamaru rhyolite, showing pessimum proportion 10%; (d) Taupo obsidian blended with greywacke, showing pessimum proportion ≥15%; (e) Tauhara dacite, showing pessimum proportion 10–25% depending on alkali content; (f) Atiamuri sand blended with Ongaroto basalt. Solid curve represents sand from which pumice had been removed.

Nevertheless, a C1260 test extended to 21 days was used in research into the potential reactivity of alluvial sand and coarse aggregate from the South Island (see also section 14.15.5 below) that had been used in concrete structures exhibiting signs of ASR/DEF. The findings showed that the small quantities of microcrystalline silica materials observed in petrographic analysis had the potential to cause expansive ASR, and thus helped to explain the incidence of ASR observed in the concrete structures examined (Freitag *et al.*, 2011).

The C1260 approach was also used when evaluating the potential reactivity of a concrete sand identified by two separate petrographic examinations as potentially reactive on the basis of a small amount of highly reactive contaminant known to exhibit pessimum proportion. Concrete prism tests produced a very small, slow but continued expansion at high alkali contents, indicating the sand was unlikely to produce a significant ASR expansion in in-situ concrete (section 14.15.5). ASTM C227 testing some 25 years earlier had produced similar results (Freitag, 1990). Consequently, C1260 tests utilising storage solutions at a range of alkali concentration were initiated to find out whether the assessment based on the petrographic examinations was appropriate. The results confirmed the sand was indeed potentially reactive, possibly 'slowly reactive', although the test specimens were not examined to identify the specific reactive constituent. It was recommended that concrete containing the sand not be subjected to temperatures exceeding 70°C during curing or in service.

Following the approach of ASTM C1567 (2013), the test has also been used to assess the effectiveness of SCMs in mitigating ASR expansion for specific products or projects. Usually, however, if SCM is used for this purpose, it is used at dosage rates widely-accepted internationally to mitigate ASR expansion.

14.15.5 Concrete prism tests (21°C, 38°C, 60°C)

Concrete expansion tests carried out by DSIR in the 1970s were limited to tests on one aggregate, an andesite, carried out at 21°C and a range of concrete alkali contents. The findings demonstrated that expansion was greater in continuously moist conditions than in a natural exposure site, and showed no obvious alkali 'threshold' below which ASR expansion would not occur, though a maximum concrete alkali content of 3.5 kg/m^3 was suggested as a practical limit for aggregate from this source (St John, 1988a).

Concrete prism tests at 38°C and 100% RH based on methodologies such as ASTM C1293 (2008) ('CPT-38') have been commissioned to assess the potential reactivity of specific aggregates. As with C1260 tests, lack of demand to establish associated acceptance criteria for New Zealand aggregates means that test programmes need to be designed with care to ensure appropriate interpretation of the results, and may involve testing concretes with a range of alkali contents. The combination of the cost and duration of this methodology means that it is only practical for major projects or aggregate resources.

One such programme involved testing a suspected potentially reactive concrete sand using a non-reactive coarse aggregate and standard mix design, at a range of alkali contents from 2.5 to 5.8 kg/m^3. At 12 months, no mixes had expanded more than 0.03 %, but at the highest alkali contents the expansions were continuing slowly and may have reached 0.03 % after two years. When interpreted according to the principles described by Nixon and Sims (2016), the results indicated that significant ASR expansion was unlikely at TR3's default alkali limit of 2.5 kg/m^3, and suggested a limit of 3.0 to 3.5 kg/m^3 would be appropriate when the sand is combined with a non-reactive coarse aggregate. These results, plus results from petrographic analyses and mortar-bar tests (see sections 14.15.1 and 14.15.4), suggested that despite very few observed cases of ASR in areas where the sand is widely used, it should not be considered completely 'non-reactive'.

Accelerated concrete prism tests at 60°C, 100% RH, and concrete alkali content 5.9 kg/m^3 ('CPT-60') were used to research the reactivity of South Island alluvial sand and aggregate. Materials that had reacted in the C1260 test (see section 14.15.4) did not react in the CPT-60 test, suggesting that observed *in situ* ASR may have been associated with concrete temperatures exceeding 60 °C when the concrete was cured, and thus explaining the limited number of cases of significant ASR damage with these aggregates (Freitag *et al.*, 2011).

The two investigations described above demonstrate how a combination of tests at different temperatures and alkali levels can be useful in identifying the risk of ASR in concrete exposed to elevated temperatures, such as in accelerated curing, and how the risk of ASR damage is related to the availability of sufficient alkali.

14.15.6 Core expansion tests (38°C)

Accelerated concrete core expansion tests at 38 °C, following the guidelines of BCA (1992), have been used to ascertain whether concrete that has undergone (or suspected to have undergone) ASR expansion *in situ* is likely to expand further.

Investigations comparing accelerated core expansions with *in situ* expansions found that core expansion behaviour after three years gave a qualitative indication of *in situ* increases in crack width observed over 10 years (Freitag, 2002). They also demonstrated the inherent variability of concrete and the importance of taking replicate samples. Similar accelerated core testing confirmed that *in situ* concrete made from greywacke that had tested as reactive in C1260 tests was unlikely to undergo further expansive ASR, and that concretes with relatively low alkali contents, but containing highly reactive rhyolite sand and showing evidence of ASR *in situ* and in petrographic examination were also unlikely to expand further (Freitag, 2002).

Comparing the expansions of cores from the same concrete stored in an alkaline solution and in moist conditions indicates whether the concrete still contains sufficient reactive aggregate and sufficient alkali to permit further *in situ* expansion if enough moisture is present. This approach has been used to inform recommendations for managing future potential ASR expansion on existing individual structures, including one of the structures that initiated the investigation into the reactivity of South Island aggregates (Freitag *et al.*, 2011).

14.15.7 Chemical analyses

Testing carried out to well-established methods sometimes produces unexpected results. They may be dismissed as 'spurious' or otherwise insignificant if the testing is simply for compliance. If the testing is part of wider research, further investigation by appropriately experienced analysts may reveal chemical interactions that explain *in situ* behaviours that are otherwise inconsistent with existing current knowledge. For example, investigations in the 1980s and 1990s into the reactivity of particular combinations of aggregate and binder, observed on individual structures, revealed that measured concrete alkali contents did not always correspond with values expected from measured cement contents and the cement alkali contents deduced from construction records and cement production records.

Construction records are not always available, or may be inaccurate. For example, a cement shortage may result in a different cement brand or type being used in one or more placements on a structure. Chemical analyses of New Zealand cements revealed chemical fingerprints related to the mineralogy of the rocks from which the cements had been produced. Analysis of these elemental ratios enabled the manufacturing plant to be identified by analysis of the binder fraction of concrete (Goguel & St John, 1993). Cement alkali contents could then be ascertained from historical analyses of the specific cement used, and concrete alkali contents then calculated from the measured cement content and the cement alkali content.

Other investigations found that the acid digestion used to extract the cement binder fraction from concrete during the determination of cement and alkali contents can also dissolve aluminosilicates within the aggregates and the binder fraction. These aluminosilicates bind alkalis, so dissolving them can result in misleadingly high results from analyses for alkali and silica content. A method of extracting the binder phase using a neutral leaching solution was thus developed for use when determining cement content and concrete alkali content (Goguel, 1995).

The ability to more accurately determine the alkali content of a concrete sample from analysis of cement composition and content led to the discovery that significant amounts of alkali can be released from particular types of rock used as aggregate (Goguel, 1995). A significant number of cases of ASR in one region were thus attributed to a nepheline basanite being used in combination with a highly reactive rhyolite sand (Goguel, 1996). Nepheline and leucite were subsequently found to be the main alkali-releasing minerals (Goguel & Milestone, 1997).

14.15.8 Inspection of structures

Although research into ASR in New Zealand started in the 1940s, the first case was not unequivocally identified until 1970 (St John, 1972). In 1983, two cores submitted to the Ministry of Works for compressive strength testing showed signs of ASR, which by chance were seen and recognised by a scientist familiar with the symptoms. This led to investigations during the 1980s and 1990s into the incidence of ASR in bridges in areas of the North Island where potentially reactive aggregates were available. Over 500 structures were inspected, and up to 10% showed evidence of ASR. Most of the damage was cosmetic or minor, and would not affect the mechanical or durability performance of the structure. Most of the affected structures were built when high alkali cement was available, including those with the more extensive damage. A few dated from the 1930s, before ASR was first recognised. A few were built after 1970, when most cement manufactured in New Zealand was low alkali.

The incidence of ASR damage was also investigated at hydro power stations in the Waikato area, where rhyolite sands had been used in concrete structures. Despite the use of low alkali cement and, in some concretes, pozzolana, ASR damage was again found to be relatively common, though the damage was not significant in terms of structure performance, durability or serviceability.

Most of the above cases would not have been identified, let alone reported, without systematic inspection of structures by concrete materials specialists supplemented by petrographic examination to ascertain whether ASR had occurred.

In contrast, bridge maintenance staff responsible for an urban motorway network were quick to recognise and identify ASR on several structures, including prestressed and precast beams, deck units, and piers. Damage includes both cracking and overall structural movement, and has ranged from minor to potentially structurally significant. These structures are critical and under intense public scrutiny, therefore the damage has been actively managed to allay public concern as much as to maintain serviceability and safety. Remedial work has included crack injection, surface coatings, externally reinforcing soffit surfaces with steel plates or carbon fibre sheets, and ventilation of enclosed cavities to facilitate evaporation of moisture from the concrete.

Concrete materials specialists have identified ASR damage in reservoirs after operations staff noted concrete deterioration and sought specialist advice. Damage thus reported ranges from minor cracking on walls and roofs, to expansion of prestressing anchor blocks, to cracking and expansion of a floor slab that needed to be replaced because of resulting leakage.

One of the more dramatic cases involved an airbase pavement constructed in the early 1960s using Waikato River sand. Initial remedial measures to counter ongoing ASR expansion of the concrete slab included reinstatement of slot drains and replacement of adjacent areas of heaved bitumen pavements. Twenty years after construction, new concrete work was laid abutting the affected area, restraining further expansion. This resulted in a spectacular case of heaving (Figure 14.64). Measured tensile strengths of cores from the affected slabs indicated a 30% reduction in strength due to the ASR. It was concluded that sufficient flexural strength remained for normal use of the slab, provided the expansion was restrained by the surrounding concrete. Nevertheless, all affected concrete was eventually replaced because of concern that spalls of concrete might be sucked into the jet engines of aircraft using the pavement. Chemical analysis of cores from the affected concrete indicated alkali contents higher than anticipated from the cement likely to have been used, suggesting additional alkalis may have been supplied from another source, although the source was not identified.

Routine inspections of three South Island bridges revealed extensive damage to precast piles, including widespread cracking, spalling and surface erosion. Subsequent investigation by concrete specialists revealed that the damage may have been caused by ASR, probably initiated by overheating of the elements during curing (Freitag et al., 2011). The combination of ASR and reinforcement corrosion meant extensive repair was required on two of these bridges to ensure they met loading requirements. On the third bridge the damage was limited to cracking, which appeared to have stabilised.

Minor damage tends to be limited to surfaces directly exposed to rain or run-off. For example, on bridges, cracking may occur only on abutment wingwalls, the ends of pier caps, or even deck soffits where runoff is not adequately directed off the structure. More extensive damage tends to affect the entire element. Thus on bridges, the entire abutment or pier cap may be affected.

Figure 14.64 This damage occurred in an airbase pavement slab when, 20 years after construction, new concrete was laid abutting one end. Five years later, the restrained expansive stress at this joint was released overnight.

To date, most inspection data are based on concrete elements above ground and readily accessible. ASR damage could be more common and/or severe in buried elements.

The key to identifying cases of ASR, and their significance, has been the early involvement of concrete materials specialists. Future identification of ASR on individual structures will depend on operations and maintenance personnel being able to recognise 'unusual' cracking, and on asset owners having processes by which such observations can be recorded and referred for further specialist investigation. Management of the risk of ASR in structures to be built in future will depend on lessons from such investigations being shared within the wider industry, rather than being treated as either purely operational or commercially sensitive.

14.16 NEW ZEALAND PRACTICES FOR MINIMISING ASR DAMAGE

14.16.1 Regulatory controls

Before 1974, the requirement for low alkali cement had to be specifically written into individual contracts. Engineers from PWD and subsequently MWD always specified the use of low alkali cement for major structures where potentially reactive aggregates were available.

In 1974, NZS 3122 introduced an option for the specifying authority to specify low alkali Portland cement where potentially reactive aggregates are to be used. As described in section 14.11.2, this option was retained in NZS 3122 until being removed in 2009, then re-introduced in 2014 as a default requirement from which purchasers or specifiers can choose to opt out.

In 1988, diagnosis of ASR as the cause of significant expansion on a motorway bridge focussed attention on ASR and led to the Cement & Concrete Association of New Zealand (CCANZ) convening a Working Party to formulate guidelines and a specification to minimise the risk of ASR causing structural damage. The first edition of CCANZ Technical Report 3 ('TR3') was published in 1991. It was based on The Concrete Society [UK] Technical Report 30 (see Chapter 6), but used appropriate New Zealand information in both the explanatory sections and the model specification clauses. Key features of the model specification clauses were:

- Potential aggregate reactivity could be assessed by petrographic examination to determine its composition, ASTM C289, field performance, or other data;
- They applied to all concrete mixes for use in permanent works in all locations unless otherwise defined;
- They required the total mass of acid soluble alkali from all sources in all concrete to be less than 2.5 kg/m^3 if potentially reactive aggregate is used. This limit was intended to restrict ASR damage to minor cracking and expansion only, not prevent ASR altogether;
- They allowed for the use of SCM as a mitigating measure;
- If ASR needed to be prevented altogether, non-reactive aggregates and a restricted concrete alkali content were needed.

In 2003, TR3 was revised to accommodate changes in regulations controlling the quality of ready mix concrete production in New Zealand (NZS 3104: 2003), as well as to introduce new standard test methodologies such as 'AMBT-80' and 'CPT-38', and the risk-based approach to managing ASR damage proposed by RILEM (see AAR-7 in Nixon & Sims, 2016) and already in use in Canada (CSA A23.1/A23.2, 2000, now 2014). Thus:

- Potential aggregate reactivity may be assessed by one or more of several means, including petrographic examination to determine its composition, mortar-bar or concrete testing, or long term field performance;
- For the majority of ready mixed concrete produced in New Zealand, defined by NZS 3104 as 'Normal Concrete', concrete alkali content is limited to no more than 2.5 kg/m^3 if the aggregate supplier and/or concrete producer cannot confirm that the proposed aggregates are non-reactive. This limit applies to all structures designed in accordance with the New Zealand Building Code, which requires the time to significant repair (defined as a proportion of capital value) to be no less than 50 years for structural elements. Under these requirements, cosmetic damage is considered acceptable;
- For concrete with specified compressive strength exceeding 50 MPa or with performance requirements in addition to compressive strength , defined by NZS 3104 as 'Special Concrete', it requires the risk of ASR on the individual structure in which it will be used to be assessed by considering the environmental

exposure conditions and the acceptable degree of damage. Risk may then be controlled via by choice of aggregate, limiting the concrete alkali content, or use of SCM;

- The risk assessment approach may also be employed when alkali contents of Normal Concrete would exceed 2.5 kg/m^3. Thus increased alkali contents are permissible if the risk to the individual structure, where the concrete will be used, is acceptable.

TR3 (2003) also acknowledged the existence of aggregates that have the potential to release alkali and recommended that they not be not be used together with potentially alkali-reactive aggregate.

In 2012, TR3 was updated to introduce provisions for the use of recycled material as concrete aggregate. It considers the potential reactivity of the recycled materials to be determined by that of the parent aggregate, and specifies that the default alkali contribution from adhering mortar be assumed to be 0.2 kg/m^3 unless the adhered mortar content and the alkali content of the parent concrete are known.

The 2012 update also acknowledges the observed alkali-reactivity of particular South Island alluvial aggregates, and stresses the need to monitor concrete temperatures during curing to ensure they do not exceed 70°C when using these or other rock types with a history of expansion related to DEF and/or ASR.

At the time of writing (2016), TR3's 2.5 kg/m^3 limit for concrete alkali content is being reviewed, along with its recommended test procedures (see 14.16.3).

New Zealand Standards for concrete design (NZS 3101), concrete production (NZS 3104) and aggregate supply (NZS 3121) cite TR3 as the means of minimising ASR damage in new structures. TR3 is also referenced in the standard for Portland cement (NZS 3122: 2009). These standards together form an acceptable solution to the New Zealand Building Code, thus TR3 has become a critical link in the durability design of concrete structures.

14.16.2 Specification and procurement practice in NZ construction industry

Until the introduction of CCANZ TR3 in 1991, the risk of ASR damage was managed via the low alkali content of cements produced in New Zealand, as required by government construction specifications. Since then, general practice has been to maintain concrete alkali levels at less than 2.5 kg/m^3.

With TR3 being cited in concrete design and production standards, the overall management of ASR has thus effectively passed from specifiers alone to the wider industry. The final responsibility for selection of concrete materials and mix designs to manage ASR lies with the concrete producers.

The indirect control of cement alkali contents via government construction specifications since the early 1940s was effective in limiting the number of significant cases of ASR damage in New Zealand, without requiring active input from specifiers, contractors, or asset owners. The disadvantage of this approach was that at times the risk of ASR being increased by other factors, such as unrecognised aggregate reactivity, alkalis from other sources, or elevated concrete temperature during curing, may have been overlooked because of a perception that 'all NZ cement is low alkali therefore ASR cannot occur'. Reliance on this belief, combined with a poor understanding of the issues, may have contributed to some apparently random incidences of ASR. For example, records indicate that in one area, potentially reactive coarse aggregate was identified, but the reactivity of a major sand source was overlooked. Thus on one bridge the prestressed beams were the only element affected, while on another bridge the precast beams were the only element not affected. Similarly, of two adjacent motorway structures built at the same time only one was significantly affected, and of several pavement slabs at the airbase site described in section 14.15.8, only those placed under one contract were affected. It may also have led to the inadvertent use of cement with elevated alkali content in conjunction with potentially reactive aggregates.

The most significant change affecting the management of ASR has been the devolvement of quality control in public infrastructure works from a central government agency to the designer and contractor. This came about with the end of major public infrastructure development projects in the 1980s, which resulted in the associated government departments becoming state-owned commercial enterprises, which were eventually privatised. In this environment, public-funded quality control of cement (and other construction materials), previously carried out by MWD and DSIR, was no longer possible. Market forces dictated that quality of materials became self-managed by the industries producing and using them.

Shifting the responsibility for control of ASR to the concrete manufacturer via the provisions of TR3 means that it is now in the hands of agencies and individuals who have a vested interest in the quality of concrete materials. Commercial interest is a strong motivator, and when combined with a good understanding of the technology, a sound and cost effective outcome can be achieved. Several features of the supply chain may, however, dilute the potential benefits.

Firstly, many designers are unfamiliar with ASR, therefore lack the confidence or willingness to take a risk-based approach to managing ASR damage on their projects. Thus some specifications unnecessarily require the use of non-reactive aggregates in addition to precautions for mitigating ASR in accordance with TR3.

Secondly, lack of demand, time, or sufficient project budget mean that potential aggregate reactivity is often assessed on the basis of generic geological composition and historical data rather than on routine examination or testing of the current aggregate supply. If the aggregate is assumed reactive, and appropriate precautions taken then the risk of ASR damage remains low, although extra cost may at times be incurred in taking unnecessary precautions. If the aggregate is incorrectly assumed non-reactive without evidence from testing or *in situ* performance, then the risk of ASR damage may be elevated, particularly if other risk factors are present. Designers on major projects are becoming more aware of these risks, and are specifying that concrete suppliers provide documented evidence of the reactivity of proposed aggregates claimed to be non-reactive.

A consequence of the lack of demand for ASR-related testing to demonstrate compliance is that independently accredited commercial testing services are not currently available for these tests within New Zealand.

Thirdly, under the terms of the New Zealand Standard for concrete production (NZS 3104), the concrete supplier is contractually responsible for ensuring that the combination of mix design and materials minimises the risk of ASR. Although TR3 offers the option of a risk based approach that considers the risk profile of individual structures, in practice a ready mixed concrete supplier rarely knows where a given batch of concrete will be used. Therefore applying the default option of maintaining concrete alkali contents under 2.5 kg/m^3 is the most widely used approach and sometimes places unnecessary limits on concrete strengths or selection of aggregates (see below).

14.16.3 Future trends and challenges

In 2015, about 8% of cement used in New Zealand was imported (including clinker). The proportion increased significantly in 2016 when one of the two major cement suppliers moved to an import model. The cement alkali limit introduced into NZS 3122 in 2014 empowers importers to demand low alkali cement from their suppliers. The 2014 amendment also introduced a new requirement for cement importers to demonstrate that the cement arriving in New Zealand conforms with NZS 3122, rather than relying on manufacturers' test results that may be based on other test methods. The onus will be on their customers to ensure they do provide such conformity confirmation.

Also over the same timeframe, concretes at the higher end of the permissible strength range for Normal Concrete (up to 50 MPa) have become more commonly used, for example to improve ease of placement, structural efficiency, or speed of construction. The cement and admixture contents of such concretes mean their alkali contents sometimes are likely to exceed the 2.5 kg/m^3 alkali limit. If an alkali reactive aggregate is being used, then alternative means of managing ASR must be applied, such as changing to a non-reactive aggregate or using SCM.

This has been a particular issue in the wider Auckland area, which is the largest centre of population, commerce and industry in New Zealand. It is also in Auckland that the highly reactive Waikato River rhyolite sand has been widely used. Most non-minor cases of ASR in New Zealand are associated with this sand. Once its potential reactivity was recognised, its use was largely discontinued because of concern about the ASR risk. In recent years, shortages of alternative sands in Auckland have led to TR3's default 2.5 kg/m^3 alkali limit being questioned by concrete suppliers wanting to use the readily available Waikato River sand. This critical aspect of ASR management is, therefore, under review, involving industry-led research commencing in 2017. The research will consider scientific, commercial, and practical aspects of managing the risk of ASR damage. The programme also seeks to demonstrate the value of current internationally-favoured AMBT-80, CPT-38, and CPT-60 test methodologies available at commercial testing laboratories, and thereby encourage their uptake by industry. Considerations will include the science behind the limit and the practicalities of managing the risk.

Whether the default concrete alkali limit is maintained or increased (and by how much) will partly depend on whether the wider industry is prepared to move from a relatively conservative, prescriptive approach, requiring little active consideration, or is willing to adopt a more performance-based approach based on risk evaluation. Previous history suggests that the extra effort and depth of understanding a risk based approach would demand of the wider industry may be a barrier to its immediate effectiveness. One way to overcome this may be for asset owners on major projects to introduce artificial financial incentives that allow a risk based approach to be put into practice and refined without negatively affecting the profitability of the contractors and concrete suppliers involved.

Although the means of identifying potentially alkali-reactive aggregate are generally well known, a practical and accurate method for identifying New Zealand aggregates that release significant amounts of alkali into the concrete pore solution is lacking. Thus the only way to manage this risk is to ban the use of suspect aggregates in combination with potentially alkali-reactive aggregates. This conservative approach may be challenged in future as the economics of aggregate supply change with time. Further research may then be required to correlate the understanding about Auckland basalts developed in the 1990s with more recent international studies such as those being undertaken under the auspices of RILEM technical committees studying ASR.

Above all, prevention of significant ASR damage in future structures will depend on maintaining an awareness of the risk throughout the wider industry, so that regulatory controls are kept up to date and applied, and that informed decisions affecting ASR are made on all projects irrespective of their value.

14.17 SUMMARY

New Zealand experience since the early 1940s has shown that successful minimisation of ASR damage relies on an awareness of the risks and appropriate regulatory control to manage them.

The first edition of this chapter (in Swamy, 1992) concluded that ASR can only be minimised, not eliminated; and that to prevent the problem requires a combination of thorough knowledge of the local aggregates and their potential reactivity as measured in the laboratory and observed in the field, combined with effective controls. Simply limiting cement alkali contents since the 1940s has successfully limited the incidence and severity of ASR damage in New Zealand. This was the basis of ASR management when the chapter was first written.

Since then, New Zealand experience has also shown that awareness of the issues must be actively maintained to ensure designers, suppliers and contractors do not overlook the risks associated with new materials or practices. Such awareness is also necessary to ensure controls are updated regularly to reflect changes in understanding, materials supply, industry regulations, procurement practices, and the wider economic and regulatory environment.

New Zealand experience has shown that ASR damage can be successfully limited without extensive compliance testing of aggregates, provided that the methods of control are conservative and followed. It has also demonstrated, however, that reliance on historical data and research, rather than routine compliance testing, can limit the availability of the experienced testing and research services needed to inform further development of the controls to meet industry's practical and economic needs.

The key to identifying cases of ASR, and their significance, has been the early involvement of concrete materials specialists. Future identification of ASR on individual structures will depend on operations and maintenance personnel being able to recognise 'unusual' cracking, and on asset owners having processes by which such observations can be recorded and referred for further specialist investigation. Management of the risk of ASR in structures to be built in future will depend on lessons from such 'operational' investigations being shared within the wider industry, as they have been in the past.

ACKNOWLEDGEMENTS

The author of this New Zealand part of the chapter acknowledges the expertise and experience of D A St John, author of the first editions of CCANZ TR3 and this chapter; the various organisations that have funded research and investigations into ASR in New Zealand; and the Cement & Concrete Association of New Zealand for co-ordinating the industry-wide Working Parties that have developed and maintained the recommended guidelines for minimising ASR damage in New Zealand structures.

REFERENCES

Abbott, J.A., & St John, D.A. (1981) *The preparation of large area thin sections in concrete.* Chemistry Division Report CD. 2311.

Abdullah, S., Shayan, A., & Al-Mahaidi, R. (2013) Assessing the mechanical properties of concrete due to alkali silica reaction. *Int J Civ Eng.*, 4 (1), January–February 2013, pp. 190–204.

Abdullah, S., Shayan, A., & Al-Mahaidi, R. (2012a) Strain monitoring of CFRP wrapped RC columns damaged by alkali aggregate reaction. In: *Proceedings of the 14th ICAAR, Austin, Texas*, USA, 20–25 May 2012.

Abdullah, S., Al-Mahaidi, R., & Shayan, A. (2012b) Experimental investigation of CFRP confined columns damaged by alkali aggregate reaction. *International Journal of Integrated Engineering.*, 4 (2), 49–52.

Ahmed, T., Burley, E., Rigden, S., & Abu-Tair, A.I. (2003) The effect of alkali reactivity on the mechanical properties of concrete. *Constr Build Mater.*, 17, 123–144.

Alderman, A.R. (1943) *A review of evidence concerning expansive reaction between aggregate and cement in concrete. Council for Scientific and Industrial Research*, Bulletin No. 161, Melbourne, Australia, 19p.

Alderman, A.R., Gaskin, A.J., Jones, R.H., & Vivian, H.E. (1947) *Studies in cement aggregate reactions: Part I. Australian aggregates and cements. Council for Scientific and Industrial Research*, Bulletin No. 229, Melbourne, Australia, pp. 7–46.

Alexander, K.M., & Davis, C.E.S. (1960) Effect of alkali on the strength of potland cement paste. *Aust J Appl Sci.*, 11, 146–156.

Alexander, K.M. (1954) Pozzolanas and their use in concretes. *Constructional Review.*, 27 (6), pp. 22–28.

Ali, M.G., & Rasheeduzzafar. (1993) Cathodic protection current accelerates alkali-silica reaction. *ACI Mater J.*, 90 (May–June), 247–252.

Alkadhimi, T.K.H., & Banfill, P.F.G. (1996) The effects of electrochemical re-alkalisation on alkali-silica expansion in concrete. In: *Proceedings of the 10th International AAR Conference*, Melbourne, Australia, pp. 637–644.

Al-Mahaidi, R., Shayan, A., & Gamage, K. (2008) Strength assessment of prestressed concrete bridge deck planks damaged by AAR. In: *Proceedings of the 23rd ARRB Conference*, Adelaide, Australia, 30 July–1 August 2008.

AS 1012.9 (2014) *Methods of testing concrete, Method 9: Compressive strength tests – Concrete, mortar and grout specimens*, SAI Global, Sydney, NSW, Australia.

AS 1012.17 (1976) *Methods of testing concrete, Method 17: Methods for the determination of the static chord modulus of elasticity and Poisson's ratio of concrete specimens*, SAI Global, Sydney, NSW, Australia.

AS 1141.38 (1975) *Methods for sampling and texting aggregates, Method 38: Alkali aggregate reactivity – Mortar bar method*. SAI Global, Sydney, NSW, Australia (discontinued since 1994).

AS 1141.39 (1975) *Methods for sampling and testing aggregates, Method 39: Alkali aggregate reactivity – Quick chemical test*. SAI Global, Sydney, NSW, Australia (discontinued since 1994).

AS 1141.60.1 (2014) *Method for sampling and testing aggregates, Method 60.1: Potential alkali aggregate reactivity – Accelerated mortar bar method*. SAI Global, Sydney, NSW, Australia.

AS 1141.60.2 (2014) *Method for sampling and testing aggregates, Method 60.2: Potential alkali-silica reactivity – Concrete prism method*. SAI Global, Sydney, NSW, Australia.

AS 1141.65 (2008) *Methods for sampling and testing aggregates, Method 65: Alkali aggregate reactivity – Qualitative petrological screening for potential alkali-silica reaction*, SAI Global, Sydney, NSW, Australia.

AS 3600 (2010) *Concrete structures*, SAI Global, Sydney, NSW, Australia.

AS 5100.5 (2004) *Bridge design: Concrete*, SAI Global, Sydney, Australia.

ASTM C1260 (2014) *Standard test method for potential alkali-reactivity of aggregates (mortar-bar method)*. American Society for Testing and Materials, West Conshohocken, USA.

ASTM C1293 (2008) *Standard method for determination of length change of concrete due to alkali-silica reactivity*. American Society for Testing and Materials, West Conshohocken, USA.

ASTM C1567 (2013) *Standard test method for determining the potential alkali-silica reactivity of combinations of cementitious materials and aggregate (mortar-bar method)*. American Society for Testing and Materials, West Conshohocken, USA.

ASTM C227 (2010) *Standard test method for potential alkali-silica reactivity of cement-aggregate combinations (mortar-bar method)*. American Society for Testing and Materials, West Conshohocken, USA.

ASTM C289 (2007, but now withdrawn) *Standard test method for potential alkali-silica reactivity of aggregates (chemical method)*. American Society for Testing and Materials, West Conshohocken, USA.

Bennett, I.C., & Vivian, H.E. (1955) Studies in cement aggregate reaction: Part XXII. The effect of fine-ground opaline material on mortar expansion. *Aust J Appl Sci.*, 6 (1), 88–93.

Berube, M.A., & Frenette, J. (1994) Testing concrete for AAR in NaOH and NaCl solutions at 38°C and 80°C. *J Cement Concrete Comp.*, 16 (3), 189–198.

Berube, M.A., & Tremblay, C. (2004) Chemistry of pore solution expressed under high pressure-Influence of various parameters and comparison with the hot water extraction method. In: *Proceedings of the 12th ICAAR*, Beijing, China, pp. 833–842.

Blaikie, N.K., Bowling, A.J., & Carse, A. (1996) The assessment and management of alkali silica reaction in the Gordon River Power Development intake tower. In: *Proceedings of the 10th International AAR Conference*, Melbourne, Australia, August 1996, pp. 500–507.

British Cement Association (1992) *The Diagnosis of Alkali-Silica Reaction, Report of a Working Party*, 2nd edition, Ref. 45.042, BCA, Wexham Springs, Slough, UK, 44p.

Carse, A., & Dux, P.F. (1990) *Development of an accelerated test on concrete prisms to determine their potential for alkali-silica reaction*, C&CR, 230, pp. 869–874.

Carse, A., & Dux, P. (2004) An assessment of the risk of using a cathodic protection system on reinforced concrete piles where very reactive aggregates have been used. In: *Proceedings of the 12th ICAAR*, Beijing, China, pp. 855–864.

Carse, A. (1988) Field survey of concrete structures suffering alkali-silica reaction distress. In: *Proceedings of the 2nd Australia/Japan Workshop on durability of reinforced concrete structures*, 28–30 November 1988, CSIRO, Highett, Victoria, pp. 2.67–2.82.

Carse, A. (1993) The identification of ASR in the concrete cooling tower infrastructure of the Tarong Power Station. *Constr Build Mater.*, 7 (2), 117–119.

Carse, A. (1996) The asset management of a long bridge structure affected by alkali-silica reaction. In: *Proceedings of the 10th IAARC*, Melbourne, Australia, pp. 1025–1032.

Cement & Concrete Association of New Zealand (2012) *Alkali Silica Reaction. Minimising the Risk of Damage to Concrete - Guidance Notes and Recommended Practice*. CCANZ Technical Report TR3. 78p.

Cement and Concrete Association of New Zealand (2011) *Best Practice Guide for the use of Recycled Aggregates in New Concrete* (CCANZ Technical Report: TR 14)

Clayton, N. (1999) *Structural Implications of Alkali-Silica Reaction: Effect of Natural Exposure and Freeze-thaw*. BRE report published by Construction Research Communications, London, UK, 23p.

Cole, W.F., Lancucki, C.J., & Sandy, M.J. (1981) Products formed in an aged concrete, *C&CR*, 11 (3), 443–454.

Concrete Society (UK) (1987) *Alkali Silica Reaction: Minimising the Risk of Damage to Concrete: Guidance notes and Model Clauses for Specifications* (3rd edition). (Concrete Society Technical Report no. 30). Berkshire: Author.

CSA A23.1-00/A23.2-00 (2000). *Concrete materials and methods of concrete construction/ Methods of Test for concrete*. Canadian Standards Association. Mississauga, Ont., Canada.

CSA A23.2-14A (2014) *Test methods and standard practices for concrete, A23.2-14A: Potential expansivity of aggregates (Procedure for length change due to alkali-aggregate reactions in concrete prisms at 38°C)*. Canadian Standards Association. Mississauga, Ont., Canada, pp. 350–362.

Davies, G., & Oberholster, R.E. (1987) Use of the NBRI accelerated test to evaluate the effectiveness of mineral admixtures in preventing the alkali-silica reaction. *Cement Concrete Res.*, 17 (1), 97–107.

Davies, M.J.S., Grace, W.R., Green, W.K., & Collins, F.G. (1996) Assessment and management of a marine structure affected by ASR. In: *Proceedings of the 10th IAARC*, Melbourne, Australia, pp. 1018–1024.

Davis, C.E.S. (1951) Studies in cement – aggregate reaction: Part XVIII. The effect of soda content and of cooling rate of Portland cement clinker on its reaction with opal in mortar. *Aust J Appl Sci.*, 2 (1), 123–131.

Davis, C.E.S. (1957) Studies in cement aggregate reactions: Part XXV. Comparison of the expansion of mortar and concrete. *Aust J Appl Sci.*, 8 (3), 222–234.

Davis, C.E.S. (1958) Studies in cement aggregate reactions: Part XXVI. Comparison of the effect of soda and potash on expansion. *Aust. J. Appl. Sci.*, 9 (1), 52–62.

Duncan, M.A.G., Swenson, E.G., Gillot, J.E., & Foran, M.R. (1973) *Alkali – aggregate reaction in Nova Scotia. Pts Ito IV. Cement Caner Res.*, 3, 55–69, 119–128, 233–244, 521–535.

Famy, C. (1999) *Expansion of heat-cured mortars*. Ph.D. Thesis, Imperial College, University of London, 256p.

Fournier, B., Chevrier, R., De Grosbois, M., Lisella, R., & Foliard, K. (2004) The accelerated concrete prism test (60°C): Variability of the test method and proposed expansion limits. In: *Proceedings of the 12th IAARC*, Beijing, China, pp. 314–323.

Freitag, S.A. (1990) *Alkali aggregate reactivity of concrete sands from the Rangitikei and Waikato Rivers*. Works Consultancy Services Central Laboratories report 90-B4207.

Freitag, S.A. (1998) *Alkali aggregate reaction: ASTM C1260 and New Zealand greywacke* (Central Laboratories Report: 98–524155.04). Wellington: Opus International Consultants.

Freitag, S.A. (2002) *Alkali aggregate reaction: research by Opus International Consultants, Central Laboratories 1994–2002* (Central Laboratories Report: 02-520913.00 2L). Wellington: Opus International Consultants.

Freitag, S.A., Bruce, S.M., & Shayan, A. (2011) *Concrete pile durability in South Island bridges.* New Zealand Transport Agency research report 454.

Freitag, S.A., St John, D.A., & Goguel, R. (2000) ASTM C1260 and the alkali reactivity of New Zealand greywackes. In: Bérubé, M.A., Fournier, B., & Durand, B., (eds.) *Alkali-aggregate reaction in concrete: Proceedings of the 11th international conference on alkali-aggregate reaction in concrete*, Quebec: ICAAR, June 11–16, pp. 305–313.

Gaskin, A.J. (1947) *Studies in cement–aggregate reaction: Part VI. The effect of carbon dioxide. Council for Scientific and Industrial Research, Bulletin No. 229*, Melbourne, Australia, pp. 78–84.

Gaskin, A.J., Jones, R.H., & Vivian, H.E. (1955) Studies in cement-aggregate reaction: Part XXI. The reactivity of various forms of silica in relation to the expansion of mortar bars. *Aust J Appl Sci.*, 6 (1), 78–87.

Gogte, B.S. (1973) An evaluation of some common Indian rocks with special reference to alkali aggregate reaction. *Eng Geol.*, 7, 135–153.

Goguel, R. (1995) Alkali release by volcanic aggregates. *Cement Concrete Res.*, 25 (4), 841–850.

Goguel, R. (1996) Selective dissolution techniques in AAR investigation : application to an example of failed concrete. In: Shayan, A. (ed.) *Alkali-Aggregate Reaction in Concrete : Proceedings of the 10th International Conference on Alkali-Aggregate Reaction in Concrete*, Melbourne, Vic.: AARC Australia, pp. 783–790.

Goguel, R., & Milestone, N.B. (1997) Auckland basalts as a source of alkali in concrete. In: Malhotra, V.M. (ed.), *CANMET/ACI International Symposium on Advances in Concrete Technology*, Farmington Hills, Mich.: American Concrete Institute, pp. 783–790.

Goguel, R., & St John, D.A. (1993a) Chemical identification of Portland cements in New Zealand concretes: I – Characteristic differences among New Zealand cements in minor and trace element chemistry. *Cement Concrete Res.*, 23 (1), 56–69.

Goguel, R., & St John, D.A. (1993b) Chemical identification of Portland cements in New Zealand concretes: II – the Ca-Sr-Mn plot in cement identification and the effects of aggregates. *Cement Concrete Res.*, 23 (2), 283–293.

Handbook HB79, (2015) *Alkali Aggregate Reaction – Guidelines on Minimising the Risk of Damage to Concrete Structures in Australia*, Standards Australia, Sydney, 80p.

Hobbs, D.W. (1988) *Alkali-silica reaction in concrete*. Thomas Telford Ltd , London, pp183.

Hutton, C.O. (1945) *The problem of reaction between aggregate materials and high alkali cements. NZ J Sci Tech.*, 26B, 191–200.

Ideker, J.H, Folliard, K.J, Fournier, B., & Thomas, M.D.A. (2006) The role of "non-reactive" aggregates in the accelerated (60°C) concrete prism test. *Marc-Andre Berube Symposium on Alkali-Aggregate Reactivity in Concrete*, held in conjunction with the 8th CANMET International Conference on Recent advances in Concrete technology, Montreal, Canada, 31 May–3 June 2006. pp. 45–70.

Idorn, G.M. (1996) Systematic ASR expertise – Australian Research 1940s to 1958. In: *Proceedings of the 10th IAARC*, Melbourne, Australia, pp. 15–26.

Institution of Structural Engineers, UK. (1992) *Structural effects of alkali-silica reaction* (D.K. Doran, Chairman of Task Group), July, 45p.

Jones, A.E.K., & Clack, L.A. (1988) The effect of ASR on the properties of concrete and the implications for assessment. *Eng Struct.*, 20 (9), 785–791.

Jones, R.H., & Vivian, H.E. (1950) *Studies in cement Aggregate reaction: Part IX. Some observations on mortar containing reactive aggregate, CSIR Bulletin 256*, Melbourne, Australia, pp. 7–12.

Shayan, A., & Morris, H. (2002) Investigation of cracking in precast, prestressed deck planks in two bridges and rehabilitation options. *ACI Mater J.*, (May), 165–172.

Shayan, A., & Morris, H. (2005) Combined deterioration problems in a coastal bridge in NSW, Australia. *Asian J of Civil Eng (Building & Housing)*, 6 (6), 477–493.

Shayan, A., Diggins, R., Ritchie, D.F., & Westgate, P. (1987) Evaluation of Western Australian aggregates for alkali-reactivity in concrete. In: *Concrete Alkali Aggregate Reactions* (P.E. Grattan-Bellew, ed.), Noyes Publications, N.J., USA, pp. 247–252.

Shayan, A., Diggins, R., & Ivanusec, I. (1996) Effectiveness of fly ash in preventing deleterious expansion due to alkali-aggregate reaction in normal and steam-cured concrete. *Cement Concrete Res.*, 26 (1), 153–164.

Shayan, A, Xu, A., & Pritchard, R. (2011) Behaviour of AAR-affected concrete piles subjected to CP currents. In: *Proceedings of the 18th International Corrosion Conference*, Perth WA, 20–24 November 2011, 14p.

Shayan, A. (1987) Alkali-aggregate reaction in Australian concrete structures. In: *Proceedings of the 13th Biennial Conference of the Concrete Institute of Australia, 1987*, Brisbane, Australia, Concrete Institute of Australia, Sydney, NSW, 5p.

Shayan, A. (1988a) Alkali aggregate reaction in a 60-year-old dam in Australia. *Int J Cement Composites and Lightweight Concrete* , 10 (4), pp. 259–266.

Shayan, A. (1988b) Developments in accelerated testing of aggregates for alkali-aggregate reaction. In: *Proceedings of the Second Australia/Japan workshop on durability of reinforced concrete structures*. 28–30 November 1988, Melbourne, pp. 2.47–2.56.

Shayan, A. (1989) Experiments with accelerated tests for predicting alkali-aggregate reaction. In: *Proceedings of the 8th International Conference on AAR*, Kyoto, pp. 321–326.

Shayan, A. (1989a) Re-examination of AAR in an old concrete. *Cem Conc Res.*, 19, 434–442.

Shayan, A. (1990) Use of fly ash and blended slag cement in prevention of alkali-aggregate reaction in concrete. In: *Proceedings of the concrete for the nineties conference, Leura, NSW*, Concrete Institute of Australia, Sydney, NSW, 17p.

Shayan, A. (1992a) Prediction of alkali-reactivity of some Australian aggregates and correlation with field performance. *ACI Mater J*, 1, 13–23.

Shayan, A. (1992b) The 'pessimum' effect in an accelerated mortar bar test using 1M NaOH at 80°C. *Cement Concrete Comp.*, 14, 249–255.

Shayan, A. (1993a) Reply to discussion on Microscopic Features of Cracked and Uncracked Concrete Railway Sleepers by M.A. Adams. *ACI Materials J*, 89, May–June 1993, 284–287.

Shayan, A. (1993b) Alkali reactivity of deformed granitic rocks: A Case Study. *Cement Concrete Res.*, 23, 1229–1236.

Shayan, A. (1994a) Alkali-aggregate reaction: A bridge management problem for Road Authorities. In: *Proceedings of the 17th ARRB Conference. Part 4*, August 94, Gold Coast, pp. 71–85.

Shayan, A. (1994b) *An illustrated guide to the identification of alkali-aggregate reaction in concrete structures*. CSIRO Tech. Report, TR94/2.

Shayan, A. (1995a) Developments in testing for AAR in Australia. In: *Proceedings of the CANMET/ACI International Workshop on Alkali-Aggregate Reaction in Concrete*, 1–4 October 1995, Dartmouth, Nova Scotia, Canada, pp. 139–152.

Shayan, A. (1995b) Behaviour of precast, prestressed concrete railway sleepers affected by AAR. Real World Concrete. In: *Proceedings of the Swamy Symposium, at 5th Internatinal CANMET/ACI Conference on Fly ash, Silica Fume, Slag and Natural Pozzolans in Concrete*. Milwaukee, Wisconsin, 4–9 June, 1995, pp. 35–56.

Shayan, A. (1997) Where is AAR heading after the 10th International Conference. *Cement Concrete Comp (UK).*, 19, 441–449.

Shayan, A. (1998) Effects of NaOH and NaCl solutions and temperature on the behaviour of specimens subjected to accelerated AAR tests. *Cement Concrete Res.*, 28 (1), 25–31.

Shayan, A. (1999) Characterisation of AAR-affected concrete from a dam structure for rehabilitation purposes. In: *Proceedings of the International Conference On Infrastructure Regeneration and Rehabilitation*, June 1999, Sheffield University, UK, pp. 777–787.

Shayan, A. (2000) Combined effects of alkali-aggregate reaction (AAR) and cathodic protection currents in reinforced concrete. In: *Proceedings of the 11th International AAR Conference. Quebec*, June 2000, Quebec City, Canada, pp. 229–238.

Shayan, A. (2001) Validity of accelerated mortar bar test methods for slowly reactive aggregates – Comparison of test results with field evidence. *Concrete in Australia*, June–August 2001, pp. 24–26.

Shayan, A. (2003) AAR in Australia and recent developments. *International seminar on road construction materials*, 2003, Kanazawa, Japan.

Shayan, A. (2004) Alkali-aggregate reaction and basalt aggregates. *International Conference on Alkali-Aggregate Reaction in Concrete, 12th, 2004*, Beijing, China, pp. 1130–1135.

Shayan, A. (2011) Aggregate selection for durability of concrete structures. *Constr MaterJ., 164*, 111–121.

Shayan, A. (2015) The Current Status of AAR in Australia and Mitigation Measures. *Concrete in Aust., 41* (2), 44–51.

Shayan, A., & Ferguson, J.A. (1996) Reactive quartz gravel from Eastern Victoria. In: *Proceedings of the 10th International AAR Conference*, Melbourne, Australia, 18–23 August 1996, pp. 703–710.

Shayan, A., & Ivanusec, I. (1989) Influence of NaOH on mechanical properties of cement paste and mortar with and without reactive aggregate. In: *Proceedings of the 8th International Conference on AAR*, Kyoto, pp. 715–720.

Shayan, A., & Ivanusec, I. (1996) An experimental verification of the association of alkali-aggregate reaction and delayed ettringite formation. *Cement Concrete Comp., 18*, 161–170.

Shayan, A., & Lancucki, C.J. (1987) Alkali-aggregate reaction in the Causeway bridge, Perth, Western Australia. In: *Concrete Alkali Aggregate Reactions* (P.E. Grattan-Bellew, editor), Noyes Publications, N.J., USA, pp. 392–397.

Shayan, A., & Morris, H. (2001) A comparison of RTA T363 and ASTM C-1260 accelerated mortar bar test methods for detecting reactive aggregates. *C&CR, 31*, 655–663.

Shayan, A., & Morris, H. (2003) Durability investigation of Deep Creek Bridge, Northern NSW. In: *Proceedings of the 21st ARRB Conference. Australia*, May 2003, Cairns, Australia, pp. 18–23.

Shayan, A., & Morris, H. (2006) A case study: Deterioration of precast, prestressed concrete piles in marine environment. *Concrete Plant & Precast Technology, BFT* (Germany), 72 (1), 38–47.

Shayan, A., & Quick, G.W. (1988) An alkali-reactive basalt from Queensland, Australia. *Int J Cement Composites & Lightweight Concrete*, 10 (4), 209–214.

Shayan, A., & Quick, G.W. (1989) Microstructure and composition of AAR products in conventional standard and new accelerated testing. In: *Proceedings of the 8th International. Conference on AAR*, Kyoto, pp. 475–482.

Shayan, A., & Quick, G.W. (1992a) Microscopic features of cracked and uncracked prestressed concrete railway sleepers. *ACI Mat J.*, (July-August), 348–361.

Shayan, A., & Quick, G.W. (1992b) Relative importance of deleterious reactions in concrete: formation of AAR products and ettringite. *Adv in Cement Res., 4* (16), 149–157.

Shayan, A., & Quick, G.W. (1994) Alkali-aggregate reaction in concrete sleepers from Finland. In: *Proceedings of the 16th ICMA Conference*, Richmond, VA., USA, pp. 69–79.

Shayan, A., & Thomas, M. (2014) Influence of fly ash in suppressing AAR expansion in a dam wall. In: *Proceedings of the 36th International Cement Microscopy Association Conference*, Milan, Italy, April 2014, 18p.

Shayan, A., & Xu, A. (2001) Some electrochemical effects of CP on concrete cracked due to AAR and contaminated by chlorideions. Corrosion & Prevention-2001, In: *Proceedings of the*

Australian Corrosion Association Conference, Newcastle, NSW, 19–21 November 2001, paper No. 65, 10p.

Shayan, A., & Xu, A. (2002) Value-added utilisation of waste glass in concrete. In: *Proceedings of the International IABSE Symposium*, Melbourne University, 11–13 September 2002, 11p.

Shayan, A. (2007) Field evidence for inability of ASTM C 1260 limits to detect slowly reactive Australian aggregates. *Aust J Civl Eng.*, *3* (1), 13–26.

Shayan, A., Al-Mahaidi, R., & Xu, A. (2008) Durability and strength assessment of AAR-affected bridge deck planks. In: *Proceedings of the 13th IAARC*, Trondheim, Norway, 16–20 June 2008, pp. 422–432.

Shayan, A., Camilleri, I., & Sadeque, A. (2008) The Structural Analysis, Material Testing and Assessment of the Como Bridge submerged piling. In: *Proceedings of the 23rd International ARRB Conference*, Adelaide, Australia, July 2008.

Shayan, A., Diggins, R.G., Ivanusec, I., & Westgate, P.L. (1988) Accelerated testing of some Australian and overseas aggregate for alkali aggregate reactivity. *Cement Concrete Res.*, *18* (6), 843–851.

Shayan, A., Green, W.K., & Collins, F.G. (1996) Alkali-aggregate reaction in Australia. In: *Proceedings of the 10th International AAR Conference*, Melbourne, Australia, 18–23 August 1996, pp. 85–92.

Shayan, A., Ivanusec, I., & Diggins, R. (1992) Comparison between two accelerated methods for determining alkali reactivity potential of aggregates. In: *Proceedings of the 9th International AAR Conference, London*, July 1992, pp. 953–957.

Shayan, A., Janetzki, D., & Witt, P. (1998) In: *Proceedings of the Australasian Corrosion Association Conference, Corrosion and Prevention 98*, pp. 188–295.

Shayan, A. & Katayama, T. (2016) Investigation of reactivity of a dolomitic aggregate from Australia. In: *Proceedings of the 15th ICAAR*, July 2016, Sao Paulo, Brazil, Paper No. 221, 10pp.

Shayan, A., Landon-Jones, I., & Nelson, P. (1997) Case study of fly ash concrete in Tallowa Dam, containing alkali-reactive aggregate. In: *Proceedings of the 4th CANMET/ACI Int. Conference on Durability of Concrete*, Sydney, Australia, August 1997, Supplementary Papers, pp. 281–293.

Shayan, A., Morris, H., & Doolan, T. (1998) Investigation Of Cracking Of Precast Concrete Bridge Piles Submerged In Tidal River Water In Northern NSW. In: *Proceedings of the 19th ARRB Conference*, November 1998, Sydney, Australia, pp. 118–141.

Shayan, A., Quick, G.W., & Lancucki, C.J. (1993) Morphological, mineralogical and chemical features of steam cured concretes containing densified silica fume and various alkali levels. *Adv Cem Res.*, *5* (20), 151–162.

Shayan, A., Quick, G.W., Lancucki, C.J., & Way, S.J. (1992) Investigation of greywacke aggregates for alkali-aggregate reactivity. In: *Proceedings of the 9th International AAR Conference*, London, July 1992, pp. 958–979.

Shayan, A., Wark, R.E., & Moulds, A. (2000) Diagnosis of AAR in Canning Dam, characterisation of the affected concrete and rehabilitation of the structure. In: *Proceedings of the 11th International AAR Conference*, June 2000, Quebec City, Canada, pp. 1383–1392.

Shayan, A., Xu, A., & Morris, H. (2008) Comparative study of the concrete prism test (CPT 60 °C, 100% RH) and other accelerated tests, In: *Proceedings of the 13th IAARC Conference*, Trondheim, Norway, 16–20 June 2008, pp. 412–421.

Shayan, A., & Xu, A. (2012) Comparison between in-situ expansion measurements on AAR-affected beams, drilled cores and large sawn sections. In: *Proceedings of the 14th ICAAR*, Austin, Texas, USA, 20–25 May 2012.

Shayan, A., Xu, A., & Andrews-Phaedonos, F. (2015) Implications of AAR for three bridges. In: Proceedings of the Combined 69th Rilem and 27th CIA *International Conferences on Concrete Durability (Concrete 2015)*, Melbourne, Australia, 31 August–3 September 2015, 12p.

Shayan, A., Xu, A., & Pritchard, R. (2012) Influence of CP impressed currents on AAR expansion of concrete containing reactive aggregates. In: *Proceedings of the 14th ICAAR*, Austin, Texas, USA, 20–25 May 2012.

Shayan, A., Xu, A., & Andrews-Phaedonos, F. (2003) Development of a performance measure for durability of concrete bridges. In: *Proceedings Of the 21st Biennial Conference of the Concrete Institute of Australia*, 18–21 July 2003, Brisbane, Australia, pp. 739–757.

Shayan, A., Xu, A., & Olasiman, R. (2008) Keynote Paper: Factors affecting the expansion and cracking of model bridge piles in seawater, and the effects of mechanical confinement. In: *Proceedings of the 13th IAARC, Trondheim, Norway*, 16–20 June 2008, pp. 1196–1202.

Shayan, A., Xu, A., & Tagnit-Hamou, A. (2004) Effects of cement composition and temperature of curing on AAR and DEF expansion in steam-cured concrete. In: *Proceedings Of the 12th, International Conference on Alkali-Aggregate Reaction in Concrete, 2004*, Beijing, China, pp. 773–788.

Sims, I., & Nixon, P.J. (2006) Assessment of aggregates for alkali-aggregate reactivity potential: Rilem international recommendations, *Marc-Andre Berube symposium on alkali-aggregate reactivity in concrete, held in conjunction with the 8th CANMET international conference on recent advances in concrete technology*, Montreal, Canada, pp. 71–91.

Smith, L.M. (1977) *Pozzolanic materials and NZS 3123. Trans. N.Z. Inst. Eng.*, 4 (1) 37–53.

St John D.A. (1972) Confirmation of Alkali Aggregate Reaction in the Te Henui Bridge. *NZ J Sci.*, *18*, 405–412.

St John D.A. (1988) *Alkali – aggregate reaction and synopsis of data on AAR. N.Z. Concr. Constr.*, 32, April 1988 pp. 7–14 and May 1988 pp. 3–11.

St John, D.A. (Ed.) (1988a) *Alkali-aggregate studies in New Zealand* (Chemistry Division Report No. C.D. 2390) p. 186. Lower Hutt: New Zealand Department of Scientific and Industrial Research.

St John, D.A., Poole, A.B., & Sims, I. (1998) *Concrete petrography: A handbook of investigative techniques*, 1st edition, Arnold, London (see Poole & Sims, 2016, for 2nd edition).

St John, D.A., Singers, W.E., & Aldous, K. (1988b) *Quarries and Minerals Database: Users' Manual.* Chemistry Division Report No. C.D. 2388.

Stanton, T.E. (1940) The influence of cement and aggregate on concrete expansion. *Eng News-Rec, 1* (February), 59.

Stanton, T.E. (1940a). Expansion of concrete through reaction between cement and aggregate. *Amer Soc Civil Eng Papers.*, December, 1781–1811.

Stanton, T.E. (1943) Studies to develop an accelerated test procedure for the detection of adversely reactive cement – aggregate combinations. In: *Proceedings* of the ASTM, Vol. 43, pp. 875–904.

Stark, D. (1992) Lithium salt admixtures- An alternative method to prevent expansive alkali silica reactivity. In: *Proceedings of the 9th ICAAR, London*, July 1992, pp. 1026–1034

Stark, J., & Seyfarth, K. (1999) Ettringite formation in hardened concrete and resulting destruction. In: Erlin, B. (ed.) *ACI international* SP-177, pp. 125–140.

Stokes, D.B. (1996) Use of lithium to combat alkali silica reactivity. In: *Proceedings of the. 10th ICAAR*, Melbourne, Australia, pp. 862–867.

Stokes, D.B., Thomas, M.D.A., & Shashiprakash, S.G. (2000) Development of lithium-based material for decreasing ASR-induced expansion in hardened concrete. In: *Proceedings of the 11th ICAAR, Quebec City*, Canada, pp. 1079–1087.

Swamy, R.N., & Al-Asali, M.M. (1988) Engineering properties of concrete affected by alkali silica reaction. *ACI Materials Journal*, 85 (5), 367–374.

Swamy, R.N. (Ed.) (1992a) *The alkali-silica reaction in concrete*. Blackie and Son Ltd. (publisher), New York, 333p.

Swamy, R.N. (Ed.) (1992b) *The alkali-silica reaction in concrete*, 1st edition, Blackie, Glasgow & London, 352p.

Thaulow. N, Holm, J., & Anderson, K.T. (1988) Petrographic examination and chemical analysis of the Lucinda Jetty prestressed concrete roadway. *In: Proceedings of the 8th ICAAR* (Eds. K. Okada *et al.*), Kyoto, Japan, pp. 573–581.

Thomas, M. (2011) The effect of supplementary cementing materials on alkali-silica reaction: A review. *Cement & Concrete Res., 41*, 1224–1231.

Thomas, M., Hooton, R.D., Rogers, C., & Fournier, B. (2012) 50 Years Old and Still Going Strong. *Concrete International*, January 2012, pp. 35–40.

Thomas, M.D.A. & Innis, F.A. (1999) Use of accelerated mortar bar test for evaluating the efficacy of mineral admixtures for controlling expansion due to alkali-silica reaction. *Cement, Concrete & Aggregates, 21* (2), 157–164.

Thomas, M.D.A. (1996a) *Review of the effect of fly ash and slag on alkali-aggregate reaction in concrete*. Building Research establishment, BR314, Garston, Watford, England, 117p.

Thomas, M.D.A. (1996b) Field studies of fly ash concrete structures containing reactive aggregates. *Mag Concrete Res., 48* (177), 265–279.

Thomas, M.D.A., Fournier, B., Folliard, K., Shehata, M., Ideker, J., & Rogers, C.A. (2007) Performance limits for evaluating supplementary cementing materials using the accelerated mortar bar test. *ACI MaterJ, 104* (2), 115–122.

Tomita, M., Miyagawa, T., & Nakano, K. (1989) Basic study for diagnosis of concrete affected by ASR using drilled concrete cores. In: *Proceedings of the 8th ICAAR Conference*, Kyoto, Japan, pp. 779–784.

Torii, K., Ishii, K., & Kawamura, M. (1997) Influence of cathodic protection on expansion and structural behaviour of RC beams containing alkali-reactive aggregates. *East Asia AAR Seminar, Tottori* (Japan), pp. 231–285.

Torii, K., Kawamura, M., Matsumoto, K., & Ishii, K. (1996) Influence of cathodic protection on cracking and expansion of beams due to alkali-silica reaction. In: *Proceedings of the 10th ICAAR*, Melbourne, Australia, pp. 653–660.

Torii, K., Kumagai, Y., Okunda, Y., Ishii, K., & Sato, K. (2000) Strengthening method for ASR affected concrete piers using prestressing steel wire. In: *Proceedings of the 11th ICAAR*, Quebec City, Canada, pp. 1225–1233.

VicRoads (2009) *Concrete coating, Standard Specification 686*, VicRoads, Kew, Vic.

VicRoads (2010) *Structural concrete, Standard Specification 610*, VicRoads, Kew, Vic.

Vivian, H.E. (1947a) Studies in cement–aggregate reaction: Part II. The effect of alkali movement in hardened mortar. *Council for Scientific and Industrial Research, Bulletin No. 229*, Melbourne, Australia, pp. 47–54.

Vivian, H.E. (1947b) Studies in cement–aggregate reaction: Part III. The effect of void space on mortar expansion. *Council for Scientific and Industrial Research, Bulletin No. 229*, Melbourne, Australia, pp. 55–66.

Vivian, H.E. (1947c) Studies in cement–aggregate reaction: Part IV. The effect of expansion on the tensile strength of mortar. *Council for Scientific and Industrial Research, Bulletin No. 229*, Melbourne, Australia, pp. 67–73.

Vivian, H.E. (1947d) Studies in cement–aggregate reaction: Part V. The effect of void space on the tensile strength changes of mortar. *Council for Scientific and Industrial Research, Bulletin No. 229*, Melbourne, Australia, pp. 74–77.

Vivian, H.E. (1947e) Studies in cement–aggregate reaction: Part VII. The effect of storage conditions on expansion and tensile strength changes of mortar. *Journal of the Council for Scientific and Industrial Research, 20* (4), Melbourne, Australia, pp. 585–594.

Vivian, H.E. (1948) Studies in cement–aggregate reaction: Part VIII. The expansion of composite mortar bars. *Journal of the Council for Scientific and Industrial Research, 21* (2), 148–159.

Vivian, H.E. (1950a) Studies in cement- Aggregate reaction: Part XII. The effect of amount of added alkalis on mortar expansion. *CSIR Bulletin 256*, Melbourne, Australia, pp. 31–47.

Vivian, H.E. (1950b) Studies in cement Aggregate reaction: Part X. The effect on mortar expansion of amount of reactive aggregate component in the aggregate, *CSIR Bulletin 256*, Melbourne, Australia, pp. 13–20.

Vivian, H.E. (1950c) Studies in cement Aggregate reaction: Part XI. The effect on mortar expansion of amount of available water in mortar, *CSIR Bulletin 256*, Melbourne, Australia, pp. 21–30.

Vivian, H.E. (1950d) Studies in cement Aggregate reaction: Part XIII. The effect of added sodium hydroxide on the tensile strength of mortar. *CSIR Bulletin 256*, Melbourne, Australia, pp. 48–52.

Vivian, H.E. (1950e) Studies in cement Aggregate reaction: Part XIV. The effect of small amounts of reactive component in the aggregate on the tensile strength of mortar. *CSIR Bulletin 256*, Melbourne, Australia, pp. 53–59.

Vivian, H.E. (1950f) Studies in cement Aggregate reaction: Part XV. The reaction product of alkalis and opal. *CSIR Bulletin 256*, Melbourne, Australia, pp. 60–78.

Vivian, H.E. (1951a) Studies in cement Aggregate reaction: Part XVI. The effect of hydroxyl ions on the reaction of opal. *Aust J Appl Sci.*, 2 (1), 108–113.

Vivian, H.E. (1951b) Studies in cement-aggregate reaction: Part XVII. Some effects of temperature on mortar expansion. *Aust J Appl Sci.*, 2 (1), 114–122.

Vivian, H.E. (1951c) Studies in cement-aggregate reaction: Part XIX. The effect on mortar expansion of the particle size of the reactive component in the aggregates. *Aust J Appl Sci.*, 2 (4), 488–494.

Vivian, H.E. (1983) Assessment of aggregate reactivity tests and significance of test results, In: *Proceedings of the. 6th International Conference on Alkalis in Concrete*, Copenhagen, Denmark, pp. 365–368.

Watters, W.A. (1969) Petrological examination of concrete aggregates. In: *Proceedings of the National Conference on Concrete Aggregates*, Hamilton, New Zealand, pp. 48–54.

Whitmore, D., & Abbott, S. (2000) Use of an applied electric field to drive lithium ions into alkali-silica reactive structures. In: *Proceedings of the 11th IAARC*, Quebec City, Canada, pp. 1089–1098.

Chapter 15

Indian Sub-Continent

Ajoy K Mullick

15.1 INTRODUCTION

Commencing in the early 1950s, post-independence India witnessed a spurt in developmental activities, which involved construction of various concrete structures. As a result, there is in India a large number of hydraulic structures such as concrete dams and bridges, which are presently more than 50 - 60 years old; well in excess of the age by which problems due to any AAR in concrete should become apparent. A majority of these structures are in satisfactory condition, barring distress due to corrosion of reinforcement in some construction. This led to confidence that the problem of distress due to AAR in concrete may not exist in India.

However, this confidence was somewhat thwarted by the report of two concrete irrigation structures in which distress during a service life of nearly 30 years was attributed to ASR (Mullick, 1987). These, coupled with similar cases in some other concrete dams or bridge substructures (which have not been fully investigated) and reports of occurrence of ASR in neighbouring Pakistan, have focused considerable attention on the problem. As a result, aggregates and cementing materials for use in most of the new hydroelectric projects are being exhaustively evaluated, and research began on the choice of appropriate cementing systems and rehabilitation techniques. This chapter describes the contemporary experience of these aspects in India and, to some extent, in the sub-continent more generally.

Since no instances of alkali-carbonate reaction (ACR) have been noticed so far, the discussion in this chapter essentially relates to ASR. The nature and reactivity of carbonate aggregate used in India are described.

15.2 DISTRESS TO CONCRETE STRUCTURES IN SERVICE

Some of the earliest references to the occurrence of ASR in concrete structures in India were made in 1962 (Gogte, 1973) however, few details were documented. The first reported cases of distress in a concrete spillway and a concrete gravity dam and the powerhouse structure can be found in Irrigation and Power Department, Government of Orissa, Expert Report (1983) and Irrigation Department, Government of Uttar Pradesh (1986) along with details of comprehensive investigations carried out, which are summarised below.

15.2.1 Hirakud Dam Spillway

The first investigation relates to concrete spillways in Hirakud dam, which is one of the longest earth dams in the world. It is a composite structure of earth, concrete and masonry. The length of the dam is 4.8 km, flanked by 21 km earthen dykes on the left and right sides, making a total length of 25.8 km. The concrete structures in the left and right bank spillways at the time of investigation were nearly 27 years old. These had suffered extensive cracking, mostly in the walls of openings like galleries, shafts and adits. Typical 'map' cracking was superimposed with longitudinal horizontal cracks. The extent of cracking had been increasing with time. In addition, malfunctioning of radial crest gates, snapping of bolts which fix the sluice gate roller tracks and guide-rails to the concrete, and deflection of side walls of the adit gallery were noticed (Irrigation and Power, Orissa, 1983).

Samples of concrete from such locations where typical signs of ASR were present, showed the unmistakable presence of ASR, as evidenced by off-white, translucent-to-opaque agglomeration of fluffy gel-type deposits in voids bordering the aggregates, or on the aggregates; the aggregates had a dark reaction rim around their edges, visible to the unaided eye (Figure 15.1).

Scanning electron microscopy (SEM) of representative samples obtained from concrete cores showed the aggregate boundaries to contain a reaction rim, altering their edges; sometimes with microcracks in the aggregate pieces, with a considerable amount of white reaction product (Figure 15.2). In many instances, such cracks in the aggregates were apparently caused by the formation of the reaction products inside the aggregates (Samuel *et al.*, 1984).

The reaction products thus formed were found to be essentially non-crystalline, gel-type material (Figure 15.3).

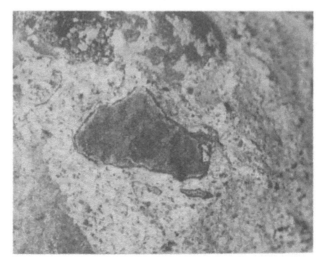

Figure 15.1 Concrete core sample showing dark reaction rim around the aggregates and typical ASR gel – Hirakud dam spillway (Irrigation and Power, Orissa, 1983).

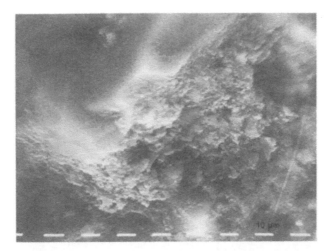

Figure 15.2 Scanning electron micrograph of the gel formation around an aggregate, altering its edges. (Samuel *et al.*, 1984).

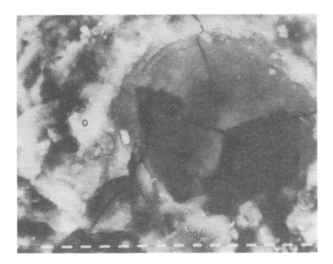

Figure 15.3 Typical gel formation and micro-cracking in the mortar phase (Samuel *et al.*, 1984).

An EDAX point scan on the aggregate and the surrounding rim showed the presence of a considerable amount of silicon, less calcium and small amounts of potassium (Figure 15.4). The presence of sodium could not be detected by EDAX. In view of the gel-type nature of the reaction products and the elemental composition, these can be described as lime-alkali-silica gel, as identified and described by Regourd and Hornain (1986), as well as Thaulow and Knudsen (1975).

In addition, the reaction products were occasionally found to be crystalline in nature (Figure 15.5); and their cornposition as detected by EDAX was similar to that

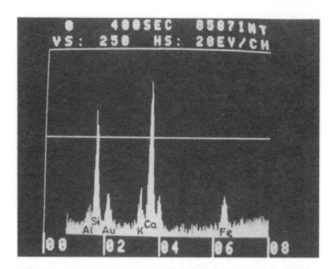

Figure 15.4 EDAX spectrum corresponding to the point marked '0' in Figure 15.3 (Samuel *et al.*, 1984).

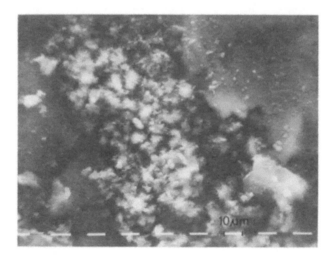

Figure 15.5 Crystalline nature of some products of ASR as seen under SEM (Mullick & Samuel, 1986).

in Figure 15.4. Away from the reaction zone, the mortar phase was generally found to be cracked and gel formation was associated with such cracking, (Samuel *et al.*, 1984).

15.2.1.1 Aggregate Types

On petrographic examination, the following three types of reactive coarse aggregates were identified by the Cement Research Institute of India report SP-147 (1983).

(a)   **Quartzite river shingles.** These consisted predominantly of crystalline quartz (β form) with grains of varying dimensions, cemented by either crystalline silica or ferruginous matter. In a number of cases, the quartz grains were cemented by near-opaque or semi-opaque – to – translucent cryptocrystalline silica. The refractive index of around 1.53 indicated the material to be chert or chalcedony (Dolar-Mantuani, 1983).

Quartz grains very often showed wavy (undulatory) extinction, determined by the procedure suggested by Dolar-Mantuani. More than 20% of the grains showed wavy extinction, the undulatory extinction (UE) angle ranging from 19° to 28°. Occasionally the river shingle consisted of dark fine-grained plagioclase, biotite mica and hornblende that showed alteration to chlorite.

(b)   **Granitic rocks.** These comprised granite, granodiorite or granite porphyry. The rock was porphyroblastic in texture and showed evidence of action of direct pressure, manifested by wavy or undulatory extinction of quartz. Large subidioblastic to xenoblastic plates of feldspar constituted orthoclase and perthitic microcline as well as plagioclase, in which the polysynthetic twin lamellae were often deformed.

The mineral analysis showed 60% alkali feldspar, 10% plagioclase feldspar, 22 – 25% quartz, 5 – 7% hornblende and biotite, and 3 – 5% accessory minerals. Fracturing and wavy extinction were common in quartz grains. The extinction of individual grains ranged from 12° to 25°. In the patchy (mottled) variety, the orthoclase feldspar occurred as large plates or laths (up to 8 mm), and this rock can be termed granite porphyry. A third type consisted of a smaller amount of quartz (15%), predominantly feldspar, and a comparatively large amount of biotite and hornblende. This type is termed granodiorite. The gradation between the three types was slight, such that they should not be considered separately. The rocks showed common and uniform alteration to chlorite (from biotite and hornblende) and sericite (from feldspar).

(c)   **Diorites.** The rock showed a predominance of plagioclase and perthitic microcline. Biotite, hornblende and quartz occurred in varying amounts. The accessories were sphene, epidote and titaniferous magnetite. A modal analysis of typical rock showed 20 – 30% plagioclase, 11 – 14% hornblende, 6 – 12% biotite, 10 – 18% perthitic or pure microcline/orthoclase and 6 – 10% quartz. The quartz grains very often showed wavy extinction and cracking, the feldspars showed bending of twin lamellae and cracks along cleavage planes.

More than 25% of quartz grains showed wavy extinction with a UE angle 15 – 30°. The rock showed conspicuous secondary alteration, which was manifested by conversion of hornblende to biotite and chlorite, feldspars to sericite and of perthitic feldspar to kaolin and other clay minerals (Cement Research Institute of India, SP-147, 1983).

Laboratory evaluation had earlier revealed the quartzite river shingles to be potentially reactive, and these were inadvertently used in some locations during the peak construction period. However, the other two types comprising crushed aggregates were considered to be 'innocuous' according to the criteria prevalent in the 1950s [*i.e.*, ASTM C227 (now 2010) and ASTM C287 (recently 2007, but now withdrawn)]. As a result, no effort was made to obtain 'low-alkali' cement, and cement used from two sources probably contained 0.8 - 1.0 total alkalis (Na_2Oeq).

15.2.2 Rihand dam and powerhouse structure

The concrete gravity dam and adjacent powerhouse of this hydroelectric project, 25 years after their construction, showed extensive distress, which was attributed to ASR (Irrigation Department, Government of Uttar Predesh, 1986). External manifestations included cracking of concrete, misalignment of hydro mechanical machinery and difficulties in the operation of gates, cranes and passenger lifts as a result of movements in concrete. In the powerhouse, the rotor assembly had sunk in relation to the stator, leading to fouling of rotor blades, and high spill current resulting in frequent tripping of the machines was reported. The rotor runner assembly had risen in relation to the speed ring. The horizontal labyrinth clearance at both the top and bottom was progressively reduced in a longitudinal direction and increased in the transverse direction. There was horizontal displacement of about 30 mm between the powerhouse crane girders in different bays, and intake gates did not seal properly. The powerhouse seemed to have tilted upstream.

Examination of the concrete samples with hand magnifying glass and visual examination on the broken surface of the concrete revealed typical white deposits associated with ASR in the voids in concrete and on aggregates, but they were not frequent. On the other hand, reaction rims around aggregates, another manifestation of ASR (Mather, 1975) were quite conspicuous. In some cases, the broken surface of the concrete showed the formation of a thin white rim around the aggregates. In most cases, however, a uniform dark band in the peripheral zone of the aggregate was observed (NCCBM [National Council for Cement and Building Materials], SP-197, 1985).

A complete scan of the sample morphology with SEM showed that the aggregates contained a reaction rim altering their borders, sometimes with microcracks either in the aggregate or in the mortar matrix, similar to Figures 15.2 and 15.3 (NCCBM, 1985). The presence of a fluffy gel-type formation with occasional crystalline deposits was observed, in which potassium was predominant. The fact that the needle-like crystal structures were not ettringite was verified with the help of EDAX, which showed the absence of sulphur. The aggregates were observed to form white fluffy gels, giving an impression that these were oozing out from the main aggregate (Visvesvaraya et al., 1986). The aggregate sample itself contained alkalis, originating from alkali feldspars. However, EDAX analysis of the reaction products showed a much larger amount of potassium.

15.2.2.1 Aggregate types

Petrographic examination of aggregates extracted from concrete samples indicated these to be mainly biotite granite, muscovite granite and mica granite (Visvesvaraya et al., 1986). In each of these rocks, the quartz content varied from 32 to 45% and alkali (sodium-potassium or sodium-calcium) feldspars such as orthoclase, microcline and plagioclase from 35 to 45%; varying amounts of biotite, muscovite and other accessories, including iron ore, chlorite and apatite were present. The average grain size of quartz varied from 0.10 to 0.20 mm, with small grains up to 0.03 mm in some cases and large grains up to 0.45 mm in others. Nearly 50 – 80% of the quartz exhibited strain effect with a UE angle varying from 25° to 30° (Figure 15.6). Potash feldspar in most of the cases was orthoclase which was found to have altered to sericite. Plagioclase

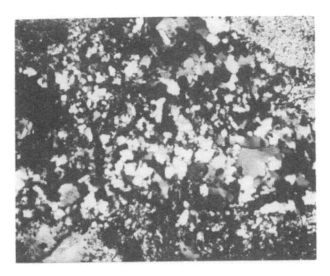

Figure 15.6 Quartz grains in granitic rocks showing undulatory extinction (crossed nicols, x 50) (NCCBM, 1985).

Figure 15.7 Alteration of feldspar to clay minerals in granitic rock aggregates (crossed nicols, x 50 (NCCBM, 1985).

feldspars were found to have altered to clay minerals (Figure 15.7). Biotite occurred in the form of laths and sometimes as sheath-like structures. These and muscovite showed the effect of bending.

The normal granitic texture of the aggregate was considerably disrupted because of the high degree of alteration of minerals. Laboratory evaluation according to the criteria existing at the commencement of the project had indicated the aggregates to be 'innocuous' and it was considered that there was no need to use low alkali cement (Irrigation Department, Government of Uttar Pradesh, 1986).

15.2.2.2 Structural interaction

The penstock gallery structure located at the toe of the concrete gravity dam comprised six blocks, corresponding to the six overflow bays. Each block had four reinforced concrete frames constructed integrally with the dam body, interconnected with the reinforced concrete beams running parallel to the axis of the dam (Figure 15.8). The major structural distress noticed in the penstock gallery and in the adjacent power-house structure, after nearly 25 years of operation, included multiple horizontal cracks on the face of columns marked 9–10–11 in Figure 15.8 (close to penstock) and extending to a height of 1 m or so from the foot of the columns, wide horizontal cracks near mid-height and snapping of main longitudinal reinforcement in column 9 – 10 - Il, spalling of concrete at the end section of beam 5 – 6, closing of 25-mm expansion joints

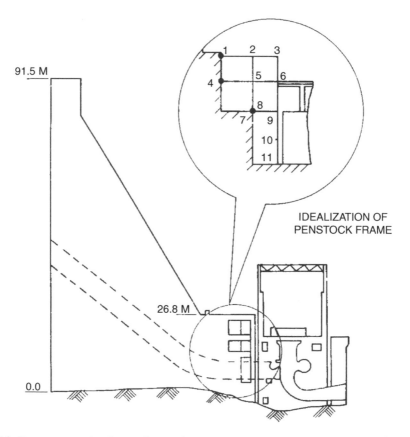

Figure 15.8 Concrete gravity dam and powerhouse structure – typical intake section (Visvesvaraya et al., 1986).

between gallery and scroll concrete, and relative horizontal shifts in the beams supporting the generator floor (Visvesvaraya *et al.*, 1986).

For such distress in concrete structures, temperature effects, deleterious chemical reactions and relative settlement of foundation merit *prima facie* consideration as probable causes. From records, it was ascertained that the concrete was pre-cooled and recorded temperatures were within the limits (Irrigation Department, Government of Uttar Pradesh, 1986). This, along with the fact that the movements were still continuing, eliminated temperature effects as a possible cause. The dam was founded on granite rock, which was considered as nearly ideal. Finite element analysis of the dam section, including a portion of the foundation rock and allowing differential modulus of elasticity of the rock to result in differential settlement revealed that this could not have caused the distress noticed in the frames (Visvesvaraya *et al.*, 1986).

Further analysis was carried out by imposing various levels of horizontal and vertical displacements of nodes 1, 4, 7 and 8 (Figure 15.8) representing the expansion in the dam body due to ASR. Because the deformation characteristics of all the frames were identical and all the frames had the same configuration and mechanical properties, a two-dimensional analysis was considered adequate. A systematic search was made to correlate the distress observed with the possible combination of vertical and horizontal movements transmitted to the joints of the frames due to volume change in the main dam.

From the analysis, it was concluded that the relative displacements due to ASR expansion could cause distress in the members of the penstock gallery frame, as observed (Visvesvaraya *et al.*, 1986). In addition, the reaction in the turbine block, which could be either passive or active, could aggravate the distress in column 9 -10 - 11. Horizontal and vertical displacements of 12.5 mm and 3 mm respectively, imposed relative to node 13 in Figure 15.8, produced the bending moment diagram shown in Figure 15.9. Visual observations of the shifts in the beams, closing of 25 mm expansion joints and further long-term observations (Mullick *et al.*, 1987) established that displacements of such magnitude were entirely feasible.

15.2.2.3 Characterisation of the reaction products

The general description of the microstructure of reaction products as observed by SEM has been mentioned already. Since these were the first reported cases of ASR in India, detailed examination of the reaction products was undertaken in order to compare them with the features reported in the literature (Mather, 1952; Thaulow & Knudsen, 1975; Regourd & Hornain, 1986).

A composite gel sample was made in both cases by carefully scooping out the gel from various locations; this was used for the chemical analyses, X-ray diffraction and optical microscopy. The chemical analysis of the gels is presented in Table 15.1. The alkali contents were determined by flame photometry. The compositions were similar to the ranges indicated by others as representative of alkali-silica gel (Mather, 1952).

15.2.3.1 Petrography

The composite gel was petrographically examined under a polarising microscope in immersion liquids. In the case of quartzite aggregates containing secondary silica minerals, the material showed the following distinct composition:

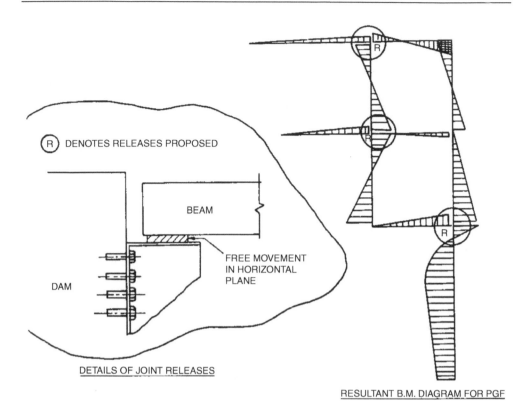

(R) DENOTES RELEASES PROPOSED

BEAM

FREE MOVEMENT
IN HORIZONTAL
PLANE

DAM

DETAILS OF JOINT RELEASES

RESULTANT B.M. DIAGRAM FOR PGF

Figure 15.9 Bending moments in penstock gallery frame corresponding to induced horizontal and vertical displacements and reaction from the turbine block (Visvesvaraya *et al.*, 1986).

Table 15.1 Chemical composition of ASR gel (Mullick & Samuel, 1986).

Sl. No.	Constituents	Quantity (%)	
		Quartzite aggregate	Granite aggregate
1	Loss on ignition	14.04	16.70
2	SiO_2	43.31	49.36
3	CaO	21.76	15.94
4	Al_2O_3	2.78	1.77
5	Fe_2O_3	0.66	0.49
6	MgO	0.83	0.49
7	alkalis		
	(a) Na_2O	3.74	3.88
	(b) K_2O	12.88	11.71

(i) amorphous gel-type matter of irregular shape with a refractive index of 1.48 – 1.50, which compared favourably with the 1.455 – 1.502 reported by Mather (1952).

(ii) distinct small grains of chert or chalcedony with a refractive index of 1.50 – 1.52 (Figure 15.10); crystalline material with patches of opaque mineral; and crystals

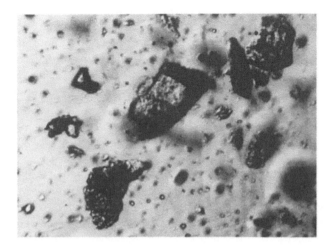

Figure 15.10 Grains of chert or chalcedony in the reaction products (polarized light, x 25) (Mullick & Samuel, 1986)

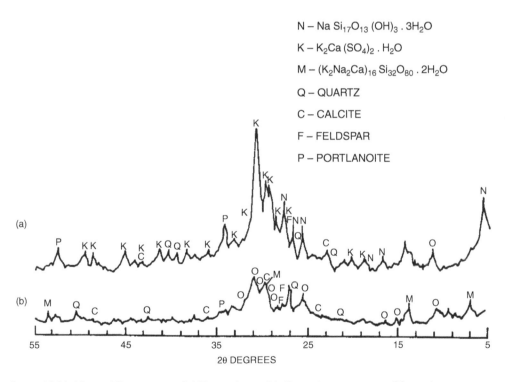

N – Na $Si_{17}O_{13}(OH)_3 . 3H_2O$

K – $K_2Ca(SO_4)_2 . H_2O$

M – $(K_2Na_2Ca)_{16} Si_{32}O_{80} . 2H_2O$

Q – QUARTZ

C – CALCITE

F – FELDSPAR

P – PORTLANOITE

Figure 15.11 X-ray diffractogram of ASR products: (a) Quartzite aggregates, (b) granite aggregates (Mullick & Samuel, 1986).

with slight anisotropy and no birefringence, presumably of the crystalline white deposits with a refractive index of 1.42 – 1.48 (Mullick & Samuel, 1986). The white material also showed occasional grains of aragonite and calcite with a refractive index of 1.65 – 1.66. In the case of granite aggregates containing strained quartz, similar features were noted, except for the presence of secondary silica such as chert or chalcedony.

15.2.3.2 X-ray diffraction analysis

Typical X-ray diffractograms of the alkali—silica gel obtained in both cases are given in Figure 15.11. The nature of the gel was predominantly amorphous, and in addition to the typical cements hydration products, some new crystalline products, believed to be due to ASR, were identified (Mullick & Samuel, 1986). In the case of quartzite pebbles, the peaks at 2Θ degrees = 5.5, 16.7 and 25.7 (Cu – Kα) were assigned to a crystalline alkali-silicate hydrate of composition $NaSi_{17}O_{13}(OH)_3.3H_2O$, and other prominent peaks at 2Θ degrees = 31, 29.8 and 29.4 to a composition, $K_2Ca(SO_4)_2.H_2O$.

In contrast, in the case of granitic aggregates, the prominent peaks at 2Θ degrees = 6.9, 13.6, 30.5 and 53.3 were ascribed to a composition of crystalline potassium-sodium-calcium-silicate hydrate $(K_2Na_2Ca)_{16}Si_{32}O_{80}.2H_2O)$. These compositions are different from these reported earlier (Cole *et al.*, 1981).

15.2.3.3 Infrared spectroscopy (IRS)

The use of IRS to study ASR has been reported before, when certain absorption bands in the 650–1600 wave number region were taken as characteristic of silica gel and calcium carbonates (Poole, 1975). Composite samples of the reaction products in both the cases discussed in this chapter were studied by recording the IRS spectra in the range 4000–200/cm, and compared with a synthetic silica gel. The samples were prepared by grinding and passing through a 45 μm sieve before drying in an oven at about 110°C for a few hours. The fine powder was made into a pellet with KBr.

The results are presented in Figure 15.12. In the case of reaction products (alkali-silica gel) obtained in the two cases, the broad band observed in the region 3000–3600/cm is due to the 0-H stretching vibrations. The bands at 1400 and 1140/cm are assigned to carbonates and feldspars respectively, while the band at 1100/cm is due to monosulphates, the bands at 680 and 590/cm are assigned to the SO_4^{2-} group. The band at 1000/cm indicates the presence of hydrated calcium silicates. Lastly, the peaks at 1030, 960, 860, 760, 460 and 440/cm are assigned to different modes of SiO_4 vibrations.

Similarly, the IRS spectrum from the synthetic silica gel contains a broad band in the range 3000–3600/cm. Most of the bands characteristic of different modes of SiO_4 vibrations, such as 1230, 1150, 1050, 940, 790 and 450/cm, are observed. The occurrence of a 'shoulder' at 590/cm, which is assigned to SO_4^{2-}, is also observed. The absence of bands at 1400, 1100, 1000 and 680/cm described earlier is quite understandable, as the alkali-silica gel was extracted from the concrete cores. In all other respects, the IRS spectra of the two materials were similar.

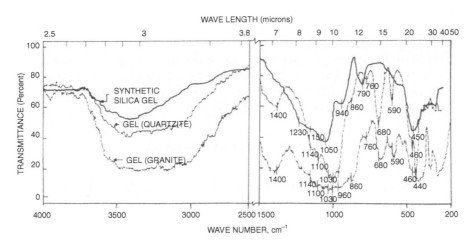

Figure 15.12 IRS spectra of products of ASR and synthetic silica gels (Mullick & Samuel, 1986).

15.2.3.4 Gel fluorescence test

By the time the above investigations were under way, a gel fluorescence test had been proposed for confirming ASR (Natesaiyer & Hover, 1989) and this was also adopted. In this test, freshly broken surfaces of the suspect concrete specimens were washed with distilled water, sprayed with uranyl acetate solution and allowed to stand for five minutes to allow adsorption of the uranyl ion by any ASR products. The specimens were then rewashed to remove any uranyl acetate solution not adsorbed, and the surfaces observed under UV light in a dark room. The presence of ASR products was indicated by characteristic greenish-yellow fluorescence of the uranyl ion.

In summary, the microstructure of the ASR products was predominantly of the amorphous gel type, with occasional crystals. In the case of metastable silica minerals, distinct reaction products seemed to be formed, whereas in the case of strained quartz the result of ASR was alteration within and of the aggregate (Mullick & Samuel, 1986).

15.3 SITUATION IN NEIGHBOURING COUNTRIES

Distress to concrete structures due to AAR in other countries in the sub-continent can reasonably be expected, wherever the geological features of the mountain ranges are similar, and the same major rivers form the sources of natural aggregate. In the north, the Himalayan region forms a border between the Indian sub-continent and the rest of Asia. Geologically, the area is complex and comprises a wide variety of igneous, metamorphic and sedimentary rocks. The prominent rivers include the Indus, Ganges and Brahmaputra, which drain into Nepal, Pakistan, Bhutan and India. Crushed rocks, gravels and river sand are mostly used as coarse and fine aggregate in concrete. Presence of strained quartz has been held as reason for slow ASR (Majid *et al.*, 2013; Bhatti *et al.*, 2003).

Sri Lanka in the south and Bangladesh in the east do not share the same mountain ranges. Documented information of occurrence of any AAR in Sri Lanka, Bangladesh or Nepal is rather limited, although the awareness of the problem is manifest. Sims & Hewson (2016) have recently described a case of severe cracking in Sri Lanka that was initially suspected to be ASR, but the charnockite aggregate was not reactive and the cracking was shown to be caused by DEF (Delayed Ettringite Formation). The information pertaining to Pakistan and Bhutan is summarised below.

15.3.1 ASR in concrete structures in Pakistan

Problems from the occurrence of ASR in the case of three dams are reported (Majid *et al.*, 2013; Bhatti *et al.*, 2003). These are the Warsak, Mangla and Tarbela dams and are discussed below.

15.3.1.1 Warsak Dam

This multi-purpose project for irrigation and power generation on the Kabul river in North West Frontier Province was commissioned in 1960. It is a 76 m high, 180 m long concrete gravity dam. The dam and powerhouse structure used coarse and fine aggregate obtained from the river bed. Cracks first appeared in 1962 and by 1982 the presence of ASR was confirmed. Petrographic analysis of concrete cores indicated that the reactive species in the coarse and fine aggregates were slate/phyllite, greywacke, schist/gneiss and quartzite having microcrystalline and strained quartz (Majid *et al.*, 2013). It was added that, *"slate/phyllite and greywacke belonged to provenance having rock type of very low grade metamorphism, where no reconstitution of minerals took place and therefore, remained harmful"* (Majid *et al.*, 2013).

15.3.1.2 Tarbela Dam

Construction of the principal structures of this multi-purpose project on the Indus river was completed in 1974. Initial assessment of gravel and sand obtained from the river bed by petrography (ASTM C295, now 2012), chemical (ASTM C289, recently 2007 but now withdrawn) and mortar-bar (ASTM C227, now 2010) tests indicated these to be non-reactive. Therefore, they were used with OPC of high-alkali content (Bhatti *et al.*, 2003). Twelve years after construction, ASR was noticed in spillway sections and irrigation tunnel structure. ASR was ascribed to the presence of strained quartz and meta-greywacke containing mineral phases that are slowly reactive (Bhatti *et al.*, 2003).

15.3.1.3 Mangla Dam

The project, located on the Jhelum river and completed in 1967, included a 13 km long embankment dam, two spillways and a powerhouse; similar to Hirakud dam discussed earlier (section 15.2.1). Both coarse and fine aggregates were obtained by processing and crushing gravel from the flood plains upstream and downstream of the main spillway. Mortar-bar tests (ASTM C227, now 2010) and petrography (ASTM C295, now 2012), at the time of initial construction, did not indicate any potential reactivity of aggregate. OPC with about 1% alkali as Na_2Oeq was used (Bhatti *et al.*, 2003).

After nearly 35 years in service, the existing spillway had to be raised for higher water level and a condition survey was carried out. Hairline cracks were present in the spillway surface. Some cracks in drainage galleries, where a humid environment exists, needed cores to be taken out for examination. Petrographic analysis revealed aggregate derived from the Jhelum riverbed to be potentially reactive due to the presence of slow and late reactive forms of silica in quartzites, volcanics, granites, quartz-wacke/grey-wacke, and slate/phyllite (Bhatti *et al.*, 2003).

Petrographic studies of cores extracted along the cracks indicated dark reaction rims around clasts. Thin sections under the polarizing microscope showed micro-cracks traversing aggregates and cement paste, and filled with secondary ASR products. Accelerated mortar-bar testing by the ASTM C1260 (now 2014) method on Jhelum riverbed gravels showed expansion of 0.25% at 16 days, indicative of deleterious reactivity (Bhatti *et al.*, 2003).

15.3.1.4 Discussion

All the three case studies indicate occurrence of ASR due to presence of strained quartz in aggregate. Most severe reaction took place in the case of the Warsak dam power-house area, followed by Tarbela dam, and Mangla dam as the least. Conventional mortar-bar and rapid chemical testing, prevalent at the time of construction, did not indicate the aggregate to be potentially reactive, and no effort was made to obtain low-alkali cement for concrete construction. The accelerated mortar-bar test (ASTM C1260, 2014) is held to be conservative (Majid *et al.*, 2013).

15.3.2 ASR Reported in concrete in Bhutan

Endowed with abundant water resource, Bhutan has potential of 30,000MW of hydroelectricity generation. One of the early hydro projects is Tala over the river Wangchu, started in 1997 and completed in 2005. It consists of a 92 m high concrete gravity dam, with a 23 km headrace tunnel leading to the powerhouse. During evaluation of the aggregate materials, accelerated mortar-bar testing indicated potential reactivity in the riverbed material. As a safeguard, Portland slag cement (with 50% GGBS content) with or without silica fume was used in different components of the Project (Singh *et al.*, 2003). No distress due to ASR has been experienced to date. Another construction is discussed in Section 15.8.2.1.

15.4 ASSESSMENT OF POTENTIAL REACTIVITY OF AGGREGATE

Various types of rocks, belonging to different classes, are used as concrete aggregate. The following is the list of trade groups of rocks used as concrete aggregate as per IS: 383 (Indian Standard Specification for Coarse and Fine aggregates from Natural Sources for Concrete):

Igneous Rocks – Granite, Gabbro, Aplite, Dolerites, Rhyolite and Basalt Groups.
Sedimentary – Sandstone and Limestone Groups,
Metamorphic – Granulite and Gneiss, Schist, Marble Groups.

Awareness among engineers in India about potential reactivity of concrete aggregates was brought about during construction of large irrigation projects in the postindependence period. Some of them were undertaken with active collaboration of agencies in other countries, where the phenomenon of ASR, due to presence of secondary silica minerals was already known. One notable example is Bhakhra Dam in the Punjab, built with collaboration of USBR in USA. Hirakud and Rihand projects, already mentioned, were contemporary constructions. Exploratory laboratory investigations were also undertaken.

Accepted test procedures, at that time comprising the 'rapid chemical' test (ASTM C289, recently 2007 but now withdrawn), the original mortar-bar expansion test (ASTM C227, now 2010) and the petrographic examination guide (ASTM C295, now 2012), were employed. Procedures in IS: 2386 (Indian Standard Methods of Test for Aggregates for Concrete, Part 7) were similar to these ASTM methods. A need for more refined test procedures was realised following the failure of these test methods to detect the potential reactivity of slowly reactive aggregates that subsequently took place. The assessments are described sequentially below.

15.4.1 Previous assessments

Bhakhra Dam in Punjab in the 1950s, which was the highest concrete gravity dam in the world when completed, used river gravels composed predominantly of quartzite, metasandstone, greywacke and sandstone, which contained cherts, chalcedonic sandstones and glassy andesites amounting to 2.3% of the total mass on an average; as well as some limestone and dolomite. Similarly, the natural sand contained less than 2% cherts. These aggregates were found to be innocuous after exhaustive laboratory evaluation and were used without any detrimental effect in service (Cement Research Institute of India, 1982). In the case of Hirakud Dam spillway, the quartzite river shingles containing cherts and chalcedony, which proved to be potentially reactive in laboratory evaluation, were used inadvertently.

A comprehensive laboratory assessment of certain Indian aggregates was made in the late 1950s (Jagus & Bawa, 1957). The following common rock types were identified as potentially reactive on the basis of their composition as well as results of the conventional ASTM C227, ASTM C289 and ASTM C295 procedures:

(1)   Fine-grained, glassy to microcrystalline basalts containing more acidic glassy phases and occurring in the Deccan Plateau, west coast, Maharashtra, Madhya Pradesh, Gujarat, Andhra Pradesh, Jammu and Kashmir, West Bengal and Bihar (see map of India (Figure 15.13) for reference).
(2)   Sandstones containing secondary silica minerals such as chalcedony, crypto- to micro-crystalline quartz, opal; and quartzite having a reactive binding matrix, found in Madhya Pradesh, West Bengal, Bihar and Delhi.
(3)   Granites and pegmatites containing opal, Rhyolite and glasses and occurring in south of India, notably Tamil Nadu and Karnataka.
(4)   Trap aggregate containing reactive constituents occurring in Jammu and Kashmir and Deccan Plateau.

Gogte had earlier postulated that the reactivity of common aggregates could also be due to mineralogical and textural features of the crystal rocks, in which the commonly

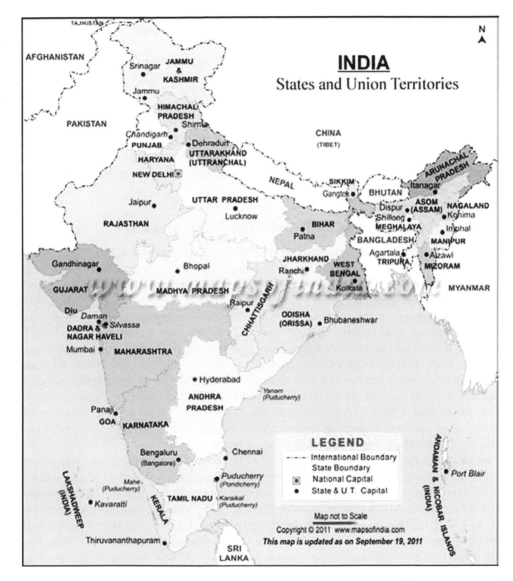

Figure 15.13 Map of India showing States and Union Territories, plus some neighbouring countries of the sub-continent.

accepted susceptible forms of glassy silica, such as opal and chalcedony, were absent (Gogte, 1973). He ascribed the reactivity of such aggregates to the presence of strained quartz. As a modification of the then existing test procedures of the Indian Standards (IS) and ASTM, Gogte had recommended that mortar-bar tests be carried out at a temperature of 50°C and suggested a criterion of mortar-bar expansion above 0.05% in 6 months as being indicative of reactivity. Accordingly, Gogte identified a number of granites, charnockites, quartzites and schistose rocks, mostly from Andhra Pradesh,

Karnataka and Tamil Nadu in south India, as well as basalts from Maharashtra and Gujarat in western India, as being potentially reactive because of the presence of strained quartz.

In most of these rocks, samples showed strongly undulatory, fractured and granulated quartz. Nearly 35 – 40% of quartz grains showed UE angle between 18° and 20°. On the other hand, rocks which contained less than 20% strained quartz, or in which most of the quartz showed uniform or only faint undulatory extinction, were considered 'innocuous'. Sandstones from Andhra Pradesh, Rajasthan, Himachal Pradesh and Madhya Pradesh owed their reactivity to the presence of cherts as a detrital constituent and sometimes as a binding matrix. Sandstones devoid of cherts but containing a few grains of strongly undulatory quartz showed expansion in mortar-bar tests within tolerable limits (Gogte, 1973).

15.4.2 Post Hirakud and Rihand – the 1980s and after

It had become clear that conventional ASTM C227 and ASTM C289 tests (similar procedures in IS: 2386 – Part 8) were inadequate for detection of potential reactivity of the aggregate types encountered in the Hirakud and Rihand dam projects in India, or those reported from Pakistan (Majid et al., 2013; Bhatti et al., 2004). Common features of these aggregate were: absence of susceptible forms of silica minerals, as in opal, chert, chalcedony etc., on the one hand, and presence of strained quartz as the reactive component, on the other. In addition, reactivity of granitic rock aggregates was also partly due to the presence of alkali feldspar, which can undergo alterations as a result of the action of hydrothermal solution and the normal process of weathering (Gogte, 1973). Laboratory experiments by Van Aardt and Visser had shown that alkali feldspars could release alkali in the presence of calcium hydroxide and water (Van Aardt & Visser, 1978), in which case they can supplement the alkali derived from cement.

In line with the suggestions of Gogte (1973), revised test methods were explored, which accorded greater acceleration to reactions by adopting either higher temperature or longer observation time. Advantage was taken of accelerated test methods proposed by NBRI in South Africa (Oberholster & Davies, 1986, which would later be developed as ASTM C1260, now 2014). A sufficiently large number of aggregate samples obtained from new constructions that were in the planning stage in the 1980s, were subjected to the modified testing protocol. A majority of these aggregate samples were of the quartzite type, while others were granitic rocks containing feldspars and mica-bearing phases in substantial quantities, as well as other varieties (Mullick, 1987; Mullick, 1994; Mullick & Wason, 1996). In addition, composite samples containing rocks of more than one type were also involved. The test procedure adopted and results are described below.

15.4.2.1 Test procedure

The starting point had to be petrographic examination. A representative summary of the petrographic details of some of the aggregates has been reported (Mullick, 1994). The description of aggregates was in terms of types (Quartzites, Granite etc.), quartz grain size, strain effect, and modal composition. The angle of undulatory extinction

(UE) in different rocks, as measured by the procedure of Dollar- Mantuani (1983), varied from 11° to 40° and the content of quartz showing strain effect varied from 15 to 90% (Mullick, 1994). Alkali feldspars in the granitic aggregates were found to have altered to clay minerals or sericite. Metastable secondary silica minerals were not detected in any of these aggregate samples. As such, any potential alkali-silica reactivity of these aggregates could be ascribed mainly to the presence of strained quartz.

For assessment of the testing regimes, the aggregate samples were subjected to rapid chemical test for durations of 24 hours, 3 days and 7 days. Also, mortar-bar expansion testing was carried out at 38°C as per IS: 2386 – Part 7 (1963), and also at 60°C, as suggested by Buck (1983). Three samples of ordinary Portland cement, variously containing 1.00, 0.57, and 0.25% alkalis (as $Na_2Oeq.$) were used in the mortar-bar tests (Mullick, 1987). The composition of the mortar-bars in both 38°C and 60°C regimes was identical, which permitted direct comparison of the results. Exploratory tests were also carried out on aggregate samples immersed in 1N KOH solution at 60°C and changes observed by SEM and IR spectroscopy.

For monitoring remaining expansion potential in concrete in service, core samples were immersed in 1N NaOH or KOH solutions at 60 °C and expansion compared to that of samples stored in water at the same temperature (Mullick *et al.*, 1987). Concrete prism tests at 60°C, without added alkali, have been used to prove mix compositions to minimize ASR activity in service (Mullick, 2001).

15.4.2.2 Results and discussions

Rapid chemical test (ASTM C289, 2007, now withdrawn) extended – Rapid chemical test ASTM C289 for 24 hours showed most of the aggregate samples to be innocuous. When the test was prolonged for storage at 3 and 7 days, the results shifted somewhat to the 'right' of the demarcation line in the case of quartzite aggregates (Figure 15.14). Many of the other aggregates continued to be in the 'innocuous' zone (Mullick, 1987).

It is not surprising that the limiting curve of Figure 15.14, developed in relation to aggregates containing secondary glassy silica minerals, should not be valid for aggregates containing strained quartz, which are relatively slowly reactive.

Mortar-bar tests at 38°C and 60°C – ASTM C227 (now 2010) mortar-bar expansion tests at 38°C and also at 60°C, with ordinary Portland cement of total alkali content more than 1% (as Na_2Oeq) were carried out to check the potential reactivity of these aggregates. In addition, most of the samples were also tested with 'low alkali' cements. Complete results with cements of different alkali contents and mortar bar expansions at both 38°C and 60°C, for periods up to 6 months are given elsewhere (Mullick, 1987).

Based on the results, the expansions at 38°C are shown in Figure 15.15 (Mullick, 1994). All except one sample satisfied the requirements of the ASTM C227 limit at 90 days (0.05%) as well as at 180 days (0.10%).

Figure 15.16 shows the ratio of expansion between 38°C and 60°C as a histogram (Mullick, 1994). The ratios varied from 2 to 5, considered to be more than in the case of usual reactive aggregate, suggesting reactivity owing to the presence of metastable silica minerals.

For slow reactive aggregates containing strained quartz, the relatively greater acceleration in expansion achieved by increase in temperature suggests that tests for such aggregate in tropical countries like India should be specified at the higher temperature (60°C).

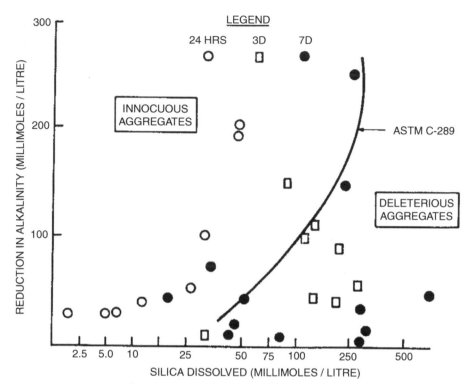

Figure 15.14 Rapid chemical test for different periods, slowly reactive aggregates (Mullick, 1987).

Sensitivity to alkali content in cement – As mentioned above, mortar-bar tests were carried out on a few aggregate samples with cements of different alkali contents. Tests were carried out both at 38°C and 60°C. Results representing different aggregate types are compared in Figure 15.17.

Expansion at 60°C after 180 days are shown. The mortar-bar expansions of the slowly reactive aggregates of various types increased with the alkali content in the cement. This is *prima facie*, proof that such aggregates are 'reactive to alkali'.

Accelerated mortar-bar tests at 80 °C in alkali solution – The procedure adopted for testing mortar-bar samples was similar to ASTM C-227 (*i.e.*, by proportioning 1 part of cement to 2.25 parts of graded aggregate by mass), however, with a fixed water/cement ratio of 0.44 for natural fine aggregate and 0.50 for coarse aggregate (Oberholster & Davies, 1986). The samples, after 24 hours of normal curing, were cured in hot water at 80°C for the next 24 hours. Finally, the specimens were stored in 1N NaOH solution at 80°C. Length change measurements are taken in the hot condition, within 20 seconds from removing the specimens from the solution. Only cement samples were used whose expansion in the autoclave (soundness) test were of the order of 0.1% or less, to rule out any possibility of the effect of the unsoundness of cement on the resultant expansion.

The same aggregate samples and the same cements as in the 38°C and 60°C tests described above were used. The results are in Figure 15.18 (Mullick & Wason, 1996).

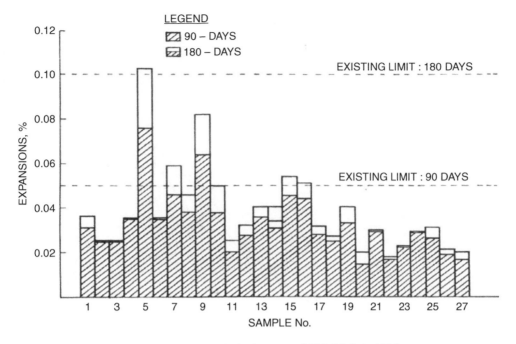

Figure 15.15 Mortar bar expansion with high alkali cement, 38°C (Mullick, 1994).

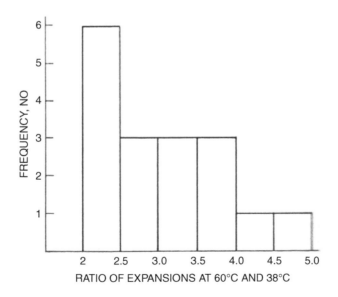

Figure 15.16 Histogram of ratio of expansion at 60°C and 38°C – 180 days, high-alkali cement (Mullick, 1994).

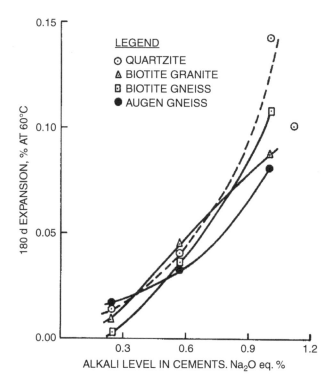

Figure 15.17 Sensitivity of mortar-bar expansion to alkali content in cements (Mullick, 1994).

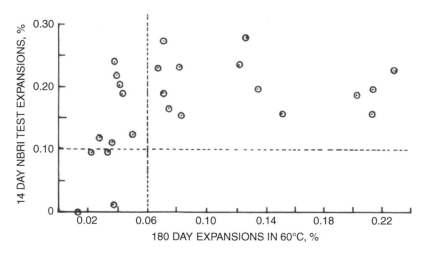

Figure 15.18 Mortar-bar expansions at 60°C and NBRI tests, slowly reactive aggregates (Mullick & Wason, 1996).

It may be seen that for 14 natural aggregate samples, which had 180 day expansions at 60°C above 0.06% and can, therefore,, be classified as 'potentially reactive' (see later), the 14 day NBRI test expansions were in the range of 0.15 to 0.28%. The expansions continued even after 14 days and increased to the order of 0.4 to 0.5% at 56 days.

Surface cracking was observed in the specimens at the age of 56 days. Presence of gel in the voids was observed in all the samples. The gel deposits were confined to the periphery of the samples at 14 days and progressed towards the interior after 56 days exposure (Mullick & Wason, 1996).

No correlation between 14 day expansions in the accelerated test and those at 180 days in the 60°C mortar-bar expansions was apparent (Figure 15.18). Another eight aggregate samples, classified as 'innocuous' in the 60°C mortar-bar tests, gave expansion above 0.10% in the accelerated test. It may be recalled that both Oberholster and Davies (1986) and ASTM C1260 (now 2014) suggest expansion less than 0.10% is indicative of 'innocuous' aggregate. Out of a total of 24, there were only three aggregate samples, which could be classified as 'innocuous' in both accelerated and 60°C mortar-bar tests (Figure 15.18). It was concluded that the accelerated test was rather severe, in that, even 'innocuous' aggregates could be classified as 'reactive'.

Concrete prism and concrete core tests – Once the onset of ASR is established, assessment of remaining potential of ASR activity is required for planning repair/remedial actions. For this purpose, concrete core samples extracted from the structure in question are exposed to the following regimes conducive to promote ASR:

I.   Water at room temperature,
II. 1N KOH solution at room temperature, and
III. 1N KOH solution at 60°C.

The dry core samples were initially immersed in water. Depending upon the relative dryness, some moisture was imbibed, resulting in absorption of the order of 0.008 to 0.09%, with accompanying gains in weight and expansion. This continued for about a week, until the length became constant. Thereafter, the initial expansion was noted and samples were exposed to the various regimes and further expansions monitored for over a year (Mullick et al., 1987).

Results of case studies relating to Rihand dam (case study I) and another concrete dam which was under construction for seven years (case II) are summarised in Figure 15.19. The initial expansion varied from 0.004 to 0.021% in case I and 0.002 to 0.016% in case II. The expansions continued after the initial phase and residual expansions were generally about 0.20% in case I and 0.3% in the case of the concrete dam nearing completion (case II) (Mullick et al., 1987). Similar observations in case of Hirakud dam spillway are presented in Section 15.7.3.

Concrete prism testing as per ASTM C1293 (now 2008), with added NaOH solution, is not commonly practiced in India. However, concrete prisms with the mix proportion recommended for construction are tested at 60°C and 100% humidity. No external alkali is added. Therefore, lower amounts of expansion than the limit of 0.04% in the ASTM test is expected to be acceptable (Mullick, 2001).

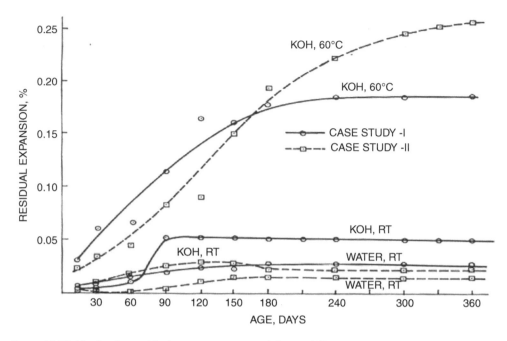

Figure 15.19 Monitoring residual expansion potential due to ASR in concrete constructions (Mullick *et al.*, 1987).

15.4.3 Assessment of limestone aggregate

The discussion so far has been on ASR due to siliceous aggregates. In India, no record of evaluation of limestone aggregates was available. Systematic evaluation for potential alkali- carbonate reactivity of three limestone aggregate samples (one dolomitic) occurring in different parts of India, which were proposed for use in concrete for construction of hydraulic structures, were carried out . Limestone 1 was a fine grained rock with abundant calcite. The average grain size of primary calcite was around 5 microns and up to 150 microns in secondary calcite. In limestone 2, calcite grains showed variation in grain size from 100 to 300 microns and that of quartz grains varied from 100 to 600 microns. A dolomite sample showed alteration of calcite and dolomite laminae. The calcite grains ranged in size from 40 to 45 microns. The constituents of the limestone samples determined by optical microscopy and the clay content for the dolomitic limestone sample determined chemically in terms of total insoluble residue, which includes silt size and larger particles also, are given in Table 15.2 (Samuel *et al.*, 1989).

15.4.3.1 Test results and discussions

Expansion Tests Expansions on concrete prisms as per CSA-A 23.2-14A (now 2014), mortar- bar tests as per ASTM C 227 (now 2010) and rock cylinder tests as per ASTM C586 (now 2011) were carried out. Additionally, mortar-bar tests were carried out at an elevated temperature regime of 60°C. Results of expansion up to 180 days are summarised in Table 15.3 (Samuel *et al.*, 1989).

Table 15.2 Constituents of limestone samples (Samuel *et al.*, 1989).

Sl. No.	Constituent	Limestone 1	Limestone 2	Dolomite
		%		
1	Primary calcite	74	40	25
2	Secondary calcite	12	–	–
3	Calcium silicates	–	25	–
4	Dolomite	4	–	70
5	Quartz	6	25	–
6	Biotite		4	–
7	Iron oxide	–	6	–
8	Accessories	4	–	5
9	Clay	–	–	6.3

Table 15.3 Results of expansion tests on limestone samples, 180 days (Samuel et al., 1989).

Sample	Concrete Prism CSA 23.2-14A			Rock Cylinder, ASTM C586	Mortar-bar, ASTM C227		
	Cement % Na_2Oeq	Temp. °C	Expansion %	Expansion %	Cement % Na_2Oeq	Temp. °C	Expansion %
LS 1	1.0	38	0.035	0.040	–	–	–
LS 2	1.13	27	0.0145	–	0.50	38	0.0146
		38	0.0337			60	0.0264
					1.13	38	0.0308
						60	0.0392
Dolomite	0.89	27	0.020	0.035	0.57	38	0.0172
		38	0.0456		1.0	38	0.0248

The net expansions in all the tests were within the permissible limits and the aggregates could therefore be held to be non expansive. These trends were expected in view of the fact that the limestone samples were not of an argillaceous nature and the clay content in the dolomitic limestone sample was rather limited *i.e.*, for a dolomite content of 70% the clay content was only 6.3%. The expansion trends of the rock cylinder test show expansion steadily without initial contraction, as noticed in some limestone samples from the Middle East (Sims & Sotiropoulos, 1983).

Rim Formation In addition to being possibly of an expansive category, limestone aggregates, depending upon their composition and formation, can also develop rims, the effect of which on the durability of concrete is not yet well understood. The composition of dolomitic limestone was such that it falls into the rim developing category (Hadley, 1964). For this purpose, the dolomitic limestone sample was also subjected to rim-development tests as suggested by Bisque and Lemish (1958). Prominent rims were observed on the longitudinally cut mortar- bars after an exposure period of two months with the exception that these rims were found to be acid insoluble. Similar white rims have also been observed around calcareous aggregates by Regourd (1983). On the other hand, the commonly reported dark reaction rims were not observed.

On examination using the scanning electron microscope, the reaction rims bordering the aggregates were found to be mostly crystalline with needle like crystals, although sometimes an amorphous nature was also detected (Samuel *et al.*, 1989). XRD of the reaction products showed the presence of dolomite and α-quartz along with the presence of calcium carbonate and calcium hydroxide. No magnesium-bearing reaction phase could be detected (Samuel *et al.*, 1989).

From the various studies discussed above, it appears that no de-dolomitization reactions, involving alkali-carbonate reactivity, which can result in either excessive expansions or in the development of dark reaction rims, had taken place. Discussion of the 'so-called alkali-carbonate reaction' can be found in Chapter 3.

15.5 DEFORMATION OF QUARTZ AND MICROSTRUCTURAL ASPECTS

The description so far brings out that ASR in Indian concrete structures made with siliceous aggregates often does not involve conventional deleterious minerals like opal, chalcedony, chert or volcanic glass etc. On the other hand, defects in the crystal lattice of the quartz grains showing strain effect, has been mentioned as a causative factor for ASR. Similar is the situation in neighbouring Pakistan (Bhatti *et al.*, 2004; Majid *et al.*, 2013). It is worthwhile to consider the related microstructural aspects of such slowly reactive rocks (Mullick, 1994).

In the context of India, Gogte (1973) was the first to propose that ASR could take place due to mineralogical and textural features of crystalline rocks like granite, quartzite and gneisses etc. This was in line with the earlier findings of Brown (1955) and Mielenz (1958) that the petrographic character of the rock could also be a factor for the initiation of alkali-aggregate reaction.

Such a phenomenon can indeed be supported from a materials science standpoint, relating to entropy and free energy of a disordered system and its chemical reactivity. When a crystalline rock is subjected to deformation during metamorphism, the strains are accommodated by individual crystals accommodating defects in the crystal lattice. According to Anderson and Thaulow, the wavy and extinction behaviour of deformed quartz, when examined under the polarizing microscope, is caused by crystallographic subdivision and disorientation of minute areas in the quartz lattice. The dislocations produced contain atoms which carry unsatisfied bonds and hence are more prone to chemical attack (Anderson & Thaulow, 1990).

Rao and Sinha examined the textural and microstructural features of four granitic aggregate types from different locations in India, and another extracted from a dam structure deteriorated due to ASR (Rao & Sinha, 1989). The granitic aggregates studied comprised biotite gneiss, augen gneiss, biotite augen gneiss, banded gneiss, and mica granite from the ASR-affected dam concrete. The textural features observed in these aggregates were myrmekites and perthites; the common microstructural features included transverse cracks, secondary mineral inclusions in quartz and feldspar porphyries and granulation of quartz.

The above petrographic studies revealed that all the aggregates were mineralogically quite similar. Their textural and microstructural features were, however, entirely different. The aggregates with highly granulated quartz, myrmekites and perthites gave mortar-bar expansions in the 60°C test greater than for the aggregates with simple gneissic texture (Rao & Sinha, 1989). Similarly, the granitic aggregate extracted from the distressed concrete showed highly granulated quartz with abundant myrmekites and perthites. On the other hand, augen gneiss and biotite augen gneiss samples showed less granulation of quartz with few myrmekites and perthites, and gave lower mortar-bar expansions; of the same order of the sample with simple gneissic texture (Rao & Sinha, 1989).

The presence of abundant myrmekitic texture mostly in the crushed zone indicates that these are formed during later deformation, in which silica migrated along skeletal defects of plagioclase feldspar (Shelley, 1964, 1967). The formation of perthites in these aggregates is closely related to the shearing stress acting upon the rock involved. According to Chayes, if quartz is undulant or granulated, perthite is commonly present in such granitic aggregates (Chayes, 1952). The occurrence of vermicular quartz in myrmekites and the blebs or tiny rods of alkali feldspar in perthites are easily susceptible to ASR, since they are structurally unstable and are developed, due to geological stress and strain, during later deformations.

In the case of concrete structures in Pakistan which have shown distress due to ASR, petrographic studies indicated slate/phyllite and greywacke in the river gravel used in the construction, deriving from a provenance having rock type of very low grade metamorphism (Majid *et al.*, 2013). Metamorphism occurred to a degree where no reconstitution of minerals took place and remained harmful. XRD of greywacke indicated absence of montmorillonite clay and the micro-crystalline matter remained as a causative factor for ASR (Majid *et al.*, 2013).

Detailed petrographic evaluation of such microstructural features of aggregate, proposed for use in construction, is common. Yet, a single parameter to indicate strain effects in quartz and potential reactivity of aggregate, is, perhaps the UE angle (Mullick, 1994). Opinion varies on whether UE is the correct measure or manifestation of the cause. The current version of the RILEM AAR-1.1 petrographic examination method (in Nixon & Sims, 2016) states (in its Annex A) *"measurement of undulatory extinction angle to predict alkali-reactivity potential has been found unreliable and now discouraged"*. French suggested that when quartzes are strained, a number of features like undulose extinction, deformation bands and deformation lamellae are produced (French, 1992). According to him, in such rocks showing reactivity, the quartz grains exhibit either strain lamellae or are cataclased, or otherwise contain microcrystalline to cryptocrystalline grains on larger grain boundaries or along fracture planes.

Grattan-Bellew stated that the presence of stress-related features, such as undulatory extinction and the formation of sub-grain boundaries in quartz grains enhance the solubility and hence the reactivity of these grains (Grattan-Bellew, 1992). He, however, is of the opinion that development of UE and microcrystalline quartz are part of the same process, which is a response of a crystal to applied stress. If the stress is high enough, development of UE will ultimately lead to the conversion of a single quartz grain to an agglomerate of microcrystalline quartz. He suggests that reactivity of such aggregate could indeed be related to the presence of microcrystalline quartz and not to strained quartz *per se* (Grattan-Bellew, 1992).

Absence of a clear correlation between the UE angle in quartz and reactivity of quartz-bearing aggregate has been commented upon by Grattan-Bellew (1992). To examine any correlation between UE angle and alkali-silica reactivity, a summary of results of ASR potential in a number of quartzite aggregate, as measured over 180 days in mortar-bar tests at 60°C, with the maximum UE angle measured as per the technique suggested by Dolar-Mantuani (1983), is shown in Figure 15.20 (Mullick, 1994).

A trend of increasing expansion with increase in UE angle can be seen. Besides UE angle, a number of other parameters like grain size, texture, proportion of quartz in the modal composition and percentage of quartz showing strain effects can influence reactivity. Microcrystalline quartz, if present, will also contribute to reactivity. Considering all these, the results shown in Figure 15.20 should be considered significant (Mullick, 1994). It is for this reason that measurement of undulose extinction in quartz grains has been adopted as a test for petrographic examination of such aggregates in the Indian Standard, as discussed later on.

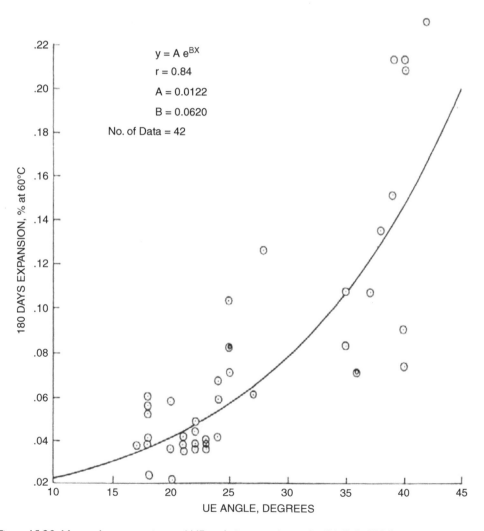

$$y = A \, e^{BX}$$
$$r = 0.84$$
$$A = 0.0122$$
$$B = 0.0620$$
No. of Data = 42

Figure 15.20 Mortar bar expansion and UE angle in quartzite rocks (Mullick, 1994).

15.6 TEST PROTOCOL AND STANDARDISATION

The relatively large database of different types of indigenous aggregates as discussed above allowed evaluation of various test methods and decision on the appropriate testing regime to be followed in India and their standardisation. These are discussed below.

15.6.1 Test protocol currently followed in India

Since the evaluation of slowly reactive aggregates was involved, the obvious choice was on tests at higher temperature or for longer durations of observation. The National Council for Cement and Building materials (NCCBM or NCB) has undertaken evaluation of almost all the hydroelectric projects for various agencies in India since the distresses in Hirakud and Rihand dams became well-known in the 1980s. According to R.C. Wason, who was involved in most of the investigations, the procedure adopted can be summed up as follows (Wason, 2013).

The aggregate samples will be examined petrographically to determine composition, texture and microstructural features, and undulose extinction. Next, accelerated mortar-bar testing at 80°C (ASTM C 1260, 2014) will be undertaken. If the expansion is less than 0.10%, the aggregate sample will be considered 'innocuous'; if the expansion exceeded 0.2%, the aggregate sample will be considered to be 'deleterious'. If the expansion was between 0.1% and 0.2%, further tests using ASTM C227 at 60°C will be carried out, before final pronouncement on the aggregate ASR potential. Criteria for a 60°C test are discussed in Section 15.6.2.

In a case in which the expansion in the accelerated test exceeded 0.10%, but sufficient time for the ASTM C227 test at 60°C was not available, the project authorities will be advised to adopt an appropriate cement system (see Section 15.9 later), treating the aggregate as potentially reactive. The concrete mix proportions with different cement systems will be evaluated in mortar-bar tests at 60°C and/or concrete prism tests (without additional alkalis) (Mullick, 2001).

15.6.2 Revised criteria of potential reactivity

Among the two tests, the criteria for the accelerated test with added alkali were available, but the criteria for the 60°C test had to be established. It may be noted that Indian Standard IS: 2386 – Part 7 (Tests for Alkali Aggregate Reactivity) has a mortar-bar test at 38°C similar to ASTM C227 (2010), but no limits of expansion to denote potential reactivity were specified. For the 60°C test, the limiting values were obtained on the basis of mortar-bar expansions with high and low alkali cements. The procedure is explained in Figures 15.21 and 15.22 for 90 and 180 days, respectively (Mullick, 1994). Use of low alkali cement is the accepted means of controlling ASR expansion. For large number of different types of aggregate tested at 60°C, the 90 days expansion was almost always less than 0.05%, and always exceeded this limit with high alkali cement (Figure 15.21). Hence, the criterion for reactivity at 90 days can be chosen as 0.05%. In a similar manner, the limit of 0.06% expansion at 180 days (Figure 15.22) was adopted (Mullick, 1994).

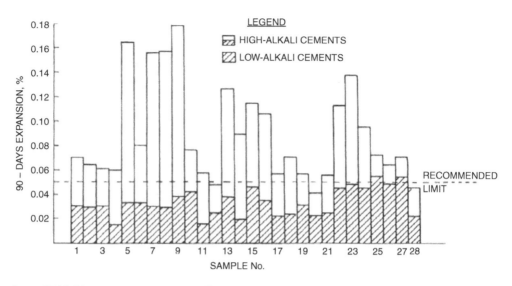

Figure 15.21 Mortar-bar expansion at 60°C with low and high alkali cements – 90 days (Mullick, 1994).

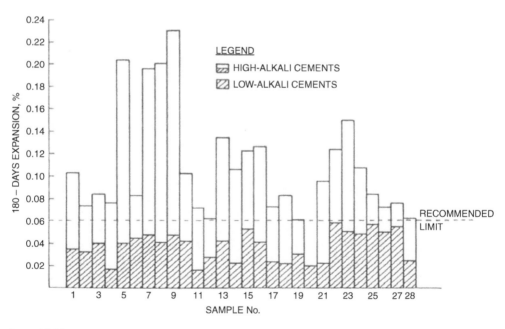

Figure 15.22 Mortar-bar expansion at 60°C with low and high alkali cements – 180 days (Mullick, 1994).

15.6.3 Standardisation

15.6.3.1 Test methods for aggregate

Indian Standard IS 383 (now 2016) prescribes quality requirements of concrete aggregate, including alkali aggregate reaction. It recognizes that certain aggregates, containing more than 20% strained quartz with undulatory extinction angle greater than 15° can be slowly reactive, for which appropriate test procedures have been prescribed. Indian Standard IS: 2386 – Methods of tests for aggregates for concrete, Parts 7 and 8 (1963 and subsequent amendments) prescribe the following:

- Part 8 – Petrographic Examination. In an amendment made in June 1999, an additional test for determination of undulatory extinction (UE) angle was prescribed. UE angle was held as a measure of quartz crystal lattice disturbance caused by geological stresses. It noted that various researchers have used UE angle as an indication of a possible indicator of alkali-reactivity of the rock.
- Part 7 – Alkali Aggregate Reactivity. In an amendment made in June 1999, mortar-bar testing at 60°C was added. This test was to be carried out on slowly reactive aggregates as defined above. It recommended that aggregates showing 180 days mortar bar expansions of greater than 0.06% using ordinary Portland cement with 1.0% or more alkali content (Na_2Oeq) be classified as potentially reactive.
- Part 7. Rapid Chemical test similar to ASTM C289 (2007, now withdrawn) and mortar bar tests at 30°C similar to ASTM C227 (now 2010) are retained, but IS 383 (now 2016) does not recommend their use in case of slowly reactive aggregates. Rapid chemical test is also not to be used in case of aggregates containing carbonates (*e.g.*, limestone) or Magnesium silicates (antigorite or serpentine)
- Part 7 Accelerated mortar bar test at 80°C using 1N NaOH (similar to ASTM C1260, 2014) has been added. IS 383 (now 2016) finds this test particularly suitable for slowly reactive aggregate. If the expansion is less than 0.10%, the aggregate sample will be considered 'innocuous'; if the expansion exceeded 0.2%, the aggregate sample will be considered to be 'deleterious'. If the expansion was between 0.1% and 0.2%, further tests using mortar bar expansion test at 60°C will be carried out, before final pronouncement on the aggregate ASR potential (see 15.6.1).

IS 383 (now 2016) states that for dolomitic and limestone aggregates, concrete prism tests shall be preferred over mortar bar tests. Until the test method is introduced into IS 2386, Part 7, specialist literature may be referred to for the test and applicable requirement.

15.6.3.2 Structural Codes

Indian Standard IS 456: (now 2000) – Code of Practice for Plain and Reinforced Concrete, is the most widely used specification for a majority of concrete constructions in the country. It had always recognised the possibility of alkali-aggregate reaction: *"some aggregates containing particular varieties of silica may be susceptible to attack by alkalis (Na_2O and K_2O) originating from cement and other sources, producing an* *expansive reaction which can cause cracking and disruption of concrete"*. By way of precautions it suggests:

- Use of non-reactive aggregate from alternate sources.
- Use of low alkali ordinary Portland cement having total alkali content not more than 0.6% (Na_2O equivalent). Further advantage can be obtained by use of fly ash conforming to IS 3812 or granulated blastfurnace slag conforming to IS 12089 as part replacement of ordinary Portland cement (having total alkali content as Na_2Oeq. not more than 0.6%), provided fly ash content is at least 20% or slag content is at least 50%.
- Measures to reduce the degree of saturation of the concrete during service such as use of impermeable membranes.
- Limiting the cement content in the concrete mix and thereby limiting total alkali content in the concrete mix. For more guidance specialist literature may be referred.

It is noteworthy that no specific guidance on the use of suitable blended cements produced commercially has been given. This aspect is discussed further in Section 15.8 later.

Concrete road bridges are built to specifications of the Indian Roads Congress (IRC). In a departure from the past, the latest revision of 'Code of Practice for Concrete Road Bridges' (IRC: 112-2011) takes note of alkali-aggregate reaction as a possible mechanism of affecting durability of concrete. The methods of assessment of reactive aggregate and recommended precautions are in line with IS: 456 discussed above.

15.7 SURVEILLANCE AND REPAIR

The procedure is explained with the help of a case study relating to distress observed in Hirakud dam spillways (Chand & Mullick, 1999). The construction had started in 1948 and been completed in 1957. The first impounding of the reservoir in 1956 was accompanied by seepage of water, mainly through joints, formed drains and gate shaft openings in the right bank spillway and the power dam. These were grouted, as were other cracks appearing up to 1973, which were considered to be normal for a concrete dam.

15.7.1 Periodic monitoring of cracks

As routine monitoring work, crack mapping is being done every year between November and December. The total length of cracks in the right spillway observed up to 1997 was 14708 metres, which included 4706 metres upstream underwater crack mapped during 1992–93 and a balance of 9372 metres in the galleries, sluice barrel and gate shaft. Cumulative crack length in the left spillway is 8328 metres, which includes 4115 metres upstream underwater cracks and a balance of 4213 metres in the galleries, gate shaft and sluice barrels. The rate of growth of total crack length in the spillways is shown in Figure 15.23, which was, however, diminishing (Chand & Mullick, 1999).

The observation of width of cracks in the right spillway has been carried out since 1974 with the help of extensometers in operation galleries and foundation galleries and since 1987 in the left spillway. Until 1989, there was an overall increase in the width of

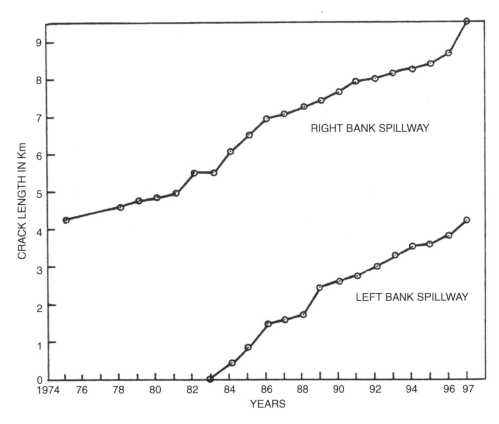

Figure 15.23 Growth of total length of cracks over years – Hirakud dam spillway (Chand & Mullick, 1999).

crack in all the observation points. In the operation gallery of right spillway, ranging from a minimum of 0.099 mm to a maximum of 0.930 mm and from 0.069 mm to 0.312 mm in the foundation gallery. In the operation gallery of the left spillway, extensometer readings from 1987 to 1989 indicated that, out of 13 points, 4 were showing crack development of minimum 0.071 mm to a maximum of 0.163 mm. other points did not show any clear opening of the cracks (Chand & Mullick, 1999).

A typical trend of increase of crack width is shown in Figure 15.24.

In spillway block-44, dial gauges in three mutually perpendicular directions along with dummy points have been fixed on the upstream face of the operation gallery, across a crack, to measure three dimensional relative movement of concrete on either side of the crack. The results showed relative movement of 0.05 mm along the dam axis on a horizontal plane and movement of 0.05 mm in cross direction up to August-1980, which then reduced to 0.01 mm up to 1982 and remaining almost constant thereafter.

15.7.2 Other *in situ* observations

The various techniques adopted were:

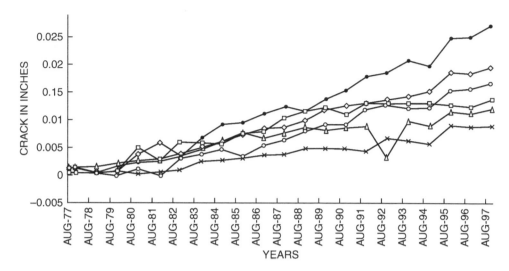

Figure 15.24 Increase in crack width over years – Hirakud dam spillway (Chand & Mullick, 1999).

15.7.2.1 Convergence by tape extensometer

The top horizontal members of the frame fixed for operating the collapsible gate provided in adit gallery in block-47 of the operation gallery of the right spillway have buckled since 1966. The difference between the curved length and straight length of the members as measured in 1982 was 24 mm. In addition, the side wall of adit at the entrance was measured to have tilted inside. To monitor progressive convergence of adit and operation gallery, measurement points have been fixed at different locations in the operation galleries in left and right spillways. The length between different points of the cross section of opening was being measured through tape extensometer. From the five sets of observations taken up to 1993, it was seen that there are maximum convergence of 2.05 mm and divergence of 3.14 mm in the right spillway. The left spillway shows the maximum convergence of 2.71 mm and divergence of 2.93 mm. The convergence and the divergence corresponded to the fluctuation of water level in the reservoir.

In addition, tilt meters, glass tell-tales and plaster tell-tales were fixed in different locations. 137 out of 276 glass tell-tales were found broken, but very few among plaster tell-tales.

15.7.2.2 Strain meters

Strain meters were embedded at the foundation level of one block of the spillway and another block of the power dam. These strain meters were to enable the measurement of three dimensional strain due to the combined effect of load, temperature and hydrostatic pressure. However, these strain meters have not been functioning for quite some time. Subsequently, five of the 300 mm long strain meters were installed across the cracks in some spillway blocks in the downstream wall of the operation gallery to measure the widening the cracks. The measured crack opening of the right spillway (0.5 mm) between 1984 to 1990 was nearly four times that of the left spillway (0.12 mm).

15.7.2.3 Strain gauges

To study the strains developed due to deformations of the steel rails and also the consequent snapping of bolts, 12 waterproof type WFLA 6-SLT 4-TML type strain gauges were installed during 1984 to 1990 in the Sluice Gate Roller Track/Side Track. Dynamic strains, which developed in the gate rails during gate operation and under full discharge through the sluices, were also measured. Dynamic strains of the order of 160 microstrain with a frequency of 1.5 cycle/second were found to develop in the roller tracks during gate operations, but under normal flow through the sluice, dynamic strain on the roller track was 30 microstrain only.

15.7.2.4 Geodetic survey

Survey of India, at Dehradun (Uttarakhand), was requested to study the movement of the crest and dam deformation in the area. For this purpose, a scheme of survey stations on either banks of Mahanadi river, towards the downstream side of the dam, was prepared in consultation with the Geological Survey of India. The scheme of observations consisted of two parts, i.e., safety zone and deformation zone. In deformation zone, 5 dam deformation stations and 3 backsight stations were established. From these stations, 18 dam targets were fixed on the downstream face of ogee spillway at three different levels by geodetic measurements (Chand & Mullick, 1999).

The scheme was first observed during 1986–87 at maximum water load and was repeated during 1987–88 both at maximum and minimum water loads, and again during 1988–89 at maximum water load of the reservoir. Results of the dam targets obtained during the last three years were analysed and tested statistically for null hypothesis. It was noticed that there were significant changes at various targets of the spillway dam. The results at maximum water load showed that there were unequal shifts or pressure on all the dam targets, however, there was no systematic trend for the changes in position of the dam targets.

15.7.2.5 Precision leveling

Precision leveling was made for monitoring the vertical movements in the area. The safety zone and deformation zone were connected by means of precision leveling. In addition, some more benchmarks were also established along the top of the dam and inside the operation gallery of the spillway.

Precision leveling was carried out near and along the dam axis and inside the dam gallery during 1986–87, 1987–88 and 1988–89 at maximum water load and during 1987–88 at maximum and minimum water loads. The changes in height between 1987–88 and 1986–87 showed that there were slight variations in the relative heights of the benchmarks, but these were not very significant. The leveling comparison between 1988–89 and 1986-87 showed that nearly all the points were approaching their original positions. Results of the year 1987–88 at maximum and minimum water loads also indicated that there were no significant changes in the height of bench marks.

15.7.2.6 Other studies

Vibration studies of Sluice Gate no. 59 were conducted in 1976. The amplitudes and particle velocities caused by the operation of the gate was found to be low and any hazard was ruled out. Static and dynamic pressure on the side walls and roofs of the sluice barrel no. 52 were measured for full sluice gate opening in 1978. The pressure was in the range of (–) 5.5 m to 1.04 m head of water.

3-D photo-elastic model studies of Hirakud Dam spillway section (for Block no. 44) were conducted in 1978. Maximum compressive tangential stress of the order of 2.6% was derived at the base. The shear stresses along the section at the region close to sluice ways were found to be higher than at other regions.

In 1991, stress analysis of the spillway section, both with and without cracking having taken place, was carried out by 3-D FEM. In the mathematical model, two cracks at RL 176.8 m and RL 167.6 m were assumed to be extending up to the line of drainage holes 5.25 m from the upstream face. The analysis showed that the stress within the body of the dam as well as foundation level were within acceptable limits, for the boundary and loading conditions normally considered (Chand & Mullick, 1999).

15.7.3 Periodic assessment of progress of ASR

Engineering properties of concrete have been periodically measured during 1983 to 1993, using core samples drilled from various locations in the spillway blocks. The compressive strength, ultrasonic pulse velocity and density were found to be adequate as per the design stipulations for the concrete mixes used. The results obtained at different periods were generally comparable. Detailed results are given elsewhere (Chand & Mullick, 1999).

The progress of alkali-silica reaction and residual expansion potential were estimated by long term measurement of concrete core samples obtained from locations where signs of ASR were prominent. The procedure has been described in Section 15.4.2.2. The core samples were stored in air, in reservoir water and potable water, all under ambient temperature condition. In addition, some samples were stored in 1N NaOH solution at 60°C. Test results are given in Table 15.4.

In the case of Hirakud Dam, expansion of the order of 0.04 to 0.06% in alkali solutions at 60°C after one year was measured. Exposure to ambient temperatures resulted in lower expansions. Similar tests on concrete core samples have been adopted by others also, to estimate residual ASR as well as long term expansion potential (Bérubé et al., 1995). Expansion rate of the order of 0.02% per year at 100% relative humidity is considered to be indicative of 'medium' rating in residual expansion.

The observed residual expansion noted in the present case can be considered to be corresponding to such 'medium' rating (Bérubé et al., 1995).

Table 15.4 ASR potential tests on concrete core samples of Hirakud Spillway (Chand & Mullick, 1999).

S. no.	Exposure	I yr. expansion, %
I	I N NaOH soln. at 60°C	0.058
2	In potable water[@]	0.02 – 0.03
3	In reservoir water[@]	0.02 – 0.03
4	In air	Nil, slight contraction
5	I N NaOH and KOH soln.[@]	0.035
6	I N KOH soln. at 60°C	0.044

@ – room temperature

From these considerations, as well as the growth of cracks depicted in Figure 15.23, it was concluded that ASR was still progressing, but at a slow rate. In, view of satisfactory strength properties of concrete as well as structural analysis carried out, it was concluded that while there was no danger to the safety of the structure, there was no scope of complacency and constant vigilance was required.

15.7.4 Treatment of cracks

Epoxy has been adopted for treatment of cracks, both surface cracks as well as underwater treatment of cracks. Exposed surface grouting of walls of sluice barrels and spillway buckets has been carried out by epoxy plastering with quartz sand and silica flour. From such treated portions, 45 core samples of 72 mm diameter and varying length were extracted. By visual inspection, the penetration of grout was found to be full and complete. On compression strength testing, it was found that at the ultimate load, failure took place in concrete and not in the epoxy treated crack plane. A further 127 core samples having 40 mm diameter and 200 mm length have been extracted from the completed surface of cracks of both spillways. Efficacy of grouting was found to be fully satisfactory, as confirmed by compressive strength tests.

The entire underwater treatment has been scanned and monitored through CCTV, which gave great confidence concerning the various treatment activities carried out underwater. Further, the examination of extracted drilled cores from the underwater treated surface confirmed satisfactory efficacy of the grouting and proper sealing of the cracks.

15.8 CHOICE OF APPROPRIATE CEMENT SYSTEMS

15.8.1 Types and quality of cements produced in India

Over a century old, the Indian cement industry has installed capacity (2015) of 386 million tonnes and 317 million tonnes of production. There are 14 different types of cement specified by the Bureau of Indian Standards (BIS), but the major production is of three varieties – ordinary Portland cement (OPC), fly ash-based Portland pozzolana cement (PPC) and Portland slag cement (PSC) (Figure 15.25). Blended cements like PPC and PSC comprise 75% of total production.

Wherever alkali-reactive aggregates are encountered, it has been customary to use cement of low-alkali content – the limit being 0.6% Na_2Oeq. The availability of low-

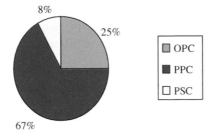

Figure 15.25 Share of different varieties of cement produced in India

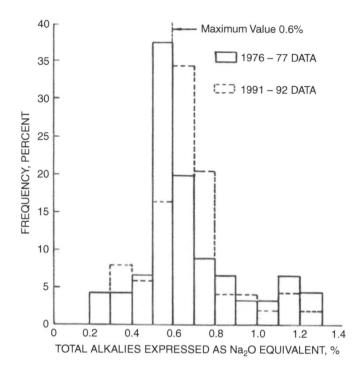

Figure 15.26 Distribution of alkali content in Indian cements.

alkali OPC in India has always been limited, and there is an increasing alkali content of cements, as more and more manufacturing plants adopt the dry process with energy conservation devices and for environmental protection. Hot exit gases containing volatiles as well as kiln dust are recirculated in the process stream and not allowed to be vented to the atmosphere.

Quality surveys of Indian cements for various periods have indicated an increasing proportion of total production having alkali content in excess of 0.6% ($Na_2Oeq.$) (Visvesvaraya & Mullick, 1986). Figure 15.26 compares the trend between two reporting periods; indicating the proportion of low-alkali cements declining. The trend is continuing even at the present, when 99% of cement capacity is based on a modern energy-efficient, environment-friendly dry process.

To meet market demand, a few cement manufacturing plants had adopted an alkali-bypass system, which increased thermal energy consumption, and, therefore, the cost of production. Low-alkali OPC produced in such manner is thus a *niche* product with a higher price. Non-availability of low-alkali OPC within economic distance for the majority of projects has forced consideration of other alternatives, as discussed below.

15.8.2 Role of blended cements and mineral admixtures

The beneficial role of mineral admixtures like fly ash or ground granulated blastfurnace slag in reducing expansion in concrete due to ASR has been known for a long time. Comprehensive investigations on the role of indigenous blended cements as well as pozzolana and slags used in commercial production of such blended cements in India, in alleviating ASR showed that, in general, blended cements are helpful, and optimum results are obtained when the amount of substitution of pozzolana or slag is relatively high, *i.e.*, of the order of 25—30% in the case of pozzolana and more than 50% in case of slag (Mullick, 2001). Indian Standard specifications permit pozzolana contents in blended cements to vary between 15 and 35%. Similarly, the slag content permitted in Portland slag cement is between 25 and 70 percent. In practice, the amounts of additions are somewhat lower, mainly to meet the requirements of early strength. In commercial blended cements with a prefixed quantity of pozzolana or slag, the flexibility to add larger doses of cement substitutes becomes somewhat restricted. The quality of the mineral admixtures to alleviate ASR are seldom evaluated in commercial cement production. In such cases, addition of the mineral admixtures in required dosage to OPC of even higher alkali content has been preferred. Typical data from one hydroelectric project is shown in Figure 15.27 (Mullick, 2001).

In this instance, aggregates from two alternate deposits (A and B) were found to be potentially reactive. OPC with a total alkali content of 1.28% was available. Replacement of OPC by fly ash up to 20% reduced the 180 days expansion in the 60°C test.

When commercial blended cements are to be used, one question that has often worried engineers is the safe limit of total alkalis in blended cements when the additives (pozzolana or slag) are not separately available for evaluation. The total alkali content in fly ash or slag is generally greater than in the cement. Depending upon the hydraulic activity of the slag or pozzolana, part of the alkalis contributed by them become available in the pore solutions. In NRMCA specifications, a limit of 0.9% total alkalis in the case of Portland slag cements in which the slag content is greater than 60% had been suggested (Mather, 1975). No such limit was, however, suggested for commercially produced Portland pozzolana cements, presumably because of not being common in USA.

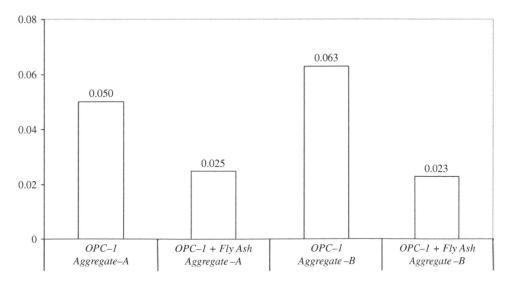

Figure 15.27 Reduction in ASR expansion by replacement of 20% fly ash to high alkali cement (values in the ordinate are expansion, %) (Mullick, 2001).

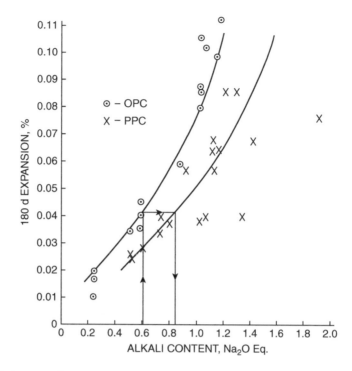

Figure 15.28 Relative performance of ordinary Portland cement and Portland pozzolana cement in mortar-bar expansion tests with slowly reactive Indian aggregates (Mullick *et al.*, 1989).

Investigation with a number of slowly reactive aggregates and commercially produced OPC and PPC showed that a limit of 0.6% total alkalis in ordinary Portland cements corresponded to a limit of the order of 0.8 – 0.9% in the case of Portland pozzolana cements, on the basis of comparable 180 days expansion in mortar-bar tests at 60°C (Figure 15.28) (Mullick *et al.*, 1989).

Nevertheless, the safe values depend upon a host of factors such as the chemical composition of the cement clinker and the pozzolana, the reactivity of the pozzolana to lime – water systems and the reactivity of the aggregates, thereby making any generalisation hazardous. It is prudent, therefore, to establish a safe aggregate – cement – pozzolana (or slag) combination by prior trials.

For many of the new constructions, in case the aggregates were classified as potentially reactive, then preventative measures, such as the provisions of IS: 456 discussed before (Section 15.6.3.2) were recommended. In case PPC was proposed to be used, the total alkali content in PPC was to be restricted to 0.8% (Wason, 2013). In case Portland slag cement was proposed to be used, it was recommended that such cement should be used after evaluation of aggregate – cement combination in mortar-bar tests, as the trend in case of such cements were not clear (Wason, 2013).

15.8.2.1 Ternary cement blends

Wanghka dam in neighbouring Bhutan required high performance concrete for the sluice glacis. The glacis is required to have high resistance to abrasion, for which high

Table 15.5 M40 grade ternary blend concrete, with resistance to ASR (Mullick, 2014).

Materials, kg/m³	Value
Cement (PPC)	425
Silica Fume	30.0
Water	164
Fine aggregate	800
Coarse aggregate, 20mm	575
Coarse aggregate, 10 mm	425
Superplasticiser (PCE) %	0.70
Water/binder ratio	0.36

strength concrete is generally specified. The aggregates were ASR-reactive and low-alkali OPC was not available. M60 concrete made with a blend of Portland slag cement with silica fume (ASTM C1240, 2015) was adopted, both for high compressive strength and adequate resistance to alkali-silica reaction (ASR) (Mullick, 2014).

For the lining of a diversion tunnel in a hydroelectric project in Sikkim presently under construction, M40 grade pumpable concrete is specified. The aggregates are alkali-reactive, for which adequate binder combination was to be adopted. Low-alkali OPC was not available. Even blending OPC with a large dose of fly ash was not possible. Only available cement was PPC, but the fly ash content could not be relied upon to combat ASR. Hence a combination of PPC with 7% silica fume was adopted, on the premise that the presence of both the fly ash in commercial PPC and silica fume (ASTM C1240, 2015) z will provide adequate resistance to ASR. The mix proportions are given in Table 15.5 (Mullick, 2014). A slump of 110 ± 10 mm and 28 days compressive strength of 54 MPa were obtained. The mix was recommended to be used after trials in mortar -bar/concrete prism tests.

15.9 CONCLUDING REMARKS

Instances of ASR in concrete structures in India have mainly been due to the presence of siliceous aggregates containing strained quartz. The potential reactivity of these slowly reactive aggregates could not be detected by the test procedures and evaluation norms existing at the time of construction. This led to modifications in the test methods and adoption of revised threshold values for mortar-bar expansion tests, according to which aggregates proposed for many new constructions are evaluated. The use of low-alkali cements, along with relatively large dosages of active pozzolana, is contemplated in such situations. Although the availability of low-alkali cements is restricted, it has been possible to meet the demand through indigenous sources and use of binary and ternary cement blends. The efficacy of the approach is borne out by the fact that no new construction in the last three decades has been reported to have exhibited alkali-aggregate reaction in concrete.

ACKNOWLEDGEMENTS

As in the previous edition, this contribution is mainly based on the research carried out at the National Council for Cement and Building Materials (NCCBM), New Delhi.

Contributions of colleagues in NCCBM namely, R.C. Wason, George Samuel, S.P. Ghosh, S.N. Ghosh, S.K. Sinha, L.H. Rao and C. Rajkumar in the research projects are acknowledged. Information received from Sanjay Pant, Director, Civil Engineering at the Bureau of Indian Standards (BIS) has been extremely valuable. Help received from Mrs Claire Bennett (RSK), Ms. Samhita Mullick and Prof. V. Sarath Babu in the preparation of the manuscript is greatly appreciated.

REFERENCES

Anderson, K. T., & Thaulow, N. (1990) The application of undulatory extinction angle (UE) as an indicator of alkali-silica reactivity of quartz, RILEM Committee Internal Report.

ASTM C227 (2010) *Standard test method for potential alkali-silica reactivity of aggregates (mortar-bar method).* West Conshohocken, USA, American Society for Testing and Materials.

ASTM C289 (2007, now withdrawn) *Standard test method for potential alkali-silica reactivity of aggregates (chemical method).* West Conshohocken, USA, American Society for Testing and Materials.

ASTM C295 (2012) *Standard guide for petrographic examination of aggregates for concrete.* West Conshohocken, USA, American Society for Testing and Materials.

ASTM C586 (2011) *Standard test method for potential alkali reactivity of carbonate rocks as concrete aggregates (rock-cylinder method).* West Conshohocken, USA, American Society for Testing and Materials.

ASTM C1240 (2015) *Standard specification for silica fume used in cementitious mixtures.* West Conshohocken, USA, American Society for Testing and Materials.

ASTM C1260 (2014) *Standard test method for potential alkali-reactivity of aggregates (mortar-bar method).* West Conshohocken, USA, American Society for Testing and Materials, West Conshohocken, USA.

Bérubé, M.A., Pedneault, A., Frenette, J., & Rivest, M. (1995) Laboratory assessment of potential for future expansion and deterioration of concrete affected by ASR, CANMET/ACI International Workshop on Alkali Aggregate Reactions in Concrete, Dartmouth, Nova Scotia, Canada (Compiled by B. Fournier), 267–291.

Bhatti, T.J., Chaudhry, M.N., & Hassan, G. (2004) Measured reaction, International Water Power Magazine, 20 December, 5p.

Bisque, R.C., & Lemish, J. (1958) Chemical characteristics of some carbonate aggregates as related to durability of concrete, *Highway Research Board Bulletin*, 196, 29–45.

Brown, L.S. (1955) Some observations on the mechanics of alkali-aggregate reaction, *ASTM Bulletin*, 205, 1–9.

Buck, A.D. (1983) Alkali reactivity of strained quartz as a constituent of concrete aggregates. *Cement Concrete Aggr.*, (ASTM), 5, 131–133.

Cement Research Institute of India (1982) Assessment of Strength of Concrete in the Training Walls and Apron Slab of the Bhakra Dam, Project Report, SP-141, 28p.

Cement Research Institute of India (1983) Assessment of Concrete in the Spillway Blocks of Hirakud Dam, Project Report, SP-147, 38p.

Chand, J.P., & Mullick, A.K. (1999) Surveillance of ASR-affected Concrete Structures: Case Study of Hirakud Dam Spillway, In: *Proceedings of the 5th International Conference on Concrete Technology for Developing Countries*, New Delhi, 2, VIII-22–31.

Chayes, F. (1952) On the association of perthitic microcline with highly undulant granular quartz in some calc – alkaline granite, *Am J Sci.*, 250 (4), 281–296.

Cole, W.F., Lancucki, C.J., & Sandy, M.J. (1981). Products formed in an aged concrete. *Cement Concrete Res.*, 11, 443–454.

CSA A23.2-14A (2014). *Test methods and standard practices for concrete, 14A, Potential expansivity of aggregates (Procedure for length change due to alkali-aggregate reactions in concrete prisms at 38°C)*, Canadian Standards Association, Mississauga, Ontario, Canada. 350–362.

Dolar-Mantuani, L. (1983) *Handbook of Concrete Aggregates, A Petrographic and Technological Evaluation.* Noyes Publication, Far Ridge, NJ, pp. 79–125.

French, W.J. (1992) The characterization of potentially reactive aggregates, In: *Proceedings of the 9th International Conference on Alkali-Aggregate Reaction in Concrete*, London, UK, pp. 338–346.

Gogte, B.S. (1973) An evaluation of some common Indian rocks with special reference to alkali – aggregate reaction. *Eng Geol.*, 7, 135–153.

Gratten-Bellew, P.E. (1992) Microcrystalline quartz, undulatory extinction and the alkali-silica reaction. In: *Proceedings of the International Conference on Alkali-Aggregate Reaction in Concrete*, London, UK, pp. 383–391.

Hadley, D.W. (1964) Alkali reactivity of Dolomitic carbonate rocks, *Highway Research Record*, 45, 1–19.

IRC: 112 (2011) Code of Practice for Concrete Road Bridges, Indian Roads Congress, Kama Koti Marg, Sector 6, RK Puram, New Delhi 110 022.

IS: 383 (2016) Coarse and Fine Aggregate for Concrete, Bureau of Indian Standards, Manak Bhawan, 9, Bahadur Shah Zafar Marg, New Delhi 110 002.

IS: 456 (2000) Plain and Reinforced Concrete – Code of Practice, Bureau of Indian Standards, Manak Bhawan, 9, Bahadur Shah Zafar Marg, New Delhi 110 002.

IS: 2386 (1963) Methods of test for aggregates for concrete – Part VII – Alkali-aggregate reactivity, Bureau of Indian Standards, Manak Bhawan, 9, Bahadur Shah Zafar Marg, New Delhi 110 002.

IS: 2386 (1999) Methods of test for aggregates for concrete – Part VIII – Petrographic examination, Bureau of Indian Standards, Manak Bhawan, 9, Bahadur Shah Zafar Marg, New Delhi 110 002.

IS: 3812 (2013) Specification for Pulverised Fuel Ash, Part I – For use as pozzolana for Cement, Cement mortar and Concrete, Bureau of Indian Standards, Manak Bhawan, 9, Bahadur Shah Zafar Marg, New Delhi 110 002.

IS: 12089 (1987) Specification for Granulated Slag for Manufacture of Portland slag cement, Bureau of Indian Standards, Manak Bhawan, 9, Bahadur Shah Zafar Marg, New Delhi 110 002.

Irrigation and Power Department, Government of Orissa, India (1983) Report of the Committee of Experts to Study and Advise Remedial Measures on Cracks in Hirakud Dam Spillway, 1, 67p.

Irrigation Department, Government of Uttar Pradesh, India (1986) Rihand Dam Experts Committee, Report, 1, 58p.

Jagus, P.J., & Bawa, N.S. (1957) Alkali – aggregate reaction in concrete construction. *Road Research Bulletin*, 3, 41–73.

Mather, B. (1952) Cracking of concrete in the Tuscaloosa lock, In: *Proceedings of the 31st Annual Meeting of Highway Research Board*, Washington, DC, pp. 218–233.

Mather, B. (1975) *New concern over alkali – aggregate reaction.* NMRCA Publication, National Ready Mixed Concrete Association, Silver Springs, MD, *149*, 20p.

Majid, C.M., Bhatti, T.J., & Chaudhry, M.N. (2013) Alkali silica reaction (ASR) potential of sand and gravels from NW-Himalayan rivers and their performance as concrete aggregate at three dams in Pakistan, *Sci. Int.* (Lahore), 25 (4), 893–899.

Mielenz, R.C. (1958) Petrographic examination of concrete aggregate to determine potential alkali reactivity, Highway Research Board, Report, 18–C, 29–35.

Mullick, A.K. (1987) *Evaluation of ASR potential of concrete aggregates containing strained quartz.* NCB Quest (New Delhi), *1*, 35–46.

Mullick, A.K. (1994) Alkali-silica reaction due to slowly reactive aggregates: Indian experience, P. K. Mehta Symposium on Durability of Concrete, Nice, France, CANMET/ACI, 175–206.

Mullick, A.K. (2001) Tests for alkali-silica reaction (ASR) in concrete and preventive measures – outline of current approach, *ICI Journal, Indian Concrete Institute, 2* (3), 31–36.

Mullick, A.K. (2014) Durability advantage of concrete with ternary cement blends and applications in India, In: *Proceedings of the 2nd International Conference on Advances in Chemically-activated Materials (CAM 2014)*, Changsha, China, pp. 320–335.

Mullick, A.K., & Samuel, G. (1986) Reaction products of alkali silica reaction – a microstructural study, In: *Proceedings of the 7th International Conference on Concrete Alkali – Aggregate Reactions*, pp. 381–385.

Mullick, A.K., Samuel, G., Sinha, S.K., & Wason, R.C. (1987) Assessment of residual expansion potential in concrete structures due to alkali – silica reaction. In: *Proceedings of the 4th International Conference on Durability of Building Materials and Components*, Singapore, II, pp. 793–800.

Mullick, A.K., & Wason, R.C. (1996) NBRI tests on aggregate containing strained quartz, In: *Proceedings of the 10th International Conference on Alkali Aggregate Reaction in Concrete*, Melbourne, Australia, pp 340–347.

Mullick, A. K., Wason, R. C., & Rajkumar, C (1989) Performance of commercial blended cements in alleviating ASR, In: *Proceedings of the 8th International Conference on Alkali Aggregate Reaction*, Kyoto, Japan, pp. 217–222.

National Council for Cement and Building Materials (1985) Assessments of Causes and Extent of Distress to Concrete in Rihand Dam and Power House Structure. Project Report, SP-197, 38p.

Natesaiyer, K., & Hover, K.C. (1989) Further study of an in-situ identification method for alkali – silica reaction products in concrete, *Cement Concrete Res.*, 19, 770–778.

Nixon, P.J., & Sims, I. (eds.) (2016) RILEM recommendations for the prevention of damage by alkali-aggregate reactions in new concrete structures. *RILEM State-of-the-Art Reports*, Volume 17, Springer, Dordrecht, ppRILEM, Paris, 168p.

Oberholster, R. E., & Davies, G. (1986) An accelerated method for testing the potential alkali reactivity of siliceous aggregates, *Cement Concrete Res.*, *16* (2), 181–189.

Poole, A.B. (1975) Alkali – silica reactivity in concrete. Symposium on Alkali – Aggregate Reaction, Preventive Measures, Reykjavik, Iceland, 101–111.

Rao, L.H., & Sinha, S.K. (1989) Textural and microstructural features of alkali reactive granitic rocks, In: *Proceedings of the 8th International Conference on Alkali Aggregate Reaction*, Kyoto, Japan, pp. 495–499.

Regourd, M. (1983) Methods of examination, In: *Proceedings of the 6th International Conference on Alkalis in Concrete*, pp. 275–289.

Regourd, M., & Hornain, H. (1986) Microstructure of reaction products.In: *Proceedings of the 7th International Conference on Concrete Alkali – Aggregate Reactions*, pp. 375–380.

Samuel, G., Mullick, A.K., Ghosh, S.P., & Wason, R.C. (1984) Alkali silica reaction in concrete – SEM and EDAX analyses. In: *Proceedings of the 6th International Conference on Cement Microscopy*, Albuquerque, New Mexico, USA, pp. 276–291.

Samuel, G., Wason, R. C., & Mullick, A. K. (1989) Evaluation of AAR Potential of Limestone Aggregates in India, In: *Proceedings of the 8th International Conference on Alkali-Aggregate Reaction*, Kyoto, Japan, 235–240.

Shelley (1964) On myrmekite, *Am Mineral.*, 49, 41–52.

Shelley (1967) Myrmekite and myrmekite like intergrowth, *Mineralogist Magazine*, 36, 491–503.

Sims, I., & Hewson, N. (2016/7) DEF in some concrete in Sri Lanka and an engineering solution. *Proceedings of the Institution of Civil Engineers, Forensic Engineering*, Themed issue: Forensic Engineering in Developing Countries, Part 2. XXX (XXX) XX–XX. The Institution of Civil Engineers, London. In preparation.

Sims, Ian & Sotiropoulos, P. (1983) Standard alkali-reactivity testing of carbonate rocks from Middle East and North Africa, In: *Proceedings of the 6th International Conference on Alkalis in Concrete*, pp. 337–350.

Singh, R., Sthapak, A.K., Goyal, D.P., & Khazanchi, R.N. (2003) Use of microsilica in concrete/shotcrete applications at Tala hydroelectric project in Bhutan Himalayas, In: *Proceedings of the Workshop on 'Use of Silica Fume Based High Performance Concrete in Hydraulic Structures'*. *CSMRS*, New Delhi, pp. 41–50.

Thaulow, N., & Knudsen, T. (1975) Quantitative microanalysis of-the reaction zone between cement paste and opal. Symposium on Alkali – Aggregate Reaction, Preventive Measures, Reykjavik, Iceland, 189–203.

Visvesvaraya, H.C., Mullick, A.K., Samuel, G., Sinha, S.K., & Wason, R.C. (1986) Alkali reactivity of granitic rock aggregates. In: *Proceedings of the 8th International Congress on the Chemistry of Cement*, Rio de Janeiro, Brasil, V, pp. 206–213.

Van Aardt, J.H.P., & Visser, S. (1978) Reaction of $Ca(OH)_2$ and of $Ca(OH)_2 + CaSO_4.2H_2O$, at various temperatures with feldspars in aggregates used for concrete making. *Cement Concr. Res.*, 8, 677–682.

Visvesvaraya, H.C., & Mullick, A.K. (1986) Quality of cements in India – results of three decadal surveys, In: Farkas, B., & Klieger, P. (eds.) Uniformity of Cement Strength, ASTM Special Technical Publication, STP 961, 66–79.

Visvesvaraya, H.C., Rajkumar, C., & Mullick, A.K. (1986) Analysis of distress due to alkali—aggregate reaction in gallery structures of a concrete dam. In: *Proceedings of the 7th International Conference on Concrete Alkali – Aggregate Reactions*, pp. 88–193.

Wason, R.C. (2013) Personal communication.

Chapter 16

Middle East & North Africa

Ted Kay, Alan B Poole & Ian Sims

16.1 INTRODUCTORY OVERVIEW

This chapter covers the geographical region of the Levant and the Arabian Peninsula as illustrated in Figure 16.1 and the North African countries which border the Mediterranean. There have been few, if any, published reports of ASR in this region, though it is known that some of the aggregate sources contain materials which are potentially reactive (French & Poole, 1976; Sims & Poole, 1977). Many specifications take this into account and require aggregates to be tested for alkali-reactivity potential and in some cases there are requirements to limit alkali content of mixes or to include materials such as pozzolanas or microsilica in order to reduce the likelihood of occurrence of a damaging reaction.

The bulk of the major concrete construction projects in this region, particularly in the Arabian Peninsula and North Africa, have been carried out since the mid 1970s. At that time, many of the countries on the peninsula had recently become independent and the development of national oil industries had given them the resources to undertake major development programmes. The construction of ports, airports and communications networks and the enlargement of centres of population and industrial facilities with their associated infrastructure required huge amounts of concrete.

In many cases, concrete production and concrete structures are exposed to environmental conditions which are much more severe than those experienced in most other regions of the world. In addition to the arid climate and high temperatures which made the production of high quality concrete with the local ingredients very challenging, many of the resulting structures were exposed to high concentrations of sulphates or chlorides in service. These conditions led to some early deterioration problems; many were the result of corrosion of reinforcement.

Amongst the underlying causes of these early deterioration problems were the lack of a concrete technology appropriate to the local conditions, the lack of an established materials supply chain and the lack of experience of the use of the local natural materials. It was not uncommon in the late 1970s and later for new quarries or borrow pits to be opened specifically to service individual construction projects. Although the use of unfamiliar and, in some cases, not entirely suitable natural aggregates played a part in the early deterioration of concrete structures, very few, if any, cases of damaging alkali-aggregate reactivity have been reported.

In the early days of concrete construction in the region, European and American Standards were applied without any modification for design and specification.

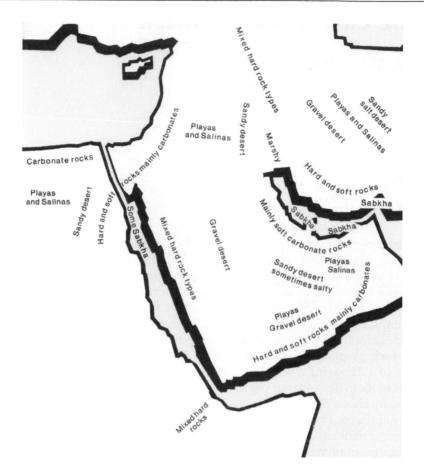

Figure 16.1 The Levant and the Arabian Peninsula after Kay *et al.* (1982) (with permission of the Concrete Society).

These standards were not particularly well focussed on durability issues, as durability of concrete structures had not, prior to that time, been a serious general concern in their countries of origin. This meant that in many cases the resulting structures were not robust when exposed to the much more severe environmental conditions in the region.

During the 1980s, soundly based concrete production and supply industries developed in most of the countries across the region. During this period also, a greater awareness began to develop of the need to use higher specification concrete in order to combat the aggressive exposures typical of the region. This led to much more focus on testing of aggregates, a general reduction in water/cement ratios, use of higher cement contents in the design of concrete mixes and greater use of admixtures. More recently, the use of cements containing fly ash and blastfurnace slag has become more widespread. During the 1980s a greater awareness of the possibility of alkali-aggregate reaction developed in the community of designers in the region and some specifications from the late 1970s onwards included requirements to test aggregates for reactivity potential.

The greater awareness of the need for more rigorous specification was promoted in publications and conferences. Individual specifying authorities, including some public or municipal bodies, produced their own documents, but to the time of writing, there are still few indigenous national standards for construction and reliance is still placed on European and American standards and consensus guidance documents.

16.2 GEOLOGICAL MATERIALS AVAILABLE FOR CONCRETE

16.2.1 The Arabian Peninsula

The geology and geomorphology of the Arabian Peninsula from an engineering point of view were first described by Fookes and Higginbottom (1980) and Owels and Bowman (1981). These papers were summarised in part of what became known as 'The CIRIA Guide' (Walker, 2002) on which the following sections are based.

For purposes of description, the Arabian Peninsula can be divided into five principal physical divisions:

1. A narrow and flat western coastal plain alongside the Red Sea,
2. A mountain range running parallel to the Red Sea coast,
3. A hilly central plateau,
4. A flat eastern coastal plain alongside the Gulf,
5. A southern coastal plain of variable width bordering the Oman Mountains and the Southern Mountains of Saudi Arabia and Yemen.

Several depressions and basins believed to be the result of either tectonic movement or erosion are superimposed on this base topography.

16.2.1.1 Underlying basement

A Precambrian complex of igneous and metamorphic rocks known as the Arabian Shield underlies approximately one-third of Saudi Arabia. The remainder is underlain by sedimentary rocks with gentle average dips to north, south or east of 1 to 2 degrees. These rocks were deposited over the basement rocks from the Cambrian to Quaternary periods with thicknesses up to several hundreds of metres.

16.2.1.2 Western coastal plain

The coast is usually protected or isolated by offshore sand bars and limestone reefs that may form lagoons. Offshore coral banks and reef limestones either outcrop or are covered with sands which are mainly carbonate.

The plain itself is covered in many areas by the salt-bearing deposits known as sabkhas and being low-lying, can be flooded as a result of the occasional rain storms. In other places there are terraces of coralline limestone bordering the Red Sea where the salt-bearing soils are absent. Elevated limestone reefs occur frequently and raised beaches, sand dunes, and locally gypsiferous sands and gravels are common. Further inland, localised gypsum outcrops and deposits of argillaceous (clayey) silts do occur.

16.2.1.3 Western mountain zone

There is a higher eastern plain which rises above the plains alongside the Red Sea. This piedmont zone, consisting of granular soils, overlies Tertiary bedrock (limestones, shales, siltstones and sandstones) or ancient strong crystalline rocks, and extends eastwards to the mountainous Arabian Shield. The higher zones of the slope tend to contain gravels and boulders, while the lower gentle slopes tend to contain finer soils. Some of the granular soils close to the coastal plain were deposited during floods that originated in the mountains to the east. Both borrow pits of fine to coarse alluvial material and bedrock quarries have been worked in this area.

Layers of salt are sometimes found in the Tertiary bedrock. For example, halite occurs at Jizan, on the coast in south-western Saudi Arabia.

16.2.1.4 Central plateau

The local bedrock basement of the Arabian Shield is formed mainly from igneous and metamorphic rock types of Precambrian origin. Examples are the granites and gneisses which occur at Khamis Mushayt and the granites, schists, amphibolites and gneisses at Taif. They are formed as domes and are often overlain by younger volcanic rocks (often basalts) or thin layers of alluvial sands and/or gravels generally less than 10m thick. Over much of the south-western part of the Shield, late Precambrian non-metamorphosed red arkose and conglomerate can occur in isolated basins or folds.

In the late Precambrian, uplift and erosion is believed to have occurred. The erosion has continued, producing the present day alluvial fans and wadi deposits. Sandstones, overlain by shales, with the older sandstones extending northwards into the marine Cambrian rocks of Jordan, cover the eroded Precambrian basement.

The basement rocks to the east of the Arabian Shield are overlain by sedimentary rocks (mostly limestone, but also sandstone and shale) of various post Precambrian ages. These extend to the Gulf. A thin soil cover, a few metres thick, resulting from either rock weathering or windblown sand, generally overlies these sedimentary rocks. In other locations there can be a thicker cover of wadi alluvium or residual soils (mostly stiff clays resulting from rock weathering) that extends to a depth of several metres.

The sedimentary shelf rocks, which dip gently to the east, have been subject to uplift and erosion, probably in the Tertiary. The more resistant or more massive rock units form west-facing escarpments. Jabal Tuwayq near Riyadh is an example. In wadis between these escarpments, sand and gravel of Quaternary age can be found, for example in Wadi Hanifa west and south of Riyadh. Erosion channels such as Wadi as Sab'ha in Al-Kharj contain deposits of Tertiary sands and gravels and lacustrian limestone (Owels & Bowman, 1981). Pits in the alluvial deposits and quarries in the bedrock have been exploited but the materials produced are often of poor or variable quality.

16.2.1.5 Eastern coastal plain

This plain alongside the Gulf is covered by Quaternary sediments. These consist of salt-bearing soils, sabkhas, some areas of dunes and outcrops of limestone and calcareous sandstone. The occurrence and layering of surface and near-surface sands in the plain and the Gulf floor alongside have their origin in Quaternary marine processes (Purser, 1973). Close to the shore, the floor of the Gulf is covered with shelly and clayey sands, corals, limestones and sandy limestones. In places, the shelly sand can be cemented to thin hard ledgerocks or marine conglomerates. A number of islands offshore from the coastal plain, Das Island off Abu Dhabi is an example, were formed as salt plugs. They are a mixture of recent sediments, evaporite minerals and exotic rocks from the underlying strata.

16.2.1.6 Southern coastal plain

The southern coastal plain is bordered to the south by the Arabian Sea and to the north by the Oman Mountains. These mountains are higher at the western (Yemen) end and the eastern (Muscat) end of the range. They are composed of igneous, metamorphic and sedimentary rocks, whereas in the central section there are only moderately high flat topped hills consisting mainly of limestone. The hills dip inland towards the Rub al Khali (Empty Quarter). At the foot of the escarpment, on the seaward side, a small pediment of eroded material, which extends across the coastal plain, occurs in many locations.

The coastal plain itself is of variable width and covered by Quaternary sediments. These consist mainly of salt-bearing soils, sabkhas, areas of sand dunes and alluvium. There are also outcrops of limestone and calcareous sandstone. The southern coastal plain has more changes in elevation than the eastern coastal plain, being intersected by many wadis of various sizes. These also cut into the escarpment of the Oman Mountains.

The near-shore sea floor consists of shelly and clayey sands, corals, limestones and sandy limestones. The shelly sands often occur with gravels that are locally derived and mainly of limestone. They may be cemented to a thin but relatively strong ledgerock (caprock) or marine conglomerate.

16.2.1.7 Geomorphology

It is probable that most of the inland landscape of the Arabian Peninsula was formed in times when the climate was much wetter than at present. For example, in the Quaternary and Late Tertiary periods, savannah conditions may have prevailed. The superficial deposits that form the modern desert surface (aeolian sand, alluvial gravel and sand and alluvial silt and clay) are of Quaternary age and were mainly laid down in the past two million years.

The advances and retreats of ice ages, though not directly affecting the region, resulted in large changes in the local sea level. These, along with the climatic changes in the Quaternary, modified the weathering and erosion patterns. The results are submarine terraces, buried coastlines, flat inland surfaces and extensive rock pavements.

The present coastline features, which include widespread coastal flats, are partly a result of the present hot, arid climate and recent sea level changes. The coastal flats are underlain by a near-surface saline water table. Evaporation from the land surface results in deposits of evaporite salts. Losses of moisture from the surface are replenished by capillary movement from below and the deposits can build up to substantial thickness. This is a hugely important consideration, as much of the modern development has occurred in the coastal plain and the evaporite salts provide a ready source of alkali metal ions in the local environment. These can easily find their way into concrete structures through moisture movement.

Inland, wind and gravity are the most typical transport mechanisms for erosion products in the lower flatter areas. Water, from occasional cloudbursts in present times or from earlier times of a wetter climate, is more important in the steeper higher locations.

16.2.2 Syria, Lebanon, Israel, Jordan

The main topographical features of these lands at the eastern end of the Mediterranean are a narrow coastal strip along the western shore, the Jordan valley running north-wards from the Gulf of Aqaba on the Red Sea along most of the west side of the region, and an elevated plateau to the east of the mountains leading to the Mesopotamian basin in Iraq in the east. The main mountainous regions are close to and run roughly parallel to the Mediterranean on either side of the Jordan valley. These include the Jebel an-Nusseiriyeh range in the north-west of Syria and the Lebanon and the Anti-Lebanon and Hermon Mountains further to the south with the Homs gap providing a route between the coast and the interior.

A low relief undulating plateau extends eastwards from the mountains to the Euphrates basin. This plain is intersected by a low chain of mountains extending north-eastward from the Jebel ad-Drouz (Jebel al Arab) near the Syria/Jordan border to the Euphrates. The barren desert region known as the Hamad lies south of this chain, while north of Jebel ar Ruwaq and east of the city of Homs is another barren area with a hard packed dirt surface known as the Homs Desert. There are also mountains to the north and north-east in the Al Jazira region of Iraq, along the border with Turkey.

Folded Precambrian rocks are exposed along the western margin between the Gulf of Aqaba and the Dead Sea. These Precambrain rocks are overlain by Palaeozoic, Mesozoic and Cenozoic sediments. Volcanic activity occurred repeatedly in the Miocene, Pliocene and Pleistocene in the Harmon and Jebel ad-Drouz regions resulting in basalt outcrops. Basalt also occurs in the southern part of the Jebel an-Nusseiriyeh region, in the area between Aleppo and the Palmyrides,

Quaternary sediments in the form of marine terraces, river terraces and alluvial fans are limited to the Mediterranena coast, the Ghab depression and inland basins.

Within Lebanon, the major geological structures of Mount Lebanon, the Bekaa Valley and the Anti-Lebanon Mountains are two very large anticlines trending north-north-east to south-south-west separated by a large syncline. Almost all the rocks in Lebanon are sedimentary rocks and most of these are pale fine grained limestones.

The region includes what is thought to be the lowest quarry in the world. The quarry at As-Sweimeh in the Jordan Valley is within sight of the Dead Sea and its highest level is 150 m below sea level. As-Sweimeh was originally opened to exploit the limestone deposits, but as it developed, marble and silica sand deposits were also found.

16.2.3 Iran and Iraq

The region of Iraq and Iran lies to the north and east of the Gulf. Iraq occupies a lowland trough which forms the flood plain and delta of the Tigris and Euphrates rivers. Iran to the immediate east is bounded by the Zagros mountain chain, which forms the physical boundary between Iraq and Iran and beyond a narrow coastal plain forms the eastern margin of the Gulf. To the east of this mountain chain is a high central plateau of desert and semi-desert (The Lut Desert). It is crossed by numerous hilly ranges trending north-west to south-east. Depressions in the plateau tend to develop salt crusts with a larger and variable area of salt marsh on the northern part of the plateau. To the north and east, beyond the east-west trending Alborz Mountain chain and coastal plain, it is bounded by the Caspian Sea.

16.2.3.1 The general geology of Iraq (after Jassim & Goff, 2006)

Iraq is formed of a lowland trough lying between asymmetrical dissimilar upland massifs. It is a region of crustal weakness and subsidence, containing Quaternary gravels and sandy sediments derived from the erosion of the Syrian-Arabian block of hard ancient granitic rocks that is overlain by Mesozoic sediments to the east and north, and from erosion of the Zagros Mountains to the east. The Zagros mountains are composed principally of younger Jurassic and Eocene sediments, including a basal layer of evaporite deposits with smaller regions of metamorphosed, intrusive and ancient hard rocks. These mountains result from the still active collision of the Arabian and Eurasian tectonic plates. They form a complex region with thrusts and folding in the north and a simpler series of sub-parallel anticlinal fold belts trending north-west to south-east in the south.

The trough contains the region's two main rivers, the Tigris and the Euphrates, which flow from north-west to south-east. They converge near Baghdad, then diverge and meet once again about 160 km (100 miles) north of the Gulf, to form the Shatt al-Arab River which drains into the Gulf through a low-lying swampy area, containing marshes, lakes and reedy waterways. Sediments in the lowland alongside the Euphrates consist of Neogene limestones and marls.

Iraq lies at the heart of the 'Fertile Crescent'. This region consists of a low central depression extending from central Syria to the Gulf. This trough which is oriented north-west to south-east contains the youngest sediments in Iraq. It has a gently inclined plateau along its west and south-west edges and a series of ridges and depressions developing into a mountainous area to its north-east. The highest point in western Iraq (936 m) lies close to the triple point where Iraq, Saudi Arabia and Jordan meet. From this high point the terrain slopes towards the Euphrates River at a gentle gradient of between 10 and 20 m/km. This plain consists of a desert divided into two distinct sectors. The Widyan Desert which lies north of latitude 32° is characterised by numerous wadis, which lie in east-west, north-east/south-west and north-south directions. The Southern Desert, lying south of the Widyan Desert and west-south-west of the Euphrates, does not have the prominent drainage which often characterises the Widyan Desert.

The central depression consists of the Plains of Mesopotamia to the south-east and Jezira in the north-west. The former runs from the Gulf at its south-east end to Wadi Thartar in the north-west. It has the Euphrates river as its western boundary and the Makhul-Hemrin-Pesh-i-Kuh range of mountains to the east. There are widespread Quaternary deposits from the Tigris and Euphrates. Older Pleistocene river fans are revealed in the elevated and dissected northern part of the Mesopotamian Plain, which has similar features on its eastern and western margins. The Jezira Plain lies between the Mesopotamian Plain and the East Syrian Euphrates depression and lies at an elevation of around 250 m.

An area with lines of hills separated by broad depressions lies to the east of the central depression. These hills and depressions are long anticlinal structures and broad synclines form the foothills of northern and north-eastern Iraq. There is an abrupt boundary between the foothills and the Mesopotamian Plain, formed by the south-west flank of the Makhul-Hemrin-Pesh-i-Kuh anticlines. Anticlines increase in frequency further to the north-east and their elevation also increases. To the east of Kirkuk and north of Mosul, high anticlinal mountains are prominent surface features, which are generally of 'whale back' character with cores of Eocene limestone and flanks of Miocene and Pliocene clastics. The mountains further to the north and north-east have Cretaceous limestone cores with Palaeogene clastics on their flanks, whilst further to the east the mountains have Upper Cretaceous and Jurassic limestone cores and flanks of Palaeogene clastics. Close to the borders with Iran and Turkey in north-east Iraq, the mountains consist of a series of thrust sheets reaching a height of 3000 m and containing metamorphic and igneous rocks.

16.2.3.2 Surface geology of Iraq

The youngest sediments in Iraq lie within the central trough and are of Quaternary and Neogene age, whilst older Palaeogene and Palaeozoic material is exposed along the flanks. In this respect, surface geology roughly reflects the morphology.

Strata which dip to the north-east at an inclination slightly greater than that of the land surface characterise the desert area south-west of the Euphrates. This is in contrast to the strata in western Iraq near Rutba, which dip to the west away from the major anticline in which rocks as old as the Permian are exposed. The crest of this anticline has formed the Ga'ara depression. The Mesopotamian depression consists of a complex of river channels with associated levees, flood plains, marshes, sabkhas and deltas. It is bordered on both sides by alluvial fans.

In the north-west, the Jezira area is dominated by the massive Tayarat uplift, which has Middle Miocene sabkha deposits exposed in its core and which is flanked to the east, north and west by Upper Miocene clastics. On the south side of the uplift, Oligocene and Lower Miocene carbonates have been exposed by the erosion of the Euphrates along anticlinal structures associated with east-west faults.

North-east of the Mesopotamian depression, the foothills are made up of narrow anticlines with Upper Miocene to Pleistocene molasse sediments, or Middle Miocene evaporites exposed in their cores. The Sinjar and Qara Chauq and similar higher amplitude anticlines have Palaeogene and sometimes Upper Cretaceous formations exposed in their cores.

To the north-east of Kirkuk, the mountainous region consists of regular folds with Cretaceous or older rocks exposed in their cores and Palaeogene and Neogene rocks forming the adjacent synclines. Along the border with Turkey, thrust faults bound tight anticlines with Palaeozoic to Cretaceous rocks exposed in their cores.

Close to the Iranian border, there are thrust sheets of sedimentary and igneous rocks which were later eroded and covered by Upper Maastrichian to Palaeogene clastics and carbonates. These in turn are often overridden by thrust sheets of Neogene volcanic and associated clastics and carbonates of Palaeogene age. Separate Cretaceous thrust sheets of metamorphic and volcanic material overly the Tertiary volcanics and have been intruded by large basic and ultrabasic plutons. The highest thrust sheets are composed of metamorphic and igneous rocks.

16.2.3.3 The general geology of Iran

Iran is bounded to the west by Iraq, Azerbaijan, the Caspian Sea and Turkmenistan in the north and the Gulf further south. Beyond a narrow coastal plain bordering the Gulf, the Zagros mountain chain forms a wide fold and thrust belt trending north-west to south-east from the Sirvan River in the north to Shiraz in the south, a distance of 900 km. These mountains were generated by the collision of the still active Eurasian and Arabian Plates. Geologically, the Zagros Mountain ranges may be divided into two major parts: an elevated region forming the north-eastern mountains, while a folded region lies to the south and west. Eastwards, the elevated Zagros region faces the inner highlands of the Zagros Mountains, also known as the Sanandej-Sirjan Zone. The folded Zagros region terminates at the Gulf in the south, and the Khuzestan plain and Iraq in the south-west and west.

The extensive folding results from deformation above a layer of rock salt in the south-eastern Zagros, but to the north-west this salt layer is missing. Subsequent erosion removed much of the pre-existing layered softer sedimentary rocks, such as mudstone and siltstone, while leaving relatively harder rocks, such as limestone and dolomite. This differential erosion has formed the linear ridges of the Zagros Mountains, which form many sub-parallel ranges 10 to 250 km wide. Salt domes and salt glaciers are a common feature of these mountains and are an important target for oil exploration, since the impermeable salt frequently traps petroleum beneath other rock layers. The Kuhrud Mountains represent one of these parallel ranges, but at a distance inland of approximately 300 km east forming the Central Iranian Mountains. They are composed of sediments and volcanics. Rivers flowing from their flanks eventually end in salt lakes.

The central Iranian plateau region is largely formed of the Gondwanan terrains between the Turan platform to the north and the Main Zagros thrust belt, which forms the suture zone between the northward moving Arabian plate and the Eurasian continent. It contains two major salt desert basins: the Dash-e Kahir in the north and the Dash-e Lut in the south-east. It is a geologically well-studied area because of general interest in continental collision zones, and because of Iran's long history of research in geology, particularly in economic geology. It extends from East Azerbaijan, north-west of Iran, to Pakistan west of the Indus River. It also includes smaller parts of the Republic of Azerbaijan and Turkmenistan. Its mountain ranges can be divided into the Alborz ranges bordering the Caspian Sea, which are composed of limestones, dolomites, volcanics and clastic sediments. On the north-western edge of the Iranian Plateau, the Turkish Pontic and Taurus Mountains converge, giving rise to rugged country with the average elevation of its peaks exceeding 3000 m. In the south-east of Iran, the barren Makran mountainous region of Balochistan continues into Pakistan to the east.

Geologically the plateau is underlain by Devonian, Carboniferous and Lower Jurassic sediments including coal seams. Some areas of these sediments were metamorphosed in the early Cretaceous. They were later covered by Eocene limestones, basic lava flows and tuffs (The Green Series) in the Elburz region, and by thick flysch deposits in the east and south-east. Further complications arise from Plio-Pleistocene volcanic deposits in Azerbaijan, Elburz, Kerman and Baluchistan. Large areas of the underlying geology of this plateau are masked by a covering of Quaternary sands and gravels. Additional detailed information concerning the geology and natural resources of Iran can be obtained from Ghorbani (2013).

16.2.4 North Africa

16.2.4.1 Geographical background

This region consists of the countries which border the Mediterranean from Morocco in the west to Egypt in the east. The dominant feature of this region is the massive sandy and sometimes gravelly Sahara desert, which stretches across almost the whole of the southern part of the region from the Atlantic Ocean to the mountain ranges along the Red Sea coast in Egypt.

A further dominant feature of the region is the Nile valley, which runs roughly south to north near the eastern edge of the region, with its massive delta where the river meets the Mediterranean in the north. Most of the population of Egypt is located in the Nile Valley or Delta.

To the north of the vast Sahara region there are intermittent mountain ranges, predominantly towards the west and in the coastal plain alongside the Mediterranean. There are also mountain ranges to the south: the Ahaggar Range in Algeria and the Tibesti south of the southern border of Libya.

In Libya, the most densely populated area is the coastal plain, which stretches for around 1750 km along the Mediterranean coast and is sometimes marshy. In the east the land rises from the coastal plain to the Jabal al Akhdar range at a height of around 900 m. In the west the land rises in a series of steps to the Plain of Jafarah and the Jabal Nafusah Plateau.

Tunisia, Algeria and Morocco, when taken together, are composed of four main geographical regions which generally run east-west:

1. A narrow discontinuous coastal strip, which generally runs along the southern shore of the Mediterranean and also extends north-south lying along the eastern side of Tunisia (the Sahel) and along the Atlantic coast of Morocco. In Tunisia and Algeria the coastal strip is known as the Tell. It consists of hills and fertile valleys and contains the main cities and most of the countries' arable land.
2. A high plateau or tableland marked by rolling hills and some steep cliffs and flat plains interspersed with large shallow basins. These basins collect water during the wet season, but dry out in the dry season sometimes forming salt flats known as chotts.
3. A region occupied by a series of mountain chains the most dominant of which, the Atlas Mountains, occupies a band stretching diagonally across the region from the south-west to the north-east starting in southern Morocco and ending in the Northern Tell and High Tell in Tunisia. Within Algeria, the Saharan Atlas is formed of three chains: Jibal Amor in the south-west, Jibal Awlad Nail in the

centre and the Monts du Zab in the north-east. There are three parallel chains in Morocco: the Moyen Atlas in the north, the Haut Atlas in the centre and the Anti-Atlas to the south.

4. The Sahara desert, a desolate flatland of gravel plains and some areas of large sand dunes with dispersed oases. The region consists of a Precambrian basin unconformably overlain by thick Palaeozoic sediment deposits (Askri *et al.*, undated).

16.2.4.2 General geology

The general geology of the North African countries is best summarized for the whole of the region. A more detailed and comprehensive guide to the geology of North Africa, country by country, is available in the Geologic Atlas of Africa (Schlèuter & Schlu'ter, 2008). This geologic summary is primarily concerned with the locations and rock types potentially suitable as concrete aggregate sources.

Populations of these countries are located principally in towns and along the coastal strip where most modern development occurs. In the very sparsely inhabited hinter-lands, the use of concrete is largely restricted to hydro-schemes in Morocco and the Nile valley, and to highways, pipelines and oil well infra-structures.

The whole region is underlain by a crystalline Precambrian basement complex, which is overlain by later sedimentary sequences of Jurassic to Pliocene age. This Precambrian basement is exposed in parts of the Atlas Mountains, a large area of the Ahaggar Mountains in southern Algeria and in the mountains bordering the Red Sea, Egypt's Eastern Desert. Large areas along the coastal strip are masked by recent river, wadi and coastal deposits of gravels and sands. Other coastal regions are made up from Miocene or Pliocene marine sands, clays, limestones and detrital continental deposits.

Areas of post-Cretaceous basalts and other volcanics form isolated mountains and long ridges in the south of Libya some 600 km south of Tripoli. Other interior parts of this region are formed of marine, freshwater and continental deposits of Cretaceous to Oligocene age. They form extensive plateau areas and dissected remnants and bold scarps. The Libyan Fezzan, an area in the south-west of the country, has been described as an ancient planation surface (Fookes & Gahir, 1995), which is filled to the north and dissected by wadi systems into a number of watershed massifs and ridges. A mountain range of Precambrian granites and metamorphic lithologies, overlain by Cambrian sandstones and cone-like remnants of volcanic domes, lies at the southern end. The margins of the mountain range are marked by extensive bajada (fan) deposits with deeply incised wadis. In the flat central areas, basalt lava flows from the Oligocene-Miocene volcanic episodes outcrop, whilst northwards the basalt occurs as fields of boulders and cobbles which have resulted from weathering and erosion of the flows.

In southern Algeria, the largely Precambrian Ahaggar Mountains and highland area, which rises to 2908 m at Mount Tahat, is a region of mountains and rocky desert formed mainly of black basalts, but granites, gneisses and volcano-clastic rocks are also present. Some areas of basic volcanics form isolated mountains and long ridges and lie within the main plateau. The mountains are fronted on the north and east by the Djanet Tassili Mountain ranges, which continue into southern Libya and are composed of hard Palaeozoic sandstones. The whole of this highland area is cut by a complex of north-south trending fault lines. The mountains often stand as isolated inselbergs or erosional pillars, surrounded by extensive scree deposits, boulder fields and areas of loose sands and dunes.

The Atlas mountain chains occupy much of Morocco, Tunisia and the northern margin of Algeria, bordering the Mediterranean and running in total for some 2500 km. The Atlas Mountains form a part of the Tertiary Alpine-Himalayan fold system. They are intensely folded and often highly metamorphosed. They are composed of a variety of rock types ranging in age from Precambrian to Miocene. Movement and uplift began at the end of the Jurassic, continuing in the Upper Cretaceous and on into the Miocene. There is evidence of earlier folding of Palaeozoic age in some parts and the existence of moraines and fluvio-glacial deposits indicate severe glaciation during the Ice Age.

These mountain chains are sub-divided into discrete ranges separated by wide valleys and basin areas. In Morocco, the ranges trend west-south-west to north-north-east with the most southerly being the Anti-Atlas of Palaeozoic age composed of crystalline Precambrian rocks and Palaeozoic sandstones and shales. The High Atlas lies to the north and east of the Anti-Atlas and is the highest range in Morocco, rising to the 4167 m peak, Toubkal. The slopes of the High Atlas are precipitous on the side facing the Atlantic, but more gradual on their southern flanks. They are largely composed of marine Jurassic rocks. In Morocco, the lower most northerly range the Middle Atlas is also composed chiefly of Jurassic sediments. The lower slopes on both flanks are wooded.

The Atlas mountain ranges continue eastwards into Algeria and Tunisia and form a boundary with the Mediterranean to the north and the Sahara to the south. The most northerly ranges in Algeria and Tunisia run parallel to the coast and are referred to as the Tell Atlas, which is largely composed of Jurassic and Tertiary sediments. They are separated by a 1000 m high plateau of the Saharan Atlas, which forms the northern boundary of the Sahara. In the East of Algeria, the two ranges converge and they extend into Tunisia as the Aurés Mountains. The high plateau region contains shallow salt water lakes and salt flats. Quaternary and recent deposits partially mask the underlying Jurassic, Cretaceous and Tertiary sediments

16.3 AGGREGATE SOURCES

16.3.1 Arabian Peninsula

In the past, both beach sand and wadi deposits and associated gravel fans were used as sources of aggregate for concrete. The use of beach sand has been discontinued in many places because of the risk of chloride contamination. Wadi deposits and the gravel fans have been worked out in many places and quarried rock and some dune sand are the common materials used in concrete production.

Limestone is the most frequently occurring sedimentary rock used for concrete aggregate, particularly in the eastern Arabian Peninsula. Chert is present in some of the limestone and has to be considered as an alkali-reactivity risk. There are vast areas of horizontally bedded young Tertiary and Quaternary limestones in the east of the region. These have been used for concrete aggregate, but are variable in character, have poor physical properties and can be contaminated with evaporite salts. There are some small areas of relatively strong Tertiary limestone and dolomite in Bahrain, Qatar and on the adjacent parts of Saudi Arabia near Dharan. The beds are usually thin and the resulting aggregates vary in physical and chemical properties. There is a widely-used source of hard good quality Jurassic limestone at the northern tip of the Oman Mountains in Ras Al Khaimah in the UAE. Limestone of this type also occurs in places in Oman. Horizons of shale or red residual clay may be present, so the rock has to be processed with care to remove this undesirable material.

Limestone in the region can be dolomitic. Some dolomites can potentially exhibit alkali-carbonate reactions. These often develop as clear de-dolomitised rims around aggregate particles, but any expansion of concrete containing such aggregates is usually ascribed to alkali-silica reaction caused by a reactive silica component contained within the carbonate aggregate. Such reactions have been reported in Cyprus and Bahrain (French & Poole, 1974 and Chapter 8, 8.11) but a series of tests on material from Attaka quarry in Egypt proved negative (Sharobim, 1995). Sims and Sotiropoulos (1983) reported finding reaction rims with dolomitic aggregates from North Africa that were not associated with any expansion. See Chapter 3 for an up-to-date account of the 'so-called alkali-carbonate reaction'.

Igneous and metamorphic rocks are won in quarries in the mountains. They tend to be harder and stronger than the limestones and range from acidic granites to basic gabbro. Each source has to be carefully assessed and worked as many occur in areas with complex geological history where there is intermixing of rock types.

Granite is found in Yemen and in the western mountains of Saudi Arabia. Rhyolite and andesite are more common in western Saudi Arabia. Some of these rocks may contain potentially alkali-reactive material. Good quality basalt can be found in the north-west and south-west of Saudi Arabia and in the Oman mountains. Dolerite and gabbro are often found in association with serpentinite. Examples occur in the Oman mountains and they are commercially exploited around the eastern Emirate of Fujeirah. Rocks containing serpentine are found in parts of the UAE, in the Oman mountains and eroded materials in the form of wadi sands and gravels have long been used as a source of aggregate for concrete. Unweathered gneiss occurs in western Saudi Arabia, Yemen and Oman. Sources of this material should be checked for ASR potential.

Dune sand is worked as a source of fine aggregate for concrete. This varies in composition from mainly carbonate shell fragments near the coast to mainly silica inland.

The boom in construction activity, particularly in the Gulf States, has meant that local sources cannot meet the demand for high quality aggregates. The concrete supply industry has had to examine the use of more distant sources, such as Iran and the southern coast of Yemen and aggregates are sometimes shipped over great distances.

16.3.2 Syria, Lebanon, Israel and Jordan

Quarrying within Lebanon has had a somewhat chequered history in the last few decades. Many unregulated quarries were opened to serve local needs in the period of reconstruction after the civil war (Anon, 2010). A study by Hamad et al. (2000) records over 500 quarry sites most of which were small, producing less than 5000 m^3 per year. Twenty of these quarries were chosen for study in more detail. These were mainly in the Mount Lebanon range and the Bekaa Valley and the products were described mainly as limestone, dolomitic limestone or marly limestone of the Cretaceous period, but with a few quarries producing material from the Jurassic or Tertiary. A range of tests was carried out on the aggregates but these did not extend to tests for alkali-reactivity. Petrographic examination revealed the presence of calcareous dolomite in a few samples and it was noted that this material is potentially alkali-reactive.

Nadu *et al.* (2003) contended that, until their time of writing, the possibility of occurrence of alkali-reactivity in Israel had been ignored and they described what they called "*the first and only Israeli investigation of AAR*". The investigations involved samples from seven sources. The materials were described as calcite-dolomite, limestone with amorphous silica, limestone-dolerite, a mixture of rock types including amorphous silica and carbonates from a river basin and quartz marine aggregate. The samples complied with the Israeli standard for mineral aggregates and the Israeli concrete code.

Samples made up with these aggregates were subjected to a range of tests including the ASTM C289: (now 2007, now withdrawn) chemical test, short and longer term mortar-bar tests, short-term autoclaved mortar-bar tests, long-term concrete prism tests and measurement of fracture-energy using the wedge splitting method. None of the aggregates showed conclusive results when tested by the 'quick chemical method' or the ASTM C227: (now 2010) mortar-bar test even after two years exposure. Limestone from Ein Harod quarry in the north of Israel was found to be reactive when tested in accordance with the AS/SR version of the long term concrete prism test described in Canadian Standard CSA A23.2-14A-M90: (1990, 1994) and the French Standard NF P18-587 (AFNOR, 1990) [this accelerated concrete prism test has now been developed by RILEM as their AAR-4.1 test, see Nixon & Sims, 2014] and also when concrete specimens with high alkali content were stored under warm humid conditions. The sample from Zeelim river basin was found to be reactive in the NBRI accelerated mortar- bar test (Canadian Standards Association, 1994 and now 2014).

It was concluded that the Ein Harod limestone was reactive with a long induction period and the Zeelim river aggregate was reactive with a short induction period. The authors speculate that damage to some Israeli structures attributed to other causes could, in fact, have been the result of alkali-reaction. In this context they mention harbours and retaining walls in contact with the Mediterranean, supporting pillars for a coal delivery system at the Hedera power station and drainage culverts under the Bersheva-Dimona highway.

Yeginobali *et al.* (1993) report a lack of high quality aggregates in Jordan. Local limestones are usually soft and produce a great deal of dust on crushing. Crushed basalt aggregates have a smooth surface resulting in poor bond with cement paste. The highly angular shape of the crushed particles also leads to a greater water requirement in concrete mixes. Wadi sands and gravels are available, but these can contain significant amounts of chert, some of which may contain reactive silica.

16.3.3 Iraq

The multiple sand and gravel terraces of the Tigris and Euphrates rivers supply the majority of the coarse and fine aggregate resources for concrete production in the north and west of the country around Baghdad. Quarries and borrow pits are located on both sides of the Tigris river. 'Black rounded gravels' are in common use in this region for both asphaltic pavement and concrete (Al Maimuri *et al.*, 2011). These are polymictic gravels contrasting with the 'white gravels' of the Kebala quarries which are principally carbonates. Other gravel quarries which lie around and north and west of Baghdad are listed as Safwan, Al-Butain, Al-Suddor, Al-Taq and Arar quarries. One of the authors recalls assessing gravels in use in concrete for a highway west of Baghdad that was originally intended to extend to Damascus in Syria, in which there was a dominance of chert considered likely substantially to exceed the 'pessimum' proportion and thus be effectively non-reactive.

There are substantial deposits of quartz sand (Jassim & Goff, 2006). Nuhr Umr and Rutba formations in the Western Desert exceed 20 m in thickness and extend for more than 100 km. Below the Ga'ara Depression floor, Lower Permian sand layers consist almost entirely of quartz derived from granite source rocks of the Arabian Shield or recycled from Palaeozoic sedimentary rocks in the west and south-west of the country. The quartz grains are generally sub-rounded to sub-angular and are from 0.1 to 0.5mm in size. Some coarse grained material occurs in the upper parts of the Ga'ara Formation, but it is generally fine grained.

The main limestone areas are the Triassic and Cretaceous units in the thrust zone close to the Iranian border, the Miocene units in the foothills of the folded zone and Eocene Miocene units in the Western and Southern Deserts. The Dammam, Ratga, Euphrates and Fatha formations contain most of the commercially economic limestone deposits. The main dolomite deposits lie in the folded zone and in the Western and Southern Deserts. In the folded zone in north-east Iraq, the Upper Eocene Pila Spi is the main limestone and dolomite bearing unit, whereas in the Western Desert, the Upper Triassic Mulussa and Zor Hauran, the Liassic Ubaid and Hussainayat, the Upper Cretaceous M'sad and Hartha, the Palaeogene Umm Er Radhuma and Dammam, and the Lower Miocene Euphrates formations contain thick beds of dolomite.

Coarse and fine aggregates for concrete usually conform to the Iraqi standard specification (IQS 45/1984: 1984). Fine aggregates are almost invariably derived from natural sands and the coarse aggregates from natural gravels, which are in some cases crushed and screened to conform with the standard. Tests for possible contamination with salts, particularly sulphates should be undertaken before approval of a new aggregate source. There appears to be no published information on aggregate testing for potential alkali-reactivity in concrete.

Several locations in a series of flat lying limestone horizons of Miocene and Eocene age 40 km west of Baghdad and running southwards to Al-Samawa City on the Euphrates have been identified by the Iraqi Government as potential sources of carbonates for cement manufacture, though there is no information concerning their exploitation. Impure crushed siliceous limestone has also been proposed as a source for road pavement material around Bahr Al-Najaf, which lies some 200 km south of Baghdad.

16.3.4 Iran

As described in Section 16.2.3.3, the geology of Iran is more complex than Iraq, with a variety of hard rock sources available for crushing as concrete aggregate in addition to the natural sands and gravels that are in common use as aggregate materials. Pumice in the vicinity of Mount Damavand 150 km east of Tehran is reported to be a source of lightweight coarse and fine aggregates. Sources of natural pozzolanas occur in the Kerman province some 300 km south-west from Tehran.

Examples of alkali-aggregate damage to concrete structures including dams have been identified in Iran (Ramezanianpour & Koloshani, 2006) and have led to the investigation of methods of avoiding potential alkali-reactivity in Iranian concrete. One such method is the addition of natural pozzolanas (see Chapter 4), which are widely available in the Kerman province. Faroughi *et al.* (2012), using natural gravel aggregate, tested four natural pozzolanas of different compositions in accelerated mortarbars according to ASTM C1260: (now 2014). They report that the effectiveness of the pozzolanas in reducing the risk of reactivity appeared to be sensitive to their calcium oxide content. Avoidance of the potential for alkali-reactivity damage appears to concentrate on the use of a low alkali cement, the use of natural pozzolanas, or other mineral additions such as silica fume, metakaolin or ground granuated blastfurnace slag. A study of the potential for alkali-aggregate reactivity in natural gravels from the Aras region of north-west Iran (Eftekar & Moghadam, 2010) reported that the gravels contained potentially reactive cherts.

In the Mazandaran province, which borders the Caspian Sea, the Alborz Mountains inland from the coastal plain are identified as a potential source of granitic, dolomitic, siliceous and limestone aggregates (Hassannejad, 2013), but no information is available relating to their alkali-silica reactivity potential. In the Lorestan Province in the central region of the Zagros Mountains, a new variable arch concrete dam, the Bakhtiari Dam, is under construction. Sources for both coarse and fine aggregates have been located near to the dam, with some material obtained from the excavation work. No details of the rock types being used are available but long term testing for the possible potential for alkali-aggregate reactivity has apparently been initiated.

16.3.5 North Africa

Typically, the coastal developments in the region use polymictic gravels as concrete aggregate, though there is hard rock inland. These are limestones of various types. Siliceous duricrust materials are also a possible source of aggregate. In the Atlas Mountains, a wide variety of sedimentary and igneous rock types is available as the source of aggregate in addition to the recent and Quaternary river and terrace gravels and sands derived from the erosion of the surrounding mountains. In Northern Egypt, sand and gravel from the Nile plain are used as concrete aggregate. The coarse aggregate from this source is well rounded and is sometimes polished. These factors can limit the maximum concrete strengths that can be achieved, as they reduce the bond between paste and aggregate. A number of different rock types including limestone and dolomite are quarried for use in concrete in the south of the country. The use of dolomite has been discouraged and possibly barred in some contracts for reasons which have not been altogether clear, but might have been associated with concerns over possible alkali-carbonate reaction (see Chapter 3). As noted above, tests on dolomitic material from Attaka quarry in Egypt proved negative (Sharobim, 1995).

During the construction of the 'Great Man-made River Project' in Libya, whereby deep seated water beneath the Sahara in the south of the country would be piped to the population centres along the northern coastal strip, large amounts of concrete were required for the sophisticated composite precast pipes and aggregate sources were assessed in detail (Stoner & Fookes, 1991; Fookes *et al.*, 1991, 1993). Pipe production in the northern sector utilised crushed carbonate rock coarse aggregate, but that in the south used aggregate processed from surface gravel deposits that sometimes included potentially alkali-silica reactive constituents, including opaline sandstone. Advised by

Professor Peter Fookes and assisted by Dr Ian Sims, the contractors undertook a rolling programme of sampling and laboratory ASR assessment in order to avoid using potentially reactive aggregates.

16.4 WATER

The majority of concrete in the region is produced using public water supplies, which are usually of good quality, but availability of supply on a day to day basis may be uncertain in some areas. Public supply is either produced by desalination of seawater at coastal sites or by pumping non-renewable groundwater. In some cases, even inland cities are supplied from distant coastal desalination plants. For example, water is pumped from the Gulf to Riyadh over a distance in excess of 450 km. Some concrete plants, particularly those set up to supply large individual projects in remote locations, are supplied from their own wells. In these cases, care has to be taken to monitor the water quality on a regular basis. In general, public supply is unlikely to contribute any significant alkali to concrete mixes. Where supply is from wells on site, the supply should be carefully monitored and the contribution of alkali to the mix taken into account where appropriate when calculating total mix alkali for control purposes.

16.5 CEMENTS, ADDITIONS & ADMIXTURES

In the 1970s and early 1980s, local cement production on the Arabian Peninsula could not keep up with the increased demand and cement was imported from a wide variety of origins including East Africa, South-East Asia and Southern and Eastern Europe. These cements, though produced to BS, ASTM or equivalent local standards, were not always in good condition by the time they were used to produce concrete. This was because of long transport times by sea, periods at anchor offshore because of port congestion and storage in unsuitable conditions in the humid local environment. Since then, cement production in the region has increased substantially and now is capable of meeting most of the local demand.

The need to produce concrete which is able to withstand the aggressive local environment has led to increased use of additions (or 'supplementary cementitious materials', SCMs), such as fly ash, microsilica and ground granulated blastfurnace slag. These materials are by-products of electricity generation by coal, semiconductor production and steel making, respectively. As these manufacturing processes are not widely undertaken in the region, the materials have to be imported. Fly ash has been imported from India and South Africa, microsilica from Europe and blastfurnace slag from India and Europe. Though there can be great variation in the quality of these sources, the inclusion of these materials in concrete is generally beneficial in reducing the potential risk of alkali-reactivity (see also Chapter 4).

16.5.1 Water reducing admixtures

Concretes with low water/cement ratios are needed to provide greater durability in the face of the high concentrations of chlorides and sulphates in the local environment. Plasticisers and high range water-reducing admixtures are necessary to give workability with these low water/cement ratios. A range of products with different chemical formulations is available. They are generally added at quite low dosage, but in some cases their alkali content needs to be taken into account when assessing the total alkali content of a concrete.

16.5.2 Waterproofing of foundations and substructures

Because of the presence of a high saline water table in many parts of the region, it has long been recognised that there are benefits to be gained from protecting foundations by a waterproofing layer or by use of waterproofing admixtures. In addition to protecting the foundations themselves, these measures also eliminate or reduce the upward movement of moisture into connected above ground members, such as walls, columns and piers. If this moisture movement takes place, the moisture tends to migrate towards the surface of the concrete above ground level, where evaporation increases the concentration of any dissolved salts. This has led to corrosion of reinforcement in a zone just above ground level in many unprotected structures. The waterproofing of underground parts of concrete structures undoubtedly reduces this effect and, concomitantly, also the risk of alkali-reactivity by reducing the moisture and additional alkalis which would otherwise be able to penetrate the concrete.

16.6 POTENTIALLY ALKALI-REACTIVE CONCRETE IN THE REGION

16.6.1 Arabian Peninsula & the Gulf

There is evidence that some of the rock types used for aggregate in the region are potentially alkali-reactive (French & Poole, 1976; Sims & Poole, 1977; Nadu et al., 2003). However, there have been very few reported cases of structures being adversely affected by alkali-reaction, even though high strength concretes with relatively high cement contents have been widely used. One self-evident reason for this could be the arid or hyper-arid nature of the climate across much of the region; there may not be sufficient moisture in the environment to permit the reaction to proceed to a damaging conclusion.

This line of reasoning, however, does not bear tighter scrutiny. Much of the development and concrete construction has taken place in the coastal plains where the humidity is often high and variable (Sims, 2006). There are frequent occasions when condensation forms on cooler concrete surfaces. One of the authors has seen map cracking on the ends of cross-heads on bridge supports in Oman. Joints in the bridge decks were situated directly above the crossheads and there was evidence of frequent water flows through the joints and down the ends of the cross-heads. These water flows resulted from early morning condensation on the bridge decks finding its way down the un-waterproofed deck joints.

Katayama (1997) also makes reference to possible ASR examples in Bahrain. Two of the authors have experience of concrete in Bahrain and are not aware of any cases of ASR, though a suspected case of alkali-carbonate reaction was subjected to early detailed analysis (French & Poole, 1974), which found reaction rim formation but no evidence of expansion.

In the late 1970s, when the construction boom in the region began, it is probable that the alkali contents of cement were not high. Durability was not immediately identified as a potential problem and so there were few requirements for high cement contents or low water/cement ratios which could have forced up cement contents in the absence of admixtures suitable for the local conditions at that time. It should also be noted that there was little experience of concrete production in the region and specified strengths were not particularly high, again reducing the need for higher cement contents. Taken together, these factors meant that total alkali contents of mixes were not high, so that the risks of alkali-reactivity in the concretes were also lower.

In the early to mid 1980s, when the durability problems with concrete, particularly reinforcement corrosion, had become widely recognised, cement contents were increased but microsilica was incorporated in many concrete mixes. This was to increase protection against sulphate attack and ingress by chloride salts, but also had the effect of reducing the risk of alkali-reactivity. In more recent times, fly ash and blastfurnace slag have been used as components of cement on many projects. This has also been to increase durability in the face of potential sulphate attack and reinforcement corrosion, but has also had the effect of reducing the likelihood of alkali-reactivity.

Tanking of foundations of concrete structures with robust membranes has been carried out to a great degree over the past thirty years or so. This was to protect against high concentrations of sulphate in the soil and groundwater in many areas, but has also had the effect of cutting off moisture supply to foundation concrete and the lower parts of superstructures which could be subject to capillary rise. This again has reduced the likelihood of damaging alkali-reactivity. Concrete piles, however, have generally not been protected in this way.

The above factors, taken in conjunction with the few reported case histories, lead to the conclusion that there is probably little potentially alkali-reactive concrete in the region. There are two caveats to this conclusion. The first is that any actual cases of alkali-reactivity may not have been widely reported, although Katayama (1997) makes brief report of possible cases of ASR in Saudi Arabia which may not have become generally known. Katayama (2012) also refers to samples he investigated from Saudi Arabia, in which dolomitic limestone aggregate developed clear reaction rims that were not expansive; he considered this was probably a case of his 'so-called alkali-carbonate reaction' (see Chapter 3), whereby expansion was caused by hitherto undetected crypto-crystalline silica within the otherwise carbonate rock. One of the authors also investigated this case and agrees that the reaction rims were not responsible for the early cracking that occurred to the thick walls in question, which was thought likely to be thermal in origin (see Figure 16.2).

The second caveat is that there were many cases of premature deterioration of structures built during the early years of the expansion of construction activity. These were mainly due to reinforcement corrosion. Many of these structures were demolished at a time before signs of damage due to alkali-reactivity would have become apparent. Another reason for early demolition was the rapid pace of development across the region. Some structures, for example bridges which restricted the increased traffic flows, became obsolete at an early stage and had to be replaced.

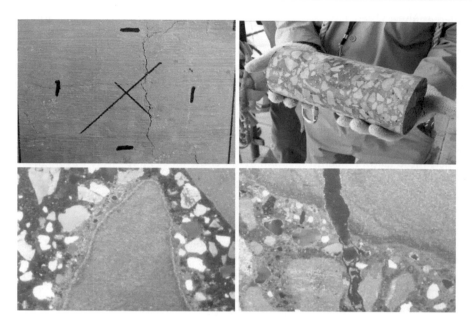

Figure 16.2 Rim formation with dolomitic limestone aggregates in a concrete in Saudi Arabia: a) cracking on the external wall surface, which appeared to penetrate through the full wall thickness, here showing a core drilling position; b) extracted core sample; c) thin-section photomicrograph in crossed polars, width of view about 4 mm, showing a complex rim formation around a dolomitic limestone particle – there is alteration to both the rock particle surface and the adjoining cement matrix; and d) another thin-section photomicrograph in crossed polars, width of view about 4 mm, showing that the crack (top to bottom) cuts across and post-dates the rim formation. Photographs courtesy of Ian Sims & RSK Environment Ltd.

16.6.2 Israel

Work by Nadu *et al.* (2003) has shown that materials described as limestone from Ein Harod quarry in the north of Israel and as a mixture of rock types including amorphous silica and carbonates from Zeelim river basin are indicated to be potentially reactive by laboratory testing. The authors also contend that that some Israeli structures may have been damaged by alkali-reaction although the deterioration was attributed to other causes. Harbours and retaining walls in contact with the Mediterranean, supporting pillars for a coal delivery system at the Hedera power station and drainage culverts under the Bersheva-Dimona highway are mentioned as possibly having been damaged by alkali-reactivity. Katayama (1997) makes brief reference to examples of ASR in Israel.

16.6.3 Alkali-aggregate reactivity in concrete structures in Iraq and Iran

As is reported above (sections 16.3.3 and 16.3.4), concrete structures utilise natural gravels and crushed rock as aggregates for concrete in Iraq and Iran. Some of these aggregates have been identified as containing material that has the potential for alkali-aggregate reactivity in concrete. Perhaps the most widely used aggregate for both coarse and fine concrete aggregate are locally sourced natural gravels and sands. Crushed rock or crushed gravels are also used for major concrete structures, but again are typically sourced local to the structure.

There appear to be no published data in Iraq concerning concrete structures damaged by alkali-aggregate reaction (AAR). In Iran, a number of structures including dams, bridges and pavements (Ramezanianpour & Koloshani, 2006) have been identified as suffering damage due to AAR.

The recognition of this potential for damage has led to extensive efforts by Iranian engineers in testing aggregates and concrete mixes prior to use and also in developing methods of avoiding the potential for alkali-aggregate reaction in new concrete structures. As already noted, a common preventative measure is to make use of mineral admixtures such as natural pozzolanas, which are widely available in the Kerman province, and ground granulated blastfurnace slag or silica fume. Although some Iranian produced cements tend towards high alkali contents, the use of a low alkali cement and the avoidance of potentially reactive aggregate types are also recognised as potential methods of avoiding potential damage associated with alkali-aggregate reaction.

16.6.4 North African and other Examples

There have been unconfirmed reports of alkali-aggregate reaction in concrete bases to transmission towers in central Libya. Other than this, the authors do not have any personal knowledge of any cases of alkali-aggregate reaction in this region and have been unable to find any references in published works.

16.7 TEST METHODS AND DIAGNOSIS

16.7.1 Testing during construction

The most widely used test for reactivity of aggregates in the region has been the 'quick chemical method', ASTM C289: (2007, but now withdrawn). In some instances, this may be the only requirement of a specification in relation to alkali-reactivity because of the long timescales involved with mortar-bar and especially concrete prism testing. If the aggregate is found to be 'innocuous' under the ASTM C289 test, there are usually no requirements for limiting the alkali content of concrete mixes. Many of the established quarries, however, will hold test result certificates for both the ASTM C289: (2007, but now withdrawn) chemical and ASTM C 227: (2010) mortar-bar methods.

More recently, the ASTM C1260: (now 2014) accelerated mortar-bar test has been specified in some contract documents in association with BRE Digest 330 (BRE, 2004). The exact implications of this may not be totally clear. ASTM C1260 is used to classify aggregates on the basis of expansion at 14 days as follows:

14 day expansion	Aggregate classification
0.00 % to 0.10 %	Non-reactive
0.10 % to 0.20 %	Potentially reactive
>0.20 %	Reactive

BRE Digest 330 provides limits on alkali contents of concrete mixes depending on the reactivity of aggregates classified in a different manner. The BRE Digest 330 aggregate classifications are 'low reactivity', 'normal reactivity' and 'high reactivity'. Lists are provided of 'low reactivity' and 'high reactivity' rock types while 'normal reactivity' aggregates are those which do not occur on these two lists. The classifications are mainly based on long experience of the use of the aggregates in British practice with some reference also to testing to BS 812-123: (1999). An up-to-date explanation of British (UK) practice is provided in Chapter 6.

The two classification systems are based on different criteria, but the interesting inference of the specifications in question appears to be that aggregates found to be 'non-reactive' under ASTM C1260 can be considered to be 'low reactivity' for the purposes of BRE Digest 330, whilst 'potentially reactive' by C1260 would relate to BRE 'normal reactivity' and C1260 'reactive' to BRE 'high reactivity'.

Concrete specifiers in Libya are generally aware of ASR and take appropriate precautions on major projects like the 'Great Man-made River' (see also section 16.3.5). Some of the gravel aggregates do contain cherts and gypsum is also reported to be common.

16.7.2 Diagnosis post construction

Initial diagnosis would be on the basis of the external appearance of structures and the obvious signs of possible reactivity, such as crack patterns, exudation of gel and general overall expansion of structures resulting in closure of movement joins and damage to adjoining elements or structures.

The initial diagnosis would be confirmed by petrographic examination to either ASTM C856: (2011) or the code of practice of the Applied Petrography Group of the Geological Society (Geological Society of London, 2008). See Chapter 5 for a detailed account of the diagnosis of any alkali-aggregate reaction in concrete.

16.8 MINIMISING THE RISK OF AAR

16.8.1 History

At the time of the start of the construction boom in the Middle East, there was little local appreciation of the potential problem of alkali-reactivity. The industry may have been protected, to a certain extent, because most cement was imported from Africa or the Far East. The cement plants in these areas used older processes which naturally resulted in lower alkali cement.

As the industry developed, there was an influx of concrete practitioners, mainly from Europe, who brought with them knowledge of the AAR problem. The big issue at that time was early deterioration of concrete structures due to corrosion of reinforcement. Internal guidelines produced by one major British consultant that was very active in the region (Kay *et al.*, 1979) contained a four page section on aggregate reactivity potential. The recommendations included petrographic exam-ination to ASTM C295: (now 2012), the rapid chemical test to ASTM C289: (now 2007, but withdrawn) and a gel-pat test to a National Building Studies Research Paper (Jones & Tarleton, 1952) and also in BS 7943: (1999). If any of these tests gave a positive result further investigation using the ASTM C227: (now 2010) mortar-bar test was recommended. It was recognised that this could be a long process and it advised that low alkali cement should be used until the aggregate was shown to be innocuous. The guidelines also note that the first and most instructive step when using an unfamiliar aggregate source is to look at existing structures which used similar aggregate to see if they showed any of the characteristic signs of alkali-reactivity.

It is of interest to note that the UK Cement and Concrete Association booklet 'Concrete Practice' (Cement and Concrete Association, 1975) bound in with the Halcrow *Notes for Guidance*, does not make any mention of alkali-reactivity. This was a time when the occurrence of AAR in the UK was only reluctantly being accepted (see Chapter 6).

The experiences of British consultants and contractors working in the Arabian Peninsula were brought together by the British research organisation CIRIA in 1984 (CIRIA, 1984). The guide notes that alkali-aggregate reaction had not been identified as a significant cause of deterioration of concrete in the Gulf region, though the reaction had been observed and some aggregates in common use showed a positive reaction when tested. The main recommendations were that:

- New sources of aggregate should be classified in accordance with ASTM C295: (2012);
- Tests should be carried out in accordance with ASTM C227: (2010) and ASTM C289: (2007, but now withdrawn);
- If the amount of potentially reactive material present in the aggregates exceeded 0.5% by weight of total aggregate, the alkali from the cement should not exceed 3 kg/m^3 of concrete.

The CIRIA guide was updated in 2002 by a joint Concrete Society-CIRIA working party with contributions from engineering societies in the Arabian Peninsula (Walker, 2002). This edition of the guidance amended some of the advice given in the initial publication, in that it stated that the ASTM C289 and C227 tests were no longer considered to be dependable. It suggested that both the intended coarse aggregate and fine aggregate together should be evaluated using an expansion test on concrete prisms, such as the tests described in BS 812-123: (1999) and ASTM C1293: (now 2008). It was noted that it could take many months for these tests to produce meaningful results and that they may not therefore be appropriate for use in specifications. An accelerated mortar-bar test, such as that set out in ASTM C1260: (now 2014), was suggested as an alternative.

A companion guide dealing with design issues has also been published by the Concrete Society (Concrete Society, 2008). The RILEM suite of AAR tests (now in Nixon & Sims, 2016) is suggested therein as the means of assessing aggregates for reactivity potential.

16.8.2 The present

It has proved difficult to obtain information on current concrete and aggregate specifications. However, it does seem that there has been little movement towards the recommendations of the more recent reports outlined above. There is still great reliance on the ASTM C227: (2010) mortar-bar and ASTM C289: (2007, but now withdrawn) chemical methods, plus also ASTM C586: (2011) 'rock cylinder' (for carbonate rocks) and ASTM C1260: (2014) accelerated mortar-bar tests when specifying aggregates. Consultants with a British connection may include reference to BRE Digest 330 Part 2 (BRE, 2004) to control the total alkali in concrete mixes. The present authors would advise that the RILEM scheme offers the most up-to-date and internationally approved approach (Nixon & Sims, 2016).

Concrete construction in Saudi Arabia has tended to follow American codes and standards. The Saudi Building Code (Al-Amoudia *et al.*, 2008) follows ACI 318: (2008) in its concrete aspects and hence requires aggregates to conform to ASTM C33: (now 2013). The recommendations in relation to reducing the risk of alkali-reactivity appear to have been adopted unaltered.

The specified use of pozzolanas and ground granulated blastfurnace slag as mineral additives with cement, together with avoidance of potentially reactive aggregate types, appear to be the usual techniques in Libya for avoiding potential AAR damage in new concrete structures.

The aggressive saline nature of the local environment means that most concrete in the region is formulated to resist the penetration of chlorides. Blastfurnace slag and fly ash have been found to be very helpful in this respect and many concrete specifications require the use of these materials. This has had the side benefit of also reducing the risk of alkali-reaction. As noted in Section 16.5.2, the widespread waterproofing of foundations has also reduced the risk of alkali-reactivity.

One of the authors is currently assisting one Gulf country with an assessment of aggregates derived from reclaiming coarse aggregates from waste residues of surface deposits previously processed to extract sand for agricultural and some building purposes (Hassan *et al.*, 2016). As well as containing gypsum that is being reduced to acceptable levels by novel processing techniques, these residues also contain some potentially reactive rock types, including rhyolite. A programme is being undertaken to assess the ASR risk and to establish the most effective and affordable preventative measures. Current trends suggest that these reclaimed aggregates will be successfully able to reduce the future dependence on imported aggregates.

16.9 METHODS OF PROTECTION OR REPAIR

There have been few publicly identified cases of alkali-reactivity and hence there is no history of protection and repair of concrete structures damaged by this process in the region. There have, however, been many cases of damage to concrete structures because of reinforcement corrosion. These structures have been dealt with by measures ranging from demolition and rebuilding to repair and cathodic protection. Undoubtedly, if any cases of damage due to alkali-reactivity do occur they will be dealt with using the procedures set out in Chapter 5 of this book. One thing to bear in mind is the effects of the local aggressive environment. The cracking that occurs in the early stages of AAR may permit ingress of moisture, salts and oxygen from the environment to reinforcement, thus accelerating reinforcement corrosion. Early decisions on protection or repair methods need to be taken in these cases.

Although the methods of repair or protection used in the region are broadly similar to those used elsewhere, there are some differences or aspects which require particular attention because of the local environment and climate. Repair procedures have been set out in a Concrete Society Guide (Walker & Kay, 2002).

As noted above, cracks resulting from AAR can in their turn lead to reinforcement corrosion. Structures which have suffered AAR cracking over a number of years can be anticipated to have a greater degree of reinforcement corrosion, when repair is undertaken, than would be anticipated in more moderate exposures.

Polymers used in concrete repair often involve exothermic reactions. That is, they give out heat as they react. The initial temperature of the reactive components influences the rate of reaction and hence their pot life and open time. There are advantages to be gained in keeping materials as cool as possible until they are needed and in shading the substrate on to which they are to be applied.

The hydration reaction between cement and water is also exothermic and the same advice applies. In this case, there is also a need to take great care over curing, so as to avoid rapid loss of moisture from repairs in the high temperatures which prevail.

16.10 ILLUSTRATIVE CASE STUDIES

Very few if any case histories of concrete structures damaged by AAR have been reported from the countries in this region.

There is anecdotal evidence that a recent case of AAR has occurred on transmission tower bases in central Libya. Also Katayama (1997) reports possible examples of ASR in structures in Bahrain, Yemen and Israel.

An anecdotal account of an unusual case of ASR gel exudations on new concrete surfaces forming within a month of removal of formwork on a structure in Saudi-Arabia was reported. Poole suggests that this unusually rapid appearance of gel at surfaces may be due to the siliceous component of the raw materials used in the original cement manufacture leaving a small proportion of unreacted small quartz particles that are altered by kiln temperatures to cristobalite and tridymite in the cement used for the concrete. These particles would react rapidly to produce the gel observed.

REFERENCES

AFNOR. (1990) *Granulats – Stabilité dimensionelle en milieu alcalin – Essai sur béton (Aggregates, dimensional stability in alkali medium, concrete test).* Paris, France, Association Francaise de Normalisation, St Denis.

Al-Amoudi, O.S.B., Maslehuddin, M., Alhozaimy, A.M., Al-Negheimish, A.I., & Khushefati, W. H. (2008) Durability and hot weather requirements in the new Saudi Building Code. In: *Proceedings of International Conference on Construction and Building Technology*, Kuala Lumpur, Malaysia. pp. 409–432. [Online]. Available from http://www.uniten.edu.my/new home/uploaded/coe/iccbt/iccbt%202008/conference%20b%20extract/UNITEN%

20ICCBT%2008%20Durability%20and%20hot%20Weather%20Requirements%20in%20The%20New%20Saudi.pdf [Accessed 5th September 2013].

Al Maimuri, N.M.L., Al-Kafaji, M.K.A., & Al Sa'adi, A.H.M. (2011) Evaluation of using a white gravel of Kerbala quarries in asphalt mixes. *Al-Taqani, 24* (3), 117–127 [Online]. Available from the Iraq Academic Journals web site www.iasj.net [Accessed 9th September 2013].

American Concrete Institute. (2011) ACI 318, *Building code requirements for structural concrete and commentary.* Farmington Hills, MI 48331 U.S.A, ACI.

American Society for Testing and Materials. (2007) ASTM C289-07. *Standard test method for potential alkali-silica reactivity of aggregates (chemical method).* West Conshohocken, PA, USA, ASTM.

American Society for Testing and Materials (2014) ASTM C1260-14. *Standard test method for potential alkali reactivity of aggregates (mortar bar method).* West Conshohocken, PA, USA, ASTM.

American Society for Testing and Materials (2008) ASTM C1293-08. *Standard test method for concrete aggregates by determination of length change due to alkali-silica reaction.* West Conshohocken, PA, USA, ASTM.

American Society for Testing and Materials (2010) ASTM C227-10. *Tests for potential alkali reactivity of cement-aggregate combinations (mortar bar test method).* West Conshohocken, PA, USA, ASTM.

American Society for Testing and Materials (2011) ASTM C856-11. *Standard practice for petrographic examination of hardened concrete.* West Conshohocken, PA, USA, ASTM.

American Society for Testing and Materials (2011) ASTM C586-11. *Standard test method for potential reactivity of carbonate rocks as concrete aggregates (rock cylinder method).* West Conshohocken, PA, USA, ASTM.

American Society for Testing and Materials (2012) ASTM C295-12. *Standard guide for examination of aggregates for concrete.* West Conshohocken, PA, USA, ASTM.

American Society for Testing and Materials (2012) ASTM C294-12. *Standard descriptive nomenclature for constituents of concrete aggregates.* West Conshohocken, PA, USA, ASTM.

American Society for Testing and Materials (2013) ASTM C33-13. *Standard specification for concrete aggregates.* West Conshohocken, PA, USA, ASTM.

Anon. (2010) Lebanon, the Quarry Mafia. *Now Lebanon* [Online] August 24. Available from: https://now.mmedia.me/lb/en/reportsfeatures/the_quarry_mafia. [Accessed 4th Septembar 2013].

Askri, H., Belmecheri, A., Benrabah, B., Boudjema, A., Boumendjel, K., Daoudi, M., Drid, M., Ghalem, T., Docca, A.M., Ghandriche, H., Ghomari, A., Guellati, N., Khennous, M., Lounici, R., Naili, H., Takherist, D., & Terkmani, M. (Undated) Geology of Algeria. *Contribution from SONATRACH exploration division, research and development centre and petroleum engineering and development division* [Online]. Available from: http://www.mem-algeria.org/fr/hydrocarbures/w1_0.pdf. [Accessed 18th September 2013].

Barnbrook, G., Dore, E., Jeffrey, A.H., Keen, R., Parkinson, J.D., Sawtell, D.L., Shacklock, B. W., & Spratt, B.H. (1975) *Concrete practice.* Slough, Cement and Concrete Association.

British Standards Institution (1999) BS 812-123:1999. *Testing aggregates: Method for determination of alkali-silica reactivity. Concrete prism method.* London, BSI.

British Standards Institution (1999) BS 7943:1999. *Guide to the interpretation of petrographical examinations for alkali-silica reactivity.* London, BSI.

Building Research Establishment (2004) *Digest 330. Alkali-silica reaction in concrete.* Watford, BRE.

Canadian Standards Association, CSA (1990 and 1994) A23.2-14A-M90. *Potential expansivity of cement-aggregate combinations, concrete prism expansion method.* Mississauga, Ontario, Canada.

Canadian Standard Association, CSA (2014) A23.2-25A. *Test method for detection of alkali-silica reactive aggregate by accelerated expansion of mortar bars*. Mississauga, Ontario, Canada.

Central Organisation for Standadization and Quality Control (1984) IQS No. 45. *Aggregates from natural sources for concrete and building*. Baghdad, COSQC.

Concrete Society (2008) *Guide to the design of concrete structures in the Arabian Peninsula*. Publication CS 163, The Concrete Society, Camberley, UK.

Construction Industry Research and Information Association (1984) *The CIRIA guide to concrete construction in the Gulf region*. London, CIRIA. Special Publication 31.

Eftekar, M.H., & Moghadam, S.R. (2010) Study of the effect of alkali-silica reaction on the properties of concrete. In: Oh, B.H. *et al.* (eds.) *Fracture mechanics and concrete structures – assessment, durability and retrofitting concrete structures*. Seoul, Korean Concrete Institute. pp. 1038–1042.

Fookes, P.G., & Higginbottom, I.E. (1980) Some problems of construction aggregates in desert areas with particular reference to the Arabian Peninsula, Part 1: Occurrence and special characteristics. *P I Civil Eng.*, Part 1, *68*, February, 39–67.

Fookes, P.G., Stoner, J.R., MacKintosh, J.H. (1991) The Libyan great man-made river project phase 1, part 2: Concrete technology of prestressed concrete pipe manufacture. *P I Civil Eng.*, *90*, 881–919. [see also Discussion (1992) in *94* (4), 501–502].

Fookes, P.G., Stoner, J.R., MacKintosh, J.H. (1993) Great man-made river project, libya, phase 1: A case study on the influence of climate and geology on concrete technology. *Q J Eng Geol.*, *26*, 25–60.

Foroughi, M., Tabatabaei, R., & Shamsadeini, M. (2012) Effect of natural pozzalans on the alkali-silica reaction of aggregates in real concrete specimens. *J Basic Appl Sci Res.*, *2* (5), 5248–54.

French, W.J., & Poole, A.B. (1974) Deleterious reactions between dolomites from Bahrain and cement paste. *Cement and Concrete Research*, *4* (6), November, 925–937.

French, W.J., & Poole, A.B. (1976) Alkali-aggregate reactions and the Middle East. *Concrete*, *10* (1), 18–20.

Ghorbani, M. (2013.) *The economic geology of Iran, mineral deposits and natural resources*. Springer, Berlin.

Hamad, B.S., Khoury, G.R., & Khawlie, M.R. (2000) Geology and location of major coarse aggregate resources in Lebanon, Eastern Mediterranean. *B Eng Geol Environ.*, *58*, Springer-Verlag, 183–190.

Hassan, K.E-G., Reid, J.M., & Al-Kuwari, M.S. (2016) Recycled aggregates in structural concrete – a Qatar case study. *Proceedings of the institution of civil engineers: Construction materials*, Themed issue on Recycled Construction Materials, *169* (CM2) April, 72–82.

Hassannejad, M., Birenjian, J., & Amin, M.J.T. (2013) The investigation on the effect of aggregate type on the modulus of elasticity and tensile strength in high strength concrete. *International Symposium on Advances in Science and Technology*. 7th SASTech. Bandar-Abbas. Paper reference 07-96-1637.

IQS 45/1984 (1984) *Aggregate from natural sources for concrete and building construction*. Central Organization of Standardization and Quality Control, Iraq.

Jassim, S.Z., & Goff, J.C. (2006) *The geology of Iraq*. Geological Society of London. 341p.

Jones, F.E., & Tarleton, R.D. (1952) *Reactions between aggregates and cement, Part II, Alkali aggregate interaction: The expansion bar test and its application to the examination of some British aggregates for possible expansive reaction with Portland cements of medium alkali content*. DSIR/BRS.National Building Studies Research Paper No 17, HMSO.

Katayama, T. (1997) *A review of alkali-aggregate reactions in Asia –Recent topics and future research*. East Asia Alkali-Aggregate Reaction seminar, Supplementary papers, A33–A43.

Katayama, T. (2012). *Rim-forming dolomitic aggregate in concrete structures in Saudi Arabia – is dedolomitization equal to the so-called alkali-carbonate reaction?* In: *Proceedings, 14th International Conference on Alkali-Aggregate Reaction in Concrete*, Austin, USA. 10p. Paper 030411-KATA-01.

Kay, E.A., Mills, M.C., Pollock, D.J., & Sharp, T.J. (1979) *Concrete practice in the Middle East – Notes for guidance*. Dubai, Halcrow International Partnership.

Kay, E.A., Pollock, D.J., & Fookes, P.G. (1982) *Concrete in the Middle East – Part 2*. Camberley, Viewpoint Publications, The Concrete Society.

Leslie, A.B., & Eden, M. (2008) *A code of practice for the petrographic examination of concrete*. Geological Society of London. APG SR1. [Online]. Available from: www.appliedpetrographygroup.com [Accessed 6th September 2013].

Nadu, M., Schieber, M., & Tschegg, E. (2003) Alkali-aggregate reaction (AAR) in concrete in Israel. In: *Proceedings of 11th International Congress on the Chemistry of Cement*, Durban, May, Vol 3, pp. 2153–2169.

Nixon, P.J., & Sims, I. (eds.) (2016) RILEM recommendations for the prevention of damage by alkali-aggregate reactions in new concrete structures. *RILEM State Art Rep.*, *17*, Springer, Dordrecht, RILEM, Paris, 168p.

Owels, I., & Bowman, J. (1981) Geotechnical considerations for construction in Saudi Arabia. *Proceedings of the American Society of Civil Engineers, The Journal of the Geotechnical Division*. Paper 16092, 107, No GT3, March, 319–338.

Purser, B.H. (ed.) (1973) *The Persian Gulf: Holocene carbonate sedimentation and diagenesis in a shallow epicontinental sea*. Springer-Verlag, Berlin.

Ramezanianpour, A.A., & Koloshani (2006) Assessment of alkali aggregate reaction of concrete in three dams of Iran. In: *Proceedings of the 7th International Conference on Civil Engineering (ICCE)*, 2nd May, Tehran, Iran.

RILEM TC 191-ARP (2003) *Recommended test method: AAR-0, Detection of potential alkali reactivity in concrete, Outline guide to the use of RILEM methods in assessments of alkali reactivity-potential*. RILEM Publications SARL, 36 (261), 472–479.

Schlèuter, T., & Schlu'ter, T. (2008) *Geologic atlas of Africa with notes on stratigraphy, economic geology, geohazards, geosites and geoscientific education of each country*. RILEM Publications SARL, Paris. Springer.

Sharobim, K.G. (1995) Potential alkali reactivity of dolomite for concrete aggregates. In: *Proceedings of the Sixth Arab Structural Engineering Conference*, 21–24 October, Damascus.

Sims, I. (2006) Selection of materials for concrete in a hot and aggressive climate, Keynote lecture. In: *Proceedings of the 8th International Concrete Conference*, Bahrain, 27–29 November 2006, 62p.

Sims, I., & Poole, A.B. (1977) Potentially alkali-reactive aggregates from the Middle East. *Concrete*, *14* (5), 27–30.

Sims, I., Sotiropoulos, P. (1983) Standard alkali-reactivity testing of carbonate rocks from the Middle East and North Africa. In: *Proceedings, 6th International Conference, 'Alkalis in Concrete - Research and Practice'*, Copenhagen, Denmark, June 1983.

Stoner, J.R., Fookes, P.G. (1991) The Libyan great man-made river project phase 1, Part 1: Manufacture of prestressed concrete cylinder pipe. *P I Civil Eng.*, *90*, 853–879. [see also Discussion (1992) in *94* (4), 501–502].

Walker, M.J. (ed.) (2002) *Guide to the construction of reinforced concrete in the Arabian Peninsula*. Concrete Society Special Publication CS136 and CIRIA publication C577.

Walker, M., & Kay, T. (2002) *Guide to evaluation and repair of concrete structures in the Arabian Peninsula*. Concrete Society Special Publication CS 137.

Yeginobali, A., Samdi, M., & Khedawi, T. (1993) Effectiveness of oil-shale ash in reducing ASR expansion. *Mater Struct.*, *26* (3), 159–164.

Index

Note: Page numbers in **bold** indicate figures and tables

T - #0455 - 071024 - C804 - 246/174/35 - PB - 9780367573331 - Gloss Lamination